REGULATORY FUNCTION OF ADENOSINE

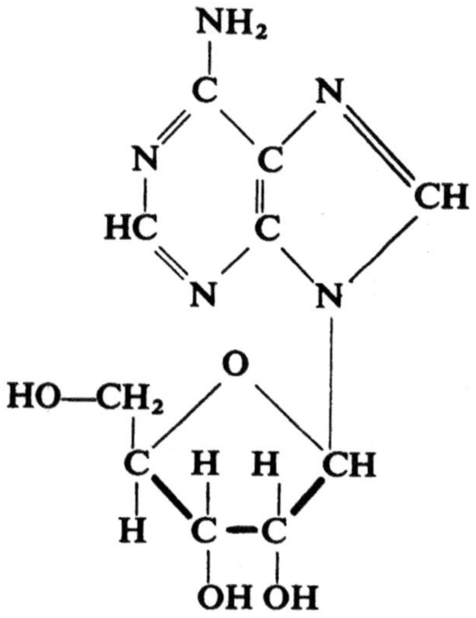

Developments in Pharmacology

Other books in the series:

J.M. Boeynaems and A.G. Herman, *Prostaglandins, Proslacyclin, and Thromboxanes Measurement*, 1980. ISBN 90 247 24171

REGULATORY FUNCTION OF ADENOSINE

PROCEEDINGS OF THE INTERNATIONAL SYMPOSIUM ON ADENOSINE, CHARLOTTESVILLE, VIRGINIA, JUNE 7–11, 1982

edited by
ROBERT M. BERNE
Department of Physiology
University of Virginia
School of Medicine
Charlottesville, Virginia, USA

THEODORE W. RALL
Department of Pharmacology
University of Virginia
School of Medicine
Charlottesville, Virginia, USA

RAFAEL RUBIO
Department of Physiology
University of Virginia
School of Medicine
Charlottesville, Virginia, USA

1983
MARTINUS NIJHOFF PUBLISHERS
BOSTON/THE HAGUE/DORDRECHT/LANCASTER

Distributors:

for the United States and Canada

Kluwer Boston, Inc.
190 Old Derby Street
Hingham, MA 02043

for all other Countries

Kluwer Academic Publishers Group
Distribution Center
P.O. Box 322
3300 AH Dordrecht
The Netherlands

Library of Congress Cataloging in Publication Data

International Symposium on Adenosine (1982:
Charlottesville, Va.)
Regulatory function of adenosine.

(Developments in pharmacology; v. 2)
Includes bibliographical references.
1. Adenosine—Congresses. 2. Metabolic regulation—Congresses. I. Berne, Robert M. II. Rall, Theodore W. III. Rubio, Rafael, 1928– . IV. Title.
QP625.A35157 1982 615'.7 82-18969
ISBN 90-247-2779-0

CONTENTS

LIST OF ABSTRACTS

CONTRIBUTING AUTHORS

Luiz Belardinelli
Division of Cardiology and
Department of Physiology
University of Virginia School of Medicine
Charlottesville, Virginia, USA

Robert M. Berne
Department of Physiology
University of Virginia School of Medicine
Charlottesville, Virginia, USA

Ruth E. Birch
Departments of Pediatrics, Pathology
and Pharmacology
Case Western Reserve University
School of Medicine
Cleveland, Ohio, USA

G. Burnstock
Department of Anatomy and Embryology
Centre for Neuroscience
University College
London, England

Carol E. Cass
Cancer Research Institute (McEachern Laboratory)
and Department of Chemistry (M.J.R.)
University of Alberta
Edmonton, Alberta
Canada

Dermot M.F. Cooper
Department of Pharmacology
University of Colorado
Health Sciences Center
Denver, Colorado, USA

Richard Crampton
Division of Cardiology and
Department of Physiology
University of Virginia School of Medicine
Charlottesville, Virginia, USA

John W. Daly
Laboratory of Bioorganic Chemistry
National Institute of Arthritis, Diabetes,
and Digestive and Kidney Diseases
The National Institutes of Health
Bethesda, Maryland, USA

James G. Dobson, Jr.
Department of Physiology
University of Massachusetts Medical School
Worcester, Massachusetts, USA

Gail S. Duncan
Department of Experimental Therapy
The Wellcome Research Laboratories
Burroughs Wellcome Company
Research Triangle Park, North Carolina, USA

Stephen W. Ely
Department of Physiology
University of Virginia School of Medicine
Charlottesville, Virginia, USA

Richard A. Fenton
Department of Physiology
University of Massachusetts Medical School
Worcester, Massachusetts, USA

Irving H. Fox
Human Purine Research Center
Department of Internal Medicine
and Biological Chemistry
University of Michigan
Ann Arbor, Michigan, USA

Bertil B. Fredholm
Department of Pharmacology
Karolinska Institutet
Stockholm, Sweden

Nai-Wu Fu
Cancer Research Unit (McEachern Laboratory)
University of Alberta
Edmonton, Alberta
Canada

E. Gerlach
Department of Physiology
University of Munich
Munich, Federal Republic of Germany

John L. Gordon
A.R.C. Institute of Animal Physiology
Babraham, Cambridge
England

Lars E. Gustafson
Department of Physiology
Karolinska Institutet
Stockholm, Sweden

Eric R. Harley
Cancer Research Institute
University of Alberta
Edmonton, Alberta
Canada

R.J. Haslam
Department of Pathology
McMaster University
Hamilton, Ontario
Canada

Per Hedqvist
Department of Physiology
Karolinska Institutet
Stockholm, Sweden

Paul G. Hellewell
A.R.C. Institute of Animal Physiology
Babraham, Cambridge
England

J. Frank Henderson
Cancer Research Unit and
Department of Biochemistry
University of Alberta
Edmonton, Alberta,
Canada

Barbara K. Henon
Division of Neurosciences
City of Hope Research Institute
Duarte, California, USA

Michael S. Hershfield
Departments of Medicine and Biochemistry
Division of Rheumatic and Genetic Diseases
Duke University Medical Center
Durham, North Carolina, USA

Ewa S. Jakobs
Cancer Research Unit (McEachern Laboratory)
University of Alberta
Edmonton, Alberta
Canada

Georg Kreutzberg
Max Planck Institute for Psychiatry
Munich, Federal Republic of Germany

Maureen T. Knabb
Department of Physiology
University of Virginia School of Medicine
Charlottesville, Virginia, USA

Robert M. Knabb
Department of Physiology
University of Virginia School of Medicine
Charlottesville, Virginia, USA

Kevin Lee
Max Planck Institute for Psychiatry
Munich, Federal Republic of Germany

B.W.G. Lemmex
Department of Pathology
McMaster University
Hamilton, Ontario
Canada

Constantine Londos
National Institute of Arthritis, Diabetes,
and Digestive and Kidney Diseases
The National Institutes of Health
Bethesda, Maryland, USA

John M. Lowenstein
Graduate Department of Biochemistry
Brandeis University
Waltham, Massachusetts, USA

D.E. Macfarlane
Department of Medicine
University of Iowa Hospital
Iowa City, Iowa, USA

Raya Mandler
Departments of Pediatrics, Pathology and Pharmacology
Case Western Reserve University
School of Medicine
Cleveland, Ohio, USA

Donald A. McAfee
Division of Neurosciences
City of Hope Research Institute
Duarte, California, USA

D.C.B. Mills
Thrombosis Research Center
Temple University Health Sciences Center
Philadelphia, Pennsylvania, USA

Yoshitsugo Naito
Graduate Department of Biochemistry
Brandeis University
Waltham, Massachusetts, USA

S. Nees
Department of Physiology
University of Munich
Munich, Federal Republic of Germany

Ray A. Olsson
Department of Internal Medicine
University of South Florida
College of Medicine
Tampa, Florida, USA

H. Osswald
Abteilung Pharmakologie
Medizinische Fakultät der RWTH
Aachen, West Germany

Alan R.P. Paterson
Cancer Research Unit (McEachern Laboratory)
University of Alberta
Edmonton, Alberta
Canada

Jeremy D. Pearson
A.R.C. Institute of Animal Physiology
Babraham, Cambridge
England

John W. Phillis
Department of Physiology
School of Medicine
Wayne State University
Detroit, Michigan, USA

Peter G.W. Plagemann
Department of Microbiology
Medical School
University of Minnesota
Minneapolis, Minnesota, USA

Stephen H. Polmar
Departments of Pediatrics, Pathology, and Pharmacology
Case Western Reserve University
School of Medicine
Cleveland, Ohio, USA

Theodore W. Rall
Department of Pharmacology
University of Virginia
School of Medicine
Charlottesville, Virginia, USA

Martin Reddington
Max Planck Institute for Psychiatry
Munich, Federal Republic of Germany

Morris J. Robins
Department of Chemistry
University of Alberta
Edmonton, Alberta
Canada

Martin Rodbell
The National Institutes of Health
Bethesda, Maryland, USA

Rafael Rubio
Department of Physiology
University of Virginia
School of Medicine
Charlottesville, Virginia, USA

Stephen A. Rudolph
Departments of Pediatrics, Pharmacology and Pathology
Case Western Reserve University
School of Medicine
Cleveland, Ohio, USA

J. Schrader
Department of Physiology
University of Munich
Munich, Federal Republic of Germany

Peter Schubert
Max Planck Institute for Psychiatry
Munich, Federal Republic of Germany

Ulrich Schwabe
Institut für Pharmakologie und Toxikologic der Universität Bonn
Bonn, Federal Republic of Germany

Alf Sollevi
Department of Pharmacology
Karolinska Institutet
Stockholm, Sweden

Trevor W. Stone
Department of Physiology
St. Georges' Hospital Medical School
University of London
London, England

Tokuma Tsukada
Department of Physiology
University of Virginia School of Medicine
Charlottesville, Virginia, USA

Per Magne Ueland
Department of Pharmacology
School of Medicine
University of Bergen
Bergen, Norway

Alexander West
Department of Physiology
University of Virginia School of Medicine
Charlottesville, Virginia, USA

H. Richard Winn
Departments of Physiology and Neurosurgery
University of Virginia School of Medicine
Charlottesville, Virginia, USA

Gerald Wohlberg
Department of Experimental Therapy
The Wellcome Research Laboratories
Burroughs Wellcome Company
Research Triangle Park, North Carolina, USA

Robert M. Wohlhueter
Department of Microbiology
Medical School
University of Minnesota
Minneapolis, Minnesota, USA

J. Wolff
National Institute of Arthritis, Diabetes, and Digestive and Kidney Diseases
The National Institutes of Health
Bethesda, Maryland, USA

Peter H. Wu
Department of Physiology
School of Medicine
Wayne State University
Detroit, Michigan, USA

Ming-kun Yu
Graduate Department of Biochemistry
Brandeis University
Waltham, Massachusetts, USA

Thomas P. Zimmerman
Department of Experimental Therapy
The Wellcome Research Laboratories
Burroughs Wellcome Co.
Research Triangle Park, North Carolina, USA

PREFACE

The principal objective of this symposium was to bring together investigators from all over the world who work in diverse areas of research but share a common interest in the physiologic and pathophysiologic functions of the nucleoside adenosine. In 1978, the "International Conference on Physiological and Regulatory Functions of Adenosine and Adenine Nucleotides," held in Banff, Canada, served as an effective catalyst in enabling scientists with different points of view to exchange information and ideas on this subject. Since 1978, the amount of research on different aspects of adenosine metabolism, mechanism of action, transport, and function in various tissues (in both normal and abnormal states) has increased exponentially, and there was an urgent need for another meeting for updating information on this burgeoning subject. The focus of the meeting was extremely sharp, but the scope of the subject material was very broad. Scientists categorized as chemists, biochemists, physiologists, pharmacologists, and immunologists, working with cells and cell fragments, with tissues such as heart, brain, muscle, fat, kidney, and with human beings having immune deficiency diseases participated in the symposium. They had an opportunity to air their differences, reinforce their areas of agreement, and, most important, they educated each other with respect to research on adenosine; it is hoped that they also enriched their own research programs from the information they gained from their fellow participants. The total number of full presentations was relatively few, so that there was considerable time available for comprehensive and provocative discussion, both invited and spontaneous.

The symposium began with consideration of mechanisms and cellular processes underlying the actions of adenosine in various systems and progressed to consideration of specific issues relating to the function of adenosine in the physiologic and pathophysiologic regulation of major organ systems. From the broad spectrum of the program, it is evident that the subject of the regulatory function of adenosine was extensively covered and updated, particularly during the discussions and in the posters, whose abstracts are included in this volume. The participants' response to the symposium was most favorable, particularly since it afforded an opportunity for scientists from distant laboratories to get together and exchange information in a congenial and informal atmosphere. Since the field will probably continue to expand at an increasing rate, it is anticipated that plans for a subsequent symposium dealing with adenosine will be undertaken in the near future.

Organizing Committee
Robert M. Berne
Eckehart Gerlach
Theodore W. Rall
Rafael Rubio
Jürgen Schrader

Editors
Robert M. Berne
Theodore W. Rall
Rafael Rubio

ACKNOWLEDGMENTS

The major support for this symposium came from the Pratt Foundation at the University of Virginia and from a grant jointly funded by the NIH (NHLBI) and the Fogarty International Center. Additional support was generously provided by the following organizations:

The American Heart Association
American Critical Care
American Cyanamid Company, Lederle Laboratories
A.H. Robbins Company
Burroughs-Wellcome Company
Hoechst Roussel Pharmaceuticals
Janssen Pharmaceutical
Karl Thomae Company
McNeil Pharmaceutical
Merck and Company, Incorporated
Pfizer Incorporated
Riker Laboratories
Schering Plough Corporation
The Upjohn Company
Warner-Lambert Company

I. ADENOSINE RECEPTORS

1. THE COMPLEX STRUCTURE AND REGULATION OF ADENYLATE CYCLASE

MARTIN RODBELL

For over twenty years the adenylate cyclase system has been investigated as a model for hormone action. Situated in the plasma membrane, the enzyme is regulated by a large number of hormones, neurotransmitters, and such "local" hormones as prostaglandins and purinergic compounds. Initially considered to be a two-component system consisting of the catalytic unit (C) and the recognition components (R) for the various hormones and hormone-like substances, a large body of evidence has accumulated that these systems are multisubunit systems composed minimally of R, C, and nucleotide regulatory components (N) that are responsible for the regulation of adenylate cyclase activity by guanine nucleotides [1, 2]. The enzyme is also regulated in opposing manner by sets of R and N units that either stimulate or inhibit the enzyme. A functional stimulatory receptor unit has been designated as RN_s, whereas the opposing inhibitory receptor unit has been designated as the RN_i unit or complex [1]. Systems displaying dual regulation contain, by definition, both types of receptor complexes. In addition, different R units are often linked structurally to both N_s and N_i in the same cell membrane.

To accommodate such regulatory and structural complexity, the hypothesis has been put forth that RN_s and RN_i complexes are present in membranes in the form of oligomeric structures in which mixtures of R, N_s, and N_i units are aggregated [1]. It was further postulated that the concerted actions of hormones and GTP on these oligomeric structures disaggregated the oligomers to

R.M. Berne, T.W. Rall, and R. Rubio (eds.), Regulatory Function of Adenosine, *pp. 3–15.*

"monomeric" structures that were capable of forming complexes with the catalytic unit.

Here I will expand and modify this theory somewhat by including recent findings on the structures of the components and on the role of ions in the regulation of adenylate cyclase activity.

THE ROLE OF CATIONS

Early studies of adenylate cyclase systems had shown that Mg^{2+} ions stimulate activity to levels approaching that observed with hormones; hormones reduced the concentrations of Mg^{2+} ions required to achieve maximal activation [3]. These findings suggested that part of the transduction process involved in hormonal activation consisted of sites for Mg^{2+} ions. Although this idea was disputed, it is now clear that regulatory sites for Mg^{2+} ions exist in a variety of cyclase systems [4, 5]. Mn^{2+} ions have generally higher affinity than Mg^{2+} ions for these sites. The idea that the divalent cation sites are linked functionally to R and N units stems in part from findings that the actions of hormones and guanine nucleotides require the divalent cations for expression. An interesting aspect of glucagon action on the hepatic adenylate cyclase system is that it promotes inhibition of the enzyme by adenosine acting at the so-called P site, which is thought to be proximal to the catalytic component; promotion is mediated by Mn^{2+} or Mg^{2+} ions with affinities resembling those involved in the activation of adenylate cyclase by glucagon and GTP [6]. Based on findings that Mn^{2+} and Mg^{2+} ions affect the binding of agonists to R units, it has also been suggested that the divalent cation sites are linked to the R units [7]. The effects are generally an enhanced affinity of R for agonists, which is opposite to the well-known effects of GTP on agonist binding to R [8]. It is unclear whether this effect of divalent cations is due to a direct action on R or to the removal of bound guanine nucleotides from N (which would result in enhanced affinity of R for agonists). Nonetheless, these findings reinforce the idea that divalent cations play an important role in the transduction process.

The components through which Mg and Mn act to stimulate adenylate cyclase remain unknown. However, the recent findings [8, 9] that the N_s component purified from rabbit liver and turkey erythrocyte is a heterodimer raise the intriguing possibility that one of the components is the site of divalent cation binding. N_s contains a subunit ranging in size from 45,000 to 55,000 daltons; these are related structurally and functionally and appear to be the major components through which hormones (through R), guanine nucleotides, cholera toxin, and fluoride ion exert their stimulatory effects [2]. The other subunit is a 35,000 dalton component that appears to be equally essential as a transduction element but is not ADP-ribosylated by the catalytic component. As illustrated schematically in Figure 1.1, the 35,000 dalton component may contain both the sites for divalent cations as well as the P site. Forskolin, a steroid-like compound that stimulates the activity of all animal cell cyclase

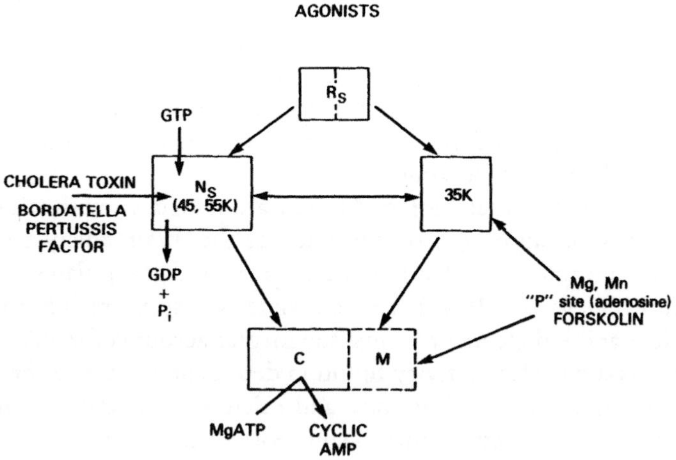

Figure 1.1. Schematic representation of the known components of systems that are stimulated by hormones and GTP. The enclosed areas are those for the hormone recognition proteins (R); the GTP-binding component (N_s) that hydrolyzes GTP to GDP and reacts with cholera toxin and a factor from *Bordatella pertussis* [30]; the 35,000 dalton component that copurifies with N_s and may be the site for divalent cation regulation, for adenosine inhibition (P site), and for the stimulatory action of forskolin. It is also possible that the latter sites are situated on a component (M) more proximal to the catalytic unit (C). For more details, see text.

systems, also acts on a component distinct from the 45,000 or 55,000 dalton unit through which GTP acts [10]. This is based on the finding that forskolin stimulates cyclase activity in membranes from a variant of S49 murine lymphoma cells that lack a functional N_s unit. Similarly, it has been reported that adenosine analogs that act through the P site also inhibit the activity of cyclase in this variant [11]. However, in light of recent information [12] that this variant contains the 35,000 dalton unit, the question remains open whether the P site, forskolin site, and sites for Mg^{2+} and Mn^{2+} ions are on the 35,000 dalton unit or possibly on another component more closely associated with the catalytic unit (indicated in Fig. 1.1 by the dashed-line region on C).

In addition to regulatory sites for Mg^{2+} and Mn^{2+} ions, it would appear that Ca^{2+} ions play a stimulatory role with certain cyclase systems. A role for Ca^{2+} ions in the stimulatory actions of ACTH was first suggested from studies of the rat adipocyte system [13]. ACTH action was abolished in the presence of EGTA, a chelator that binds Mg^{2+} ions much less effectively than Ca^{2+} ions. Action by ACTH was restored by the addition of as little as $10^{-8}M$ free calcium ions. By contrast, the stimulatory actions of glucagon, epinephrine, and several other stimulatory hormones were unaffected by the chelator or the addition of Ca^{2+} ions. Similar effects of the chelator and Ca^{2+} ions have been reported with cyclase systems stimulated by oxytocin and vasopressin [14, 15]. In view of the selective effects of Ca^{2+} ions and the different structures of the hormones involved in Ca^{2+} ion regulation, it seems unlikely that the R unit is the site of calcium ion regulation. ACTH, in addition to its involvement with Ca^{2+} ions, promotes activation by Mg^{2+} ions. Therefore, it is possible that certain hormones promote activation through sites or components for GTP, Mg^{2+} or Mn^{2+} ions, and for Ca^{2+} ions. Perhaps two or more metal ion components (M) operate in tandem.

The idea that distinct metal ion or M units are associated with stimulation of adenylate cyclase is supported by the findings that brain adenylate cyclase is activated by calmodulin, a calcium-binding protein that regulates a number of enzymes (for review, see [16]). Brain adenylate cyclase exists in both calmodulin-dependent and independent forms; the former accounts for 20% to 50% of the total adenylate cyclase activity found in detergent extracts of brain [17]. In such extracts, addition of calmodulin and calcium ions activates the enzyme independently of the effects of guanine nucleotides exerted through the N_s unit [18]. It has not been established that hormones or neurotransmitters regulate adenylate cyclase through the calmodulin-stimulated process. By the same token, it is not evident that calmodulin is involved in the stimulatory actions of ACTH, vasopressin, and oxytocin. Clearly, these are problems worthy of resolution. Nonetheless, there is sufficient evidence to warrant the suggestion that M components exist independently of the N_s components and that divalent cations regulate adenylate cyclase through at least two distinct types of M units, one for Mg or Mn, the other for Ca^{2+} ions.

A role for monovalent cations in the regulation of adenylate cyclase activity was first suggested from findings that Na ion affected both the binding of opiate agonists to receptors and the ability of opiates to inhibit cyclase activity in neuroblastoma cells (reviewed in [19]). GTP, through N_i, inhibits cyclase by a process that is independent of Na^+ ion but like sodium it affects the binding of agonists to opiate receptors as well as other receptors known to affect inhibition of cyclase systems [1]. Arguments that N_i and sodium regulation are independent processes stem in part from findings that GTP and Na^+ ion act synergistically but not dependently. Recent findings [20, 21] point to different structural components for Na regulation. RN_i complexes have been extracted with detergents in forms that retain the ability of agonist binding to be sensitive to GTP but not to Na^+ ion, suggesting that the Na component was released during detergent solubilization of RN_i.

The available information suggests that regulation of cyclase activity by monovalent cations (Li^+ is also effective) is restricted to systems that are inhibitory. By analogy to the stimulatory systems, perhaps the systems containing RN_i units also contain an M unit that uniquely reacts with Na^+ or Li^+ ions, as is schematically represented in Figure 1.2. Mn^{2+} and Mg^{2+} ions may also react with this M component since, in systems displaying inhibition by GTP, the divalent cations (Mn > Mg) block inhibition. However, it is also possible that the blocking effects are related to actions at the M sites through

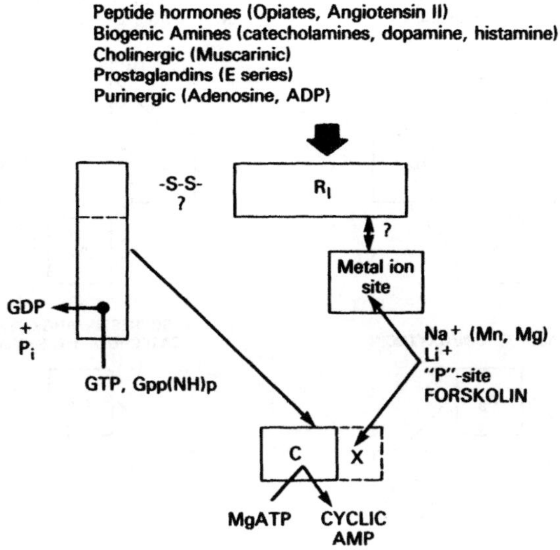

Figure 1.2. Schematic representation of components in hormone-inhibitable forms of adenylate cyclase systems. Enclosed areas are R_i for hormones and neurotransmitters, and a GTP-regulatory unit that inhibits the enzyme (N_i) and hydrolyzes GTP but not Gpp(NH)p. The putative metal ion unit (M) that reacts with Na^+ or Li^+ ions is possibly linked to R_i and/or to a component (X) associated with the catalytic unit (C).

which the cations stimulate adenylate cyclase (the N_s system depicted in Fig. 1.1).

HETEROGENEITY OF NUCLEOTIDE REGULATORY COMPONENTS

In addition to sites for monovalent cations, there are major differences in the structures of N_i and N_s systems. An important distinction is that the GTP-inhibitory process is abolished when cells containing both N_s and N_i components are treated with proteases; N_s survives whereas the functions of N_i are abolished [22]. Another difference is that the binding of opiates and adenosine to inhibitory R units and the effects of GTP on agonist binding are preserved following detergent extraction of receptors. These findings infer that N_i remains associated with the R units during solubilization and form stable RN_i complexes. This is not the case for the glucagon and beta adrenergic RN_s systems. Stabilization of RN_s complexes requires binding of agonists to R prior to solubilization with detergents [1, 7]. Based on these differences, it would appear from the few examples in hand that inhibitory R units form stable, possibly covalent attachments with N_i (perhaps through disulfide bridges). The attachments may be at the outer surface of the membrane through components susceptible to trypsin-degradation (hence the dashed area on N_i). This would imply that N_i units span the plasma membrane, whereas it

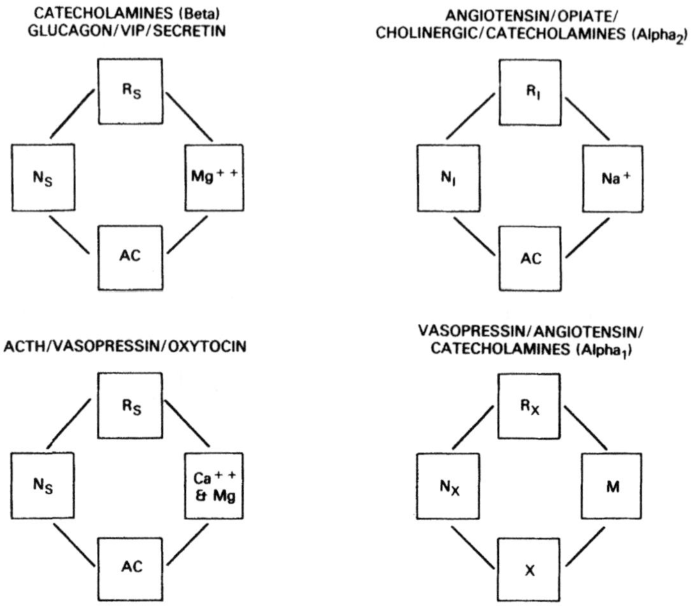

Figure 1.3. Schematic representation of different regulatory forms of adenylate cyclase systems. Each type of system contains hormone recognition units (R), a GTP-regulatory protein (N), a metal ion component for monovalent or divalent cations (M), and the catalytic unit. A hypothetical signal-generating system responding to various hormones that do not regulate adenylate cyclase activity directly is also shown.

is thought that the N_s unit is associated only with the inner aspect of the membrane and is not an intrinsic membrane protein.

Structural differences between RN_s and RN_i systems have also been indicated from differences in sizes, as revealed by target analysis [23]. In addition, N_i is considerably more sensitive to sulfhydryl alkylating agents than is N_s [24].

Collectively, the information gained thus far on the role of cations argues for selective types of metal ion components (M) associated differentially with adenylate cyclase systems depending on the types of R and N units comprising the receptors. Three such "families" of cyclase systems are depicted schematically in Figure 1.3. Also included is a fourth category of signal generating systems that may contain R, N, and M units that regulate enzymes or processes other than adenylate cyclase. Hints that such systems may exist stem from findings that guanine nucleotide affects the binding of hormones to receptors thought not to be involved in cyclic AMP production [1].

STRUCTURE OF MULTIRECEPTOR SYSTEMS

It is rare to find a cell that contains only one type of receptor for the regulation of adenylate cyclase. Most cells contain several types of R units associated with both RN_s and RN_i units in the same membrane, the classical example being the rat adipocyte. In this system, addition of maximally stimulating concentrations of the various stimulatory hormones does not lead to additive responses, indicating that each receptor type has access to a common C unit. The M units must vary since ACTH and vasopressin action involve Ca^{2+} ions, whereas the actions of glucagon, VIP, secretin, and catecholamines do not. Moreover, the inhibitory agents (adenosine, prostaglandins, and nicotinic acid) that act through RN_i require Na^+ ion and thus separate type of M unit. The latter agents also do not act in additive fashion. Moreover, these agents whether combined or not fail to inhibit adenylate cyclase to an extent greater than about 50% [19], a phenomenon seen with all systems displaying both stimulatory and inhibitory circuits. One explanation is that adenylate cyclase is a family of enzymes, some linked to inhibitory RN_i units, others not. If this is correct, then the various populations of the enzyme could be in discrete, nonoverlapping domains that have quite different regulatory features. Data taken from studies of the actions of Gpp(NH)p on the adipocyte system [29] are consistent with discrete populations of cyclase systems that are differentially susceptible to inhibition and stimulation by guanine nucleotides.

Shown in Figure 1.4 is the time course of basal activity of the adipocyte enzyme in the absence and presence of 0.1 mM Gpp(NH)p. Note that the initial basal rate is high and then declines to a much lower rate after 1 min incubation. Gpp(NH)p causes an immediate, profound (greater than 80%) inhibition of the initial basal rate. Inhibition is followed by progressive increase in rate until finally the rate is essentially equivalent to the high rate given by the basal state. These findings are consistent with the existence of two noninterconvertible populations of the adipocyte enzyme, one containing an N_i unit through which Gpp(NH)p inhibits, the other containing N_s through which the nucleotide activates. The recent report that forksolin acts in additive fashion with Gpp(NH)p on the adipocyte system [10] is also consistent with two or more populations of the enzyme coexisting in the same cell membrane.

Given the large and varied number of components involved in the regulation of adenylate cyclase systems, the organization of these components in the membrane seems a formidable problem. It has been suggested that the components are initially separate in the membrane and that hormones and guanine nucleotides, in stepwise fashion, induce physical coupling of the units [7]. An alternative [1] is an integrated, multicomponent structure in which the various effector ligands induce structural rearrangements of the organized components. In the light of what is known of many complex regulatory systems in biology, including that of the recently purified acetylcholine receptor [25], an organized structure would seem the most efficacious for prompt and highly coordinate regulation. In any event, without prior knowledge of structure, it is not possible to differentiate clearly between uncoupled and precoupled forms of the enzyme, as some have argued.

When applied judiciously [28], analysis of functional size by high-energy

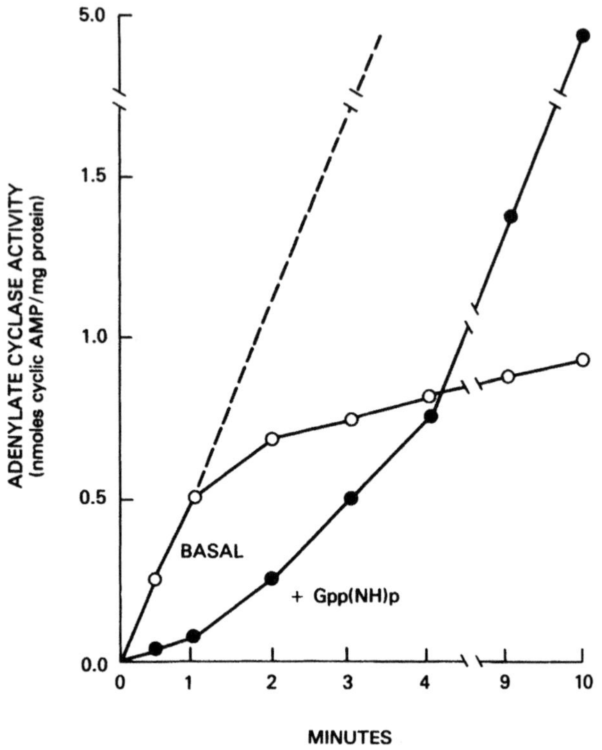

Figure 1.4. Time course of cyclic AMP production by adenylate cyclase in rat adipocyte membranes in the absence (○) and presence (●) of Gpp(NH)p. Membranes (5 μg protein) were incubated at 30° in 0.35 ml medium consisting of 25 mM Tris-HCl, pH 7.4; 5 mM $MgCl_2$/1 mM dithiothreitol; an ATP-regenerating system; 0.5 mM ATP/0.1% bovine serum albumin; and, when present, 0.1 mM Gpp(NH)p. Data are from ref. 29.

irradiation (target analysis) is the only procedure currently available for evaluating some aspect of membrane-bound structures. This procedure has been applied to the ground-state and activated states of the enzyme systems in rat hepatocytes and adipocytes [23]. The ground-state or precursor sizes in both cell membranes approximated 1.2×10^6 daltons. Analysis of the size of the hormone-activated states yielded about 350,000 daltons, or about one-quarter that of the ground-state (Table 1.1). Assuming that both the precursor and final activated structures contain R, N_s, and C, then the precursor is an oligomer consisting of four units of the complex formed between R, N, and C. Such assumptions are, of course, subject to debate. Nonetheless, at least this methodology supplies data for interpretation. Moreover, target analysis has given some insight, particularly when taken in conjunction with reported sizes of the subunits of R, N, and C, regarding the structures of the units that are required for activation of adenylate cyclase in its membrane environment.

Table 1.1 includes recent data on the subunit sizes reported for the glucagon

Table 1.1. Subunit and functional sizes of adenylate cyclase components in rat hepatocyte

Component	Subunit Analysis M_r (daltons)		Reference
Catalytic (C)	50,000		[27]
GTP-regulatory (N)	35,000, 45,000, and 55,000*		[9]
Glucagon-receptor (R)	58,000		[26]

Function assayed	Functional Sizes (target analysis)** Target size (daltons)	Possible target	Target based on subunit sizes
MnATP as substrate	~100,000	C	C_2
F^- or Gpp(NH)p-activated	~250,000	NC	$[NC]_2$
Glucagon + GTP-activated	~350,000	RNC	$[RNC]_2$
Ground-state precursor	~1,300,000	$[RNC]_n$ or $[RN]_n$	$([RNC]_2)_4$ or $[RN]_{10}$

*45,000 and 55,000 dalton units are structurally and functionally homologous.
**Data taken from ref. 29.

receptor [26], the purified N_s unit [9], and the estimated size of C when liberated in soluble form from membranes by proteases [Stengel et al., unpublished observations]. Note that the target sizes of the units are approximately twice that of the subunits for R, N, and C. Again assuming a complex between these units, such findings suggest that the R unit for glucagon is linked to C via a dimer of N_s which, as discussed previously, is a heterodimer. From the standpoint of the known regulatory properties of the glucagon-sensitive system, this is an interesting structure. For example, guanine nucleotides induce different states of the glucagon receptor from that of the resting state and, under the same conditions, promote activation of C in the presence of the hormone. Analysis of the actions of guanine nucleotides on hormone binding and activation of C gave rise to the conclusion that the sites of guanine nucleotide action on these processes were different, and were designated N_1 and N_2 [27]. If, indeed, a dimer of N_s links R with C, a relatively straightforward explanation for N_1 and N_2 can be surmised—namely, that one unit of the dimer interacts with R whereas the other interacts with C. In effect, each unit of the assembled N_s dimer can be considered structurally and functionally as distinct N units with differing interacting forces for R and C. With the possibility of an asymmetrical arrangement of the 45,000 and 35,000 dalton subunits in the dimeric structure and of differing articulation with the dimers of R and C, the holoenzyme could take a number of configurations. These could satisfy differing states of R for agonist binding, different activity states of N for the binding and actions of GTP and divalent cations, and different states of the catalytic unit that converts MgATP to cyclic AMP. Finally, even though the various subunits comprising the holoenzyme are structurally "leashed" through intersubunit forces, the system could display kinetics as though they functioned as separate units and then coupled, in stepwise fashion, in response to the effector ligands.

CONCLUSIONS

It is evident from this brief discussion of adenylate cyclase systems that regulation of cyclic AMP production by hormones, neurotransmitters, and local hormones can take different pathways, some convergent, others divergent. From the standpoint of structural features, notable achievements have been made in purifying N_s, in identifying and isolating R units, and in gaining some information on structure with the target size technique. The nature of N_i and its attachments to R need to be further clarified. There are a number of functional similarities in the behavior of N_s and N_i to warrant the suggestion that they are likely to be structural homologues. Since so many hormones and neurotransmitters act in common through N_s or N_i units, one might also argue that the R units for each hormone are structural homologues, the variant regions of each being the sites of binding of the hormones or neurotransmitters. However, since the R units appear to interact in rather select manner with N_s, N_i, and the putative M units for cations, one must also consider that R units diverge in structure and function not only with respect to the specific hormone-binding regions but also with respect to the regions for interaction with the N and M components. Target analysis suggests that R units function as dimers. These could consist both of homologous and heterologous subunits, each type designed for selective interactions with the other subunits that comprise what I term the functional receptor (the RN complexes). The major point to be emphasized is that receptor specificity, long held to be primarily the function of the hormone-binding sites, is actually the combined result of all of the subunits comprising the receptor, including R, N, M and whatever other subunits are eventually discovered to be part of the regulatory process. It is already evident from studies with the purified acetylcholine receptor that much greater insights are to be gained by understanding the structure of all components, not simply those responsible for binding hormones.

REFERENCES

1. Rodbell M: The role of hormone receptors and GTP-regulatory proteins in membrane transduction. *Nature* 284:17–22, 1980.
2. Ross EM, Gilman AG: Biochemical properties of hormone-sensitive adenylate cyclase. *Annu Rev Biochem* 49:533–564, 1980.
3. Birnbaumer L, Pohl SL, Rodbell M: Adenyl cyclase in fat cells: I. Properties and the effects of adrenocorticotropin and fluoride. *J Biol Chem* 244:3468–3476, 1969.
4. Londos C, Rodbell M: Multiple inhibitory and activating effects of nucleotides and magnesium on adrenal adenylate cyclase. *J Biol Chem* 250:3459–3465, 1975.
5. Londos C, Preston MS: Activation of the hepatic adenylate cyclase system by divalent cations: A reassessment. *J Biol Chem* 252:5957–5961, 1977.

6. Londos C, Preston MS: Regulation by glucagon and divalent cations of inhibition of hepatic adenylate cyclase by adenosine. *J Biol Chem* 252:5951–5956, 1977.
7. Stadel JM, DeLean A, Lefkowitz RJ: Molecular mechanisms of coupling in hormone-receptor-adenylate cyclase systems. *Adv Enzymol* 53:1–43, 1982.
8. Hanski E, Sternweis PC, Northrup JK, Dromerick AW, Gilman AG: The regulatory component of adenylate cyclase: Purification and properties of the turkey erythrocyte protein. *J Biol Chem* 256:12911-12919, 1981.
9. Sternweis PC, Northrup JK, Smigel MD, Gilman AG: The regulatory component of adenylate cyclase: Purification and properties. *J Biol Chem* 256:11517-11526, 1981.
10. Seamon KB, Daly JW: Activation of adenylate cyclase by the diterpene forskolin does not require the guanine nucleotide regulatory protein. *J Biol Chem* 256:9799–9801, 1981.
11. Florio VA, Ross EM: Direct inhibition of the catalytic protein of adenylate cyclase at the adenosin P site. *Fed Proc* (abstr), p. 1407, 1982.
12. Ferguson KM, Northrup JK, Gilman AG: Goat antibodies to the regulatory component of adenylate cyclase. *Fed Proc* (abstr), p. 1407, 1982.
13. Birnbaumer L, Rodbell M: Adenyl cyclase in fat cells: II. Hormone receptors. *J Biol Chem* 244:3477–3482, 1969.
14. Bar HP, Hechter O: Adenyl cyclase and hormone action: III. Calcium requirement for ACTH stimulation of adenylate cyclase. *Biochem Biophys Res Commun* 35:681–686, 1969.
15. Campbell BJ, Woodward G, Borber V: Calcium-mediated interactions between the antidiuretic hormone and renal plasma membranes. *J Biol Chem* 247:6167–6173, 1972.
16. Bradham LS, Cheung WY: Calmodulin-dependent adenylate cyclase, in Cheung WY (ed): *Calcium and Cell Function*. New York, Academic Press, 1980, pp 109–126.
17. Westcott KR, LaPorte DC, Storm DR: Resolution of adenylate cyclase sensitive and insensitive to Ca^{2+} and calcium-dependent regulatory protein (CDR) by CDR-sepharose affinity chromatography. *Proc Natl Acad Sci USA* 76:204–208, 1979.
18. Heideman W, Wierman BM, Storm DR: GTP is not required for calmodulin stimulation of bovine brain adenylate cyclase. *Proc Natl Acad Sci USA* 79:1462–1465, 1982.
19. Jakobs KH, Aktories K, Schultz G: Inhibition of adenylate cyclase by hormones and neurotransmitters. *Adv Cyclic Nucleotide Res* 14:173–187, 1981.
20. Simonds, WF, Koski G, Streaty RA, Hjelmeland LM, Klee WA: Solubilization of active opiate receptors. *Proc Natl Acad Sci USA* 77:6423–6427, 1980.
21. Gavish M, Goodman RR, Snyder SH: Solubilized adenosine receptors in the brain: Regulation by guanine nucleotides. *Science* 215:1633–1634, 1982.
22. Evain D, Anderson WB: Inhibitory effect of guanyl nucleotides toward adenylate cyclase of Chinese hamster ovary cell membranes activated in vitro by cholera toxin. *J Biol Chem* 254:8726–8729, 1979.
23. Rodbell M, Lad PM, Nielsen TB, Cooper DMF, Schlegel W, Preston MS, Londos C, Kempner ES: The structure of adenylate cyclase systems. *Adv Cyclic Nucleotide Res* 14:3–14, 1981.
24. Jakobs KH, Lasch P, Minuth M, Aktories K, Schultz G: Uncoupling of alpha-adrenoceptor-mediated inhibition of human platelet adenylate cyclase by N-ethylmaleimide. *J Biol Chem* 257:2829–2833, 1982.
25. Wastek GJ, Yamamura HI: Acetylcholine receptors, in Yamamura HI, Enna SJ (eds), *Neurotransmitter Receptors,* part 2. London, Chapman and Hall, 1981, pp 103–128.
26. Johnson GL, MacAndrew VI, Pilch PF: Identification of the glucagon receptor in rat liver membranes by photoaffinity crosslinking. *Proc Natl Acad Sci USA* 78:875–878, 1981.
27. Lad PM, Welton AF, Rodbell M: Evidence for distinct guanine nucleotide sites in the regulation of the glucagon receptor and of adenylate cyclase activity. *J Biol Chem* 252:5942–5946, 1977.
28. Schlegel W, Kempner ES, Rodbell M: Activation of adenylate cyclase in hepatic membranes involves interactions of the catalytic unit with multimeric complexes of regulatory proteins. *J Biol Chem* 254:5168–5176, 1979.
29. Rodbell M: On the mechanism of activation of fat cell adenylate cyclase by guanine nucleotides: An explanation for biphasic inhibitory and stimulatory effects of the nucleotides and the role of hormones. *J Biol Chem* 250:5826–5834, 1975.
30. Katada T, Amano T, Ui M: Modulation by islet-activating protein of adenylate cyclase activity in C_6 Glioma cells. *J Biol Chem* 257:3739–3746, 1982.

DISCUSSION

DR. A.C. NEWBY: The idea of separate families of cyclase in the fat cell seems inconsistent with the lack of additivity of the effect of hormones. How can you reconcile this?

DR. RODBELL: The question of additivity should be reevaluated in the light of what is now understood about the complex structure of adenylate cyclase systems.

DR. T.W. RALL: I was struck by your proposal regarding the assembly of dimeric complexes of the various components of the cyclase system and by the suggestion that different receptors share homologous structural features. Could you comment on the possibility that "heterologous" receptors could dimerize or associate in order to account for interactions between heterologous agonists?

DR. RODBELL: This is a plausible solution to account for synergistic actions of hormones on cyclase systems. Moreover, by inserting different R units into functional dimers, one could also account for nonadditivity of response to combined hormones. Finally, heterologous R units incorporated into dimers or higher orders of oligomers offer an explanation for why a given hormone might display different efficacy and potency of action on cells depending solely on how many of its R units are homologous or heterologous dimers.

DR. R.A. OLSSON: As I understand your model, there is a particular N subunit for each type of hormone receptor. Is it possible that there is but one type of N subunit with one interaction site for hormones and that a common feature of the various receptors is an identical site for interaction with N?

DR. RODBELL: I suggest that N_s, for example, is the same for all R_s units linked to this component. However, each R_s and N_s unit may interact differently (however slightly) depending on the species of R_s, i.e., there may be a spectrum of complementary structures. This would require, of course, that N and R have been sites for interaction. When R and N are purified from a variety of sources, this possibility will be amenable for testing.

DR. J. LARNER: How do you visualize the role of phospholipids in controlling the activity of the adenylate cyclase enzyme system?

DR. RODBELL: I view them as playing predominantly a structural role—a type of "lubricating oil" to maintain the protein components in a functional state. This is not to suggest that lipids don't play a specific role in some functional aspect of the system.

DR. H.P. BAER: How do you feel about the role or future of "genetic analysis" to probe the existence of multiple components of the cyclase system or multiple forces of one component? (One of the early suggestions by the late Gordon Tomkins, of course, was that there is only one cyclase catalytic unit associated with different hormone receptors.)

DR. RODBELL: Clearly, "Genetic analysis" is an important tool, as evidenced by the intensive studies with the S49 murine lymphoma cell line developed and propagated by Tomkins and co-workers. However, I would stress at this stage that such analysis be complemented with quantitative analytical tools that assess the levels of the various components rather than relying solely on the measurement of activity. Reliance on the latter has tended to confuse some of the issues regarding what is missing or modified in the variants that have been tested.

DR. J. WOLFF: It should be pointed out that lipids or phospholipids interact in more complicated ways than as a structural support. It can be shown that phospholipids activate soluble (*B. pertussis*) adenylate cyclase at concentrations well below the critical micelle concentration. This virtually requires assuming that they act through hydrophobic binding to the catalytic protein.

DR. RODBELL: I don't disagree with your contention. We demonstrated a long time ago that hydrolysis of a small percentage of the acidic phospholipids can profoundly alter the ability of glucagon to bind and activate the cyclase system in liver. At the moment, however, I don't see any evidence for a specific regulatory role of lipids.

2. ADENOSINE RECEPTORS AND ADENYLATE CYCLASE INTERACTIONS

CONSTANTINE LONDOS, J. WOLFF, DERMOT M.F. COOPER

For many years it has been known that adenosine, at low concentrations, initiates numerous metabolic actions in a variety of cells. Although adenosine effects may result from altered purine metabolism [1, 2], some persist under conditions where the nucleoside is prevented from entering cells. Thus, a number of metabolic effects of adenosine persist or are enhanced in the presence of adenosine transport inhibitors, such as dipyridamole and nitrobenzylthioguanosine. These include inhibition of lipolysis in fat cells [3], inhibition of platelet aggregation [4], increased steroid production [5], potentiation of histamine release from mast cells [6], inhibition of vas deferens contraction [7], and coronary vasodilation [8]. Moreover, conjugates of adenosine with large molecules that are thought not to enter cells mimic the actions of free adenosine, the potency of adenosine in the conjugates being similar to that of the free nucleoside [3, 9–12]. Such findings gave birth to the notion that cells contain surface receptors for adenosine.

One action mediated by the external action of adenosine in a wide variety of cells is a change in cellular cyclic AMP concentrations. The early observations of Sattin and Rall [13] and of Shimizu and Daly [14] that adenosine increases cyclic AMP in brain slice preparations have now been extended to numerous types of cells [3, 15–18]. However, several cells from different sources contain adenosine receptors that mediate a *decrease* in cellular cyclic AMP concentra-

The authors gratefully appreciate the expert secretarial assistance of Bonnie Richards.

R.M. Berne, T.W. Rall, and R. Rubio (eds.), Regulatory Function of Adenosine, *pp. 17–32.*

tions [3, 15–18]. With human platelets, the adenosine analogues that increase cyclic AMP concentrations also inhibit platelet aggregation [19]. Analogue studies in adipocytes have revealed a correlation in potency between decreased cyclic AMP production and inhibition of lipolysis [3, 15–18, 20]. Although none of the foregoing studies provides a direct link between cellular cyclic AMP and the metabolic reaction observed, they all suggest that the cell surface sites through which nucleosides mediates cyclic nucleotide concentrations and other metabolic effects are similar.

A key to confirming the existence of adenosine receptors was the discovery that adenosine, at low concentrations, stimulates adenylate cyclase activity in isolated membranes. This finding, first reported with platelet membranes [21] and subsequently with membranes from numerous types of cells [3, 15–18], indicated that the adenylate cyclase system is a major target for adenosine action. Adenylate cyclase is a multicomponent enzyme system located on the plasma membrane, composed of a catalytic component on the inner cell surface, receptors for hormones on the outer cell surface, and a complex of recently identified proteins known as N or G/F components [22]. The nucleotide regulatory components (N, G/F) react with GTP and are necessary for transmission of the hormone signal at the receptor to the catalytic component. Numerous studies on the GTP regulation of receptor-mediated adenylate cyclase activity have provided a set of criteria for determining whether or not a ligand acts via a receptor [23]. In this chapter we review the evidence for adenylate cyclase-linked adenosine receptors.

Two functionally and pharmacologically distinct adenosine receptors have been identified [24, 25] as being associated with adenylate cyclase (Table 2.1). One receptor mediates stimulation of activity; the other mediates inhibition. In addition to the receptors, adenylate cyclases contain another site for adenosine action, termed P site (Table 2.1), which mediates inhibition of activity and is not a receptor [15–17, 26]. The existence of the P site presents a problem with the use of adenosine for examining receptors, since the nucleoside will act at both the receptor and the P site. Thus, with membranes that contain the stimulatory form of the receptor, adenosine will both stimulate activity by reacting with the receptor and inhibit activity via the P site. With membranes that contain the inhibitory receptor, adenosine will inhibit activity by reacting with the receptor or the P site. However, a number of analogues exist that are selective for the receptors, and most of the receptor studies discussed below have been performed with such analogues.

NOMENCLATURE FOR SITES OF ADENOSINE ACTION

Several nomenclatures have been proposed for the different sites of adenosine actions (Table 2.1). Burnstock [27] suggested P_1 for adenosine receptors and P_2 for purine nucleotide receptors. Van Calker et al. [24] proposed A_1 and A_2 for the adenosine receptors that mediate, respectively, inhibition and stimulation of adenylate cyclase. Based on their different chemical specificities, we pro-

Table 2.1. Sites of adenosine reaction with adenylate cyclase

Property	Receptors (R)		
	Stimulatory receptor (R_a)	Inhibitory receptor (R_i)	P Site
1. Location	Outer cell surface	Outer cell surface	Cytoplasmic surface of plasma membrane
2. Effect on activity	Stimulation	Inhibition	Inhibition
3. Methyl xanthines	Antagonists	Antagonists	No effect
4. Purine-modified adenosine analogues	Agonists	Agonists	Poor effectors
5. Ribose-modified[a] adenosine analogues	Poor effectors	Poor effectors	Agonists
6. GTP	Required	Required	Not required
7. Relative potency: PIA and NECA	NECA > PIA	PIA > NECA	Not effectors
8. NECA: half-maximal effects	20 n*M* – 2 μ*M*	100–500 n*M*	No effect
9. PIA: half-maximal effects	200 n*M* – 100 μ*M*	5–10 n*M*	No effect
10. Divalent cations	Little effect	Reduce expression	Increase expression
Nomenclatures:			
Burnstock [27]	P_1	P_1	None
Van Calker et al. [24]	A_2	A_1	None

[a] 5′-N-ethylcarboxamide-adenosine, in contrast to most ribose-modified analogues, is a potent agonist at the stimulatory receptors.

posed R and P for two adenylate cyclase-associated sites [26]. One site, which recognized analogues with intact purine moieties, was designated P, and the other, which seems to require an intact ribose, was termed R. Subsequently, it became clear that the R site is a bona fide receptor, whereas the P site exhibits no receptor characteristics. We extended the R nomenclature to include R_a and R_i, the subscripts indicating the change in adenylate cyclase activity promoted by the receptor—R_a leading to activation and R_i to inhibition [25]. It should be emphasized that these functional designations are restricted to the adenylate cyclase reaction.

If one anticipates that refinements in adenosine receptor research will reveal additional major receptor subclasses or demand a further subclassification of the known receptors, the existing nomenclatures could pose problems. Addition of more subscripts, such as A_{1a} or R_{i1}, might prove cumbersome. Some improvement, perhaps based on the α and β system for adrenergic receptors, or omission of subscripts (Ado 1, Ado 2, etc.) should be considered for adenosine receptor nomenclature.

In this chapter we shall use the terms R_a and R_i to designate receptors and P to indicate the nonreceptor inhibitory site for adenosine action.

MEASUREMENT OF ADENOSINE EFFECTS ON ADENYLATE CYCLASE

One major difficulty in measuring adenosine receptor-mediated effects on adenylate cyclase is that the enzyme is associated with plasma membranes that contain nucleotide phosphohydrolases that metabolize the substrate, ATP, to adenosine. Also, membranes may contain sufficient contaminating endogenous adenosine to obscure the actions of added adenosine receptor agonists. Adenosine deaminase converts adenosine to inosine, which is not a receptor agonist [3, 15–18]. Thus, deaminase can be used to metabolize unwanted adenosine in the adenylate cyclase assay medium. Adenosine receptors may then be examined in the presence of adenosine deaminase with the use of analogues that are not substrates for the deaminase [15–17]. However, pharmacological studies of adenosine analogues in the presence of deaminase should take into account the possibility that the compounds may produce effects on adenylate cyclase indirectly by inhibiting the deaminase and thus preserving endogenous adenosine.

The use of dATP as substrate provides a second method for examining adenosine effects on adenylate cyclase [28]. Metabolism of dATP is not a problem, since 2′-deoxyadenosine is a weak receptor and P site agonist [15–17]. With adenylate cyclases from several cells it has been shown that the V_{max} with dATP is equivalent to or better than that seen with ATP as the substrate [28]. Barring significant contamination of membranes with endogenous adenosine, the use of dATP provides a means for examining the action of adenosine per se, as long as it is established that the membranes contain no other enzyme that metabolizes the nucleoside. In some cases, however, accurate assessments of the concentration-dependency of adenosine at its receptor are difficult to obtain because, in addition to receptor action, the nucleoside may act at the inhibitory P site.

STIMULATORY RECEPTORS: R_a

Rodbell and colleagues demonstrated the GTP requirement for glucagon stimulation of hepatic adenylate cyclase [29]. Subsequently, the requirement for GTP has been found to be a universal feature of hormone-stimulated adenylate cyclase [23]. In the absence of hormone, the onset of activation of adenylate cyclase by the relatively phosphohydrolase-resistant GTP analogue, Gpp(NH)p, is preceded by a lag period. Hormone binding to receptors characteristically shortens the lag period for activation by Gpp(NH)p [30].

As with other receptor-mediated adenylate cyclases, those stimulated by adenosine require GTP. Table 2.2 shows the GTP dependence for adenosine stimulation of the cyclase in purified hepatic plasma membranes. The enzyme from several other tissues has been shown to be enhanced by or dependent on guanine nucleotides [3, 15–17]. Moreover, as with other ligands that stimulate adenylate cyclases by receptors, adenosine reduces the lag period for activation by Gpp(NH)p [31, 32]. Thus, by such elementary tests the adenosine receptors appear to function in a manner similar to other hormone receptors.

The turkey erythrocyte adenylate cyclase, which is stimulated both by adenosine action at an R_a receptor [31, 32] and by catecholamine action at a β-adrenergic receptor, possesses unique characteristics that have permitted a dissection of the mechanism whereby hormones and guanine nucleotides lead to cyclase activation. Turkey erythrocyte membranes, upon purification, are thought to have GDP bound to the nucleotide binding site on the nucleotide regulatory component. Since a nucleotide triphosphate is required for activity, the enzyme with GDP bound is relatively inactive. Hormones appear to act by promoting the release of GDP, thus permitting GTP to enter the site [33]. The sequence of events in the foregoing scheme can be followed by a series of incubations and washings to remove ligands prior to exposure to new ligands

Table 2.2. GTP requirement for adenosine receptor-mediated stimulation of hepatic adenylate cyclase

Additions to assay	Adenylate cyclase activity[a]
None	12
GTP	13
NECA	11
GTP + NECA	46

Assays were performed with dATP as the substrate [28].
[a]pmol cyclic dAMP formed/mg prot/min.

Table 2.3. Outline of adenylate cyclase exchange reaction

Membrane treatment	Adenylate cyclase configuration	Activity state
None	GDP R·Ṅ·C	Low
Receptor agonist plus GMP	R·Ṅ·C	Low
Receptor agonist plus GMP, followed by Gpp(NH)p	GPP(NH)p R·Ṅ·C	High

Scheme as described in [34] for the actions of isoproterenol on the β-adrenergic receptor of turkey erythrocyte adenylate cyclase system. N represents the guanine nucleotide regulatory component, R the receptor, and C the catalytic unit.

Table 2.4. Adenosine receptor-mediated exchange reaction in turkey erythrocyte membranes

Membrane treatment	Adenylate cyclase activity[a]
None	2
NECA	4
GMP	2
NECA + GMP	26

Membranes were pretreated and assayed as described previously [31].
[a]pmol cyclic AMP formed/mg prot/min.

(Table 2.3). As worked out initially for the mechanism of activation by β-adrenergic agonists, isoproterenol plus GMP are incubated with membranes to remove the bound endogenous GDP. In this so-called exchange reaction, GMP is required; incubation with hormone or GMP alone does not clear GDP from the regulatory component. Upon washing membranes free of isoproterenol and GMP, the nucleotide site is empty and the enzyme exhibits low activity. However, high activity is seen upon introduction of Gpp(NH)p into the nucleotide binding site [34].

The data in Table 2-4 reveal similarities between the β-adrenergic and adenosine receptors in the turkey erythrocyte adenylate cyclase system [31]. NECA, the adenosine receptor agonist, promotes activation of the enzyme by Gpp(NH)p only upon preincubation with GMP. Treatment of membranes with either NECA or GMP alone has no effect. Thus, by a variety of criteria including GTP dependence, the agonist plus GMP exchange reaction, and shortening of Gpp(NH)p lag in activation, it would appear that adenosine receptor-mediated stimulation of adenylate cyclase proceeds by a mechanism similar if not identical to the mechanism employed by other hormone receptors.

INHIBITORY RECEPTORS: R_i

In recent years receptor-mediated inhibition of adenylate cyclase has been an active area of research. Among hormones known to inhibit activity of adenylate cyclases in isolated membrane preparations are α-adrenergic agonists, muscarinic cholinergic agonists, opiates, and prostaglandins [23, 35–37]. Inhibitory adenosine receptors have been demonstrated in purified fat cell membranes [38] and in crude membrane pellets from brain cortex [39].

Although receptor-mediated stimulation and inhibition of adenylate cyclases share a requirement for GTP, the nucleotide-regulated processes for the two different reactions are functionally distinct. As reviewed recently [35–37], lower concentrations of GTP are required for hormone-mediated stimulation than for hormone-mediated inhibition of activity. Moreover, Gpp(NH)p, while supporting the action of stimulatory hormones, does not permit observation of inhibitory hormone action. Similarly, high concentrations of Mg^{2+} and low concentrations of Mn^{2+} mask the effects of inhibitory hormones but not of stimulatory hormones. Sodium ions in the range of 100 mM generally enhance the magnitude of inhibitory hormone action while having little or no effect on stimulatory hormone action. Thus, criteria have been established for determining if a ligand inhibits adenylate cyclase via a receptor-mediated reaction.

Inhibition of adenylate cyclases in brain cortex and fat cell membranes by appropriate adenosine analogues satisfies the foregoing criteria for receptor-mediated processes. Inhibition of both enzymes is GTP-dependent. Inhibition of adipocyte adenylate cyclase is eliminated by high Mg^{2+} and low Mn^{2+} concentrations and by inclusion of Gpp(NH)p in the assay medium [40]. Sodium ions increase the magnitude of adenosine inhibition of both fat cell and brain adenylate cyclases [25, 39]. In addition to adenosine, the fat cell cyclase is inhibited by prostaglandins and nicotinic acid [35]. Other than acting at separate receptors, no kinetic differences are observed in the cyclase inhibition by these three agonists. Thus, in all respects, adenosine receptor agonists inhibit the fat cell and brain adenylate cyclase by a mechanism similar if not identical to that employed by other hormones and neurotransmitters in their inhibition of adenylate cyclases from a variety of cells.

SUBCLASSES OF ADENOSINE RECEPTORS

Based on studies with selected antagonists and agonists, a number of examples exist in which a given hormone has been shown to act at pharmacologically or functionally different receptors. This receptor heterogeneity has been demonstrated for histamine, dopamine, adrenergic, and cholinergic receptors [23]. For example, catecholamines act at β receptors to stimulate adenylate cyclase and at α receptors to inhibit activity. Similarly, the evidence suggests that at

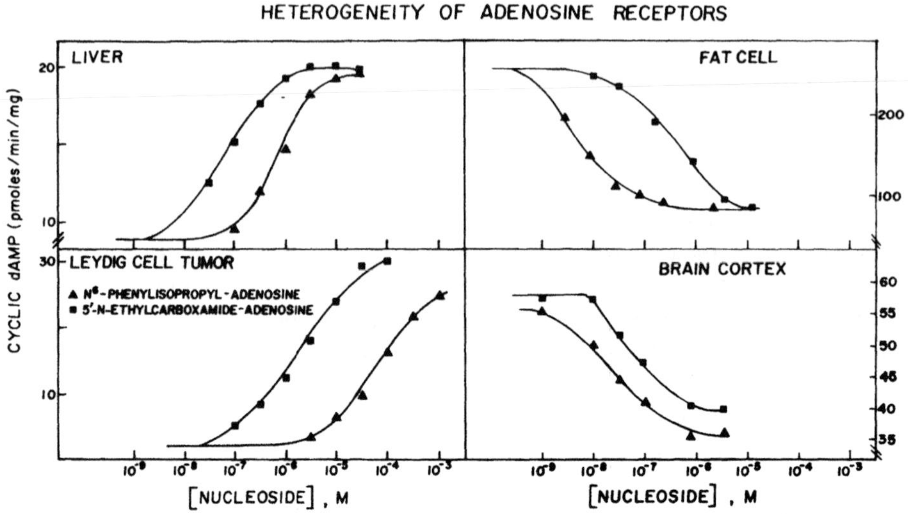

Figure 2.1. Comparison of pharmacological and functional differences between adenosine receptor subclasses. Adenylate cyclase activities were measured with dATP as the substrate, as described previously [28]. Activities in fat cell and brain cortical membranes were determined in the presence of NaCl [25, 39]; otherwise, assay conditions were identical for the four membranes.

least two major classes of adenosine receptors are found in cell membranes [15–18, 24, 25].

Figure 2.1 compares the actions of two analogs, NECA and PIA, on four different adenylate cyclase systems: adipocyte, brain cortex, liver, and Leydig cell tumor [16, 25, 39]. In the two membranes stimulated by the receptor agonists, Leydig cell tumor and liver, NECA is more potent than PIA. Also, NECA is more potent than PIA in stimulating the adenylate cyclase from PC12 cells, a line derived from a pheochromocytoma [41]. A similar potency order is found in turkey erythrocyte membranes [31]. By contrast, with those membranes in which adenosine receptors mediate inhibition of adenylate cyclase, brain and fat cell, it is clear that PIA is more potent than NECA. In those cases where an attempt has been made to compare adenosine to NECA and PIA, the potency of the natural nucleoside appears to fall between the potencies of the analogues [25]. In separate studies, the membranes tested in Figure 2.1 were shown not to metabolize adenosine [25]. Thus, the differences in potencies cannot be attributed to differential effects of the analogues on metabolism of endogenous adenosine. Moreover, the studies were performed with dATP as substrate and in the absence of adenosine deaminase, thus eliminating artefacts that might arise from protection by analogues against metabolism of endogenous adenosine. Thus, one can conclude that the different potency orders exhibited in Figure 2.1 reflect different subclasses of adenosine receptors.

It can be seen in Figure 2.1 that discrimination between the two receptor subclasses with NECA and PIA depends primarily on the great potency differences of PIA in the various test systems. The potency of NECA is somewhat similar with the four enzymes; the analogue concentration required for half-maximal effects ranges from 20 n*M* to 2 μ*M*. However, the IC_{50} for PIA in inhibiting the fat cell and brain cyclases is 5 to 10 n*M*, whereas the EC_{50} for this analog in stimulating the liver enzyme is approximately 1 μ*M*, and nearly 100 μ*M* with the Leydig cell tumor cyclase. The ability to discriminate between the receptor subclasses is seen also with the L and D steroisomers of PIA. L-PIA is nearly 100-fold more potent than D-PIA in inhibiting the fat cell enzyme, but only 5-fold more potent than D-PIA in stimulating the liver enzyme [unpublished observation]. Discrimination of different adenosine receptor subclasses with the use of PIA stereoisomers has been accomplished in binding studies as well [18]. Thus, it would appear that the differences between R_a and R_i receptors might be reflected most acutely in the region that binds N^6 amino group of the purine moiety.

Within each subclass of adenosine receptor, kinetic differences are observed among the various cell types tested (Fig. 2.1). Note that the potency differences between PIA and NECA are greater in the fat cell than in the brain cortex, yet both contain R_i receptors. Similarly, differences are seen in R_a receptors from different sources. NECA is approximately 100-fold more potent than PIA in stimulating Leydig tumor cell cyclase, 40-fold greater with the liver enzyme, and only 10-fold greater with the PC12 cyclase [41]. Such differences might reflect subpopulations of adenosine receptors within the major subclasses. However, they could reflect differences in the character of the nucleotide regulatory components within the various membranes, since affinity of most hormone receptors [23], including those for adenosine [18], are known to be regulated by GTP. Whatever the explanation, it is clear that receptors within a given subclass exhibit subtle differences. As in the case of adrenergic receptors, the postulated differences among adenosine receptors, between and within the major subclasses, are based on kinetic evidence. Whether or not these differences represent different gene products remains to be determined.

A word of caution is in order on the designations "high-affinity" and "low-affinity" receptors to describe the major adenosine receptor subclasses. As noted earlier, results with NECA would not permit such qualitative statements on the apparent ability of the different receptors to bind nucleosides. However, data obtained with PIA suggest that one subclass, R_a or A_1, is a high-affinity receptor, whereas the other subclass, R_a or A_2, is a low-affinity receptor. However, the separation of effective PIA concentration ranges between R_a and R_i receptors may become less distinct as additional receptors are examined. Consider, for example, the PC12 cells [41], which contain an R_a receptor that exhibits an EC_{50} of 200 n*M* for PIA. This value is merely 20-fold greater than the IC_{50} for PIA at the brain R_i receptor. Although measurements of effective ligand concentration ranges for modifying adenylate cyclase are not true affinity determinations, it can be assumed that such data reasonably reflect receptor affinity. One should assume that R_i receptors occur that bind PIA more weakly than the brain receptors, and that R_a receptors will be found that bind that analog more tightly than the PC12 R_a receptors. With such receptors, knowledge of the absolute binding affinity would give no information on the pharmacological and functional subclass. In contrast to several frequently used analogues, data obtained with adenosine per se are limited. Should adenosine prove to be less discriminating than PIA, considerable overlap might occur between affinity ranges for R_a and R_i receptors.

METHYLXANTHINES

As first suggested by the data of Sattin and Rall [13] and subsequently confirmed in numerous studies [3, 15–18], methylxanthines, such as 1-methyl-3-isobutylxanthine (MIX), theophylline, and caffeine, are competitive antagonists of adenosine receptors. Previously, most effects seen with these compounds were attributed to their ability to inhibit cyclic nucleotide phosphodiesterases. However, the widespread occurrence of adenosine receptors has forced a reevaluation of the pharmacological basis for the methylxanthine effects [42], especially since they may antagonize adenosine receptors at concentrations far lower than those required for inhibition of the diesterases [43]. As reported by Smellie et al. [43], some alkylxanthines, such as 8-phenyltheophylline and 1,3-dibutylxanthine, are potent adenosine receptor antagonists with relatively weak phosphodiesterase actions. Conversely, potent phosphodiesterase inhibitors with weak adenosine receptor effects are known [43]. Substitution of these more discriminating compounds for MIX, theophylline, and caffeine would facilitate the interpretation of results obtained with complex physiological systems.

Despite their general usefulness as tools in adenosine receptor research, the methylxanthines do not discriminate between the different adenosine receptor subclasses. The potency order MIX > theophylline > caffeine is seen with either R_a or R_i receptors [15, 18]. Moreover, the apparent K_i values for the methylxanthines are similar with both receptors [15–18, 39, 41]. Great benefit would derive from the development of receptor-specific antagonists.

THE P SITE

The P site is a site that recognizes adenosine and mediates inhibition of adenylate cyclase. In contrast to adenosine receptors, this site is located on the inner cell surface [44], is not blocked by methylxanthines, does not require GTP, and survives solubilization of membranes with detergents [15–19, 26, 44, 45). Also the P site can be distinguished from receptors by analogues, especially those modified at the ribose moiety. For example, 2′,5′-dideoxyadenosine (DDA), arabinofuranosyladenosine, xylofuranosyladenosine, and 2′-deoxyadenosine-3′-monophosphate, compounds with no receptor agonist properties, are agonists at the P site, whereas NECA, PIA, and a variety of purine-modified analogues that are receptor agonists have little or no P site activity [15–19, 26, 44]. The P site poses problems for the use of adenosine for examining receptors, since in some membranes there is considerable overlap in the concentration range of the nucleoside over which it displays both P site and receptor agonism. Thus, for platelet [19, 21] and liver [15] adenylate cyclases, both of which contain R_a receptors, the concentration dependency for adenosine at the receptor cannot be determined accurately; stimulation by low adenosine concentrations at the receptor and inhibition by higher concentrations at the P site produce biphasic curves. Unfortunately, no selective antagonists of the P site are known.

Currently available evidence suggests that the P site is located on the catalytic subunit of the adenylate cyclase system. Based on their strong preference for MnATP over MgATP as substrate, several preparations are thought to represent models for the catalytic component. These include the naturally soluble testicular enzyme, the enzyme found in a mutant SV49 lymphoma cell line, and the cyclase solubilized from hepatic membranes not treated with guanine nucleotides prior to extraction with detergents [16]. These three cyclases are relatively resistant to inhibition by adenosine or DDA. Millimolar concentrations of DDA inhibit the testicular and mutant lymphoma cyclase by only 20% to 30%, whereas low micromolar concentrations of DDA abolish activity of most membrane-bound adenylate cyclases that contain a normal complement of nucleotide-regulatory proteins [15–17]. When the mutant lymphoma cyclase is complemented with nucleotide-regulatory proteins from human erythrocyte membranes, reactivity with DDA is restored [16]. This finding not only showed the importance of guanine nucleotide-regulatory components in P site reactivity but also suggested that the regulatory proteins might contain the P site. However, if one considers the inhibition, albeit small, of the MnATP-dependent cyclases by DDA, and the reports by Neer and Salter [46] and by Premont et al. [47] showing inhibition by P site agonists of soluble adenylate cyclases that are apparently free of associated nucleotide-regulatory components, it is clear that the P site resides on the cyclase catalytic component.

One of the interesting characteristics of the P site is the apparent change in affinity for agonists upon stimulation of cyclases with hormones [15–17]. Whether or not this represents a physiological regulatory mechanism remains to be determined, as does the identity of the natural P site agonist, should it be a compound other than adenosine.

CONCLUDING REMARKS

The list of cells containing adenosine receptors has undergone an acute growth spurt within the past few years. In most cases, adenosine receptors are revealed by detection of physiological phenomena expressed upon application of receptor agonists. Considered less often, however, is the possibility that test systems contain cryptic adenosine sufficient for receptor stimulation. Both

Schwabe and colleagues and Fain and co-workers demonstrated the importance of adenosine in fat cell suspension media (see [3] for review). Green [48] has shown that cultured neuroblastoma cells release adenosine in proportion to cell density and that at high densities the nucleoside is sufficient to stimulate adenylate cyclase. Such findings may have important implications for the interpretation of cell density-dependent phenomena in cultured cells.

Other situations in which adenosine receptors are not readily revealed are those in which the nucleoside exerts little or no effect unless tested in combination with a second agent. Thus, adenosine modifies histamine release from mast cells only in combination with other release stimulants [49]. The pheochromocytoma cell line, PC12, serves as a bioassay system for nerve growth factor, which stimulates neurite outgrowth. Adenosine receptor agonists alone have no effect on PC12 cell morphology but markedly stimulate the rate of neurite outgrowth in the presence of nerve growth factor [41]. In contrast to the frequently studied acute effects of adenosine, this trophic action on PC12 cells requires at least 24 hours to become manifest.

It is likely that many more systems exist that contain adenosine receptors but, as in the examples cited, the potential role for the nucleoside in regulation is not readily evident. Thus, the full scope of biological regulation by adenosine receptors has yet to be realized.

REFERENCES

1. Fox IH, Kelly WN: The role of adenosine and 2′-deoxy-adenosine in mammalian cells. *Annu Rev Biochem* 47:655–686, 1978.
2. Arch JRS, Newsholme EA: The control of the metabolism and the hormonal role of adenosine. *Essays Biochem* 14:82–123, 1978.
3. Fain JN, Malbon CC: Regulation of adenylate cyclase by adenosine. *Mol Cell Biochem* 23:1–27, 1979.
4. Haslam RJ, Rosson GM: Effects of adenosine on levels of adenosine cyclic 3′,5′-monophosphate in human blood platelets in relation to adenosine incorporation and platelet aggregation. *Mol Pharmacol* 11:528–544, 1975.
5. Wolff J, Cook GH: Activation of steroidogenesis and adenylate cyclase by adenosine in adrenal and Leydig tumor cells. *J Biol Chem* 252:687–693, 1977.
6. Marquardt DL, Parker CE, Sullivan TJ: Potentiation of mast cell mediator release by adenosine. *J Immunol* 120:871–878, 1978.
7. Muller MJ, Paton DM: Presynaptic inhibitory actions of 2-substituted adenosine derivative on neurotransmission in rat vas deferens: Effects of inhibitors of adenosine uptake and deamination. *Arch Pharmacol* 306:23–28, 1979.
8. Olsson RA, Snow JA, Gentry MK, Erick GP: Adenosine uptake by canine heart. *Circ Res* 31:767–778, 1972.
9. Olsson RA, Davis CJ, Khouri EM, Patterson RE: Evidence for an adenosine receptor on the surface of dog coronary myocytes. *Circ Res* 39:93–98, 1976.
10. Schrader J, Nees S, Gerlach E: Evidence for a cell surface adenosine receptor on coronary myocytes and atrial muscle cells. *Pfluegers Arch* 369:251–257, 1977.
11. Olsson RA, Davis CC, Khouri EM: Coronary vasoactivity of adenosine covalently linked to polylysine. *Life Sci* 21:1343–1350, 1977.
12. Okwuasaba FK, Hamilton JT, Cook MA: Evidence for the cell surface locus of presynaptic purine nucleotide receptors in the guinea pig ileum. *J Pharmacol Exp Ther* 207:779–786, 1978.
13. Sattin A, Rall TW: The effect of adenosine and adenine nucleotides on the cyclic adenosine 3′,5′-phosphate content of guinea pig cerebral cortex slices. *Mol Pharmacol* 6:13–23, 1970.

14. Shimizu H, Daly JW: Formation of cyclic adenosine 3',5'-monophosphate from adenosine in brain slices. *Biochim Biophys Acta* 222:465–473, 1970.
15. Londos C, Wolff J, Cooper DMF: Adenosine as a regulator of adenylate cyclase, in Burnstock G (ed): *Purinergic Receptors*. London, Chapman and Hall, 1981, pp 287–323.
16. Wolff J, Londos C, Cooper DMF: Adenosine receptors and the regulation of adenylate cyclase. *Adv Cyclic Nucleotide Res* 14:199–214, 1981.
17. Londos C, Wolff J, Cooper DMF: Action of adenosine on adenylate cyclase, in Baer HP, Drummond GI (eds): *Physiological and Regulatory Functions of Adenosine and Adenine Nucleotides*. New York, Raven Press, 1979, pp 271–281.
18. Daly JW: Adenosine receptors: Targets for future drugs. *J Med Chem* 25:197-207, 1982.
19. Haslam RJ, Davidson MML, Lemmex BWG, Desjardins JV, McCarry BE: Adenosine receptors of the blood platelet: Interaction with adenylate cyclase, in Baer HP, Drummond GI (eds): *Physiological and Regulatory Functions of Adenosine and Adenine Nucleotides*. New York, Raven Press, 1979, pp 189–204.
20. Trost T, Stock K: Effects of adenosine derivatives on cAMP accumulation and lipolysis in rat adipocytes and on adenylate cyclase in adipocyte plasma membranes. *Arch Pharmacol* 299:33–40, 1977.
21. Haslam RJ, Lynham JA: Activation and inhibition of blood platelet adenylate cyclase by adenosine or by 2-chloroadenosine. *Life Sci* 11:1143–1154, 1972.
22. Sternweis PC, Northrup JK, Hanski E, Schleifer LS, Smigel MD, Gilman AG: Puriciation and properties of the regulatory component (G/F) of adenylate cyclase. *Adv Cyclic Nucleotide Res* 14:23–36, 1981.
23. Rodbell M: The role of hormone receptors and GTP-regulatory proteins in hormone transduction. *Nature* 284:17–22, 1980.
24. Van Calker D, Muller M, Hamprecht B: Adenosine regulates via two different types of receptors, the accumulation of cyclic AMP in cultured brain cells. *J Neurochem* 33:999–1005, 1979.
25. Londos C, Cooper DMF, Wolff J: Subclasses of adenosine receptors. *Proc Natl Acad Sci USA* 77:2551–2554, 1980.
26. Londos C, Wolff, J: Two distinct adenosine-sensitive sites on adenylate cyclase. *Proc Natl Acad Sci USA* 74:5482–5485, 1977.
27. Burnstock G: Past and current evidence for the purinergic nerve hypotheses, in Baer HP, Drummond GI (eds): *Physiological and Regulatory Functions of Adenosine and Adenine Nucleotide*. New York, Raven Press, 1979, pp 3–32.
28. Cooper DMF, Londos C: Evaluation of the effects of adenosine on hepatic and adipocyte adenylate cyclase under conditions where adenosine is not generated endogenously. *J Cyclic Nucleotide Res* 5:289–302, 1979.
29. Rodbell M, Birnbaumer L, Pohl SL, Krans HM: The glucagon-sensitive adenyl cyclase system in plasma membranes of rat liver: An obligatory role of guanyl nucleotides in glucagon action. *J Biol Chem* 246:1877–1882, 1971.
30. Londos C, Salomon Y, Lin MC, Harwood JP, Schramm M, Wolff J, Rodbell M: 5'-Guanylylimidodiphosphate, a potent activator of adenylate cyclase in eukaryotic cells. *Proc Natl Acad Sci USA* 71:3087–3090, 1974.
31. Lad PM, Nielsen TB, Londos C, Preston MS, Rodbell M: Independent mechanisms of adenosine activation and inhibition of the turkey erythrocyte adenylate cyclase system. *J Biol Chem* 255:10841–10846, 1980.
32. Sevilla N, Tolkovsky AM, Levitzki A: Activation of turkey erythrocyte adenylate cyclase by two receptors: Adenosine and catecholamines. *FEBS Lett* 81:339–341, 1977.
33. Selinger Z, Cassel D: Role of guanine nucleotides in hormonal activation of adenylate cyclase. *Adv Cyclic Nucleotide Res* 14:15–22, 1981.
34. Lad PM, Nielsen TB, Preston MS, Rodbell M: The role of the β-adrenergic receptor and in the actions of catecholamines and cholera toxin on adenylate cyclase in turkey erythrocyte membranes. *J Biol Chem* 255:988–995, 1980.
35. Londos C, Cooper DMF, Rodbell M: Receptor-mediated stimulation and inhibition of adenylate cyclases: The fat cell as a model system. *Adv Cyclic Nucleotide Res* 14:163–172, 1981.
36. Cooper DMF, Londos C: GTP-dependent stimulation and inhibition of adenylate cyclase, in Kohn LD (ed): *Hormone Receptors*. London, John Wiley, 1982, pp 309–333.

37. Jakobs KH, Aktories K, Schultz G: Inhibition of adenylate cyclase by hormones and neurotransmitters. *Adv Cyclic Nucleotide Res* 14:173–188, 1981.
38. Londos C, Cooper DMF, Schlegel W, Rodbell M: Adenosine analogs inhibit adipocyte adenylase cyclase by a GTP-dependent process: Basis for actions of adenosine and methylxanthines on cyclic AMP production and lipolysis. *Proc Natl Acad Sci USA* 75:5362–5366, 1978.
39. Cooper DMF, Londos C, Rodbell M: Adenosine receptor-mediated inhibition of rat cerebral cortical adenylate cyclase by a GTP-dependent process. *Mol Pharmacol* 18:598–601, 1980.
40. Cooper DMF, Schlegel W, Lin MC, Rodbell M: The fat cell adenylate cyclase system: Characterization and manipulation of its bimodal regulation by GTP. *J Biol Chem* 254:8927–8931, 1979.
41. Guroff G, Dickens G, End D, Londos C: The action of adenosine analogs on PC12 cells. *J Neurochem* 37:1431–1439, 1981.
42. Fredholm BB: Are methylxanthine effects due to antagonism of endogenous adenosine? *Trends Pharmacol Sci* 1:129–132, 1980.
43. Smellie FW, Davis CW, Daly JW, Wells JN: Alkylxanthines: Inhibition of adenosine-elicited accumulation of cyclic AMP in brain slices and of brain phosphodiesterase activity. *Life Sci* 24:2475–2482, 1979.
44. Haslam RJ, Davidson MML, Desjardins JV: Inhibition of adenylate cyclase by adenosine analogues in preparation of broken and intact human platelets. *Biochem J* 176:83–95, 1978.
45. Wolff J, Londos C, Cook GH: Adenosine interactions with thyroid adenylate cyclase. *Arch Biochem Biophys* 191:161–168, 1978.
46. Neer EJ, Salter RS: Reconstituted adenylate cyclase from bovine brain: Functions of the subunits. *J Biol Chem* 256:10102–12107, 1981.
47. Premont J, Guillon G, Bockaert J: Specific Mg^{2+} and adenosine sites involved in a bireactant mechanism for adenylate cyclase inhibition and their probable localization on this enzyme's catalytic component. *Biochem Biophys Res Commun* 90:513–519, 1979.
48. Green RD: Effects of exogenous adenosine deaminase and erythro-9-(2-Hydroxy-3-nonyl)adenine on intracellular cyclic AMP levels in C1300 murine neuroblastoma in tissue culture. *Life Sci* 26:399–406, 1980.
49. Holgate ST, Lewis RA, Austen KF: Role of adenylate cyclase in immunologic release of mediators from rat mast cells: Agonist and antagonist effects of purine- and ribose-modified adenosine analogs. *Proc Natl Acad Sci USA* 77:6800–6804, 1980.

DISCUSSION

DR. M.G. COLLIS: How do you interpret the wide variation in the relative potencies of PIA and NECA of R_i sites?

DR. LONDOS: Such variations suggest that the R_i receptor might represent a family of similar but not identical receptors. They could reflect either different gene products or differing behaviors of a single gene product imposed by regulatory proteins within the membrane.

DR. H. McILWAIN: Did you suggest that inhibition by adenosine at nanomolar concentrations was observed only under limited experimental conditions?—so that if one wishes to invoke such inhibition under physiological conditions, does one have to postulate regions with particular [Na^+] or [GTP] concentrations?

DR. LONDOS: I do not doubt the existence of high-affinity receptors that respond *in vivo* to nanomolar adenosine concentrations. I merely suggested that we avoid linking affinity with receptor subclass, since it is entirely possible that "high-affinity" R_a/A_2 and "low-affinity" R_i/A_1 receptors await discovery. Note that PC12 cell adenylate cyclase responds to PIA with an EC_{50} of $2 \times 10^{-7} M$. Should these cells respond to adenosine at similar or lower concentrations, one would have an example of a "moderately high-affinity" receptor that exhibits the pharmacological profile of what has been termed a "low-affinity" receptor.

DR. A.C. NEWBY: For technical reasons it is always most difficult to measure cyclase activation or receptor binding with adenosine itself. What do you think are the best estimates of adenosine's affinity for A_1 and A_2 receptors? Intact cells offer some advantages for such studies since extracellular adenosine levels can be measured and correlated directly with, for instance, cAMP rise. This seems to reveal large differences in potency for adenosine in inhibiting, for example, fat cell cAMP as opposed to stimulating platelet cAMP rise.

DR. LONDOS: I would agree that for the very few currently available examples of adenosine potency estimates produced under conditions of limited adenosine metabolism, R_i/A_1 receptors appear to have a higher affinity for the nucleoside than R_a/A_2 receptors. I merely suggested that we avoid the "high" versus "low" affinity designations, which do seem to apply with certain N^6-substituted analogues but not with the 5′-carboxamide derivatives. We should be prepared to encounter high-affinity receptors that both mediate adenylate cyclase stimulation and show a pharmacological profile similar to receptors currently designated as "low-affinity" receptors.

DR. J.E. SEEGMILLER: Could you comment on similarities and differences, if any, in cyclic AMP and cyclic dAMP which you measured interchangeably in your assay systems, and does cyclic dAMP serve a physiological role *in vivo*?

DR. LONDOS: I do not know if physiologically significant levels of cyclic dAMP are formed *in vivo*.

DR. R. RUBIO: (1) In intact cells, do changes in extracellular [Na^+] lead to changes in cyclase activity or changes in sensitivity to adenosine modulation? (2) Does tetrodotoxin simulate in the absence of extracellular Na^+?

DR. LONDOS: We have not examined the effects of Na^+ on the response in intact cells. However, Blume and colleagues have published data on the neuroblastoma × glioma hybrid cell suggesting that monovalent cations in the extracellular medium support the actions of inhibitory hormones. We have not tested tetrodotoxin.

DR. J.P. MANFREDI: Does the cyclase exhibit desensitization with either adenosine or an adenosine analogue as agonist?

DR. LONDOS: I am not aware of any study designed to demonstrate desensitization of an adenylate cyclase-linked adenosine receptor. However, one could infer from Dr. Richard Green's studies on cells in tissue culture that desensitization does occur.

DR. H.J. LEIGHTON: I am concerned about the use of apparent K_A and/or K_i values to describe high- and low-affinity adenosine sites and the implication that these sites represent different adenosine receptors. Apparent K_A is a function of intrinsic efficacy and therefore not a true estimate of K_A. True K_A values would allow one to distinguish receptor subtypes; apparent K_A's do not allow one this luxury. Would you care to comment?

DR. LONDOS: This depends on the criteria one establishes for distinguishing one receptor subclass from another. Imagine if all adenosine recognition sites on receptors were carried on the same protein but the ability to discriminate among agonists and to produce different functional responses with adenylate cyclase was determined not by the receptor but by receptor-associated regulatory proteins. In this situation, a functional response accompanied by a unique potency series for selected agonists would be far more valuable than knowledge of "true" affinities. The lack of selective adenosine receptor antagonists notwithstanding, the argument for adenosine receptor subclasses is no less valid than the case for different (alpha and beta) adrenergic receptors.

DR. A. SATTIN: Please comment on the interchangeability of Li^+ with Na^+ on the adenosine inhibition of the cyclase, and the relative potency of these ions. Are there instances of analogous effects of monovalent cations on stimulatory cyclases?

DR. LONDOS: At low concentrations, Li^+ is more effective than Na^+ in promoting inhibition. However, at higher concentrations, Li^+, but not Na^+, abolishes the cyclase response to stimulatory hormones. I know of no examples of regulation of stimulatory hormone responses by monovalent cations.

DR. S.J. MUSTAFA: What is the physiological function of the P site receptor?

DR. LONDOS: First, the P site should not be thought of as a receptor, since, unlike the R_a and R_i adenosine receptors, it has little in common with other hormone receptors. Second, although there is no known physiological function for the P site, the fact that hormonal stimulation of adenylate cyclases increases reactivity of adenosine at the site, one should not discount its potential regulatory function. However, the natural ligand for the P site might be a molecule other than adenosine.

3. ADENOSINE RECEPTORS ON VASCULAR SMOOTH MUSCLE

RAY A. OLSSON

The first account of the cardiovascular actions of adenosine included a description of its coronary vasoactivity [1]. Evidence that adenosine operates through specific receptors came in 1970 with the discovery that adenosine stimulates accumulation of cyclic AMP in brain and that theophylline antagonizes this effect [2]. Simultaneously, theophylline was reported to antagonize the coronary vasoactivity of adenosine [3] in a manner later shown to obey competitive kinetics [4]. This agonist–competitive antagonist relationship between adenosine and theophylline is strong evidence for a specific receptor. Experiments in many laboratories have now confirmed the existence of adenosine receptors in many tissues, have described three receptor subtypes, and have demonstrated that these receptors usually, perhaps invariably, operate through adenylate cyclase.

Adenosine is a vasodilator in all vascular beds studied to date save the kidney, in which adenosine causes intense vasoconstriction [5]. This review focuses on the coronary circulation, because the information about this circulation is the most detailed, and addresses two unresolved questions. The first concerns the structure of the vascular adenosine receptor and the second the mechanism whereby receptor occupancy initiates coronary relaxation.

STRUCTURE OF THE CORONARY ADENOSINE RECEPTOR

A number of studies [6–18] have shown that certain chemical features of the adenosine molecule promote vasoactivity, further supporting the notion of a

R.M. Berne, T.W. Rall, and R. Rubio (eds.), Regulatory Function of Adenosine, *pp. 33–47.*

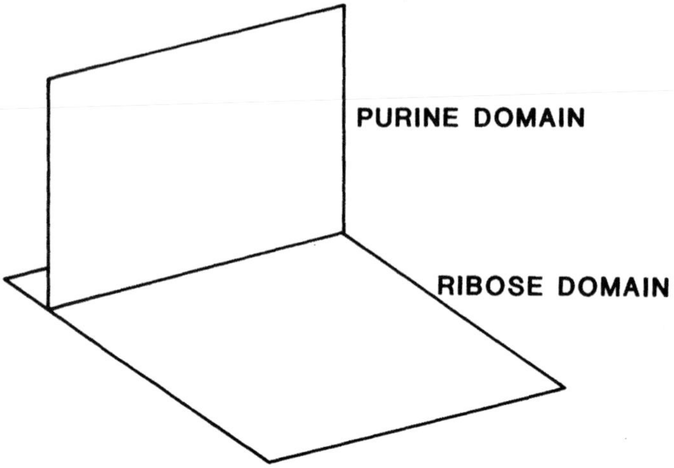

Figure 3.1. Generalized model of the coronary adenosine receptor, represented here as separate, antiplanar purine and ribose domains.

specific receptor. The most prominent determinants of adenosine's vasoactivity seem to be its six-amino group, the β-anomeric configuration of the ribose, and the ribofuranose C2′ and C3′ hydroxyl groups. We have used the results of these studies and knowledge of the chemistry of adenosine to develop our model of this receptor.

Adenosine is a complex molecule that consists of a flat, relatively hydrophobic heterocyclic base and a nonplanar, hydrophobic sugar. Recognizing that the two parts of the molecule are chemically dissimilar and that both are required for vasoactivity (neither adenine nor ribose is vasoactive), we posit separate purine and ribose domains, each with its own determinants of binding affinity. Due to the angle of the N9-C1′ glycosylic bond, the base and sugar moieties are antiplanar; thus, a simple way to picture the purine and ribose domains of the receptor is as two intersecting planes (Fig. 3.1).

Determinants of binding to the purine domain

Data from more than 100 base-modified ribofuranosides, whose coronary vasoactivity ranges from some that are inactive to others whose activity is an order of magnitude more than adenosine, constitute an important source of information about the factors governing binding to the purine domain. This group includes nucleosides having exocyclic substituents at N1, C2, C6, N7, and C8 as well as several purine ring isosteres. General structure-activity rules for coronary vasoactivity include the following:

1. N1 alkylation or oxidation reduces potency.
2. Several 2-haloadenosines are more potent vasodilators than adenosine.

The vasoactivity of adenosines substituted at C2 with an alkoxy, aryloxy, alkylthio, arylthio, alkylamino, or arylamino group vary widely, tending to increase as a function of the size—and hydrophobicity—of the alkyl or aryl group. This suggests the presence of a hydrophobic region in the vicinity of C2.

3. Alteration of the C6 substituent usually reduces vasoactivity. For examples, the N^6-methyladenosines and the halo, alkoxy, and alkylthioinosines are two or more orders of magnitude less potent than adenosine. However, certain hydrophobic N^6 substituents preserve or enhance vasoactivity, which suggests the presence of a hydrophobic region in the vicinity of C6. Important differences in the vasoactivity of adenosine whose C6 substituents are optically active amines strengthen the case for stereospecific binding to such a region. Thus, N^6-[*R*(−)-1-methyl-2-phenethyl]adenosine (L-PIA)[1] is an order of magnitude more potent than its isomer derived from the *S*(+) amine [14]. Similarly, N^6-[*R*(+)-1-phenethyl]adenosine is more vasoactive than the derivative of the *S*(−) amine [19], confirming that in addition to the hydrophobic contribution of the phenyl group, binding depends on the absolute configuration at the chiral center.

The goal of our structure-activity studies is to identify the fundamental determinants of the affinity of adenosine for its receptor and thereby to gain some insight into the structure of the receptor. The quantitative structure-activity relationship (QSAR) technique [20] is one approach to this kind of information. QSAR posits that biological activity (C) is linearly related to the sum of hydrophobic (π), electronic (σ), and steric (s) contributions to the free energy of ligand (or drug) binding to a receptor:

$$\log (1/C) = k_, \pi + k_2\sigma + k_3 s$$

Our previous analysis [17] showed that the Hammett σ_m substituent parameter, an index of electronic structure, accounted reasonably well for the vasoactivity of C2-substituted adenosines. However, there was essentially no relationship between this parameter and the vasoactivity of C6- and C8-substituted analogues. That neither hydrophobic nor steric factors seemed to contribute to vasoactivity was additionally disconcerting. Accordingly, we considered other ways of representing the electronic structure of these nucleosides and, additionally, tested nucleosides that had hydrophobic substituents.

The purine dipole moment appears to be the major electronic determinant of binding to the purine domain. The purine dipole results from atom-to-atom differences in electron density, i.e., atomic charge, within the electron-rich pyrimidine and electron-deficient imidazole rings [21]. Dipoles are vectors, having both magnitude, which is measured in Debye (D—1D $= 10^{-10}$ Å · esu), and direction, which is conventionally [22] measured as angular displacement from a line whose origin is the center of mass of the purine base and

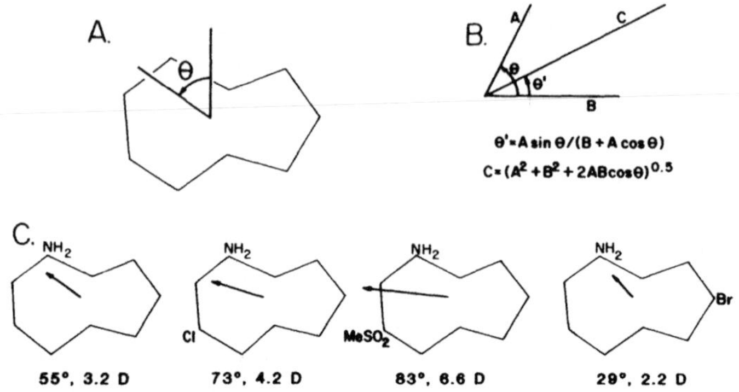

Figure 3.2. Purine dipole moment. A: Conventional method for measuring angular displacement [8]. B: Calculation of dipole direction and magnitude by vector addition. C: Purine dipole moments of several adenosines; from left to right: adenosine, MPR 1.0; 2-chloradenosine, MPR 13; adenosine-2-methyl-sulfone, inactive; and 8-bromoadenosine, inactive.

which is directed parallel to the C4 → C5 bond (Figure 3.2). One employs vector addition to calculate changes in dipole moment that result from the addition of substituent groups. Our calculations of the dipole moments of adenosine analogues substituted at N1, C2, C6, and C8 employ a literature value for the dipole moment of 9(*H*) purine of 4.3 D, 49° [23], X-ray crystallographic data on the geometry of the adenine base [24], and group dipole moments of various substituents [25]. In accord with the work of others [6, 9, 12, 13], we express agonist potency as a molar potency ratio (MPR), the quotient of the ED_{50} of adenosine divided by that of a test analogue. Because by this convention the potency of adenosine is always unity, one can substitute log (MPR) directly for log (1/*C*) in the QSAR equation.

This analysis shows that the only nucleosides having appreciable coronary vasoactivity are those whose purine dipole strength is between 3 D and 4.5 D and whose angle ranges between 55° and 75° (Fig. 3.3). Most of these analogues are C2-substituted adenosines. The only other vasoactive nucleosides are analogues of L-PIA, which probably owe their vasoactivity to the hydrophobicity of the N6 substituent rather than to electronic factors. The C8-modified adenosines studied to date include only nucleosides whose C8 substituent is strongly electronegative, e.g., sulfur, bromine or oxygen. Such substituents strongly shift the purine dipole moment clockwise and reduce its strength to $\geqslant$ 2.5 D. Thus, although the conventional explanation for the inactivity of these nucleosides is that the steric bulk of these C8 substituents may promote a *syn* conformation ill-suited for binding, the important shift in dipole moment is an equally plausible reason.

The purine dipole hypothesis successfully accounts for the binding of the adenosine antagonist theophylline. Calculation of the dipole moment of this purine based on that of 7(*H*) xanthosine [21] yields a value of 4.3 D, 102°. The

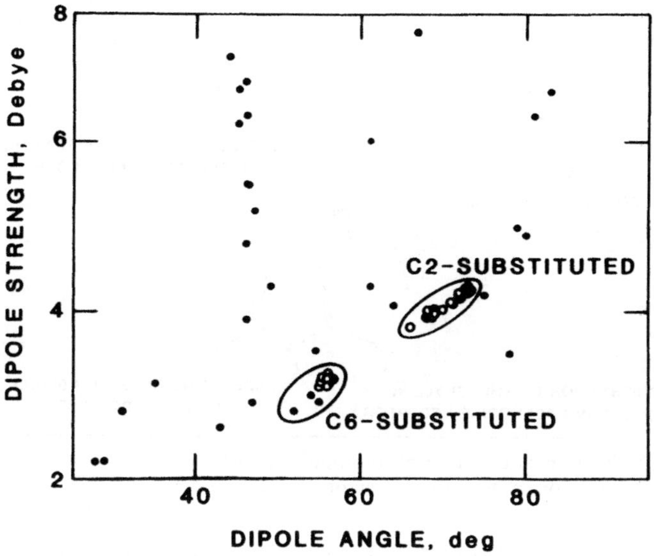

Figure 3.3. Relationship of coronary vasoactivity of adenosine analogues to purine dipole moment. Nucleosides having molar potency ratios ≥ 0.1 relative to adenosine are denoted by open circles, those less active than this limit by filled circles. Note that those having substantial vasoactivity are localized in two discrete regions. The region labeled "C6-SUBSTITUTED" includes adenosine and adenosines having hydrophobic substituents at C6. Those in the region labeled "C2-SUBSTITUTED" are mainly 2-halo-, 2-alkylthio-, 2-alkoxy-, or 2-arylaminoadenosines.

angular displacement of this dipole seems to lie far outside the range favored for binding to the receptor. However, a 180° rotation of this molecule around its long axis so that the "back" side is now "front" leads to a dipole angle of 180° − 102° = 78°, which is only slightly outside the "favored" range (Fig. 3.4). This angle is still suboptimal, but this could be the reason why the antagonist potency of theophylline, IC_{50} 5 × $10^{-5}M$ [4], is two orders of magnitude lower than the agonist potency of adenosine, ED_{50} 5 × $10^{-7}M$, in a similar experimental preparation [26]. Alternatively, the orientation of theophylline bound to the receptor is not constrained by the simultaneous binding of a sugar to the ribose domain. Thus, the molecule could rotate slightly in the clockwise direction to properly align its dipole with that of the receptor.

We have just begun to synthesize and assay the vasoactivity of nucleosides designed to test the predictive power of the purine dipole hypothesis. For example, the hypothesis predicts that N1 methylation of isoguanosine (2-hydroxyadenosine, dipole moment 4.2 D, 72°), should reduce vasoactivity. This is indeed the case; 1-methylisoguanosine (dipole moment 6.3 D, 81°) is two orders of magnitude less potent than the parent nucleoside. Similarly, the hypothesis predicts that the vasoactivity of adenosine-2-methyl-sulfone should

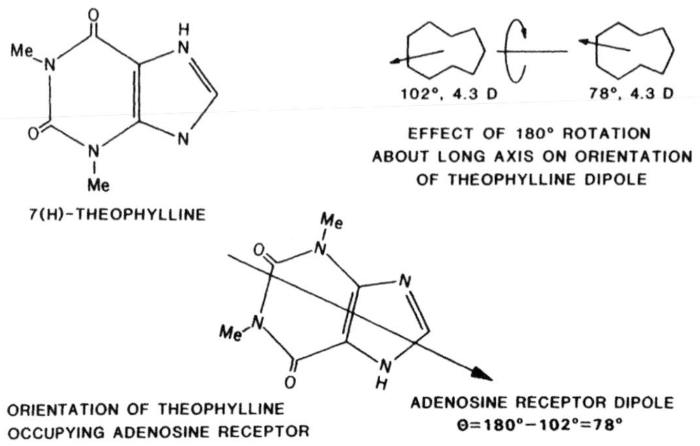

Figure 3.4. Explanation based on the purine dipole moment hypothesis for the binding of theophylline to the coronary adenosine receptor. See text for details.

be low because the $-SO_2CH_3$ substituent (group dipole moment -3.85 D) rotates the purine dipole past the optimum range to a value of 6.6 D, 83°. Experiment bears out this prediction as well; adenosine-2-sulfone is almost devoid of coronary vasoactivity.

Despite these successes, validation of the QSAR approach is incomplete. Estimates of the hydrophobicity (2-butanol: water distribution coefficient) of the analogues whose vasoactivity was studied by other investigators [6, 9, 12, 13] are not yet available. Coronary vasoactivity data on ring isosteres of adenosine such as formycin or 2-aza adenosine are uninterpretable without information about dipole moment, which requires molecular orbital calculations beyond our competence. Further, base dipole moments are strongly influenced by rotation of the ribose around the glycosylic bond [27]; thus, the dipole moment of the ligand bound to the receptor may differ systematically from our estimate. Until the conformation of bound ligand is known, there is no way to assess the size of this systematic error. Additionally, the strength and direction of a dipole moment are to some extent covariant, so it is difficult to decide which factor contributes more to binding affinity. Above all, coronary vasoactivity in the intact heart is only approximately related to a ligand's affinity for the adenosine receptor. Additional factors, such as hydrophobicity and whether a nucleoside is a substrate for the adenosine transport carrier or for enzymes such as adenosine kinase, will affect ligand concentration in the vicinity of the receptor. The vasoactivity of certain nucleosides, e.g., inosine and N^6-(4-nitrobenzyl) thioguanosine, appears to reflect, at least in part, their inhibition of the adenosine transport carrier and, consequently, local accumulation of adenosine [17]. Nevertheless, the dipole moment hypothesis serves as a useful guide for further experimental challenge, ultimately at the level of isolated membranes and purified adenosine receptors.

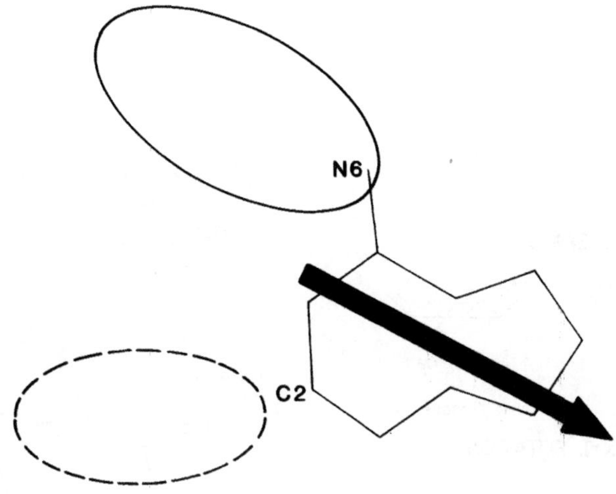

Figure 3.5. Topography of the purine domain of the coronary adenosine receptor. The arrow indicates the orientation of the receptor dipole 180° away from the average dipole of vasoactive adenosines. The ellipses represent the hydrophobic regions believed to lie near C2 and N6.

Figure 3.5 summarizes our provisional model of the purine domain of the coronary adenosine receptor. The main feature of this model is a dipole moment oriented at 250°, opposite the alignment of that of the average base dipole of the most vasoactive adenosine analogues. This model includes a hydrophobic region near C6 and a second hydrophobic region near C2. The latter region is indicated in the diagram by dashed lines, since experimental support for its existence is less certain. Note that nonlocalized, i.e., "bulk," properties of the base—namely, dipole moment and hydrophobicity—are primary determinants of binding affinity. This suggests that the chemical properties of a specific substituent, e.g., the ability of an amino group to participate in hydrogen bonding, may be less important than properties that are distributed over the entire purine base.

Determinants of binding to the ribose domain

The *ribo* configuration of the C2′ and C3′ hydroxyl groups is unquestionably the prime determinants of binding to the ribose domain of the coronary adenosine receptor. The types of forces mediating this interaction are uncertain. Reduction (2′- and 3′-deoxyadenosine), etherification (*O*-methyl ethers), or steric hindrance exerted by a bulky vicinal substituent (psicofuranine, adenine-α-L-lyxoside, and 4′-*O*-methyladenosine) greatly reduce coronary vasoactivity, while epimerization (adenine arabinoside and adenine xyloside) abolishes activity altogether. Many C5′-substituted adenosines are vasodilators whose potency is only slightly less than that of adenosine, which suggests little if any interaction of 5′ hydroxyl with the receptor. However,

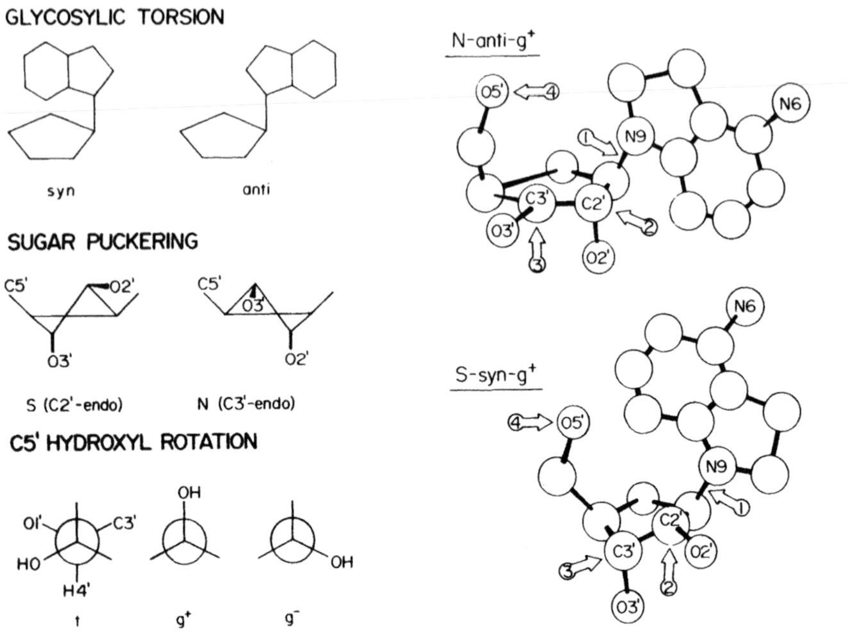

Figure 3.6. *Left:* Factors influencing the conformation of nucleoside sugar binding to the ribose domain of the adenosine receptor. *Right:* Structures of the two most common solution conformers of adenosine, *N-anti-g*$^+$ and *S-syn-g*$^+$, exhibiting concerted changes in *syn* ⇌ *anti* and *N* ⇌ *S* equilibria. Numbered arrows refer to, respectively: 1, glycosylic bond; 2, C2′ and its hydroxyl group; 3, C3′ and its hydroxyl group, 4, 05′. Note the important changes in the location of 02′ and 03′ as a consequence of the N → S transition.

polarizability of the bond between C5′ and its substituent importantly influences vasoactivity, perhaps reflecting an electrostatic interaction with N3 or C8 of the purine [18, 28].

The flexibility of adenosine is a serious obstacle to using structure-activity relationships to describe the spatial relationship of the purine and ribose domains. As shown in Figure 3.6, rotation of the base around the axis of the glycosylic bond generates *syn* and *anti* rotamers. C-C bond strain in a cyclopentane ring causes the ring to pucker such that two atoms lie outside the plane formed by the other three. X-ray crystallography shows that this is also the case for the furanose ring of ribonucleosides; C2′ and C3′ are usually displaced on opposite sides of the ring. Depending on whether C2′ or C3′ is "up," i.e., on the same side of the ring as C5′, one has either an *N* or an *S* conformer [29]. Additionally, rotation of C5′ around the axis of the C4′-C5′ bond orients the 5′-OH in g^-, g^+ or *t* conformations. Proton NMR studies demonstrate rapid (nanosecond) interconversion of these various conformations, so that adenosine in solution is an equilibrium mixture of all possible combinations [30]. Recent theoretical [31] and experimental [32] studies suggest that a transition from *syn* to *anti* causes a change from *S* to *N*, explaining why adenosine solutions seem to consist almost entirely of *S-syn* and *N-anti* conformers.

The consequences of these conformational changes are not trivial. The poise of the *syn* ⇌ *anti* equilibrium determines whether the base and ribose moieties can fit into their respective binding sites. Because the $S \rightarrow N$ transition rotates both the C2′ and C3′ hydroxyls over a distance > 1 Å, which is approximately the distance over which binding forces operate, e.g., hydrogen bonding, the poise of the $S \rightleftharpoons N$ equilibrium strongly influences how well these important functional groups interact with the receptor. Because the 5′-OH seems to contribute little to vasoactivity, rotation of this functional group is relatively unimportant. Further, solutions of most nucleosides consist mainly (≥ 70%) of the g^+ rotamer, which suggests that this factor is not likely to account for individual differences in ligand affinity.

Available experimental evidence concerning which adenosine conformer binds to the receptor is meager and highly ambiguous. Cycloadenosines held rigidly in *anti* conformation by -O- or -S- bonds between purine C8 and ribose C2′, C3′, or C5′ have no coronary vasoactivity [17]. This could mean that the receptor accommodates only *syn* nucleosides. However, oxygen and sulfur are strongly electronegative and thus doubtless alter the electronic structure of the adenine base. Additionally, this kind of base-to-sugar bond importantly changes the structure of the sugar. Either factor could by itself explain the loss of coronary vasoactivity. Prasad et al. [18] have suggested that the extremely powerful coronary vasoactivity of adenosine 5′-uronic acid carboxamides [16] is due to a hydrogen bond between the carboxamide N and purine N3, which stabilizes these nucleosides in the *syn* conformation. These authors support this hypothesis with the observation that dialkyl carboxamides, which lack an amide hydrogen to participate in hydrogen bonding, have no coronary vasoactivity. Finally, proton NMR studies of monoalkyl carboxamides showed selective nuclear Overhauser enhancement of the ribose H1′ signal upon irradiation at the frequency of the purine H8 resonance. This evidence for the proximity of C8 and H1′ suggests that *syn* conformers may have constituted the major species in solution. It does not necessarily mean, however, that this is the species bound to the receptor. Moreover, this hypothesis fails to account for the coronary vasodilator potency of the lower alkyl esters of adenosine-5′-uronic acid [15]. These esters lack the ability to form hydrogen bonds with adenine N3, yet they are only slightly less potent than the homologous carboxamides. Recently we have synthesized and assayed the coronary vasoactivity of adenosine-5′-uronic acid β-*unsym*-dimethyl hydrazide. This nucleoside is an analogue of the inactive dimethyl carboxamide but has a hydrogen on the hydrazine β N, which is available for hydrogen bonding. The Prasad hypothesis predicts that it should be vasoactive, but our experiments show that it is more than two orders of magnitude less active than adenosine.

Information about the influence of the $S \rightleftharpoons N$ equilibrium on binding affinity is completely lacking. Because the information about the influence of the *syn* ⇌ *anti* equilibrium is ambiguous and information about the *S* ⇌ *N* equilibrium is nonexistent, a description of the topography of the ribose domain seems remote at this time.

MECHANISM OF CORONARY RELAXATION BY ADENOSINE

There are currently two hypotheses to explain how adenosine initiates coronary relaxation. The first is that adenosine is a Ca^{2+} entry blocker. Following the lead of earlier studies showing that adenosine inhibits the slow inward current of the atrial action potential via surface receptors [33, 34], Harder et al. [35] showed that adenosine in micromolar concentrations likewise inhibits the slow inward current in coronary myocytes. Additionally, adenosine inhibits ^{45}Ca uptake by both intact and cultured vascular smooth muscle cells [36]. The calcium entry blocker hypothesis predicts that coronary relaxation by adenosine and by conventional Ca^{2+} entry blockers should be additive. Consequently, the report that nifedipine, one of the Ca^{2+} entry blockers, antagonizes the coronary vasoactivity of adenosine [37] is of great interest. Ca^{2+} entry blockers are thought to inhibit voltage-operated Ca^{2+} channels, whereas adenosine presumably acts through so-called receptor-operated Ca^{2+} channels. Why blockade of one type of Ca^{2+} channel should antagonize blockade of the other type is an unexplained paradox. Nifedipine bears no obvious structural resemblance to adenosine or to adenosine analogues, so antagonism through binding to the adenosine receptor seems an unlikely explanation.

The second hypothesis, that adenylate cyclase mediates coronary relaxation by adenosine, is based on the observation that adenosine-induced relaxation parallels cyclic AMP accumulation in bovine epicardial coronary rings [38]. This hypothesis has substantial support, not the least of which is that all adenosine receptors described to date seem to operate through adenylate cyclase. Table 3.1 summarizes additional experimental evidence that shows that

Table 3.1. Comparison of coronary adenosine and adenylate cyclase-associated stimulatory (R_a) receptors

Characteristic	Adenylate cyclase R_a receptor[a]	Coronary adenosine receptor	Reference
Cell surface	Yes	Yes	[34, 42]
Effect on adenylate cyclase	Stimulatory	Stimulatory	[38]
Theophylline antagonism	Yes	Yes	[3]
GTP requirement	Absolute	Untested	
Analogue potency, NECA > L-PIA	Yes	Yes, 19:1[b]	
ED_{50} of adenosine	0.5–25 μM	0.5 μM	[17, 26]

Abbreviations: NECA, adenosine-5′-uronic acid-*N*-ethyl carboxamide; L-PIA, N^6-[*R*(−)-1-methyl-2-phenethyl]adenosine.
[a]From [41].
[b]This ratio applies to the N-cyclopropyl analogue of NECA, which is a powerful stimulant of adenylate cyclase [43].

the coronary adenosine receptor exhibits most of the characteristics by which one recognizes the adenylate cyclase-associated stimulatory (R_a or A_2) adenosine receptor.

If adenosine relaxation employs cyclic AMP as a second messenger, one might ask what the substrate is for the protein kinase that cyclic AMP presumably activates. Phosphorylation of plasma membrane proteins—perhaps the Ca^{2+} channels themselves [39]—is one possibility, which is supported by a report of the phosphorylation of sarcolemmal proteins by intramembranous cyclic AMP-dependent protein kinases [40]. Such a notion is consonant with Harder's electrophysiological observations, which clearly demonstrate adenosine effects at the level of the coronary myocyte sarcolemma.

Indeed, proof that the membrane electrical effects of adenosine are secondary to a cyclic AMP-dependent mechanism would unify the two contending hypotheses that now seek to explain vascular relaxation by adenosine.

ADENOSINE RECEPTORS IN OTHER VASCULAR BEDS

The foregoing description of the coronary adenosine receptor may apply to other vascular beds that are dilated by adenosine, such as those of brain and skeletal muscle. This model may not describe the vascular receptors for adenosine in the renal circulation, which undergoes intense vasoconstriction in response to adenosine administration (see Chapter 25).

NOTE

1. The ambiguous trivial name "phenylisopropyladenosine" is highly confusing because it refers to a derivative of any of six isomeric amines, two of which exist as optical isomers by virtue of a chiral center. These six amines are N-phenyl-2-propylamine; 2-, 3-, or 4-(2-propyl) aniline; 1-methyl-2-phenethylamine (either *R* or *S*); and 2-methyl-2-phenethylamine (either *R* or *S*). The confusion arising from this trivial name extends to the marketplace. The tritiated PIA offered by Amersham as "N^6-L-2-phenylisopropyladenosine" is a derivative of 2-(2-propyl) aniline, whereas that offered by New England Nuclear is a derivative of L(−)-1-methyl 2-phenethylamine. The "N^6-(L-2-phenyl-isopropyl)-adenosine" sold by Boehringer Mannheim is derived from L(−)-1-methyl-2-phenethylamine, has the *R* configuration [personal communication], and is identical to L-PIA studied by Westermann's group [14]. In this chapter, L-PIA refers to N^6-[*R*(−)-1-methyl-2-phenethyl]adenosine.

REFERENCES

1. Drury AN, Szent-György A: The physiological activity of adenine compounds with especial reference to their action upon the mammalian heart. *J Physiol (Lond)* 28:213–237, 1929.
2. Sattin A, Rall TW: The effect of adenosine and adenine nucleotides on the cyclic adenosine 3′, 5′-monophosphate content of guinea pig cerebral cortex slices. *Mol Pharmacol* 6:13–23, 1970.
3. Afonso S: Inhibition of coronary vasodilating action of dipyridamole and adenosine by aminophylline in the dog. *Circ Res* 26:743–752, 1970.
4. Bünger R, Haddy FJ, Gerlach E: Coronary responses to dilating substances and competitive inhibition by theophylline in the isolated perfused guinea pig heart. *Pfluegers Arch* 358:213–224, 1975.
5. Thurau K: Renal hemodynamics. *Am J Med* 36:698–719, 1964.
6. Gough G, Maguire MH: 2-Trifluoromethyladenosine. *J Med Chem* 8:866–867, 1965.
7. Jahn W: Kreislaufwirkungen 5′-substituierter adenosinderivate. *Naunyn Schmiedebergs Arch Pharmacol* 251:95–104, 1965.
8. Jahn W: N^6-[Naphthyl-(1)]-methyl-adenosin, ein Adenosin-Derivat mit lang anhaltender Coronarwirkung. *Arzneim Forsch* 19:701–704, 1969.

9. Angus JA, Cobbin LB, Einstein R, Maguire MH: Cardiovascular actions of substituted adenosine analogues. *Brit J Pharmacol* 41:592–599, 1971.
10. Shimizu B, Kaneko M, Saito A, Nishino H, Mizuno H, Oshima T, Nakayama K, Koike H: An alternate synthesis of N^6 substituted adenosines and their coronary dilator activities. *Ann Sankyo Res Lab* 23:117–123, 1971.
11. Scholtholt J, Nitze RE, Schraven E: On the mechanism of the antagonistic action of xanthine derivatives against adenosine and coronary vasodilators. *Arzneim Forsch* 22:1255–1259, 1972.
12. Cobbin LB, Einstein R, Maguire MH: Studies on the coronary dilator actions of some adenosine analogues. *Brit J Pharmacol* 50:25–33, 1974.
13. Marumoto M, Yoshioka Y, Miyashita O, Shima S, Imai K-I, Kawazoe K, Honjo M: Synthesis and coronary vasoactivity of 2-substituted adenosines. *Chem Pharm Bull* 23:759–774, 1975.
14. Vapaatalo H, Onken D, Neuvonene PJ, Westermann E: Stereospecificity in some central and circulatory effects of phenylisopropyl-adenosine (PIA). *Arzneim Forsch* 25:407–410, 1975.
15. Prasad RN, Fung A, Tietje K, Stein HH, Brondyk HD: Modification of the 5′ position of purine nucleosides: 1. Synthesis and biological properties of alkyl adenosine-5′-carboxylates. *J Med Chem* 19:1180–1186, 1976.
16. Raberger G, Schütz W, Kraup O: Coronary dilatory action of adenosine analogues: A comparative study. *Arch Int Pharmacodyn Ther* 230:140–149, 1977.
17. Olsson RA, Khouri EM, Bedynek JL Jr, McLean J: Coronary vasoactivity of adenosine in the conscious dog. *Circ Res* 45:468–478, 1979.
18. Prasad RN, Bariana DS, Fung A, Savic M, Tietje K, Stein HH, Brondyk H, Egan R: Modification of the 5′ position of purine nucleosides: 2. Synthesis and some cardiovascular properties of adenosine-5′-(*N*-substituted) carboxamides. *J Med Chem* 23:313–319, 1980.
19. Olsson RA, Kusachi S: Adenosine (Ado) initiates coronary relaxation via adenylate cyclase-associated stimulatory (R*a*) receptors. (abst) *Proc Intl Symp Adenosine,* 1982.
20. Hansch C: A quantitative approach to biochemical structure-activity relationships. *Accts Chem Res* 2:232–239, 1969.
21. Pullman B, Berthod H, Bergmann F, Neiman Z, Weiler-Feilchenfeld H, Bergmann ED: Electronic structure of purine tautomers. *Tetrahedron* 26:1483–1491, 1970.
22. DeVoe H, Tinoco I Jr: The stability of helical polynucleotides: Base contributions. *J Mol Biol* 4:500–517, 1962.
23. Bergmann, ED, Weiler-Feilchenfeld H: The dipole moments of purine, in Bergmann ED, Pullman B (eds): *The purines: Theory and experiment.* Jerusalem, Israel Acad Sci Human, 1972, pp 21–28.
24. Rao ST, Sundaralingam M: Stereochemistry of nucleic acids and their constituents: XIII. The crystal and molecular structure of 3′-*O*-acetyladenosine. *J Am Chem Soc* 92:4963–4970, 1970.
25. Minkin VI, Isipov OA, Zhdanov YA: *Dipole Moments in Organic Chemistry.* New York, Plenum Press, 1970, p 91.
26. Schrader J, Haddy FJ, Gerlach E: Release of adenosine, inosine and hypoxanthine from isolated guinea pig heart during hypoxia, flow-autoregulation and reactive hyperemia. *Pfluegers Arch* 369:1–6, 1977.
27. Berthod H, Pullman B: A molecular orbital investigation of the *syn* and *anti* conformers of purine and pyrimidine nucleosides. *Biochem Biophys Res Commun* 46:125–129, 1972.
28. Follman H, Gremels G: Adenine nucleosides in solution: Stabilization of the *anti*-conformation by C5′ substituents. *Eur J Biochem* 47:187–197, 1974.
29. Altona C, Sundaralingam M: Nucleotides: A new description using the concept of pseudorotation. *J Am Chem Soc* 94:8205–8212, 1972.
30. Saenger W: Nucleotide conformations: Rules and exceptions, in Walker RT, Le Clercg E, Eckstein F (eds): *Nucleoside Analogues: Chemistry, Biology and Medical Applications.* New York, Plenum Press, 1979, pp 1–13.
31. Kitamura K, Wakahara A, Mizuno H, Baba Y, Tomita K-I: Conformationally "concerted" changes in nucleotide structures: A new description using circular correlation and regression analyses. *J Am Chem Soc* 103:3899–3904, 1981.
32. Klimke G, Lüdemann H-D, Townsend LB: Further evidence for a S-syn correlation in the purine (β) ribosides: The solution conformation of two tricyclic analogs of adenosine and guanosine. *Z Naturforsch [C]* 34:653–657, 1979.
33. Schrader J, Rubio R, Berne RM: Inhibition of slow action potentials of guinea pig atrial muscle by adenosine: A possible effect on Ca^{2+} influx. *J Mol Cel Cardiol* 7:427–433, 1975.

34. Schrader J, Nees S, Gerlach E: Evidence for a cell surface receptor on coronary myocytes and atrial muscle cells. *Pfluegers Arch* 369:251–257, 1977.
35. Harder DR, Bellardinelli L, Sperelakis N, Rubio R, Berne RM: Differential effects of adenosine and nitroglycerin on the action potentials of large and small coronary arteries. *Circ Res* 44:176–182, 1979.
36. Fenton RA, Bruttig SP, Rubio R, Berne RM: Effect of adenosine on calcium uptake by intact and cultured vascular smooth muscle. *Am J Physiol* 242:H797–H804, 1982.
37. Young MA, Merrill GF: Nifedipine inhibition of adenosine dilation and reactive hyperemia in the coronary vasculature of dogs (abst) *Fed Proc* 40:691, 1981.
38. Kukovetz WR, Wurm A, Holzmann S, Pöch G, Rinner I: Evidence for an adenylate cyclase-linked adenosine receptor mediating coronary relaxation, in Baer HP, Drummond GI (eds): *Physiological and Regulating Functions of Adenosine and Adenine Nucleotides*. New York, Raven Press, 1979, pp 205–213.
39. Katz AM, Messineo FC, Herbette L: Ion channels in membranes. *Circulation* 65 (Suppl I): I2–I10, 1982.
40. Walsh DA, Clippinger MS, Srirmakrishnan S, McCullough TE: Cyclic adenosine monophosphate dependent and independent phosphorylation of sarcolemma membrane proteins in perfused rat heart. *Biochemistry* 18:871–877, 1979.
41. Wolff J, Londos C, Cooper DMF: Adenosine receptors and the regulation of adenylate cyclase. *Adv Cyclic Nucleotide Res* 14:199–214, 1981.
42. Olsson RA, Davis CJ, Khouri EM, Patterson RE: Evidence for an adenosine receptor on the surface of dog coronary myocytes. *Circ Res* 39:93–98, 1976.
43. Daly JW, Nimitkitpaisan Y, Pons F, Bruns RF, Smellie F, Skolnick P: Binding sites for adenosine analogs: Possible relationship to cyclic AMP-generating systems in brain tissue. (abst) *Pharmacologist* 21:253, 1979.

DISCUSSION

DR. H.V. SPARKS: What are the interactions between pharmacologic Ca^{2+} antagonists and adenosine?

DR. OLSSON: A report in abstract (Young and Merrill, *Fed Proc* 40:691, 1981) describes antagonism of adenosine coronary vasodilation by nifedipine, one of the Ca^{2+} entry blockers. This result bemuses me. Ca^{2+} entry blockers are said to act on "voltage-operated" Ca^{2+} channels. With the exception of D-600 and one of its congeners, verapamil, the drugs in this class bear no structural similarity to one another or to adenosine. Contrariwise, the coronary vasodilation by adenosine analogues is so highly stereospecific that one must conclude that if adenosine acts directly on Ca^{2+} channels, these must be the "receptor-operated" variety. I find it paradoxical that blockade of a "voltage-operated" channel should antagonize blockade of a "receptor-operated" channel. One possible explanation for this dilemma lies in recognizing that a "channel" is an operational concept, i.e., a channel is defined by its behavior. Since Ca^{2+} channels have not been isolated and chemically characterized, one cannot dismiss the possibility that "voltage-" and "receptor-operated" channels are but different functional manifestations of a single physical entity.

DR. S.J. MUSTAFA: In reference to Dr. Sparks's question as to the relationship of adenosine to calcium channel blockers, we have recently reported at the last Federation Meeting in New Orleans that the adenosine-induced relaxation of bovine coronary arteries is not affected by D-600, verapamil, nifedipine, and diltiazem. However, lidoflazine, another calcium blocker, which also inhibits the uptake of adenosine, potentiates the relaxing activity of adenosine several-fold. This would suggest that the site of action of adenosine in relation to calcium is different from calcium blockers. Also, different calcium blockers may have different mechanisms of action.

DR. R. RUBIO: Adenosine and La^{3+} block Ca^{2+} action potentials through a different mechanism. Adenosine acts independently of the extracellular Ca^{2+}, whereas La^{2+} acts by competing with Ca^{2+}. Various Ca^{2+} blockers act probably through different mechanisms.

DR. H. NAWRATH: I should like to comment on the interaction between calcium antagonists and adenosine. A possible mechanism of the action of adenosine is the activation of potassium conductance at the cell membrane, which leads indirectly to a diminished calcium influx. Calcium antagonists, however, are well able to inhibit potassium currents. This may be the basis for the nonspecific antagonism between adenosine and calcium antagonists.

DR. D.A. MCAFEE: Does adenosine cause a change (decrease) in input impedance in coronary vessels and thus short-circuit the Ca^{2+} spike?

DR. OLSSON: This refers to Dr. Harder's experiments done in Dr. Berne's laboratory. I can't answer this question, but Dr. Berne replied that adenosine does not alter input impedance nor short-circuit the Ca^{2+} spike.

DR. R.F. BRUNS: My studies on the A_2 receptor in human fibroblasts have shown that the *anti* conformation is required for affinity. Cycloadenosines that are locked in the *anti* conformation by bridges between the 5′ and 8 positions are adenosine antagonists, whereas those bridged between the 8 position and the 2′ or 3′ position were inactive.

DR. OLSSON: I am aware of your careful studies and the possible bearing they have on our work with the coronary adenosine receptor. In our hands, the 8,2′-, 8,3′-, and 8,5′-cyclonucleosides to which you refer had no coronary vasoactivity. Unfortunately, the quantities of these nucleosides were not sufficient to test antagonism. Your work is important to our model in another respect. If the coronary adenosine receptor turns out to be an R_a receptor, your experiments with C8-substituted adenosines predict a third hydrophobic patch, this one in the vicinity of the site occupied by the purine C8.

4. A COMPARISON OF RECEPTORS FOR ADENOSINE AND ADENINE NUCLEOTIDES

G. BURNSTOCK

Evidence was presented in 1972 to support the hypothesis that a purine nucleotide, probably adenosine 5′-triphosphate (ATP), was the transmitter in nonadrenergic, noncholinergic nerves supplying the smooth muscle of the gut and bladder, and the term *purinergic* was coined [1]. Implicit in the hypothesis was the existence of purinergic receptors on the postjunctional muscle membranes [2].

In 1978, I proposed that two types of purinergic receptors could be distinguished [3]. Their subclassification into P_1 and P_2 purinoceptors was based on four criteria: the relative potency of the adenine compounds, selective antagonism, activation of adenylate cyclase by adenosine, and induction of prostaglandin synthesis by ATP and ADP (see Table 4.1). This was derived from a combination of experimental data from our own laboratory and a detailed analysis of the voluminous literature on the actions of purine nucleotides and nucleosides on a wide variety of tissues. For example, the log concentration response curves for the inhibitory effects of ATP and ADP on the guinea pig taenia coli were found to be nonparallel to those for AMP and adenosine [4]; 2,2′-pyridylisatogen tosylate was shown to block the inhibitory responses of the taenia coli to ATP and ADP but not to adenosine [5]; adenosine and ATP were reported to induce opposite responses on the renal vascular bed [6]; and theophylline was shown to antagonize adenosine actions in the cerebral cortex [7]. Thus, it was proposed that P_1 purinoceptors are more responsive to adenosine and AMP than to ATP and ADP, that methylxanthines such as

R.M. Berne, T.W. Rall, and R. Rubio (eds.), Regulatory Function of Adenosine, *pp. 49–62.*

Table 4.1. A classification of extracellular receptors for adenosine and adenine nucleotides into P_1 and P_2 purinoceptors

	Antagonists	Agonist potencies	cAMP production
P_1	Methylxanthines Quinidine 2-substituted imidazolines	AD$\geqslant$AMP>ADP$\geqslant$ATP	Yes
P_2	2'2-pyridilisatogen Apamin Arylazido aminopropionyl ATP	ATP$\geqslant$ADP>AMP$\geqslant$AD	No

Source: Modified from Burnstock [3].
Note: Each purinoceptor subtype exhibits a different rank order of potency for adenosine and its related nucleotides. The P_1 purinoceptor is generally competitively antagonized by the methylxanthines, and occupation of that receptor often leads to the production of cyclic AMP. Although no drug can presently be considered as a competitive antagonist of the effects of ATP at the P_2 purinoceptor, a variety of compounds have been used in the past to antagonize ATP-induced responses. More recently, the bee venom toxin apamin (which blocks some potassium channels) and the photoaffinity analogue arylazido aminopropionyl ATP ($ANAPP_3$) have been used as antagonists of ATP-induced responses.

theophylline and caffeine are selective competitive antagonists, and that occupation of these receptors leads to activation of an adenylate cyclase system with resultant changes in levels of intracellular cyclic AMP; whereas the P_2 purinoceptors are more responsive to ATP and ADP than to AMP and adenosine, are not antagonized by methylxanthines, do not act via an adenylate cyclase system, and their occupation may lead to prostaglandin synthesis. It was recognized, however, that some of the actions of ATP and ADP might be mediated via P_1 purinoceptors following their rapid breakdown by ectoenzymes to AMP and adenosine.

Since 1978, evidence has accrued to support the P_1/P_2 purinoceptor hypothesis [8]. For example, in the guinea pig bladder, P_1 purinoceptors mediate relaxation, while P_2 purinoceptors mediate contraction [9, 10]; in rat ileum, both P_1 and P_2 purinoceptors mediate relaxation, but only the P_1 purinoceptors are blocked by theophylline [11]. P_2 purinoceptors directly mediate contraction of the smooth muscle of dog saphenous vein, while P_1 purinoceptors are present on adrenergic nerve terminals to mediate inhibition of norepinephrine release [12]; in the frog atrium, P_2 purinoceptors mediate excitation, whereas P_1 purinoceptors are inhibitory, although guinea pig atria appear to possess only inhibitory P_1 purinoceptors [13]. Theophylline and apamin have been shown to differentially block the responses of the taenia coli to adenosine and ATP, respectively [14], and separate receptors for adenosine and ATP in these preparations were also indicated on the basis of studies with slowly degradable analogues of ATP [15]. P_2 purinoceptors have been detected in cultured renal epithelial cells [16]; P_1, but not P_2, purinoceptors have been seen on cholinergic nerve terminals in the guinea pig ileum [17]; and P_2 purinoceptors have been identified on parotid acinar cells [18].

P_1 PURINOCEPTORS

Stereospecificity

Evidence for stereospecificity of the P_1 purinoceptor has been presented recently [19]. The potency of the D-enantiomers for producing inhibitory effects on the driven left atria, the trachea and cholinergic excitatory (twitch) responses of the ileum of the guinea pig was 5'-N-ethylcarboxamidoadenosine (NECA) > 2-chloroadenosine ≥ 2-azidoadenosine > adenosine. The L-enantiomers of adenosine and these analogues did not produce inhibitory responses in the atrium and transmurally stimulated ileum, and they were considerably less potent in the trachea.

Antagonists

Whereas theophylline and caffeine are effective competitive antagonists to P_1 purinoceptors in a wide variety of tissues [8, 20], 8-phenyltheophylline has been shown to be substantially more potent than theophylline in antagonizing adenosine-induced accumulation of cyclic AMP in slices of guinea pig cerebral cortex [21] and in human fibroblasts [22], as well as antagonizing the responses elicited by adenosine of the guinea pig atrium, rabbit basilar artery, cholinergic nerve terminals in the guinea pig ileum [23], and adrenergic terminals in the rat vas deferens [24].

Autoradiographic localization

Subclassifications of the P_1 purinoceptors into R_i and R_a or A_1 and A_2 have been proposed largely on the basis of the outcome of receptor occupation (decreased versus increased) in cyclic AMP production and/or in the relative potencies of adenosine analogues such as N^6-cyclohexyladenosine (CHA), NECA, and N^6-phenylisopropyladenosine (PIA) [25–29]; however, the validity of these subclasses is still being debated. One approach that might help to resolve this question is the ingenious technique developed by Young and Kuhar [30] that employs high-affinity radioligands to give autoradiographic localization of receptors. Using this technique, P_1 purinoceptors have been localized in the brain using ^{3}H-CHA [31, 32]. In my own laboratory, Noel Buckley has recently demonstrated binding sites in the gut using ^{3}H-NECA; the autoradiographic grains are displaced by 2-chloroadenosine but not by dipyridamole or dialazep, indicating that binding is not at adenosine uptake sites and suggesting that these grains are marking P_1 purinoceptors.

P_2 PURINOCEPTORS

Stereospecificity

D-methylthio ATP has been shown recently by Ian MacKenzie, Perry Meghji, and Judith Hills in my laboratory to be about 200 times more potent than ATP and 700 times more potent than the L-methylthio ATP enantiomer in producing relaxation of the guinea pig taenia coli, suggesting that the inhibitory P_2

purinoceptors are probably stereospecific. However, significant differences in the excitatory responses of the guinea pig bladder and frog ventricle to the D- and L-enantiomers were not found, which may indicate that a different receptor is involved. Shuba and Vladimirova [33] have also suggested that there are subclasses of P_2 purinoceptors based on the actions of apamin; apamin blocks the inhibitory actions of ATP on various gut preparations but does not block the excitatory actions of ATP in bladder and various blood vessels. More recently, block of the excitatory but not the inhibitory actions of ATP on guinea pig vas deferens and stomach, respectively, by arylazido aminopropionyl ATP ($ANAPP_3$) led to a similar view [34]. The inhibitory actions of ATP on the intestinal smooth muscle result from a specific increase in K^+ conductivity [35, 36], whereas excitatory actions of ATP may result from activation of Ca^{2+} channels [37–39].

Responses to ultraviolet (UV) light

The inhibitory responses of the guinea pig taenia coli to ATP and to nonadrenergic, noncholinergic (purinergic) nerve stimulation are precisely mimicked by exposure to UV light (maximum response produced by wavelengths of 340–380 nm) [40]. This remarkable effect is not due to release of ATP through initiation of impulses in nerves, since the response to UV light is not affected by tetrodotoxin, or to direct release from nerve terminals, since ATP was not detected in perfusates during the response. It may be possible to use this approach to characterize the chemical nature of the P_2 purinoceptors.

Antagonists

Several compounds, including quinidine, 2-substituted imidazolamines and 2,2′pyridylisatogen tosylate have been shown during the past 10 years to antagonize responses to ATP in various preparations, but none of these compounds are competitive antagonists and all have nonspecific actions [41]. When low concentrations of apamin were shown by Shuba and Vladimirova [33] to block ATP but not adenosine, the possibility that this substance acted as a P_2 purinoceptor antagonist was considered. However, it was subsequently shown to act by blocking K^+ channels rather than as a competitive ATP antagonist [42]; nevertheless, it is a useful agent to distinguish P_1 and P_2 purinoceptor inhibitory actions in the gut [14].

The most promising development has appeared recently, with the claim that $ANAPP_3$ is a specific P_2 antagonist. However, it is a difficult compound to use, because it must be incubated with the tissues in the dark before exposure to light in order to be effective [43]; further, it may be more effective against the excitatory than the inhibitory actions of ATP [34, 44].

ATP-induced prostaglandin synthesis

Prostaglandins are often involved in the actions of ATP on smooth muscle. ATP and ADP are potent inducers of prostaglandin synthesis in a wide variety of tissues, whereas adenosine is not [45]. The "rebound contractions" that typically follow the ATP-induced inhibitory responses of the guinea pig taenia coli have been shown to be blocked by the prostaglandin synthesis inhibitor, indomethacin [46], and more recently by the prostaglandin receptor antagonist, phenylphosphonate N-0164 [47]. The ATP-induced responses of a number of other tissues have since been shown to be reduced by indomethacin, including guinea pig trachea [48], intestine [49], bladder [50–52], anococcygeus [53], and uterus [54]. Purinergic stimulation of prostaglandin synthesis has also been demonstrated in rabbit kidney and heart [55]. The relaxation of tracheal strips in response to exogenous ATP are entirely blocked by indomethacin, and the slowly degradable analogues of ATP, α, β-methylene ATP, and β,γ-methylene ATP, which do not induce prostaglandin synthesis, have no effect on this preparation [56]. It has also been suggested that during the axon reflex, ATP is released from sensory collaterals, probably together with substance P, and in addition to producing direct vasodilation of skin vessels, induces prostaglandin synthesis, which may contribute to the inflammatory reaction [57].

PHYSIOLOGICAL ROLES OF P_1 AND P_2 PURINOCEPTORS

Evidence has been presented for physiological roles mediated by P_1 and P_2 purinoceptors in a variety of organs, whether the source of adenyl compounds is from neuronal or nonneuronal cells. Holton [58] showed that ATP release accompanied vasodilation of rabbit ear vessels during antidromic stimulation of the great auricular nerve, and Berne [59] proposed that adenosine was the physiological mediator of coronary vasodilation during hypoxia. Although it was assumed that the source of the adenosine was nonneuronal, the possibility has been raised that some, if not all, of the adenosine may arise from ATP released from nerves [60, 61].

The evidence that ATP is the principal transmitter released from intramural nonadrenergic, noncholinergic neurones supplying the nonsphincteric smooth muscle of the gastrointestinal tract, urinary bladder, and rabbit portal vein is well documented and will not be repeated here [1, 62–64]. In addition to purinergic nerves, evidence has been presented in more recent years for the coexistence of ATP with the classical neurotransmitters acetylcholine (Ach) and norepinephrine (NE) in sympathetic and parasympathetic autonomic nerve terminals and in the motor end-plate of skeletal muscle, and for its release during nerve activation [65–74]. The physiological role of ATP released from adrenergic and cholinergic nerves is becoming clearer. In the general model presented in Figure 4.1, ATP released from sympathetic nerve terminals may act as a genuine *cotransmitter* on postjunctional P_2 purinoceptors, resulting in contractions that parallel those produced by NE acting on α_1-adrenoceptors. ATP may play a role as a *postjunctional neuromodulator,* since it potentiates the excitatory actions of NE [75], and postjunctional enhancement of the excitatory action of Ach by ATP at the motor end-plate has also been demonstrated [76]. This rapid mutual enhancement of the transmission process is followed by a phase of *prejunctional neuromodulation,*

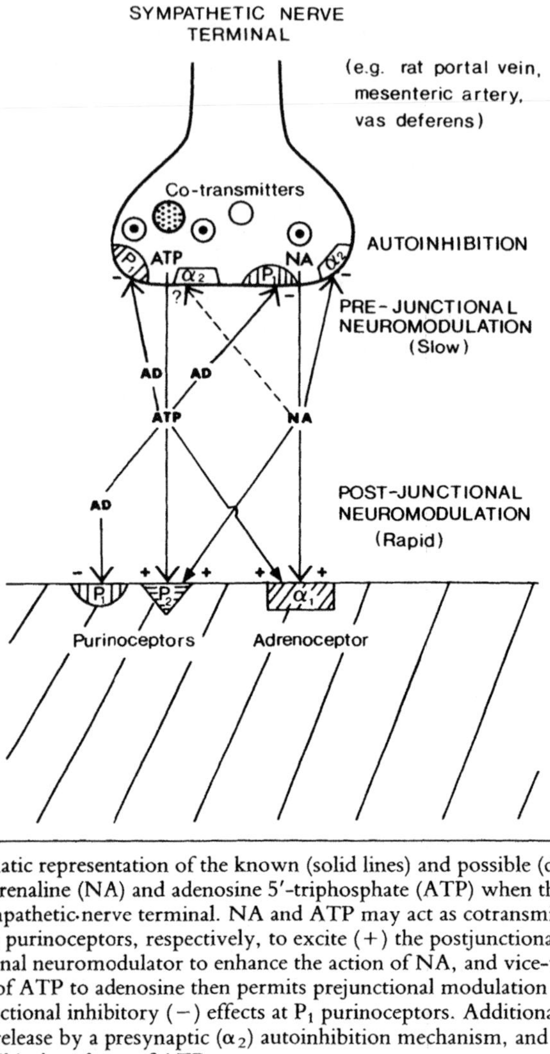

Figure 4.1. Schematic representation of the known (solid lines) and possible (dashed line) interactions of noradrenaline (NA) and adenosine 5′-triphosphate (ATP) when they are released from the same sympathetic nerve terminal. NA and ATP may act as cotransmitters on α_1 adrenoceptors and P_2 purinoceptors, respectively, to excite (+) the postjunctional cell. ATP may act as a postjunctional neuromodulator to enhance the action of NA, and vice-versa. The subsequent breakdown of ATP to adenosine then permits prejunctional modulation (−) of the release of NA and postjunctional inhibitory (−) effects at P_1 purinoceptors. Additionally, NA may modulate its own release by a presynaptic (α_2) autoinhibition mechanism, and similarly, adenosine may inhibit the release of ATP.

which facilitates the inactivation of the transmission process; i.e., adenosine resulting from the breakdown of ATP released from the nerve acts on prejunctional P_1 purinoceptors to reduce the release of NE [77–80]. By analogy, it is possible that NE may act prejunctionally to reduce the release of ATP, but this has yet to be tested. In addition to pre- and postjunctional modulation of NE release and activity, *autoinhibition* has been demonstrated, whereby release of ATP is inhibited via a negative feedback mechanism involving prejunctional P_1 purinoceptors [73], and release of NE is inhibited via presynaptic α_2-adrenoceptors [77]; however, it seems unlikely that this autoinhibitory mechanism plays a major role in the cotransmitter situation, as it would in a

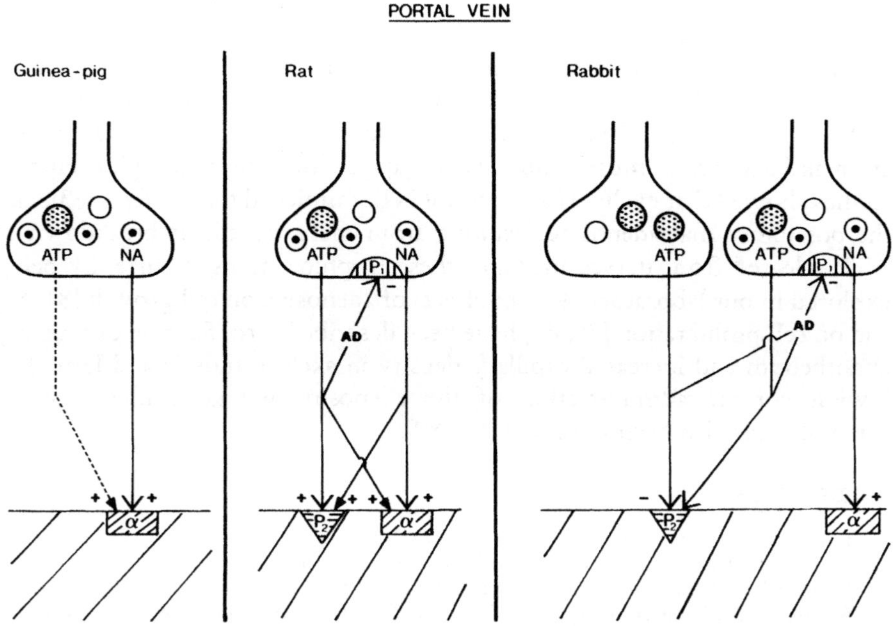

Figure 4.2. Schematic representation of the release of noradrenaline (NA) and adenosine 5'-triphosphate (ATP) from nerves supplying the portal veins of guinea pig, rat, and rabbit. The scheme is taken to support the view that prejunctional inhibition of NA release by adenosine acting on P_1 purinoceptors is a physiological event, since it takes place in the rat portal vein (where substantial ATP is released together with NA from sympathetic nerves) and the rabbit portal vein (where ATP is released as the principal transmitter from purinergic nerves as well as co-transmitter with NA in sympathetic nerves), but not in the guinea pig portal vein (where little ATP appears to be stored with NA in sympathetic nerves). Postjunctional neuromodulation occurs in the rat, and possibly guinea pig, portal veins, where both NA and ATP produce vasoconstriction, but not in the rabbit portal vein, where NA and ATP have opposite actions.

nerve terminal releasing a single (or dominant) transmitter. Postjunctional P_1 purinoceptors mediating relaxation would also operate to speed up the termination of excitatory neurotransmission.

The view that prejunctional P_1 purinoceptor inhibition of NE release from adrenergic nerves is a physiological process is supported by experiments on the portal veins of three species (see Fig. 4.2) [Török & Burnstock, unpublished observations]. In the portal vein of the guinea pig, where only low levels of ATP appear to coexist with NE in perivascular sympathetic nerves, there is no evidence for prejunctional modulation of NE release; ATP is virtually inactive on smooth muscle, although there may be some postjunctional enhancement of NE contractions. In the rat portal vein, ATP released from sympathetic nerves plays a major role as a cotransmitter and postjunctional modulator, and prejunctional modulation of NE release is prominent. Finally, in the rabbit portal vein, where there is evidence for release of ATP from purinergic inhibitory nerves as well as from sympathetic nerves together with NE [71, 81], prejunctional inhibition of NE release is again prominent.

TROPHIC ACTIONS OF ADENOSINE

The possibility that adenyl compounds may play a role in long-term "trophic" interactions between nerves and muscle, as well as acting as neurotransmitters and neuromodulators, is beginning to be explored. For example, in tissue culture, the presence of sympathetic nerves delays smooth muscle dedifferentiation and accelerates muscle bundle and gap-junction formation [82]. Further studies showed that cyclic AMP, but not NE, mimicked these effects [83], and the possibility that adenosine resulting from the extracellular breakdown of ATP released from nerves mediates these trophic actions is currently being explored in our laboratory. Potent effects of adenosine on cell growth [84, 85] and on cell proliferation [86, 87] have been described. Proliferation of capillary endothelium and increased capillary density in skeletal muscle and heart following chronic administration of the adenosine uptake inhibitor, dipyridamole, has also been observed [88, 89].

REFERENCES

1. Burnstock G: Purinergic nerves. *Pharmacol Rev* 24:509–581, 1972.
2. Burnstock G: Purinergic receptors. *J Theoret Biol* 62:491–503, 1976.
3. Burnstock G: A basis for distinguishing two types of purinergic receptor, in Bolis L, Straub RW (eds): *Cell Membrane Receptors for Drugs and Hormones*. New York, Raven Press, 1978, pp 107–118.
4. Satchell D, Burnstock G: Comparison of the inhibitory effects on guinea pig taenia coli of adenine nucleotides and adenosine in the presence and absence of dipyridamole. *Eur J Pharmacol* 32:324–328, 1975.
5. Spedding M, Weetman DF: Identification of separate receptors for adenosine and adenosine 5′-triphosphate in causing relaxations of the isolated taenia of the guinea-pig caecum. *Br J Pharmacol* 57:305–310, 1976.
6. Haddy FJ, Scott JB: Metabolically linked vasoactive chemicals in local regulation of blood flow. *Physiol Rev* 48:688–707, 1968.
7. Sattin A, Rall TW: The effect of adenosine and adenine nucleotides on the cyclic adenosine 3′,5′-phosphate content of guinea pig cerebral cortex slices. *Mol Pharmacol* 6:13–23, 1970.
8. Burnstock G (ed): *Purinergic Receptors*. Receptors and Recognition, Series B, Vol. 12, London, Chapman & Hall, 1981.
9. Burnstock G, Cocks T, Crowe R, Kasakov L: Purinergic innervation of the guinea pig urinary bladder. *Br J Pharmacol* 63:125–138, 1978.
10. Brown C, Burnstock G, Cocks T: Effects of adenosine 5′-triphosphate (ATP) and beta-gamma methylene ATP on the rat urinary bladder. *Br J Pharmacol* 65:97–102, 1979.
11. Bartlett V, Stewart RR, Nakatsu K: Evidence for 2 adenine derivative receptors in rat ileum which are not involved in the non-adrenergic, non-cholinergic response. *Can J Physiol Pharmacol* 57:1130–1137, 1979.
12. De Mey J, Burnstock G, Vanhoutte PM: Modulation of the evoked release of noradrenaline in canine saphenous vein via presynaptic receptors for adenosine but not ATP. *Eur J Pharmacol* 55:401–405, 1979.
13. Burnstock G, Meghji P: Distribution of P_1 and P_2 purinoceptors in the guinea pig and frog heart. *Br J Pharmacol* 73:879–885, 1981.
14. Brown C, Burnstock G: Evidence in support of the P_1/P_2 purinoceptor hypothesis in the guinea pig taenia coli. *Br J Pharmacol* 73:617–624, 1981.
15. Maguire MH, Satchell DG: The contribution of adenosine to the inhibitory actions of adenine nucleotides on the guinea pig taenia coli: Studies with phosphate-modified adenine nucleotide analogs and dipyridamole. *J Pharmacol Exp Ther* 211:626–631, 1979.
16. Simmons NL: Identification of a purine (P_2) receptor linked to ion transport in a cultured renal (MDCK) epithelium. *Br J Pharmacol* 73:379–384, 1981.

17. Moody C, Burnstock G: Evidence for the presence of P_1-purinoceptors on cholinergic nerve terminals in the guinea pig ileum. *Eur J Pharmacol* 77:1–9, 1982.
18. Gallacher DV: Are there purinergic receptors on parotid acinar cells? *Nature* 296:83–86, 1982.
19. Brown C, Burnstock G, Cusack NJ, Meghji P, Moody C: Evidence for stereospecificity of the P_1-purinoceptor. *Br J Pharmacol* 75:101–107, 1982.
20. Daly JW: Adenosine receptors: Targets for future drugs. *J Med Chem* 25:197–207, 1982.
21. Smellie FW, Davis CW, Daly JW, Wells JN: Alkylxanthines: Inhibition of adenosine-elicited accumulation of cyclic-AMP in brain slices and of brain phosphodiesterase activity. *Life Sci* 24:2475–2481, 1979.
22. Bruns RF: Adenosine antagonism by purines, pteridines and benzopteridines in human fibroblasts. *Biochem Pharmacol* 30:325–333, 1981.
23. Griffith S, Meghji P, Moody CJ, Burnstock G: 8-phenyltheophylline: A potent P_1-purinoceptor antagonist. *Eur J Pharmacol* 75:61–64, 1981.
24. Clanachan AS: Antagonism of presynaptic adenosine receptors by theophylline 9-beta-D-riboside and 8-phenyltheophylline. *Can J Physiol Pharmacol* 59:603–606, 1981.
25. Londos C, Wolff J: Two distinct adenosine-sensitive sites on adenylate cyclase. *Proc Natl Acad Sci USA* 74:5482–5486, 1977.
26. Van Calker D, Müller M, Hamprecht B: Adenosine regulates via 2 different types of receptors, the accumulation of cyclic-AMP in cultured brain cells. *J Neurochem* 33:999–1005, 1979.
27. Fain JN, Malbon CC: Regulation of adenylate cyclase by adenosine. *Mol Cell Biochem* 25:143–169, 1979.
28. Bruns RF, Daly JW, Snyder SH: Adenosine receptors in brain membranes: Binding of N^6-cyclohexyl[^{3}H]adenosine and 1,3′-diethyl-8-[^{3}H]phenylxanthine. *Proc Natl Acad Sci USA* 77:5547–5551, 1980.
29. Schwabe U: Direct binding studies of adenosine receptors. *Trends Pharmacol Sci* 2:299–303, 1981.
30. Young WS, Kuhar MJ: A new method for receptor autoradiography: [^{3}H] opioid receptors in rat brain. *Brain Res* 179:255–270, 1979.
31. Goodman RR, Snyder SH: Autoradiographic localization of adenosine receptors in rat brain using ^{3}H-cyclohexyladenosine. *J Neurosci* 2:1230–1241, 1982.
32. Lewis ME, Patel J, Moon Edly S, Marangos PJ: Autoradiographic visualization of rat brain adenosine receptors using N^6-cyclohexyl[^{3}H]adenosine. *Eur J Pharmacol* 73:109–110, 1981.
33. Shuba MF, Vladimirova IA: Effect of apamin on the electrical responses of smooth muscle to adenosine 5′-triphosphate and to non-adrenergic, non-cholinergic nerve stimulation. *Neuroscience* 4:853–859, 1980.
34. Frew R, Lundy PM: Effect of amylazido aminopropionyl ATP ($ANAPP_3$), a putative ATP antagonist, on ATP responses of isolated guinea pig smooth muscle. *Life Sci* 30:259–267, 1982.
35. Tomita T, Watanabe H: A comparison of the effects of adenosine triphosphate with noradrenaline and with the inhibitory potential of the guinea pig taenia coli. *J Physiol (Lond)* 231:167–177, 1973.
36. Den Hertog A: Calcium and the action of adrenaline, adenosine triphosphate and carbachol on guinea pig taenia caeci. *J Physiol (Lond)* 325:423–439, 1982.
37. Sjögren C, Andersson K-E: Effect of cholinoceptor blocking drugs, adrenoceptor stimulants, and calcium antagonists on the transmurally stimulated guinea pig urinary bladder *in vitro* and *in vivo*. *Acta Pharmacol Toxicol* 44:228–234, 1979.
38. Stone TW: Differential blockade of ATP, noradrenaline and electrically evoked contractions of the rat vas deferens by nifedipine. *Eur J Pharmacol* 74:373–376, 1981.
39. Yatani A, Tsuda Y, Akaike N, Brown AM: Nanomolar concentrations of extracellular ATP activate membrane Ca channels in snail neurones. *Nature* 296:169–171, 1982.
40. Burnstock G, Wong H: Comparison of the effects of UV light and purinergic nerve stimulation on the guinea pig taenia coli. *Br J Pharmacol* 62:293–302, 1978.
41. Burnstock G, Brown C: An introduction to purinergic receptors: History, classification and future developments, in Burnstock G (ed): *Purinergic Receptors*. Receptors and Recognition, Series B, Vol. 12. London, Chapman & Hall, 1981, pp 1–45.
42. Banks BEC, Brown C, Burgess GM, Burnstock G, Claret M, Cocks T, Jenkinson, DH: Apamin blocks certain neurotransmitter-induced increases in potassium permeability. *Nature* 282:415–417, 1979.

43. Hogaboom GK, O'Donnell JP, Fedan JS: Purinergic receptors: Photoaffinity analog of adenosine triphosphate is a specific adenosine triphosphate antagonist. *Science* 208:1273–1276, 1980.
44. Belis JA, Colby JE, Adlestein LB, Westfall DP: Characterization of neurotransmission in the bladder using arylazido-amino-propionyl ATP ($ANAPP_3$). *Surg Forum* 32:625–627, 1981.
45. Needleman P, Minkes MS, Douglas JR: Stimulation of prostaglandin biosynthesis by adenine nucleotides: Profile of prostaglandin release by perfused organs. *Circ Res* 34:455–460, 1974.
46. Burnstock G, Cocks T, Paddle BM, Staszewska-Barczak J: Evidence that prostaglandin is responsible for the 'rebound contraction' following stimulation of non-adrenergic, non-cholinergic ('purinergic') inhibitory nerves. *Eur J Pharmacol* 31:360–362, 1975.
47. Maas AJJ, Den Hertog A: The effect of the phenyl phosphonate N-0164 on prostaglandin action and on post inhibitory excitation in the taenia of guinea pig caecum. *Eur J Pharmacol* 62:157–166, 1980.
48. Kamikawa Y, Shimo Y: Mediation of prostaglandin-E2 in biphasic response to ATP of isolated tracheal muscle of guinea pigs. *J Pharm Pharmacol* 28:294–297, 1976.
49. Kamikawa Y, Serizawa K, Shimo Y: Some possibilities for prostaglandin mediation in contractile response to ATP of guinea pig digestive tract. *Eur J Pharmacol* 45:199–203, 1977.
50. Johns A, Paton DM: Effect of indomethacin on atropine resistant transmission in rabbit and monkey urinary bladder: Evidence for involvement of prostaglandins in transmission. *Prostaglandins* 13:245–254, 1977.
51. Choo LK, Mitchelson, F: The effect of indomethacin and adenosine 5′-triphosphate on the excitatory innervation of the rat urinary bladder. *Can J Physiol Pharmacol* 58:1042–1048, 1980.
52. Anderson GF: Evidence for a prostaglandin link in the purinergic activation of rabbit bladder smooth muscle. *J Pharmacol Exp Ther* 220:347–352, 1982.
53. Burnstock G, Cocks T, Crowe R: Evidence for purinergic innervation of the anococcygeus muscle. *Br J Pharmacol* 64:13–20, 1978.
54. Moritoki H, Takei M, Kasai T, Matsumuri Y, Ishida Y: Possible involvement of prostaglandins in the action of ATP on guinea pig uterus. *J Pharmacol Exp Ther* 211:104–111, 1979.
55. Schwartzman M, Pinkas R, Raz A: Evidence for different purinergic receptors for ATP and ADP in rabbit kidney and heart. *Eur J Pharmacol* 74:167–173, 1981.
56. Brown CM, Burnstock G: The structural conformation of the polyphosphate chain of the ATP molecule is critical for its promotion of prostaglandin biosynthesis. *Eur J Pharmacol* 69:81–86, 1981.
57. Burnstock G: Autonomic neuroeffector junctions: Reflex vasodilatation of the skin. *J Invest Dermatol* 69:47–57, 1977.
58. Holton P: The liberation of adenosine triphosphate on antidromic stimulation of sensory nerves. *J Physiol (Lond)* 145:494–504, 1959.
59. Berne RM: Cardiac nucleotides in hypoxia: Possible role in regulation of coronary blood flow. *Am J Physiol* 204:317–322, 1963.
60. Burnstock G: Cholinergic and purinergic regulation of blood vessels, in Bohr DF, Somly AD, Sparks HW, Geiger SR (eds): *Handbook of Physiology, Section 2: The Cardiovascular System, Vol. II, Vascular Smooth Muscle.* Amer Physiol Soc., Baltimore, Waverly Press, 1980, pp 567–612.
61. Burnstock G: Adenine nucleotides and nucleosides in cerebral hypoxia, in Rose FC, Amery WK (eds): *Cerebral Hypoxia in the Pathogenesis of Migraine Attacks.* Bath, Pitman Press, 1982, pp 77–81.
62. Burnstock G: Purinergic transmission, in Iversen LL, Iversen SD, Snyder SH (eds): *Handbook of Psychopharmacology, Vol. 5. Synaptic Modulators.* New York, Plenum Press, 1975, pp 131–194.
63. Burnstock G: The ultrastructure of autonomic cholinergic nerves and junctions. *Prog Brain Res* 49:3–21, 1979.
64. Burnstock G: Neurotransmitters and trophic factors in the autonomic nervous system. *J Physiol (Lond)* 313:1–35, 1981.
65. Su C, Bevan JA, Burnstock G: (^3H) adenosine triphosphate: Release during stimulation of enteric nerves. *Science* 173:337–339, 1971.
66. Burnstock G: Do some nerve cells release more than one transmitter? *Neuroscience* 1:239–248, 1976.
67. Burnstock G: Purinergic modulation of cholinergic transmission. *Gen Pharmacol* 11:15–18, 1980.

68. Burnstock G: The co-transmitter hypothesis, with special reference to the storage and release of ATP with noradrenaline and acetylcholine, in Cuello AC (ed): *Co-transmission*. London, Macmillan, 1982.
69. Su C: Purinergic inhibition of adrenergic transmission in rabbit blood vessels. *J Pharmacol Exp Ther* 204:351–361, 1978.
70. Ribeiro JA: Purinergic modulation of transmitter release. *J Theor Biol* 80:259–270, 1979.
71. Fedan JS, Hogaboom GK, O'Donnell JP, Colby J, Westfall DP: Contribution by purines to the neurogenic response of the vas deferens of the guinea pig. *Eur J Pharmacol* 69:41–53, 1981.
72. Potter DD, Furshpan EJ, Londis SC: Multiple-transmitter status and "Dale's Principle." *Neurosci Commentaries* 1:1–9, 1981.
73. Katsuragi T, Su C: Augmentation by theophylline of [^{3}H] purine release from vascular adrenergic nerves: Evidence for presynaptic autoinhibition. *J Pharmacol Exp Ther* 220:152–156, 1982.
74. MacKenzie I, Burnstock G, Dolly JO: The effects of purified botulinum neurotoxin A on cholinergic, adrenergic and non-adrenergic, atropine-resistant autonomic neuromuscular transmission. *Neuroscience* 7:997–1006, 1982.
75. Kažić T, Milosavljević D: Interaction between adenosine triphosphate and noradrenaline in the isolated vas deferens of the guinea pig. *Br J Pharmacol* 71:93–98, 1980.
76. Akasu T, Hirai K, Koketsu K: Increase of acetylcholine-receptor sensitivity by adenosine triphosphate: A novel action of ATP on ACh-sensitivity. *Br J Pharmacol* 74:505–507, 1981.
77. Vizi ES: Presynaptic modulation of neurochemical transmission. *Prog Neurobiol* 12:181–290, 1979.
78. Vizi ES: Non-synaptic modulation of transmitter release: Pharmacological implication. *Trends Pharmacol Sci* 1:172–175, 1980.
79. Fredholm BB, Hedqvist P: Modulation of neurotransmission by purine nucleotides and nucleosides. *Biochem Pharmacol* 29:1635–1643, 1980.
80. Husted S, Nedergaard OA: Inhibition of adrenergic neuroeffector transmission in rabbit pulmonary artery and aorta by adenosine and adenine nucleotides. *Acta Pharmacol Toxicol* 49:334–353, 1981.
81. Burnstock G, Crowe R, Wong HK: Comparative pharmacological and histochemical evidence for purinergic inhibitory innervation of the portal vein of the rabbit, but not guinea pig. *Br J Pharmacol* 65:377–388, 1979.
82. Chamley J, Campbell GR, Burnstock G: Dedifferentiation, redifferentiation and bundle formation of smooth muscle cells in tissue culture: Influence of cell number and nerve fibres. *J Embryol Exp Morphol* 32:297–323, 1974.
83. Chamley JH, Campbell GR: Trophic influences of sympathetic nerves and cyclic AMP on differentiation and proliferation of isolated smooth muscle cells in culture. *Cell Tissue Res* 161:497–510, 1975.
84. Henderson JF, Scott FW: Inhibition of animal and invertebrate cell growth by naturally occurring purine bases and ribonucleosides. *Pharmacol Ther* 8:539–571, 1980.
85. Phillis JW, Wu PH: Adenosine may regulate the vascular supply and thus the growth and spread of neoplastic tissues: A proposal. *Gen Pharmacol* 12:309–310, 1981.
86. Ichikawa A, Esumi K, Takagi M, Yatsunami K, Negishi M, Yokoyama K, Tomota K: Effect of adenosine and adenosine 5′-monophosphate on cell division of cultured mastocytoma P-815 cells. *J Pharm Dyn* 3:123–135, 1980.
87. Sandberg G, Fredholm BB: Regulation of thymocyte proliferation: Effects of L-alanine, adenosine and cyclic AMP *in vitro*. *Thymus* 3:63–75, 1981.
88. Tornling G, Unge G, Ljungqvist L, Carlsson S: Dipyridamole and capillary proliferation: A preliminary report. *Acta Pathol Microbiol Scand [A]* 86:82, 1978 (abst.).
89. Wright AJA, Hudlicka O, Tyler KR, Ziada A: The effect of vasoactive drugs on capillary density and performance in skeletal muscles (11th European Conference Microcirculation), in P Gaehtgens (ed): Bibliotheca Anatomica, No. 20. Basel, S. Karger, 1981, pp 362–365.

DISCUSSION

DR S.P. BRUTTIG: For years, we've been plagued with the problem of differentiation of smooth muscle cells in culture. I was intrigued with your comment that adenosine may prevent dedifferentiation or potentiate redifferentiation. In our hands, with primary or subcultured, enzyme-dissociated vascular smooth muscle cells of rat and pig aorta and rat mesenteric artery, these cells produce large amounts of adenosine in culture, yet they remain dedifferentiated.

DR. BURNSTOCK: In our experience, the crucial factors for differentiation of smooth muscle to proceed in culture is that confluence (i.e., close contact between cells) occurs following cell division. If the initial suspension of dissociated muscle cells is not sufficiently dense to lead to confluence even after several cell divisions, then differentiation will not occur. In the experiments I described in my talk, a trophic factor, perhaps adenosine resulting from ATP released together with norepinephrine from sympathetic nerves, accelerates the differentiation of smooth muscle bundles. However, it cannot accelerate muscle differentiation unless confluence has first taken place. The same trophic factor delays differentiation of individual smooth muscle cells, through reducing cell proliferation, confluence and subsequent differentiation. If adenosine is the trophic factor involved, then your interesting result is consistent with our findings.

DR. D.G. SATCHELL: We have recently studied the effects of adrenergic and nonadrenergic inhibitory nerve stimulation in tracheal preparations from guinea pig, rat, and rabbit. We find that both sympathetic nerves and nonadrenergic inhibitory nerves cause relaxation of the trachealis muscle of guinea pig but only the nonadrenergic nerves cause relaxation in the trachealis muscle from the rat and rabbit preparations. These findings parallel your experiments on portal vein in the three species.

DR. J.P. MANFREDI: Is there evidence to suggest that adenosine exhibits prejunctional inhibition of acetylcholine release from motor end-plates?

DR. BURNSTOCK: Yes, ATP and adenosine have been shown to act on presynaptic receptors on motor nerve terminals to reduce release of Ach in the rat diaphragm [Ginsborg & Hirst, *J Physiol* 224:629, 1972; Ribeiro & Walker, *Br J Pharm* 34:213, 1975], frog sartorious [Ribeiro & Dominguez, *J Physiol (Paris)* 74:491, 1978], and pectoris [Silinsky, *Br J Pharm* 71:191, 1980], and fish electric organ [Israel et al., *J Neurochem* 34:923, 1980]. These responses are blocked by methylxanthines, so are likely to be mediated by P_1 purinoceptors.

DR. H.P. BAER: As you know, apamin in our hands blocks adenosine or AMP action as well as ATP and nonadrenergic inhibitory nerve action. These are differences in experimental design: we used taenia coli not stimulated with acetylcholine and, in particular, we used hydroxynitrobenzylthioguanosine (adenosine uptake blocker) when studying adenosine-mediated relaxation. Could this explain your failure to see apamine effects on adenosine action? Apart from this particular question, we would suggest that adenosine uptake blockers must be included in studies dealing with kinetic aspects of adenosine effects and functions, otherwise results or conclusions can be ambiguous.

DR. BURNSTOCK: Apamin has been shown by other groups [e.g., Shuba & Vladimirova, *Neuroscience* 5:583, 1980] to block the effects of ATP but *not* adenosine, in guinea pig intestinal muscle. This agrees with our results. As far as experimental design is concerned, the response of the taenia coli to ATP and adenosine depends on the tone of the preparation. Thus, we controlled this parameter before and after apamin treatment by contracting the tissue to the same extent with carbachol. Under these conditions the response to adenosine was not attenuated. In your experiments, there was no control over the developed tone in the preparation, and as apamin increases the tone we feel this could give rise to a physiological rather than pharmacological antagonism of adenosine responses. We did not include adenosine uptake blockers in our protocol, but cannot see why this should lead to the discrepancy in the two sets of results.

DR. J.M. LOWENSTEIN: (1) In your presentation you seem to lump adenosine and AMP together. Does this imply that you believe AMP can act as a vasodilator without being hydrolyzed? (2) We have tested the compound 5′-deoxy-5′-methylphosphate, an analog of AMP in which the 5′-oxygen is replaced by -CH_2-, as a vasodilator. This compound cannot be hydrolyzed by 5′-nucleotidase because it is a phosphonate. It showed no activity as a vasodilator, although it is active in a number of reactions involving AMP.

DR. BURNSTOCK: We have usually assumed that AMP can act directly on P_1 purinoceptors, although without excluding the likelihood that some of its action was due to adenosine following hydrolysis. However, now that you raise the issue, I realize that the evidence for a direct action of AMP is limited, particularly in vascular smooth muscle, as you pointed out. Perhaps the best piece of evidence for some direct AMP action on P_1 purinoceptors in guinea pig atrium and also on cholinergic nerve terminals in the guinea pig ileum (there are no P_2 purinoceptors in these preparations) is that adenosine deaminase only partially blocks the responses to AMP, but a combination of 5′-nucleotidase and adenosine deaminase causes a substantial reduction, suggesting that AMP can act per se on P_1 purinoceptors [Meghji & Moody, personal communication].

DR. G. RABERGER: I am very surprised about your comment on indomethacin abolishing reactive hyperemia in the heart. We did experiments and could show that indomethacin abolishes occlusion-induced prostaglandin release but does not influence reactive hyperemia.

DR. BURNSTOCK: My comment was based on a publication by Alexander and his colleagues [*J Clin Invest* 55:1174–1181, 1975] that indomethacin significantly reduced the reactive hyperemia that occurred following coronary occlusion in the dog heart. However, I gather from what you say that this finding is controversial.

DR. C. LONDOS: On the matter of lumping AMP and adenosine: We find that AMP actions on adenylate cyclase-linked receptors are abolished by adenosine deaminase, yet the deaminase metabolized none of the AMP (determined chromatographically using [^{3}H]-AMP).

DR. L. BELARDINELLI: You stated that there is no adenosine receptor in the ventricle. How do you reconcile such a statement with the following findings: (a) adenosine is known to antagonize the inotropic effects of catecholamines in ventricular muscle; (b) adenosine antagonizes the isoproterenol-induced prolongation and displacement of the action potential plateau to more positive levels in multicellular ventricular preparations as well as in isolated ventricular myocytes (bovine and guinea pig); (c) adenosine also depresses pacemaker activity in His-bundle and Purkinje fibers. Thus, why cannot these effects of adenosine be attributed to adenosine receptors? Wouldn't this be similar to what is known about acetylcholine receptors in the ventricle?

DR. BURNSTOCK: My immediate reaction to your question was that since adenosine has no direct action on the ventricle, there is no receptor present. However, on thinking it over, I must agree that since adenosine is capable of modulating the actions of catecholamines, it is likely to be acting through a receptor, just as the prejunctional modulatory actions of adenosine on adrenergic and cholinergic nerve terminals are mediated by P_1 purinoceptors.

5. GENERAL ASPECTS OF LIGAND-BINDING PHENOMENA

THEODORE W. RALL

With the advent of radioactive compounds with high affinity and high specific activity, the investigation of ligand-binding properties of various receptor systems has proceeded with increasing intensity over the past decade. These studies have sometimes provided new insights into the nature and complex relationships of components of hormone receptor-effector systems. In addition, the experimental approach can rapidly provide extensive information on structure-activity relationships of compounds that has been most useful in endeavors ranging from the identification of endogenous active substances to the development of new medicinal agents. Adenosine receptor systems have recently become a target of such investigations and undoubtedly will be explored even more extensively in the future. This experimental approach has become so commonplace that research publications often do not provide sufficient information on theoretical and practical issues to allow many readers to understand the implications of the data presented. This chapter represents a consumer's view of this experimental approach, which is intended to increase the comprehension of the information it provides.

Investigation of ligand-binding phenomena generally proceeds in two stages. First, it is necessary to characterize the properties of a particular radioactive ligand or set of ligands in order to validate the proposition that the binding sites detected are the recognition sites of functional receptor complexes. At present, these data are restricted principally to a body of correlative information on structure-activity relationships; in some cases the kinetics of

R.M. Berne, T.W. Rall, and R. Rubio (eds.), Regulatory Function of Adenosine, *pp. 63–76.*

binding are compared with the rates of onset or offset of receptor function. In the second stage, the characterized ligands are used to investigate the properties of the receptor in question, such as the anatomical location and relative concentration of receptors or the detection of regulatory ligands. Paradoxically, apparently anomalous ligand-binding behavior makes it very difficult to achieve the first objective but sometimes provides important clues regarding the complexities of receptor systems. One general point to be made in this presentation is that it is important to be able to characterize the binding of both antagonist and agonist ligands in order to extract the maximum amount of information from this type of investigation.

GENERAL THEORETICAL CONSIDERATIONS

The underlying theory is both straightforward and familiar. Data are acquired and analyzed using the simple model of a biomolecular, reversible reaction between a ligand (L) and a single population of binding sites (symbolized presumptively as R in the following discussion):

$$R + L \underset{k_{-1}}{\overset{k_1}{\rightleftharpoons}} RL \tag{5.1}$$

The symbols k_1 and k_{-1} refer to the rate constants of the forward and reverse reactions, respectively. At equilibrium, the rates of formation and decomposition of the complex RL are equal. Thus,

$$(R)(L)k_1 = (RL)k_{-1} \tag{5.2}$$

The equilibrium constant (K_d) can therefore be defined as follows:

$$K_d = \frac{(R)(L)}{(RL)} = \frac{k_{-1}}{k_1} \tag{5.3}$$

Another useful equation follows from the obvious relationship that the total number of binding sites in the system represents the sum of liganded and unliganded sites. Hence,

$$(R_t) = (RL) + (R) \tag{5.4}$$

From equation (5.3) we know that

$$(R) = K_d \frac{(RL)}{(L)} \tag{5.5}$$

and by substituting in equation (5.4) we obtain

$$(R_t) = (RL) + K_d\frac{(RL)}{(L)} = (RL)\left(1 + \frac{K_d}{(L)}\right) \tag{5.6}$$

Rearrangement of equation (5.6) yields the equation for the familiar double-reciprocal plot utilized in the analysis of the kinetics of enzyme-catalyzed reactions:

$$\frac{1}{(RL)} = \frac{1}{(R_t)} + \frac{1}{(L)}\left(\frac{K_d}{(R_t)}\right) \tag{5.7}$$

Rearrangement of equation (5.6) in another fashion yields the equation for the Scatchard plot [1]:

$$\frac{(R_t) - (RL)}{K_d} = \frac{(RL)}{(L)} = -\frac{1}{K_d}(RL - R_t) \tag{5.8}$$

Virtually every publication will include one or more graphs of data wherein "bound over free" ligand concentration is plotted against "bound" ligand concentration. A straight-line relationship is consistent with the assumptions made in equations (5.1) and (5.2)—namely, a system containing a single population of binding sites at equilibrium with a ligand—and the equilibrium constant and the concentration of binding sites can be calculated from the slope and intercept on the abscissa, respectively (Fig. 5.1). A linear Scatchard plot would also be obtained if the system contained several populations of binding sites with approximately equal equilibrium constants. Such a complexity might be revealed in experiments dealing with structure-activity or kinetic measurements.

In the presence of a nonradioactive ligand capable of binding to the same site as the radioactive ligand, a second equilibrium is established. Using the symbol I for this second ligand, we see that

$$RL \underset{K_i}{\rightleftharpoons} I + R + L \overset{K_d}{\rightleftharpoons} RL \tag{5.9}$$

The equilibrium constant for the binding of I is defined as follows:

$$K_i = \frac{(R)(I)}{(RI)} \tag{5.10}$$

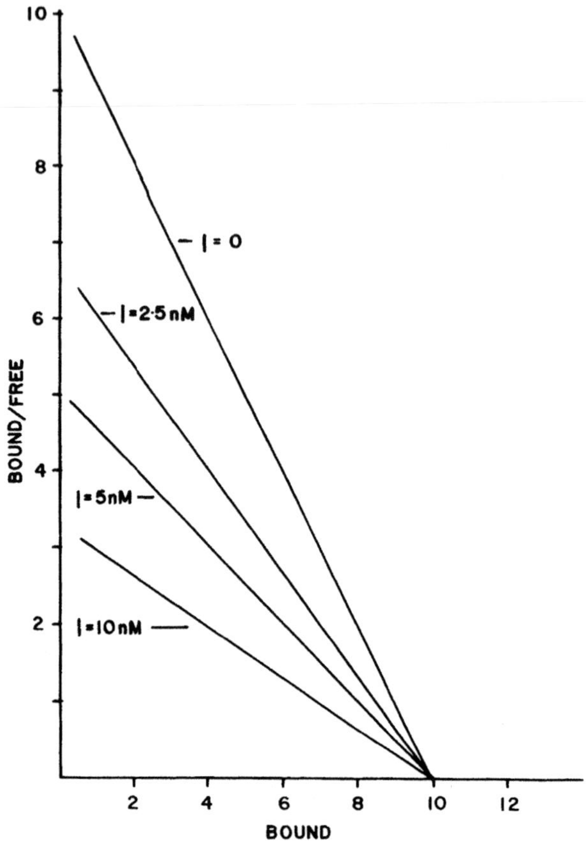

Figure 5.1. Scatchard analysis of binding data. Data are plotted for a hypothetical system containing preparation of membranes (100 μgm protein per reaction vessel) with a capacity of 100 fmol per mg protein for a radioactive ligand (specific activity = 1000 cpm per fmol). Using equation (5.14), the binding of the ligand (K_d = 1 nM) has been calculated for ligand concentrations between 0.05 nM and 20 nM in the absence and in the presence of a nonradioactive competitive inhibitor of binding (I) (K_i = 5 nM) at concentrations of 2.5, 5, and 10 nM. The system is assumed to be at equilibrium, and nonspecific binding has been neglected.

By substituting equation (5.5) into this expression and rearranging, we can derive the relationship

$$(RI) = (RL)\left(\frac{K_d}{(L)}\frac{(I)}{K_i}\right) \qquad (5.11)$$

Equation (5.4) must be expanded to include the species RI:

$$(R_t) = (RL) + (R) + (RI) \qquad (5.12)$$

By substituting equations (5.5) and (5.11) into equation (5.12), we obtain

$$(R_t) = (RL)\left(1 + \frac{K_d}{(L)} + \frac{K_d(I)}{(L)K_i}\right) \tag{5.13}$$

This expression can be rearranged into the form of the Scatchard equation:

$$\frac{(RL)}{(L)} = -\frac{1}{K_d\left(1 + \frac{(I)}{K_i}\right)}(RL - R_t) \tag{5.14}$$

Thus, Scatchard plots at various concentrations of (I) will yield a family of straight lines with the same intercept on the abscissa, and the value for K_i can be calculated from the slopes in the presence and absence of I (Fig. 5.1). Although such behavior would provide evidence that the two ligands were indeed competing for the same binding site, it is rare that data are gathered or presented in this fashion. Instead, the reduction in bound radioactivity as a function of the concentration of nonradioactive ligand is determined using a fixed concentration of radioactive ligand. This yields a "competition curve" from which is estimated an IC_{50} value, the concentration of nonradioactive ligand at which the amount of bound radioactive ligand is reduced by 50% (Fig. 5.2). Hence,

$$(RL)_{I=0} = 2(RL)_{IC_{50}} \tag{5.15}$$

Rearrangement of equation (5.13) yields

$$(RL) = \frac{R_t}{1 + \frac{K_d}{(L)} + \frac{K_d}{(L)}\frac{(I)}{K_i}} \tag{5.16}$$

Substituting in equation (5.15) yields

$$\frac{R_t}{1 + \frac{K_d}{(L)}} = \frac{2R_t}{1 + \frac{K_d}{(L)} + \frac{K_d}{(L)}\frac{(IC_{50})}{K_i}} \tag{5.17}$$

Solving for K_i generates the relationship

$$K_i = (IC_{50})\frac{K_d}{(L) + K_d} \tag{5.18}$$

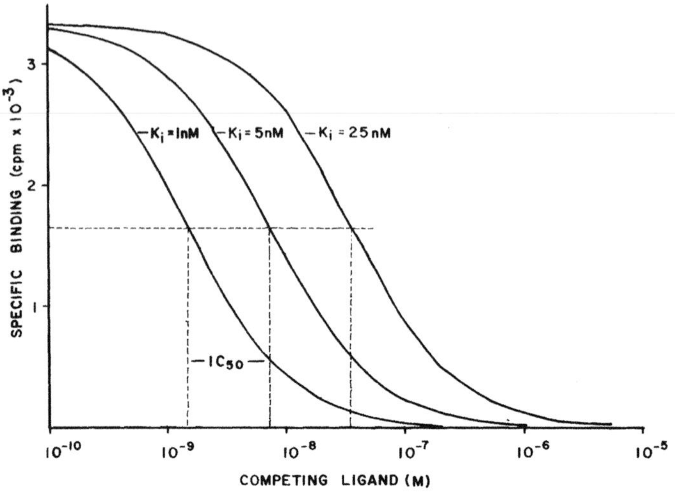

Figure 5.2. Competition curves. Data calculated according to equation (5.16) are plotted for a hypothetical system containing a preparation of membranes (100 μgm protein per reaction vessel) with a capacity of 100 fmol per mg protein for a radioactive ligand (specific activity = 1000 cpm per fmol). Each reaction vessel contains 0.5 nM of radioactive ligand (K_d = 1 nM) and various concentrations of three different nonradioactive competing ligands with K_i's of 1, 5, and 25 nM, respectively. The corresponding IC_{50} values (calculated from equation [5.18]) are 1.5, 7.5, and 37.5 nM, respectively. The system is assumed to be at equilibrium, and nonspecific binding has been neglected.

If the concentration of radioactive ligand used is equal to the K_d for that ligand, then the K_i for the nonradioactive ligand is equal to 50% of the IC_{50}. By this means, the binding of a host of substances can be evaluated, including those whose affinities are much too low to be determined directly. However, it must be remembered that the K_i values so determined have meaning as equilibrium constants only in the presence of evidence that the ligands behave according to the simple model expressed in equation (5.9). Nonparallel competition curves would signal some kind of complexity; and even though IC_{50} values can be estimated, any K_i values calculated from such data would have indeterminate meaning.

GENERAL EXPERIMENTAL CONSIDERATIONS

Except for experiments utilizing equilibrium dialysis, discussed later, the concentration of R_t will almost always be very much smaller than the concentration of radioactive ligand used, and the reduction in (L) due to the formation of (RL) usually can be neglected. Therefore, unless there is significant metabolism or decomposition of the ligand during the experiment, (L) is equal to the concentration of the ligand added to the system and the crucial measurement is the determination of (RL). The number of hormone-binding sites in target tissues has usually been found to be quite small, on the order of 10^3 sites per cell [2]. Most binding studies are performed with relatively crude membrane fractions derived from homogenates of cellularly heterogeneous tissues, and the binding capacity (R_t) observed for most hormones (including adenosine) in such preparations rarely falls outside the range of 10 to 100 fmol per mg protein [2]. Thus, it is easy to understand why ligand-binding experiments could not be done until the advent of radioactive ligands with very high specific activity. Even so, the determination of (RL) is sometimes difficult and requires the use of relatively large amounts of tissue. For example, a tritium-labeled ligand with a specific activity of 30 Ci/mmol will permit the recovery of about 130 to 1300 cpm per mg protein when used at a concentration equal to the K_d in systems with a capacity of 10 to 100 fmol per mg protein (assuming a counting efficiency of 40%). It is also easy to understand why iodinated polypeptide hormones were among the first ligands to be used in such experiments in view of the ability to achieve specific activities of more than 1000 Ci/mmol with ^{125}I.

In addition to the utilization of equilibrium conditions and Scatchard analysis, the K_d of radioactive ligands can also be measured by determining the ratio of reaction rate constants (Eq. 5.3). Unfortunately, this is often impossible to do for a number of reasons, including the usually rapid rate of association. In general, the range of values for k_1 has been found to be much smaller than the range of values for K_d; this means that the relative K_d values for a group of substances reflects principally the relative k_{-1} values [2]. It is possible to estimate a priori that a ligand with a K_d of 10 nM and possessing the most frequently observed k_1 value (of the order of 10^8 L/mol/min at 22° to 37°) would generate complexes with a half-time of dissociation of around 40 sec. This can be calculated from equation (5.3) ($k_{-1} = K_d \cdot k_1$) and from the first-order rate equation where $T_{1/2} = 0.693/k_{-1}$. Thus, the method of separation of bound from free ligand will determine the range of affinities for binding sites that can be dealt with accurately. Since hormone binding sites are membrane-bound (except for steroid hormones), filtration through glass fiber filters is usually the best and most convenient method for achieving this separation. Even using cold temperatures to slow the dissociation rate and utilizing no more than 10 to 20 sec for dilution, transfer, filtration, and rinsing procedures, binding sites with a K_d of greater than 10 to 100 nM for a radioactive ligand cannot usually be examined successfully by this method. Ligand complexes with more rapid rates of dissociation can sometimes be recovered by centrifugation followed by superficial rinsing of the pellet. However, the trapped solute that helps maintain equilibrium conditions during centrifugation becomes the source of a relatively large background of free ligand, which must be corrected for and which reduces the precision of measurements. Theoretically, equilibrium dialysis can be used to analyze the binding of ligands with a wide range of affinities. However, this procedure depends on the measurement of free ligand on one side of the dialysis membrane and of free plus bound ligand on the other side; the difference between these two values represents the concentration of.bound ligand. Thus, it is necessary to use amounts of binding sites that are of the same order of magnitude as the concentration of ligand; furthermore, as the K_d for the ligand becomes larger, so must the concentration of binding sites be raised in order to detect bound ligand. Unless one is dealing with very rich sources of binding sites, such as the electric organ for cholinergic receptors, this method is not very useful.

In theory, as the concentration of radioactive ligand is increased, the amount of bound radioactivity should reach a maximum determined by the number of binding sites in the system. In practice, the amount of radioactivity recovered by filtration or centrifugation is observed to increase indefinitely. Furthermore, using a fixed concentration of radioactive ligand, the recovery of radioactivity is found to approach a limiting value greater than zero as the concentration of nonradioactive ligand included in the system is increased. This limiting value is referred to as nonspecific binding, and when subtracted from the total radioactivity recovered yields a calculated value for specific binding (Fig. 5.3). Nonspecific binding results from a number of circumstances, including the trapping of solvent with its dissolved ligand, the partitioning of ligand into the lipids of membrane preparations, and the possible presence of other binding sites with much higher capacity and lower affinity than the sites under consideration. The calculated values for specific binding are those that are analyzed further. The term *specific binding* has a seductive ring to it, but it must be remembered that it simply means "binding prevented by a large excess of a nonradioactive ligand." The substance and its concentration used in this operational definition is a matter for experimentation. It seems obvious that the larger the proportion of nonspecific binding observed, the greater the error in estimating specific binding becomes.

If the Scatchard plot of specific binding data yields a straight line, then only a limited number of interpretations is possible, including the conclusion that a single population of binding sites is present. Curvilinear Scatchard plots can indicate a large number of possibilities, including experimental problems such as nonequilibrium conditions. Most frequently, curvilinear plots are observed to be concave-up; this would be consistent with the presence of more than one population of binding sites or with "negative cooperativity" between members of a single population of binding sites. It is sometimes helpful to examine the rate of disappearance of specific binding following dilutions of mixtures at equilibrium in both the presence and absence of excess nonradioactive ligand. Acceleration of dissociation in the presence of nonradioactive ligand would be consistent with negative cooperativity but would not prove it. Curvilinear, concave-up Scatchard plots could also signify other complexities, including the interaction of other components of a receptor system with the ligand-binding component [3]. In addition to hypotheses, analyses of adenylate cyclase systems have revealed the functional interaction of several membrane components with hormone-binding sites involved in the regulation of this enzyme (see Chapter 1). In this event, the simple model depicted in equation

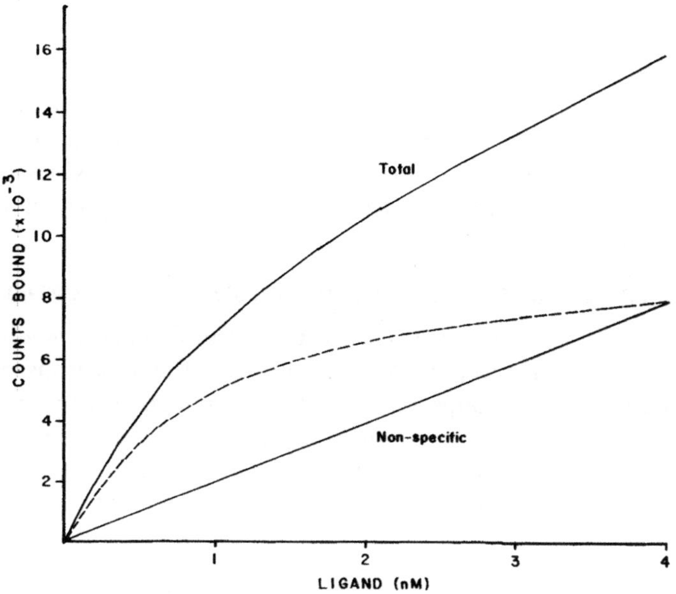

Figure 5.3. Specific and nonspecific binding as a function of ligand concentration. Data are plotted for a hypothetical system containing a preparation of membranes (100 μgm protein per reaction vessel) with a "specific" (high-affinity) binding capacity of 100 fmol per mg protein and a "nonspecific" (low-affinity) binding capacity of 100 pmol per mg protein for a radioactive ligand (specific activity = 1000 cpm per fmol). The high-affinity site was assigned a K_d of 1 n*M*, and the low-affinity site was assigned a K_d of 5 μ*M*. The radioactivity bound at each site was calculated and plotted as a function of the concentration of radioactive ligand (labeled "Specific" and "non-specific," respectively). The sum of the radioactivity bound at both sites is plotted in the uppermost curve (labeled "Total"). Addition of 100 n*M* or 1 μ*M* of the same ligand in a nonradioactive form would lead to 2079 cpm or 1678 cpm of total binding at 1 n*M* radioactive ligand; this would result in estimates of specific binding that were 98.4% and 106.4% of the true value, respectively. At 2 n*M* radioactive ligand, these maneuvers would lead to estimates of specific binding that were 97.1% and 109.6% of the true values, respectively. The use of 200 n*M* nonradioactive ligand would result in estimates of specific binding that were 100.4% and 100.2% of the true values at 1 n*M* and 2 n*M* radioactive ligand, respectively. Thus, the proper amount of nonradioactive ligand to use for the definition of specific binding will depend on the behavior of the nonspecific binding and will require optimization by experimentation.

(5.1) would not be applicable, and additional equilibria would have to be proposed defining the combination of the several components with each other as well as with various ligands. In this context, it may be helpful to consider the difference between the behavior of agonist and antagonist ligands. Functionally, agonists set in motion some sequence of events leading to an observable response, while antagonists have no effect other than to prevent the response to an agonist. It seems reasonable to expect that, compared to an agonist, the less complicated consequences of binding of an antagonist should be reflected in less complex binding behavior of an antagonist. This expectation has been borne out in a number of instances, especially in the case of β-adrenergic ligands [2]. One generalization that follows from these considerations is that examination of the binding of candidate antagonists has the greatest chance of yielding a radioactive ligand that displays simple and therefore quantifiable binding behavior and that can be used to characterize and identify a given ligand-binding site. However, examination of the binding of agonist ligands is more apt to reveal some of the complexities of receptor function. Thus, it is ultimately desirable to have both types of radioactive ligands available.

EVIDENCE FOR RELATIONSHIP OF BINDING SITES TO RECEPTORS

The principal category of evidence used to identify ligand-binding sites as entities functioning in receptor systems consists of comparisons of the rank order of potencies of groups of agonists and antagonists in producing or preventing biological responses with the rank order of K_d values observed directly or indirectly (via IC_{50} estimates) in binding experiments. In addition to qualitative information, these data usually provide quantitative comparisons on the ratio of potencies of key index substances, including pairs of stereoisomers. By selection of appropriate substances or conditions, these data can also provide negative evidence regarding the possible participation of other macromolecular entities in the ligand-binding phenomena. These entities include enzymes capable of combining with and metabolizing hormones, and putative carrier sites, functioning in facilitated transport systems. The presence of transport systems may be of particular importance for those neurotransmitter substances whose termination of physiologic action depends substantially on uptake processes. Membrane fragments containing transport carriers have been known to form vesicles encasing trapped solvent; facilitated diffusion into such compartments can sometimes be confused with binding. However, evaluation of the impact of substances known to influence uptake but not receptor processes as well as of mechanical disruption can provide important evidence or suggest useful conditions for conducting binding experiments.

The absolute values for the K_d of agonist ligands has usually not been found to correspond to the K_{act} (ED_{50}) values for biological activity even in adenylate cyclase systems in which the vagaries of "spare receptors" can be avoided and in which binding and activity potencies can be determined under identical conditions [2, 4]. Discrepancies in both directions have been observed; that is, K_d values have been found that are either much larger or much smaller than the corresponding K_{act} values. However, when carefully analyzed, the K_d values for antagonist ligands have usually been found to correspond very closely to the K_i values for these substances in biological systems [2, 4]. Furthermore, it is possible to estimate equilibrium constants for antagonists from data gathered in complex, intact-cell systems by use of the approach developed by Schild [5]. Receptor theory postulates that, for a given agonist, equal fractional receptor occupancy leads to equal biological responses, regardless of how complex the relationship between receptor occupancy and response may be.

Rearrangement of equation (5.6) leads to an expression for fractional receptor occupancy:

$$\frac{(RL)}{(R_t)} = \frac{(L)}{(L) + K_d} \tag{5.19}$$

Dose-response relationships are determined for an agonist (L) in the absence and presence of various concentrations of an antagonist (I). A level of response is chosen for comparison (e.g., 50% of maximal response) and the concentration of agonist (L') required to achieve this response in the presence of a given concentration of antagonist is estimated. According to the postulates, the fractional receptor occupancy by agonist in the two circumstances will be equal. Hence,

$$\frac{(L)}{(L) + K_d} = \frac{(L')}{(L') + K_d\left(1 + \frac{(I)}{K_i}\right)} \tag{5.20}$$

This reduces to the relationship

$$\frac{(I)}{K_i} = \frac{(L')}{(L)} - 1 \tag{5.21}$$

This expression no longer contains the term for the equilibrium constant for the agonist (K_d). In its logarithmic form, this becomes the equation for a Schild plot:

$$\log\left(\frac{L'}{L} - 1\right) = \log(I) - \log\,(K_i) \tag{5.22}$$

The equilibrium constant (K_i) can be calculated from the intercept on the abscissa, and the slope should be equal to one if the substance is a competitive antagonist. It is instructive that, in one case, virtually identical K_i values for propranolol have been derived from data measuring ligand binding, antagonism of isoproterenol-induced increase in contractile force in isolated atria, and antagonism of isoproterenol-induced stimulation of cardiac adenylate cyclase preparations [4]. The potential ability to acquire quantitative comparisons of equilibrium constants for antagonists under diverse experimental conditions is yet another reason to use radioactive antagonist ligands if at all possible in the attempt to characterize and identify ligand-binding sites.

In a few cases, restricted to adenylate cyclase systems, attempts have been made to compare the kinetics of ligand binding with the kinetics of receptor activation by agonists under identical conditions [2, 4]. Although it is obvious that the rate of association of agonist ligands should not be slower than the rate of activation of the receptor, it is not clear what criteria can be established a priori for the correspondence between the decay of receptor activation and the rate of dissociation of agonist ligands. It is possible that an active state persists for a finite period following agonist dissociation or that refractory, inactive agonist-receptor complexes may outlast the duration of receptor activation upon dilution or addition of an antagonist. Therefore, kinetic comparisons may be of more value in detecting complexities of receptor function than in establishing the identity of ligand-binding sites.

CONCLUDING REMARKS

It is hoped that the foregoing discussion will have dispelled any mystery connected with investigations using ligand-binding approaches. Such experiments are deceptively simple, both conceptually and in their execution. Although the basic assumptions are not complicated—namely, that systems represented by equations (5.1) and (5.9) are being dealt with at equilibrium—these assumptions must be kept in mind in examining data and experimental conditions. Furthermore, the data are analyzed using relationships that are derived from these basic assumptions by straightforward algebraic manipulations. It should be clear that Scatchard plots of data that fit straight lines are consistent with the original model but do not prove its validity and that curvilinear plots can signify a large number of possibilities.

The opinion has been expressed that antagonists make superior radioactive ligands for binding studies. This conclusion has been based on arguments and observations that antagonists are less apt to exhibit complex binding behavior and that they offer greater opportunity for quantitative correlations with receptor phenomenology. To this can be added the generalization that, except for most polypeptide hormones, antagonists are apt to display greater affinity and greater selectivity than agonists. These considerations, together with the practical issues concerning the separation of bound from free ligand and the minimizing of nonspecific binding, begin to provide a description of an ideal radioactive ligand for use as a probe in binding studies. First, the compound obviously should have high selectivity for the receptor in question. Second, it should have high affinity either with a K_d no greater than 1 nM or with a half-time of dissociation of the bound ligand no less than 5 to 10 min at room temperature. This would permit the use of filtration and thorough rinsing to reduce nonspecific binding without danger of appreciable loss of specific binding. Third, it should have high specific radioactivity, perhaps by the introduction of ^{125}I; for example, a ligand with a K_d of 1 nM and a specific activity of 2000 Ci/mol would generate about 1000 cpm of bound ligand at 0.5 nM in a system containing 100 μg of a preparation with a capacity of 10 fmol per mg protein. This sensitivity would permit competition studies to be conducted at concentrations of radioactive ligand where the contribution of nonspecific binding would be relatively low. Fourth, it should display simple binding behavior and have other properties that would permit unambiguous identification of the binding site; this might be another way of saying that it should be an antagonist. Finally, it should have a constellation of convenient chemical and physical properties. These would include reasonable radiochemical stability and a low propensity to being adsorbed to the surfaces of reaction vessels, pipettes, or filters. Obviously, few substances will measure up to all these criteria, and a great deal of useful information can be acquired with less than ideal reagents. However, it is important to be aware of the difficulties or compromises imposed by the properties of available ligands.

As will become apparent in other discussions in this volume, the generalization concerning the relative selectivity and affinity of antagonists does not hold true for the ligands that have been developed to date for use in the study of adenosine receptors. This state of affairs does not diminish the desirability of developing antagonist ligands for adenosine systems. Imperfect as they may be, the currently available adenosine agonist ligands could be used to screen a large number of chemical substances with structures not necessarily confined to derivatives of the methylxanthines. Ligand-binding protocols are much more suited to this type of enterprise than are experiments involving biochemical or physiologic measurements. Such an effort might be rewarded with the discovery of substances useful both in the investigation of adenosine receptors and in the treatment of human disease.

REFERENCES

1. Scatchard G: The attractions of proteins for small molecules and ions. *Ann NY Acad Sci* 51:660–672, 1949.
2. Maguire ME, Ross EM, Gilman AG: β-adrenergic receptor: Ligand binding properties and the interaction with adenylyl cyclase. *Adv Cyclic Nucleotide Res* 8:1–83, 1977.
3. Boevnaems JM, Dumont JE: Quantitative analysis of the binding of ligands to their receptors. *J Cyclic Nucleotide Res* 1:123–142, 1975.
4. Birnbaumer L, Pohl SL, Kauman AJ: Receptors and acceptors: A necessary distinction in hormone binding studies. *Adv Cyclic Nucleotide Res* 4:239–281, 1974.
5. Schild HO: pA_x and competitive drug antagonism. *Br J Pharmacol* 4:277–280, 1949.

DISCUSSION

DR. R.D. GREEN: While I fully agree that antagonist ligands are preferable to agonist ligands because they usually provide simpler binding data, I would like to interject a word of caution. We tend to extend what is known about ligand binding to β-adrenergic receptors to other systems. It is now clear that guanine nucleotides affect antagonist binding to D-2 dopamine receptors and to muscarinic receptors. We also find that Gpp(NH)p affects ^{3}H-DPX binding to rat hippocampus. In each of these cases, the guanine nucleotides increase antagonist binding. It is intriguing that these receptors all mediate inhibitions of adenylate cyclase activity.

DR. RALL: These are interesting observations and may be telling us something about the complexities of the organization of the components of some receptor systems. For this presentation, I have not considered the issues raised by the general phenomenon of "partial agonists." Various proposals to account for these data (and some opiate and α-adrenergic ligand-binding data) have to do with the possibility that the "receptor" exists in two interconvertible forms having differing affinities for agonists and antagonists. If we superimpose another reactant (e.g., Ni) in both its GTP-liganded and unliganded form, we probably could explain almost any observation by assigning various binding constants to the large number of possible complexes.

DR. M. WILLIAMS: Demonstration of pharmacology and kinetics of a particular radioligand does not per se indicate a pharmacologically relevant recognition site. Anomalous binding is a problem. Zukin showed ^{3}H-PCP (Phencyclidine) binding with all the relevant pharmacology and kinetics in brain. Subsequent examination of these phenomena by Weinstein showed that ^{3}H-PCP binding was to the glass fiber filters and was not relevant to tissue present in a nondenatured state. Studies to eliminate such anomalies are as critical as establishing kinetics and pharmacology to the elucidation of a relevant acceptor site.

DR. RALL: I agree heartily. I hope I made it clear that ligand-binding data must be carefully correlated with physiologic and pharmacologic information before interpretation is possible. Of course, one control included in binding studies would simply be to omit the membrane preparation. There are other similar horror stories of apparently specific binding displayed by filters or support materials (e.g., opioid ligands).

6. GENERAL ASPECTS OF BINDING OF LIGANDS TO ADENOSINE RECEPTORS

ULRICH SCHWABE

The concept of adenosine receptors has been developed from several independent lines of research. A first biochemical basis for the study of the molecular mechanisms of adenosine action was discovered when Sattin and Rall [1] and Shimizu et al. [2] described the potent stimulatory effects of adenosine and adenine nucleotides on cyclic AMP formation in brain slices. These authors proposed the existence of an external adenosine receptor that stimulates adenylate cyclase in a hormone-like manner and that is competitively inhibited by several methylxanthines. In other tissues, adenosine was found to inhibit adenylate cyclase-dependent reactions [3–5]. Based on these observations, two types of adenosine receptors have been classified by the use of selected adenosine analogues [6]. The inhibitory adenosine receptor has a high affinity for adenosine and has been designated R_i. The stimulatory adenosine receptor has a lower affinity for adenosine and has been designated R_a. An alternative nomenclature, using A_1 for the inhibitory receptor and A_2 for the stimulatory receptor, has been proposed [5]. Moreover, nearly all cyclases contain an additional inhibitory P site located at the internal side of the cell membrane [7].

The adenosine receptor concept was extended by Burnstock [8], who proposed two classes of purinergic receptors for the contrasting effects of adenosine and ATP on various physiologic systems. A purinergic adenosine receptor linked to adenylate cyclase was termed P_1, and a purinergic receptor for adenine nucleotides was termed P_2. Although many effects of adenosine appear to involve adenylate cyclase, some may be mediated independently

R.M. Berne, T.W. Rall, and R. Rubio (eds.), Regulatory Function of Adenosine, *pp. 77–96.*

from cyclic AMP [9]. The mechanism whereby P_2 receptor stimulation is transformed into physiologic responses still remains to be elucidated.

During the last five years, radioligand binding techniques have been employed for a further characterization of adenosine receptors. Most studies were performed to label sites of the R_i or R_a type in membrane preparations containing an adenosine-sensitive adenylate cyclase, whereas the inhibitory P site (inhibitory to adenylate cyclase) receptors and the purinergic P_2 receptors for ATP have not been studied as extensively by radioligand binding.

CHARACTERIZATION OF [^{3}H]ADENOSINE BINDING

The earliest attempts to identify adenosine receptors by radioligand binding studies have been performed with radioactively labeled adenosine in fat cell membranes [10]. Although the binding was rapid and reversible, it appeared to be nonsaturable and did not have the characteristics one would expect of binding to physiologically important adenosine receptors. Subsequently, several investigators attempted to use [^{3}H]adenosine and [^{14}C]adenosine for binding studies in membrane preparations of brain [11–13], heart [14], and carotid arteries [15, 16]. Some of the results are listed in Table 6.1. For several reasons, the binding sites labeled in these studies did not have the binding characteristics expected of adenosine receptors, as shown by low affinities and unusually large numbers of binding sites.

In addition, the displacement curves of several analogues and metabolites of adenosine revealed unexpected features of [^{3}H]adenosine binding (Table 6.2). In most tissues, the k_i values for N^6-phenylisopropyladenosine (PIA) were above 100 μM and [^{3}H]adenosine was not stereospecifically displaced by the (−) and (+) isomers of PIA. Similar high k_i values were obtained for 2-chloroadenosine. However, the potent P site agonist 2′,5′-dideoxyadenosine and the adenosine metabolite inosine, both of which are devoid of adenosine-like biological activity, significantly inhibited binding, with k_i values ranging from 14 to 586 μM. Thus, it became increasingly apparent that the sites labeled by [^{3}H]adenosine in these studies were not R site adenosine receptors. Compo-

Table 6.1. Binding of radioactively labeled adenosine in various tissues

Membrane preparation	K_d (μM)	B_{max} (pmol/mg protein)	Reference
Rat fat cell	9.5; 950	28; 1700	[10]
Rat brain	1.7; 13.6	31; 165	[11]
Dog heart	0.6	152	[14]
Dog carotid arteries	1.3	140	[15]
Guinea pig brain	0.7; 5.4	4.5; 28	[12]
Rat brain cortex	0.3; 10	2.8; 10.5	[13]
Hog carotid arteries	0.2; 4.3	1.7; 22	[16]
Rat liver	0.08; 2.6	13	[36]

Table 6.2. Specificity of [^{3}H]adenosine binding in various membrane preparations

Compound	Rat brain[a]	Rat brain synaptosomes[b]	Hog carotid arteries[c]
		K_i (μM)	
Adenosine	4	11	4
(−)N^{6}-phenylisopropyladenosine	>100	10	>100
2-chloroadenosine	83	10	100
2′,5′-dideoxyadenosine	142	108	111
Inosine	586	14	342
Dipyridamole	>100	40	>100

Source: [a]Schwabe et al. [11]; [b]Newman and Levitzki [13]; [c]Schütz and Brugger [16].

Table 6.3. Identification of radioactivity after binding of [^{3}H]adenosine to rat brain membranes

	% of total bound radioactivity			
	Control		0.1 μM Deoxycoformycin	
Adenosine metabolites	37°C	0°C	37°C	0°C
ATP, ADP	2	1	3	1
5′-AMP	52	13	47	10
Hypoxanthine	2	1	2	1
Inosine	17	25	5	7
Adenosine	20	52	40	75
Not identified	7	8	3	6

Source: Data from Schwabe et al. [11].
Values are the mean of two experiments.

nents of the adenosine uptake system were not consistently labeled by [^{3}H]adenosine since in two of these studies the uptake blocker dipyridamole at concentrations up to 100 μM had no significant effect on binding.

A further difficulty emerged from the extensive metabolism of [^{3}H]adenosine under the conditions of receptor binding studies. Table 6.3 shows results of experiments in which the identity of radioactivity bound to rat brain membranes was determined after elution from the receptor sites. After incubation at 37°C the bulk of bound radioactivity was found in the fraction of phosphorylated adenosine metabolites, 17% was deaminated to inosine, and only 20% was recovered as the unchanged radioligand. The metabolism of [^{3}H]adenosine was considerably diminished at reduced temperatures and in the presence of the adenosine deaminase inhibitor deoxycoformycin. Despite these precautions, 25% of the radioligand was metabolized during a relatively short incubation time of 12 min. Similar results have been obtained by other authors [10, 12, 13, 16].

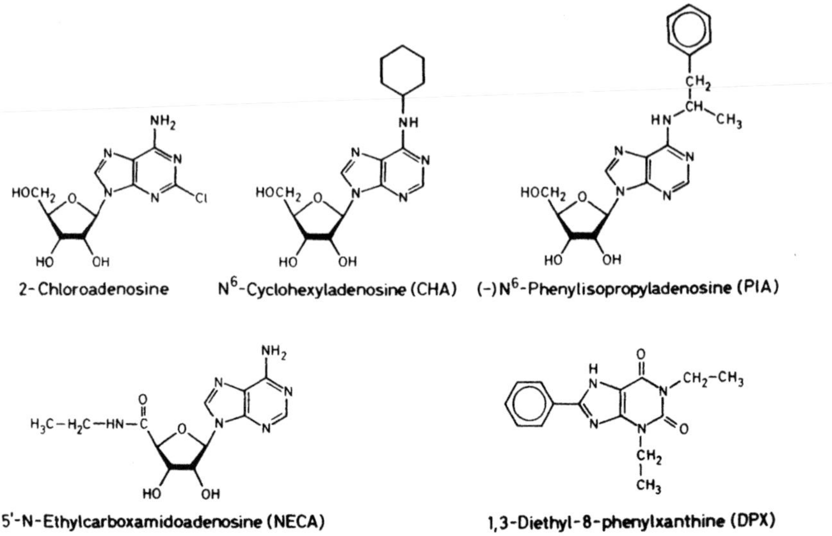

Figure 6.1. Structures of derivatives of adenosine and xanthine.

From these results it has been concluded that it is necessary to use radiolabeled adenosine analogues that are resistant to adenosine deaminase and other adenosine metabolizing enzymes. Due to these considerations [^{3}H]adenosine-cyclopropylcarboxamide [17] and [^{3}H]5′-adenylylimidodiphosphate [18] have been examined as potential ligands for adenosine receptors. However, the use of these metabolically stable analogues of adenosine did not permit characterization of the adenosine receptors. Similar to the experiments with [^{3}H]adenosine, low affinities, low specific binding, and large numbers of binding sites made these attempts unsatisfactory. According to these results, metabolic stability might be an important but not the most significant property of radioligands for adenosine receptors.

GENERAL ASPECTS OF NEWLY DEVELOPED ADENOSINE RADIOLIGANDS

In 1980, four groups independently developed methods that allowed the direct study of physiologically important adenosine receptors [19–22]. At that time, a number of R site-selective adenosine analogues had been radioactively labeled. The structures of these analogues and of a xanthine derivative are depicted in Figure 6.1.

All adenosine analogues are agonists and have been introduced because of their enhanced and prolonged activity compared to adenosine. 2-Chloroadenosine had been developed in 1952 as a compound with potent vasodepressor properties [23]. The increased potency of this analogue compared to adenosine has been attributed to its resistance to adenosine deaminase and adenosine kinase. Among the N^6-substituted analogues, PIA was first described as an antilipolytic agent [24] and subsequently characterized as a potent adenosine agonist in many adenosine-sensitive systems [3, 5, 7, 25–27]. 5′-N-Ethylcarboxamidoadenosine (NECA) has been shown to possess the highest potency in a series of amidated 5′-carboxyl analogues of adenosine [28, 29].

PIA and NECA gained additional interest among these adenosine analogues when Londos et al. [6] found that PIA is a more potent agonist at R_i adenosine receptors than is NECA and that the reverse order of potency is seen at stimulatory R_a adenosine receptors. 1,3-Diethyl-8-phenylxanthine (DPX) is a potent xanthine antagonist that was directly developed specifically for use as an antagonist ligand for adenosine receptor studies [19].

All these compounds are stable against enzymes of the catabolic pathways of adenosine and thus fulfill a general obligatory condition for radioligands. This property appears to be important for an additional reason. In adenylate cyclase studies of R_i receptor-selective cells, it has been observed that a full expression of enzyme activity was only obtained in the presence of adenosine deaminase [26]. It was concluded that the enzyme degraded endogenous adenosine to inosine and thereby prevented its binding to the inhibitory adenosine receptors.

Similar effects were observed in radioligand binding studies of R_i adenosine receptors. Table 6.4 gives an example of the effects of adenosine deaminase on [^{3}H]PIA binding in membrane preparations of rat fat cells and rat brain. In membranes pretreated with adenosine deaminase during the preparation, specific binding was 3- to 4-fold higher than in untreated membranes. The effect of adenosine deaminase treatment is probably due to the removal of endogenous adenosine still present in the repeatedly washed membrane preparations. The effect of the enzyme was further checked during the binding assay by the addition of deoxycoformycin, a potent adenosine deaminase inhibitor. Deoxycoformycin partially reversed the effect in adenosine deaminase-treated membranes but had no effect in untreated membranes. These results indicate

Table 6.4. Effect of adenosine deaminase treatment on [^{3}H]PIA binding

	Specific binding of [^{3}H]PIA (fmol/mg protein)	
Membranes	Fat cell[a]	Rat brain[b]
Untreated	259 ± 44	90 ± 21
+ 0.1 μM deoxycoformycin	232 ± 30	
Adenosine deaminase-treated[c]	966 ± 138	272 ± 30
+ 0.01 μM deoxycoformycin	555 ± 106	
+ 0.1 μM deoxycoformycin	488 ± 86	
+ 1 μM deoxycoformycin	489 ± 82	

Source: Data from [a]Trost and Schwabe [30]; [b]Schwabe and Trost [20].
[c]Membranes were pretreated 10 min at 30°C with adenosine deaminase (1 μg/ml).
Values are means ± SEM of three experiments.

an additional formation of adenosine during the incubation and show the necessity for the continuous presence of the enzyme. Furthermore, the partial reversal makes it unlikely that the increase in specific binding is due to an association of the radioligand to the added enzyme.

Similar results on the importance of adenosine deaminase pretreatment have been obtained with other radioligands for adenosine receptors [19, 21]. They confirm that in membrane preparations used for radioligand binding assays, 50% to 70% of the adenosine receptors appear to be occupied by small amounts of endogenous adenosine that obviously cannot be removed by washing the membranes. This condition is an additional reason for the failure to label physiologically important adenosine receptors with [^{3}H]adenosine. In the absence of exogenous adenosine deaminase, erroneous results may be obtained even with metabolically stable adenosine analogs. Under such conditions the dissociation constant for 2-chloro[^{3}H]adenosine was relatively high, and radioligand binding was displaced by compounds such as inosine and adenine [22].

Binding data obtained with five newly developed radioligands are listed in Table 6.5. Both the dissociation constants and the maximal number of binding sites are several orders of magnitude lower than the corresponding values previously reported for radioactively labeled adenosine. The high-affinity agonists N^6-cyclohexyl[^{3}H]adenosine ([^{3}H]CHA) and [^{3}H]PIA have been the most useful ligands for the study of R_i adenosine receptors. In most tissues studied, the dissociation constants for these two ligands are in the low nanomolar

Table 6.5. Radioligands used for identification of adenosine receptors

Radioligand	Membrane preparation	K_d (nM)	B_{max} (fmol/mg protein)	Reference
2-chloro-[^{3}H]adenosine	Rat brain	23	476	[22]
		1.3; 16	207; 380	[21]
N^6-cyclohexyl-[^{3}H]adenosine	Bovine brain	0.3; 1.8	340^a; 200^a	[19]
		0.4	340	[31]
		0.5	410	[32]
	Guinea pig brain	6	370^a	[19]
		5	206	[32]
	Rat brain	0.7;2.4	230; 120	[33]
	Rat testes	2	200^a	[34]
	Rat liver	168; 2,400	1,050; 14,400	[36]
(−)N^6-phenylisopropyl-[^{3}H]adenosine	Rat brain	5	810	[20]
		1.4	700	[35]
	Rat fat cell	6	1,900	[30]
5′-ethylcarboxamido-[^{3}H]adenosine	Rat liver	84; 4200	1,030; 29,400	[36]
1-diethyl-3-phenyl-[^{3}H]xanthine	Bovine brain	5	1,000^a	[19]
	Guinea pig brain	70	500^a	[19]

$^a B_{max}$ recalculated (1 gm tissue = 100 mg protein).

range. The only exception is rat liver membranes, with a K_d of 168 nM [36]. It has therefore been assumed that in this tissue R_a adenosine receptors are labeled by [^{3}H]CHA, but a high level of nonreceptor binding did not permit determination of the appropriate structure-activity profile of adenosine analogs.

2-Chloro[^{3}H]adenosine appears to be another suitable ligand for high-affinity R_i adenosine receptors if the membrane preparations are adequately pretreated with deaminase [21]. For this radioligand, and in some cases for [^{3}H]CHA, two sites have been described in brain tissues, but whether these sites may be attributed to different subclasses of adenosine receptors has not yet been explored.

[^{3}H]NECA has recently been studied as a ligand for R_a adenosine receptors in liver membranes [36]. Although the dissociation constant of the high-affinity site was in the same range as the K_a for the stimulatory effect of NECA on the hepatic adenylate cyclase, appropriate structure-activity correlations for other adenosine analogues were not obtained.

[^{3}H]DPX is the first antagonist ligand for adenosine receptors and appears to label both R_i and R_a adenosine receptors [19]. In bovine brain membranes, a high-affinity site with a K_d of 5 nM and further characteristics suggestive of R_i adenosine receptors were obtained. In guinea pig brain membranes, the binding sites have a lower affinity and additional properties suggesting interactions with R_a adenosine receptors. The [^{3}H]DPX binding data agree with the previous finding that xanthine adenosine antagonists distinguish between P site and R site adenosine receptors but are not selective for R_i and R_a receptor subtypes [6].

BINDING OF RADIOLIGANDS TO R_i ADENOSINE RECEPTORS

A desirable approach to delineate the selectivity of ligands for adenosine receptors is to study radioligand binding in cells that have previously been characterized by physiological or pharmacological studies to possess a single receptor subtype. For the R_i receptor, the most suitable candidate probably is the fat cell, since this cell type has a high affinity for PIA and similar N^6-substituted adenosine analogues [24, 26]. Furthermore, it is considered R_i-selective on the basis of cyclase studies with several adenosine analogs [6]. Thus, fat cells offer the advantage of the ability to study R_i receptors without interference by R_a receptors, whereas in brain tissue both receptor subtypes may contribute to overall radioligand binding. [^{3}H]PIA was selected for use as the radioligand, because the pharmacological properties of this adenosine analogue have been extensively investigated in adenylate cyclase systems [26, 27].

First, we have determined the kinetic properties of [^{3}H]PIA binding in rat fat cell membranes. Figure 6.2 shows the time course measured at 37°C. Specific binding occurred rapidly and achieved equilibrium after approximately 15 min of incubation. The association constant k_1 has been determined from the pseudo-first-order plot in the time shown in the inset to Figure 6.2A and is $19.5 \times 10^6 M^{-1}\text{min}^{-1}$. The dissociation of [^{3}H]PIA was relatively slow but

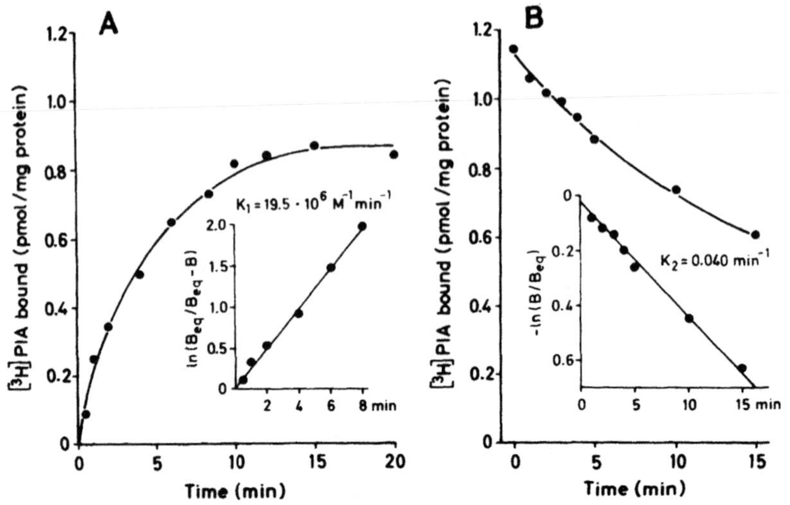

Figure 6.2. Time course of [^{3}H]PIA binding to rat fat cell membranes. Association (A) and dissociation (B) of [^{3}H]PIA (10 nM) were measured at 37°C. Values are means of three experiments. Source: Data from Trost and Schwabe [30].

constant over the entire time measured, yielding a rate constant (k_2) of 0.040 min^{-1}, which is equivalent to a $t_{1/2}$ of about 17 min (Fig. 6.2B). The ratio k_2/k_1 of the rate constants is 2.1 nM and provides a first estimate of the equilibrium dissociation constant K_d.

As shown in Figure 6.3, [^{3}H]PIA binding is saturable in rat fat cell membranes with a dissociation constant (K_d) of 6.1 nM and a maximal number of binding sites of 1.89 pmol per mg of protein. Scatchard analysis of these binding data, as is shown in the inset to Figure 6.3A, yields a linear plot suggesting a single class of binding sites. This is confirmed by a Hill coefficient (n_H) of approximately 1.0 (Fig. 6.3B). The equilibrium dissociation constant determined by Scatchard analysis is in good agreement with that determined kinetically.

The structural specificity of [^{3}H]PIA binding was determined by competition experiments with adenosine analogues, xanthine derivatives, and several other agents. In Table 6.6, the data for the inhibition of [^{3}H]PIA binding and of adenylate cyclase by various compounds using membranes from fat cells and rat brain are compared. In all cases studied, the ability to compete for [^{3}H]PIA binding is closely related to the potency as agonist or antagonist in the adenylate cyclase system. (−)PIA was the most effective adenosine analogue, as shown by k_i values of 2 and 5 nM for [^{3}H]PIA binding. The binding sites are highly stereospecific, as was demonstrated by a more than 10-fold lower k_i value for (−)PIA than for (+)PIA. With the exception of 2′,5′-dideoxyadenosine, all other adenosine analogues have k_i values in the nanomolar range. These data confirm that [^{3}H]PIA binding in both membrane prepa-

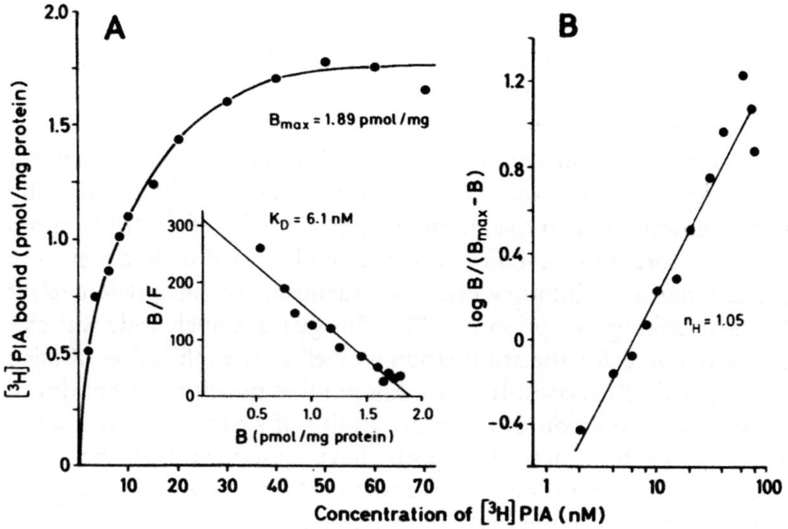

Figure 6.3. A: Saturation of [^{3}H]PIA binding to rat fat cell membranes. Each value is the mean of two to three experiments. Inset: Scatchard plot of the same data. B: Hill plot of the same data. Source: Data from Trost and Schwabe [30].

Table 6.6. Structure-activity relationship for the inhibition of [^{3}H]PIA binding and adenylate cyclase by adenosine analogues and related compounds

	k_i(nM)			
	[^{3}H]PIA binding		Adenylate cyclase	
Compound	Fat cell[a]	Rat brain[b]	Fat cell[c]	Rat brain[d]
(−)N^6-phenylisopropyladenosine	5	2	20	10
N^6-cyclohexyladenosine	7	2		10
5′-N-ethylcarboxamidoadenosine		7		100
2-chloroadenosine	9	5	10	
(+)N^6-phenylisopropyladenosine	54	31		
2′,5′-dideoxyadenosine	11,900	17,000		
3-isobutyl-1-methylxanthine	21,000	3,500	7,000	450
Theophylline	23,300	10,200	7,000	
Caffeine	57,000	20,700	30,000	
Dipyridamole	>100,000	21,000		
Inosine	>100,000	>100,000		
GTP	>100,000[e]	260		

Source: Data from [a]Trost and Schwabe [30]; [b]Lenschow and Schwabe [35]; [c]Londos et al. [26]; [d]Cooper et al. [27], [e]Grandt et al. [37].

rations occurs at the high-affinity R_i adenosine receptors. The k_i values for the xanthine antagonists are between 3 and 57 μM and thus are comparable to those obtained in adenylate cyclase studies. As would be expected, the adenosine uptake blocker dipyridamole in fat cell membranes does not inhibit [^{3}H]PIA binding, indicating that adenosine uptake sites are not labeled. In rat brain, the k_i value for dipyridamole is 21 μM. Agents devoid of adenosine-like activity, such as inosine and adenine, do not compete for the binding sites. These data emphasize that the pharmacological profile of [^{3}H]PIA binding sites in fat cell and brain membranes closely resembles that of R_i adenosine receptors, mediating the inhibitory effects of adenosine on adenylate cyclase.

The only striking exception is GTP. This guanine nucleotide was examined since it is required for the stimulation as well as the inhibition of adenosine-sensitive adenylate cyclases. It is also known that guanine nucleotides cause an agonist-specific reduction of receptor affinity of several hormone and neurotransmitter receptors. Recent reports have described that the binding of adenosine agonist ligands but not of antagonist ligands is decreased by guanine nucleotides in brain membranes [19, 31, 32]. We have confirmed that [^{3}H]PIA binding is displaced in this tissue by GTP with a k_i value of 260 nM [35]. By contrast, GTP does not inhibit [^{3}H]PIA binding in fat cell membranes. Under certain conditions, GTP has even the opposite effect and reverses the reduced binding in the presence of divalent cations [37]. These data show that radioligand binding to R_i adenosine receptors with identical pharmacological profiles is differently modulated by GTP. It remains an interesting question whether these apparently discrepant results may be explained by additional subtypes of the R_i receptor or by different forms of regulatory protein (e.g., N_i) in the two tissues.

STUDY OF R_a ADENOSINE RECEPTORS BY RADIOLIGAND BINDING

The identification of R_a adenosine receptors by radioligand binding studies still lags behind that of the R_i receptors, although the stimulatory effects of adenosine on cyclic AMP formation in the brain, which led to the initial proposal of adenylate cyclase-linked adenosine receptors, were actually described first [1]. This delay may be attributed to problems in finding highly specific radioligands and to the relatively low affinity of these sites for various adenosine analogues. Adenosine 5′-cyclopropylcarboxamide, a potent R_a receptor agonist was the first agent used for radioligand binding studies of the R_a receptors [17]. The radioactively labeled compound showed saturable high-affinity binding to brain membranes, but the displacement profile by adenosine agonists and antagonists was not consonant with selective binding to either R_a receptors or P sites. The results obtained with [^{3}H]NECA in rat liver membranes [36] and with [^{3}H]DPX in guinea pig brain membranes [19] have already been mentioned (Table 6.5). In both studies, labeling of R_a receptors was only partially achieved, and several discrepancies in the structure-activity profile were observed.

Some of the difficulties may arise from the common occurrence of both

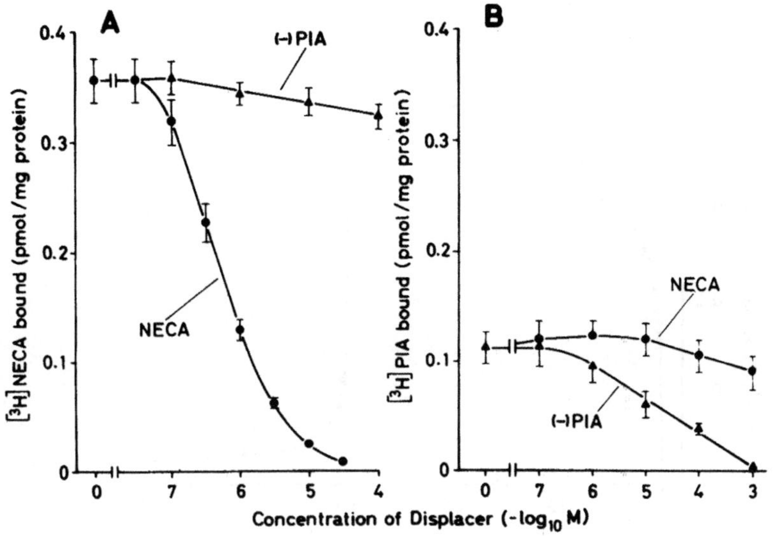

Figure 6.4. Binding of [^{3}H]NECA and [^{3}H]PIA to human platelet membranes. A: Displacement of specific binding of [^{3}H]NECA (10 n*M*) at 0°C. B: Displacement of specific binding of [^{3}H]PIA (10 n*M*) at 0°C. Values are means ± SEM of five experiments. Source: Data from Lenschow et al. [40].

receptor subtypes in these tissues. Therefore, we have used a similar approach for the characterization of R_i adenosine receptors and attempted to study radioligand binding of R_a receptors in cells that did not display R_i receptor function. Human platelets were chosen because they have been previously classified as possessing only R_a adenosine receptors [38]. [^{3}H]NECA appeared to be a suitable ligand for R_a receptors, since this adenosine analogue has been shown to be 5- to 10-fold more potent than adenosine in stimulating adenylate cyclase in platelets [39].

We first compared the binding of [^{3}H]NECA and [^{3}H]PIA to human platelet membranes. Figure 6.4 shows the competition curves of unlabeled NECA and PIA versus each of their radioactive counterparts used at 10 n*M*. For several reasons, [^{3}H]NECA appeared to be a more suitable ligand than [^{3}H]PIA. Specific binding of [^{3}H]NECA was 4-fold higher than that of [^{3}H]PIA. Nonspecific binding of [^{3}H]NECA was only 4%, compared to 50% for that of [^{3}H]PIA (data not shown). Furthermore, NECA was found to be a potent displacer at [^{3}H]NECA sites with an IC_{50} value of 0.5 μ*M*, but it had only weak effects on the binding of [^{3}H]PIA, even at 1000-fold higher concentrations. However, PIA was a relatively weak displacer of [^{3}H]PIA, with an IC_{50} value of about 30 μ*M* and thus was about 60-fold less potent than NECA at [^{3}H]NECA sites. Unexpectedly, PIA did not affect [^{3}H]NECA binding. This finding does not correlate with the known pharmacological potency of PIA on platelet aggregation and adenylate cyclase.

The study of the kinetics and the saturation of binding revealed some addi-

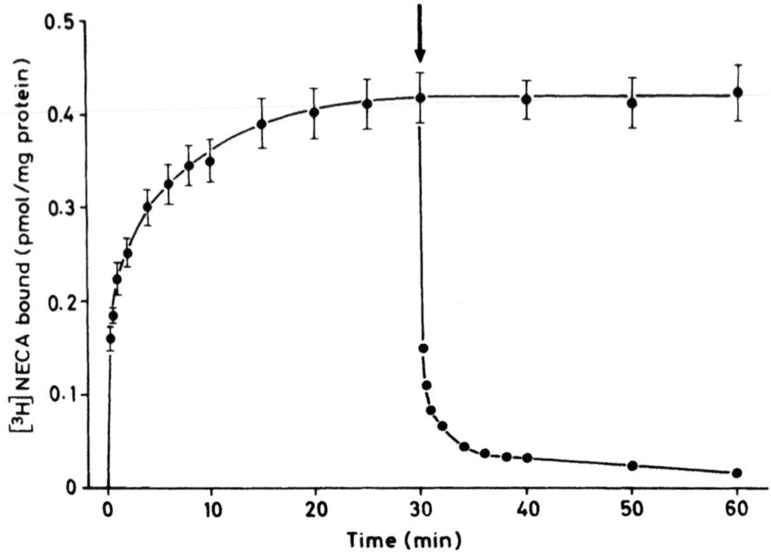

Figure 6.5. Time course of [^{3}H]NECA binding to human platelet membranes. Association and dissociation of [^{3}H]NECA (10 n*M*) were measured at 0°C. Values are means ± SEM of five experiments. Data from Lenschow et al. [40].

tional difficulties associated with the use of [^{3}H]NECA. Figure 6.5 shows the time course of binding at 0°C, using a radioligand concentration of 10 n*M*. Binding was rapid and reversible, but dissociation induced by addition of an excess of unlabeled ligand was faster than association. More than 60% of the bound ligand was dissociated within 20 sec. Therefore, it was not possible to calculate a kinetically determined dissociation constant.

The specific binding appeared to be saturable at ligand concentrations between 0.25 and 8 μ*M*; at higher concentrations, additional binding occurred (Fig. 6.6). Scatchard analysis of the data revealed a curvilinear plot suggesting the presence of two binding sites. For the high-affinity site, a K_d of 0.16 μ*M* and a B_{max} of 8.4 pmol per mg protein and for the low-affinity site a K_d of 2.9 μ*M* and a B_{max} of 33.67 pmol per mg protein were calculated. As would be expected for R_a receptors, the affinity of the high-affinity site is about 100-fold lower than that of R_i adenosine receptors.

The [^{3}H]NECA binding was further characterized by competition experiments with various ligands. In Table 6.7 the k_i values for human platelets are listed and compared to those obtained for calf thymocytes, another R_a-selective cell type. Data for [^{3}H]DPX binding in guinea pig brain membranes and for [^{3}H]CHA binding in rat liver membranes are also included in Table 6.7. The [^{3}H]NECA sites in platelets and thymocytes have the same substrate specificity. NECA, the most potent adenosine analogue, is about 10 times more potent than 2-chloroadenosine and 25- to 50-fold more potent than

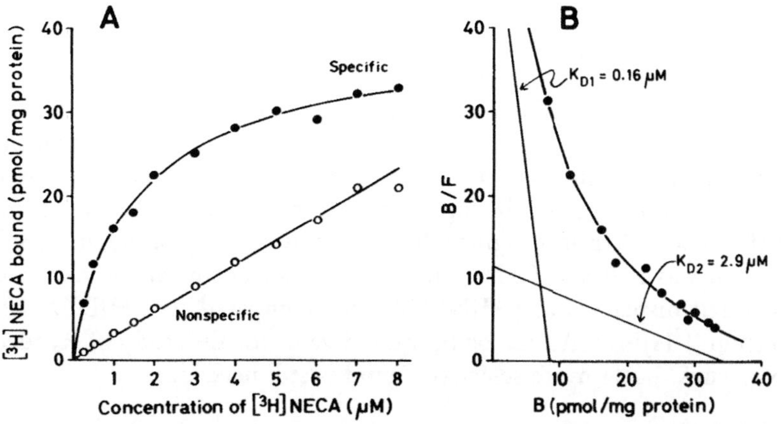

Figure 6.6. A: Saturation of [^{3}H]NECA binding to human platelet membranes. Each value is the mean of four experiments. B: Scatchard plot of the same data. Source: Data from Lenschow et al. [40].

Table 6.7. Specificity of radioligands used for R_a adenosine receptor studies

	k_i(μM)			
Compound	[^{3}H]NECA, human platelets[a]	[^{3}H]NECA, calf thymocytes[b]	[^{3}H]DPX, guinea pig brain[c]	[^{3}H]CHA, rat liver[d]
5′-N-ethylcarboxamidoadenosine	0.5	0.3	93[e]	108
2-chloroadenosine	6	2	2	
Adenosine	12	14		9
(−)N^6-phenylisopropyladenosine	>1000	1400	140	3
N^6-cyclohexyladenosine	>1000	>1000	75	1
(+)N^6-phenylisopropyladenosine	>1000	>1000	93	5
3-isobutyl-1-methylxanthine	92	32	3	195
Theophylline	770	325	19	
Caffeine	5300	9500	187	
2′,5′-dideoxyadenosine	14	78	28	24
Adenine	86	190	187	>500
Dipyridamole	> 100	> 100		
Inosine	>1000	>1000	>1000	232

Source: Data from [a]Lenschow et al. [40]; [b]Ukena et al. [41]; [c]Bruns et al. [19]; [d]Schütz et al. [36].
[e]Value of adenosine 5′-cyclopropylcarboxamide.

adenosine. Similar structure-activity profiles have been obtained with these three ligands for the stimulation of adenylate cyclase [6, 39]. As already mentioned, PIA and other N^6-substituted adenosine analogues did not affect [^{3}H]NECA binding at concentrations up to 1 mM. The xanthine derivatives displaced radioligand binding, but their k_i values were about 10-fold higher than those determined at R_i receptor sites and in adenylate cyclase studies.

Relatively low k_i values were obtained with 2′,5′-dideoxyadenosine and adenine, whereas dipyridamole and inosine were completely inactive.

The comparison of the [^{3}H]NECA binding data with those of the two other ligands reveals that [^{3}H]DPX and [^{3}H]CHA sites are not able to discriminate between R_a and R_i selective agonists, since NECA and adenosine 5′-cyclopropylcarboxamide are only weakly effective. These typical R_a-selective agonists are much better distinguished by the [^{3}H]NECA sites. However, [^{3}H]NECA sites do not recognize the residual R_a activity of potent R_i-selective ligands such as PIA and CHA. Furthermore, adenosine antagonists are relatively weak displacers at [^{3}H]NECA sites compared to [^{3}H]DPX sites. In conclusion, [^{3}H]NECA may be of limited value for the study of R_a adenosine receptor sites, but a more selective ligand is still needed.

CONCLUSIONS

The data presented show that a considerable amount of information has been generated by radioligand binding studies of adenosine receptors. Some of the general aspects are summarized in Table 6.8. It has been established beyond doubt that the N^6-substituted adenosine analogues [^{3}H]CHA and [^{3}H]PIA are at present the most suitable ligands for labeling of R_i adenosine receptors. They exhibit high subtype selectivity, and [^{3}H]PIA does not significantly associate with R_a receptors of subtype-selective cells. With regard to R_a receptors, some progress has been achieved by the development of [^{3}H]NECA, but a more specific ligand with less binding to very low affinity sites is still needed.

The availability of [^{3}H]DPX as the first antagonist ligand reveals additional possibilities of investigating adenosine receptors. Furthermore, a new direct approach to developing subtype-selective antagonists is offered. These compounds not only will be of general interest for a more detailed characterization

Table 6.8. Comparison of radioligand binding to adenosine receptors

Characteristics	R_i receptors	R_a receptors
Radioligands		
Agonists	[^{3}H]CHA, [^{3}H]PIA	[^{3}H]NECA (?)
Antagonists	[^{3}H]DPX	[^{3}H]DPX (?)
Affinity range (K_d)	0.3–6 n*M*	70–160 n*M*
Binding kinetics		
Equilibrium	[^{3}H]PIA, 15 min [^{3}H]CHA, 100 min	[^{3}H]NECA, 30 min
Dissociation ($t_{1/2}$)	20 min at 25–37°C	<20 sec at 0°C
Modulators		
Adenosine deaminase	3- to 4-fold increase	No effect
GTP	Brain, inhibition Fat cell, no inhibition	No effect
Divalent cations	Brain, increase Fat cell, inhibition	Platelets, inhibition Thymocytes, increase

of receptor subtypes but they could also have far-reaching implications for drug therapy.

The affinity of radioligands for R_i receptors is in the low nanomolar range, whereas R_a receptors have 10- to 100-fold lower affinities. These differences have important effects on receptor kinetics and the performance of binding assays. Whereas R_i receptor ligands dissociate slowly and can be analyzed at 25–37°C, R_a ligands rapidly dissociate even at 0°C and reach the limit of the usual filtration separation.

Finally, some interesting aspects are provided by endogenous modulators of adenosine receptors. Adenosine deaminase treatment has been revealed as an important condition for the binding studies of R_i but not of R_a receptors. This again is probably due to the high affinity of R_i receptors. It does not seem unlikely that this property reflects the basic condition of the usual physiological state of the R_i receptor, that is, to be occupied by the endogenous ligand. The R_a receptor, however, is occupied only to a small extent at low physiological concentrations of adenosine and is capable of responding to increased levels of endogenous adenosine.

The effects of GTP and cations indicate that the seemingly uniform adenosine receptor subtypes respond quite differently to these modulators of adenylate cyclase. It may be taken as another promising prospect for future studies that radioligand binding may help to clarify the interaction of possibly different GTP regulatory components with adenosine receptors linked to the adenylate cyclase in various cell types.

REFERENCES

1. Sattin A, Rall TW: The effect of adenosine and adenine nucleotides on the cyclic adenosine 3′,5′-phosphate content of guinea pig cerebral cortex slices. *Mol Pharmacol* 6:13–23, 1970.
2. Shimizu H, Creveling CR, Daly J: Stimulated formation of adenosine 3′,5′-cyclic phosphate in cerebral cortex: Synergism between electrical activity and biogenic amines. *Proc Natl Acad Sci USA* 65:1033–1040, 1970.
3. Fain JN, Pointer RH, Ward WF: Effects of adenosine nucleosides on adenylate cyclase, phosphodiesterase, cyclic adenosine monophosphate accumulation, and lipolysis in fat cells. *J Biol Chem* 247:6866–6872, 1972.
4. Schwabe U, Ebert R, Erbler HC: Adenosine release from isolated fat cells and its significance for the effects of hormones on cyclic 3′,5′-AMP levels and lipolysis. *Naunyn Schmiedebergs Arch Pharmacol* 276:133–148, 1973.
5. Van Calker D, Müller M, Hamprecht B: Adenosine inhibits the accumulation of cyclic AMP in cultured brain cells. *Nature* 276:839–841, 1978.
6. Londos C, Cooper DMF, Wolff J: Subclasses of external adenosine receptors. *Proc Natl Acad Sci USA* 77:2551–2554, 1980.
7. Londos C, Wolff J: Two distinct adenosine-sensitive sites on adenylate cyclase. *Proc Natl Acad Sci USA* 74:5482–5486, 1977.
8. Burnstock G: Past and current evidence for the purinergic nerve hypothesis, in Baer HP, Drummond GI (eds): *Physiological and Regulatory Functions of Adenosine and Adenine Nucleotides*. New York, Raven Press, 1979, pp 3–32.
9. Schütz W, Tuisl E: Evidence against adenylate cyclase-coupled adenosine receptors in the guinea pig heart. *Eur J Pharmacol* 76:285–288, 1981.
10. Malbon CC, Hert RC, Fain JN: Characterization of [^{3}H]adenosine binding to fat cell membranes. *J Biol Chem* 253:3114–3122, 1978.

11. Schwabe U, Kiffe H, Puchstein C, Trost T: Specific binding of [^{3}H]adenosine to rat brain membranes. *Naunyn Schmiedebergs Arch Pharmacol* 310:59–67, 1979.
12. Newman ME, Patel J, McIlwain H: The binding of [^{3}H]adenosine to synaptosomal and other preparations from the mammalian brain. *Biochem J* 194:611–620, 1981.
13. Newman M, Levitzki A: Characteristics of high-affinity [^{3}H]adenosine binding to rat brain synaptosomes and turkey erythrocyte membranes. *Biochim Biophys Acta* 685:129–136, 1982.
14. Dutta P, Mustafa SJ: Saturable binding of adenosine to the dog heart microsomal fraction: competitive inhibition by aminophylline. *J Pharmacol Exp Ther* 211:496–501, 1979.
15. Dutta P, Mustafa SJ: Binding of adenosine to the crude plasma membrane fraction isolated from dog coronary and carotid arteries. *J Pharmacol Exp Ther* 214:496–502, 1980.
16. Schütz W, Brugger G: Characterization of [^{3}H]-adenosine binding to media membranes of hog carotid arteries. Pharmacology 24:26–34, 1982.
17. Daly JW, Nimitkitpaisan Y, Pons F, Bruns RF, Smellie F, Skolnick P: Binding sites for adenosine analogs: possible relationship to cyclic AMP-generating systems in brain tissue. *Pharmacologist* 21:253, 1979 (abst).
18. Williams M, Risley EA: Binding of ^{3}H-adenyl-5′-imidodiphosphate (AppNHp) to rat brain synaptic membranes. *Fed Proc Fed Am Soc Exp Biol* 39:1009, 1980 (abst).
19. Bruns RF, Daly JW, Snyder SH: Adenosine receptors in brain membranes: Binding of N^6-cyclohexyl[^{3}H]adenosine and 1,3-diethyl-8-[^{3}H]phenylxanthine. *Proc Natl Acad Sci USA* 77:5547–5551, 1980.
20. Schwabe U, Trost T: Characterization of adenosine receptors in rat brain by (−)[^{3}H]N^6-phenylisopropyladenosine. *Naunyn Schmiedebergs Arch Pharmacol* 313:179–187, 1980.
21. Williams M, Risley EA: Biochemical characterization of putative central purinergic receptors by using 2-chloro[^{3}H]adenosine, a stable analog of adenosine. *Proc Natl Acad Sci USA* 77:6892–6896, 1980.
22. Wu PH, Phillis JW, Balls K, Rinaldi B: Specific binding of 2-[^{3}H]chloroadenosine to rat brain cortical membranes. *Can J Physiol Pharmacol* 58:576–579, 1980.
23. Clarke DA, Davoll J, Philips FS, Brown GB: Enzymatic deamination and vasodepressor effects of adenosine analogs. *J Pharmacol Exp Ther* 106:291–302, 1952.
24. Westermann E, Stock K: Inhibitors of lipolysis: Potency and mode of action of α- and β-adrenolytics, methoxamine derivatives, prostaglandin E_1 and phenylisopropyl adenosine, in Jeanrenaud B, Hepp D (eds): *Adipose Tissue Regulation and Metabolic Functions*. Stuttgart, Georg Thieme Verlag, 1970, pp 47–54.
25. Vapaatalo H, Onken D, Neuvonen PJ, Westermann E: Stereospecificity in some central and circulatory effects of phenylisopropyl-adenosine (PIA). *Arzneim Forsch* 25:407–410, 1975.
26. Londos C, Cooper DMF, Schlegel W, Rodbell M: Adenosine analogs inhibit adipocyte adenylate cyclase by a GTP-dependent process: Basis for actions of adenosine and methylxanthines on cyclic AMP production and lipolysis. *Proc Natl Acad Sci USA* 75:5362–5366, 1978.
27. Cooper DMF, Londos C, Rodbell M: Adenosine receptor-mediated inhibition of rat cerebral cortical adenylate cyclase by a GTP-dependent process. *Mol Pharmacol* 18:598–601, 1980.
28. Stein HH, Somani P, Prasad RN: Cardiovascular effects of nucleoside analogs. *Ann NY Acad Sci* 255:380–389, 1975.
29. Raberger G, Schütz W, Kraupp O: Coronary dilatory action of adenosine analogues: A comparative study. *Arch Int Pharmacodyn Ther* 230:140–149, 1977.
30. Trost T, Schwabe U: Adenosine receptors in fat cells: Identification by (−)-N^6-[^{3}H]phenylisopropyladenosine binding. *Mol Pharmacol* 19:228–235, 1981.
31. Gavish M, Goodman RR, Snyder SH: Solubilized adenosine receptors in the brain: Regulation by guanine nucleotides. *Science* 215:1633–1635, 1982.
32. Goodman RR, Cooper MJ, Gavish M, Snyder SH: Guanine nucleotide and cation regulation of the binding of [^{3}H]cyclohexyladenosine and [^{3}H]diethylphenylxanthine to adenosine A_1 receptors in brain membranes. *Mol Pharmacol* 21:329–335, 1982.
33. Patel J, Marangos PJ, Stivers J, Goodwin K: Characterization of adenosine receptors in brain using N^6-cyclohexyl[^{3}H]adenosine. *Brain Res* 237:203–214, 1982.
34. Murphy KM, Snyder SH: Adenosine receptors in rat testes: Labeling with ^{3}H-cyclohexyladenosine. *Life Sci* 28:917–920, 1981.
35. Lenschow V, Schwabe U: Effects of guanine nucleotides on adenosine receptors in rat brain. *Naunyn Schmiedebergs Arch Pharmacol* 319 (Suppl):R6, 1982 (abst).
36. Schütz W, Tuisl E, Kraupp O: Adenosine receptor agonists: Binding and adenylate cyclase

stimulation in rat liver plasma membranes. *Naunyn Schmiedebergs Arch Pharmacol* 319:34–39, 1982.

37. Grandt R, Gabel G, Jakobs KH: Guanine nucleotides and monovalent cations increase affinities of adipocyte prostaglandin E_2 and adenosine receptors for agonists. *Naunyn Schmiedebergs Arch Pharmacol* 319 (Suppl):R30 1982 (abst).
38. Haslam RJ, Cusack NJ: Blood platelet receptor for ADP and for adenosine, in Burnstock G (ed): *Purinergic Receptors.* London, Chapman & Hall, 1981, pp 223–285.
39. Cusack NJ, Hourani SMO: 5'-N-Ethylcarboxamidoadenosine: A potent inhibitor of human platelet aggregation. *Br J Pharmacol* 72:443–447, 1981.
40. Lenschow, V., Hüttemann, E., Ukena, D., Schwabe, U.: Study of R_a adenosine receptors in human platelets by radioligand binding. Naunyn-Schmiedeberg's Arch Pharmacol 321 (Suppl): R 31, 1982 (abst).
41. Ukena, D., Martens, D., Schwabe, U.: Specific binding of 5'-N-ethylcarboxamido[^{3}H]adenosine to calf thymocyte membranes. Naunyn Schmiedeberg's Arch Pharmacol 321 (Suppl): R 38, 1982 (abst).

DISCUSSION

DR. H.J. LEIGHTON: The K_d's that you report for R_i agonists and DPX are much lower than what one would predict on the basis of isolated smooth muscle effects. Why do you think this difference occurs?

DR. SCHWABE: Hitherto, we have not performed [^{3}H]PIA binding in smooth muscle preparations and therefore the K_d values obtained in brain and fat cell membranes are not directly comparable to the physiological data from other tissues. In addition, it does not seem unlikely that adenosine receptors in at least some of these tissues may be of the R_a type.

DR. D.M. PATON: Would you comment on the relative potencies of L-PIA and NECA as inhibitors of [^{3}H]PIA binding and how this relates to the classification of R_i and R_a adenosine receptors?

DR. SCHWABE: The relative potencies of (−)PIA and NECA as displacers of [^{3}H]PIA binding have only been determined in rat brain membranes in which a 3- to 4-fold higher affinity for the (−)PIA was observed. This is not a very marked difference, but it is very similar to the ratio obtained in adenylate cyclase studies by Londos and co-workers. However, the information on the relative potencies would come out much clearer in a subtype-selective cell type, such as the fat cell, in which [^{3}H]NECA is not supposed to bind to R_a receptors.

DR. J. WOLFF: How do you deal with the discrepancy in the binding *rates* and the much more rapid attainment of a steady catalytic rate of the adenylate cyclase reaction? Could this be due to the ATP, Mg^{2+}, etc., in the assay or are there sites of equal affinity *not* coupled to adenylate cyclase and acting on some other response?

DR. SCHWABE: The radioligand binding assay did not contain all constituents of the usual adenylate cyclase assay (no ATP, GTP, $MgCl_2$, etc.), since some of these agents inhibit binding of adenosine radioligands. Furthermore, we have not yet attempted to develop an experimental design that aims especially to achieve a fast equilibration of ligand binding. Nevertheless, [^{3}H]PIA binding reaches equilibrium within 10–15 min, whereas [^{3}H]CHA binding requires 100–120 min. According to Dr. Bruns, this difference might not be due to the lower temperature (25°C) of [^{3}H]CHA binding assays. In intact fat cells, equilibrium is reached within one minute.

DR. W. SCHÜTZ: The K_d of [^{3}H]NECA binding to platelet membranes is very high, i.e., above 100 nM. Hence, did you also perform the centrifugation method for separation of bound and free ligand?

DR. SCHWABE: The K_d for [^{3}H]NECA in platelets is 0.16 μM but it cannot be designated as very high, since it reasonably agrees with the K_a for NECA obtained for the stimulation of human platelet adenylate cyclase. Filtration was used as a method of separation since nonspecific binding is much lower (about 4%) compared to separation by microcentrifugation.

DR. B.B. FREDHOLM: Could there be agonist-induced changes in adenosine receptors, as reported for both muscarinic and nicotinic receptors? If this occurs, could it perhaps explain the relatively slow rate of association and the low potency of xanthines as displacers of adenosine analogues once bound?

DR. SCHWABE: It is not unlikely that agonist-induced desensitization occurs during radioligand binding studies of R_i adenosine receptors. In many of our experiments, the concentration of agonist radioligands is far above the K_d value of the ligand. The easiest way to exclude this kind of change might be to utilize antagonist ligands such as [^{3}H]DPX; we have not yet studied this ligand.

DR. G. MISTRY: Have the investigators of [^{3}H]PIA or [^{3}H]NECA binding studies looked at the effect of dithiothreitol (DTT)? DTT is known to inhibit binding of [^{3}H]DHA to β_2 receptors of skeletal muscle membranes by about 50%. How does one correlate adenylate cyclase studies in which DTT is used with the binding studies in which DTT is omitted?

DR. SCHWABE: The effects of dithiothreitol (DTT) on binding of radioactively labeled adenosine analogues have not been studied; however, this should not be an issue, since studies on the effects of adenosine and adenosine analogues on adenylate cyclase in rat fat cell membranes and rat brain membranes have been performed without DTT.

DR. I.H. FOX: Your human platelet data are similar to our observations in human placenta. The ligand used was [^{3}H]chloroadenosine. The K_d by Scatchard analysis was 56 nM and was 17 nM from the ratio of k_2 to k_1. There is very rapid dissociation with more than 50% dissociation in 20 sec. L-PIA and CHA do not inhibit binding up to 10 μM. In contrast, NECA causes 50% inhibition at 400–800 nM. There is 50% inhibition with 30 μM IBMX and 200 μM theophylline. The characterization of adenylate cyclase properties related to this receptor has not yet been made.

DR. SCHWABE: From your data it may be suggested that [^{3}H]2-chloroadenosine labels a similar type of adenosine receptor in human placenta as [^{3}H]NECA in human platelets. Although some problems in relating the data to physiological responses are raised by the inability of (−)PIA and CHA to displace radioligand binding, both radioligands seem to recognize a common property of the R_a sites in these two tissues. Perhaps these data indicate another subclass of R_a receptors, but the better way to solve the problem would be to look for more selective radioligands.

DR. A.C. NEWBY: Despite all the technical problems, can you give any kinetic data for the binding of adenosine itself to the R_i receptor? Given inhibitors of both adenosine deaminase and adenosine kinase, could not displacement of [^{3}H]PIA by adenosine be studied?

DR. SCHWABE: We have studied the displacement of [^{3}H]PIA by adenosine in membrane preparations that had not been pretreated by adenosine deaminase. The k_i value for adenosine was about 0.6 μM, but it may not be correct because under these conditions the k_i value for (−)PIA was almost 10-fold higher than in adenosine deaminase-treated membranes. The use of inhibitors of adenosine deaminase and adenosine kinase may create similar problems, since the membrane preparations used for radioligand binding studies still continue to release endogenous adenosine, as shown by the effects of deoxycoformycin in adenosine deaminase-treated membranes.

DR. PATON: Dr. Schwabe, the rates of achieving equilibrium and dissociation you have reported for [^{3}H]PIA binding are very similar to findings we have observed in pharmacological studies using L-PIA and CHA in isolated preparations (rat vas deferens, guinea pig ileum, rat aorta). In these studies, the maximal effects of a given concentration of such analogues develop gradually and are washed out slowly, taking many minutes to return to the control situation.

DR. SCHWABE: We have not worked with smooth muscle tissue, but I know from several other groups that similar slow time courses for (−)PIA but not for NECA have been observed in isolated guinea pig heart preparations.

DR. R.D. GREEN: Helena Yeung and I have been studying [^{3}H]NECA binding to preparations of rat hippocampus and striatum. Our binding studies were performed at 37°C and suggest that [^{3}H]NECA binds to R_i sites in hippocampus and to both R_i and R_a receptors in striatum. Interestingly, agonist ([^{3}H]CHA and [^{3}H]NECA) binding to R_i sites is blocked by treatment with N-ethylmaleimide (NEM), whereas antagonist binding ([^{3}H]DPX) is increased by this treatment. As the binding of [^{3}H]NECA to R_a sites is unaffected by NEM, [^{3}H]NECA binding to R_a sites can be quantitated in NEM-pretreated preparations. It thus appears that although [^{3}H]NECA is not an ideal ligand for studying R_a receptors, it will be useful for such studies until a better ligand is discovered.

7. BINDING OF RADIOACTIVE LIGANDS TO ADENOSINE RECEPTORS IN THE CENTRAL NERVOUS SYSTEM

JOHN W. DALY

FUNCTIONAL ROLES OF ADENOSINE IN THE CENTRAL NERVOUS SYSTEM

Adenosine and analogues can inhibit spontaneous activity of a very large percentage of the neurons in brain and can markedly inhibit transsynaptic potentials in certain but not all neuronal pathways of the central nervous system [1]. Whether adenosine has other physiological functions in the central nervous system beyond serving as an inhibitory modulator of neuronal activity is unknown. However, the ability of adenosine to activate cyclic AMP-generating systems in all brain regions suggests a role for adenosine in the control of central metabolic function. Behaviorally, adenosine and analogues have marked sedative effects [2–4]. The behavioral effects of adenosine and analogues are opposite to those of adenosine antagonists, such as theophylline and caffeine. The latter compounds are central stimulants and cause marked increases in behavioral activity. An understanding of the behavioral effects of adenosine and of the xanthines theophylline and caffeine must come from further definition of the biochemical basis for the action of these compounds in specific loci of brain. Such studies have in the past focused on a definition of the nature of the adenosine receptors involved in the control of adenylate cyclase in different brain regions. Now, with the advent of binding assays for adenosine receptors, research is focusing on the localization and properties of such adenosine receptors and their significance to the function of the central nervous system.

R.M. Berne, T.W. Rall, and R. Rubio (eds.), Regulatory Function of Adenosine, *pp. 97–113.*

INTERACTION OF ADENOSINE WITH CYCLIC AMP-GENERATING SYSTEMS

A significant advance in research on adenosine occurred in the late 1970s, when it became clear that adenosine affected adenylate cyclase through two distinct classes of receptors with markedly different affinities for adenosine and markedly different profiles for interactions with adenosine analogues [5, 6]. A "low-affinity" class of adenosine receptor is stimulatory, while a "high-affinity" class of adenosine receptor is inhibitory to adenylate cyclase. In addition, it was realized that adenosine and certain analogues could inhibit nearly all adenylate cyclases via an intracellular low-affinity site, which to differentiate it from the two classes of adenosine receptors was called the P site [5]. In cells from fetal mouse brain, both classes of adenosine receptors are present: the low-affinity stimulatory receptor was termed the A_2 adenosine receptor, and the high-affinity inhibitory receptor was termed the A_1 adenosine receptor [7]; these are the terms used in this review. It is not yet clear whether both A_1 and A_2 receptors are ever present in the same cell type.

The potencies and profile of interaction with adenosine analogues for the two classes of adenosine receptors are quite different [2, 5–8]. The most potent adenosine analogues at the A_1 receptor are N^6-substituted compounds. N^6-cyclohexyladenosine, N^6-phenyladenosine, and N^6-L-phenylisopropyl adenosine are very potent, while N^6-methyladenosine, N^6-benzyladenosine, and the D-diastereomer of N^6-phenylisopropyladenosine are considerably less potent. Thus, the nature of the N^6 substitution has remarkable effects on the potency at the A_1 receptor. 2-Chloroadenosine and adenosine are quite potent at the A_1 receptor, while 5′-carboxamides of adenosine such as the 5′-ethylcarboxamide are somewhat less potent. The most potent adenosine analogues at the A_2 receptors are the 5′-ethyl- and 5′-cyclopropyl-carboxamides. Adenosine and 2-chloroadenosine are nearly 10-fold less potent. N^6 substitution decreases the potency of adenosine at this class of receptors even further. Thus, the N^6-cyclohexyl, N^6-phenylisopropyl, N^6-phenyl, N^6-benzyl, and N^6-methyladenosines have relatively low potency. Furthermore, the nature of the N^6 substituent has no remarkable effect on potency at A_2 receptors, unlike the situation at A_1 receptors. The D-diastereomer is only a few-fold less potent than N^6-L-phenylisopropyladenosine at A_2 receptors. The N^6-phenyladenosine is no more potent than its N^6-benzyl analogue. 8-bromoadenosine and 2′,5′-dideoxyadenosine are virtually inactive at both A_1 and A_2 receptors.

The differences in the rank order of potency of various analogues at so-called A_1 and A_2 receptors would seem to provide a means of classifying a physiological response as being linked to inhibition (A_1) or stimulation (A_2) of adenylate cyclase. However, data with adenylate cyclases originating from different peripheral tissues indicate that there may be subclasses of A_2 receptors, some showing relatively high affinity and a somewhat differing profile of interactions with analogues [6]. Furthermore, although the adenosine receptors involved in regulation of neurotransmitter release from peripheral sympathetic and parasympathetic neurons exhibit profiles with N^6-substituted adenosines consonant with an A_1 receptor classification, adenosine 5′-ethylcarboxamide shows a remarkable high potency [9]. Indeed, the carboxamide is the most potent adenosine analogue. It should be mentioned that not all adenosine receptors may ultimately prove to be linked to adenylate cyclase.

The major class of antagonists at adenosine receptors are the alkylxanthines such as theophylline, caffeine, and 3-isobutyl-1-methylxanthine. A large number of xanthines have been evaluated as antagonists at A_2 receptors stimulatory to adenylate cyclase [10], while only a limited number of xanthines have been evaluated as antagonists of A_1 receptors inhibitory to adenylate cyclase [11]. It would appear that none of the xanthine antagonists are truly selective for blockade of either A_1 or A_2 receptors [10–12]. An 8-phenyl substituent confers high activity as an adenosine antagonist to 1,3-dialkylxanthines [10]. The nature of the alkyl substituents is also important. Potency is greatest with propyl or butyl substituents at the 1 and 3 positions.

Most adenosine analogues that are inactive as agonists have no activity as adenosine antagonists [8]. One notable exception is a naturally occurring nucleoside, 5′-methylthioadenosine. This analogue is a competitive antagonist at A_2 receptors [8] and probably has antagonist activity at A_1 receptors [12].

The inhibitory adenosine P site exhibits an entirely different profile for interaction with adenosine analogues than do the adenosine receptors [5, 13] (see Table 7.1). The most potent analog is 2′,5′-dideoxyadenosine. The 5′-carboxamides and the N^6-substituted adenosine are inactive either as inhibitory agonists at the P site or as antagonists. 5′-methylthioadenosine has no action either as agonist or antagonist. Adenosine has low affinity and 2-chloroadenosine is nearly inactive. 2-fluoroadenosine, however, is more active than adenosine [13]. There are no known antagonists for the adenosine P site.

DEVELOPMENT OF BINDING ASSAYS FOR THE CHARACTERIZATION OF ADENOSINE RECEPTORS

Although a wide range of adenosine analogues and xanthine derivatives were developed during the past decade for the characterization of physiological roles for adenosine, until 1980 there were no satisfactory methods for the investigation of binding of radioactive ligands to any of the adenosine receptors or to the P site.

Adenosine

Initial efforts to develop a binding protocol for adenosine receptors had utilized radioactive adenosine. The earliest study on adenosine binding was with fat cell membranes [14], in which both high-affinity A_1 receptors and low-affinity adenosine P sites might be expected. A filtration assay was employed. The standard conditions consisted of incubations at 4° for 20 min with 30 n*M* [^{3}H]adenosine in Tris buffer with 5 m*M* magnesium ions. An inhibitor of adenosine deaminase was used to prevent deamination of [^{3}H]adenosine.

Table 7.1. Adenosine analogues and xanthines: Effects on cyclic AMP generation and on binding of radioactive ligands in rat brain

	EC_{50} or IC_{50} (μM), Cyclic AMP generation in striatum			IC_{50} (μM), Binding in brain membranes	
	A_2 receptor		P site		
Agent	Membrane	Slice	Membrane	[^{3}H]N^6-cyclohexyl-adenosine	[^{3}H]2′,5′-dideoxy-adenosine
Agonists					
N^6-cyclohexyladenosine	25	—	Inactive	0.0011	—
N^6-L-phenylisopropyladenosine	35	100	Inactive	0.0011	>100
N^6-D-phenylisopropyladenosine	100	800	Inactive	0.090	>1000
2-chloroadenosine	0.5	20	Inactive	0.005	>100
Adenosine 5′-cyclopropylcarboxamide	0.15	3	Inactive	0.005	1000
2′,5′-dideoxyadenosine	Inactive	Inactive	10	100	1.0
Antagonists					
5′-methylthioadenosine	20	40	Inactive	0.30	0.8
8-phenyltheophylline	0.8	3	Inactive	0.16	—
Isobutylmethylxanthine	20	40	Inactive	4	—
Theophylline	25	40	Inactive	16	>100
Caffeine	80	90	Inactive	50	—
Theobromine	>500	>500	Inactive	>100	—

Source: Data from [12, 13, 29, 31]. (Data for A_2 receptors in rat striatum are unpublished.)

Only 50% of the total binding of 30 nM [^{3}H]adenosine was displaced by 100 μM adenosine. Kinetic analysis indicated one low-affinity site with a K_d of 9 μM and a density of 28 pmol per mg protein. In addition, there was a very low affinity binding site with a 50-fold higher density. Adenine was more effective than adenosine in antagonizing binding. 2′,5′-dideoxyadenosine, ADP, and ATP were effective antagonists. Xanthines such as theophylline and isobutylmethylxanthine partially antagonized binding. N^6-phenylisopropyladenosine was ineffective. Clearly, a major portion of the binding was to sites other than A_1 receptors. A later study on binding of [^{14}C]adenosine to arterial membranes gave evidence for a binding site with an affinity constant of 1.3 μM and a high density of 140 pmol per mg protein [15, 16]. Adenosine, 2-chloroadenosine, and 5′-AMP were effective antagonists, but theophylline had little effect except at very high concentrations. Thus, the binding site was not typical of an adenosine receptor.

The characteristics of binding of [^{3}H]adenosine to brain membranes were also not entirely consonant with those expected of an adenosine receptor. In one study, with rat brain membranes, a centrifugation assay was employed, since binding and release of [^{3}H]adenosine were very rapid [17]. Standard incubations were at 0° for 6 min with 10 nM ligand in Tris buffer with 1 mM magnesium ions. An adenosine deaminase inhibitor was used to prevent deamination of [^{3}H]adenosine. Specific binding represented 85% of total binding, as defined by antagonism with 1 mM adenosine. Kinetic analysis indicated two binding sites. One site had an affinity constant of 1.7 μM and a density of 30 pmol per mg protein. A lower affinity site had an affinity constant of 14 μM and a density of 170 pmol per mg protein. Adenosine, 5′-AMP, and cyclic AMP inhibited specific binding by 50% at 4 to 24 μM. Higher concentrations (40 – 80 μM) of ATP, S-adenosylhomocysteine, 2′-deoxyadenosine, 2-chloroadenosine, and adenine were required for 50% inhibition of binding. 2′,5′-dideoxyadenosine inhibited binding by 50% at 140 μM. At 100 μM both D- and L-diastereomers of N^6-phenylisopropyladenosine inhibited binding by only 30%. Isobutylmethylxanthine, theophylline, and caffeine were ineffective except at millimolar concentrations. Clearly, a major portion of the sites did not have characteristics of adenosine receptors. Another study with rat cerebral cortical membranes employed a filtration assay [18]. Incubations were at 0° for 7 min with 4 nM [^{3}H]adenosine in Tris buffer. Specific binding as defined by antagonism with 1 mM adenosine was 75% of total binding. The affinity constant was about 2 μM and the density about 9 pmol per mg protein. Binding was blocked equally effectively by adenosine and S-adenosylhomocysteine. A more recent study on binding of [^{3}H]adenosine to rat cerebral cortical or striatal synaptosomal membranes also afforded results inconsistent with selective binding to adenosine receptors [19]. A filtration assay was employed. Standard incubations were with 100 nM [^{3}H]adenosine at 0′ for 60 min in Tris buffer with 5 mM magnesium ions. An adenosine deaminase inhibitor had no effect on binding of [^{3}H]adenosine. Specific binding as defined with 1 mM adenosine represented about 60% of total binding. Kinetic analysis for cerebral cortical membranes indicated two sites—one with an affinity constant of 0.3 μM and a density of 3 pmol per mg protein and the other with an affinity constant of 11 μM and a density of 11 pmol per mg protein. N^6-methyladenosine, N^6-L-phenylisopropyladenosine, and 2-chloroadenosine inhibited specific binding by 50% at 6 to 10 μM. However, inosine, a nucleoside with no activity at adenosine receptors, also blocked binding by 50% at 14 μM. 2′-deoxyadenosine at 30 μM and 2′,5′-dideoxyadenosine at about 100 μM blocked 50% of the binding to cortical membranes. Theophylline was a potent antagonist, with an IC_{50} of about 12 μM. However, the potent adenosine receptor antagonist isobutylmethylxanthine was much less potent with an IC_{50} of about 200 μM. Inhibition curves with N^6-phenylisopropyladenosine, 2-chloroadenosine, 2′-deoxyadenosine, inosine, theophylline, and isobutylmethylxanthine were indicative of at least two binding sites. Similar results were obtained with rat striatal membranes. Earlier studies with guinea pig cerebral cortical membranes and nearly the same protocol afforded results also suggestive of multiple binding sites for [^{3}H]adenosine, with characteristics not entirely consonant with those expected for adenosine receptors [20, 21]. A filtration assay and a standard incubation at 0′ for 90 min with 100 nM [^{3}H]adenosine in Tris buffer with 5 mM magnesium ions were employed. Binding was relatively slow. Specific binding as defined by antagonism with 100 μM adenosine was 80% of total binding. Kinetic analysis indicated the presence of two binding sites—one with an affinity constant of 1.7 μM and a density of 5 pmol per mg protein and the other with a constant of 5 μM and a density of 30 pmol per mg protein. The density of binding sites was higher in guinea pig than in rat or bovine brain membranes. 2-chloroadenosine was a potent antagonist. 5′-AMP, cyclic AMP, and inosine appeared to be only partial antagonists. Theophylline with an IC_{50} of about 5 μM appeared to be a partial antagonist. Clearly, binding of [^{3}H]adenosine to brain membranes occurs to multiple sites, limiting the usefulness of this ligand for characterization of adenosine receptors. Some of these sites probably correspond to adenosine receptors, some to the P sites, and others to sites of unknown significance.

2′,5′-Dideoxyadenosine

The binding of the specific P site agonist [^{3}H]2′,5′-dideoxyadenosine to rat cerebral cortical and striatal membranes was recently compared to binding of another P site agonist, [^{3}H]adenine arabinoside, and to binding of [^{3}H]adenosine [13]. A filtration assay was employed. Binding and release of [^{3}H]2′,5′-dideoxyadenosine, [^{3}H]adenine arabinoside, and [^{3}H]adenosine were relatively slow. Standard incubations were at 25° for 40 min with 100 nM ligand in Tris buffer. For [^{3}H]adenosine, specific binding as defined by antagonism with 100 μM adenosine was 65% of total binding. For [^{3}H]2′,5′-dideoxyadenosine, specific binding as defined with 100 μM adenine arabinoside was 30% to 60% of total binding. Preincubated membranes showed the highest degree of specific binding. Kinetic analysis suggested that [^{3}H]adenosine bound to only one class of sites in rat cerebral cortical membranes, with an affinity constant of about 70 n*M* and a density of 7 pmol per mg protein. However, inhibition of binding by certain adenosine analogues was clearly indicative of more than one binding site. Kinetic analysis suggested that [^{3}H]2′,5′-dideoxyadenosine bound to only one class of sites in rat cerebral cortical membranes, with an affinity constant of about 80 n*M* and a density of 1.8 pmol per mg protein. Similar results were obtained for [^{3}H]adenosine arabinoside. A variety of compounds were tested as antagonists of binding of [^{3}H]2′,5′-dideoxyadenosine. The IC_{50} values showed no correlation with the potencies of the compounds as agonists at the P site [13] (Table 7.1). For example, 5′-methylthioadenosine was a potent antagonist of binding of [^{3}H]2′,5′-dideoxyadenosine but had no activity as a P site agonist or antagonist. Furthermore, displacement curves with various compounds were indicative of the presence of at least two binding sites for [^{3}H]2′,5′-dideoxyadenosine. The nature of the various sites that bind [^{3}H]adenosine and [^{3}H]2′,5′-dideoxyadenosine in brain membranes remains speculative. One possibility is that some of these sites correspond to intracellular binding sites recently proposed to be involved in facilitated uptake of adenosine [22].

It was clear that successful development of ligands for adenosine receptors required the use of adenosine analogues that were much more specific in their sites of action. In 1980, four laboratories reported the development of successful protocols for the characterization of binding of radioactive adenosine analogues to brain membranes. The analogues were [^{3}H]2-chloroadenosine [23–25], [^{3}H]N^6-cyclohexyladenosine [12], and [^{3}H]N^6-L-phenylisopropyladenosine [26]. All these compounds appear useful as ligands. The profiles of antagonism of binding by adenosine analogues are consonant with binding primarily to A_1 adenosine receptors. Pretreatment of membranes with adenosine deaminase appeared necessary to remove endogenous adenosine.

2-Chloroadenosine

[^{3}H]2-chloroadenosine was developed independently as a ligand for adenosine receptors by two laboratories [23–25]. 2-chloroadenosine is a potent agonist at A_1 receptors but is also a relatively potent agonist at A_2 receptors. It has little affinity for P sites. Thus, high-affinity binding would be expected at A_1 receptors, but a significant contribution might be expected from putative subclasses of A_2 receptors that have relatively high affinity for adenosine [6]. The protocols used by the two laboratories for the binding assays differed somewhat, as did the results. Filtration assays were used.

In one protocol, a pretreatment of rat brain membranes with adenosine deaminase was found essential [25, 27]. Standard incubations were at 23° for 60 min with 1 n*M* [^{3}H]2-chloroadenosine in Tris buffer. Specific binding as defined by antagonism with 10 μM 2-chloroadenosine was 90% of total binding. Kinetic analysis indicated the presence of two binding sites for [^{3}H]2-chloroadenosine in rat brain membranes. One had an affinity constant of 1.4 n*M* and a density of 0.2 pmol per mg protein, while the other had an affinity constant of 12 n*M* and a density of 0.3 pmol per mg protein. The IC_{50} values of some adenosine analogues as antagonists of binding were as follows: N^6-L-phenylisopropyladenosine 1 n*M* > 2-chloroadenosine 2.6 n*M* > adenosine 5′-cyclopropylcarboxamide 11 n*M* > N^6-D-phenylisopropyladenosine 30 n*M* > adenosine 5′-ethylcarboxamide 40 n*M*. Inosine and 8-bromoadenosine were ineffective. 5′-methylthioadenosine had an IC_{50} of 70 n*M*. Xanthines had IC_{50} values consonant with those expected from their effects on adenosine-receptor mediated modulation of adenylate cyclase. Etazolate (IC_{50} 6 μ*M*) was a rather potent blocker. The binding of [^{3}H]2-chloroadenosine was reduced by GppNHp or GTP. The density of binding sites for [^{3}H]2-chloroadenosine was much higher in rat brain than in peripheral tissues such as liver, kidney or heart. Regionally in rat brain there was only about a two-fold range in density of binding sites, with the highest density in caudate, hippocampus, cerebellum, and thalamus. The binding sites were enriched in synaptosomal fractions. This laboratory did not obtain satisfactory results with subsequent samples of [^{3}H]2-chloroadenosine [28].

In the other protocol for [^{3}H]2-chloroadenosine binding to brain membranes, a treatment with adenosine deaminase was not employed [23, 24]. Standard incubations were at 0° for 60 min with 33 n*M* [^{3}H]2-chloroadenosine in Tris buffer with 25 m*M* magnesium ions and 25 m*M* calcium ions. Specific binding as defined by antagonism with 250 μ*M* 2-chloroadenosine was 60% of total binding. Kinetic analysis suggested the presence in rat cerebral cortical membranes of only one site with an affinity constant of 24 n*M* and a density of 0.5 pmol per mg protein. The profile of inhibition by the few compounds tested was not entirely consonant with an adenosine receptor, but further studies with a more extensive series of analogues will be required to clarify this point. Thus, adenosine exhibited an IC_{50} of 120 n*M* versus binding of [^{3}H]2-chloroadenosine, but inosine and adenine, two compounds that have no affinity for adenosine receptors, were also remarkably potent, with IC_{50} values of 400 to 450 n*M*. Xanthines such as isobutylmethylxanthine and theophylline were about 10-fold more potent in this protocol as antagonists of binding of [^{3}H]2-chloroadenosine than would be expected from their effects on adenosine receptor-mediated modulation of adenylate cyclase.

N^6-Cyclohexyladenosine

N^6-cyclohexyladenosine is an extremely potent agonist at A_1 receptors, whereas, like most N^6-substituted adenosine analogues, it is a relatively weak agonist at A_2 receptors and has no effect on the adenosine P site. Thus, it appeared to be an ideal ligand for selective labeling of A_1 receptors in brain membranes. Initial studies were with guinea pig brain membranes [12]. A filtration assay was used. The binding and release was relatively slow. At 0° the specific binding was greatly reduced. Standard incubations were at 25° for 2 hr with 1 μ*M* [^{3}H]N^6-cyclohexyladenosine in Tris buffer. Membranes were pretreated with adenosine deaminase. Specific binding as defined by antagonism with 10 μ*M* N^6-L-phenylisopropyladenosine was 90% of total binding. Kinetic analysis of binding to guinea pig brain membranes indicated the presence of one site with an affinity constant of 6 n*M* and a density of 0.4 pmol per mg protein [12]. In bovine brain membranes, two sites appeared present—one with an affinity constant of 0.3 n*M* and a density of 0.3 pmol per mg protein, the other with a constant of 2 n*M* and a density of 0.2 pmol per mg protein. In rat brain membranes, one site with an affinity constant of 1.6 n*M* and a density of 0.1 pmol per mg protein appeared to be present [29]. Another laboratory using the same protocol has recently reported [30] the presence of two binding sites in rat brain synaptosomal membranes—one with an affinity constant of 0.7 n*M* and a density of 0.23 pmol per mg protein, the other with a constant of 2.4 n*M* and a density of 0.12 pmol per mg protein. The profile of inhibition of binding by adenosine analogues in guinea pig brain membranes [12] was consonant with that expected of an A_1 receptor. The IC_{50} values were as follows: N^6-L-phenylisopropyladenosine 4 n*M;* 2-chloroadenosine 10 n*M;* adenosine 5′-cyclopropylcarboxamide 10 n*M;* N^6-D-phenylisopropyladenosine 150 n*M.* 2′,5′-dideoxyadenosine, adenine, 8-bromoadenosine, and inosine were virtually inactive. 5′-methylthioadenosine had an IC_{50} value of 300 n*M.* Similar profiles were obtained with bovine brain and rat brain membranes [12, 29–31] (Table 7.1). However, all compounds were more potent in these species than in guinea pig. Displacement curves were not indicative of multiple binding sites. Guanyl nucleotides inhibited the binding of [^{3}H]N^6-cyclohexyladenosine [12, 32]. In bovine brain membranes, the affinity constant for [^{3}H]N^6-cyclohexyladenosine could be reduced five-fold by GppNHp. A series of xanthine antagonists showed profiles for inhibition of binding [13] consonant with their ability to antagonize adenosine receptor-mediated modulation of adenylate cyclase. The potency of xanthines as displacers was unaffected by guanyl nucleotides, while the potency of adenosine analogues was decreased [32].

[^{3}H]N^6-cyclohexyladenosine has now been employed extensively as a ligand from A_1 adenosine receptors. It has been investigated in brain membranes from mouse [4], rat [28–32], guinea pig [12], rabbit [31], cattle [12, 31–33], and human [31]. In all cases binding appears to be to a site with characteristics expected of an A_1 receptor. There apparently were some significant species differences in the potencies of agonists and antagonists for the binding sites. Binding sites have been solubilized from bovine brain membranes [33]. Such solubilized sites display unaltered affinity for the various adenosine analogues and for xanthines. Guanyl nucleotides still inhibited binding. In rat brain, the density of binding sites for [^{3}H]N^6-cyclohexyladenosine was 9 pmol per gram wet weight, whereas it was 2 pmol in testes and less than 0.3 pmol in heart, lung, kidney, adrenal, stomach, small intestine, fat, thyroid, muscle, vas deferens, ovary, and submandibular gland [29]. Regionally in rat brain, density of binding sites varied no more than two-fold, with highest densities in cerebellum, hippocampus, and striatum and lower values in hypothalamus-thalamus, cortex, midbrain, and brain stem [29, 31]. Another laboratory reported densities in rat hippocampus > striatum, cerebellum, cortex > hypothalamus > medulla, spinal cord [30]. Densities in pineal were only about one-fourth those in brain, and densities in pituitary were even lower. In rabbit brain, highest densities were in hippocampus and lowest in midbrain, cerebellum, and brain stem [31]. In bovine brain, highest densities were in hippocampus and lowest in hypothalamus-thalamus and brain stem. Binding sites were enriched in synaptosome fractions [30]. The regulation of binding of [^{3}H]N^6-cyclohexyladenosine by guanyl nucleotides and ions deserves further investigation. Sodium ions decrease binding to brain membranes, while divalent cations such as manganese, magnesium, and calcium can increase binding 1.5-fold in EDTA-treated membranes [32]. EDTA treatment appears to decrease binding. Another group also reported enhancement of binding of [^{3}H]N^6-cyclohexyladenosine by calcium and magnesium in rat brain membranes [30]. Recently, it has been found that in fat cell membranes, GTP can have opposite effects on binding of [^{3}H]N^6-cyclohexyladenosine, depending on the concentration of magnesium [D.M.F. Cooper, personal communication]. In the absence of magnesium, GTP decreases binding, whereas in the presence of 0.2 to 20 m*M* magnesium, GTP increases binding. Magnesium alone decreases binding with maximal inhibitory effects at 1 to 2 m*M*. All studies on effects of guanyl nucleotides on binding of [^{3}H]N^6-cyclohexyladenosine to brain membranes have been done in magnesium-free media.

N^6-L-Phenylisopropyladenosine

N^6-L-phenylisopropyladenosine represents another candidate for a selective high-affinity ligand for A_1 receptors since it has low affinity at A_2 receptors and is inactive at adenosine P sites. Initial studies on binding were with rat brain membranes [26]. A filtration assay was employed. Binding and release occurred at a moderate rate. Standard incubations were at 37° for 15 min with 14 n*M* [^{3}H]N^6-L-phenylisopropyladenosine in Tris buffer with 1 m*M* magnesium ions. Adenosine deaminase was present. The specific binding as defined by antagonism with 100 μ*M* N^6-L-phenylisopropyladenosine was 60% of total binding. Kinetic analysis indicated one binding site with an affinity constant of 5 n*M* and a density of 0.8 pmol per mg protein. The IC_{50} values for certain adenosine analogs at 0°C were as follows: N^6-L-phenylisopropyladenosine 90 n*M;* N^6-phenyladenosine 340 n*M;* N^6-benzyladenosine 3000 n*M;* N^6-D-phenylisopropyladenosine 3600 n*M;* and 2-chloroadenosine 3600 n*M*. 2′,5′-dideoxyadenosine, 5′-AMP, and cyclic AMP were inactive. Xanthines such as isobutylmethylxanthine, theophylline, and caffeine showed potencies as displacers consonant with their activity as antagonists of adenosine receptors linked to adenylate cyclase. The low activity of the

N^6-benzyladenosine and N^6-D-phenylisopropyladenosine is consonant with that expected of an A_1 receptor. The low activity of 2-chloroadenosine is probably an effect of temperature. In a subsequent study on binding of [^{3}H]N^6-L-phenylisopropyladenosine to rat fat cell membranes at 37°C, 2-chloroadenosine was quite potent as displacer [34]. Xanthines were much less potent as displacers in fat cell membranes at 37°C than in brain membranes at 0°C.

Adenosine 5′-carboxamides

The availability of radioactive ligands for the A_1 receptor makes possible studies correlating binding parameters at this receptor with both function and localization of these receptors in brain. However, a ligand for the A_2 receptor is lacking. This receptor has a relatively low affinity for most adenosine analogues, making the task of developing a ligand difficult. The most attractive candidates are either adenosine 5′-cyclopropylcarboxamide or adenosine 5′-ethylcarboxamide, since these compounds have the highest potencies among adenosine analogues at A_2 receptors. A preliminary report suggested that binding of [^{3}H]adenosine 5′-ethylcarboxamide to rat striatal membranes occurs at both A_1 and A_2 receptors [35].

1,3-Diethyl-8-phenylxanthine

An alternate approach to labeling of A_1 and A_2 receptors is to use 8-phenyl-1,3-dialkylxanthine. Such compounds have very high potencies as antagonists at adenosine receptors that modulate adenylate cyclase activity. The binding of [^{3}H]1,3-diethyl-8-phenylxanthine to brain membranes was therefore investigated [12]. A filtration assay was used. Binding and release to brain membranes was rapid. With bovine brain membranes, standard incubations were at 0° for 4 hr with 1 n*M* [^{3}H]1,3-diethyl-8-phenylxanthine in Tris buffer. The specific binding as defined by antagonism with 10 μM N^6-L-phenylisopropyladenosine was 90% of total binding. Kinetic analysis indicated the presence of one site in bovine brain membranes with an affinity constant of 5 n*M* and a density of 1.0 pmol per mg protein. The profile of inhibition of binding by adenosine analogues was consonant with that expected of an A_1 receptor. Thus, IC_{50} values were as follows: N^6-L-phenylisopropyladenosine 0.8 n*M*; N^6-cyclohexyladenosine 3n*M*; N^6-D-phenylisopropyladenosine 20 n*M*. Xanthine analogues showed a profile as displacers consonant with that expected at adenosine receptors coupled to adenylate cyclase. The potencies of adenosine analogues as inhibitors of binding of [^{3}H]1,3-diethyl-8-phenylxanthine was reduced by GppNHp and GTP [32]. The binding parameters in guinea pig brain membranes differed significantly from those in bovine brain membranes. The affinity was much lower, and adenosine analogues were very weak inhibitors. Standard incubations were at 0° for 1 hr with 5 n*M* [^{3}H]1,3-diethyl-8-phenylxanthine. The specific binding as defined by antagonism with 1 m*M* 2-chloroadenosine or with theophylline was only 50% of total binding. Kinetic analysis indicated one site with an affinity constant of 70 n*M* and a density of 0.5 pmol per mg protein. 2-chloroadenosine was the most potent analogue, versus binding in guinea pig brain membranes with an IC_{50} of 2 μM. The profile of inhibition by other analogues was not consonant with that expected at either A_1 or A_2 receptors. The IC_{50} for certain analogues were as follows: 8-bromoadenosine 50 μM; N^6-cyclohexyladenosine 80 μM; adenosine 5′-cyclopropylcarboxamide 100 μM; N^6-D-phenylisopropyladenosine 100 μM; N^6-L-phenylisopropyladenosine 140 μM. Xanthines antagonized binding with a profile consonant with their potencies as adenosine antagonists. Thus, although [^{3}H]1,3-diethyl-8-phenylxanthine appears to label primarily A_1 receptors in bovine brain membranes and in rat and rabbit cerebral cortical membranes, the binding sites in guinea pig brain membranes and in human brain membranes do not have characteristics expected of adenosine receptors [12, 31, 32]. The significance of the binding sites in guinea pig and human brain membranes is unknown. The sites might be an antagonist state of an adenosine receptor or a site unrelated to adenosine receptor. However, such unexpected results serve as a warning that binding studies with [^{3}H]1,3-diethyl-8-phenylxanthine in different species and tissues must be evaluated carefully. In a recent comparison of [^{3}H]N^6-cyclohexyladenosine and [^{3}H]1,3-diethyl-8-phenylxanthine binding, it was found that adenosine derivatives have enhanced affinity for these sites at higher temperatures, while xanthines have higher affinities at lower temperatures [31].

CORRELATION OF BINDING WITH FUNCTIONAL ROLES FOR ADENOSINE

It is apparent that 1980 was a turning point in research on physiological functions of adenosine in the central nervous system. [^{3}H]N^6-cyclohexyladenosine has been already used as a ligand for radioautographic localization of A_1 receptors in rat brain [36, 37]. Binding sites were found to be particularly dense in the superficial layer of superior colliculus and in the molecular layer of the cerebellum. There were laminar and regional differences in binding sites in cerebral cortex and laminar labeling in dorsal hippocampus [37]. Such heterogeneous distribution suggests that adenosine receptors may be selectively associated with specific neuronal systems and perhaps functions. Indeed, many of the receptors in cerebellum were associated with the excitatory granule cells, particularly the parallel fibers and terminals of the molecular layer [36]. The relative potencies of N^6-cyclohexyladenosine > N^6-L-phenylisopropyladenosine > 2-chloroadenosine > N^6-D-phenylisopropyladenosine in (1) inhibiting evoked potentials in rat hippocampal slices and (2) inhibiting binding of [^{3}H]N^6-cyclohexyladenosine to rat brain membranes showed an excellent correlation [38]. The ability of parenteral N^6-cyclohexyladenosine, N^6-L-phenylisopropyladenosine, and N^6-D-phenylisopropyladenosine to cause a profound behavioral depression of locomotor activity in mice correlates with the ability of these compounds to inhibit binding of [^{3}H]N^6-cyclohexyladenosine to brain membranes [4]. The behavioral depressant effects of N^6-L-phenylisopropyladenosine are not due to peripheral hypotension, since the polar adenosine antagonist 8-(p-sulfophenyl)theophylline has no effect on the behavioral depression [4] but can reverse the hypotensive effects of N^6-L-phenylisopropyladenosine [P. Skolnick, personal communication]. Other xanthines that penetrate into brain do reverse behavioral depression caused by parenteral N^6-L-phenylisopropyladenosine [4]. Xanthines are behavioral stimulants, and their stimulatory activity correlates well with their ability to antagonize binding of $[^3H]N^6$-cyclohexyladenosine to mouse brain membranes [4]. Surprisingly, xanthines such as caffeine and theophylline are much more potent and effective central stimulants when administered with a depressant dose of L-phenylisopropyladenosine. Thus, theophylline plus the adenosine analogue enhanced locomotor activity more than theophylline alone [4]. Indeed, isobutylmethylxanthine, which is a behavioral depressant alone, perhaps due to its high activity as a phosphodiesterase inhibitor, became a central stimulant when administered with N^6-L-phenylisopropyladenosine. The mechanism involved in these unusual interactions of adenosine agonists and antagonists is unknown. The effects of parenteral N^6-cyclohexyladenosine and 2-chloroadenosine on spontaneous motor activity, exploratory activity, motor coordination, and pain threshold have been compared [4]. Both analogues reversed diazepam-induced increases of exploratory behavior in mice. 8-phenyltheophylline, theophylline, and caffeine, but not isobutylmethylxanthine, reversed the analgesic effects of intracisternal 2-chloroadenosine and the 5′-carboxamides [39]. The carboxamide was more potent as an analgesic than 2-chloroadenosine. Rats rendered tolerant to the behavioral effects of N^6-L-phenylisopropyladenosine showed no change in the properties or levels of central receptors that bind $[^3H]N^6$-cyclohexyladenosine [R.F. Bruns, J.M. Carney, personal communication]. After pentylenetetrazole-induced convulsions, levels of binding sites for $[^3H]$cyclohexyladenosine in rat cerebellar membranes decreased significantly after one hour [40]. There was no change in cerebral cortical levels.

An understanding of behavioral effects of adenosine analogues and xanthine derivatives will require a much more intimate knowledge of the nature and function of adenosine receptors in specific pathways and regions of the central nervous system. Striatum appears to be one region that should be more thoroughly studied. Not only is this an important region with respect to a variety of behaviors, but it is also a somewhat unique region with respect to adenosine receptors linked to adenylate cyclase. In brain slices from all regions, adenosine elicits a marked accumulation of cyclic AMP through interaction with a low-affinity A_2 receptor [2]. In contrast, it has proved difficult to detect A_1 receptor-mediated inhibition of cyclic AMP in brain slices [unpublished results]. After disrupting the cells, the situation reverses. Thus, in membranes from rat cerebral cortex, an A_1 receptor-mediated inhibition of adenylate cyclase can be detected [41], but it has proved difficult to detect an A_2 receptor-mediated stimulation of adenylate cyclase in membranes from any brain region except in striatum and other limbic structures [42]. Adenosine analogues activate adenylate cyclase in striatal membranes, elicit accumulations of cyclic AMP in striatal slices, and inhibit binding [^{3}H]N^6-cyclohexyladenosine to striatal membranes (Table 7.1). The A_2 receptor that activates adenylate cyclase in membranes appears to be a much higher affinity receptor than that which stimulates accumulation of cyclic AMP in slices. The data suggest that the limbic system contains both the usual low-affinity A_2 receptor system whose coupling is lost in membranes and a high-affinity A_2 receptor system whose coupling to adenylate cyclase is maintained in membranes. It would follow that the high-affinity system must contribute only marginally to total accumulations of cyclic AMP in striatal slices. Fortunately, the affinity of N^6-substituted adenosines for both membrane and slice A_2 receptor systems in striatum is very low, so that binding of [^{3}H]cyclohexyladenosine to A_2 receptors remains insignificant. However, for other ligands, such as the 5′-carboxamides, one might expect significant binding to the putative high-affinity A_2 receptors in striatal membrane but perhaps not to the more ubiquitous low-affinity A_2 receptors of brain. Clearly, our insights into the central role of adenosine receptors remain limited. Challenges for the future include the creative exploitation of the A_1 receptor ligands, the development of ligands for A_2 receptors, and the development of selective antagonists for the different subclasses of adenosine receptors.

REFERENCES

1. Stone TW: Physiological roles for adenosine and adenosine 5′-triphosphate in the nervous system. *Neuroscience* 6:523–556, 1981.
2. Daly JW, Bruns RF, Snyder, SH: Adenosine receptors in the central nervous system: Relationship to the central actions of methylxanthines. *Life Sci* 28:2083–2097, 1981.
3. Snyder SH, Katims JJ, Annau Z, Bruns RF, Daly JW: Adenosine receptors and behavioral actions of methylxanthines. *Proc Natl Acad Sci USA* 78:3260–3264, 1981.
4. Crawley JN, Patel J, Marangos PJ: Behavioral characterization of two long-lasting adenosine analogs: Sedative properties and interaction with diazepam. *Life Sci* 29:2623–2630, 1981.
5. Londos C, Wolff J: Two distinct adenosine-sensitive sites on adenylate cyclase. *Proc Natl Acad Sci USA* 74:5482–5486, 1977.
6. Londos C, Cooper DMF, Wolff J: Subclasses of external adenosine receptors. *Proc Natl Acad Sci USA* 77:2551–2554, 1980.
7. Van Calker D, Müller M, Hamprecht B: Adenosine regulates via two different types of receptors, the accumulation of cyclic AMP in cultured brain cells. *J Neurochem* 33:999–1005, 1979.
8. Bruns RF: Adenosine receptor activation in human fibroblasts: Nucleoside agonists and antagonists. *Can J Physiol Pharmacol* 58:673–691, 1980.
9. Paton DM: Structure-activity relations for presynaptic inhibition of noradrenergic and cholinergic transmission by adenosine: Evidence for action on A_1 receptors. *J Autonomic Pharmacol* 1:287–291, 1981.
10. Bruns RF: Adenosine antagonism by purines, pteridines and benzopteridines in human fibroblasts. *Biochem Pharmacol* 30:325–333, 1981.
11. Londos C, Cooper DMF, Schlegel W, Rodbell M: Adenosine analogs inhibit adipocyte adenylate cyclase by a GTP-dependent process: Basis for actions of adenosine and methylxanthines on cyclic AMP production and lipolysis. *Proc Natl Acad Sci USA* 75:5362–5366, 1978.
12. Bruns RF, Daly JW, Snyder SH: Adenosine receptors in brain membranes: Binding of N^6-cyclohexyl[^{3}H]adenosine and 1,3-diethyl-8-^{3}H]phenylxanthine. *Proc Natl Acad Sci USA* 77:5547–5552, 1980.

13. Nimit Y, Law J, Daly JW: Binding of 2′,5′-dideoxyadenosine to brain membranes: Comparison to P-site inhibition of adenylate cyclase. *Biochem Pharmacol* 31:3279–3287, 1982.
14. Malbon CC, Hert RC, Fain JN: Characterization of [^{3}H]adenosine binding to fat cell membranes. *J Biol Chem* 253:3114–3112, 1978.
15. Dutta P, Mustafa SJ: Saturable binding of adenosine to the dog heart microsomal fraction: Competitive inhibition by aminophylline. *J Pharmacol Exp Ther* 211:496–501, 1979.
16. Dutta P, Mustafa SJ: Binding of adenosine to the crude plasma membrane fraction isolated from dog coronary and carotid arteries. *J Pharmacol Exp Ther* 214:496–502, 1980.
17. Schwabe U, Kiffe H, Puchstein C, Trost T: Specific binding of ^{3}H-adenosine to rat brain membranes. *Naunyn Schmiedebergs Arch Pharmacol* 310:59–67, 1979.
18. Fonlupt P, Rey C, Pacheco H: *In vitro* and *in vivo* binding of S-adenosyl-L-homocysteine to membranes from rat cerebral cortex. *J Neurochem* 36:165–170, 1981.
19. Newman M, Levitzki A: Characteristics of high-affinity]^{3}H]adenosine binding to rat brain synaptosomes and turkey erythrocyte membranes. *Biochim Biophys Acta* 685:129–136, 1982.
20. Newman ME, De Lucia R, Patel J, McIlwain H: Adenosine-binding to cerebral preparations in interpretation of adenosine activation of adenosine 3′:5′-cyclic monophosphate formation. *Biochem Soc Trans* 8:141–142, 1980.
21. Newman ME, Patel J, McIlwain H: The binding of [^{3}H]adenosine to synaptosomal and other preparations from the mammalian brain. *Biochem J* 194:611–620, 1981.
22. Green RD: Adenosine transport by a variant of C1300 murine neuroblastoma cells deficient in adenosine kinase. *Biochim Biophys Acta* 598:366–474, 1980.
23. Wu PH, Phillis JW: Specific binding of ^{3}H-2-chloroadenosine to rat brain cortical membranes: Adenosine receptors. *Can J Neurol Sci* 7:239, 1980.
24. Wu PH, Phillis JW, Balls K, Rinaldi B: Specific binding of 2-[^{3}H]chloroadenosine to rat brain cortical membranes. *Can J Physiol Pharmac* 58:576–579, 1980.
25. Williams M, Risley EA: Biochemical characterization of putative central purinergic receptors by using 2-chloro[^{3}H]adenosine, a stable analog of adenosine. *Proc Natl Acad Sci USA* 77:6892–6896, 1980.
26. Schwabe U, Trost T: Characterization of adenosine receptors in rat brain by (−)[^{3}H]N^6-phenylisopropyladenosine. *Naunyn Schmiedebergs Arch Pharmacol* 313:179–188, 1980.
27. Williams M, Risley EA, Huff JR: Interaction of putative anxiolytic agents with central adenosine receptors. *Can J Physiol Pharmacol* 59:897–900, 1981.
28. Haubrich DR, Williams M, Yarbrough GG: 2-chloroadenosine inhibits brain acetylcholine turnover *in vivo*. *Can J Physiol Pharmacol* 59:1196–1198, 1981.
29. Murphy KMM, Snyder SH: Adenosine receptors in rat testes: Labeling with ^{3}H-cyclohexyladenosine. *Life Sci* 28:917–920, 1981.
30. Patel J, Marangos PJ, Stivers J, Goodwin FK: Characterization of adenosine receptors in brain using N^6-cyclohexyl[^{3}H]adenosine. *Brain Res* 237:203–214, 1982.
31. Murphy KMM, Snyder SH: Heterogeneity of adenosine A_1 receptor binding in brain tissue. *Mol Pharmacol* 22:250–257, 1982.
32. Goodman RR, Cooper MJ, Gavish M, Snyder SH: Guanine nucleotide and cation regulation of the binding of [^{3}H]cyclohexyladenosine and [^{3}H]diethylphenylxanthine to adenosine A_1 receptors in brain membranes. *Mol Pharmacol* 21:329–335, 1982.
33. Gavish M, Goodman RR, Snyder SH: Solubilized adenosine receptors in the brain: Regulation by guanine nucleotides. *Science* 215:1633–1635, 1982.
34. Trost T, Schwabe U: Adenosine receptors in fat cells: Identification by (−)-N^6-[^{3}H]phenylisopropyladenosine binding. *Mol Pharmacol* 19:228–235, 1981.
35. Yeung S-M, Green RD: Binding of 5′-N-ethylcarboxamide [^{3}H]adenosine (^{3}H-NECA) to adenosine receptors in rat striatum. *Pharmacologist* 23:184, 1981 (abst).
36. Goodman RR, Snyder SH: Autoradiographic localization of adenosine receptors in rat brain using ^{3}H-cyclohexyladenosine. *J Neuroscience* 2:1230–1241, 1982.
37. Lewis ME, Patel J, Edley SM, Marangos PJ: Autoradiographic visualization of rat brain adenosine receptors using N^6-cyclohexyl[^{3}H]adenosine. *Eur J Pharmacol* 73:109–110, 1981.
38. Reddington M, Lee KS, Schubert P: An A_1-adenosine receptor, characterized by [^{3}H]cyclohexyladenosine binding, mediates the depression on evoked potentials in a rat hippocampal slice preparation. *Neurosci Lett* 28:275–280, 1982.
39. Yarbrough GG, McGuffin-Clineschmidt JC: *In vivo* behavioral assessment of central nervous system purinergic receptor. *Eur J Pharmacol* 76:137–144, 1981.

40. Wybenga MP, Murphy MG, Robertson HA: Rapid changes in cerebellar adenosine receptors following experimental seizures. *Eur J Pharmacol* 75:79–80, 1981.
41. Cooper DMF, Londos C, Rodbell M: Adenosine receptor-mediated inhibition of rat cerebral cortical adenylate cyclase by a GTP-dependent process. *Mol Pharmacol* 18:598–601, 1980.
42. Premont J, Perez M, Blanc G, Tassin J-P, Thierry A-M, Herve D, Bockaert J: Adenosine-sensitive adenylate cyclase in rat brain homogenates: Kinetic characteristics, specificity, topographical, subcellular and cellular distribution. *Mol Pharmacol* 16:790–804, 1979.

DISCUSSION

DR. A. SATTIN: Could not the "sedative" effect of parenteral L-PIA in mice be a consequence of vasomotor changes?

DR. DALY: There is no doubt that animals treated with parenteral L-PIA are in a severe hypotensive state. Undoubtedly, this has behavioral consequences. In our initial report, we found that 8-p-sulfophenyltheophylline, a polar adenosine antagonist that should not penetrate well into brain, did not reverse the sedative effects of parenteral L-PIA, while theophylline in an equivalent dose did reverse the sedation. 8-p-sulfophenyltheophylline and theophylline are of comparable potency as adenosine antagonists. This was, however, an incomplete study since blood pressure was not measured. Recently, Dr. Phil Skolnick and colleagues have shown that doses of 8-p-sulfophenyltheophylline that nearly completely reverse the hypotensive effects of L-PIA do not affect the behavioral effects. Further studies are needed, but at present it appears that the hypotension is not a major determinant of the behavioral effects of L-PIA.

DR. J.S. FEDAN: Several speakers have noted the diversity in nomenclature regarding purine receptors. When should a uniform nomenclature be decided?

DR. DALY: Later.

II. ADENOSINE METABOLISM AND TRANSPORT

8. REGULATION OF ADENOSINE METABOLISM BY 5'-NUCLEOTIDASES

JOHN M. LOWENSTEIN, MING-KUN YU, YOSHITSUGO NAITO

When an organ is subjected to anoxia or an increased work load, its rates of ATP utilization and synthesis change. Such changes are accompanied by adjustments in the concentrations of many intermediates of energy metabolism, including ATP, ADP, AMP, creatine phosphate, and creatine. However, changes in the concentration of free ATP are relatively small. This is a consequence of the equilibrium of the creatine kinase reaction, which, as written, lies far to the right:

creatine phosphate + ADP + H^+ ↔ creatine + ATP

$K_{eq} = 1.51 \times 10^8 M^{-1}$

In other words, if the activity of creatine kinase is high enough to maintain equilibrium under all conditions, an increase in ATP utilization will chiefly result in a decrease in creatine phosphate and in a relatively small decrease in ATP. The higher the concentration of creatine phosphate, the greater is the ATP-buffering capacity of this system. Although the ATP decrease will be relatively small, it will be reflected by an increase in ADP, and this in turn will lead to a rise in AMP through the action of adenylate kinase:

2 ADP ↔ ATP + AMP

The work described in this chapter was supported by NIH grant GM-07261.

R.M. Berne, T.W. Rall, and R. Rubio (eds.), Regulatory Function of Adenosine, *pp. 117–131.*

Metabolite measurements on skeletal muscle [1, 2] and heart muscle [3] show that an increased work load causes relatively little change in the total contents of ADP and AMP. Muscle actin contains one tightly bound molecule of ADP per actin monomer that is not available to the pool of free ADP. The content of actin monomer in skeletal muscle is about 0.6 μmol per gm fresh weight [4], a value similar to the total ADP content. The methods of deproteinizing tissue samples for the purpose of metabolite measurements release the bound ADP. As a result, bound and free ADP are measured together. The ADP bound to actin exceeds free ADP by a wide margin. Hence it is difficult, if not impossible, to measure free ADP directly. However, the free ADP content can be calculated from the creatine phosphate, creatine, and adenine mononucleotide contents if it is assumed that the creatine kinase and adenylate kinase reactions are in equilibrium [4]. This assumption is justified because of the very high activities of these enzymes in heart and skeletal muscle.

Measured contents of AMP and ADP and calculated contents of free AMP and ADP in skeletal muscle during rest and exercise are shown in Table 8.1, part A. Similar values for heart muscle under different work loads are shown in Table 8.1, part B. The measured amounts of AMP and ADP are much higher than the calculated amounts of free AMP and ADP. As has already been mentioned, the large difference between measured and calculated ADP can be explained largely or wholly in terms of actin-bound ADP. However, there is no similar protein capable of binding 50 to 150 nmol AMP per gm fresh weight. Thus the metabolite measurements indicate that there is a pool of AMP in skeletal muscle and heart that is not in equilibrium with ADP and ATP through the adenylate kinase reaction.

We showed previously that in the perfused rat heart, radioactive adenosine is incorporated into a nucleotide pool in which the specific activity of AMP, ADP, and ATP becomes equilibrated in 20 min or less, whereas radioactive AMP enters a pool where the specific activities of AMP, ADP, and ATP do not equilibrate during 20 min of perfusion followed by a washout period of 3 min [5]. We thought at the time that the radioactive AMP in the perfusate enters a compartment that is not in equilibrium with the rest of the adenine nucleotide pool. Although this evidence is not as strong as the comparison of calculated free AMP with measured AMP made by Nishiki et al. [3], it points in the same direction, namely, a separate pool of AMP.

If the adenylate kinase reaction remains in equilibrium, the AMP concentration will rise in proportion to the square of the ADP concentration. ADP is a signal of energy deficit, and a rise in ADP increases the rate of oxidative phosphorylation and hence of oxygen consumption. The extent of this increase is determined by the K_m for ATP synthetase of oxidative phosphorylation and by competition between ADP and ATP for the mitochondrial adenine nucleotide translocase [6]. The rise in AMP is also a signal of energy deficit. AMP is an activator of phosphorylase *b* and phosphofructokinase and an inhibitor of fructose diphosphatase. In other words, AMP serves as a signal for

Table 8.1. Measured and calculated levels of free ADP and AMP in skeletal muscle and heart of rat under various conditions

A. Skeletal Muscle [1]

		Measured		Calculated	
Conditions	Nucleotide	Rest	Exercise	Rest	Exercise
	AMP	81	144		
	ADP	792	864		
Assumed pH					
7.2	AMP			0.31	6.5
	ADP			29	116
7.0	AMP			0.12	2.6
	ADP			19	75
6.8	AMP			0.054	1.2
	ADP			13	51
6.6	AMP			0.026	0.56
	ADP			8.9	36
6.4	AMP			0.012	0.27
	ADP			6.4	26

B. Heart [3]

Conditions	Nucleotide	Measured	Calculated
80 cm water	AMP	74	3.6
	ADP	1000	219
120 cm water	AMP	152	28
	ADP	910	464
3 μM epinephrine	AMP	121	24
	ADP	1080	497
20 mM K^+	AMP	119	2.6
	ADP	900	178
Ca^{2+} free	AMP	24	0.6
	ADP	710	115

Values are in nanomoles per gram fresh weight. The calculations are based on the assumption that the creatine kinase and myokinase reactions are in equilibrium and were made according to the method of McGilvery and Murray [4].

increasing the rates of glycogen mobilization and glycolysis and for inhibiting fructose diphosphatase. AMP is thus an intracellular signal of energy deficit. Many organs are capable of converting this intracellular signal into an extracellular signal by hydrolyzing some of the AMP to adenosine. A specific transport protein facilitates diffusion of adenosine from the cell. Adenosine then acts as an intercellular signal; in this capacity it has a variety of functions. In its role of vasodilator, adenosine increases the blood supply to the organ and thus helps to correct the energy deficit by increasing the organ's supply of oxygen and metabolic fuels.

Table 8.2. Comparison of 5′-nucleotidase activity and rate of adenosine production by rat heart (Frick & Lowenstein, ref. 8)

	Rate (μmol/gm wet weight/min)
Perfused heart	
Hydrolysis of AMP added to perfusate (V_{max})	7–9
Heart homogenate	
Hydrolysis of AMP (V_{max})	4–6
Perfused heart	
Adenosine production from endogenous precursors during anoxia	0.08–0.1

In particular, the heart can regulate its own blood supply by producing adenosine, which then acts as a coronary vasodilator [7]. The reaction chiefly responsible for producing adenosine is that catalyzed by 5′-nucleotidase. Rat heart possesses a membrane-bound 5′-nucleotidase with a maximum reaction capacity of about 9 μmol AMP hydrolyzed per gm wet weight per minute, which is about 100 times greater than the maximum observed rates of adenosine production by the perfused heart (Table 8.2). In the mammalian heart, the membrane-bound enzyme occurs mainly on the outside of the plasma membrane and in structures contiguous with the outside of the cell surface [9–11]. The perfused heart hydrolyzes externally added AMP [12–14]. The rate of hydrolysis of externally added AMP by the perfused rat heart is about the same as the rate of AMP hydrolysis by heart homogenates (Table 8.2) [8]. In the case of cultured fibroblasts, only 40% of the 5′-nucleotidase activity is expressed externally [15]. In murine plasmocytoma cells, 5′-nucleotidase occurs largely on the inner face of the plasma membrane [16].

Sucrose density gradient centrifugation of rat heart homogenates shows that the highest specific activities of 5′-nucleotidase are observed in the F_1 fraction (mainly plasma membranes) and the F_2 fraction (the so-called microsomal fraction) (Fig. 8.1). However, 5′-nucleotidase activity is also observed in the F_3 fraction. In heart of avian species such as chicken, a large portion of 5′-nucleotidase is found in the cytosolic fraction [18].

The outward-facing orientation of the membrane-bound 5′-nucleotidase raises the question of whether this enzyme produces the adenosine used for vasodilation. Several possibilities may be considered. First, not all of the enzyme faces outward. If as little as 1% of the rat heart membrane enzyme faces the cytosol, this would provide sufficient activity to account for the maximum observed rates of adenosine production (Table 8.2). Second, although the heart plasma membrane enzyme faces outward, its active site may be accessible to intracellular AMP, either because the enzyme acts as its own translocase for

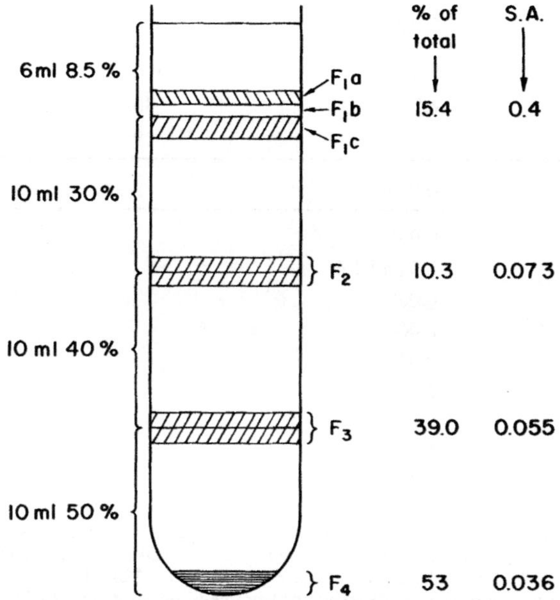

Figure 8.1. Distribution of 5′-nucleotidase in particulate fractions of rat heart. A homogenate was prepared from 20 gm of rat hearts, as described under Methods. The particulate fraction obtained after centrifuging the homogenate at 78,000*g* for 30 min was resuspended in 60 ml of 50% sucrose (w/v) containing 10 m*M* Tris-HCl buffer, pH 7.5, using a Potter-Elvehjem homogenizer. The resulting suspension was pipetted into six centrifuge tubes so that each contained 10 ml. The tubes were then filled with layers of different densities of sucrose, as shown. The tubes were centrifuged in a preparative ultracentrifuge (Spinco model L2-75B), using a swinging bucket rotor (SW27) at 100,000*g* for 17 hours. The layers observed visually at the end of the centrifugation were sketched and are labeled F_1–F_4. F_1 is rich in plasma membranes; it consists of three closely spaced bands, labeled a, b, and c. F_1a is faint and turbid; F_1b is clear, with some white material floating in it; and F_1c is strongly turbid. F_2 is the so-called microsomal fraction, F_3 contains mitochondria, and F_4 contains contractile proteins, cell debris, and nuclei. At the end of the centrifugation, a hole was punched in the bottom of the tube and 1 ml fractions were collected and analyzed for 5′-nucleotidase activity and protein. Recovery of 5′-nucleotidase activity was 108%. Specific activity is given in μmol/mg protein/min. The fractions containing activity were diluted with 4 volumes of 10 m*M* Tris-HCl buffer, pH 7.5, and centrifuged at 78,000*g* for 30 min. The membrane fraction so obtained was resuspended in 400 m*M* sucrose. The centrifugation method is a modification of that described by Kidwai et al. [17].

AMP or because there is a separate translocase for AMP with one end at or near the active site of the enzyme. Third, in liver and cultured fibroblasts, there is a continuous process of internalization of part of the 5′-nucleotidase of the plasma membrane [15, 19]; some of the enzyme so internalized may be accessible to cytosolic substrates. A similar type of internalization may occur in heart. Finally, an intracellular enzyme, distinct from the membrane-bound 5′-nucleotidase, is responsible for producing the adenosine needed for vasodilation.

Table 8.3. K_m and K_i values for 5′-nucleotidase from rat heart

Substrate or inhibitor	K_m (μM)	V_{max} (%)	Relative V_{max}/K_m
AMP	4.7	100	21
ϵAMP	6.5	45	6.9
AMPS	2.3	6	2.6
	K_I (μM)		
ADP	0.081		
ATP	1.5 and 4.4		
AOPCP	0.006		
AOPCPOP	0.066		
AOPOPCP	0.54		
ACP	45		
ACPOP	0.067		
ACPOPOP	1.3		
ϵADP	0.47		

Abbreviations: ϵAMP, 1,N^6-ethenoadenosine 5′-monophosphate; ϵADP, 1,N^6-ethenoadenosine 5′-diphosphate; AOPCP, α,β-methylene adenosine 5′-diphosphate; AOPCPOP, α,β-methylene adenosine 5′-triphosphate; AOPOPCP, β,γ-methylene adenosine 5′-triphosphate; ACP, 5′-deoxyadenosine 5′-methylene phosphonate (alternative designation, 6′-deoxyhomoadenosine-6′-phosphonate); ACPOP, the ADP analogue corresponding to ACP; ACPOPOP, the ATP analogue corresponding to ACP.
Measurements were in 50 mM Tris-HCl buffer, pH 7.5, at 30°. Assay 1 was used, except in experiments with ACP, ATP, and adenosine, for which Assay 3 was used [23]. Values of K_I were calculated from Dixon plots for three AMP concentrations. Inhibitor concentrations were used in the following ranges: ADP, 0.025–0.4 μM; AOPCP, 2.5–30 nM; AOPCPOP, 0.04–0.24 μM; AOPOPCP, 0.2–2.0 μM; ACP, 20–100 μM; ACPOP, 0.02–0.2 μM; ACPOPOP, 0.4–10 μM; ϵADP, 0.2–1.0 μM.

Are the regulatory properties of the membrane-bound 5′-nucleotidase consistent with its involvement in the production of adenosine for the purpose of regulating coronary blood flow? At one point it was believed that this is the case, because the enzyme is inhibited by ATP. An increase in energy demand was thought to result in lower levels of ATP and hence in a partial release of the ATP inhibition of the enzyme [14, 20]. However, it was shown that ADP is a much more powerful inhibitor of the enzyme than ATP [10, 21] and this cast doubt on the hypothesis that the enzyme is regulated by changes in ATP concentration. The pure enzyme isolated from rat heart plasma membranes has K_I values for ADP and ATP of 0.08 μM and 5 μM, respectively (Table 8.3) [unpublished observations]. If conditions of energy deficit lead to a rise in ADP, then the plasma membrane enzyme would be expected to become more inhibited. It may be argued that the location of the enzyme in the cell membrane is such as to limit access of ADP; if this were the case, then access to AMP should also be limited. In this connection it is worth noting that in the perfused heart, the membrane-bound enzyme shows a K_m for external AMP of 1.3 to 1.6 mM [8]. Perhaps AMP has limited access to the enzyme in the intact membrane and this is responsible for a higher apparent K_m. It should not be overlooked that inhibition of the enzyme by ADP might be overcome by an activator which rises during anoxia and conditions of energy deficit; so far we have failed to find such a substance.

Considerations of this kind led us to examine rat hearts for a 5′-nucleotidase that is not inhibited by ADP. A 5′-nucleotidase has been found to occur in the soluble fraction of heart homogenates. In this chapter, we describe some of the properties of this enzyme.

METHODS

Preparation of rat heart homogenates

Rat hearts, 45 gm wet weight, were homogenized in 180 ml of ice-cold 10 m*M* Tris-HCl buffer, pH 7.5, using a Polytron model PCU-2-110 (Brinkman Instruments, Westbury, NY) in three bursts at half-maximum speed for 30 seconds. Each burst of homogenization was followed by a one-minute rest. The mixture was kept on ice during this treatment. The homogenate was centrifuged at 5,000*g* for 60 min. The residue was homogenized in 180 ml of the same buffer and the centrifugation was repeated. This treatment was repeated once more. The supernatants were combined and centrifuged at 78,000*g* for 30 min. This yielded a *homogenate supernatant* and a *homogenate residue*.

Other methods

Membrane-bound 5′-nucleotidase was purified from rat hearts as described previously [22]. The enzyme was assayed using Assay 1 or 3 of Naito and Lowenstein [22]. Soluble 5′-nucleotidase and acid phosphatase were assayed by measuring the orthophosphate liberated from various substrates by the method of Sanui [23].

RESULTS

Supernatants were prepared from homogenates of rat hearts and examined for 5′-nucleotidase activity using AMP as substrate. Enzyme activity was determined by measuring the orthophosphate formed in the reaction. Antibody to the pure enzyme from plasma membranes of rat heart completely inhibits the plasma membrane enzyme; however, it inhibits only about 10% of the activity found in the high-speed supernatant. The soluble AMP phosphatase activity may be due entirely to a nonspecific phosphatase, or to a 5′-nucleotidase, or both. These possibilities can best be distinguished by enzyme purification. The AMP-hydrolyzing activity in the high-speed supernatant obtained from rat heart homogenates was purified with the object of separating it from nonspecific phosphatase activities. Table 8.4 shows that during the purification, the soluble 5′-nucleotidase activity was enriched 19-fold over the nonspecific phosphatase activity.

The soluble 5′-nucleotidase is activated by ATP and, to a lesser extent, by ADP (Fig. 8.2). In contrast, the plasma membrane enzyme is inhibited very strongly by ADP and less strongly by ATP (Table 8.3). In the presence of 5

Table 8.4 Purification of 5′-nucleotidase from rat heart cytosol

Fraction	Protein (mg)	Specific activity			Activity Ratio	
		AMP	IMP	p-$NO_2\phi P$	IMP/AMP	p-$NO_2\phi P$/IMP
Homogenate supernatant	1600	30.5	30.4	64.4	1.0	2.1
$(NH_4)_2SO_4$ supernatant	939	4.5	2.1	18.3	0.47	8.7
$(NH_4)_2SO_4$ precipitate	564	43.9	44.0	36	1.0	0.82
pH 5.6 supernatant	356	13.1	9.7	37	0.74	3.8
pH 5.6 precipitate	173	50.6	56.8	14.1	1.1	0.25
DEAE peak	39.5	28.6	69.8	7.6	2.4	0.11

Homogenate supernatant was prepared from 45 gm of rat heart, as described under Methods. Solid ammonium sulfate (30.3 gm/100 ml) was added to the supernatant slowly with stirring. The mixture was allowed to stand on ice for 30 min and was then centrifuged at 9,000 *g* for 10 min to yield $(NH_4)_2SO_4$ supernatant and precipitate. The ammonium sulfate precipitate was dissolved in 60 ml of 50 m*M* Tris-HCl buffer, pH 7.5, and was dialyzed against 61 of the same buffer, with one change of buffer, for 16 hr. A small precipitate that formed during the dialysis was removed by centrifugation. The resulting solution is referred to as $(NH_4)_2SO_4$ precipitate. It was adjusted to pH 5.6 with 2 *M* sodium acetate buffer, pH 5.0. The mixture was allowed to stand for 30 min and was then centrifuged at 9,000 *g* for 10 min to yield pH 5.6 supernatant and precipitate. The pH 5.6 precipitate was dissolved in 30 ml of 100 m*M* Tris-HCl buffer, pH 7.5, and a small amount of insoluble residue was removed by centrifugation at 108,000 *g* for 30 min. The resulting solution is referred to as pH 5.6 precipitate. The solution was applied to a DEAE-cellulose column (2 cm diameter × 20 cm long) that had been equilibrated with 100 m*M* Tris-HCl buffer, pH 7.5. The column was then eluted with 100 ml of the same buffer, containing 50 m*M* NaCl. This was followed by a linear gradient from 50 m*M* to 400 m*M* NaCl over 500 ml. Fractions of 6 ml were collected. The fractions with the highest 5′-nucleotidase activity were pooled (66 ml) and concentrated to 4 ml, using a pressure concentrator with a YM30 filtration membrane (Amicon, Danvers, Mass.). The resulting solution is referred to as DEAE peak. p-$NO_2\phi P$ stands for p-nitrophenylphosphate.

m*M* ATP and 6 m*M* $MgCl_2$, the soluble enzyme shows a K_{app} for AMP of 0.5 m*M* (Fig. 8.3). In the absence of ATP, the K_m value for AMP is about 3 m*M*. Thus ATP increases the apparent affinity of the enzyme for AMP. Similar effects are produced by ADP, except that ATP is more effective.

The soluble 5′-nucleotidase shows little activity in the absence of magnesium ions. With 0.5 m*M* AMP, the enzyme shows a slight activation by magnesium. However, in the presence of 5 m*M* ATP and 6 m*M* magnesium ions, the enzyme exhibits a 42-fold activation (Fig. 8.4). In the presence of ATP and without added magnesium ions, the enzyme shows only a 3-fold activation (not illustrated). In other words, the combination of ATP and magnesium ions shows the greatest activating effects.

COMMENTS

The soluble enzyme is activated by both ATP and ADP, and this is as puzzling as the inhibition of the membrane-bound 5′-nucleotidase by ADP and ATP. Perhaps the true regulation of the soluble enzyme occurs by inhibition of the activated enzyme. A search for other regulators of the enzyme is currently being pursued. Activation of the soluble enzyme from liver by ATP was

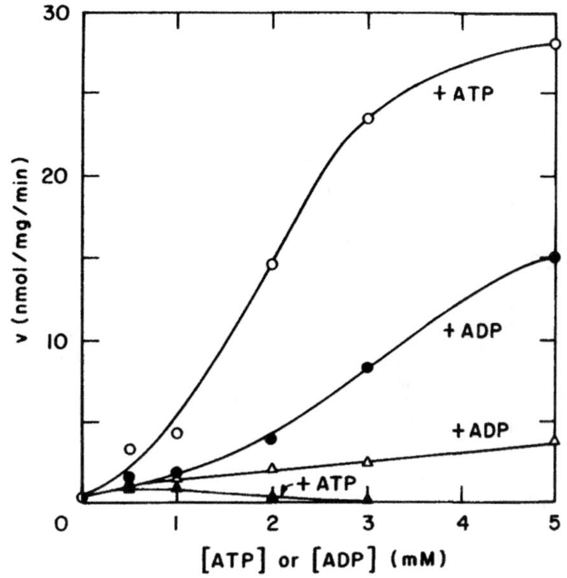

Figure 8.2. Effects of ATP and ADP on 5′-nucleotidase and acid phosphatase activities in fractions prepared from rat heart cytosol. The reaction mixture contained 0.5 m*M* AMP, 5 m*M* $MgCl_2$, 50 m*M* maleate-Tris buffer, pH 7.0, ADP or ATP at the indicated concentrations, and either 5′-nucleotidase (Table 8.4, DEAE peak), 54 μgm protein (●, ○), or acid phosphatase (Table 8.4, pH 5.6 supernatant), 556 μgm protein (▲, △). The reaction was started by adding enzyme and was run at 37° for 60 min.

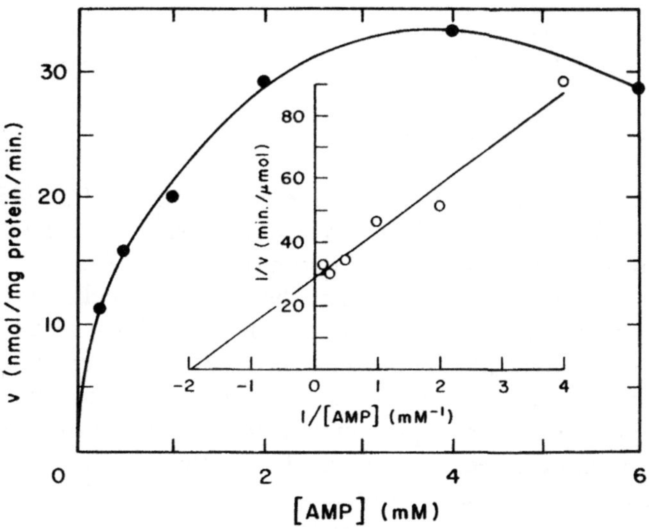

Figure 8.3. Effect of AMP concentration on activity of 5′-nucleotidase from soluble fraction of heart. Reaction mixtures contained AMP as indicated, 100 m*M* maleate-Tris buffer, pH 7.0, 5 m*M* ATP, 6 m*M* $MgCl_2$, and enzyme, 54 μgm protein. The reaction was run at 37° for 60 min.

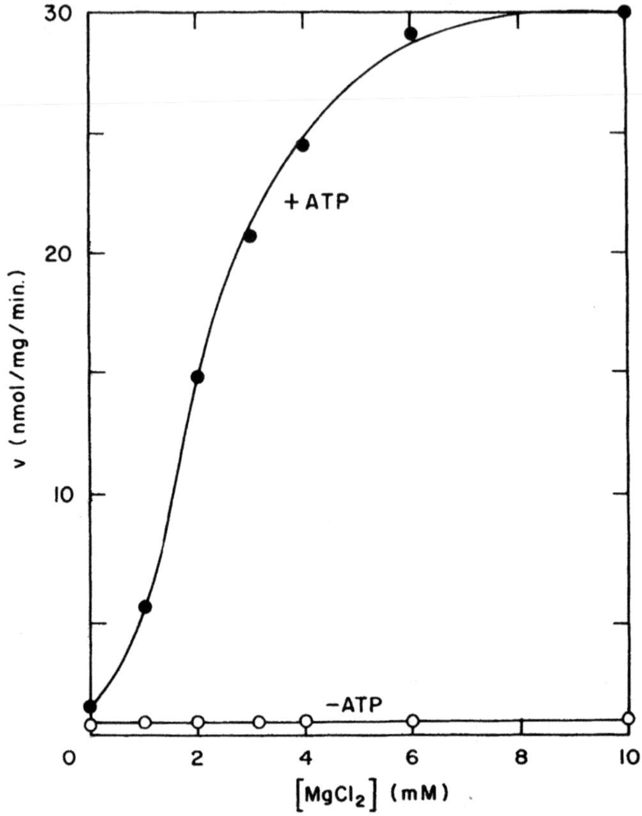

Figure 8.4. Effect of magnesium chloride and ATP on activity of 5′-nucleotidase from soluble fraction of rat heart. Reaction mixtures contained 0.5 m*M* AMP, 5 m*M* ATP, 100 m*M* maleate-Tris buffer, pH 7.0, and $MgCl_2$, as indicated in a final volume of 0.5 ml. The reaction was started by adding enzyme, 54 μgm protein, and was run at 37° for 60 min.

reported previously by Van den Berghe et al. [24] and Itoh [25]. Like the enzyme from liver [26, 27], the soluble enzyme from heart shows a requirement for magnesium ions and a pH optimum of about 6.5.

Additional work is required to define the optimum conditions of activating the soluble heart enzyme, but it is clear that the high concentrations of free magnesium ions used previously with enzyme from liver [26, 28] are not required for the soluble 5′-nucleotidase from rat heart. Note that at 5m*M* ATP and 6 m*M* magnesium ion, approximately the levels expected in heart cytosol, the activation has reached a maximum. Under these conditions, the reaction mixture contains about 4.2 m*M* $MgATP^{2-}$ and 1 m*M* free Mg^{2+} [29]. In other words, the kinetics of the enzyme appear geared for physiological concentrations of AMP, MgATP, and free Mg^{2+}.

Before we can link the soluble form of 5′-nucleotidase with adenosine production as a signal for vasodilation, we require a more detailed knowledge of the regulatory properties of the enzyme. Activation of the enzyme by physiological levels of ATP does not by itself make regulatory sense. Various possibilities are under investigation, including the hypothesis that changes in AMP are the primary determinant of activity, the idea that the enzyme may be regulated by an inhibitor that declines during anoxia or increased work load, and the possibility that there is some additional activator, which is better than ATP or which acts synergistically with ATP.

Our studies do not reveal the cell type from which the soluble 5′-nucleotidase is derived. A particular cell type may be rich in the enzyme and act as a signaling cell, whereas other cell types may have a paucity of enzyme. The acid phosphatase activity is located mainly in the coronary endothelial cells [S. Nees, personal communication].

Changes in the AMP concentration of the heart may be the primary determinant of adenosine production. Adenosine then acts as an extracellular signal of the intracellular level of AMP [30]. The effects of acetate on coronary flow serve as an example. Acetate is converted to acetyl-CoA by acetyl-CoA synthetase (formerly acetate thiokinase): acetate + ATP + CoA → acetyl-CoA + AMP + pyrophosphate. In the perfused rat heart, AMP levels rise from 0.35 μmol per gm dry weight in controls to 0.70 μmol per gm dry weight in hearts perfused with 10 m*M* acetate [31]. This is equivalent to a change from 0.14 to 0.28 m*M* if it is assumed that AMP is evenly distributed in the intracellular water. In contrast to acetate, addition of 10 m*M* pyruvate to the perfusate causes no increase in AMP.

In the anaesthetized dog, infusion of as little as 0.038 mmol acetate/kg/min causes an 88% increase in coronary blood flow, whereas a similar rate of pyruvate infusion has no effect (Table 8.5) [32]. Infusion of 0.075 and 0.15 mmol acetate/kg/min increases the coronary flow rate by 150% and 190%, respectively. These changes are accompanied by a doubling and a tripling, respectively, of the hypoxanthine concentration in coronary sinus plasma. Moreover, the adenosine content of the myocardium is significantly higher

Table 8.5. Effects of acetate and pyruvate infusions on coronary blood flow in dogs

	Coronary Blood Flow	
Infusion rate (mmol/kg/min)	Acetate infusion (ml/100 gm/min)	Pyruvate infusion (ml/100 gm/min)
Control	90 ± 6	107 ± 12
0.0375	169 ± 13*	100 ± 12
0.075	227 ± 22*	112 ± 14
0.15	258 ± 13*	107 ± 14

Coronary blood flow was measured by the radioactive microsphere method. The number of dogs for the acetate and pyruvate infusions was 7 and 6, respectively. Asterisks (*) indicate values significantly different from the preinfusion controls, with $p < 0.05$ [32].

after infusing acetate (2.85 ± 0.68 nmol per gm wet weight) than after infusing pyruvate (0.68 ± 0.11 nmol per gm wet weight). Acetate infusion increases blood flow in many other organs as well. An analysis of skeletal muscle shows that the AMP content almost doubles upon acetate infusion, whereas the ADP levels increase by about 40% [32].

REFERENCES

1. Goodman MN, Lowenstein JM: The purine nucleotide cycle: Studies of ammonia production by skeletal muscle *in situ* and in perfused preparations. *J Biol Chem* 252:5054–5060, 1977.
2. Lowenstein JM, Goodman MN: The purine nucleotide cycle in skeletal muscle. *Fed Proc* 37:2308–2312, 1978.
3. Nishiki K, Erecińska M, Wilson DF: Energy relationships between cytosolic metabolism and mitochondrial respiration in rat heart. *Am J Physiol* 234:C73–C81, 1978.
4. McGilvery RW, Murray TW: Calculated equilibria of phosphocreatine and adenosine phosphates during utilization of high energy phosphate by muscle. *J Biol Chem* 249:5845–5850, 1974.
5. Frick GP, Lowenstein JM: Vectorial production of adenosine by 5′-nucleotidase in the perfused rat heart. *J Biol Chem* 253:1240–1244, 1978.
6. Klingenberg M: The ADP-ATP translocation in mitochondria, a membrane potential controlled transport. *J Membr Biol* 56:97–105, 1980.
7. Berne RM, Foley DH, Watkinson WP, Miller WL, Winn HR, Rubio R: Role of adenosine as a mediator of metabolic vasodilation in the heart and brain: A brief overview and update, in Baer HP, Drummond GI (eds): *Physiological and Regulatory Functions of Adenosine and Adenine Nucleotides*. New York, Raven Press, 1979, pp 117–126.
8. Frick GP, Lowenstein JM: Studies of 5′-nucleotidase in the perfused rat heart: Including measurements of the enzyme in perfused skeletal muscle and liver. *J Biol Chem* 251:6372–6378, 1976.
9. Rubio R, Berne RM, Dobson JG Jr: Site of adenosine production in cardiac and skeletal muscle. *Am J Physiol* 225:938–953, 1973.
10. Burger R, Lowenstein JM: Preparation and properties of 5′-nucleotidase from smooth muscle of small intestine. *J Biol Chem* 245:6274–6280, 1970.
11. Arch JRS, Newsholme EA: The control of metabolism and the hormonal role of adenosine. *Essays Biochem* 14:82–123, 1978.
12. Hoffman PC, Okita GT: Penetration of ATP into the myocardium. *Proc Soc Exp Biol Med* 119:573–576, 1965.
13. Williamson JR, DiPietro DL: Evidence for extracellular enzymic activity of the isolated perfused rat heart. *Biochem J* 95:226–232, 1965.
14. Baer HP, Drummond GI: Catabolism of adenine nucleotides by the isolated perfused rat heart. *Proc Soc Exp Biol Med* 127:33–36, 1968.
15. Widnell CC, Schneider Y-J, Pierre B, Baudhuin P, Trouet A: Evidence for a continual exchange of 5′-nucleotidase between the cell surface and cytoplasmic membranes in cultured rat fibroblasts. *Cell* 28:61–70, 1982.
16. Zachowski A, Aubry J, Jonkman-Bark G, Lelievre L: Localization of the catalytic site of 5′-nucleotidase at the inner surface of murine plasmocytoma plasma membranes. *FEBS Lett* 75:197–200, 1977.
17. Kidwai AM, Radcliffe MA, Duchon G, Daniel EE: Isolation of plasma membrane from cardiac muscle. *Biochem Biophys Res Commun* 45:901–910, 1971.
18. Gibson WB, Drummond GI: Properties of 5′-nucleotidase from avian heart. *Biochemistry* 11:223–229, 1972.
19. Little JS, Widnell CC: Evidence for the translocation of 5′-nucleotidase across hepatic membranes *in vivo*. *Proc Natl Acad Sci USA* 72:4013–4017, 1975.
20. Baer H-P, Drummond GI, Duncan EL: Formation and deamination of adenosine by cardiac muscle enzymes. *Mol Pharmacol* 2:67–76, 1966.
21. Sullivan JM, Alpers JB: *In vitro* regulation of rat heart 5′-nucleotidase by adenine nucleotides and magnesium. *J Biol Chem* 246:3057–3063, 1971.
22. Naito Y, Lowenstein JM: 5′-Nucleotidase from rat heart. *Biochemistry* 20:5188–5194, 1981.

23. Sanui H: Measurement of inorganic orthophosphate in biological materials: Extraction properties of butyl acetate. *Anal Biochem* 60:489–504, 1974.
24. Van Den Berghe G, Van Pottelsberghe C, Hers H-G: A kinetic study of the soluble 5'-nucleotidase of rat liver. *Biochem J* 162:611–616, 1977.
25. Itoh R: Purification and some properties of cytosol 5'-nucleotidase from rat liver. *Biochim Biophys Acta* 657:402–410, 1981.
26. Fritzson P: Nucleotidase activities in the soluble fraction of rat liver homogenate: Partial purification and properties of a 5'-nucleotidase with pH optimum 6.3. *Biochim Biophys Acta* 178:534–541, 1969.
27. Naito Y, Tsushima K: Cytosol 5'-nucleotidase from chicken liver: Purification and some properties. *Biochim Biophys Acta* 438:159–168, 1976.
28. Naito Y, Itoh R, Tsushima K: 5'-Nucleotidase of chicken liver: A comparison of soluble 5'-nucleotidase activities in chicken and rat livers. *Int J Biochem* 5:807–810, 1974.
29. Storer AC, Cornish-Bowden A: Concentration of $MgATP^{2-}$ and other ions in solution: Calculation of the true concentrations of species present in mixtures of associating ions. *Biochem J* 159:1–5, 1976.
30. Schultz V, Lowenstein JM: The purine nucleotide cycle: Studies of ammonia production and interconversion of adenine and hypoxanthine nucleotides and nucleosides by rat brain *in situ*. *J Biol Chem* 253:1938–1943, 1978.
31. Williamson JR: Glycolytic control mechanisms: I. Inhibition of glycolysis by acetate and pyruvate in the isolated, perfused rat heart. *J Biol Chem* 240:2308–2321, 1965.
32. Liang C-s, Lowenstein JM: Metabolic control of the circulation. Effects of acetate and pyruvate. *J Clin Invest* 62:1029–1038, 1978.

DISCUSSION

DR. A.C. NEWBY: Is your soluble 5′-nucleotidase identical with the nucleotidase described in the liver by Fritzson?

DR. LOWENSTEIN: Fritzson reported no regulation of the soluble liver enzyme by ATP or ADP. He reported a high K_m for AMP and a high requirement for free magnesium ions. The enzyme we have found in heart has a low K_m for AMP and a low requirement for magnesium ions in the presence of ATP. Nevertheless, the enzymes from heart and liver may be similar. The enzymes from heart and liver require further characterization before a more definitive answer can be given to your question.

DR. M.S. HERSHFIELD: Aldo Bagnara and I have recently reported on a cytoplasmic nucleotidase present in human lymphocytes that is also activated by both ATP and dATP. We believe that, when dATP accumulates as a result of deoxyadenosine phosphorylation in patients treated with deoxycoformycin, this cytoplasmic nucleotidase plays a role in causing net catabolism of adenine ribonucleotides. However, in this situation IMP, not AMP, is the substrate for the nucleotidase, since AMP deaminase in lymphocytes is also activated by dAMP. Do you think that in the heart, under conditions when ATP conversion to AMP occurs, that the activity of the cytoplasmic nucleotidase toward AMP would diminish to any significant extent?

DR. LOWENSTEIN: No, because the ATP levels do not drop sufficiently during an increase in the work load. Moreover, ADP also activates the enzyme, so the loss in ATP activation is partially compensated for by the ADP activation.

DR. R. VON BORSTEL: What is the effect of inhibitors of the transport of adenosine across cell membranes (such as dipyridamole) upon the release of adenosine from heart?

DR. LOWENSTEIN: A number of different research groups have reported that dipyridamole inhibits adenosine efflux.

DR. VON BORSTEL: Dipyridamole potentiates the release of adenosine from depolarized brain slices, and yet studies with 5′-nucleotidase inhibitors such as AOPCP suggest that released nucleosides do not derive from released nucleotides. How might these intracellularly formed nucleosides be crossing cell membranes in the presence of dipyridamole?

DR. LOWENSTEIN: Perhaps some transport systems are insensitive to dipyridamole. I do not know how it might potentiate adenosine release.

DR. H.P. BAER: The K_m values for AMP and K_i values for inhibitors ATP and AOPCP are one to two orders of magnitude lower than those reported originally for a partially purified rat heart 5′-nucleotidase. Can you explain this?

DR. LOWENSTEIN: The pure membrane enzyme exhibits K_m and K_i values that are substantially lower than the values observed before the enzyme is extracted from membrane fragments. This may be due to limited access of the substrate to the enzyme in the membrane fragments, or to the presence in the membrane of regulatory proteins that are lost upon purification, to mention but two possibilities.

DR. BAER: You reported substrate levels and enzyme levels measured in, or purified from, whole heart homogenates. But there may be different cellular sources for these, which of course would reflect very much on any interpretations regarding physiological and functional consequences. How did you, or could you, deal with this problem?

DR. LOWENSTEIN: Metabolite measurements show substantially higher levels of free AMP and ADP in heart and skeletal muscle than are predicted by calculations based on the assumption that the creatine kinase and the myokinase reactions are in equilibrium. This suggests that either these reactions are not in equilibrium or that the nucleotides are in two or more compartments. The compartments may be different cell types, or cytosol and mitochondria, to mention only two possibilities. These considerations suggest how the problems you mention might be approached.

DR. I.H. FOX: What is the pH profile of your soluble 5′-nucleotidase? We have a highly purified cytoplasmic 5′-nucleotidase from human placenta that has a pH plateau from 7.5 and up. The enzyme requires Mg^{2+} absolutely, has about equal activity with AMP and IMP, and about three-fold higher activity with CMP. The K_m for AMP is below 1 m*M* with no ATP present.

DR. LOWENSTEIN: The soluble 5′-nucleotidase from rat heart has a broad optimum around pH 6.5; it has about 75% of optimum activity at pH 7. However, the soluble enzyme is a regulated enzyme and it is not uncommon that such enzymes show a different pH optimum in the presence of activators or inhibitors.

DR. J.P. MANFREDI: Mathematical models of nucleotide metabolism in isolated rat hearts by Garfinkel's lab indicate that the pool of AMP not available to myokinase is within mitochondria. What do you think of this?

DR. LOWENSTEIN: To the best of my recollection, this model did not rule out other compartments, or cell types other than myocytes, in which the AMP pool might be exceptionally large.

DR. MANFREDI: There have been studies that indicate that stimulation of slow-twitch red muscle is accompanied by an *increase* in tissue ATP content. I cannot think of a mechanism by which this increase in ATP can occur. Can you?

DR. LOWENSTEIN: No.

9. METABOLISM OF ADENOSINE AND SITES OF PRODUCTION IN THE HEART

J. SCHRADER

Adenosine formed by the heart has been proposed to participate in (1) the metabolic regulation of coronary flow [1–4], (2) the attenuation of several hemodynamic and metabolic effects of catecholamines by a pre- and post-synaptic mechanism [5–8], and (3) the inhibition of impulse conduction in the atrioventricular node during hypoxia [9]. Although the cardiac effects of adenosine involve a variety of different structures, the action of adenosine is always inhibitory, and the basic mechanism appears to be similar in all cases [4]. Adenosine is likely to act via specific receptors, located on the outer surface of the cell membrane of smooth muscle, cardiac muscle, and nodal cells [10, 11]. The adenosine receptors are linked to the adenylate cyclase system [5–7] and receptor activation can inhibit transmembrane Ca^{2+} influx [12, 13].

The ways in which synthesis of adenosine is regulated in the heart are rather complex. Only recently, it has become appreciated that this complexity is due only in part to the different kinetic properties of the adenosine-metabolizing enzymes and their modulation [14, 15]. Additional difficulties arise from the fact that cells other than the myocardium, such as smooth muscle and endothelial cells, exhibit an active nucleotide metabolism and are also capable of producing adenosine [16–18]. Furthermore, there is good evidence that adenosine formed by the heart is derived from at least two different pools of adenine nucleotides [19]. Whether this compartmentation occurs at the subcellular or cellular level, however, remains to be elucidated.

Several aspects of the metabolism of adenosine in different tissues have been

R.M. Berne, T.W. Rall, and R. Rubio (eds.), Regulatory Function of Adenosine, *pp. 133–156.*

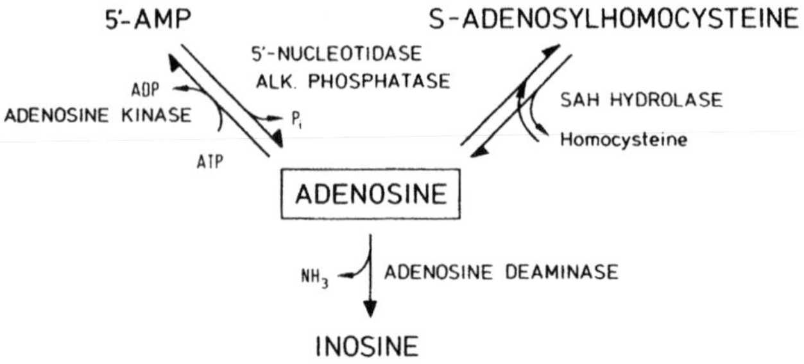

Figure 9.1. Pathways of adenosine metabolism in the heart.

reviewed [20–24]. In this chapter, some more recent findings of myocardial adenosine metabolism shall be discussed. Special emphasis will be given to those factors that may control the metabolism of endogenously formed adenosine under the dynamic conditions of the beating heart.

OVERVIEW OF ADENOSINE METABOLISM

The principal pathways of adenosine metabolism are illustrated in Figure 9.1. Adenosine can be formed either from AMP by action of 5′-nucleotidase (EC: 3.1.3.5) or alkaline and acid phosphatase (EC: 3.1.3.1 and 3.1.3.2) or from S-adenosylhomocysteine (SAH) by means of SAH-hydrolase (EC: 3.3.1.1). However, adenosine is removed by phosphorylation to AMP (adenosine kinase, EC: 2.7.1.20) and/or deamination to inosine (adenosine deaminase, EC: 3.5.4.6). Although the equilibrium constant of SAH-hydrolase favors SAH formation [25], the reaction will function in the hydrolytic direction under physiological conditions [26]. However, when tissue L-homocysteine and adenosine levels are high, adenosine was shown to be preferentially converted into SAH [15].

5′-nucleotidase was shown by histochemical studies to be associated with the myocardial cell membrane, including the T-tubular system [27], but was also reported to be preferentially associated with the endothelial cells of capillaries [28]. Consistent with this latter observation is the finding that cultured coronary endothelial cells from guinea pig hearts exhibit a very high 5′-nucleotidase activity [29]. Membrane-bound 5′-nucleotidase, which is an ecto-enzyme, has also been characterized by functional studies in the intact heart [13, 30–32]. A soluble 5′-nucleotidase was described in rat liver [33, 34], chicken liver [35], and human erythrocytes [34] and has been shown to exhibit higher affinities for IMP and GMP than for AMP [33, 36]. Similarly, we found in the guinea pig heart the apparent K_m for IMP to be 9.4 μM, while that for AMP was 23.9 μM.

In order to further define the sites of adenosine formation in the heart, we have determined the subcellular distribution of all enzymes involved in the metabolism of adenosine [14, 15]. For this purpose, the kinetic parameters of

Table 9.1. Subcellular distribution and kinetic parameters of enzymes involved in adenosine metabolism of guinea pig hearts

	Membrane fraction		Cytosolic fraction	
	K_m	V_{max}	K_m	V_{max}
5′-nucleotidase	14.0 ± 1.3	10.0 ± 0.4	23.9 ± 6.7	2.0 ± 0.3
Alkaline phosphatase	441.0 ± 12.0	9.1 ± 1.1	269.0 ± 71.0	1.0 ± 0.1
SAH hydrolase	—		2.9 ± 0.6	1.3 ± 0.2
SAH synthetase[a]	—		0.4 ± 0.03	1.9 ± 0.1
Adenosine kinase	—		0.9 ± 0.2	1.3 ± 0.1
Adenosine deaminase	—		90.3 ± 19.7	37.6 ± 5.1

Source: From [14, 15].
$K_m = \mu M$; V_{max} = nmol · min^{-1}.mg $protein^{-1}$. $n = 5$, ±SEM.
[a]L-homocysteine: 2 mM.

the different adenosine-metabolizing enzymes were measured in a crude membrane preparation and in a 200,000 × *g* cytosolic fraction from guinea pig ventricular muscle. From the results summarized in Table 9.1, it is evident that 5′-nucleotidase is predominantly present in the membrane fraction. About 20% of 5′-nucleotidase activity, however, was also found in the cytosolic fraction. There are several reasons to assume that the cytosolic 5′-nucleotidase constitutes a truly soluble enzyme, which differs from membrane-bound 5′-nucleotidase by substrate specificity and kinetic parameters [14]. In addition, separation of cytosolic proteins on a Con A-Sepharose-column suggests that soluble 5′-nucleotidase of the heart is not fully glucosylated [unpublished observation]. Thus, the heart possesses the capability to dephosphorylate mononucleotides both by intracellular and extracellular nucleotidases. Under which functional conditions these nucleotidases may come into play shall be discussed later.

A similar distribution as for 5′-nucleotidase was also obtained for the alkaline phosphatase (Table 9.1). As in other tissues, the K_m value for the alkaline phosphatase is high, so that this enzyme is not likely to be an important catalyst for the formation of adenosine under physiological conditions [21, 22, 24].

SAH-hydrolase, adenosine deaminase, and adenosine kinase proved to be exclusively present in the cardiac cytosol (Table 9.1). The K_m values for these enzymes are similar to those reported for other tissues [21, 22, 24]. It should be noted that SAH-hydrolase and adenosine kinase exhibit the highest affinities for adenosine of all adenosine-metabolizing enzymes. The activity of SAH-hydrolase, however, differs greatly in hearts of different species. As is shown in Table 9.2, highest activities were found in the guinea pig heart, whereas SAH-hydrolase of the human heart was considerably lower. Unfortunately, no firm conclusion can be drawn from these data as to the *in vivo* activity of the

Table 9.2. S-adenosylhomocysteine hydrolase in mammalian hearts of different species measured in the direction of S-adenosylhomocysteine hydrolysis

	n (±SEM)	S-adenosylhomocysteine hydrolase activity (nmol/min/mg protein)
Guinea pig	5	1.29 ± 0.17
Cat	5	0.27 ± 0.05
Rat	7	0.21 ± 0.04
Human[a]	2	0.14
Rabbit	2	0.092
Dog	3	0.079 ± 0.011

Source: From [15].
[a]Papillary muscle.

SAM → SAH → adenosine pathway, because respective substrate concentrations are not known in the tissues studied, and intracellular factors may significantly modulate enzymatic activity.

REGULATION OF ADENOSINE FORMATION

It is generally assumed that myocardial adenosine is formed by dephosphorylation of AMP [37, 38]. 5′-nucleotidase is known to be strongly inhibited by ATP, ADP, and creatine phosphate, whereas Mg^{2+} can reverse the creatine phosphate-induced inhibition [37–41]. These results from *in vitro* experiments suggested that in the heart *in vivo*, 5′-nucleotidase normally is also almost fully inhibited. Based on these findings, an enhanced production of adenosine was postulated [38] to involve the following sequence of events: An increase in heart work leads to an enhanced energy utilization, which results in an increase in AMP and a decrease in creatine phosphate and energy charge. Thus, both an increase in substrate concentration and deinhibition of 5′-nucleotidase may be responsible for leading to an accelerated adenosine production [38].

Can this model be supported by measurements of respective metabolites and ions in the intact heart? Recent measurements on the influence of increased heart work on the subcellular distribution of ATP, ADP, and CrP in guinea pig hearts appear to be relevant to this question [42]. Using density gradient centrifugation of lyophilized myocardial homogenates in nonaqueous media, it was found that the concentration of ATP in the cytosol was higher than in the mitochondria, whereas the opposite was true for ADP. Thus, the ratio of free ATP to ADP in the cytosol was 10- to 40-fold higher than mitochondrial ATP/ADP ratios. From the mass action equation for adenylate kinase, the concentration of free AMP was calculated to be only about 1 μM [43]. This value is about a hundred times higher than the AMP levels measured in whole heart extracts [19] and is in the same order of magnitude as adenosine. Most likely, part of the cytosolic AMP is bound to proteins [43, 44]. When heart work and thus oxygen consumption were increased with norepinephrine (8 × $10^{-8}M$, however, the adenylate energy charge remained unchanged [43]. However, cytosolic creatine phosphate was slightly decreased, whereas the calculated free cytosolic AMP was doubled [42, 43].

In an attempt to relate the foregoing findings to the rate of adenosine production, we have repeated the experiments of Soboll and Bünger and measured the release of adenosine into the coronary effluent perfusate in the same heart preparation (guinea pig) under identical experimental conditions. In the presence of 8 × $10^{-8}M$ norepinephrine, we found the adenosine release to increase from a control value of 96.3 ± 24.3 pmol/min/gm (± SEM, n = 9) to 1380 ± 239 pmol min/gm (± SEM, n = 9). Obviously this 14-fold increase in adenosine production cannot be explained solely by the changes in the substrate concentration of AMP. Also, the changes in cytosolic levels of adenine nucleotides and CrP are rather small, and it is questionable whether this can result in substantial activation of 5′-nucleotidase.

Increasing Mg^{2+} concentrations was shown to antagonize the inhibitory effect of creatine phosphate on 5′-nucleotidase *in vitro* [38]. Therefore, increasing free cellular Mg^{2+}, possibly by breakdown of MgATP, may lead to an enhanced production of adenosine. It has recently been shown, however, that *in vivo* the free cardiac Mg^{2+}, as measured by NMR [45] and ion-sensitive microelectrodes [46], is rather stable and does not change under a variety of experimental conditions, including ischemia, hypoxia, and metabolic poisoning [45, 46].

Another interesting model for regulation of adenosine production was suggested by Arch and Newsholme [21]. It is assumed that adenosine kinase, because of its low K_m value, approaches saturation with substrate under basal conditions. Since 5′-nucleotidase and adenosine kinase produce a substrate cycle between AMP and adenosine, any increase in activity of 5′-nucleotidase leads to a proportionally larger change in the concentration of adenosine, thereby improving sensitivity [21].

The limitation of both these models is that they do not incorporate the possibility of adenosine being formed from SAH by action of low K_m SAH-hydrolase. In order to determine whether SAH-hydrolase activity can be demonstrated in the intact heart, we attempted to saturate the enzyme by its substrates and perfused isolated guinea pig hearts with L-homocysteine and adenosine [15]. The rate of formation of SAH is then a direct measure of SAH-hydrolase activity *in vivo*. From the data given in Figure 9.2, it is evident that SAH drastically increased, reaching 1400 nmol/gm after 30 min of perfusion. Continuation of perfusion can elevate SAH to the level of ATP [15]. It is also shown in Figure 9.2 that after cessation of L-homocysteine and adenosine infusion, SAH levels slowly declined, which would be predicted from the kinetic parameters of this enzyme (Table 9.1). The initial rate of rise of SAH was determined to be 62 nmol/min/gm wet weight. This value compares well with the respective value calculated from the kinetic properties of SAH-

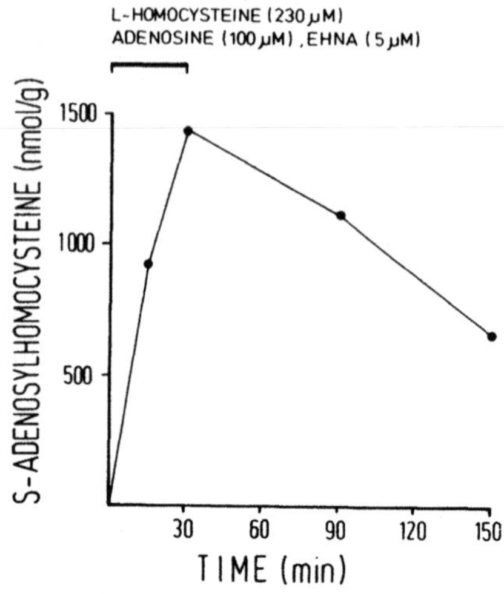

Figure 9.2. Increase in tissue S-adenosylhomocysteine in the presence of L-homocysteine thiolactone (0.23 m*M*), adenosine (0.1 m*M*), and EHNA (5 µ*M*). Isolated guinea pig hearts were perfused with a modified Krebs-Henseleit solution at constant coronary flow (10 ml/min) and were electrically paced (270 beats/min).

hydrolase and known substrate concentrations. Assuming simple Michaelis-Menten kinetics

$$V = \frac{V_{max} \cdot S}{K_m + S}$$

Taking V_{max} and K_m values given in Table 9.1 (V_{max} = 1.96 nmol/min/mg protein, K_m = 0.39 μM) and substrate concentrations used in the experiment depicted in Figure 9.2 (100 μM adenosine), the rate of SAH formation by the whole heart (29 mg protein/gm wet weight) can be calculated to be 57 nmol/min/gm. The similarity of measured and calculated rates of SAH synthesis strongly suggests that SAH-hydrolase in the guinea pig heart is not greatly inhibited by its cytosolic environment. Conversely, it can be expected that under physiological conditions (low adenosine, low L-homocysteine), adenosine is continuously formed by hydrolysis of SAH.

In an attempt to determine more directly the rate of adenosine formation from SAH, the turnover of SAH was measured by perfusion of isolated guinea pig hearts under steady-state conditions with ^{35}S-methionine (54 μM) for up to 75 min. The specific activities of SAM and SAH measured at different time intervals were found to be identical in all experiments, even after perfusion for only 1.5 min [unpublished observation]. Since a certain period is required for exogenously applied methionine to mix with its intracellular pool, the turnover of SAH can be assumed to be shorter than 1 min. These findings suggest that at the given concentration of SAH in the heart (0.568 ± 0.048 nmol/gm wet weight; SEM ± 13), the rate of SAH turnover and thus adenosine production is at least 600 pmol/gm/min, or higher.

A similar conclusion was reached when using a different experimental approach to assess the flux through the SAM → SAH pathway. This is based on the fact that the rate of methylation in a tissue is equal to the rate at which adenosine is produced from SAH. Histones, DNA, RNA, phospholipids, proteins, amino acids, and other small molecular weight molecules are known to be methylated [26]. Among the macromolecular compounds of cardiac tissue, the major fraction being methylated is proteins [unpublished observation]. We therefore determined, again under steady-state conditions, the rate of protein methylation by perfusion of isolated hearts with ^{14}C-methyl-labeled methionine. After proper corrections for protein synthesis were made, the rate of posttranscriptional modification of proteins by methyl groups was calculated to be 655 pmol methyl groups/min/gm tissue [unpublished observation]. This value certainly is a minimal value for total methyl group transfer, not only because compounds other than proteins are methylated, but also because protein degradation and recycling of the label may lead to underestimation of total cellular methylation rate.

The foregoing findings indicate that under control conditions, the SAM → SAH pathway is very active, leading to a substantial production of intracellular adenosine (> 600 pmol/min/gm). It is interesting to note that the basal release of adenosine into the coronary effluent perfusate of isolated hearts amounts only to about 30 to 50 pmol/min/gm. This strongly suggests that most of the adenosine formed from SAH is reincorporated into adenine nucleotides by phosphorylation to AMP. Thus, intracellular salvage of adenosine appears to be substantial leading to a continuous cycling of the adenosine moiety through AMP → ADP → ATP → SAM → SAH.

REGULATION OF ADENOSINE REMOVAL

The kinetic properties of adenosine kinase have recently been determined in highly purified enzyme preparations from rat heart [47–49], human placenta [50, 51], and rabbit liver [52–54]. A complex relationship was shown to exist between enzymatic activity, the concentration of adenosine, ATP, magnesium ion, and pH [51, 52, 54]. Inhibition of adenosine kinase by adenosine was observed at concentrations greater than 2.5 $\mu M/l$ [51]. The enzyme was also found to be inhibited by free ATP and free magnesium as well as by ADP, AMP, and S-adenosylhomocysteine [51, 52, 54].

Similar to studies in other tissues, the K_m value of cardiac adenosine kinase is more than one order of magnitude higher than that of adenosine deaminase (Table 9.1) [55]. These kinetic data suggest that at low adenosine concentrations, phosphorylation will be favored over deamination. Considerable evidence suggests that this view also pertains to the physiological condition *in vivo* [55].

Provided that intracellular flux from adenosine to AMP normally is relatively high, as suggested by the rate of cardiac SAH hydrolysis, then inhibition of adenosine kinase should lead to accumulation and release of adenosine. In order to test this hypothesis, we have perfused isolated guinea pig hearts with 5′-amino-5′-deoxyadenosine [52], which is also a potent inhibitor of cardiac adenosine kinase *in vivo* and *in vitro* [unpublished observation]. Under the influence of 5′-amino-5′-deoxyadenosine ($10^{-5}M$), and in the presence of EHNA ($5 \times 10^{-6}M$) to inhibit deamination, the release of endogenously formed adenosine increased about 15-fold and reached 800 pmol/min/gm. EHNA when applied alone only marginally increased the release of adenosine without altering the release of inosine. These findings suggest that inosine released by the guinea pig heart may not be formed solely by deamination of adenosine but was generated by the pathway leading from AMP via IMP to inosine. Furthermore, our results provide additional evidence that in the well-oxygenated heart, intracellular rephosphorylation of adenosine is substantial and predominates over deamination. In addition, it now appears possible that modulation of the activity of adenosine kinase could be an additional mechanism whereby adenosine production may also be regulated *in vivo*.

Another interesting finding concerns the fact that parallel to the 5′-amino-5′-deoxyadenosine-induced rise of adenosine release, coronary flow also increased. The time course of adenosine release and its concentrations in the effluent perfusate suggest that the observed change in coronary flow was not due to a primary dilator action of the nucleoside analogue but was mediated by the secondary release of endogenous adenosine. The implication of this finding is that the biological effect of adenosine derivatives cannot necessarily be taken as a measure of the affinity of these compounds to the adenosine receptor when metabolic actions predominate.

From the foregoing discussion of the pathways participating in the metabolism of adenosine, the following hypothesis of control of adenosine formation and release by cardiac tissue is proposed:

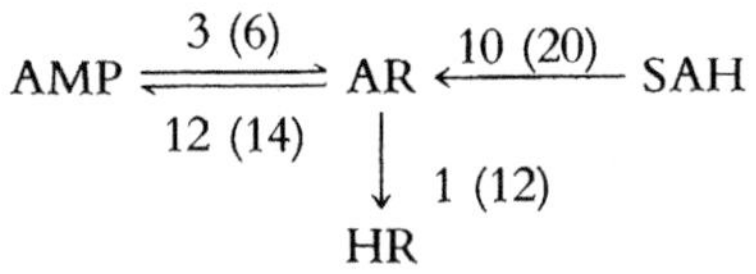

The numbers have been arbitrarily chosen to signify relative flux rates through the different reactions. It is proposed that there is a continuous cycling of the adenosine moiety leading from SAH via adenosine to AMP and ATP, as already discussed. Under control conditions, 5′-nucleotidase activity is greatly inhibited, so that only comparatively little adenosine is formed from AMP. As has been pointed out by Arch and Newsholme [21], low K_m adenosine kinase can be assumed normally to operate close to saturation with its substrate, whereas only a little adenosine is degraded by adenosine deaminase because of the high K_m of the enzyme. Thus, there are two substrate cycles, one between AMP and adenosine and another that extends via AMP → ATP → SAM → SAH to adenosine. Such substrate cycles can provide regulatory sensitivity with respect to the control of adenosine formation [55, 56]. By this mechanism, for example, a doubling in the rate of adenosine formation by 5′-nucleotidase and SAH-hydrolase (numbers in parentheses) leads to a 12-fold increase in the formation of adenosine, which subsequently is released from the cell or is deaminated. The amplification can be even further enhanced when, in addition, adenosine kinase is inhibited.

FUNCTIONAL IMPLICATIONS

Assessment of the role of adenosine in coronary flow regulation, cardiac contractility, and impulse conduction would ideally require measurement of adenosine in the interstitial fluid in close vicinity to respective adenosine receptors. Present techniques, however, only allow quantitation of either total tissue content of adenosine [19, 57, 58] or the release and washout of adenosine into the coronary system [59, 60, 61] or pericardial sac [62]. Both approaches have important limitations.

Tissue content of adenosine has been reported to be 0.9 to 2.0 nmol per gm wet weight in hearts of different species [3]. If all the adenosine were located within the extracellular space, as initially suggested by Rubio et al. [27], the calculated adenosine concentration would be in the micromolar range [19]. This adenosine concentration, however, is known to elicit maximal coronary dilation [59, 63]. These considerations already suggested that adenosine cannot be distributed solely in the extracellular space [19].

More recently, it was shown that SAH-hydrolase tightly binds adenosine [64] and that this enzyme is present in appreciable amounts in the cytosolic fraction of mammalian hearts [14, 15]. From the kinetic properties of a highly purified SAH-hydrolase and the reported binding constants of this enzyme, it was calculated that about 37% of the total myocardial adenosine is bound to SAH-hydrolase [15]. Similar values were reported for binding of adenosine in crude extracts from cardiac tissue [65]. From these findings, it appears that the amount of SAH-hydrolase in each tissue may be a major factor determining total tissue content of adenosine. Furthermore, additional binding proteins, as yet undefined, have been reported for the dog heart [66].

In an attempt to determine more directly the adenosine-binding capacity of cardiac tissue, we have used the following approach. Isolated guinea pig hearts perfused under normoxic conditions with a saline medium were stop-frozen, and ventricular tissue was pulverized to a fine powder in a mortar under liquid nitrogen. Cardiac tissue was then extracted with a small volume of 5 ml 150 m*M* phosphate-buffer pH 7.2 for 5 sec by sonication, and whole tissue extract was immediately placed on a short sephadex G-10 column which separated high molecular proteins from free adenosine within 10 min (−2°C). Extraction of the protein fraction with PCA revealed that 0.9 nmol adenosine per gm wet weight were protein-bound. These findings provide direct evidence that a substantial fraction of adenosine in the heart is protein-bound. Currently, however, no differential extraction procedures or other techniques are available to estimate only the free, biologically active adenosine in tissue samples.

The other possibility for assessing the role of adenosine experimentally is to determine its release and washout into the coronary circulation. Since the site of adenosine action is locally and temporarily distinct from the site of measurement—usually the coronary venous outflow—there are also some limitations with this approach. Once adenosine is formed, there are several mechanisms that tend to rapidly decrease its concentration in the extracellular space. It is well known that cells lining the vascular system, such as endothelium, smooth muscle, and myocardial cells, rapidly take up adenosine [3]. Furthermore, erythrocytes avidly incorporate adenosine [67], and plasma contains significant amounts of adenosine deaminase [68]. By reuptake, subsequent rephosphorylation, and deamination, as much as 80% to 90% of the adenosine can be removed during a simple passage through the heart [69, 70]. It therefore can be assumed that a great portion of the adenosine formed by the heart is also removed prior to its measurement in the coronary venous outflow.

INTRACELLULAR VERSUS EXTRACELLULAR SITE OF ADENOSINE FORMATION

The site of adenosine formation in the heart and in other tissues is generally assumed to be the plasma membrane, mainly because 5′-nucleotidase is a membrane-associated ecto-enzyme [27, 30, 31, 71]. This hypothesis implies that intracellular AMP can cross the sarcolemma membrane to reach the active site of the enzyme, which is extracellular. By which mechanism this proposed transfer of negatively charged AMP is brought about is not fully understood at present. However, fractionation studies in cardiac muscle (see Table 9.1) and other tissue [33–37] revealed 5′-nucleotidase to be present in both the particulate and the cytosolic fractions. Thus, adenosine may also be formed from AMP within the myocardial cell. Due to the presence of SAH-hydrolase, an exclusively cytosolic enzyme, additional adenosine can be generated by this latter pathway.

To determine whether adenosine is formed intracellularly or extracellularly in the intact heart and under physiological conditions, we made use of specific and potent inhibitors of 5′-nucleotidase (α,β-methylene adenosine 5′-diphosphate (AOPCP) [38]) and transmembranal nucleoside transport (4-nitrobenzylthioinosine (NBMPR) [72]).

To assess ecto-5′-nucleotidase activity, we perfused isolated guinea pig hearts with $10^{-4} M$ AMP, and the rate of extracellular dephosphorylation was 150 nmol/min/gm, which amounts to about 20% of the AMP added [14]. Addition of 50 μM AOPCP inhibited ecto-5′-nucleotidase activity by about 85% within 2 to 3 min, and this effect was reversible. If ecto-5-nucleotidase is

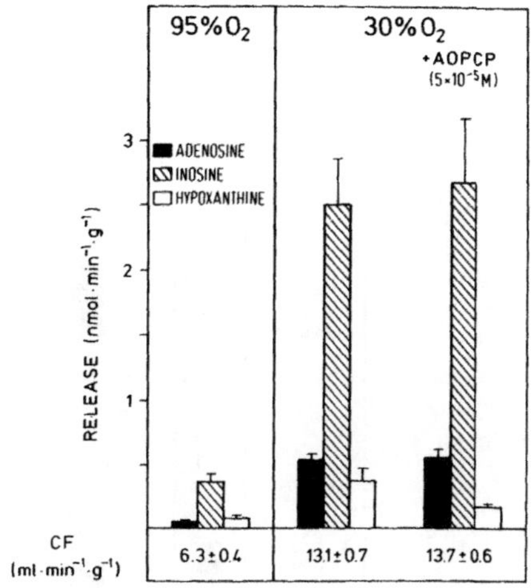

Figure 9.3. Changes in coronary flow (CF) and release of adenosine, inosine, and hypoxanthine from isolated perfused guinea pig heart during hypoxia, in the absence and presence of 50 μM AOPCP. ($n = 5$; $x = \pm$SE)

involved in the formation of endogenous adenosine, AOPCP should lead to a greatly diminished release of this nucleoside from the heart. Figure 9.3 clearly demonstrates that the hypoxia-induced release of adenosine remained unaltered despite effective inhibition of ecto-5′-nucleotidase. These findings suggest that membrane-bound 5′-nucleotidase may not be involved in adenosine production by the heart. Similarly, it was recently shown for rat polymorphonuclear leukocytes that ectophosphatases do not participate in cyanide- and 2-deoxy-glucose-induced catabolism of intracellular AMP [73, 74].

NBMPR has been demonstrated to inhibit nucleoside transport effectively in various tissues and cells [72, 73] and also to greatly depress adenosine uptake by the isolated guinea pig heart [14]. The effect of 5 μM NBMPR on the tissue content and release of endogenously formed adenosine is shown in Figure 9.4. Obviously, NBMPR potentiated the hypoxia-induced increase of tissue adenosine. This effect was particularly pronounced in the inosine fraction, indicating that intracellular deamination dominates at elevated adenosine levels. It is interesting that hypoxanthine remained almost unchanged with NBMPR. This may be explained by the fact that nucleoside phosphorylase is only present in the vessel wall [76], to which inosine did not have access in these experiments. In contrast to findings in cardiac tissue, less adenosine and inosine were released into the effluent perfusate with NBMPR (Fig. 9.4), again documenting the potency of the transport inhibitor. More importantly, however, these findings demonstrate that adenosine formation most likely occurred within the myocardial cell. A similar conclusion was recently drawn from experiments on polymorphonuclear leukocytes, using inhibitors of adenosine kinase and adenosine deaminase [74].

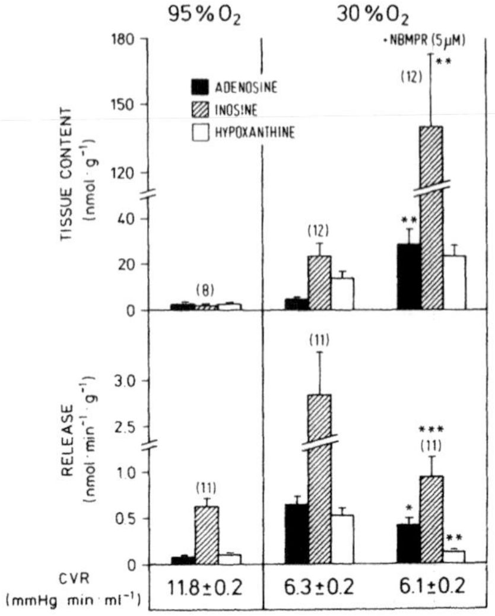

Figure 9.4. Influence of NBMPR (4-nitrobenzylthioinosine) on the hypoxia-induced changes in the formation of adenosine, inosine, and hypoxanthine, and coronary vascular resistance (CVR). (n = 8 − 12; x = ± SD) Source: From [14].

Additional support for a predominantly intracellular site of adenosine formation was derived from experiments in which use was made of the kinetic properties and intracellular distribution of SAH-hydrolase [15]. The rationale of this approach was the following: Because of the equilibrium constant of SAH-hydrolase [24], increasing L-homocysteine results in condensation of adenosine with L-homocysteine. Thus, L-homocysteine should lead to the accumulation of S-adenosylhomocysteine in proportion to the intracellular concentration of free adenosine. As is shown in Figure 9.5, perfusion of isolated hearts with L-homocysteinethiolactone significantly increased myocardial SAH levels, even under normoxic perfusion conditions, suggesting the presence of intracellular free adenosine in the normoxic heart. However, when endogenous adenosine production was accelerated by hypoxia, SAH increased 18-fold within 15 min. Also note that, under these latter conditions, adenosine levels were significantly reduced.

One might argue that the experimental evidence presented so far does not rule out the possibility that adenosine is first released into the extracellular space and then, during reuptake by the heart, is trapped by L-homocysteine. However, inhibition of nucleoside transport by NBMPR in L-homocysteine-treated hearts further augmented the increase in SAH (Fig. 9.5), thus ruling out this possibility.

Similar to the results obtained with NBMPR, L-homocysteine also greatly diminished the hypoxia-induced release of adenosine and its degradatives in the effluent perfusate (Table 9.3). The smaller total amounts of purines re-

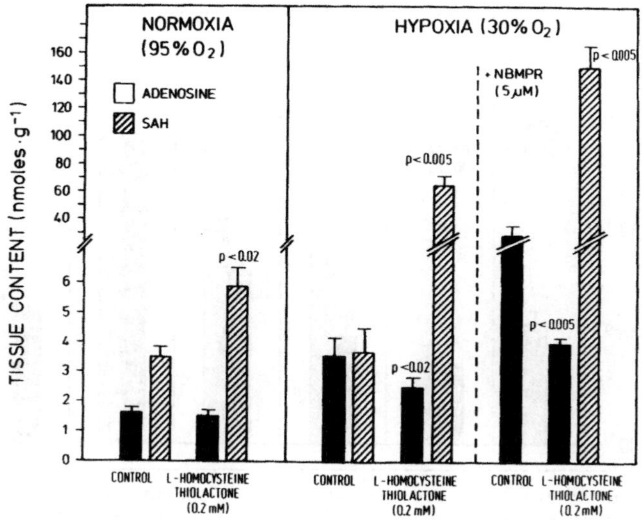

Figure 9.5. Influence of L-homocysteine thiolactone and NBMPR (4-nitrobenzylthioinosine) on tissue content of S-adenosylhomocysteine and adenosine in isolated perfused guinea pig hearts. (n = 5–12; x = ±SEM)

Table 9.3. Effect of 0.2 m*M* L-homocysteine thiolactone (homocysteine) on the hypoxia-induced release of adenosine, inosine, and hypoxanthine into the coronary effluent perfusate

	Adenosine (pmol · $min^{-1}.gm^{-1}$)	Inosine (pmol · $min^{-1}.gm^{-1}$)	Hypoxanthine (pmol · $min^{-1}.gm^{-1}$)
Normoxia (90% O_2)	86 ± 12	750 ± 178	188 ± 41
Hypoxia (30% O_2)	1209 ± 353	5295 ± 1840	2292 ± 706
Hypoxia (30% O_2) + homocysteine	439 ± 110*	2514 ± 789*	1575 ± 532*

Source: From [15].
n = 7, ±SEM
*$p < 0.005$.

leased with L-homocysteine (64 nmol/15 min/gm) compare favorably with the tissue content of SAH accumulated over the same period of time (Fig. 9.5).

The foregoing discussion provides strong evidence that the site of the hypoxia-induced formation of adenosine in the guinea pig heart is intracellular. Most probably, adenosine and inosine reach the extracellular space by diffusion across the sarcolemma membrane along their concentration gradient.

When considering the different sites of adenosine formation in the heart, the question arises as to the function of ecto-5′-nucleotidase and the source of external nucleotides. It is well documented that adenine nucleotides can be released into the extracellular space as a result of a cotransport with neuro-

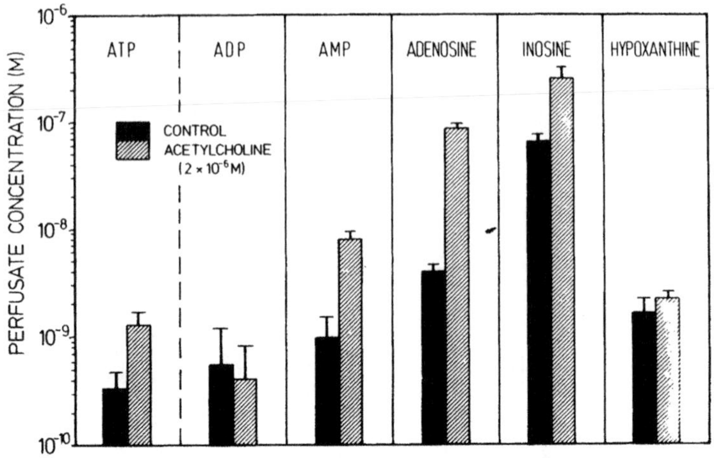

Figure 9.6. Concentration of various purine compounds in the effluent of isolated guinea pig hearts in the absence and presence of acetylcholine. (N = 6–12; x = ±SEM) Source: From [93].

transmitters [77] or in the course of the platelet release reaction [78]. ATP was also demonstrated to be released from nerve [79], skeletal muscle [80], neuromuscular junction [81], endothelial cells [17], and possibly purinergic nerves [82]. It is also likely that because of cell damage or death, small quantities of nucleotides are continuously released into the circulation. Thus, ectophosphatases may serve as effective means of removing biologically active nucleotides by degradation to adenosine, which in turn is inactivated by cellular reuptake or deamination.

Whether adenine nucleotides are normally released from the heart and can serve as a signal for vasodilation is a matter of debate [3, 4, 83]. An enhanced release of ATP has been reported during anoxic perfusion of the heart [84] and with increased work load [85]. Whether the amounts released are quantitatively sufficient to be involved in a functional response (e.g., increase in coronary flow) remains unclear. However, extracellular degradation of ATP is extremely rapid [32, 84], and most of the ATP may be dephosphorylated before analysis can be performed. In the isolated guinea pig heart, we found that about 99% of infused ATP ($10^{-7}M$) was degraded during a single passage through the heart [86]. Our experiments with AOPCP, however, exclude extracellular nucleotide breakdown as a major source for the hypoxia-induced adenosine formation.

In view of these findings, we were intrigued by the observation that $10^{-6}M$ acetylcholine considerably increased the release of ATP, AMP, and adenosine into the coronary circulation of well-oxygenated hearts (Fig. 9.6). Acetylcholine also markedly increased coronary flow, while heart rate and intraventricular pressure decreased. Consequently, myocardial oxygen consumption and lactate release were significantly diminished [86].

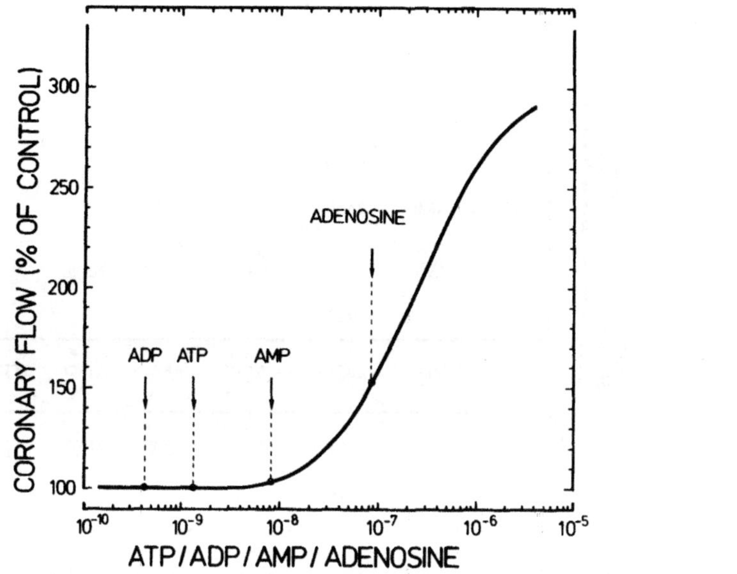

Figure 9.7. Comparison of the dose-response curve to exogenously applied ATP, ADP, AMP, and adenosine, with the concentration of each purine compound measured in the effluent under the influence of acetylcholine. Source: Data taken from Fig. 9.6.

The functional condition of purine release by acetylcholine appears to be fundamentally different from those previously described. Acetylcholine decreased myocardial oxygen requirements and also increased oxygen supply via the coronary system. Thus, the observed purine release, which can be blocked by atropine [86], cannot be due to myocardial hypoxia but is most likely independent of energy metabolism of the heart.

No firm conclusion can be drawn at present as to the molecular mechanism underlying the acetylcholine-mediated ATP and adenosine release. It is possible that ATP was released from adrenergic nerves together with norepinephrine as a cotransmitter or was derived from intramural purinergic nerves [84]. More recent evidence suggests that endothelial cells may be involved. Muscarinic cholinergic receptors on endothelial cells have been shown to play an obligatory role in the acetylcholine-induced smooth muscle relaxation [87]. Furthermore, coronary endothelial cells contain high levels of adenine nucleotides [16]; they selectively release adenine nucleotides in response to various stimuli [17] and dephosphorylate adenine nucleotides by ecto-enzymes [88].

When comparing the sensitivity of the coronary vessels to adenosine with the concentration changes of adenosine induced by acetylcholine (Fig. 9.7), it is obvious that endogenously formed adenosine falls within the coronary vasodilator range. Thus, a great portion of the coronary dilating effect of acetylcholine may be mediated by adenosine or a nucleotide. This conclusion

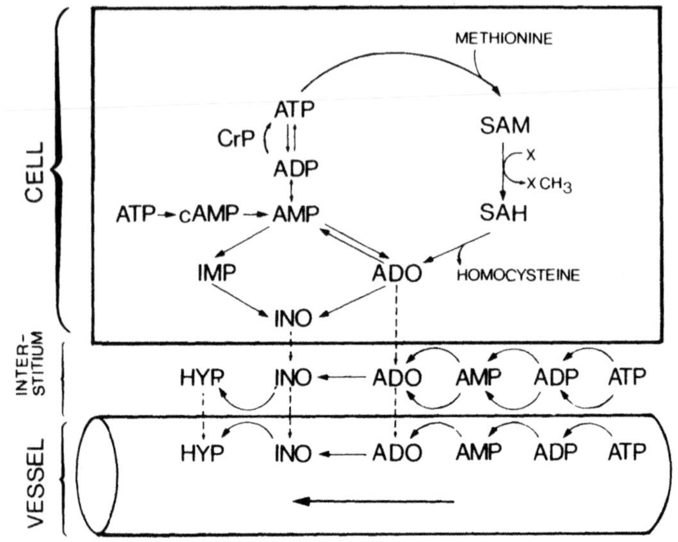

Figure 9.8. Schematic representation of the metabolism of adenosine and its cellular and subcellular distribution.

is supported by the observation that 2×10^{-5} theophylline, known to inhibit the cardiac actions of ATP and adenosine [89], also attenuated by 42% ($n = 4$) the coronary flow response to $10^{-6}M$ acetylcholine.

SUMMARY

Evidence has been presented that the site of adenosine formation in the heart may be intracellular and extracellular, depending on the experimental condition (Fig. 9.8). During hypoxia, adenosine is produced predominantly intracellularly and reaches the extracellular space by a transport process across the sarcolemma membrane. However, when hearts are treated with acetylcholine, adenine nucleotides are released, and by action of ecto-phosphatases on coronary myocytes and endothelial cells they are rapidly degraded to adenosine. Whereas the extracellular formation of adenosine is catalyzed by ecto-5′-nucleotidase, the situation is more complex within the myocardial cell. Most likely, intracellular adenosine is turning over at a rapid rate, which under control conditions is determined by the rate of cellular methylation processes. Due to the rapid cycling of the adenosine moiety through the pathway leading from SAH → ADO → AMP, modulation of the activity of adenosine kinase can considerably influence the rate at which adenosine accumulates in the heart. Furthermore, regulation of intracellular adenosine formation involves the activity of a soluble 5′-nucleotidase. The presence of two substrate cycles (AMP ⇄ ADO and ADO ⇄ SAH) can provide additional sensitivity and may explain why only moderate changes in the rate of adenosine formation from AMP or SAH can lead to substantial changes in the rate of adenosine release.

Little information is available at present regarding whether adenosine can be formed in the heart by the pathway leading from cyclic AMP via AMP to adenosine (Fig. 9.8). Furthermore, it is not known to what extent the methylation pathway in the heart can be accelerated by physiological stimuli. Therefore, the precise contribution of AMP versus SAH to the formation of adenosine remains to be elucidated. Another important problem is the fact that, in addition to the myocardial cell, coronary smooth muscle and endothelium may also produce adenosine by an intracellular mechanism. Thus, the situation in the intact heart may be more complex, as is outlined in Figure 9.8. Elucidation of the relative contribution of these various cells to the formation of adenosine as related to the regulation of physiological function remains a major challenge for future studies.

REFERENCES

1. Berne RM: Cardiac nucleotides in hypoxia: Possible role in regulation of coronary blood flow. *Am J Physiol* 204:317–322, 1963.
2. Gerlach E, Deuticke B, Dreisbach RH: Der Nucleotid-Abbau im Herzmuskel bei Sauerstoffmangel und seine mögliche Bedeutung für die Koronardurchblutung. *Naturwissenschaften* 50:228–229, 1963.
3. Berne RM, Rubio R: Coronary circulation, in Berne RM, Sperelakis N (eds): *Handbook of Physiology: The Cardiovascular System I.* 1979, pp 873–952.
4. Schrader J: Sites of action and production of adenosine in the heart, in Burnstock G (ed): *Purinergic Receptors.* London, Chapman & Hall, 1981, pp 121–162.
5. Schrader J, Baumann G, and Gerlach E: Adenosine as inhibitor of myocardial effects of catecholamines. *Pfluegers Arch* 372:29–35, 1977.
6. Baumann G, Schrader J, Gerlach E: Inhibitory action of adenosine on histamine- and dopamine-stimulated cardiac contractility and adenylate cyclase in guinea pigs. *Circ Res* 48:259–26, 1981.
7. Dobson JG: Reduction by adenosine of the isoproterenol-induced increase in cyclic 3′,5′-monophosphate. *Circ Res* 43:785–793, 1978.
8. Hedqvist P, Fredholm B: Inhibitory effect of adenosine on adrenergic neuroeffector transmission in the rabbit heart. *Acta Physiol Scand* 105:120–122, 1979.
9. Belardinelli L, Belloni FL, Rubio R, Berne RM: Atrioventricular conduction disturbances during hypoxia: Possible role of adenosine in rabbit and guinea pig heart. *Circ Res* 47:684–691, 1980.
10. Schrader J, Nees S, Gerlach E: Evidence for a cell surface adenosine receptor on coronary myocytes and atrial muscle cells. *Pfluegers Arch* 369:251–257, 1977.
11. Olsson RA, Davis CJ, Khouri EM, Patterson R: Evidence for an adenosine receptor on the surface of dog coronary myocytes. *Circ Res* 39:93–98, 1976.
12. Schrader J, Rubio R, Berne RM: Inhibition of slow action potentials of guinea pig atrial muscle by adenosine: A possible effect on Ca^{++} influx. *J Mol Cell Cardiol* 7:427–433, 1975.
13. Belardinelli L, Rubio R, Berne RM: Blockade of Ca^{++}-dependent rat atrial slow action potentials by adenosine and lanthanum. *Pfluegers Arch* 380:19–27, 1979.
14. Schütz W, Schrader J, Gerlach E: Different sites of adenosine formation in the heart. *Am J Physiol* 240:H963–H970, 1981.
15. Schrader J, Schütz W, Bardenheuer H: Role of S-adenosylhomocysteine hydrolase in adenosine metabolism in mammalian heart. *Biochem J* 196:65–70, 1981.
16. Nees S, Gerbes AL, Willershausen-Zönnchen B, Gerlach E: Purine metabolism in cultured coronary endothelial cells, in *Purine Metabolism in Man III. Adv Exp Med Biol* 122B:25–30, 1980.
17. Pearson JD, Gordon JL: Vascular endothelium and smooth muscle cells in culture selectively release adenine nucleotides. Nature 281:384–386, 1979.
18. van Harn GL, Rubio R, Berne RM: Formation of adenosine nucleotide derivatives in isolated hog carotid artery strips. *Am J Physiol* 233(2):H299–H304, 1977.

19. Schrader J, Gerlach E: Compartmentation of cardiac adenine nucleotides and formation of adenosine. *Pfluegers Arch* 367:129–131, 1976.
20. Berne RM, Rubio R: Adenine nucleotide metabolism in the heart. *Circ Res* (Suppl. III) 34–35:109–120, 1974.
21. Arch JRS, Newsholme EA: The control of the metabolism and the hormonal role of adenosine. *Essays Biochem* 14:82–123, 1978.
22. Fox IH, Kelly WN: The role of adenosine and 2′-deoxyadenosine in mammalian cells. *Annu Rev Biochem* 47:655–686, 1978.
23. Martin DW, Gelfand EW: Biochemistry of diseases of immunodevelopment. *Annu Rev Biochem* 50:845–877, 1981.
24. Henderson, JF: Regulation of adenosine metabolism, in Baer HP, Drummond GI (eds): *Physiological and Regulatory Functions of Adenosine and Adenine Nucleotides*. New York, Raven Press, 1979, pp 845–322.
25. De La Haba G, Cantoni GL: The enzymatic synthesis of S-adenosyl-L-homocysteine from adenosine and homocysteine. *J Biol Chem* 234:603–608, 1959.
26. Kredich N, Hershfield MS, Johnston JM: Role of adenosine metabolism in transmethylation, in Usdin E, Borchardt RT, Creveling CR (eds): *Transmethylation*. Amsterdam, Elsevier North-Holland, 1979, pp 225–230.
27. Rubio R, Berne RM, Dobson JG: Sites of adenosine production in cardiac and skeletal muscle. *Am J Physiol* 225:938–953, 1973.
28. Nakatsu K, Drummond GI: Adenylate metabolism and adenosine formation in the heart. *Am J Physiol* 223:1119–1127, 1972.
29. Nees S, Gerbes AL, Gerlach E: Isolation, identification, and continuous culture of coronary endothelial cells from guinea pig hearts. *Eur J Cell Biol* 24:287–297, 1981.
30. Frick GP, Lowenstein JM: Studies of 5′-nucleotidase in the perfused rat heart. J Biol Chem 251:6372–6378, 1976.
31. Frick GP, Lowenstein JM: Vectorial production of adenosine by 5′-nucleotidase in the perfused rat heart. *J Biol Chem* 253:1240–1244, 1978.
32. Baer HP, Drummond GI: Catabolism of adenine nucleotides by isolated perfused rat heart. *Proc Soc Exp Biol Med* 127:33–36, 1968.
33. Itoh R, Usami C, Nishino T, Tsushima K: Kinetic properties of cytosol 5′-nucleotidase from chicken liver. Biochim Biophys Acta 526:154–162, 1978.
34. Paglia DE, Valentine WN: Characteristics of a pyridine-specific 5′-nucleotidase in human erythrocytes. J Biol Chem 250:7973–7979, 1975.
35. Itoh R, Usami C, Nishino T, Tsushima K: Kinetic properties of cytosol 5′-nucleotidase from chicken liver. *Biochim Biophys Acta* 526:154–162, 1978.
36. Fritzon P: Regulation of nucleotidase activities in animal tissues. *Adv Enzyme Regul* 16:43–61, 1978.
37. Olsson RA, Gentry MK, Townsend RS: Adenosine metabolism: Properties of dog heart microsomal 5′-nucleotidase, in Bloor CM, Olsson RA (eds): *Current Topics in Coronary Research*. New York, Plenum Press, 1973, 27–39.
38. Rubio R, Belardinelli L, Thompson CI, Berne RM: Cardiac adenosine: Electrophysiological effects, possible significance in cell function, and mechanisms controlling its release, in Baer HP, Drummond GI (eds): *Physiological and Regulatory Functions of Adenosine and Adenine Nucleotides*. Raven Press, 1979, pp 167–182.
39. Baer HP, Drummond GI, Duncan EL: Formation and deamination of adenosine by cardiac muscle enzymes. *Mol Pharmacol* 2:67–76, 1966.
40. Sullivan JM, Alpers JB: In vitro regulation of rat heart 5′-nucleotidase by adenine nucleotides and magnesium. J Biol Chem 246:3057–3063, 1971.
41. Edwards MJ, Maguire MH: Purification and properties of rat heart 5′-nucleotidase. *Mol Pharmacol* 6:641–648, 1970.
42. Soboll S, Bünger R: Compartmentation of adenine nucleotides in the isolated working guinea pig heart stimulated by noradrenaline. *Hoppe Seylers Z Physiol Chem* 362:125–132, 1981.
43. Bünger R, Soboll S, Permanetter B: *Effects of Norepinephrine on Coronary Flow, Myocardial Substrate Utilization and Subcellular Adenylates*. Munich & Baltimore, Urban u. Schwarzenberg, 1982 (forthcoming).
44. Hiltunen JK, Hassinen IE: Energy-linked regulation of glucose and pyruvate oxidation in isolated perfused rat heart. *Biochim Biophys Acta* 440:377–390, 1976.

45. Wu ST, Pieper GM, Salhany IM, Eliot RS: Measurement of free magnesium in perfused and ischemic-arrested heart muscle: A quantitative phosphorus-31 nuclear magnetic resonance and multiequilibria analysis. *Biochemistry* 20:7399–7403, 1981.
46. Hess P, Metzger P, Weingart R: Free magnesium in striated muscle at rest measured with ion-selective microelectrodes. *J Physiol,* 1982 (in press).
47. De Jong JW: Partial purification and properties of rat heart adenosine kinase. *Arch Int Physiol Biochem* 85:557–569, 1977.
48. De Jong JW, Keijzer E, Uitendaal MP, Harmsen E: Further purification of adenosine kinase from rat heart using affinity and ion-exchange chromatography. *Anal Biochem* 101:407–412, 1980.
49. De Jong JW, Keijzer E: Dipyridamole affects myocardial adenosine kinase activity. *Arch Int Physiol Biochem* 87:525–532, 1979.
50. Andres CM, Fox IH: Purification and properties of human placental adenosine kinase. *J Biol Chem* 254:11388–11393, 1979.
51. Palella TD, Andres, CM, Fox IH: Human placental adenosine kinase: Kinetic mechanism and inhibition. *J Biol Chem* 255:5264–5269, 1980.
52. Miller RL, Adamczyk DL, Miller WH: Adenosine kinase from rabbit liver: I. Purification by affinity chromatography and properties. *J Biol Chem* 254:2339–2345, 1978.
53. Miller RL, Adamczyk DL, Miller WH, Koszalka GW, Rideout JL, Beacham LM, Chao EY, Haggerty JJ, Krenitsky TA, Elion GB: Adenosine kinase from rabbit liver: II. Substrate and inhibitor specificity. *J Biol Chem* 254:2346–2352, 1978.
54. Miller RL, Adamczyk DL: Adenosine kinase: Regulation by substrates, magnesium and pH, in *Purine Metabolism in Man III. Adv Exp Med Biol* 122B: 145–149, 1980.
55. Arch JRS, Newsholme EA: Activities and some properties of 5′-nucleotidase, adenosine kinase and adenosine deaminase in tissues from vertebrates and invertebrates in relation to the control of the concentration and the physiological role of adenosine. *Biochem J* 174:965–977, 1978.
56. Newsholme EA, Start C: *Regulation in Metabolism*. London, New York, Toronto, John Wiley and Sons, 1973.
57. McKenzie JE, McCoy FP, Bockman EL: Myocardial adenosine and coronary resistance during increased cardiac performance. *Am J Physiol* 239:H509–H515, 1980.
58. Saito D, Nixon DG, Vomacka RB, Olsson RA: Relationship of cardiac oxygen usage, adenosine content, and coronary resistance in dogs. *Circ Res* 47:875–882, 1980.
59. Schrader J, Haddy, FJ, Gerlach E: Release of adenosine, inosine and hypoxanthine from the isolated guinea pig heart during hypoxia, flow-autoregulation and reactive hyperemia. *Pfluegers Arch* 369:251–257, 1977.
60. Wiedmeier VT, Spell LH: Effects of catecholamines, histamine, and nitroglycerin on flow, oxygen utilization, and adenosine production in the perfused guinea pig heart. *Circ Res* 41:503–508, 1977.
61. Rubio R, Wiedmeier VT, Berne RM: Relationship between coronary blood flow and adenosine production and release. *J Mol Cell Cardiol* 6:561–566, 1974.
62. Miller WL, Belardinelli L, Bachhus A, Foley DH, Rubio R, Berne RM: Canine myocardial adenosine and lactate production, oxygen consumption, and coronary blood flow during stellate ganglion stimulation. *Circ Res* 45:708–718, 1979.
63. Olsson RA, Khouri EM, Bedynek JL, McLean J: Coronary vasoactivity of adenosine in the conscious dog. *Circ Res* 45:468–478, 1979.
64. Hershfield MS, Kredich NM: S-adenosylhomocysteine-hydrolase is an adenosine-binding protein: A target for adenosine toxicity. *Science* 202:757–760, 1978.
65. Ueland PM, Saebø J: Sequestration of adenosine in crude extract from mouse liver and other tissue. *Biochim Biophys Acta* 587:341–352, 1979.
66. Olsson RA, Vomacka RB, Nixon DG: Adenosine-binding proteins in dog hearts, in Baer HP, Drummond GI (eds): *Physiological and Regulatory Functions of Adenosine and Adenine Nucleotides*. New York, Raven Press, 1979, pp 297–403.
67. Schrader J, Berne RM, Rubio R: Uptake and metabolism of adenosine by human erythrocyte ghosts. *Am J Physiol* 223:159–166, 1972.
68. Belle VW: Uptake and deamination of adenosine by blood. *Biochim Biophys Acta* 192:124–132, 1969.
69. Olsson RA, Snow JA, Gentry MG, Frick GP: Adenosine uptake by canine heart. *Circ Res* 21:767–778, 1972.

70. Wiedmeier VT, Rubio R, Berne RM: Incorporation and turnover of adenosine-U-^{14}C in perfused guinea pig myocardium. *Am J Physiol* 223:51–54, 1972.
71. Newby AC, Luzio JP, Hales CN: The properties and extracellular location of 5′-nucleotidase of the rat fat cell plasma membrane. *Biochem J* 146:625–633, 1975.
72. Paterson ARP, Babb LR, Paran JH, Cass CE: Inhibition by nitrobenzylthioinosine of adenosine uptake by asynchronous HeLa cells. *Mol Pharmacol* 13:1147–1158, 1977.
73. Newby AC: Role of adenosine deaminase, ecto-(5′-nucleotidase) and ecto-(nonspecific phosphatase) in cyanide-induced adenosine monophosphate catabolism in rat polymorphonuclear leucocytes. *Biochem J* 186:907–918, 1980.
74. Newby AC, Holmquist CA: Adenosine production inside rat polymorphonuclear leucocytes. *Biochem J* 200:399–403, 1981.
75. Paterson ARP: Adenosine transport, in Baer HP, Drummond GI (eds): *Physiological and Regulatory Functions of Adenosine and Adenine Nucleotides*. New York, Raven Press, 1979, pp 305–313.
76. Rubio R, Wiedmeier VT, Berne RM: Nucleoside phosphorylase: Localization and role in the myocardial distribution of purines. *Am J Physiol* 222:550–555, 1972.
77. Kuperman AS, Volpert WA, Okamoto M: Release of adenine nucleotide from nerve axons. *Nature* 204:1000–1001, 1964.
78. Holmsen H, Weiss HJ: Secretable storage pools in platelets. *Annu Rev Med* 30:119–134, 1979.
79. White TD: Direct detection of depolarisation-induced release of ATP from a synaptosomal preparation. *Nature* 267:67–68, 1977.
80. Forrester T: An estimate of adenosine triphosphate release into the venous effluent from exercising human forearm muscle. *J Physiol* 244:611–628, 1972.
81. Silinsky EJ, Hubbard JI: Release of ATP from rat motor nerve terminals. *Nature* 243:404–405, 1973.
82. Burnstock G: Purinergic nerves. *Pharmacol Rev* 24:509–581, 1972.
83. Forrester T: Adenosine or adenosine triphosphate? in Vanhoutte PM, Leusen I (eds): *Vasodilation*. New York, Raven Press, 1981, pp 205–229.
84. Paddle BM, Burnştock G: Release of ATP from perfused heart during coronary vasodilation. *Blood Vessels* 11:110–119, 1974.
85. Clemens MG, Forrester T: Appearance of adenosine triphosphate in the coronary sinus effluent from isolated working rat heart in response to hypoxia. *J Physiol* 312:143–158, 1980.
86. Schrader J, Thompson CI, Hiendlmayer G, Gerlach E: Role of purines in acetylcholine-induced coronary vasodilation. *J Cell Mol Cardiol* 14:427–430, 1982.
87. Furchgott RF, Zawadzki IV: The obligatory role of endothelial cells in relaxation of arterial smooth muscle by acetylcholine. *Nature* 288:373–376, 1980.
88. Pearson JD, Carleton JS, Gordon JL: Metabolism of adenine nucleotides by ectoenzymes of vascular endothelial and smooth muscle cells in culture. *Biochem J* 190:421–429, 1980.
89. Bünger R, Haddy FJ, Gerlach E: Coronary responses to dilating substances and competitive inhibition by theophylline in the isolated perfused guinea pig heart. *Pfluegers Arch* 359:213–224, 1975.

DISCUSSION

DR. J.P. MANFREDI: Interpretation of intracellular AMP content in terms of adenosine release can be misleading. It is the concentration difference across the cell membrane which determines release. So, it is possible that a doubling of intracellular adenosine concentration (as a result of an approximate doubling of intracellular AMP) may be associated with a manyfold increase in the concentration difference of adenosine across the membrane, and consequently a manyfold increase in adenosine release.

DR. SCHRADER: Precise estimates of the extent to which changes of intracellular adenosine can influence its concentration gradient and release would require measurements of interstitial adenosine in the heart. However, this is technically not feasible. In addition, the intracellular free adenosine cannot be measured directly. Thus, your suggestion is an interesting theoretical possibility but cannot be studied experimentally at present.

DR. H.V. SPARKS: When relating AMP levels to adenosine release, one should be aware of the transient nature of adenosine release in response to catecholamines. Release is very high after 4 minutes, but much lower 8 minutes later. It is not safe to assume that there is a steady state during the first 15 minutes of catecholamine stimulation.

DR. SCHRADER: It would be interesting to know whether the transient nature of adenosine release is paralleled by similar changes in AMP levels. However, it is also possible that the phenomenon you describe is due to hormonal stimulation of the transmethylation pathway.

DR. R. BÜNGER: In our hands, 0.08 μM norepinephrine (NE) increased adenosine release from the working guinea pig heart up to approximately 4.5-fold when measured after 12 to 16 min of NE infusion. This increase is more than that expected from the 2- to 3-fold change in calculated free cytosolic AMP (1–3 μM) as measured by the non-aqueous fractionation technique under similar conditions. We therefore tend to assume that there is still some allosteric inhibition of the 5′-nucleotidase under our experimental conditions even when NE decreases the CrP/Cr ratio and the cytosolic ATP potential falls because of increased cardiac work performance and pressure development.

DR. R. RUBIO: Are the levels of homocysteine you use physiological?

DR. SCHRADER: The concentration of L-homocysteine in plasma of humans was recently determined by us to be about 10 μM. To what extent and under which conditions, other than homocysteinuria, this basal level can change is not known.

DR. RUBIO: Is your inhibitor of adenosine uptake, 5′-deoxy-5-aminoadenosine, also an inhibitor of transport of pyrimidines and purines that share the carrier of adenosine transport?

DR. SCHRADER: I do not have data to answer your question directly. However, I would like to remind you that 5′-deoxy-5-aminoadenosine greatly augmented adenosine release. The opposite would be expected when 5′-deoxy-5-aminoadenosine inhibited adenosine transport.

DR. A.C. NEWBY: At least in my studies on the polymorphonuclear leukocyte, there was no effect of 5′-deoxy-5-aminoadenosine on adenosine transport. It appears to get in by another mechanism.

DR. R. VON BORSTEL: Your studies with nitrobenzylthioinosine answered my question from Dr. Lowenstein's talk on the influence of transport inhibitors on cardiac adenosine release. Does nitrobenzylthioinosine (or dipyridamole) reduce coronary vasodilation in hypoxic hearts? If not, do these transport inhibitors have vasodilator actions not related to inhibition of adenosine transport?

DR. SCHRADER: Nitrobenzylthioinosine, in the concentration used in our studies, greatly dilated coronary vessels. The rate of onset of its action, however, was very slow, and this may be a direct effect unrelated to inhibition of adenosine transport.

DR. B.B. FREDHOLM: Do you think that the *in vitro* conditions you use influence the rate of methylation and the rate of AMP dephosphorylation and that the quantitative aspects of adenosine production may be substantially different *in vivo*?

DR. SCHRADER: The isolated guinea pig heart we used in our study is metabolically very stable and exhibits all functional features (autoregulation of coronary flow, reactive hyperemia, metabolic vasodilation) characteristic of the heart *in situ*. Provided that adenosine is an important metabolite involved in the regulation of these hemodynamic phenomena, the quantitative aspect of its metabolism should not be substantially different from the *in vivo* situation. In fact, the release of adenosine from the blood-perfused heart *in situ* is similar to that found in the isolated saline-perfused heart preparation.

DR. V. CHAGOYA DE SANCHEZ: I would like to comment about the importance of the activity of S-adenosylhomocysteine (SAH) hydrolase in relation to adenosine concentration. We find a good correlation of the activity of SAH-hydrolase and adenosine concentration in the liver, brain, and heart.

DR. SCHRADER: Similar to your observation, we could establish a good correlation between activity of SAH-hydrolase and adenosine in hearts of different species.

DR. R.A. OLSSON: The *in vitro* rate of the dissociation of the Ado.SAH complex (an intermediate product of the hydrolysis of S-adenosylhomocysteine) [Ado.Hcy] is quite slow; we estimate a $t_{1/2}$ of about 2.5 hours. As I understand your model, adenosine kinase serves as an "adenosine trap," thus driving the reaction in the direction of Ado.Hcy hydrolysis. During hypoxia and similar physiological stimuli, adenosine release is greatly stimulated. Does this mean that the adenosine carrier mediating the export of adenosine now competes with adenosine kinase? If so, what accounts for this change in the behavior of the carrier?

DR. SCHRADER: Whereas the rate of dissociation of the SAH-adenosine complex is a slow process, adenosine formation by hydrolysis of SAH is rather rapid and normally produces significant quantities of adenosine. In most animal cells, the adenosine carrier is characterized by a high K_m and high capacity, whereas adenosine kinase shows a low K_m and low capacity. Under control conditions—low cellular adenosine—most of the adenosine is rephosphorylated and only a little is transported, constituting basal cellular release. When cellular adenosine is elevated, its concentration may approach saturation of the kinase, so that adenosine is either directly transported out of the cell or is deaminated to inosine.

DR. R.M. BERNE: You speak of free adenosine within the myocardial cell. How can there be free adenosine intracellularly in the presence of cytosolic adenosine deaminase and adenosine kinase? Furthermore, adenosine is readily bound to protein (SAH-hydrolase), and its release is very slow. Finally, studies from our laboratory have failed to find free adenosine within cardiomyocytes or vascular smooth muscle cells. I would like you to explain how you account for free intracellular adenosine and, if it is present, what is its concentration?

DR. SCHRADER: Since there is a constant flux of adenosine through the kinase and deaminase pathway (both cytosolic enzymes in the guinea pig heart) even under noroxic conditions, there must be a certain substrate concentration of adenosine in the cytosol for these enzymes to be active. Under control conditions, the intracellular free adenosine concentration must be assumed to be rather low, but it increases during hypoxia. It is unlikely that the hypoxia-induced formation of adenosine is derived from adenosine released from the SAH-hydrolase complex. If adenosine is produced by this pathway, it is only by hydrolysis of SAH. Measurement of free adenosine would require the subtraction of the adenosine bound to SAH-hydrolase (including other cellular binding sites) from total tissue adenosine.

DR. E. GERLACH: What bothers me is that we are thinking in terms of only one cell compartment, namely, the cardiomyocytes. We should be aware that not only cardiomyocytes but also endothelial cells have to be taken into account because of their very active adenine nucleotide and adenosine metabolism. I feel it is really important not to ignore the significance of the vascular endothelial compartment, which in our opinion can contribute extensively to various steps of adenine nucleotide and adenosine metabolism in the heart.

DR. SCHRADER: Cardiomyocytes certainly constitute by far the major fraction of cardiac tissue, and thus most of the adenine nucleotides of the heart can be assumed to be associated with cardiomyocytes. That endothelial cells may be important functionally is illustrated by our experiments with acetylcholine. Whether endothelial cells have a more far-reaching function may become clearer as the conference progresses. Certainly, future studies on the site and metabolism of adenosine in the heart must include endothelial and also smooth muscle cells. I would also like to draw your attention to the fact that besides compartmentation at the cellular level, adenosine within the cell is present in free and protein-bound form. Similarly, there is evidence that only a small fraction of the measured total AMP serves as precursor for the formation of adenosine. Perhaps Dr. Bünger would like to comment on this problem.

DR. BÜNGER: Using whole tissue data, various investigators have calculated the distribution of adenylates between cytosol and mitochondria. Our approach was quite different. We applied the creatine kinase mass action ratio and adenylate kinase equilibrium relations, respectively, only *after* fractionating whole myocardium into six to eight portions, each of which contained enriched mitochondria or enriched cytosolic components. It was of particular interest to estimate the free concentrations of ADP and AMP, respectively, using data obtained from the fractionated myocardium. The adenylate kinase equilibrium constant was taken as 0.95, a value that holds for free Mg^{2+} concentrations between 0.4 and 0.8 m*M*, pH ≅ 7.0 [*Proc Natl Acad Sci USA* 61:1079, 1968]. (The presumptive free [Mg^{2+}] in the cytosol appears to be near 0.5 m*M* [*Circ Res* 33:367, 1973; *J Mol Cell Cardiol* 10:617, 1978]). We obtained a cytosolic concentration of free AMP ([AMP]) that was on the order of a few micromolar, i.e., it was three to ten times lower than the lowest K_m values for AMP of cardiac 5′-nucleotidase, according to recent literature [*J Biol Chem* 246:3057, 1971; *Adv Exp Med*

Biol 39:27, 1973; *Biochemistry* 20:5188, 1981]. [AMP] in the cytosol only doubled or tripled during infusion of 0.08 μ*M* norepinephrine (poster abstract), whereas adenosine release increased 4- to 6-fold under identical conditions of perfusion and heart work [unpublished data]. From these findings, it appears conceivable that (1) the availability of AMP in the cytosol may influence 5′-nucleotidase activity and thus adenosine formation, (2) because the Michaelis-Menten characteristics of the cardiac adenylate kinase seem to predict adenosine release rates higher than those actually measured, it appears that allosteric regulator(s), e.g., creatine phosphate, are present that might restrain the enzyme's activity and therefore the formation of vasodilator adenosine (especially in control hearts). Furthermore, the extremely low [AMP] in the cytosol suggests that most of whole myocardium AMP is not a substrate of the cytosolic adenylate kinase; hence, it is likely compartmentalized and perhaps located within the mitochondria [see, e.g., *Biochem Biophys Res Commun* 79:1155, 1977; *Hoppe Seylers Z Physiol Chem* 358:647, 1977].

DR. J.M. LOWENSTEIN: I don't think there is any disagreement among us as to what regulates adenosine production and that, clearly, if the AMP goes up, that is going to make a difference and it is not enough by itself, so there is some additional regulation.

DR. NEWBY: We have studied adenine nucleotide catabolism and adenosine formation in an intact isolated cell preparation, rat polymorphonuclear leukocytes, as a simple biochemical model of these processes in complex multicellular organs. Adenosine formation in response to 2-deoxyglucose poisoning was shown to take place within the cells by a pathway independent of the cell membrane ecto-5′-nucleotidase [Newby AC, Holmquist CA: *Biochem J* 200:399–403, 1981]. The use of inhibitors of both adenosine kinase and adenosine deaminase allowed us to quantitate the rate of adenosine synthesis precisely. We have since isolated a cytoplasmic 5′-nucleotidase from polymorphs. The enzyme is indistinguishable kinetically and electrophoretically from the liver enzyme discovered by Fritzon and studied extensively by Itoh and Van den Berghe and their colleagues. The following table compares the maximal rate of adenosine formation in 2-deoxyglucose-poisoned cells with the activity of the cytoplasmic 5′-nucleotidase determined *in vitro* at the appropriate AMP concentration.

Adenosine formation	0.14 nmol/10^7 cells/min
AMP concentration	1.3 nmol/10^7 cells
Cytoplasmic water space (3-0-methylglucose space)	1.23 μl/10^7 cells
AMP concentration	1.06 m*M*
Cellular content of cytosolic 5′-nucleotidase	3.6 nmol/10^7 cells
AMP'ase activity at 1.06 m*M*	0.23 nmol/10^7 cells/min

We conclude that the cytosolic 5′-nucleotidase is sufficiently active to account for adenosine synthesis in this cell type.

DR. MANFREDI: It should also be pointed out that the 2-deoxyglucose intervention induces a drastic reduction in the cell's energy state. The conclusion that intracellular adenosine production is attributable to flux through 5′-nucleotidase may be accurate only for such energy-depleting interventions.

10. INTERACTION OF ADENOSINE WITH ADENOSINE-BINDING PROTEIN, S-ADENOSYLHOMOCYSTEINE HYDROLASE

PER MAGNE UELAND

Soluble adenosine-binding proteins were first described by Yuh and Tao in rabbit erythrocytes [1], and similar proteins were later isolated from various eukaryotes [2–8]. The physiological role of these binding proteins for adenosine (Ado) remained obscure until Hershfield [9] and Hershfield and Kredich [10] demonstrated that such proteins from human placenta, spleen, and lymphoblasts were identical to S-adenosylhomocysteine (AdoHcy) hydrolase (EC: 3.3.1.1), the enzyme responsible for the metabolic degradation of the endogenous transmethylase inhibitor, AdoHcy [11]. This finding has been confirmed by others [12–14]. This chapter is a brief review of the properties of AdoHcy hydrolase, with emphasis on the interaction of the enzyme with Ado.

PROPERTIES AND OCCURRENCE OF AdoHcy HYDROLASE

AdoHcy hydrolase catalyzes the reversible hydrolysis of AdoHcy to Ado and L-homocysteine (Hcy) (Fig. 10.1). At high concentrations of substrates (>1 μM), the AdoHcy hydrolase reaction favors synthesis of AdoHcy [15]. This can be more precisely expressed by the equilibrium constant (K_{eq}) of the reaction, defined by the following equation [15, 16]:

$$K_{eq} = \frac{[\text{Ado}] \times [\text{Hcy}]}{[\text{AdoHcy}]} = 10^{-6}\,M$$

R.M. Berne, T.W. Rall, and R. Rubio (eds.), Regulatory Function of Adenosine, *pp. 157–170.*

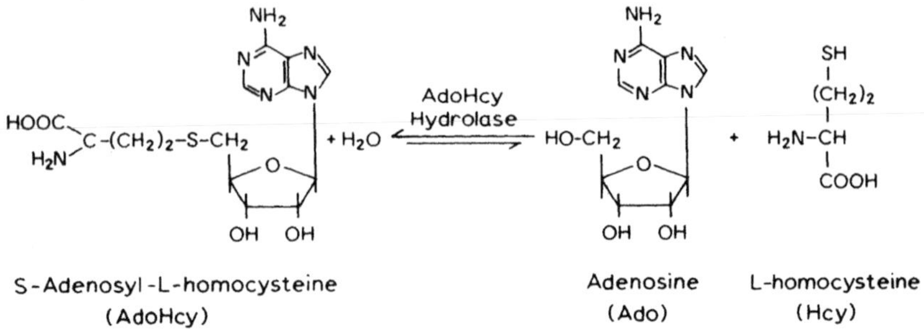

Figure 10.1. The reaction catalyzed by AdoHCy hydrolase.

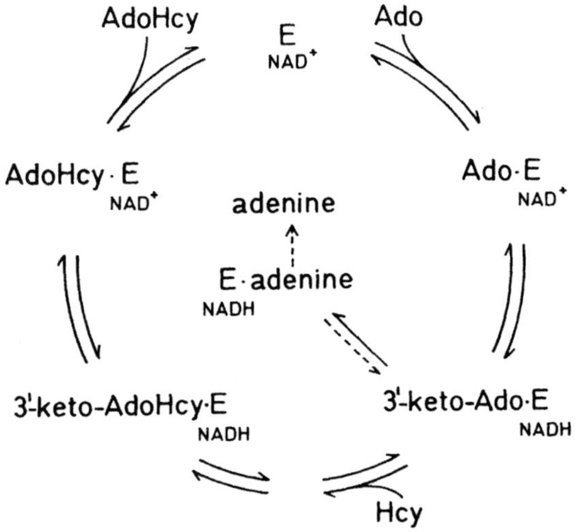

Figure 10.2. The catalytic cycle of AdoHcy hydrolase.

Ado and, to a lesser degree, Hcy are inhibitors of hydrolysis of AdoHcy, and the reaction can be directed toward hydrolysis by removal of Ado or Hcy [15].

Knowledge of the catalytic mechanism of AdoHcy hydrolase has been provided by the work of Palmer and Abeles [17]. They found that the calf liver enzyme contains tightly bound NAD^+, which participates in the catalytic cycle. Synthesis and hydrolysis of AdoHcy involve oxidation of Ado and AdoHcy, respectively, to their 3′-keto derivatives, with a concomitant reduction of NAD^+ to NADH (Fig. 10.2; for details, see [17]).

Data on physiochemical and kinetic properties of AdoHcy hydrolase purified to apparent homogeneity from various sources are summarized in Table 10.1. Because somewhat inconsistent data on this item exist (Table 10.1), a study was undertaken in my laboratory to compare the properties of AdoHcy hydrolase purified to homogeneity from three different sources, i.e.,

Table 10.1. Physiochemical and kinetic properties of AdoHcy hydrolase purified to homogeneity from various sources

		Subunits			K_m			
Source	M_r	No.	M_r	pI	Ado (μM)	AdoHcy (μM)	Hcy (mM)	Reference
Rabbit erythrocytes	240,000	5	48,000					[1]
Beef liver	192,000	4	48,000		45	10.5		[18]
Mouse liver	185,000	4	45,000–46,500	5.7	0.2	0.75	0.15	[4, 12, 19]
Yellow lupin seeds	110,000	2	55,000	4.9	2.3	12.0	4.6	[20]
Calf liver	237,000	4	50,000–60,000	5.8	420	63		[21]
Human placenta	190,000	4			0.5			[9]
Rat brain	180,000	4	48,000	5.6		36.6		[22]
Rat liver	220,000	5	47,000	6.6	0.6	0.90	0.06	[23]
Rat liver	188,000	4	47,000	5.7	1.05	15.2	0.155	[24]
Rat liver			47,000–57,500		0.94	12.3	0.164	[25]
Bovine adrenal cortex	185,000	4	45,000–46,000	5.35	0.2	0.75		[15]
Bovine liver	185,000	4	45,000–46,000	5.35	0.2	0.75		[15]

mouse liver, bovine liver, and bovine adrenal cortex [14]. Similar kinetic behavior was observed for these enzymes (K_m for Ado and AdoHcy in the micromolar range), and all enzymes were quadrimers with M_r of about 190,000 [14] (Table 10.1).

AdoHcy hydrolase is widely distributed in vertebrates, yeast, and plants, but bacteria are devoid of AdoHcy hydrolase activity [25]. In vertebrates, highest activity is found in liver and pancreas, but kidney is also rich in this enzyme. Smaller amounts are found in spleen, brain, skeletal muscle, heart, and testes, but some species variations have been observed [25–28]. This has been most clearly demonstrated in mammalian heart, where the enzyme activity is about 20-fold higher in guinea pig than in dog [29].

AdoHcy hydrolase from liver [26, 30], brain [21, 31], heart [29], and probably other tissues is a soluble protein localized to the cytosol fraction of the tissue homogenate.

ADENINE ANALOGUE-BINDING PROTEINS

In 1971, Chambaut and co-workers demonstrated in rat liver the presence of a cyclic AMP-binding protein devoid of protein kinase activity [32]. Similar proteins were later described in various cells and tissues (Table 10.2). Some of these proteins bind Ado with high affinity [1, 3, 4, 6, 7] and also interact with adenine nucleotides other than cyclic AMP [3, 36]. Sugden and Corbin [3] suggested the name adenine analogue-binding proteins (AABP) for such proteins from rat and bovine tissues. The finding that Ado-binding proteins are associated with AdoHcy hydrolase activity [9, 10, 12, 14] raises the question of whether AdoHcy hydrolase may account for all soluble Ado-binding factors.

Table 10.2. Properties of adenine analogue binding proteins from various sources

Source	Molecular size	Affinity (K_d) for Ado (μM)	Affinity (K_d) for cAMP (μM)	Reference
Rat liver				[33]
Rabbit erythrocytes	240,000, 9S	0.1	0.3	[1]
Mouse liver	185,000, 9S	0.2	0.15	[4, 19]
Drosophila melanogaster	200,000		0.1	[34]
Rat liver		0.24		[2]
Rat liver	180,500, 10.1 S 370,500, 14.2 S			[3]
Bovine liver	185,000, 9.8 S			[3]
Rabbit erythrocytes		0.0053	0.009	[7]
Human erythrocytes	230,000			[6]
Jerusalem artichoke rhizome tissues	240,000		0.23	[35]
Trypanosoma gambiense	150,000		0.6	[36]
Rat heart	176,000			[38]

Binding proteins for Ado have been described in dog, cow [8], and rat heart [37], and these proteins are not associated with AdoHcy synthase or hydrolase activity. A cyclic AMP-Ado-binding protein from rat heart has been separated from AdoHcy hydrolase by affinity chromatography [37]. However, these reports do not definitely prove the existence *in vivo* of Ado-binding proteins not related to AdoHcy hydrolase. The enzyme is inactivated in the presence of various naturally occurring purines, including 2′-deoxyadenosine, Ado, inosine [38], and adenine nucleotides [39].

INTERACTION OF ADENOSINE ANALOGUES WITH AdoHcy HYDROLASE

Numerous analogues of Ado interact with AdoHcy hydrolase (Table 10.3). Some Ado analogues serve as substrate for the enzyme and may thereby function as a metabolic precursor for the corresponding AdoHcy analogue. 3-deazaadenosine is an even better substrate for AdoHcy hydrolase than Ado itself, and this compound has been widely used as a pharmacological tool for studying methyltransfer reactions [11, and references therein]. Other analogues of Ado are reversible inhibitors of the enzyme. These compounds may induce a buildup of cellular AdoHcy derived from S-adenosyl-L-methionine (AdoMet). Carbocyclic Ado seems to be particularly effective [13]. Irreversible inactivation of AdoHcy hydrolase was first demonstrated with 9-β-D-arabinofuranosyladenine and 2′-deoxyadenosine [40], and these compounds inactivate the enzyme *in vitro* by a mechanism that involves irreversible reduction of enzyme-bound NAD^+ [38, 41, 42]. However, the intracellular enzyme exposed to 9-β-D-arabinofuranosyladenine [43] is slowly reactivated [44]. The existence of such a mechanism in the intact cell might be expected in the light

Table 10.3. Adenosine analogues as substrates, inhibitors, and inactivators of AdoHcy hydrolase

Adenosine analogue	Substrate	Inhibitor	Inactivator	Reference
Adenosine	+++	+++	(+)	[13, 23, 46, 47]
3-deazaadenosine	+++	+++	(+)	[11, 13, 46, 47]
Nebularin	++	(+)	++	[13, 46, 47]
Formycin	+	+	(+)	[13, 46, 47]
N^6-methyladenosine	+	+	(+)	[13, 46–48]
8-azaadenosine	+	+	(+)	[13, 46, 47]
2-chloroadenosine[a]			+++	[13, 46]
Adeninearabinoside	(+)	++	+++	[13, 39, 41, 43, 44]
Carbocyclic adenosine	(+)	+++	+++	[13, 47]
2′-deoxyadenosine			+	[13, 41, 42]
Adenosine dialdehyde			+++	[49, 50]

[a]2-Chloroadenosine seem to be a potent inactivator of the isolated enzyme [13] but induces only a moderate increase in AdoHcy content in intact cells [46].

of the fact that naturally occurring purines, including Ado, are inactivators of AdoHcy hydrolase [38].

Ado analogues that are reported to function as substrates, inhibitors, or inactivators of isolated AdoHcy hydrolase and the enzyme in intact cells are listed in Table 10.3.

INTERACTION OF ADENOSINE WITH AdoHcy HYDROLASE *IN VITRO*

Ado forms a stable complex with AdoHcy hydrolase [4, 9, 10, 13, 50]. Compared with the site interacting with cyclic AMP, which accepts various adenine derivatives, the Ado-binding site is rather restrictive [4, 7, 36]. The ado-binding site may be identical to the catalytic site of the enzyme [16]. The Ado-enzyme complex formed after prolonged incubation shows a long half-life [50, 51], and the bound Ado is not displaced by excess unlabeled Ado [9, 50] and is not available for deamination catalyzed by adenosine deaminase [51–53].

A fraction of Ado bound to AdoHcy hydrolase is converted to adenine or a substance liberating adenine [52], and a significant amount of adenine is dissociated from the enzyme [52]. Tightly bound adenine (derived from Ado) is reconverted back to Ado under certain conditions [51], but synthesis of Ado from added adenine could not be demonstrated [52]. Adenine formation may be explained by liberation of the purine base from 3′-ketoadenosine [13, 53], a proposed intermediate in the catalytic cycle [17] (Fig. 10.2).

High concentrations of Ado in the absence of Hcy induce an irreversible inactivation of isolated AdoHcy hydrolase [22, 38]. The kinetics of the inactivation process suggest a suicide or K_{cat} mechanism of action of Ado, which implies that the inactivation proceeds from a reversible Ado-enzyme complex [38]. A mechanism of action of Ado has been proposed [13] that involves irreversible reduction of enzyme-bound NAD^+ and liberation of adenine from 3′-ketoadenosine (Fig. 10.2), i.e., a mechanism of action similar to that of such inactivators as 2′-deoxyadenosine [41] and 9-β-D-arabinofuranosyladenine [42].

Tight binding of Ado to AdoHcy hydrolase, formation of adenine, and inactivation of the enzyme by Ado are probably closely related processes [53].

AdoHcy hydrolase from plant is a dimer (Table 10.1) and binds two molecules of Ado per molecule of enzyme [54]. The binding of the first Ado molecule is fast, whereas the binding of the second molecule is a slow process. This suggests negative cooperativity among binding sites. The Ado-enzyme complex reacts slowly to form adenine, ribose, and active enzyme. Thus, in contrast to mammalian AdoHcy hydrolase, Ado does not inactivate the plant enzyme. The half-life of the newly formed complex is short, whereas the old complex dissociates slowly [54].

Ado added to concentrated crude extracts from various tissues, including liver, kidney, adrenal cortex, brain, heart, skeletal muscle, and uterus, is not completely metabolized. A fraction of Ado added to the extracts is not available for deamination catalyzed by adenosine deaminase [28]. This phenome-

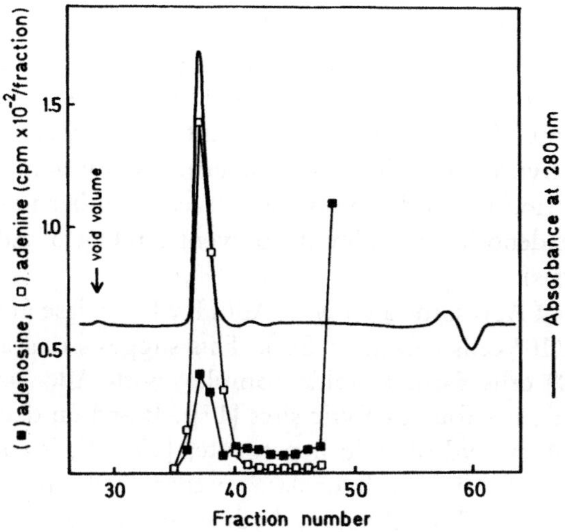

Figure 10.3. HPLC gel-filtration of purified AdoHcy hydrolase (—) and cytosol from rat hepatocytes incubated with [^{14}C]adenosine (■,□). Adenine (□), adenosine (■).

non, termed sequestration of Ado, is explained by formation of a stable Ado-AdoHcy hydrolase complex. The ability of various tissues to sequester Ado parallels the tissue level of AdoHcy hydrolase [28].

The possibility existed that tight binding of Ado to AdoHcy hydrolase and formation of adenine are *in vitro* phenomena related to the absence of the other substrates of the enzyme, namely, Hcy or AdoHcy. Therefore, synthesis and hydrolysis of AdoHcy were determined at assumedly physiological concentrations of enzyme (about 10 μM [28]) and its substrates. Under these conditions, both sequestration of Ado and formation of adenine could be demonstrated. Furthermore, sequestered Ado is not available for synthesis of AdoHcy [53].

BINDING OF ADENOSINE TO INTRACELLULAR AdoHcy HYDROLASE

In the light of the data given in the preceding paragraph, a study was carried out in my laboratory to investigate whether AdoHcy hydrolase functions as an Ado-binding protein in intact cells. Isolated rat hepatocytes were chosen as the experimental system because these cells are rich in AdoHcy hydrolase [25]. Hepatocytes incubated with [^{14}C]Ado were extracted in a buffer containing high concentrations of unlabeled Ado (10 mM) to prevent binding of radioactive Ado to the enzyme after cellular lysis. Extract from these cells was analyzed on a high-performance liquid chromatography (HPLC) protein column, and radioactive adenine and Ado eluted as a peak that co-chromatographed exactly with AdoHcy hydrolase (Fig. 10.3). Formation of this peak was inhibited by exposure of the cells to compounds (9-β-D-arabinofuranosyladenine, 3-deazaadenosine, or homocysteine) interacting with the catalytic site of AdoHcy hydrolase. On this basis, it was concluded that labeled Ado forms a stable complex with intracellular AdoHcy hydrolase [55].

It has been reported that a fraction of endogenous Ado in rat hepatocytes is not mobilized by treating the cells with extracellular adenosine deaminase [44, 56]. This amount of Ado equals the amount of endogenous Ado associated with AdoHcy hydrolase, which was determined by a modification of the HPLC gel-filtration method, involving extraction of hepatocytes in the presence of 3-deazaadenosine and Hcy (to prevent binding of Ado to the enzyme after cellular lysis).

The amount of Ado associated with AdoHcy hydrolase in rat hepatocytes is 37 ± 15 pmol/10^6 cells (mean ± SD). This suggests that about 10 pmol of enzyme per 10^6 cells form a stable complex with Ado (assuming that the enzyme molecule has four catalytic sites [17]). Based on data on the catalytic activity of AdoHcy hydrolase in hepatocytes [43, 44], it could be calculated that about 50% of AdoHcy hydrolase in these cells may exist as a stable complex with Ado. Thus, AdoHcy hydrolase functions as an Ado-binding protein in rat hepatocytes.

POSSIBLE ROLE OF AdoHcy HYDROLASE IN ADENOSINE METABOLISM

AdoHcy hydrolase catalyzes the reversible hydrolysis of AdoHcy to Ado and Hcy. Although the reaction favors synthesis of AdoHcy, the metabolic flow is probably in the hydrolytic direction, because both Ado and Hcy are continuously trapped in the intact cell [11, 15]. Ado is either deaminated or phosphorylated [11] (Fig. 10.4). The cellular level of Hcy is low [57], suggesting no synthesis of AdoHcy under physiological conditions.

Apart from the work of Schütz et al. [58] investigating the site of production of Ado in the heart, no attention has been given to the possible role of the AdoHcy hydrolase reaction as a source of Ado *in vivo*. This question obviously deserves further attention.

Conversion of Ado to adenine is a side reaction of AdoHcy hydrolase, but this reaction is ineffective [52, 53]. Only small amounts of adenine are formed from Ado in isolated rat liver [56]. Thus, it seems that formation of adenine from Ado catalyzed by AdoHcy hydrolase is not a quantitatively important source of cellular adenine, relative to other metabolic pathways.

Addition of exogenous Ado leads to accumulation of AdoHcy, which is greatly enhanced by Hcy. This has been demonstrated with intact cells [59], perfused liver [57, 60], and heart [29] and in whole animals [61]. Thus, administration of these compounds is a convenient procedure for obtaining an elevated cellular level of AdoHcy. The mechanisms underlying this metabolic effect of Ado and Hcy involve inhibiting AdoHcy hydrolase or directing the AdoHcy hydrolase reaction toward synthesis of AdoHcy. Data provided by Duerre and Briske-Anderson [57], who perfused isolated rat liver with labeled Ado, suggest that reversal of the AdoHcy hydrolase reaction takes place even in the absence of exogenous Hcy. Because the cellular level of Hcy is far below

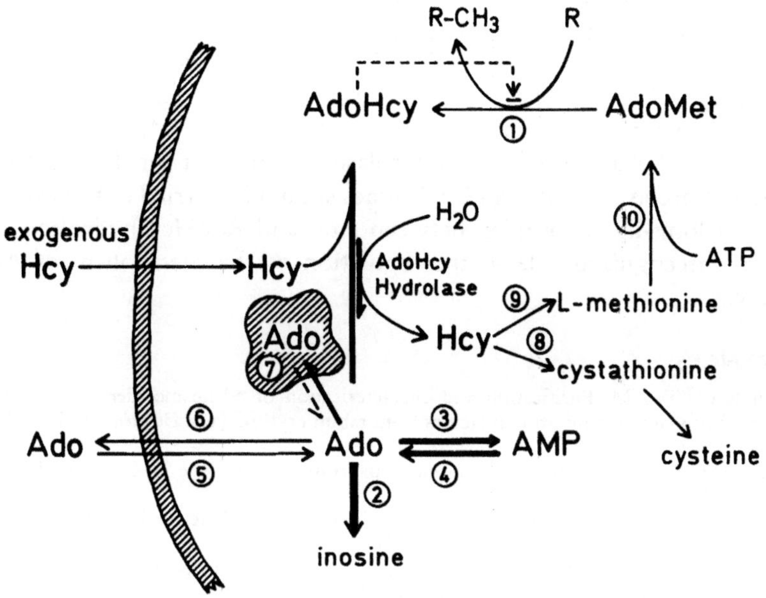

Figure 10.4. Role of AdoHcy hydrolase in intermediary metabolism. (1) AdoMet-dependent methyltransferases; (2) adenosine aminohydrolase (adenosine deaminase); (3) adenosine kinase; (4) adenosine 5′-monophosphate nucleotidase; (5) adenosine uptake; (6) adenosine release; (7) adenosine tightly bound to AdoHcy hydrolase; (8) cystathionine-β-synthase; (9) 5′-methyltetrahydrofolate: L-homocysteine methyltransferase and betaine: L-homocysteine S-methyltransferase; (10) ATP: L-methionine S-adenosyltransferase.

the apparent K_m value for Hcy, the existence of a cellular pool of Hcy has been postulated to explain this observation [57]. However, under this condition (i.e., in the absence of exogenous Hcy), the bulk of AdoHcy may be derived from various methyltransfer reactions. When the liver is perfused with both Ado and Hcy, a significant amount of AdoHcy is formed via synthesis [57].

It seems reasonable to suggest that intracellular AdoHcy hydrolase functions as an Ado-binding protein in various tissues and cells in addition to rat hepatocytes. Green [62] has recently reported that Ado is taken up by murine neuroblastoma cells deficient in adenosine kinase by a concentrative and saturable process, and binding of Ado to intracellular protein such as AdoHcy hydrolase is suggested as an explanation for this finding. Furthermore, injection of mice with Ado decreases the AdoHcy hydrolase activity in brain extracts. This effect could not be reversed by dialysis and may be related to the formation of a stable Ado-AdoHcy hydrolase complex [61].

The fraction of Ado bound to AdoHcy hydrolase is probably not available for other enzymes participating in the metabolic handling of Ado [28] and for synthesis of AdoHcy [53]. Biological effects of Ado that are not mediated by AdoHcy hydrolase are probably induced by the free fraction of the nucleoside.

In my opinion, it is unlikely that the biological implications of the existence of a stable Ado-AdoHcy hydrolase complex are limited to a functional compartmentation of Ado. The size of a cellular pool of Ado existing as an Ado-AdoHcy hydrolase complex seems to be small relative to the metabolic capacity of other pathways leading to Ado formation (Fig. 10.4). It is conceivable that binding of Ado to AdoHcy hydrolase is important for the regulation of AdoHcy hydrolase activity *in vivo*. Studies should be carried out to investigate whether a long-lived complex between Ado and AdoHcy hydrolase plays a hitherto unrecognized role in the regulation of the metabolism of Ado or AdoHcy.

REFERENCES

1. Yuh K-C, Tao M: Purification and characterization of adenosine-adenosine cyclic 3′,5′-monophosphate binding protein factors from rabbit erythrocytes. *Biochemistry* 13:5220–5226, 1974.
2. Hsu HHT: Specific adenosine binding proteins from rat liver. *Proc Soc Exp Biol Med* 149:698–701, 1975.
3. Sugden PH, Corbin JD: Adenosine 3′:5′-cyclic monophosphate-binding proteins in bovine and rat tissues. *Biochem J* 159:423–437, 1976.
4. Ueland PM, Døskeland SO: An adenosine 3′:5′-monophosphate-adenosine binding protein from mouse liver. *J Biol Chem* 252:677–686, 1977.
5. Døskeland SO, Ueland PM: Binding proteins for adenosine 3′:5′-cyclic monophosphate in bovine adrenal cortex. *Biochem J* 165:561–573, 1977.
6. Kumar R, Yuh K-C, Tao M: Human erythrocyte proteins associated with adenosine 3′,5′-cyclic monophosphate action. *Enzymes* 23:73–83, 1978.
7. Olsson RA: Ligand binding to adenine analogue binding protein of the rabbit erythrocytes. *Biochemistry* 17:367–375, 1978.
8. Olsson RA, Vomacka RB, Nixon DG: Adenosine-binding proteins in dog heart, in Baer HP, Drummond GI (eds): *Physiological and Regulatory Functions of Adenosine and Adenine Nucleotides*. New York, Raven Press, 1979, pp 297–304.
9. Hershfield MS: Human cytoplasmic adenosine binding protein: Identification as S-adenosylhomocysteinase. *Fed Proc* 37:1466, 1978 (abst).
10. Hershfield MS, Kredich NM: S-adenosylhomocysteine hydrolase is an adenosine binding protein: A target for adenosine toxicity. *Science* 202:757–760, 1978.
11. Cantoni GL, Chiang PK: The role of S-adenosylhomocysteine and S-adenosylhomocysteine hydrolase in the control of biological methylations, in Cavallini D, Gaull GE, Zappia V (eds): *Natural Sulfur Compounds: Novel Biochemical and Structural Aspects*. New York, Plenum Press, 1980, pp 67–80.
12. Saebø J, Ueland PM: An adenosine 3′:5′-monophosphate adenosine-binding protein from mouse liver. *FEBS Lett* 96:125–128, 1978.
13. Chiang PK, Guranowski A, Segall JE: Irreversible inhibition of S-adenosylhomocysteine hydrolase by nucleoside analogs. *Arch Biochem Biophys* 207:175–184, 1981.
14. Døskeland SO, Ueland PM: Comparison of some physicochemical and kinetic properties of S-adenosylhomocysteine hydrolase from bovine liver, bovine adrenal cortex and mouse liver. Biochim. Biophys. Acta 708:185–193, 1982.
15. De La Haba G, Cantoni GL: The enzymatic synthesis of S-adenosyl-L-homocysteine from adenosine and homocysteine. *J Biol Chem* 234:603–608, 1959.
16. Ueland PM, Saebø J: S-adenosylhomocysteinase from mouse liver. *Biochemistry* 18:4130–4135, 1979.
17. Palmer JL, Abeles RH: The mechanism of action of S-adenosylhomocysteinase. *J Biol Chem* 254:1217–1226, 1979.
18. Ueland PM, Skotland T, Døskeland SO, Flatmark T: An adenosine 3′:5′-monophosphate-adenosine binding protein from mouse liver: Some physiochemical properties. *Biochim Biophys Acta* 533:57–65, 1978.

19. Guranowski A, Pawelkiewicz J: Adenosylhomocysteinase from yellow lupin seeds. *Eur J Biochem* 80:517–523, 1977.
20. Richards HH, Chiang PK, Cantoni GL: Adenosylhomocysteine hydrolase. *J Biol Chem* 253:4476–4480, 1978.
21. Schatz RA, Vunnam CR, Sellinger OZ: S-adenosyl-L-homocysteine hydrolase from rat brain: Purification and some properties, in Usdin E, Borchardt RT, Creveling CR (eds): *Transmethylation*.New York, Elsevier North-Holland, 1979, pp 143–153.
22. Kajander EO, Raina AM: Affinity-chromatographic purification of S-adenosyl-L-homocysteine hydrolase. *Biochem J* 193:503–512, 1981.
23. Fujioka M, Takata Y: S-adenosylhomocysteine hydrolase from rat liver. *J Biol Chem* 256:1631–1635, 1981.
24. Briske-Anderson M, Duerre JA: S-adenosylhomocysteine hydrolase from rat liver. *Can J Biochem* 60:118–123, 1982.
25. Walker RD, Duerre JA: S-adenosylhomocysteine metabolism in various species. *Can J Biochem* 53:312–319, 1975.
26. Finkelstein JD, Harris B: Methionine metabolism in mammals: Synthesis of S-adenosylhomocysteine in rat tissues. *Arch Biochem Biophys* 159:160–165, 1973.
27. Eloranta TO: Tissue distribution of S-adenosylmethionine and S-adenosylhomocysteine in the rat. *Biochem J* 166:521–529, 1977.
28. Ueland PM, Saebø J: Sequestration of adenosine in crude extract from mouse liver and other tissues. *Biochim Biophys Acta* 587:341–352, 1979.
29. Schrader J, Schütz W, Bardenheuer H: Role of S-adenosylhomocysteine hydrolase in adenosine metabolism in mammalian heart. *Biochem J* 196:65–70, 1981.
30. Ueland PM, Berge RK, Saebø J, Farstad M: S-adenosylhomocysteine hydrolase in human and rat liver is localized to the cytosol fraction of the tissue homogenate. *FEBS Lett* 101:184–186, 1979.
31. Broch OJ, Ueland PM: Regional and subcellular distribution of S-adenosylhomocysteine hydrolase in adult rat brain. *J Neurochem* 35:484–488, 1980.
32. Chambaut AM, Leray F, Hanoune J: Relationship between cyclic AMP dependent protein kinase(s) and cyclic AMP binding protein(s) in rat liver. *FEBS Lett* 15:328–334, 1971.
33. Tsuzuki J, Kiger JA Jr: Cyclic AMP binding proteins in early embryos of *Drosophilia melanogaster*. *Biochim Biophys Acta* 393:225–235, 1975.
34. Giannattasio M, Carratu G, Tucci GF, Carafa AM: Cyclic AMP binding protein from Jerusalem artichoke rhizome tissues. *Phytochemistry* 18:1–4, 1979.
35. Walter RD: Adenosine 3′,5′-cyclic monophosphate-binding proteins from *Trypanosoma gambiense*. *Hoppe Seylers Z Physiol Chem* 359:607–612, 1978.
36. Ueland PM: An adenosine 3′:5′-monophosphate/adenosine binding protein from mouse liver. *Eur J Biochem* 86:27–34, 1978.
37. Endrizzi A, Hrelia S, Lucacchini A, Rossi CA: Isolation of a cyclic AMP-adenosine binding protein from rat heart. *FEBS Lett* 108:395–398, 1979.
38. Hershfield MS et al. Inactivation of S-adenosylhomocysteine hydrolase: Mechanism and occurrence in vivo in disorders of purine nucleoside catabolism, in Borchardt RT, Creveling CR, Usdin E (eds): *The Biochemistry of S-adenosylmethionine and Related Compounds*. London, Macmillan Publishers, 1982, pp. 657–665.
39. Ueland PM: S-adenosylhomocysteine hydrolase from mouse liver. *Int J Biochem* 14:207–213, 1982.
40. Hershfield MS: Apparent suicide inactivation of human lymphoblast S-adenosylhomocysteine hydrolase by 2′-deoxyadenosine and adenine arabinoside. *J Biol Chem* 254:22–25, 1979.
41. Abeles RH, Tashjian AH Jr, Fish S: The mechanism of inactivation of S-adenosylhomocysteinase by 2′-deoxyadenosine. *Biochem Biophys Res Commun* 95:612–617, 1980.
42. Helland S, Ueland PM: The relation between the functions of 9-β-D-arabinofuranosyladenine as inactivator and substrate of S-adenosylhomocysteine hydrolase. *J Pharmacol Exp Ther* 218:758–763, 1981.
43. Helland S, Ueland PM: Inactivation of S-adenosylhomocysteine hydrolase by 9-β-D-arabinofuranosyladenine in intact cells. *Cancer Res* 42:1130–1136, 1982.
44. Helland S, Ueland PM: Reactivation of S-adenosylhomocysteine hydrolase activity in cells exposed to 9-β-D-arabinofuranosyladenine. *Cancer Res* 42:2861–2866, 1982.
45. Zimmerman TP, Wolberg G, Duncan GS, Elion GB: Adenosine analogues as substrates and

inhibitors of S-adenosylhomocysteine hydrolase in intact lymphocytes. *Biochemistry* 19:2252–2259, 1980.
46. Guranowski A, Montgomery JA, Cantoni GL, Chiang PK: Adenosine analogues as substrates and inhibitors of S-adenosylhomocysteine hydrolase. *Biochemistry* 20:110–115, 1981.
47. Hoffman JL: Biosynthesis of S-N^6-methyladenosylhomocysteine, an inhibitor of RNA methyltransferases. *J Biol Chem* 253:2905–2907, 1978.
48. Hoffman JL: The rate of transmethylation in mouse liver measured by trapping S-adenosylhomocysteine. *Arch Biochem Biophys* 205:132–135, 1980.
49. Borchardt RT, Patel UG, Bartel RL: Adenosine dialdehyde: A potent inhibitor of S-adenosylhomocysteine hydrolase, in Borchardt RT, Creveling CR, Usdin E (eds): *The Biochemistry of S-adenosylmethionine and Related Compounds*. London, Macmillan Publishers, 1982, pp. 645–652.
50. Ueland PM, Døskeland SO: An adenosine 3′:5′-monophosphate-adenosine binding protein from mouse liver. *J Biol Chem* 253:1667–1676, 1978.
51. Saebø J, Ueland PM: A study on the sequestration of adenosine and its conversion to adenine by the cyclic AMP-adenosine binding protein/S-adenosylhomocysteinase from mouse liver. *Biochim Biophys Acta* 587:333–340, 1979.
52. Ueland PM, Saebø J: Cyclic AMP-adenosine binding protein/S-adenosylhomocysteinase from mouse liver. *Biochim Biophys Acta* 585:512–526, 1979.
53. Ueland PM, Helland S: S-adenosylhomocysteinase from mouse liver. *J Biol Chem* 255:7722–7727, 1980.
54. Jakubowski H, Guranowski A: S-adenosylhomocysteinase from yellow lupin seeds: Stoichiometry and reactions of the enzyme adenosine complex. *Biochemistry* 20:6877–6881, 1981.
55. Ueland PM: Intracellular sequestration of adenosine in rat hepatocytes. *Acta Pharmacol Toxicol* 49(Suppl. 1):80, 1981.
56. Berne RM, Belloni L, Rubio R: Adenosine associated with dispersed hepatocytes. *Basic Res Cardiol* 76:377–379, 1981.
57. Duerre JA, Briske-Anderson M: Effect of adenosine metabolites on methyltransferase reactions in isolated rat liver. *Biochim Biophys Acta* 678:275–282, 1981.
58. Schütz W, Schrader J, Gerlach E: Different sites of adenosine formation in the heart. *Am J Physiol* 240:H963–H970, 1981.
59. Kredich NM, Hershfield MS, Johnston JM: Role of adenosine metabolism in methylation, in Usdin E, Borchardt RT, Creveling CR (eds): *Transmethylation*. New York, Elsevier North-Holland, 1979, pp 225–230.
60. Hoffman DR, Marion DW, Cornatzer WE, Duerre JA: S-adenosylmethionine and S-adenosylhomocysteine metabolism in isolated rat liver. *J Biol Chem* 255:10822–10827, 1980.
61. Schatz RA, Wilens TE, Sellinger OZ: Decreased transmethylation of biogenic amines after in vivo elevation of brain S-adenosyl-L-homocysteine. *J Neurochem* 36:1739–1748, 1981.
62. Green RD: Adenosine transport by a variant of C1300 murine neuroblastoma cells deficient in adenosine kinase. *Biochim Biophys Acta* 598:366–374, 1980.

DISCUSSION

DR. H. McILWAIN: In relation to further effects consequential to S-adenosylhomocysteine formation, we have found recently (Newman, McIlwain, and Reason) that addition of homocysteine to incubating neocortical tissues diminishes their cyclic AMP content. Preincubation of such tissues with both adenosine and homocysteine, however, permits unusually large responses to the later addition of other neurohumoral agents, including norepinephrine and histamine. We judge that, in the first case, adenosine is being sequestered; in the second, a reserve of AdoHcy has been formed and is supplying adenosine for activation of adenylate cyclase.

DR. UELAND: I think that the answer to your question is as follows: Addition of homocysteine (Hcy) alone will decrease adenosine content. The mechanism involves the condensation of adenosine with Hcy, through the action of AdoHcy hydrolase. When adenosine is added together with Hcy, this would lead to a large accumulation of AdoHcy. Dr. Zimmerman and co-workers have reported that high levels of AdoHcy inhibit cyclic AMP phosphodiesterase. We have recently obtained similar data [*FEBS Lett* 135:165, 1982].

DR. J. SCHRADER: Do you have recent estimates of the amount of free adenosine versus SAH-bound adenosine?

DR. UELAND: I would like to refer to our poster presented at this meeting. Briefly, the amount of adenosine associated with AdoHcy hydrolase (about 40 pmol/10^6 cells) was of the same order of magnitude as the amount of cellular adenosine not mobilized by extracellular adenosine deaminase. There is a problem here because adenosine bound to AdoHcy hydrolase liberates adenine when the complex is denatured. The conversion to adenosine depends on the denaturation procedure used, i.e., a large amount of adenine is formed in the presence of high concentrations of acid and small amounts when mild denaturation procedures were used. However, under the latter conditions, interfering enzymes, such as adenosine deaminase, are not inactivated. Despite these difficulties, it seems that the major fraction of adenosine in hepatocytes exposed to extracellular adenosine deaminase is associated with AdoHcy hydrolase.

DR. M.S. HERSHFIELD: Is the conversion of enzyme-bound adenosine to adenine reversible? Will ^{14}C adenosine exchange into enzyme-bound adenosine?

DR. UELAND: With isolated enzyme, we have demonstrated that ^{14}C adenine tightly bound to the enzyme, or more likely its precursor, is reconverted to ^{14}C adenosine. However, ^{14}C adenine added to the enzyme is not converted to ^{14}C adenosine.

DR. H. OSSWALD: I would like to make a comment on the intracellular binding of adenosine. According to your data shown in Table 10.3, you have a maximal sequestration of 25 pmol adenosine per mg protein of kidney tissue, which is 25 nmol per gm protein or 5 nmol per gm wet weight. This is very close to the tissue content we find in rat kidneys [Osswald et al., *Pfluegers Arch* 371:45, 1977]. In addition, we find an adenosine-binding protein in the dog kidney with a K_d of 11 nM at 0°C for adenosine [Sacher R, Osswald H: *Naunyn Schmidedebergs Arch Pharmacol* (Suppl.) 311:K49, 1980] that corresponds to your values. Therefore, it seems likely that more than 95% of the tissue adenosine is bound to the adenosine-binding proteins.

DR. C.I. THOMPSON: Is it possible that the stable adenosine-SAH hydrolase complex maintains significant hydrolytic activity? If not, how does one account for the apparent steady-state activity of the hydrolytic pathway under control conditions?

DR. UELAND: Yes, I feel it might be possible. There are data suggesting that the enzyme has four catalytic sites. Besides, there is no definite evidence that the sites sequestering adenosine are identical to the catalytic site.

DR. V. CHAGOYA DE SANCHEZ: Do you know the role of NAD in the stability of the complex? Is it possible that the redox state of NAD^+, NADH, is implicated in the liberation of adenosine?

DR. UELAND: The formation of the complex is probably associated with reduction of tightly bound NAD^+. I feel that further studies into the fate of NADH and adenosine bound to the enzyme should be carried out by workers well trained in chemistry. I am aware of no data on the possible relation between the redox state of NAD^+ and the sequestration of adenosine.

DR. R. BÜNGER: In considering a possible physiological role of the stable SAH-adenosine complex, it would seem of critical importance to know the association constants of the complex as well as the $t_{1/2}$ of the system. Is it known whether the adenosine that is sequestered as SAH-adenosine complex can dissociate quickly enough (say, within 30 to 60 sec) in sufficient amounts to contribute substantially to metabolic vasoregulation in, for example, heart or liver? Perhaps Dr. Schrader would also comment on this question.

DR. UELAND: The adenosine-AdoHcy hydrolase complex in isolated rat hepatocytes seems to have a $t_{1/2}$ of several hours. The turnover of this complex is an important question related to the possible biological implication of this phenomenon, and the question has recently been addressed by Belloni et al. and in our laboratory. The long turnover of the intracellular complex may be related to the experimental conditions used. It may be possible that complexes with different $t_{1/2}$ exist, as was recently shown for adenosine binding to purified AdoHcy hydrolase from plant. The HPLC gel-filtration technique used by us may be limited to the detection of complexes characterized by long $t_{1/2}$.

DR. SCHRADER: The reported *in vitro* rate of dissociation of the adenosine-SAH complex is rather slow ($t_{1/2}$ = 2.5 hr). It therefore seems rather unlikely that adenosine released from this complex can contribute to metabolic control of flow on a short-term basis.

11. S-ADENOSYLHOMOCYSTEINE HYDROLASE AS A TARGET IN GENETIC AND DRUG-INDUCED DEFICIENCY OF ADENOSINE DEAMINASE

MICHAEL S. HERSHFIELD

Several participants in this conference will discuss the physiological regulatory roles that adenosine (Ado) may serve, beyond its status as an intermediate in the interconversion of purine compounds. Along with others at this symposium, I have been primarily concerned with the nonphysiologic, cytotoxic effects of Ado and 2′-deoxyAdo (dAdo) that occur *in vivo* when their major route of metabolism, deamination, is blocked. Interest in the cytotoxic, and particularly the lymphocytotoxic, effects of these nucleosides stems from the discovery, made 10 years ago by Giblett and her colleagues [1], that genetic deficiency of Ado deaminase (ADA) is associated with selective depletion of lymphoid tissue, manifest clinically as severe combined immunodeficiency disease (SCID). In addition, there is now also considerable interest in the use of inhibitors of ADA, alone and in combination with Ado analogues such as adenine arabinoside (Ara-A), as rational therapies for lymphoproliferative diseases and as immunosuppressive agents.

METABOLISM OF Ado AND dAdo

Ado and dAdo arise as transient intermediates in the catabolism or interconversion of adenine nucleotides and nucleic acids. (Dr. Nicholas Kredich and I have recently reviewed genetic ADA deficiency [2]; that review may be used as a source of references for statements made here regarding Ado and dAdo metabolism). Ado is also derived from S-adenosylmethionine (AdoMet) after its conversion in transmethylation reactions to S-adenosylhomocysteine

R.M. Berne, T.W. Rall, and R. Rubio (eds.), Regulatory Function of Adenosine, *pp. 171–178.*

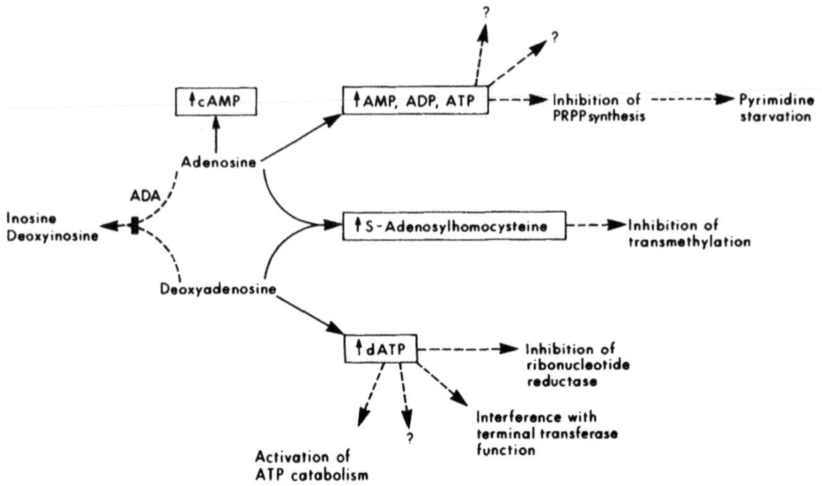

Figure 11.1. Potential mechanisms of cytotoxicity in adenosine deaminase deficiency. Source: adapted from [2].

(AdoHcy). Hydrolysis of AdoHcy, catalyzed by AdoHcy hydrolase, which is estimated to produce 14 to 23 mmol of Ado daily in adults [3], may be a major metabolic source of Ado and is the only source of homocysteine. Under normal conditions, the plasma concentration of Ado is about 0.3 to 0.5 μM, and dAdo no higher than one-tenth of this level. At these concentrations there is probably appreciable rephosphorylation of Ado to AMP, catalyzed by Ado kinase, which has a K_m for Ado of around 0.5 to 1 μM. ADA has a higher K_m for Ado, 24 to 75 μM. Nevertheless, there is also likely to be considerable deamination of Ado, since in many tissues, and especially in lymphoid cells, there is far more ADA than Ado kinase activity. In contrast, dAdo is probably metabolized exclusively by deamination in normal individuals. In heritable ADA deficiency plasma levels of Ado are elevated to the 2 to 7 μM range; dAdo is rarely higher than 1 μM, and the highest value reported has been about 2 μM. It is these levels that produce the strikingly selective depletion of all lymphoid cells observed in the 85% of affected children who have the most complete enzyme deficiency. In patients who have a few percent of normal ADA activity, there is a severe defect in cell-mediated (T cell) immunity, but B cells and humoral immune function are relatively spared. A few children with slightly higher residual ADA activity have been reported to have completely normal immune function.

ACCUMULATION OF dATP AND dAdo TOXICITY TO T AND B LYMPHOBLASTS

A number of specific actions of Ado and dAdo have been identified over the past decade that may contribute to their cytotoxicity (Fig. 11.1). The most potent cytotoxicity is attributable to expansion of the intracellular dATP pool [4–6]. Two enzymes, Ado kinase and deoxycytidine kinase, are capable of phosphorylating dAdo, although much less efficiently than the substrates for which they are named. We have attempted to determine the relative contributions of these enzymes to dAdo nucleotide accumulation in intact lymphoblasts by examining the effects of mutational loss of these activities [7–9]. In the case of the CEM human T cell line [9], we found that deoxycytidine kinase was responsible for dAdo phosphorylation at the lowest concentrations of dAdo that inhibit growth (0.5–2 μM). Loss of Ado kinase or deoxycytidine kinase diminished dAdo toxicity to CEM cells by <2- and 3-fold, respectively. Loss of both kinase activities abolished dATP accumulation and diminished sensitivity to dAdo by 100-fold. The importance of the accumulation of dATP as a cause of toxicity to T lymphoblasts *in vivo* was also made apparent during a recent study of a patient with refractory T cell leukemia who underwent treatment with the potent ADA inhibitor dCF [10]. In the early phase of treatment, this patient's circulating malignant cells accumulated large amounts of dATP and were sensitive to therapy. A remission was induced, during which dCF was administered every three weeks. After three such treatments, leukemic cells reappeared and were relatively resistant to treatment. The resistant cells accumulated <10% as much dATP *in vivo* than was originally observed.

We have also examined the consequences of loss of deoxycytidine kinase and Ado kinase to dATP accumulation in the WI-L2 human B lymphoblastoid cell line [7, 8]. As with B cell lines in general, WI-L2 is far less efficient than T cell lines in trapping dAdo and is thus relatively resistant to dAdo toxicity. WI-L2 cell extracts contain both deoxycytidine and Ado kinases, and in about the same relative amounts as CEM [9]. However, with WI-L2, loss of deoxycytidine kinase had virtually no effect on dATP accumulation from dAdo or on dAdo toxicity; loss of Ado kinase essentially eliminated dATP accumulation and diminished dAdo toxicity by 3- to 4-fold, to about the same extent as loss of both deoxycytidine kinase and Ado kinase [7, 8]. These results are consistent with the operation in WI-L2 of an inhibitor of dAdo phosphorylation by deoxycytidine or, as has been suggested [11–14], an active dAdo nucleotide catabolizing enzyme.

AdoHcy HYDROLASE AS A TARGET IN ADA DEFICIENCY

In contrast to the high degree of resistance that is conferred by loss of ability to phosphorylate dAdo, we observed several years ago that loss of Ado kinase had little if any effect on the toxicity of Ado to ADA-inhibited WI-L2 lymphoblasts [15]. To account for this observation, we suggested that the "nucleotide-independent" toxicity might be mediated by a cytoplasmic protein with high affinity for Ado. About this time Kredich and Martin [16] had suggested that some of the toxicity of Ado to ADA-deficient cells might arise from inhibition of transmethylation because of the ability of Ado to inhibit the hydrolysis of AdoHcy, which is a consequence of the unfavorable equilibrium for this reaction [17]. Dr. Kredich and I found that the major cytoplasmic Ado-binding protein from human placenta copurified with AdoHcy hydrolase [18]. In addition to the affinity of AdoHcy hydrolase for stable Ado binding (K_d of $2.5 \times 10^{-7} M$), its K_m for Ado in the AdoHcy synthesis reaction was about 1 μM, comparable to that of Ado kinase. In other studies we showed that accumulation of AdoHcy, causing inhibition of nucleic acid methylation, could account for the toxicity of Ado to Ado kinase-deficient human lymphoblasts [19]. When ADA was inhibited, homocysteine in the absence of exogenous Ado was selectively toxic to lymphoblasts that lacked Ado kinase. This toxicity was caused by AdoHcy accumulation due to the condensation of homocysteine with endogenously produced Ado.

A second property of AdoHcy hydrolase also makes this enzyme a unique target in ADA deficiency. We found that the enzyme is rapidly inactivated by dAdo and by its analogues, notably, adenine arabinoside [20]. The inactivation of the purified enzyme by these nucleosides was irreversible, with kinetics that suggested a suicide-like mechanism. Inactivation is associated with conversion of the enzyme-bound cofactor, NAD [21], to NADH, with concomitant cleavage of the inactivating nucleoside at the purine-glycosidic bond, releasing adenine [22–25]. Evidence of inactivation by dAdo *in vivo* is the finding that erythrocytes of three children with ADA deficiency had $<2\%$ of the normal activity of AdoHcy hydrolase [26]. The enzyme also becomes extensively inactivated in erythrocytes and lymphoblasts of patients undergoing treatment for lymphoid malignancies with dCF or with dCF plus Ara-A [10, 27]. Low levels of red cell AdoHcy hydrolase occur in children with genetic deficiency of purine nucleoside phosphorylase [28, 29]. Inosine, which accumulates to levels as high as 100 μM in the plasma of these children, is also a weak inactivator of AdoHcy hydrolase, which probably explains their secondary deficiency of erythrocyte AdoHcy hydrolase [29].

Although AdoHcy hydrolase inactivation of $>98\%$ did cause accumulation of AdoHcy to toxic levels in cultured WI-L2 lymphoid cells, hydrolase inactivation was only a minor contributor to the toxicity of dAdo to these cells [7]. *In vivo,* however, the inactivating effect of dAdo (or Ara-A) would be synergistic with the inhibitory effect of Ado in causing AdoHcy accumulation. Continued studies of the biochemical consequences of treatment with dCF, alone or with Ara-A, should provide an opportunity for examining the combined effects of these actions on cellular AdoHcy levels and on specific transmethylation reactions that might be critical to cell viability or function.

We have recently discovered another intriguing link between AdoHcy hydrolase and ADA, namely, that the human genes for these enzymes both reside on the same chromosome [30]. We mapped the hydrolase gene to chromosome 20 by the human–Chinese hamster cell hybrid technique, using a mouse monoclonal antibody to human AdoHcy hydrolase to identify the product of the human gene. Certain observations suggest that the coincidence of the

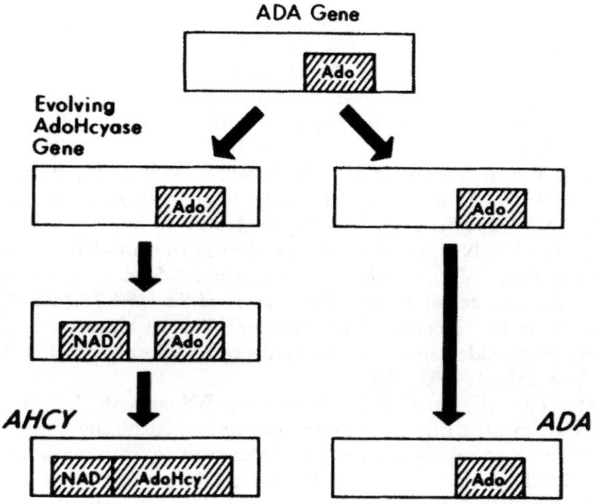

Figure 11.2. The gene for AdoHcy hydrolase (*AHCY*) may have evolved from the gene for adenosine deaminase (*ADA*). The hatched boxes indicate nucleotide sequences that code for amino acids of the binding domains for adenosine (ADO), NAD^+ (the tightly bound cofactor for AdoHcy hydrolase) [21], and AdoHcy.

AdoHcy hydrolase and ADA genes on the same chromosome may be nonrandom. First, it is likely that AdoHcy hydrolase arose later in evolution than ADA, since the hydrolase is only found in eukaryotic cells, whereas ADA occurs in both prokaryotes and eukaryotes. Second, because of the effects of Ado and dAdo, it is unlikely that AdoHcy hydrolase activity could have arisen in a cell that lacked ADA activity. Third, both AdoHcy hydrolase and ADA have binding domains for Ado. Finally, the subunit MW's for the two enzymes are similar. Linkage of the ADA and evolving AdoHcy hydrolase could have ensured the presence of ADA activity during the long time required for evolution of a catalytically efficient hydrolase. It is also possible that the hydrolase gene arose from the ADA gene, with conservation of nucleotide sequences that code for the Ado-binding domain of ADA (Fig. 11.2). To evaluate these possibilities further will require direct comparisons of amino acid sequences for these proteins and ultimately of the nucleotide sequences of their structural genes.

REFERENCES

1. Giblett ER, Anderson JE, Cohen F, Pollara B, Meuwissen HJ: Adenosine deaminase deficiency in two patients with severely impaired cellular immunity. *Lancet* 2:1067–1069, 1972.
2. Kredich NM, Hershfield MS: Immunodeficiency diseases associated with adenosine deaminase deficiency and purine nucleoside phosphorylase deficiency, in Stanbury JB, Wyngaarden JB, Fredrickson DS, Goldstein J, Brown M (eds): *The Metabolic Basis of Inherited Disease,* 5th ed. New York, McGraw-Hill, pp 1157–1183, 1983.
3. Mudd HS, Poole JR: Labile methyl balances for normal humans on various dietary regimens. *Metabolism* 24:721–732, 1975.

4. Coleman MS, Donofrio J, Hutton JJ, Daoud A, Lampkin B, Dyminski, J: Abnormal concentrations of deoxynucleotides in adenosine deaminase (ADA) deficiency and severe combined immunodeficiency disease (SCID). *Blood* 50(Suppl.1):292, 1977 (abst).
5. Coleman MS, Donofrio J, Hutton JJ, Hahn L, Daoud A, Lampkin B, Dyminski J: Identification and quantification of adenine deoxynucleotides in erythrocytes of a patient with adenosine deaminase deficiency and severe combined immunodeficiency. *J Biol Chem* 253:1619–1626, 1978.
6. Cohen A, Hirschhorn R, Horowitz SD, Rubinstein A, Polmar SH, Hong R, Martin DW Jr: Deoxyadenosine triphosphate as a potentially toxic metabolite in adenosine deaminase deficiency. *Proc Natl Acad Sci USA* 75:472–476, 1978.
7. Hershfield MS, Kredich NM: Resistance of an adenosine kinase-deficient human lymphoblastoid cell line to effects of deoxyadenosine on growth, S-adenosylhomocysteine hydrolase inactivation, and dATP accumulation. *Proc Natl Acad Sci USA* 77:4292–4296, 1980.
8. Ullman B, Levinson BB, Hershfield MS, Martin DW Jr: A biochemical genetic study of the role of specific nucleoside kinases in deoxyadenosine phosphorylation by cultured human cells. *J Biol Chem* 256:848–852, 1981.
9. Hershfield MS, Fetter JE, Small WC, Bagnara AS, Williams SR, Ullman B, Martin DW Jr, Wasson DB, Carson DA: Effects of mutational loss of adenosine kinase and deoxycytidine kinase on deoxyATP accumulation and deoxyadenosine toxicity in cultured CEM human T-lymphoblastoid cells. *J Biol Chem* 257:6380–6386, 1982.
10. Hershfield MS, Kredich NM, Falletta JM, Kinney TR, Mitchell BS, Koller CA. Effects of deoxyadenosine and adenine arabinoside during deoxycoformycin therapy. *Proc Am Assoc Cancer Res* 22:157, 1981 (abst).
11. Carson DA, Kaye J, Matsumoto S, Seegmiller JE, Thompson L: Biochemical basis for the enhanced toxicity of deoxyribonucleosides toward malignant human T cell lines. *Proc Natl Acad Sci USA* 76:2430–2433, 1979.
12. Wortmann RL, Mitchell BS, Edwards NL, Fox IH: Biochemical basis for differential deoxyadenosine toxicity to T and B lymphoblasts: Role for 5′nucleotidase. *Proc Natl Acad Sci USA* 76:2434–2437, 1979.
13. Carson DA, Kaye J, Wasson DB: The potential importance of soluble deoxynucleotidase activity in mediating deoxyadenosine toxicity in human lymphoblasts. *J Immunol* 126:348–352, 1981.
14. Fox RM, Piddington SK, Tripp EH, Tattersall MHN: Ecto-ATPase deficiency in cultured human T and null leukemic lymphocytes: A biochemical basis for thymidine sensitivity. *J Clin Invest* 68:544–552, 1981.
15. Hershfield MS, Snyder FF, Seegmiller JE: Adenine and adenosine are toxic to human lymphoblast mutants defective in purine salvage enzymes. *Science* 197:1284–1287, 1977.
16. Kredich NM, Martin DW Jr: Role of S-adenosylhomocysteine in adenosine-mediated toxicity in cultured mouse T lymphoma cells. *Cell* 12:931–938, 1977.
17. De La Haba G, Cantoni GL: The enzymatic synthesis of S-adenosyl-L-homocysteine from adenosine and homocysteine. *J Biol Chem* 234:603–608, 1959.
18. Hershfield MS, Kredich NM: S-adenosylhomocysteine hydrolase is an adenosine-binding protein: A target for adenosine toxicity. *Science* 202:757–760, 1978.
19. Kredich NM, Hershfield MS: S-adenosylhomocysteine toxicity in normal and adenosine kinase-deficient lymphoblasts of human origin. *Proc Natl Acad Sci USA* 76:2450–2454, 1979.
20. Hershfield MS: Apparent suicide inactivation of human lymphoblast S-adenosylhomocysteine hydrolase by 2′-deoxyadenosine and adenine arabinoside: A basis for direct toxic effects of analogs of adenosine. *J Biol Chem* 254:22–25, 1979.
21. Palmer JL, Abeles RH: Mechanism for enzymatic thioether formation: Mechanism of action of S-adenosylhomocysteinease. *J Biol Chem* 251:5817–5819, 1976.
22. Hershfield MS: Alternate reactions for S-adenosylhomocysteine hydrolase. *Fed Proc* 39:1858, 1980 (abst).
23. Abeles RH, Tashjian AH, Fish S: The mechanism of inactivation of S-adenosylhomocysteinease by 2′-deoxyadenosine. *Biochem Biophys Res Commun* 95:612–617, 1980.
24. Helland S, Ueland PM: The relation between the functions of 9-β-D-arabinofuranosyladenine as inactivator and substrate of S-adenosylhomocysteine hydrolase. *J Pharmacol Exp Ther* 218:758–763, 1981.
25. Hershfield MS, Small WC, Premakumar R, Bagnara AS, Fetter JE: Inactivation of S-

adenosylhomocysteine hydrolase: Mechanism and occurrence in vivo in disorders of purine nucleoside catabolism, in Borchardt RT, Usdin E, Creveling CR (eds): *The Biochemistry of S-adenosylmethionine and Related Compounds*. New York, Macmillan, 1982 (forthcoming).

26. Hershfield MS, Kredich NM, Ownby DR, Ownby H, Buckley R: In vivo inactivation of erythrocyte S-adenosylhomocysteine hydrolase by 2′deoxyadenosine in adenosine deaminase-deficient patients. *J Clin Invest* 63:807–811, 1979.
27. Mitchell BS, Koller CA, Heyn R: Inhibition of adenosine deaminase activity results in cytotoxicity to T lymphoblasts in vivo. *Blood* 56:556–559, 1980.
28. Kaminska JE, Fox IH: Decreased S-adenosylhomocysteine hydrolase in inborn errors of purine metabolism. *J Lab Clin Med* 96:141–147, 1980.
29. Hershfield MS: Proposed explanation for S-adenosylhomocysteine hydrolase deficiency in purine nucleoside phosphorylase and hypoxanthine guanine phosphoribosyltransferase deficient patients. *J Clin Invest* 67:696–701, 1981.
30. Hershfield MS, Francke U: The human genes for S-adenosylhomocysteine hydrolase and adenosine deaminase are syntenic on chromosome 20. *Science* 216:739–742, 1982.

DISCUSSION

DR. J. P. R. M. VAN LAARHOVEN: You have shown us the experiments in T cell lines in which the AdoMet/AdoHcy ratio dropped when the cells were cultured with deoxyadenosine. Also, the mutants were derived from a T cell line. Do you have information on similar experiments performed in B cell lines? I wonder whether the mechanism discussed is responsible or at least contributes to the differential sensitivity that exists between T and B cells toward deoxyadenosine.

DR. HERSHFIELD: Yes, we have obtained similar kinds of results with B cell lines. To what extent these cell lines resemble nonmalignant or nontransformed T and B cells is always a question. Nevertheless, the T cell defect in ADA deficiency is probably best explained by the toxic effects of dATP accumulation. Cultured B cell lines do not trap dAdo as dATP very efficiently. Whether or not impaired catabolism of AdoHcy contributes to the immunologic defect *in vivo* is difficult to say. There may be differences among cells in their sensitivity to Ado- and dAdo-induced AdoHcy hydrolase inhibition or to the effects of AdoHcy accumulation. For example, cells with low AdoMet pools could be more sensitive to a rise in AdoHcy concentration. There may be a more stringent dependence in some cells for a particular transmethylation reaction that is very sensitive to inhibition when the ratio of AdoMet/AdoHcy falls. However, at present there is no evidence that this is the case. I do suspect that the hepatic and central nervous system toxicities seen in some patients undergoing treatment with dCF, or with dCF plus Ara-A, may be due to inhibition of methylation reactions. It is more likely that the information needed to evaluate these possibilities will be gained through study of dCF-treated patients than through study of children with the rare genetic ADA deficiency, who lack lymphocytes at the time of diagnosis.

DR. J.E. SEEGMILLER: Have you been able to identify the basis for the resistance to deoxycoformycin that developed in your patient?

DR. HERSHFIELD: No, we have not. We have excluded several possibilities, including a nucleoside (dAdo, dCF) transport defect, alteration in ADA activity (diminished sensitivity to dCF, or ADA overproduction), or any alteration in the activities of deoxycytidine kinase and adenosine kinase, the enzymes that phosphorylate dAdo.

DR. H. OSSWALD: I have a question concerning the renal function of your patients treated with deoxycoformycin. Did you see any change in glomerular filtration rate or in plasma renin activity?

DR. HERSHFIELD: Renal function was not impaired by treatment with dCF in the patient that we studied. However, renal toxicity has occurred during dCF therapy in some individuals. This complication can be very dangerous, since Ado and dAdo are excreted very efficiently by the kidney, and loss of renal function can elevate plasma levels of these nucleosides to the point where generalized toxicity may occur, particularly involving the liver and the central nervous system.

12. NUCLEOSIDE TRANSPORT IN MAMMALIAN CELLS AND INTERACTION WITH INTRACELLULAR METABOLISM

PETER G.W. PLAGEMANN, ROBERT M. WOHLHUETER

The salvage pathways for nucleosides in animal cells in their simplest form consist of two steps: facilitated diffusion into the cell and subsequent intracellular phosphorylation (Fig. 12.1). Phosphorylation effectively activates and traps the nucleoside intracellularly. This simple pathway describes the salvage of uridine and thymidine in most cultured cells, because these cells generally lack thymidine phosphorylase activity and uridine phosphorolysis is slow relative to the rates of transport and phosphorylation, if the cells possess uridine phosphorylase at all.

Uridine and thymidine kinase-deficient variants have been isolated from a number of established animal cell lines, which made it possible to analyze the transport of uridine and thymidine unimpeded by intracellular metabolism. The transport of these nucleosides in cultured cells has also been measured in cells in which phosphorylation was blocked by depleting the cells of ATP and in human erythrocytes that lack kinases and phosphorylases for uridine and thymidine [1]. The availability of these cell systems has led to the use of uridine and thymidine as the main model substrates for investigating the kinetic properties of the nucleoside transporter and the interaction between the

We thank John Erbe for excellent technical assistance, Colleen O'Neill for competent secretarial work, and Dr. H. E. Machamer, Warner/Lambert, Parke-Davis Co., for a gift of dCF (Pentostatin). The experimental work reported was supported by USPHS Research Grant GM24468.

R.M. Berne, T.W. Rall, and R. Rubio (eds.), Regulatory Function of Adenosine, *pp. 179–201.*

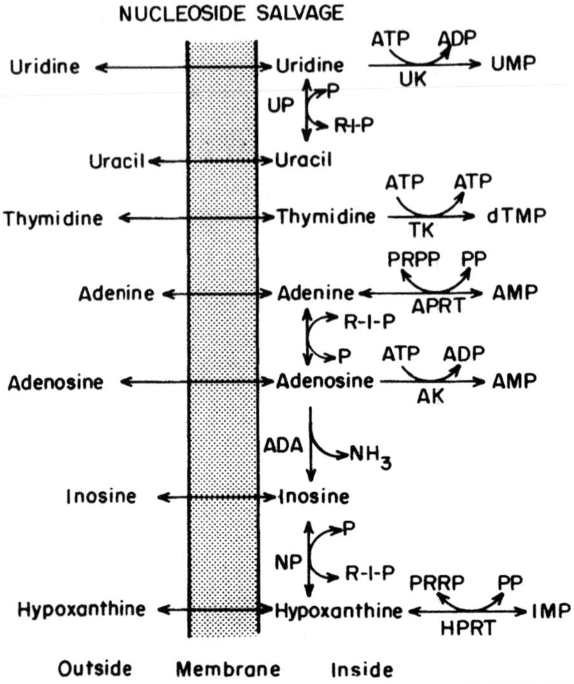

Figure 12.1. Salvage pathways for uridine, thymidine, and adenosine in mammalian cells. UK, uridine kinase; UP, uridine phosphorylase; TK, thymidine kinase; AK, adenosine kinase; APRT, adenine phosphoribosyltransferase; ADA, adenosine deaminase; NP, nucleoside phosphorylase; HPRT, hypoxanthine phosphoribosyltransferase; PRPP, phosphoribosyl pyrophosphate; R-1-P, ribose-1-phosphate.

transport and phosphorylation steps in salvage (for a detailed presentation, see [2–5]).

Similar studies with adenosine, however, have been complicated by the multitude of enzymes involved in its salvage (Fig. 12.1). Besides being directly phosphorylated, adenosine is rapidly deaminated to inosine, and inosine is phosphorolyzed to hypoxanthine, which in turn is phosphoribosylated. In addition, adenosine seems also to be a substrate, though an inefficient one, for a phosphorylase, perhaps 5′-methylthioadenosine phosphorylase (see later discussion). Furthermore, all unphosphorylated intermediates are subject to exit transport. Nevertheless, all information available at present indicates that adenosine is transported into cells by the same carrier that transports uridine, thymidine, and other ribo- and deoxyribonucleosides.

In this chapter, we will first discuss methodologies for measuring nucleoside transport and the kinetic and other properties of the nucleoside transporter of animal cells and then briefly examine the interrelationship between transport and intracellular metabolism in the salvage of nucleosides, with specific emphasis on uridine and adenosine.

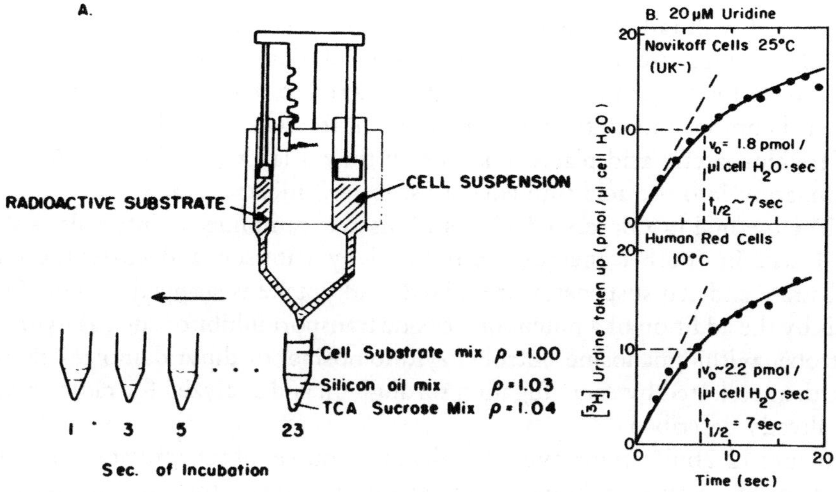

Figure 12.2. A: Schematic view of the rapid mixing/sampling procedure described in the text. The longest time sample is dispensed first; the countdown ends when the centrifuge is started at 0 sec. The trichloroacetic acid phase (TCA sucrose) is optional where termination of cellular metabolism is necessary. B: Typical initial time courses of accumulation of 20 μM [^{3}H]uridine in the zero-*trans* mode by UK-deficient Novikoff cells and human erythrocytes (for details, see [2]).

TRANSPORT METHODOLOGY: RAPID KINETIC TECHNIQUES

Initial studies with human erythrocytes and cultured cells in which uridine and thymidine phosphorylation was blocked genetically or by ATP depletion demonstrated that the uridine and thymidine permeation is energy-independent, nonconcentrative, and saturable; hence, transport occurred by facilitated diffusion. Transport was found to be very rapid; at concentrations in the micromolar range, uridine and thymidine equilibrate across the plasma membrane within less than 1 min at 20°–25°C. This rapidity of transport complicates its kinetic analysis and necessitates the development of methods that allow the measurement of intracellular or extracellular concentrations of substrate over time periods of a few seconds. We have developed one such method (Fig. 12.2A), which is based on dual-syringe devices used in stopped-flow enzyme kinetics [6]. The method consists of mixing constant aliquots of a solution of radiolabeled substrate and of a cell suspension by means of a hand-operated, dual-syringe apparatus, according to an appropriate time schedule. The effluent from the mixing chamber is directly dispensed serially into 12 centrifuge tubes that have been placed in a microcentrifuge and that contain an oil mixture somewhat denser than water. In influx measurements, the substrate solution is placed in the smaller syringe and the cell suspension in the larger syringe. After the last mixture is dispensed, the centrifuge is started, and within 2 sec entry is effectively stopped as the cells are separated from the extracellular fluid by sedimentation beneath the oil. The cells are collected in a pellet and are analyzed for radioactivity (for additional details, see [6]). Substrate efflux can be measured in a similar manner, except that the cells are preloaded with radiolabeled substrate, and the preloaded cells, placed into the smaller syringe, are then mixed with buffer solution that is devoid of substrate present in the larger syringe. When it is desired to fractionate and analyze the labeled components of the intracellular pool, the cells are collected into a layer of trichloroacetic acid placed underneath the oil layer (see Fig. 12.2A). Centrifugation into the acid immediately stops cell metabolism.

The method just described allows 12 discrete samplings at intervals as short as 1 sec. In another method, introduced by Paterson and co-workers [5], substrate and cell suspension are mixed, and uptake is stopped at short intervals by the addition of a potent nucleoside transport inhibitor such as hydroxynitrobenzylthioguanosine, nitrobenzylthioinosine, or dipyridamole. The cells are then collected by centrifugation through oil and analyzed for radioactivity, as already described.

Figure 12.2B illustrates typical initial time courses of accumulation of 20 μM uridine to transmembrane equilibrium obtained by the dual-syringe methodology. The results are for a mutant of Novikoff rat hepatoma cells deficient in uridine kinase and for human red cells, at 25°C and 10°C, respectively. These time courses illustrate the rapidity of nucleoside transport in mammalian cells; the half-time for attainment of transmembrane equilibrium under these conditions is about 7 sec. In other words, by 7 sec the entry velocity has decreased by 50%. This fact limits the accuracy of graphical estimation of initial transport velocities. The initial linear phase is so short that graphical analysis tends to underestimate the true initial velocity. We have approached this problem by estimating initial velocities through analysis of the entire time course of attainment of transmembrane equilibrium in terms of integrated rate equations based on various transport models [2]. We have fitted such equations by computer and have calculated initial transport velocities as the zero time slope of the fitted curves, as illustrated in Figure 12.2B.

THE SIMPLE CARRIER MODEL

Analysis of numerous time courses of nucleoside accumulation by and release from both cultured cells and human erythrocytes indicates that these time courses are satisfactorily described by the simple carrier model, which is summarized in Figure 12.3, along with two crucial experimental protocols needed to test the applicability of the model. The simple carrier model has been analyzed mathematically in a number of ways (for reviews, see [7, 8]). We prefer and have used the nomenclature and rate equations developed by Stein and collaborators [8] because of their relative simplicity and adaptability to our experimental protocols. The integrated forms of the rate equations developed by these investigators and details of the procedures for fitting these equations to time courses of substrate accumulation and release have been described previously [2, 6]. We have arbitrarily designated the outside and inside faces of the plasma membrane as sides 1 and 2, respectively.

The simple carrier is defined by five fundamental constants: four resistance

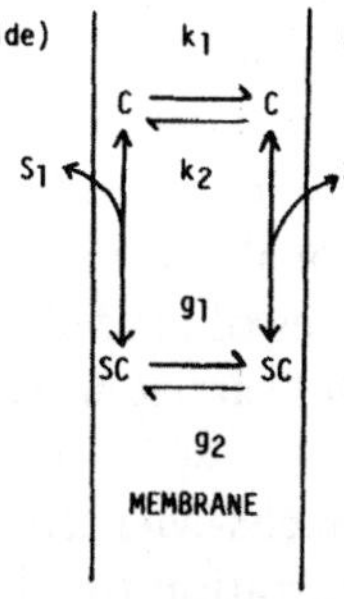

Protocol	Direction	Face 1 (outside)	Face 2 (inside)
Zero-trans (zt)	12	S_1^* varied	At $t = 0$, $S_2 = 0$
	21	At $t = 0$, $S_1 = 0$	S_2^* varied (cell preloaded)
Equilibrium exchange (ee)	12	S_1^* varied	S_2 varied as S_1^* (cells preloaded)
	21	S_1 varied as S_2^*	S_2^* varied (cell preloaded)

	Direction 12	21	
$R_{12} = 1/V^{zt}$	Loaded	Empty	$K^{zt} = (R_{oo}/R_{12})K$
$R_{21} = 1/V^{zt}$	Empty	Loaded	$K^{zt} = (R_{oo}/R_{21})K$
$R_{ee} = 1/V^{ee}$	Loaded	Loaded	$K^{ee} = (R_{oo}/R_{ee})K$
R_{oo}	Empty	Empty	$R_{oo} + R_{ee} = R_{12} + R_{21}$

K = Intrinsic dissociation constant

$R_{21} = R_{12}$ Directional symmetry

$R_{oo} = R_{ee}$ Equal mobility of loaded and empty carrier

$R_{oo} > R_{ee}$ Loaded carrier moves more rapidly than empty carrier

Figure 12.3. The simple carrier model and its assessment in various experimental protocols. The nomenclature is that of Stein [8]. S_1, extracellular concentration of substrate; S_2, intracellular concentration of substrate; S*, radiolabeled substrate; C, carrier; k_1, k_2, and g_1, g_2 are the rate constants for the "movement" of the empty and substrate-loaded carrier, respectively. K^{ee} and K^{zt} are the Michaelis-Menten constants for equilibrium exchange and zero-*trans* flux, respectively, and V^{ee} and V^{zt} denote the corresponding maximum velocities. R = resistance factor, as discussed in the text.

terms, R_{12}, R_{21}, R_{ee}, and R_{oo}, and an intrinsic dissociation constant, K [2, 8]. The R terms are reciprocally related to the rate constants associated with various elements of the model (k_1, k_2, g_1, g_2) and might be understood in simple terms as the resistance encountered by the carrier when moving empty (C) or substrate-loaded (CS) from outside-in or inside-out, where carrier mobility connotes some macromolecular movement, if only a conformational shift, and not necessarily carrier translocation. In other words, the R terms are proportional to the time it takes for a round trip of the carrier in four modes: (1) loaded on the inbound trip and empty on the outbound trip (R_{12}); (2) empty on the inbound trip and loaded on the outbound trip (R_{21}); (3) loaded in both directions (R_{ee}); and (4) empty in both directions (R_{oo}). The R terms are reciprocals of the maximum velocities measurable in three experimental protocols: zero-*trans* entry, zero-*trans* exit, and equilibrium exchange (Fig. 12.3). In zero-*trans* entry, the accumulation of substrate by the cells is measured at various extracellular concentrations (S_1) under conditions in which the intracellular concentrations of substrate (S_2) at $t = 0$ is zero. In zero-*trans* exit, the cells are preloaded with various concentrations of substrate, and its loss into the medium initially free of substrate is measured. In equilibrium exchange, the equilibration of radiolabeled substrate is measured in either direction, with the absolute concentration of substrate equal on both faces of the membrane. R_{oo} cannot be measured but is determined by the other three R terms ($R_{12} + R_{21} = R_{ee} + R_{oo}$). If substrate influx and efflux are equal, $R_{12} = R_{21}$, and the carrier is said to exhibit directional symmetry. If equilibrium exchange flux equals zero-*trans* flux, the mobilities of loaded and empty carrier are equal ($R_{oo} = R_{ee}$). $R_{oo} > R_{ee}$ implies that the carrier moves faster when loaded with substrate than when empty.

K is the limit Michaelis-Menten constant of the carrier, which expresses the half-saturation concentration for the hypothetical situation in which substrate is present on neither face of the membrane. It is related to the Michaelis-Menten constants measurable in the zero-*trans* and equilibrium exchange protocols, as shown in Figure 12.3.

All nucleoside transport data available at present for both cultured mammalian cells and human erythrocytes are in compliance with the simple carrier model. Furthermore, as documented in the next section, the nucleoside transporter of both types of cells exhibits directional symmetry, but they differ in their relative mobility when loaded and empty. For cultured cells, $R_{oo} = R_{ee}$. In contrast, $R_{oo} > R_{ee}$ for the erythocyte nucleoside transporter, which indicates that the loaded carrier moves more rapidly than the empty carrier, that is, the zero-*trans* entry or exit of nucleoside is limited by the rate of return of the empty carrier.

COMPARISON OF ZERO-*TRANS* INFLUX AND EFFLUX AND OF ZERO-*TRANS* AND EQUILIBRIUM EXCHANGE FLUX, AND KINETICS OF TRANSPORT

Zero-*trans* influx and efflux of uridine (and of other nucleosides) in both cultured cells and human erythrocytes is about equal; thus, $R_{12} = R_{21}$ (Fig. 12.4) [9, unpublished observations]. The transporters of both types of cells therefore exhibit directional symmetry. Our results for human red cells do not agree with a previous report by Cabantchik and Ginsburg [10], which indicated that uridine efflux exceeded influx severalfold, but the reasons for this discrepancy have not been resolved. The transporters of cultured cells and erythrocytes, however, clearly differ in their relative mobility when loaded and empty. This conclusion is indicated by comparing zero-*trans* and equilibrium exchange flux at concentrations well above the Michaelis-Menten constant for transport (Fig. 12.4). The initial velocities of zero-*trans* entry and inward equilibrium exchange of uridine in cultured Novikoff cells are about the same, i.e., $R_{oo} = R_{ee}$ (Fig. 12.4A), and equality of zero-*trans* and equilibrium exchange fluxes has also been demonstrated for other types of cultured cells [2, 9]. In contrast, the equilibrium exchange in red blood cells is much more rapid than zero-*trans* flux (Fig. 12.4B). This effect of the presence of substrate on both sides of the membrane on uridine flux in red cells, also known as *trans*-stimulation, was first described by Paterson and his collaborators [11, 12] and has also been confirmed by Cabantchik and Ginsburg [10]. As previously explained, these findings indicate that the movement of the nucleoside carrier of cultured cells

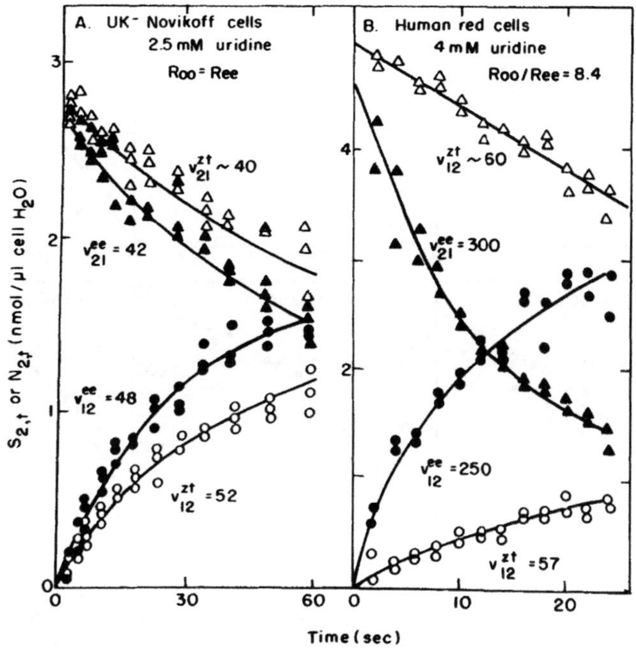

Figure 12.4. Zero-*trans* influx (○-○) and efflux (△-△) and inward (●-●) and outward (▲-▲) equilibrium exchange of uridine in UK-deficient Novikoff rat hepatoma cells (A) and human erythrocytes (B) at 24–25°C. Data for human red cells are from Plagemann et al. J Biol Chem 257:12069–12074, 1982. The experiment with Novikoff cells was conducted in the same manner [previously unpublished results]. The initial velocities for zero-*trans* influx (ν_{12}^{zt}) and efflux (ν_{21}^{zt}) and equilibrium exchange (ν^{ee}) are in pmol/µl cell $H_2O \cdot$ sec.

is the same whether or not it is loaded with substrate, i.e., $R_{oo} = R_{ee}$, whereas the red cell transporter moves more rapidly when loaded than when empty. The ratio R_{oo}/R_{ee} quantitates the differential mobility of loaded and empty carrier; this ratio was 8.4 for the experiment shown in Figure 12.4B.

The time courses of [^{3}H]uridine accumulation in the zero-*trans* and equilibrium exchange modes in Novikoff cells differ markedly (Fig. 12.4A) and might have suggested *trans*-stimulation. These different time courses, however, are entirely predictable by the appropriate rate equations, which have been fitted to these time courses for the simple carrier model. Before we discuss this point further, however, we will present examples for the kinetic characterization of the transporters.

In routine analysis, we measure the accumulation of radiolabeled nucleoside to transmembrane equilibrium at six to eight nucleoside concentrations. Figure 12.5 illustrates six representative initial time courses of zero-*trans* entry of uridine by uridine kinase-deficient Novikoff cells ranging from 20 to 640 μM uridine as well as the theoretical curves resulting from the fitting of the appropriate integrated rate equation for the simple carrier to the data pooled for all the concentrations. The closeness of the fits indicates that the time courses are

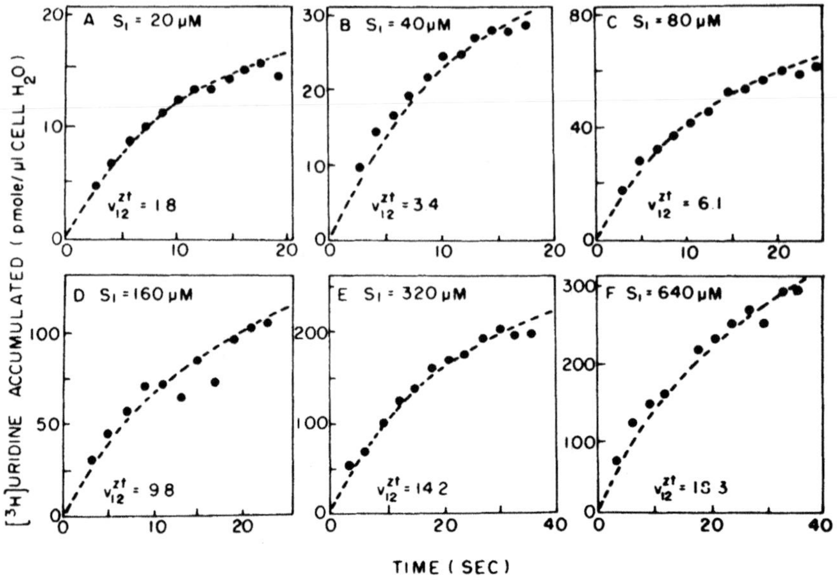

Figure 12.5. Kinetics of zero-*trans* entry of uridine in UK-deficient Novikoff hepatoma cells at 24°C. The experiment has been described in detail in Plagemann and Wohlhueter [2]. The appropriate integrated rate equation was fitted to the time courses pooled for seven uridine concentrations (20–1280 µ*M*) with R_{12} held equal to R_{21} and R_{ee} (= R) as justified by directional symmetry of the transporter and equal mobility whether empty or loaded. The best fitting parameters were $K = 261 \pm 12$ µ*M* and $V = 25.8 \pm 0.5$ pmol/µl cell H_2O · sec ($1/R$). The initial velocities were calculated from the fitted parameters for the given S_1 as the slopes of the theoretical curves at $t = 0$.

well described by the equation. The fit yields estimates of the Michaelis-Menten constant and the maximum velocity, which in this instance were 261 µ*M* and 26 pmol/µl cell H_2O · sec, respectively.

Table 12.1 summarizes the average kinetic parameters for the transport of various nucleosides by Novikoff cells. Similar values have been obtained for the transport of these nucleosides by P388 mouse leukemia cells, Chinese hamster ovary cells, sarcoma 180 cells, L929 cells, and HeLa cells [9], and values for adenosine have been tabulated. Transport of uridine, thymidine, and adenosine has been measured in both kinase-deficient and ATP-depleted cells with similar results [2]. ATP depletion per se has no significant effect on the nucleoside transport capacity of cells or on the kinetic properties of the transporter. The transport of the other nucleosides was measured in ATP-depleted cells only, and in the case of adenosine the cells were pretreated with deoxycoformycin (dCF) to prevent the deamination of adenosine [2, unpublished data].

Experimental evidence available at present indicates that all deoxy- and ribonucleosides are transported by a single carrier with relatively low affinity but high capacity. The affinity of the transporter for various nucleosides dif-

Table 12.1. Kinetic parameters for the nucleoside transporters of various cell culture lines and human erythrocytes at 24°–25°C

Cells	Nucleoside	K		K^{zt}	K^{ee}		V	V^{zt}	V^{ee}	R_{oo}/R_{ee}
Novikoff hepatoma	Adenosine	124 ± 17	(5)		130 ± 24	(7)	17 ± 3		12 ± 4	1
	Tubercidin	87 ± 19	(3)				19 ± 2			
	Tricyclic nucleoside	156 ± 26					10 ± 1			
	Inosine	150 ± 9					13 ± 0.3			
	Thymidine	228 ± 14	(9)		273 ± 42	(4)	27 ± 5		46 ± 9	
	Uridine	250 ± 13	(12)		241 ± 22		25 ± 4		35 ± 1	
	Deoxycytidine	626 ± 52	(4)				41 ± 6			
	Cytidine	2425 ± 492	(4)				59 ± 2			
P388	Adenosine	129 ± 9					17 ± 1			
Chinese hamster ovary	Adenosine	89 ± 24	(5)		128 ± 29	(4)	24 ± 2		19 ± 7	
Sarcoma 180	Adenosine	111 ± 7					16 ± 3			
Human erythrocytes	Uridine	40–140	(5)	70–210	440–700			15–240	140–480	3–18

K, K^{zt}, K^{ee} = μM; V, V^{zt}, V^{ee} = pmol/μl cell $H_2O \cdot$ sec.
The experiments were conducted as illustrated in Fig. 12.5, except that in equilibrium exchange experiments, the cells were preloaded with substrate at the same concentration as radiolabeled substrate was used (see Fig. 12.3). Most of the values for cultured cell lines have been summarized previously [2]. Those for human erythrocytes are from Plagemann et al. [10]. Zero-*trans* entry and inward equilibrium exchange were measured in the same population of cells. The appropriate integrated rate equations were fitted to the time courses of substrate equilibration across the membrane pooled for seven or eight concentrations. In the integrated rate analysis of the zero-*trans* data, R_{ee} was fixed at the value experimentally determined in the companion equilibrium exchange run and R_{12} was held equal to R_{21}. Values are either means of the number of experiments indicated in parentheses ± SEM or from single experiments where the kinetic parameters are stated ± SE of estimate.

Table 12.2. Kinetic parameters for transport and metabolic reactions involved in adenosine uptake by mammalian cells

Process	K_m (μ*M*)	V_{max} [a] (pmol/μl cell $H_2O \cdot$ sec)
Adenosine (inosine) transport	50–150[b]	30–100
Adenosine kinase	0.4–6.0[c]	0.5–2.0
Adenosine deaminase	35–400[c]	30–220
Nucleoside phosphorylase	15–30[c]	20–60
Hypoxanthine phosphoribosyltransferase	2–6[c]	0.5–5
Hypoxanthine transport	350–1500[b]	20–150

[a] V_{max} values encompass ranges for P388, CHO, and Novikoff cells at 37°C [2, and unpublished data].
[b] From [2] and Table 12.1.
[c] Ranges of K_m values with respect to the purine or purine riboside reported for different mammalian cell systems, as tabulated previously [2].

fers, however; its affinity seems to be highest for purine nucleosides and lowest for cytosine nucleosides.

In its broad substrate specificity, the nucleoside transporter of human erythrocytes resembles that from cultured cells [12], and its kinetic parameters are also of the same order of magnitude, though somewhat higher (Table 12.1). The main difference concerns the differential mobility of loaded and empty carrier, which necessitated modifications in the integrated rate analysis of zero-*trans* data for red cells (see legend to table 12.1) and which is reflected in $V^{ee} > V^{zt}$, and $K^{ee} > K^{zt} > K$. We have observed marked differences in the uridine and thymidine transport capacity of different batches of erythrocytes, which generally have been obtained from different individuals. The extent of differential mobility of loaded and empty carrier also varied over a severalfold range. Table 12.1 summarizes the ranges of kinetic parameters observed with different batches of cells. The differential mobility of loaded and empty carrier, as quantitated by the R_{oo}/R_{ee} ratio, ranged from 3 to 18 for different batches of red cells. It should be pointed out, however, that the first-order rate constants for equilibrium exchange (V^{ee}/K^{ee}) and zero-*trans* (V^{zt}/K^{zt}) flux were about the same for each batch of cells, as is implicit in the simple carrier model, and that the loaded carrier always moved more rapidly than the empty carrier, as judged by an R_{oo}/R_{ee} ratio > 1. A high R_{oo}/R_{ee} ratio, nevertheless, was invariably associated with a low V^{zt}, indicating that the main cause for a high differential mobility of loaded and empty carrier is a low mobility of the empty carrier, the rate-limiting step in zero-*trans* flux. These differences in the kinetic behavior of the nucleoside transporter of erythrocytes from different individuals are intriguing, but they need to be confirmed by a more detailed analysis of red cells from each individual of a group of individuals over an extended time span. The differential mobility of loaded and empty carrier and the apparent variability in the mobility of the empty carrier are not understood

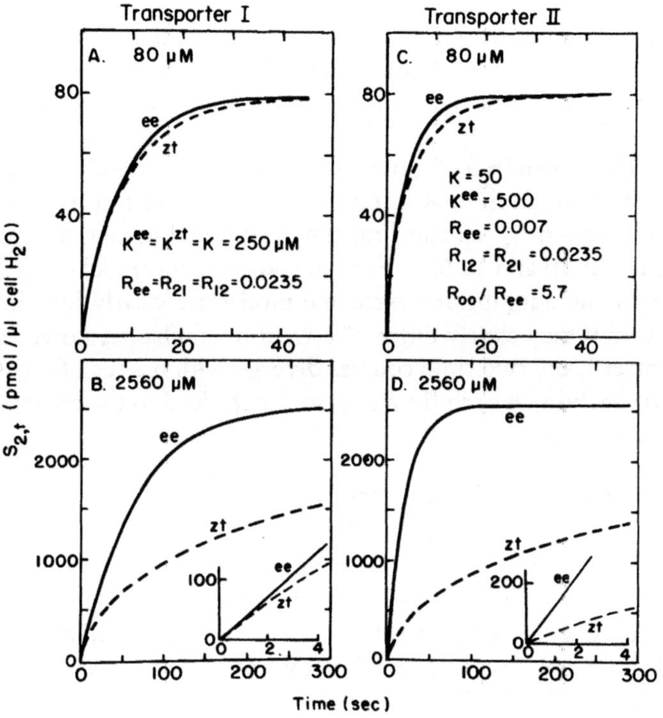

Figure 12.6. Simulated time courses of zero-*trans* entry and inward equilibrium exchange for simple transporters with directional symmetry and equal mobility of loaded and empty carrier (Transporter I) or a 5.7-fold greater mobility of loaded than empty carrier (Transporter II). The progress curves for the intracellular concentrations of radiolabeled substrate ($S_{2,t}$ or $N_{2,t}$, see [2]) were generated for the indicated substrate concentrations by numerical solution of the appropriate integrated rate equations [2] with the K and R values indicated for each transporter.

at the molecular level but may entail important physiological effects on the efficiency of nucleoside salvage (see later discussion).

Let us briefly return to the differences in zero-*trans* and equilibrium exchange time courses at substrate concentrations $\gg K$ (Fig. 12.4), which are entirely consistent with the simple carrier model. This contention is documented in Figure 12.6, which illustrates computer simulations of zero-*trans* entry and equilibrium exchange time courses at substrate concentrations of 80 μM and 2560 μM for two transporters with kinetic properties approximating those from cultured cells (I) and erythrocytes (II). For transporter I, $R_{ee} = R_{12} = R_{21}$, whereas for transporter II, $R_{ee}/R_{oo} = 5.7$. Note that the simple carrier model predicts that at substrate concentrations $\ll K^{ee}$, the zero-*trans* and equilibrium exchange time courses are very similar, but that at concentrations $\gg K^{ee}$ the two time courses differ greatly for either transporter, just as was observed experimentally (Fig. 12.4). The reasons for the differences in zero-*trans* and equilibrium exchange time courses, however, differ for the two transporters. The difference for transporter I is strictly related to the fact that in inward equilibrium exchange, the radiolabeled substrate entering the cell will become immediately diluted with unlabeled substrate present in the cells and, since the carrier is saturated at 2560 μM substrate, the reexit of radiolabeled substrate will be effectively slowed down; this is not the case under zero-*trans* conditions. For transporter II, however, the differential mobility of loaded and empty carrier is an additional important contributing factor. This difference can be best assessed by scrutinizing the very initial uptake time courses (see inserts in Fig. 12.6B and 12.6D). For transporter I, the initial entry velocities in the zero-*trans* and equilibrium exchange modes are clearly identical, whereas for transporter II they clearly differ. We want to emphasize, however, that even for transporter I, the two time courses diverge within 2 sec of incubation, and a superficial analysis of such time courses might lead to the erroneous conclusion of a *trans*-effect.

INTERRELATIONSHIP BETWEEN URIDINE TRANSPORT AND INTRACELLULAR PHOSPHORYLATION

Uridine and thymidine have also served as model substrates for probing the interaction between transport and phosphorylation because of their relatively simple metabolism in cultured cells (Fig. 12.1). Figure 12.7A compares time courses of uptake of 1 μM [^{3}H]uridine by untreated and ATP-depleted wild-type Novikoff cells. ATP-depleted cells are incapable of phosphorylating uridine, and the uptake time course, therefore, is typical for the accumulation of unmodified uridine to transmembrane equilibrium (indicated by broken lines) in the zero-*trans* mode. In untreated cells, however, uptake deviates upward from this time course within 5 sec of incubation and enters a second linear phase, which might last for several minutes. The uptake rate estimated from the slope of the line is significantly lower than the rate of transport. Figure 12.7B illustrates an experiment in which the intracellular labeled components were fractionated; the results show that the linear phase of uptake by untreated cells reflects the accumulation of labeled uracil nucleotides, mainly UTP [2, 13], and not the rate of uridine transport, as was assumed in earlier work [14]. Intracellular free uridine, however, attains a steady-state level within a few seconds of incubation. Figure 12.7C illustrates time courses of uptake of radiolabeled uridine as a function of its concentration. In this experiment, samples of a cell suspension were supplemented with a constant concentration of radiolabeled uridine and increasing concentrations of unlabeled uridine. Thus, a decrease in the rate of radioactivity uptake with an increase in concentration reflects saturation of the uptake process. The slopes of the lines give an estimate of the rate of intracellular phosphorylation of uridine from which an apparent $K_m = 6.5\ \mu M$ and a $V_{max} = 0.38$ pmol/μl cell $H_2O \cdot$ sec have been computed. It is obvious that the K_m must be somewhat overestimated, since it is calculated on the basis of the extracellular uridine concentration and, as indicated by the data in Figure 12.7B, the intracellular steady-state

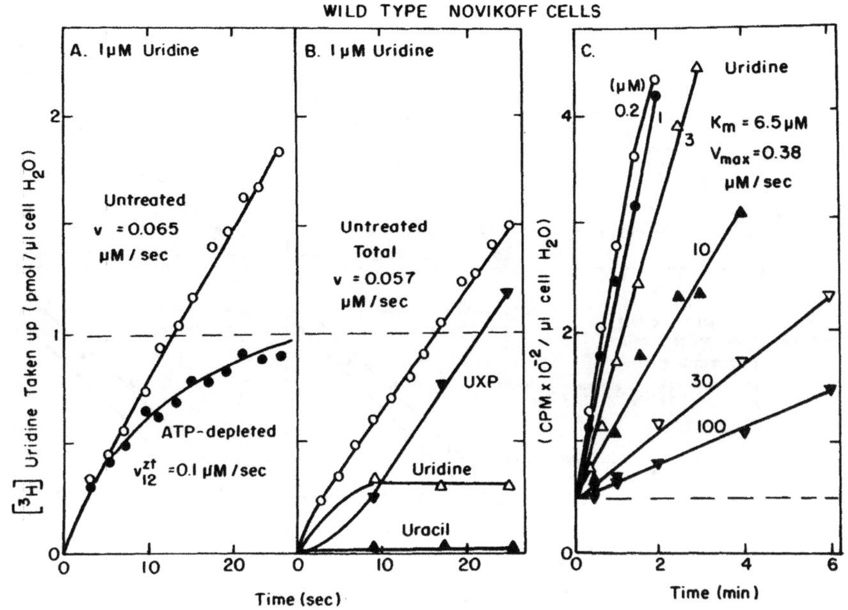

Figure 12.7. Uptake of uridine by wild-type Novikoff rat hepatoma cells at 25°C. A: The uptake of 1 μM [^{3}H-5]uridine by untreated and ATP-depleted cells was measured by the rapid kinetic technique, as illustrated in Fig. 12.2. B: The experiment was conducted in the same manner, except that at some time points, samples of cells were collected into an acid layer and the acid extracts were fractionated chromatographically. Source: Data are from Plagemann et al. [13]. C: Samples of a suspension of about 2×10^7 cells/ml were supplemented with the indicated concentrations of [^{3}H]uridine (570 cpm/μl, irrespective of concentration). At the indicated times, samples of cells were collected by centrifugation through oil and analyzed for radioactivity [previously unpublished data].

concentration of free uridine is significantly lower than that found extracellularly, particularly at lower uridine concentrations.

A more appropriate approach is to treat uptake as a simple two-component system composed of symmetrical carrier-mediated, nonconcentrative transport across the membrane coupled with a simple, irreversible Michaelian phosphorylation within the cells (Fig. 12.8, insert). Equations describing uptake as a function of the extracellular concentration and the kinetic behavior of the two individual steps have been described by Heichel et al. [15] and ourselves [3] and have been applied here to analyze time courses of uridine uptake by wild-type Novikoff hepatoma cells in the following manner. The zero-*trans* entry of uridine is measured in triplicate under transport conditions (Fig. 12.8A), and the maximum transport velocity (V) is estimated by integrated rate analysis, with K fixed at 250 μM (see Table 12.1). The uptake of various concentrations of [^{3}H]uridine is measured in the same cell population over a longer time span (Fig. 12.8B–12.8E), and the equation for coupled transport and phosphorylation is fitted to the compiled data with the kinetic parameters

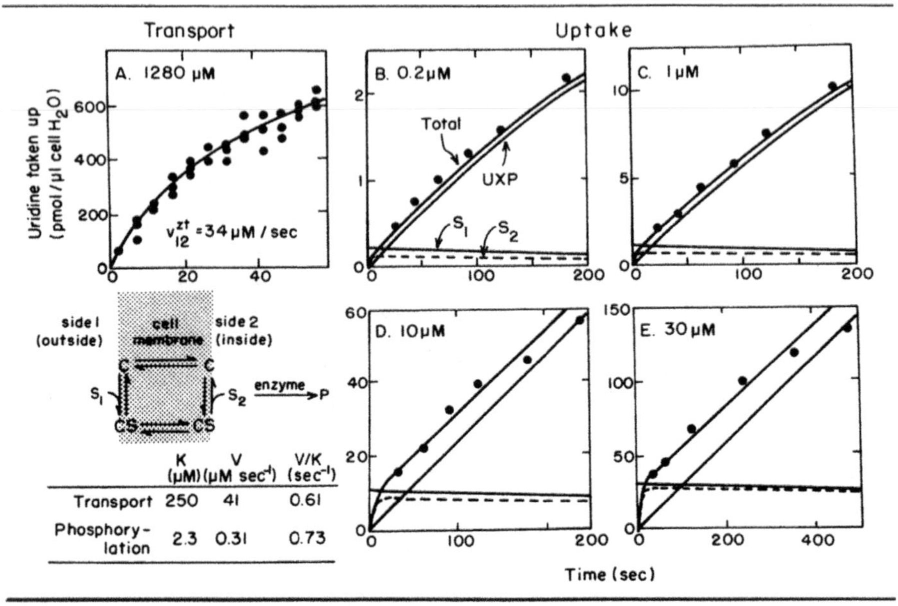

Figure 12.8. Simple model for transport and phosphorylation in tandem (insert) and application to time courses of uptake of uridine by wild-type Novikoff hepatoma cells at 25°C. A: The zero-*trans* accumulation of 1280 μ*M* [^{3}H]uridine was measured in triplicate by the rapid kinetic technique, as illustrated in Fig. 12.2. The appropriate integrated rate equation was fitted to the pooled data with the Michaelis-Menten constant for transport (K_t) fixed at 250 μ*M* (see Table 12.1) and all *R* terms held equal. The fit yields an estimate for the maximum transport velocity (V_t, see table in insert). The uridine uptake time courses (B–E) are for the same population of cells and are the same as illustrated in Fig. 12.7C. The Michaelis-Menten constant for intracellular phosphorylation (K_p) and the maximum velocity (V_p) as well as S_2 (the intracellular concentration of unmodified uridine) were estimated by solving the following equation for the pooled data, with K_t and V_t fixed at the experimentally determined values and substituting appropriate values for S_1 [3]:

$$-(V_tK_t + V_pK_t + V_pS_1)S_2^2 + (K_tV_tS_1 - K_pK_tV_p - K_tV_pS_1 - K_t^2V_p)S_2 + K_tK_pV_tS_1 = 0$$

for transport fixed at the experimentally determined values. The computer fit yields best fitting curves for uptake of substrate into total cell material, for formation of phosphorylated products (UXP), for extracellular (S_1) and intracellular (S_2) concentrations of unmodified substrate and estimates of the Michaelis-Menten constant and maximum velocity for the *in situ* phosphorylation of the substrate, which in this experiment were 2.3 μ*M* and 0.31 pmol/μl cell H_2O · sec, respectively. We have found in repeated experiments that the fitted curves for total uridine uptake by Novikoff cells coincided quite well with the experimental values, particularly at low substrate concentrations (0.2 and 1 μ*M*). At higher concentrations of uridine (10 and 30 μ*M*) the rate of uptake tends to decrease with time below that expected from the kinetic properties of transport and phosphorylation (Fig. 12.3, Fig. 12.8E). We attribute this decrease to the operation of some kind of feedback inhibition, most likely at the level of phosphorylation [16].

The Michaelis-Menten constant and maximum velocity for the *in situ* phosphorylation of uridine by Novikoff cells computed by the method illustrated in Figure 12.8 are, as expected, slightly lower than the values estimated by the graphical approach illustrated in Figure 12.7B. Note that kinetic parameters for uridine transport are about 100-fold higher than those for its phosphorylation, but that the first-order rate constants for the two steps are about the same (Fig. 12.8, insert). Therefore, the salvage pathway for uridine operates very efficiently at low, physiologically relevant extracellular concentrations of uridine, as, for example, at 0.2 μM. At this concentration, uridine is phosphorylated inside the cell and effectively trapped at a rate about 70% of that of its entry into the cell, and one cannot point to a rate-limiting step in uptake. Both steps are codeterminants in the overall rate of uptake. The cells will very efficiently remove the substrate from the extracellular fluid and, in fact, will almost completely deplete the medium of substrate. At concentrations of 10 μM or higher, however, *in situ* uridine phosphorylation becomes rapidly saturated, intracellular steady-state levels of free uridine approach those in the medium, and phosphorylation is clearly the rate-determining step in uptake. The biphasic nature of the uptake time course is clearly evident at these concentrations. The second phase corresponds to the steady state, whereas the initial phase corresponds to the zero-*trans* entry of substrate. With decrease of the uridine concentration below the K_m for intracellular phosphorylation, the extent and duration of the initial phase decreases (Fig. 12.8B and 12.7C).

RELATION BETWEEN ADENOSINE TRANSPORT AND INTRACELLULAR METABOLISM

The relationship between adenosine transport and phosphorylation can only be scrutinized critically if adenosine deamination is blocked, because, with increase in extracellular adenosine concentration, deamination becomes the predominant intracellular metabolic step. Figure 12.9 illustrates this point. Figure 12.9A shows the time course of uptake of 100 μM [^{3}H]adenosine into total cell material by CHO cells, whereas Figure 12.9B shows the time courses of adenosine disappearance from the medium and of the accumulation of labeled inosine and hypoxanthine in the culture fluid. A comparison of the data shows that adenosine disappears about 10 times more rapidly from the medium than can be accounted for by the intracellular accumulation of nucleotides. The remainder is accounted for by inosine and hypoxanthine accumulating in the medium. Since deamination by whole cells is inhibited by inhibitors of nucleoside transport at least as much as the conversion to intracellular nucleotides, deamination seems to occur mainly, if not exclusively, intracellularly. Similar results have been reported previously for other types of cultured cells [2], and the events can be readily understood on the basis of the kinetic

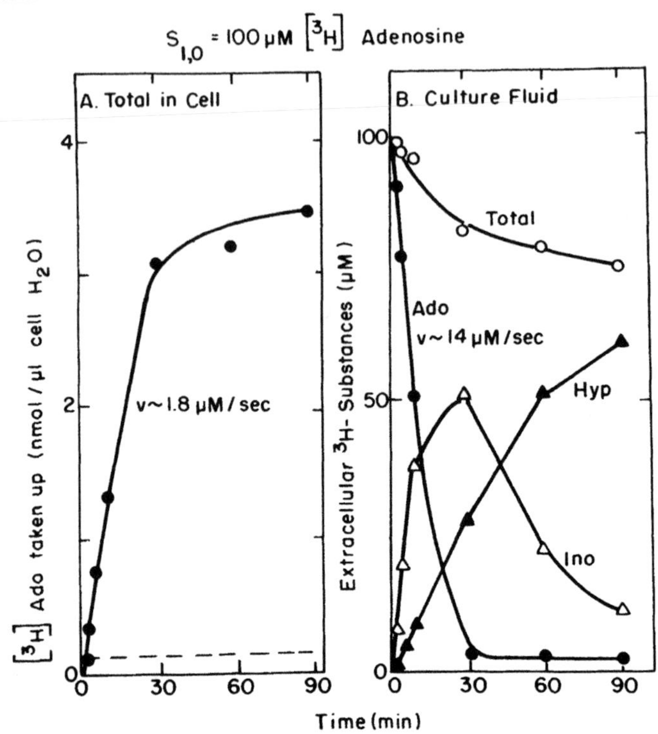

Figure 12.9. Uptake and metabolism of adenosine by wild-type CHO cells at 25°C. A suspension of 4.8×10^6 cells/ml of serum-free, basal medium was supplemented (0 time) with 100 μM [^{3}H]adenosine (8 cpm/pmol). At various times, the cells from 0.5 ml samples of suspension were collected by centrifugation through oil and analyzed for radioactivity. All values represent averages of duplicate samples. B: Adenosine, inosine, and hypoxanthine in the cell-free supernatant were separated chromatographically [17]. The velocities of uptake into the cells (A) and of the disappearance of adenosine from the medium (B) were estimated graphically from the linear portions of the curves. The broken line in (A) indicates the intracellular concentration of radioactivity equivalent to that in the medium at 0 time.

properties of the various processes involved (Table 12.2). It is apparent that the K_m and V_{max} values for the anabolic reactions involved in adenosine salvage, adenosine kinase (AK), and HPRT are one to two orders of magnitude lower than those for the catabolic reactions as well as those for nucleoside and hypoxanthine transport, but that the first-order rate constants (V_{max}/K_m) are very similar—approximately 1 sec^{-1}. Under these conditions, the intracellular steady-state concentration of free adenosine will remain relatively low, regardless of the absolute concentration of extracellular adenosine. At low physiological concentrations of adenosine ($\leq 1\ \mu M$) adenosine kinase and deaminase will compete about equally for the adenosine transported into the cells. Obviously, if the first-order rate constant for phosphorylation is higher than that for deamination, phosphorylation is favored over deamination; the opposite is true if the first-order rate constant for phosphorylation is lower than that for

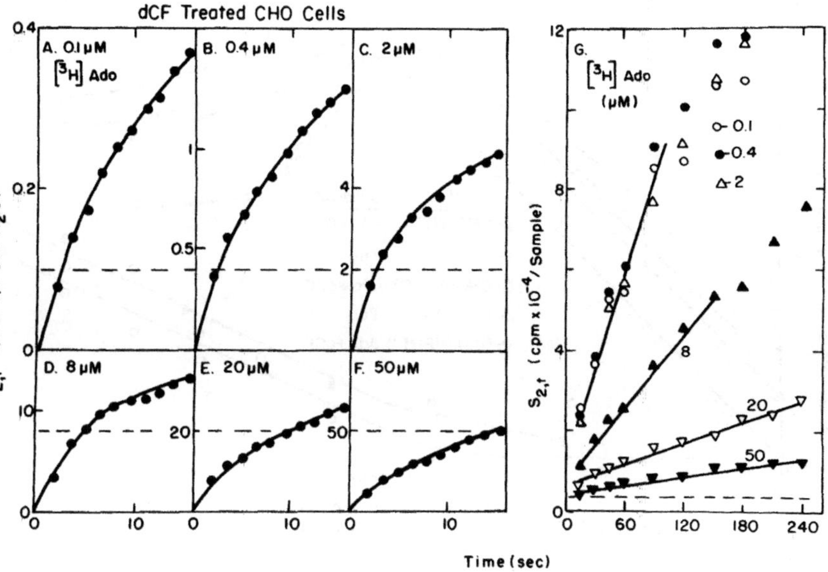

Figure 12.10. Time courses of uptake of adenosine at various concentrations by dCF-treated CHO cells at 25°C. A suspension of 1.2 × 10^7 CHO cells/ml of basal medium was supplemented with 25 μ*M* dCF and incubated for at least 5 min. Then the uptake of [^{3}H]adenosine at the indicated concentrations (530 cpm/μl, irrespective of concentration) was measured by the rapid kinetic technique at two time scales, (A–F and G). Uptake velocities were estimated from the linear portions of the uptake curves in G and are plotted as a function of adenosine concentration in Fig. 12.11 (▲-▲).

deamination. There exists a critical balance between the anabolic and catabolic reactions for adenosine, which seem to differ slightly for different types of cells. At a concentration of 100 μ*M* adenosine, however, adenosine kinase will be saturated and deamination will predominate. Adenosine will be converted to inosine, inosine in turn will be phosphorolyzed to hypoxanthine, and both will rapidly equilibrate with the medium (see Fig. 12.1).

The interaction between adenosine transport and phosphorylation has been further assessed in cells in which deamination was inhibited by pretreatment of the cells with high concentrations of deoxycoformycin [17, 18]. Figure 12.10 illustrates early time courses of uptake of various concentrations of [^{3}H]adenosine by dCF-treated CHO cells and demonstrates the rapidity of the uptake system. At a concentration of 2 μ*M* and lower, the intracellular level of adenosine equivalents exceeds that in the medium (indicated by broken lines in the figure) within 3 sec of incubation at 25°C, and analysis of the intracellular contents shows that over 80% of the radioactivity is present in the form of nucleotides. Uptake clearly becomes saturated between 2 and 50 μ*M* adenosine, but this saturation cannot be construed as a saturation of transport, since it coincides with a decrease in the conversion of adenosine to nucleotides. Furthermore, the initial uptake velocities for 0.1, 0.4, and 2 μ*M* adenosine

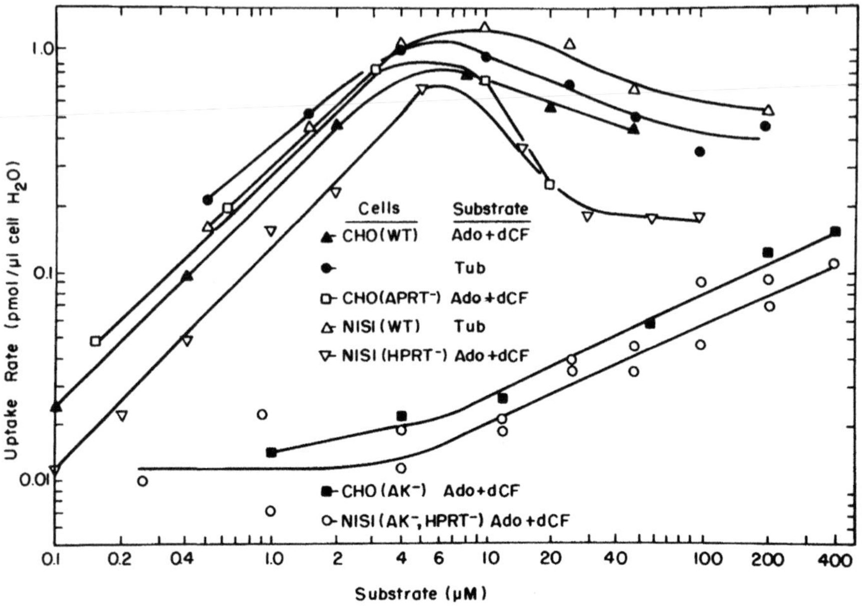

Figure 12.11. Velocities of uptake of adenosine and tubercidin by wild-type CHO and Novikoff (N1S1) cells and various enzyme-deficient variants thereof at 25°C. The experiments were conducted in a similar manner as described in the legend to Fig. 12.10.

estimated from the first 2 sec time points are entirely consistent with those predicted for adenosine transport on the basis of the kinetic properties of the adenosine transporter of CHO cells (see Table 12.1). It needs to be emphasized, however, that accurate initial transport velocities cannot be estimated from such uptake time courses, because the strong curvature apparent within the first few seconds reflects a combination of saturation of adenosine kinase and substrate backflow, the contributions of which differ for different adenosine concentrations. Appropriate equations describing adenosine uptake time courses have not been developed. The equation for coupled transport and simple Michaelian phosphorylation (Fig. 12.8) is not applicable because of substrate inhibition of adenosine kinase, which will be discussed later.

Similarly, as discussed for uridine, adenosine uptake by dCF-treated cells is biphasic, and the second phase, followed over a 4 min period in Figure 12.10G, reflects the rate of accumulation of phosphorylated products. Figure 12.11 is a log/log plot of the rate of phosphorylation estimated from uptake slopes such as those illustrated in Figure 12.10G as a function of the extracellular adenosine concentration. It summarizes results for wild-type CHO and Novikoff cells and a number of enzyme deficient variants thereof and includes data for adenosine and tubercidin as substrates. It is apparent that the rates of phosphorylation of both adenosine and tubercidin are maximal at a concentration of 5 to 10 μM and then decrease with further increase in concentration. This

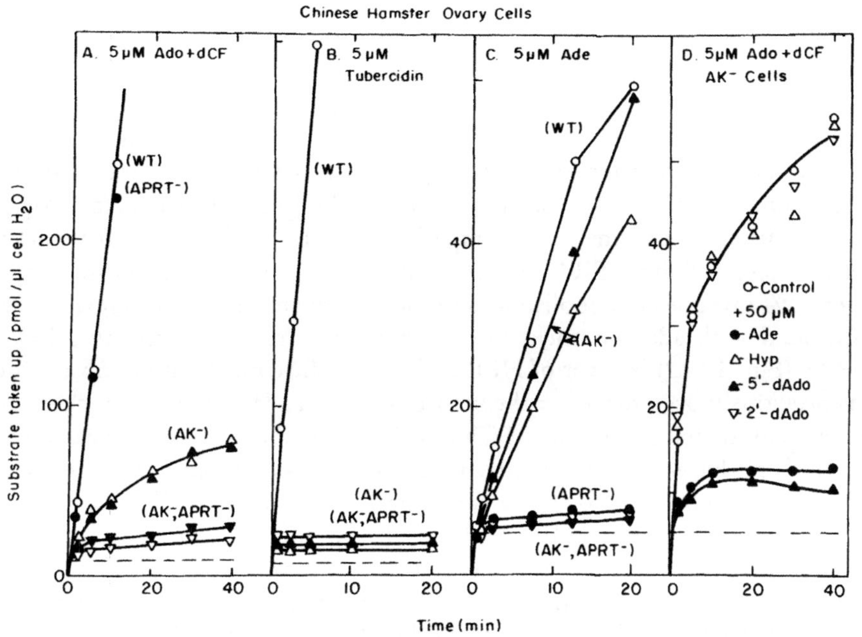

Figure 12.12. Uptake of adenosine, tubercidin, and adenine by wild-type CHO cells and various enzyme-deficient variants thereof at 25°C. Samples of suspensions of 6–12 × 10^6 cells of the various strains per ml of basal medium were supplemented (0 time) with 5 μM [^{3}H]adenosine, [^{3}H]adenine, or [^{3}H]tubercidin and where indicated in frame D, with 50 μM unlabeled adenine (Ade), hypoxanthine (Hyp), 5′-deoxyadenosine (5′-dAdo), or 2′-deoxyadenosine (2′-dAdo). At various times, the cells from 0.5 ml samples of suspension were collected by centrifugation and analyzed for radioactivity. All points represent averages of duplicate samples. Whenever adenosine was the substrate, the cells were preincubated with 50 μM dCF for 5 min and dCF was present during the labeling period. The data are for two independently derived AK-deficient variants (△-△, ▲-▲) and two independently derived AK- and APRT-deficient variants (▽-▽, ▼-▼). The broken lines indicate the intracellular concentration of substrate equivalents equal to those in medium at 0 time.

decrease in uptake with increase in adenosine concentration above 10 μM seems to reflect substrate inhibition of adenosine kinase, since it is also observed with cell-free preparations of adenosine kinase (AK) obtained from various sources [18–21]. Substrate inhibition of the kinase seems a second safeguard, after deamination, to protect against excessive phosphorylation of adenosine from extracellular or intracellular sources.

The dose response curves for adenosine phosphorylation are similar for wild-type cells and variants deficient in APRT or HPRT but differ greatly for AK-deficient variants. As documented in more detail in Figure 12.12, AK-deficient cells are still capable of converting adenosine to nucleotides, even when treated with dCF and deficient in HPRT, but the uptake rate is low at low adenosine concentrations and is not substrate-inhibited. On the contrary, there is no evidence of saturation up to a concentration of 400 μM adenosine, so that at these concentrations, AK-deficient cells take up adenosine almost as rapidly as the substrate-inhibited wild-type cells. A similar observation has been made with wild-type and AK-deficient Ehrlich ascites cells [22].

Figure 12.12A–12.12C compares the uptake of adenosine, tubercidin, and adenine, all at a concentration of 5 μM, by wild-type CHO cells and various enzyme-deficient variants thereof; similar results have been obtained with corresponding strains of Novikoff hepatoma cells [unpublished data]. As discussed earlier, adenosine is very rapidly taken up by dCF-treated wild-type cells or cells lacking APRT. A lack of AK activity reduces adenosine uptake over 90% but does not abolish it, and most of the intracellular radioactivity is associated with adenine nucleotides. Uptake is further reduced by a lack of APRT (Fig. 12.12) but not of HPRT (data not shown). A similar observation has previously been made for the incorporation of adenosine into nucleic acids by enzyme-deficient variants of Chinese hamster fibroblasts by Debatisse and Buttin [23]. However, AK-deficient cells fail to take up adenosine analogues that are also substrates for adenosine kinase, such as tubercidin (Fig. 12.12B) and tricyclic nucleoside [24], and little, if any, intracellular phosphorylation occurs. Furthermore, cell-free extracts from these cells phosphorylate adenosine, tubercidin, or tricyclic nucleoside at less than 1% of the rate observed with extracts from wild-type cells [24, and unpublished data].

The results suggest that adenosine is converted to nucleotides in AK-deficient cells via adenine, that is, that adenosine is subject to phosphorolysis. An enzyme possibly responsible for the degradation of adenosine to adenine is 5′-methylthioadenosine phosphorylase. This hypothesis is supported by the findings that adenosine uptake by the dCF-treated, AK-deficient cells is strongly inhibited by adenine, a strong product inhibitor of 5′-methylthioadenosine phosphorylase (K_i = 11 μM) [25–27], and by 5′-deoxyadenosine, an efficient substrate for the phosphorylase [25], but not by 2′-deoxyadenosine or hypoxanthine (Fig. 12.12D). However, the affinity of this enzyme (if it is responsible) for adenosine seems to be very low, since no saturation of adenosine uptake by AK-deficient cells is observed up to concentrations of 400 μM (Fig. 12.11), whereas the K_m's of the enzyme from various sources for their natural substrate and for 5′-deoxyadenosine fall below 0.4 and 25 μM [25–28].

The double mutants lacking AK and APRT, in which adenosine is converted very slowly to nucleotides when pretreated with dCF, have also been used to assess the kinetics of adenosine transport, and tubercidin transport has been similarly studied in various AK-deficient strains. The Michaelis-Menten constants and maximum velocities for adenosine and tubercidin transport in these cells are similar to those previously determined for ATP-depleted Novikoff and P388 cells (summarized in Table 12.1). There is no evidence for the operation of a second adenosine transporter with a K_m in the low micromolar range. It seems likely that in previous studies [5, 14], saturation of *in situ* conversion of adenosine to nucleotides by whole cells was erroneously interpreted as evidence for the operation of a second adenosine-specific, high-affinity transporter.

Thus, we contend that the uptake of adenosine is distinguished from that of other nucleosides only in the diversity of the metabolic routes available to it. Its transport into the cell is accomplished by the same broadly specific, facilitated diffusion system responsible for the transport of its metabolically simpler cousins, uridine and thymidine.

REFERENCES

1. Henderson JF, Paterson ARP: *Nucleotide Metabolism: An Introduction*. New York, Academic Press, 1973.
2. Plagemann PGW, Wohlhueter RM: Permeation of nucleosides, nucleic acid bases, and nucleotides in animal cells. *Curr Top Membr Transp* 14:225–330, 1980.
3. Wohlhueter RM, Plagemann PGW: The roles of transport and phosphorylation in nutrient uptake in cultured animal cells. *Int Rev Cytol* 64:171–240, 1980.
4. Koren R: The relevance of the state of growth and transformation of cells to their patterns of metabolic uptake. *Int Rev Cytol* 68:127–172, 1980.
5. Paterson ARP, Kolassa N, Cass CE: Transport of nucleoside drugs in animal cells. *Pharmacol Ther* 12:515–536, 1981.
6. Wohlhueter RM, Marz R, Graff JC, Plagemann PGW: A rapid-mixing technique to measure transport in suspended animal cells: Applications to nucleoside transport in Novikoff rat hepatoma cells. *Methods Cell Biol* 20:211–236, 1978.
7. Widdas WF: The asymmetry of the hexose transfer system of the human red cell membrane. *Curr Top Membr Transp* 14:165–223, 1980.
8. Stein WD: Concepts of mediated transport, in Bonting SL, dePont JJHM (eds): *Membrane Transport*. New York, Elsevier, 1981, pp 123–157.
9. Wohlhueter RM, Plagemann PGW: On the functional symmetry of nucleoside transport in mammalian cells. *Biochim Biophys Acta* 689:249–260, 1982.
10. Cabantchik ZI, Ginsburg H: Transport in human red blood cells: Demonstration of a simple carrier mediated process. *J Gen Physiol* 69:75–96, 1977.
11. Pickard MA, Paterson ARP: Nucleoside transport in human erythrocytes: Equilibrium exchange diffusion of uridine. *Can J Biochem* 50:704–705, 1972.
12. Cass CE, Paterson ARP: Mediated transport of nucleosides in human erythrocytes: Accelerated exchange diffusion of uridine and thymidine and specificity towards pyrimidine nucleosides as permeants. *J Biol Chem* 247:3314–3320, 1972.
13. Plagemann PGW, Marz R, Wohlhueter RM: Uridine transport in Novikoff rat hepatoma cells and other cell lines and its relationship to uridine phosphorylation and phosphorolysis. *J Cell Physiol* 97:49–72, 1978.
14. Plagemann PGW, Richey DP: Transport of nucleosides, nucleic acid bases, choline and glucose by animal cells in culture. *Biochim Biophys Acta* 344:263–305, 1974.
15. Heichal O, Ish-Shalom D, Koren R, Stein WD: The kinetic dissection of transport from metabolic trapping during substrate uptake by intact cells: Uridine uptake by quiescent and serum activated Nil 8 hamster cells and their MSV-transformed counterparts. *Biochim Biophys Acta* 551:169–186, 1979.
16. Vidar C, Rubin H: Evaluation of Mg^{2+} as an intracellular regulator of uridine uptake. *J Cell Physiol* 108:317–325, 1981.
17. Lum CT, Marz R, Plagemann PGW, Wohlhueter RM: Adenosine transport and metabolism in mouse leukemia cells and in canine thymocytes and peripheral blood leukocytes. *J Cell Physiol* 101:173–200, 1979.
18. Plagemann PGW, Wohlhueter RM: 2-deoxycoformycin inhibition of intracellular phosphorylation of adenosine in Novikoff rat hepatoma cells. *Biochem Pharmacol* 30:417–426, 1981.
19. Divekar AY, Fleysher MH, Slocum HK, Kenny LN, Hakala MT: Changes in sarcoma 180 cells associated with drug-induced resistance to adenosine analogs. *Cancer Res* 32:2530–2537, 1972.

20. Miller RL, Adamczyk DL, Miller WH: Adenosine kinase from rabbit liver: I. Purification by affinity chromatography and properties. *J Biol Chem* 254:2339–2345, 1979.
21. Palleda TD, Andres CM, Fox IH: Human placental adenosine kinase: Kinetic mechanism and inhibition. *J Biol Chem* 255:5264–5269, 1980.
22. Lomax CA, Henderson JF: Phosphorylation of adenosine and deoxyadenosine in Ehrlich ascites carcinoma cells resistant to 6-(methylmercapto)purine ribonucleosides. *Can J Biochem* 50:423–427, 1972.
23. Debatisse M, Buttin G: The control of cell proliferation by preformed purines: A genetic study. I. Isolation and preliminary characterization of Chinese hamster lines with single and multiple defects in purine "salvage" pathways. *Somatic Cell Genet* 3:497–511, 1977.
24. Plagemann PGW: Transport, phosphorylation, and toxicity of a tricyclic nucleoside in cultured Novikoff rat hepatoma cells and other cell lines and release of its monophosphate by the cells. *J Natl Cancer Inst* 57:1283–1295, 1976.
25. Savarese TM, Crabtree GW, Parks RE Jr: 5′-methylthioadenosine phosphorylase: I. Substrate activity of 5′-deoxyadenosine with the enzyme from sarcoma 180 cells. *Biochem Pharmacol* 30:189–199, 1981.
26. Shugart L, Mahoney L, Chastain B: Kinetic studies of *Drosophila melanogaster* methylthioadenosine nucleoside phosphorylase. *Int J Biochem* 13:559–564, 1981.
27. Guranowski AB, Chiang PK, Cantoni GL: 5′-methylthioadenosine nucleotidase: Purification and characterization of the enzyme from *Lupinus luteus* seeds. *Eur J Biochem* 114:293–299, 1981.
28. Ferro AJJ, Wrobel NC, Nicolette JA: 5-methylthioribose-1-phosphate: A product of partially purified rat liver 5′-methylthioadenosine phosphorylase activity. *Biochim Biophys Acta* 570:65–73, 1979.

DISCUSSION

DR. J.D. PEARSON: Dr. Plagemann has demonstrated that nucleoside carriers efficiently coupled to intracellular enzymes keep the intracellular concentration of adenosine extremely low under normal conditions. Additionally, Newby et al. and Vincent et al. have shown that the predominant rate of nucleotide catabolism during hypoxia is via deamination of AMP to IMP. How, then, can adenosine be produced in sufficient quantities intracellularly to be exported during reactive hyperemia following hypoxia, as would seem to occur (according to the evidence presented by Dr. Schrader)?

DR. PLAGEMANN: Whether adenosine is released from cells or taken up by cells depends on its relative intracellular and extracellular concentration, since transport is nonconcentrative and exhibits directional symmetry. If adenosine is generated intracellularly and its concentration exceeds that in the extracellular compartment, there will be a rapid net export. Since transport is very rapid, I would expect that the intracellular and extracellular concentrations of free adenosine will not differ greatly.

13. INHIBITION OF NUCLEOSIDE TRANSPORT

ALAN R.P. PATERSON, EWA S. JAKOBS, ERIC R. HARLEY,
NAI-WU FU, MORRIS J. ROBINS, CAROL E. CASS

In many types of animal cells, the movement of nucleoside molecules across the plasma membrane is a specific, transporter-mediated process [1, 2]. The existence of the nucleoside transport mechanism has been recognized from the kinetic properties of nucleoside fluxes, but nucleoside transporter elements of the plasma membrane have not yet been visualized or isolated, and their properties have only been studied indirectly. The potent, tightly bound inhibitor of nucleoside transport, nitrobenzylthioinosine (NBMPR),[1] has been a valuable probe of transporter function and biology, enabling, for example, quantitation of transporter elements on cells [3] and providing an approach to their isolation from plasma membrane preparations [4]. This chapter discusses the nucleoside transport-inhibitory properties of NBMPR, N^6-nitrobenzyldeoxyadenosine, and certain of their congeners, and of dilazep, a coronary vasodilator.

NUCLEOSIDE TRANSPORT IN ANIMAL CELLS

In some cell types, the reversible, transporter-mediated influx of nucleoside molecules may be more rapid than the enzymatic transformations to which the

The work reported here was supported by the National Cancer Institute of Canada and the Medical Research Council of Canada. N.-W. Fu was on leave from the Department of Pharmacology, Cancer Institute, Chinese Academy of Medical Sciences, Beijing, Peoples' Republic of China, with support from the Alberta Cancer Board and the Alberta Heritage Foundation for Medical Research. We acknowledge the very competent assistance of Jean Wrobel, Christine Panasiuk, and Patricia Roth. Dilazep was provided through the generosity of Hoffmann La Roche & Co., Basel.

R.M. Berne, T.W. Rall, and R. Rubio (eds.), Regulatory Function of Adenosine, *pp. 203–220.*

influent nucleoside molecules are subject, a circumstance that results in complex time courses of net nucleoside uptake [5, 6]. The transport step of the nucleoside uptake process has been studied independently of the subsequent enzymatic events by measuring fluxes of unmetabolized nucleoside analogues [6–8] or of physiological nucleosides in (1) cells intrinsically incapable of metabolizing the permeant [8–11] or (2) cells in which nucleoside phosphorylation has been impaired by mutational deletion of a kinase activity or by ATP depletion [12, 13]. Under such conditions, intracellular permeant metabolism is suppressed, and permeant fluxes in either direction may be studied to characterize transporter function. With rapid sampling methods, fluxes in either direction across the membrane have been measured, and *trans* effects such as counter-transport have been demonstrated, signifying involvement of a transport mechanism [1, 13].

With this experimental approach, uridine and thymidine permeation in human erythrocytes (neither permeant is metabolized in these cells [8]), and in cultured cells of several types under the condition of impaired permeant phosphorylation, have been recognized as nonconcentrative, reversible processes that are kinetically symmetrical[2] [1, 10, 13] and apparently obey the predictions of a simple carrier model [1, 10, 13, 14]. Although the passive, facilitated diffusion character of nucleoside transport in nucleated cells is generally accepted, we note that this characterization derives from observations with only a few cell-permeant combinations.

Rapid sampling methods have been required to define the time-dependent changes in cell content of nucleoside permeants. With such methods, Plagemann and colleagues [1, 13] have extracted kinetic parameters of transport by fitting time course data to an integrated rate equation based on a simple, symmetrical carrier model for the transport process. Others [15–17], including ourselves [18–20], have measured transport rates directly, employing the concept that the initial rate of a cellular nucleoside uptake process measures unidirectional influx and is intrinsically the rate of the first step in the uptake process, transport. Thus, rates of inward transport of nucleosides may be obtained from definitive time courses of cellular uptake, regardless of whether transport only or transport plus subsequent metabolic conversions of the permeating nucleoside are operative. In this direct approach, the potent inhibitors of nucleoside transport have provided methods for (1) experimentally determining time-zero origins of time courses [18–20] and (2) exposing populations of suspended cells for brief intervals to nucleoside permeants [20].

The nucleoside transport mechanism is of broad specificity with respect to the aglycone portion of the nucleosides transported, as the following findings have shown: (1) various nucleosides induce counter-transport of other nucleosides [12, 13]; (2) binding of the nucleoside transport inhibitor, NBMPR, to homogeneous, high-affinity plasma membrane sites blocks cellular uptake of many nucleosides [2, 21]; and (3) a mutationally induced transporter defect imparted resistance to a variety of nucleoside analogues [22].

NITROBENZYLTHIOINOSINE (NBMPR), A TIGHTLY BOUND INHIBITOR OF NUCLEOSIDE TRANSPORT

NBMPR is the best studied member of a family of S^6-substituted 6-thiopurine nucleosides that interact strongly with the nucleoside transport mechanism of animal cells. NBMPR and certain congeners are very potent, tightly bound inhibitors of nucleoside transport in a wide variety of cell types [2]. NBMPR binds reversibly to high-affinity sites on nucleoside transporter elements of erythrocytes and cultured cells, and occupancy of these sites on human erythrocytes correlates directly with inhibition of uridine transport [3]. The relationship between site occupancy and inhibition of transporter function may be complex in some cell types [19, 23]. The high-affinity sites appear to be of a single type [24], and dissociation constants of 0.1 to 1 n*M* for site-bound NBMPR have been reported (for examples, see [3, 18, 24]).

The saturation number (B_{max}) of cellular NBMPR binding sites estimated in equilibrium binding studies has been regarded as a measure of transporter abundance on cells of different types. As reported elsewhere in this book [25], NBMPR binding-site density on erythrocytes differs markedly from one animal species to another, ranging from no detectable sites on dog erythrocytes to about 10^4 sites per cell on human erythrocytes. In that study, it was apparent that nucleoside transporter elements on erythrocytes from different species had similar uridine translocation capacities. The inference in these and other studies [17] that cellular, site-specific binding of NBMPR is a quantitative indicator of the presence of functional nucleoside transporter sites is supported by the absence of NBMPR binding sites on AE_1 cells [18], a nucleoside transport-defective clone derived from the S49 lymphoma line [22]. B_{max} values of about 10^5 sites per cell for several types of cultured cells have been reported [18, 19, 23, 24].

NBMPR binding sites are accessible from the outer face of the plasma membrane, and occupancy affects both influx and efflux processes. Jarvis and Young [26] have reported that NBMPR effects on zero-*trans* influx of uridine in fresh "nucleoside-permeable type" sheep erythrocytes were competitive, and they were noncompetitive when zero-*trans* efflux was measured. These authors have also proposed a model for the nucleoside transport mechanism in which the NBMPR binding sites are located on the outward aspect of the transporter [27] and they consider the possibility that NBMPR binds at the permeation sites. Apparently competitive inhibition of nucleoside transport by NBMPR has been reported [27], but whether this kinetic result may be interpreted in terms of actual competition between inhibitor and permeant at the permeation site is not yet clear, because nonnucleoside inhibitors (e.g., dipyridamole) also have apparently competitive inhibitory effects on nucleoside transport rates [26] and on NBMPR binding [23]. Furthermore, the apparent purine specificity of the NBMPR sites [19] does not correlate with the broad substrate specificity of the transport mechanism and argues for a distinction between the transport and permeation sites.

NBMPR AS A PROBE OF NUCLEOSIDE TRANSPORTER FUNCTION

The presence of NBMPR in the culture medium of RPM1 6410 cells protects those cells against the antiproliferative effects of a variety of cytotoxic nucleosides through NBMPR impairment of nucleoside entry [21]. Protection experiments have shown that the transporter is insensitive to drastic structural variations in the base portion of its substrates. For example, ribofuranosides as structurally diverse as showdomycin, azomycin, and formycin[3] are accepted as substrates for the transporter in this cell line [18, 21, 28]. The remarkable indifference of the nucleoside transport mechanism toward the base portion of its substrates indicates that the transporter provides a means for the introduction of unphysiological substances into cells [28].

HOMOLOGUES OF NBMPR

Various S^6-substituted 6-thiopurine nucleosides inhibit nucleoside transport, as we have reported previously [29, 30]. The hydrophobic nature of the S^6-substituent contributes importantly to the interaction between the inhibitor and the transport-inhibitory site [31]. The pentofuranosyl moiety of the inhibitor is also an important determinant of this interaction, although the effects of structural variations in the glycosyl substituent have not been explored, apart from some variations at the 2′ position, which have indicated that 2′-hydroxyls are not involved in the high-affinity binding of these compounds (Table 13.1) [29, 30]. The experiments summarized in Figure 13.1 are consistent with the idea that the arabinofuranosyl and ribofuranosyl derivatives of nitrobenzylthioinosine compete at the high-affinity, transport-inhibitory sites on cultured S49 cells. These data (Table 13.1, Fig. 13.1) demonstrate that various homologues of NBMPR interact strongly with the nucleoside transport mechanism and, accordingly, that one might expect common effects in this series related to inhibition of nucleoside transport. For example, if the hemodynamic effects of NBMPR-P[4] derive ultimately from NBMPR effects on adenosine concentrations in the vascular system, then other potent inhibitors of nucleoside transport may have similar effects.

N^6-SUBSTITUTED ADENINE NUCLEOSIDES

Table 13.2 lists dissociation constants at the NBMPR binding site on S49 cells for several adenine nucleosides bearing variously substituted benzyl groups on the purine 6-amino group. These values were determined by competitive inhibition of [G-^{3}H]NBMPR binding. Several compounds in this series interacted strongly with the NBMPR binding site; for example, nitrobenzyladenosine was bound at the high-affinity site about 100-fold less tightly than the S-isostere, NBMPR (Table 13.1). The N^6-adenine derivatives that interacted strongly with the NBMPR binding site were potent inhibitors of adenosine uptake by cultured cells, as the experiment data in Table 13.3 show. The assay employed (cellular uptake of 15 μM adenosine during 10 sec) does

Table 13.1. Dissociation constants for the site-specific binding of S^6-derivatives of 6-thiopurine nucleosides to S49 lymphoma cells

No.	Nonisotopic ligand	K_d (nM)
218	S^6-(4-nitrobenzyl)-6-thioguanosine, expt. a	0.057 ± 0.012
	expt. b	0.036 ± 0.007
844	S^6-(4-nitrobenzyl)-6-thio-2′-deoxyinosine	0.26 ± 0.02
947	S^6-(4-nitrobenzyl)-6-thiopurinearabinoside	0.49 ± 0.07
555	S^6-(2-hydroxy-5-nitrobenzyl)-6-thioguanosine	0.56 ± 0.08

K_d values at the NBMPR binding site of the compounds listed were determined from their inhibitory effects on the equilibrium binding of [G-^{3}H]NBMPR to S49 cells. S49 cells were incubated under previously described conditions [18] with graded concentrations of [G-^{3}H]NBMPR in the presence and absence of various concentrations of nonisotopic test inhibitor. Incubation mixtures (1.0 ml, 2 × 10^6 cells, duplicates) were contained in 1.5 ml plastic centrifuge tubes and layered over 100 μl of an oil mixture [20]. After 30 min at 22°, cells were pelleted under the oil and the [^{3}H]contents of medium samples and cell pellets were determined. Nonspecific retention of [G-^{3}H]NBMPR, defined as the cellular [^{3}H]content in the presence of 3 μM NBTGR, was deducted. Binding data were subjected to mass law analysis by means of the Feldman equation [32] to obtain K_d values ± SE (listed). B_{max} values were between 0.7 and 1.2 × 10^5 sites/cell. For NBMPR, the average K_d value ± SE (n = 4) in these experiments was 0.096 ± 0.009 nM.

not determine adenosine transport rates, but the process measured (cellular retention of adenosine) is transport-dependent and sensitive to inhibitors of nucleoside transport. The inhibitory activity of the N^6-adenosine derivatives in this assay was related generally to NBMPR binding site affinity (Table 13.2).

DILAZEP

Several coronary vasodilators are well-recognized inhibitors of adenosine transport (dilazep, dipyridamole, hexobendine, lidoflazine). We have previously reported the interaction of dipyridamole and lidoflazine with high-affinity, nucleoside transport-inhibitory sites [24], and we now describe some aspects of the interaction of dilazep with the nucleoside transport mechanism. The potent inhibition of adenosine permeation by dilazep recognized previously [33] was also apparent in the experiment summarized in Figure 13.2. Under the conditions of that experiment, the dilazep concentration that reduced adenosine uptake rates in L5178Y/wt cells by 50% was about 40 nM. We have demonstrated that dilazep will displace bound NBMPR from S49 cells (data not known).

The experiment summarized in Figure 13.3 demonstrated that very rapid blockade of adenosine permeation may be achieved with dilazep. Illustrated is a time course of adenosine uptake by L5178Y/wt cells obtained by exposure of cells to ^{3}H-adenosine for graded intervals that were ended by addition of dilazep (final concentration 333 μM). This time course met virtually at time zero, with the time course for adenosine uptake by cells exposed simultaneously to adenosine and dilazep. The intersection of these time courses at time zero indicates that permeation of adenosine was blocked "instantaneously," as in an earlier use of this method, which demonstrated the similarly rapid blockade of adenosine transport by NBMPR [20].

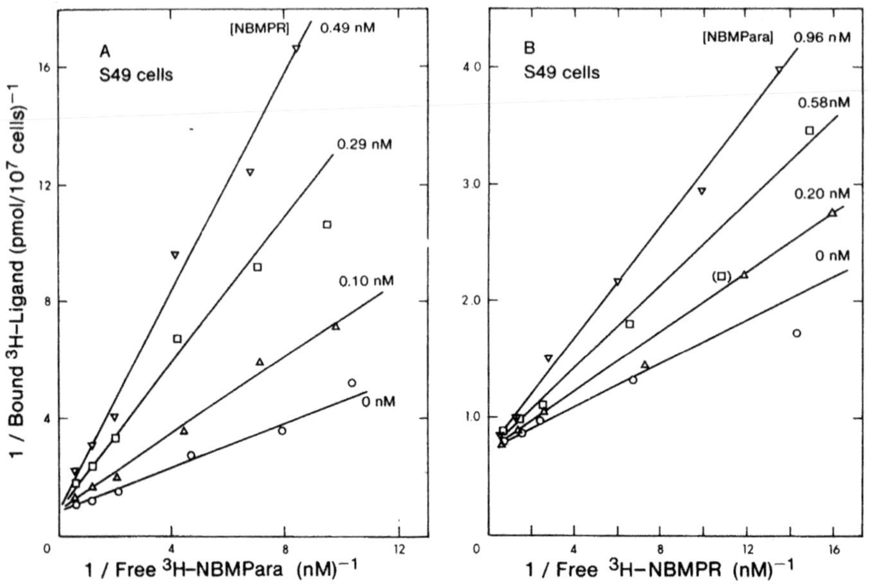

Figure 13.1. Site-specific binding of NBMPR and NBMPara to S49 cells. A: S49 cells were incubated with graded concentrations of [G-^{3}H]NBMPR in the presence and absence of the indicated concentrations of NBMPara and site-specific binding of [G-^{3}H]NBMPR was determined, as in Table 13.1. Binding data were subjected to mass law analysis, as in Table 13.1. B: The influence of NBMPR on the site-specific binding of [G-^{3}H]NBMPara was determined, as in frame A. Binding constants (± SE) determined (1) directly from observed concentrations of bound and free ^{3}H ligand and (2) from inhibitory effects of the competing nonisotopic ligand, as in Table 13.1, were as follows:

	K_d (nM)		
Ligand	(1) Direct	(2) Competition	B_{max} (sites per cell) × 10^{-5}
NBMPR	0.11 ± 0.009	0.074 ± 0.015	0.80 ± 0.02
NBMPara	0.60 ± 0.10	0.49 ± 0.07	0.86 ± 0.10

Table 13.2. Dissociation constants for the site-specific binding of N^6-derivatives of adenine nucleosides to S49 lymphoma cells

No.	Inhibitor[a]	K_d (nM)
1015	N^6-(4-nitrobenzyl)-2′-deoxyadenosine	2.4 ± 0.2
562	N^6-(4-nitrobenzyl)adenosine	8.7 ± 0.7
2003	N^6-(4-azidobenzyl)adenosine	37 ± 6
558	N^6-(benzyl)adenosine	407 ± 84
1016	N^6-(4-aminobenzyl)adenosine	631 ± 63
1014	N^6-(4-nitrobenzyl)-2′,3′-0-isopropylideneadenosine	273 ± 26

K_d values ± SE for the compounds listed were determined from their inhibitory effects on the equilibrium binding of [G-^{3}H]NBMPR to S49 cells, assuming competition of the two ligands for the NBMPR binding site. Experimental conditions and data analysis were as described in Table 13.1.

[a]Prepared by M. Kaneko, S. Shibuya, and M.J. Robins, unpublished results.

Table 13.3. Inhibition of adenosine uptake by N^6-derivatives of adenine nucleosides

No.	Compound	Inhibition of adenosine uptake IC_{50} (nM)
218	S^6-(4-nitrobenzyl)thioguanosine	0.1
554	S^6-(4-nitrobenzyl)thioinosine	1.7
1015	N^6-(4-nitrobenzyl)-2′-deoxyadenosine	4.0
562	N^6-(4-nitrobenzyl)adenosine	12.0
2003	N^6-(4-azidobenzyl)adenosine	50.0
558	N^6-(benzyl)adenosine	350.0
	Dilazep	8.0

The uptake of [2-^{3}H]adenosine by cultured S49 cells during intervals of 10 sec was measured as follows [21]. Intervals of permeant uptake were initiated by adding cell suspension (100 μl, 4 × 10^7 cells) to 100 μl medium containing 15 μM [2-^{3}H]adenosine (0.1 μCi) with and without graded concentrations of test inhibitor; incubation mixtures were layered over 100 μl of oil in 1.5-ml plastic centrifuge tubes. Incubation intervals (10 sec) were ended by adding 200 μl of 30 μM NBMPR, and cells were immediately pelleted under the oil. Incubation mixtures were prepared in triplicate for each condition, and IC_{50} values were estimated from plots of cellular [^{3}H] content versus log inhibitor concentration. Adenosine uptake rates in the absence of inhibitors were about 2.5 pmol/μl cell water.

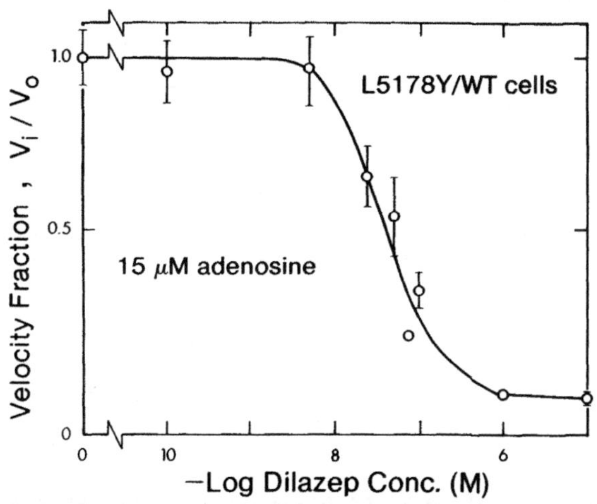

Figure 13.2. Dilazep inhibition of adenosine permeation in L5178Y/WT cells. Cultured L5178Y/WT cells were incubated (4 × 10^7 cells in 1.0 ml serum-free, HEPES-buffered culture medium [20]) for 10 min at 22°C with graded concentrations of dilazep. Cells were then assayed for their ability to take up [2-^{3}H]adenosine (15 μM) from their own dilazep-containing incubation medium during 3 sec intervals at 22° that were ended by the addition of 200 μl dilazep (1 mM) by the method described in Table 13.3. Assays were conducted in triplicate, means ± SD are shown. The rate of adenosine uptake was essentially constant during this uptake interval. The rate of adenosine uptake in cells incubated without dilazep was 7.8 ± 0.6 pmol/μl cell water/sec. The plot indicates an IC_{50} value of 40 nM for the dilazep inhibition of adenosine permeation under these conditions.

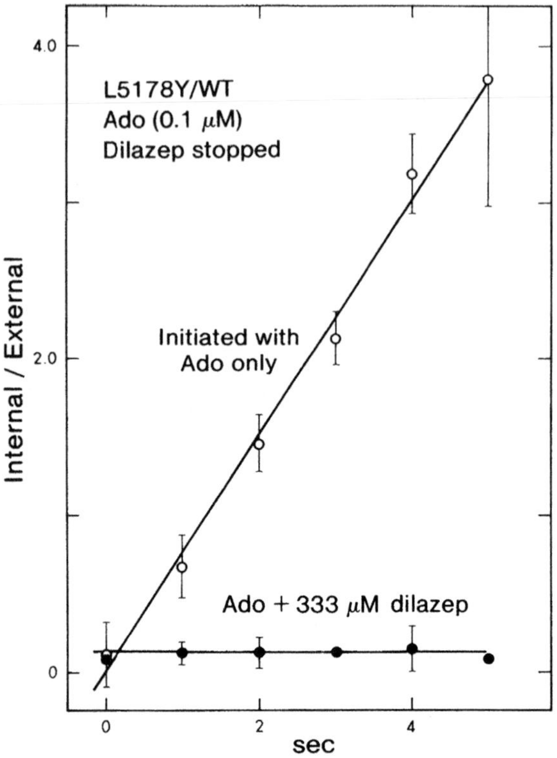

Figure 13.3. Rapid stoppage by dilazep of adenosine permeation in L5178Y/WT cells. To obtain time courses of adenosine uptake, cells were exposed to 0.1 μ*M* [2-^{3}H]adenosine in absence (○) or presence (●) of 333 μ*M* dilazep for the intervals indicated, as described in Table 13.3. Intervals of exposure were ended by addition of 200 μl of medium containing 666 μ*M* dilazep to incubation mixtures. Cells were pelleted under oil, and cell content of [^{3}H]adenosine equivalents was determined as in Table 13.3. Plotted are means of triplicates ± SD. The lines are fitted using a weighted, least-squares method, as before [20]. The calculated abscissa value for the intersection of the time courses was 0.007 ± 0.05 sec.

A QUENCHED-FLOW METHOD EMPLOYING DILAZEP FOR MEASURING CELLULAR UPTAKE OF ADENOSINE DURING BRIEF INTERVALS

Because of the potent and rapid inhibition of adenosine permeation by dilazep, we have employed dilazep as the "quencher" in a quenched-flow method for measuring rates of cellular uptake of adenosine during time intervals of less than 1 sec. In this method, separate streams of cell suspension and of [^{3}H]adenosine-containing medium meet in a mixing chamber to initiate an interval of adenosine uptake (Fig. 13.4). The combined streams then flow in a reaction line before meeting a stream of medium containing the transport quencher, dilazep, in a second mixing chamber to end the uptake interval. Cells from the quenched mixture are then immediately pelleted under oil for analysis of permeant content.

During passage through the reaction line, cells were exposed to ^{3}H-

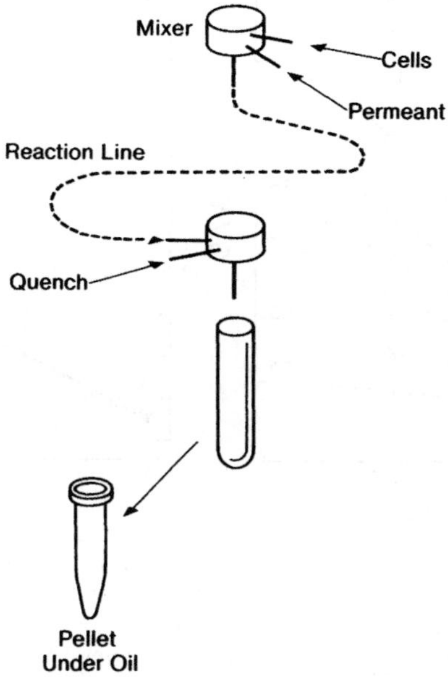

Figure 13.4. Quenched-flow system for measuring cellular uptake of adenosine during brief intervals. Cell suspension, dilazep (1000 μM)-containing medium and [^{3}H]adenosine-containing medium flowed from identical syringes operated by a Harvard 940 syringe pump. Intervals of cell exposure to adenosine began when their solutions met in the first mixer and ended when the reaction stream met the dilazep solution in the second mixer. The total flow rate from the system was 21.6 ml/min.

adenosine for intervals determined by the flow velocity and the dimensions of the reaction line. The three reactant solutions were contained in identical syringes operated by a Hamilton Model 940 syringe pump. Intervals of cellular exposure to [^{3}H]adenosine between 50 and 500 msec were varied by using reaction lines of graded length. In the mixing devices, two jets (21 gauge, thin-walled needle tubing) directed parallel streams tangentially to opposite sides of a cylindrical (1.6 mm diameter) mixing cavity. The mixing-quenching time was less than 16 msec.

A kinetic study of adenosine transport in L5178Y/MMPR cells (an adenosine kinase-deficient line) employing the dilazep quenched-flow method is shown in Figure 13.5. Uptake rates were determined from the slopes of lines fitted to the time course data shown in Figure 13.5A. These rates saturated and yielded the following values ($\pm$ S.E.) for the kinetic constants: K_m, 39 $\pm$ 6 μM; V_{max}, 42 $\pm$ 4 pmol/μl cell water/sec (Fig. 13.5B). An additional example of a kinetic study with the quenched-flow method is shown in Figure 13.6, which summarizes an experiment on adenosine transport in mouse erythrocytes. In this experiment, transport rates were determined from cellular uptake during intervals of 500 msec; the kinetic constants obtained were K_m, 41 $\pm$ 3 μM, and V_{max}, 40 $\pm$ 2 pmol/10^8 cells/sec.

The dilazep quenched-flow method has yielded kinetic constants for

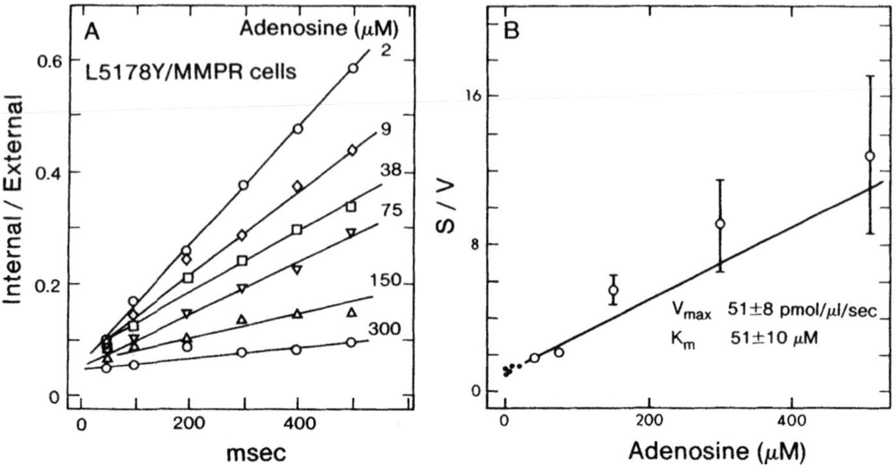

Figure 13.5. Quenched-flow study of adenosine transport in L5178Y/MMPR cells. The quenched-flow system was flow- and volume-calibrated and equipped with reaction lines of such lengths that the passage of cells from one mixer to the other took the time intervals indicated. Serum-free, HEPES-buffered culture medium [18] was employed for suspending cells (2.6 × 10^7 cells/ml) and for solutions of [2-^{3}H]adenosine (13 μCi/ml) and dilazep (1000 μ*M*). These reactant solutions were contained in 10-ml Hamilton syringes pumped simultaneously at a fixed speed. Each experimental condition (time and adenosine concentration) required a separate operation of the system. A: The quenched reaction mixture was sampled after a flow of about 1 ml and cells from triplicate 300 μl portions were pelleted under oil. The [^{3}H] content of the cell samples was determined as before [18]. Water and sucrose spaces were determined in a parallel experiment by conventional methods [18]. Straight lines were fitted to the data by a weighted, least-squares method for the determination of rates of cellular uptake of adenosine [18]. B: Rate-concentration relationship in data from frame A.

adenosine transport that generally confirm those obtained in a previous study [20], in which (1) transport rates were determined as initial rates of uptake calculated from time courses of adenosine uptake, (2) those time courses included experimentally determined values for uptake at time zero, and (3) intervals of permeant uptake were ended by addition of NBMPR to incubation mixtures.

In this study, we have presumed two time factors to be negligible in the context of uptake intervals of 50 to 500 msec: (1) differences between the time courses of the mixing processes in the initiating and quenching mixing devices and (2) onset of dilazep inhibition. These factors must be addressed in further development of this technique. A quenched-flow system developed for studies of ion fluxes in membrane vesicles has been described recently [35].

IN VIVO EFFECTS AND POSSIBLE THERAPEUTIC APPLICATIONS OF THE NUCLEOSIDE TRANSPORT INHIBITORS

We have reported that NBMPR and NBMPR-P (the 5′-monophosphate of NBMPR) protected mice against otherwise lethal treatment protocols with cytotoxic nucleoside analogues [36–39], and we have suggested that such pro-

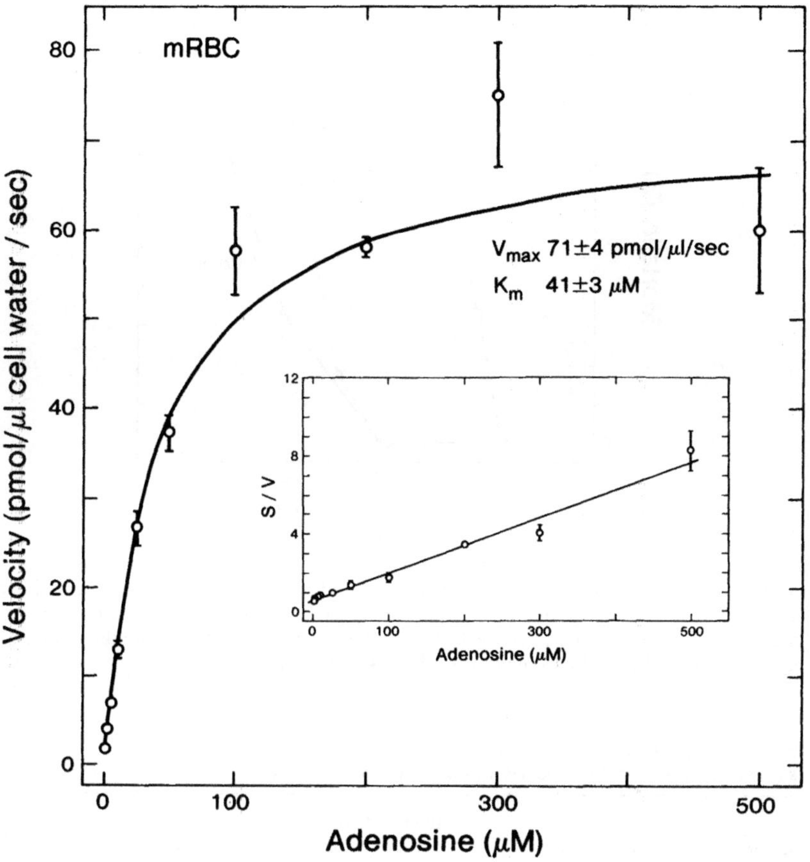

Figure 13.6. Quenched-flow study of the inward transport of adenosine in mouse erythrocytes. Rates of adenosine transport by washed mouse erythrocytes were determined from the cellular uptake of [^{3}H]adenosine equivalents from medium containing graded concentrations of [2-^{3}H]adenosine during intervals of 500 msec. The quenched-flow procedure was employed. The erythrocyte concentration in the reaction line was 5 × 10^7 cells/ml. The method of Young [34] was employed to measure cellular content of [^{3}H] in samples of dilazep-quenched reaction mixtures.

tection is probably due to inhibitor-reduced access of the toxic agents to vital tissues. The strong interactions with the nucleoside transport mechanism shown to occur with dilazep and with nitrobenzyldeoxyadenosine and congeners indicated that these agents might also influence the toxicology of nucleoside analogues. Figure 13.7 illustrates protection of mice against otherwise lethal intraperitoneal treatment schedules with nebularine (purine riboside) by the intraperitoneal administration of nitrobenzyldeoxyadenosine or dilazep prior to that of the lethal agent; these data show that the tolerance of mice to nebularine was increased two- to three-fold by prior treatment with the transport inhibitors.

The *in vivo* protection demonstrated in these and in previous experiments

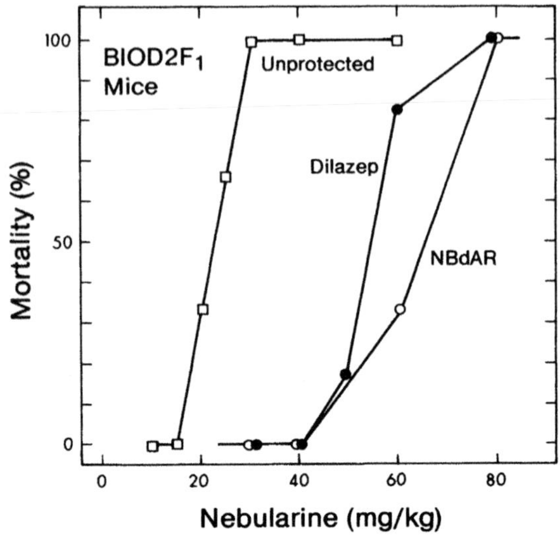

Figure 13.7. Protection of mice against lethality of intraperitoneally administered nebularine. Female B10D2F$_1$ mice in groups of six received a total of 5 i.p. treatments with nebularine at 24-hr intervals. The protecting agents, dilazep (40 mg/kg) and nitrobenzyldeoxyadenosine (7 mg/kg), were administered by the i.p. route 30 sec and 60 sec, respectively, before the toxic agent. Deaths were recorded daily for 30 days after the first treatment.

Table 13.4. Daily treatment of mice bearing the Ehrlich ascites (EA) carcinoma with NBdAR and high doses of nebularine

Nebularine (mg/kg)	NBdAR (mg/kg)	Survival time of mice that died (days)	ILS[a] (%)	Mice alive on Day 50
0	0	12.9 ± 1.0[b]		0/12
20	0	19.0 ± 5.5	47	2/6
0	10	13.6 ± 1.1	5	0/6
37	8	19.5 ± 5.5[c]	51	4/6
37	10	14	9	5/6

Female B10D2F$_1$ mice were implanted intraperitoneally with 10^6 EA carcinoma cells and 24 hr later received the first of 5 i.p. treatments (specified in table) at 24 hr intervals. NBdAR was given 1 min before nebularine.
[a]ILS = increase in life span (difference between mean survival times of treated and untreated control animals that died, expressed as a percentage of control group survival time).
[b]Mean ± SD, except as noted.
[c]Mean ± average deviation from the mean.

[36–39] by several different agents indicates that the observed protection is related to the property common to these agents, inhibition of nucleoside transport.

We have previously reported that high, potentially lethal doses of either nebularine or tubercidin may be employed in chemotherapy of several mouse leukemias, provided that host-protective doses of NBMPR-P were coadminis-

Table 13.5. Treatment of mouse leukemia L1210/TG8 with dilazep and high doses of nebularine

Nebularine (mg/kg)	Dilazep (mg/kg)	Survival time (days)	ILS[a] (%)	Mice alive on Day 45
0	0	7.8 ± 2.1[b]		0/12
37	50	11.3 ± 0.7	45	0/6
37	40	13.1 ± 3.8	68	0/6
37	30	11.8 ± 1.6	51	0/6
20	0	8.8 ± 0.3	13	0/6
0	50	9.0 ± 0.8	15	0/6

Female BD10D2F$_1$ mice were implanted intraperitoneally with 10^6 leukemia L1210/TG8 cells and 24 hr later received the first of 5 i.p. treatments (specified in table). Dilazep was administered 30 sec before nebularine.
[a]ILS = increase in life span (see Table 13.4).
[b]Mean ± SD

tered [38, 39]. In those experiments, NBMPR-P protected vital host tissues against the toxic agents, yet substantial kill of leukemic cells resulted, as indicated by the production of long-term survivors ("cures") in the treated, leukemic mice. This result suggested some apparent selectivity for normal tissues of the NBMPR-P protection, an effect that we are unable to explain. The experiments reported here in Tables 13.4 and 13.5 demonstrate that nitrobenzyldeoxyadenosine and dilazep also provided host protection in experiments in which mice implanted with the Ehrlich ascites carcinoma and with leukemia L1210/TG8 were treated with high doses of nebularine. Again, the inhibitors provided host protection without eliminating therapeutic effects. This was evident in the occurrence of long-term survivors and increased survival times among treated mice bearing Ehrlich ascites carcinoma and leukemia L1210/TG8 cells, respectively. Thus, a variety of compounds appear to be able to serve as modulators of nucleoside analogue toxicity and, presumably, of the *in vivo* disposition of those agents. It remains to be seen whether it will be possible to optimize these effects through different combinations of transport inhibitors and nucleoside analogues and whether undesirable effects of inhibitors (such as apparent central nervous system effects of NBMPR-P[5]) may be avoided through selection of inhibitor-drug combinations.

NOTES

1. Abbreviations used herein are as follows: NBMPR, S^6-(4-nitrobenzyl)-6-thioinosine; NBMPR-P, the 5′-monophosphate of NBMPR; NBTGR, S^6-(4-nitrobenzyl)-6-thioguanosine; NBMPara, S^6-(4-nitrobenzyl)-9-β-D-arabinofuranosyl-6-thiopurine; NBAR, N^6-(nitrobenzyl)adenosine; NBdAR, N^6-4(nitrobenzyl)-2′-deoxyadenosine.
2. Uridine permeation in stored erythrocytes of the human is kinetically asymmetrical [9, 10], but in fresh erythrocytes it is symmetrical [10].
3. Showdomycin, azomycin, and formycin are β-D-ribofuranosides of maleimide, 2-nitroimidazole, and 7-amino-pyrazolo[4,3 *d*]pyrimidine, respectively.
4. NBMPR-P administered by intravenous bolus injection at doses of 10 mg/kg and above causes marked lowering of mean blood pressure in conscious dogs [N.W. Holfeld, P.M. Venner, A.S. Clanachan, and A.R.P. Paterson, unpublished results].

5. NBMPR markedly potentiates barbiturate anesthesia in mice and dogs [N.W. Holfeld, P.M. Venner, A.S. Clanachan, and A.R.P. Paterson, unpublished results].

REFERENCES

1. Plagemann PGW, Wohlhueter RM: Permeation of nucleosides and nucleic acid bases and nucleotides in animal cells. *Curr Top Membr Transp* 14:225–330, 1980.
2. Paterson ARP, Kolassa N, Cass CE: Transport of nucleoside drugs in animal cells. *Pharmacol Ther* 12:515–536, 1981.
3. Cass CE, Gaudette LA, Paterson ARP: Mediated transport of nucleosides in human erythrocytes: Specific binding of the inhibitor nitrobenzylthioinosine to nucleoside transport sites in the erythrocyte membrane. *Biochim Biophys Acta* 345:1–10, 1974.
4. Jarvis SM, Young JD: Extraction and partial purification of the nucleoside transport systems from human erythrocytes based on the assay of nitrobenzylthioinosine binding activity. *Biochem J* 194:331–339, 1981.
5. Wohlhueter RM, Plagemann PGW: The roles of transport and phosphorylation in nutrient uptake in cultured animal cells. *Int Rev Cytol* 64:171–240, 1980.
6. Koren R, Shohami E, Yeroushalmi S: A kinetic analysis of the uptake of cytosine-β-D-arabinoside by rat B77 cells: Differentiation between transport and phosphorylation. *Eur J Biochem* 95:333–339, 1979.
7. Kessel D: Transport of a nonphosphorylated nucleoside, 5′-deoxyadenosine, by murine leukemia L1210 cells. *J Biol Chem* 253:400–403, 1978.
8. Oliver JM, Paterson ARP: Nucleoside transport: 1. A mediated process in human erythrocytes. *Can J Biochem* 49:262–270, 1971.
9. Cass CE, Paterson ARP: Mediated transport of nucleosides in human erythrocytes: Accelerative exchange diffusion of uridine and thymidine and specificity toward pyrimidine nucleosides as permeants. *J Biol Chem* 247:3314–3320, 1972.
10. Cabantchik ZI, Ginsburg H: Transport of uridine in human red blood cells: Demonstration of a simple carrier mechanism. *J Gen Physiol* 69:75–96, 1977.
11. Jarvis SM, Hammond JR, Paterson ARP, Clanachan AS: Nucleoside transport differs in erythrocytes from fresh and outdated human blood, in Berne RM, Rall TW, Rubio R (eds): *The Regulatory Function of Adenosine*. The Hague, Martinus Nijhoff Medical Division, 1983 (abst).
12. Plagemann PGW, Marz R, Wohlhueter RM: Uridine transport in Novikoff rat hepatoma cells and other cell lines and its relationship to uridine phosphorylation and phosphorolysis. *J Cell Physiol* 97:49–72, 1978.
13. Wohlhueter RM, Marz R, Plagemann PGW: Thymidine transport in cultured mammalian cells: Kinetic analysis, temperature dependence and specificity of the transport system. *Biochim Biophys Acta* 553:261–268, 1979.
14. Lieb WR, Stein WD: Testing and characterizing the simple carrier. *Biochim Biophys Acta* 373:178–196, 1974.
15. Bowen D, Diasio RB, Goldman ID: Distinguishing between membrane transport and intracellular metabolism of fluorodeoxyuridine in Ehrlich ascites cells by application of kinetic and high performance liquid chromatographic techniques. *J Biol Chem* 254:5333–5339, 1979.
16. Kolassa N, Plank B, Turnheim K: pH and temperature dependence of adenosine uptake in human erythrocytes. *Eur J Pharmacol* 52:345–351, 1978.
17. Jarvis SM, Young JD: Nucleoside transport in human and sheep erythrocytes: Evidence that nitrobenzylthioinosine binds specifically to functional nucleoside transport sites. *Biochem J* 190:377–383, 1980.
18. Cass CE, Kolassa N, Uehara Y, Dahlig-Harley E, Harley ER, Paterson ARP: Absence of binding sites for the transport inhibitor nitrobenzylthioinosine on nucleoside transport-deficient mouse lymphoma cells. *Biochim Biophys Acta* 649:769–777, 1981.
19. Dahlig-Harley E, Eilam Y, Paterson ARP, Cass CE: Binding of nitrobenzylthioinosine to high affinity sites on the nucleoside transport mechanism of HeLa cells. *Biochem J* 200:295–305, 1981.
20. Harley ER, Paterson ARP, Cass CE: Initial rate kinetics of the transport of adenosine and 4-amino-7-(β-D-ribofuranosyl)pyrrolo[2,3-*d*]pyrimidine (tubercidin) in cultured cells. *Cancer Res* 42:1289–1295, 1982.

21. Paterson ARP, Yang S-E, Lau EY, Cass CE: Low specificity of the nucleoside transport mechanism of RPMI 6410 cells. *Mol Pharmacol* 16:900–908, 1979.
22. Cohen A, Ullman B, Martin DW Jr: Characterization of a mutant mouse lymphoma cell with deficient transport of purine and pyrimidine nucleosides. *J Biol Chem* 254:112–116, 1979.
23. Paterson ARP, Lau EY, Dahlig E, Cass CE: A common basis for inhibition of nucleoside transport by dipyridamole and nitrobenzylthioinosine? *Mol Pharmacol* 18:40–44, 1980.
24. Lauzon GJ, Paterson ARP: Binding of the nucleoside transport inhibitor nitrobenzylthioinosine to HeLa cells. *Mol Pharmacol* 13:883–891, 1977.
25. Clanachan AS, Paterson ARP, Hammond JR, Jarvis SM: Species differences in nucleoside transport by mammalian erythrocytes, in Berne RM, Rall TW, Rubio R (eds): *Regulatory Function of Adenosine*. The Hague, Martinus Nijhoff Medical Division, 1983 (abst).
26. Jarvis SM, McBride D, Young JD: Erythrocyte nucleoside transport: Asymmetric binding of nitrobenzylthioinosine to nucleoside permeation sites. *J Physiol* 324:31–46, 1982.
27. Jarvis SM, Young JD: Nucleoside translocation in sheep reticulocytes and fetal erythrocytes: A proposed model for the nucleoside transporter. *J Physiol* 324:47–66, 1982.
28. Jarvis SM, Chapman JD, Ngan-Lee J, Rutledge KA, Barr PJ, Paterson ARP: Azomycin riboside: A sugar homologue of misonidazole with favorable radiosensitizing properties. *Cancer Res* 42:4358–4363, 1982.
29. Paterson ARP, Naik SR, Cass CE: Inhibition of uridine uptake in HeLa cells by nitrobenzylthioinosine and related compounds. *Mol Pharmacol* 13:1014–1023, 1977.
30. Paterson ARP, Babb LR, Paran JH, Cass CE: Inhibition by nitrobenzylthioinosine of adenosine uptake by asynchronous HeLa cells. *Mol Pharmacol* 13:1147–1158, 1977.
31. Paul B, Chen MF, Paterson ARP: Inhibitors of nucleoside transport: A structure-activity study. *J Med Chem* 18:968–973, 1975.
32. Feldman HA: Mathematical theory of complex ligand binding systems at equilibrium: Some methods for parameter fitting. *Anal Biochem* 48:317–338, 1972.
33. Pohl J, Brock N: Vergleichende untersuchungen zur hemmung des adenosinabbaus in vitro durch dilazep. *Arzneim Forsch* 24:1901–1905, 1974.
34. Young JD: Nucleoside transport in sheep erythrocytes: Genetically controlled transport variation and its influence on erythrocyte ATP concentrations. *J Physiol* 277:325–339, 1978.
35. Cash DJ, Hess GP: Quenched flow technique with plasma membrane vesicles: Acetylcholine receptor mediated transmembrane ion flux. *Anal Biochem* 112:39–51, 1981.
36. Paterson ARP, Paran JH, Yang S-E, Lynch TP: Protection of mice against lethal dosages of nebularine by nitrobenzylthioinosine, an inhibitor of nucleoside transport. *Cancer Res* 39:3607–3611, 1979.
37. Paterson ARP, Jakobs ES, Lauzon GJ, Weinstein WM: Drug sequence-dependent toxicity and small bowel mucosal injury in mice treated with low doses of 3-deazauridine and 1-β-D-arabinofuranosylcytosine. *Cancer Res* 39:2216–2219, 1979.
38. Lynch TP, Jakobs ES, Paran JH, Paterson ARP: Treatment of mouse neoplasms with high doses of tubercidin. *Cancer Res* 41:3200–3204, 1981.
39. Lynch TP, Paran JH, Paterson ARP: Therapy of mouse leukemia L1210 with combinations of nebularine and nitrobenzylthioinosine 5′-monophosphate. *Cancer Res* 41:560–565, 1981.

DISCUSSION

DR. T. SPECTOR: (1) Do the transport inhibitors have any antitumor activity when tested with tumor-bearing mice? (2) Can you suggest a reason (or mechanism) why the transport inhibitors do not also protect the tumor cells against tubercidin or nebularine?

DR. PATERSON: (1) As single agents, NBMPR, NBMPR-P, and NBdAR, and dilazep have not shown significant antineoplastic activity against several transplanted mouse neoplasms, including leukemia L1210, in our hands. (2) In chemotherapy experiments with mice bearing the Ehrlich ascites carcinoma or leukemia L1210/TG8 [38], we have shown that at optimized dosages, NBMPR-P protection against tubercidin cytotoxicity favored vital host tissues over the neoplastic cells. Substantial kill of neoplastic cells ("cures") was achieved at optimal doses of NBMPR-P; however, at supraoptimal doses of NBMPR-P, neoplastic cells were also protected against tubercidin, and therapeutic failures (neoplastic deaths) resulted. The basis of the differential NBMPR-P protective effect favoring host tissues has not been explored further.

DR. S.M. JARVIS: (1) I notice that Dr. Paterson reported apparent K_m values for adenosine transport of approximately 17 to 50 μM, whereas those reported by Dr. Plagemann are about 100 μM. Are these values similar, or do they represent real differences? (2) The translocation capacity for the nucleoside erythrocyte transport system is very similar for many mammalian species. Are the changes in V_{max} for nucleoside transport from cell batch to cell batch due to changes in the translocation capacity of a constant number of functional nucleoside transport systems, or is the translocation capacity constant and the number of transport sites, as determined by nitrobenzylthioinosine binding changing from cell batch to cell batch?

DR. P.G.W. PLAGEMANN: (1) I do not consider these differences significant. They are inherent in the way transport velocities were estimated. Dr. Paterson estimated transport velocities by linear regression of or fitting parabolic equations to initial time courses of uptake of adenosine by cells that were not blocked in adenosine metabolism. We have estimated transport velocities from fits of integrated rate equations based on the simple carrier model (zero-*trans* and equilibrium exchange) to overall time courses of transmembrane equilibration of adenosine in cells in which both phosphorylation and deamination were blocked by prior depletion of the cells of ATP and incubation with deoxycoformycin, respectively. (2) We do not know. We have not yet conducted any studies of the number of nitrobenzylthioinosine binding sites in red cells.

DR. PATERSON: In the examples we have presented, K_m values for the inward transport of adenosine determined by the quenched-flow method ranged between 17 and 39 μM. With the qualification that our experience with this method is very limited, these results appear consistent with values previously reported from this laboratory by Harley et al. [20]. Although our particular rapid-sampling methods may be yielding K_m values for adenosine transport in several cultured cell lines somewhat lower than the values reported by Dr. Plagemann and colleagues (Lum, C.T., Marz, R., Plagemann, P.G.W., Wohlhueter, R.M., *J. Cell. Physiol.* 101:173–200, 1979), the aggregate experience in this area is still rather limited.

DR. T.W. RALL: At least one transport antagonist (N^6-benzyladenosine) also has been used as our adenosine agonist (IC_{50} for transport approximately 350 nM, and according to Dr. Bruns this substance has K_d or K_{act} for A_1 receptors of about 150 μM and about 100 μM for A_2 receptors). My question is: Is this substance (or other adenosine analogues that act as agonists) transported? Further, is it possible that some of these agonist analogues can "release" or "preserve" adenosine in intact cell preparations and account for some of their apparent activity?

DR. PATERSON: N^6-benzyladenosine interacts with the nucleoside transport mechanism as a weak (relative to NBMPR) inhibitor of NBMPR binding and of adenosine transport, but we do not know if this compound enters cells via the nucleoside transport mechanism. Although the specificity of the transport mechanism is markedly broad with respect to the nucleobase portion of the ribosides accepted as substrates, this compound has not been studied as a permeant. To my knowledge, the effects on adenosine efflux of the potent agonists at adenosine receptor sites have not been evaluated.

DR. D.G. SATCHELL: Can the sedative properties of the compounds mentioned earlier be related to specific actions of compounds or to structural characteristics?

DR. PATERSON: The sedative and other pharmacological effects of the nucleoside transport inhibitors have not been explored systematically. Our cage-side impressions that nucleoside transport inhibitor activities do not correlate with sedative effects in mice remain to be substantiated in quantitative studies.

DR. R.A. OLSSON: 6-phenylthio- and 6-benzylthiopurine ribosides have been used as affinity probes of the AMP allosteric regulatory site of phosphofructokinase. In this instance, ^{35}S labeling shows that the N^6 substituent is covalently bound to the allosteric site. Does this occur to any extent in the case of the adenosine transport?

DR. PATERSON: Although NBMPR is bound tightly to specific, transporter-associated sites in plasma membranes in erythrocytes and cultured cells, that binding is reversible and noncovalent. This is evident in the displaceability of site-bound, labeled NBMPR by NBMPR congeners. Further, site-bound isotopic NBMPR has been recovered unaltered from erythrocytes. Nonspecifically bound NBMPR may be accounted for by simple partition into membrane lipids. In short, there is no evidence for covalent binding of NBMPR to the nucleoside transport mechanism.

DR. R. VON BORSTEL: I have found that 10 mg/kg NBMPR-P intraarterially induces a 40% drop in blood pressure within a few seconds, and this hypotension is reversed by caffeine. Do you think that the sedative effect of this compound is secondary to its effect on blood pressure (i.e., have you found any related compounds with an apparent vasoactive effect but not sedative effect)?

DR. PATERSON: We agree that NBMPR-P is a potent hypotensive agent. Our observations on the sedative effects of NBMPR-P are very limited. Although systematic studies of vasoactivity and other pharmacological effects remain to be done in the array of known inhibitors of nucleoside transport, it is our impression that sedative effects may not correlate with inhibition of nucleoside transport. Sedative effects have not been apparent in our use of BNdAR and dilazep as host-protective agents against nebularine toxicity.

DR. G. MISTRY: (1) Has anyone characterized the adenosine transport protein? (2) In the AE_1 cells, Dr. Paterson's data show lack of binding of [^{3}H]NBMPR. Does lack of binding of [^{3}H]NBMPR indicate lack of the adenosine transport protein, or is the protein present but not functional?

DR. PATERSON: (1) Characterization of the nucleoside transport protein (or complex) is presently at a very early stage. Jarvis and Young (*Biochem. J.* 194:373–376, 1981) have achieved a 13-fold purification of the NBMPR binding activity, which may have an in situ molecular weight of about 122,000, as indicated by radiation inactivation studies (Jarvis, S.M., Young, J.D., Ellory, J.C., *Biochem. J.* 190:373–376, 1980). Although it has not been demonstrated that the NBMPR binding site is an intrinsic part of the transporter, this is a reasonable assumption at present. (2) We know only that cells of the AE_1 clone lack nucleoside transport capability and NBMPR binding sites. It is conceivable that nonfunctional components of the nucleoside transporter may be present in AE_1 cells, but the means of their detection does not yet exist.

DR. D. GERMAN: (1) I noticed an increase in long-term survival in mice treated with transport inhibitors. This increase was prominent in the Ehrlich ascites carcinoma but, except for one case, was absent in leukemia cells. Could you comment on this? (2) Data were shown using dilazep with leukemic mice with no long-term survivors. Have you used dilazep with the Ehrlich ascites carcinoma?

DR. PATERSON: (1) Transplantable neoplasms of mice differ markedly in sensitivity to treatment with antineoplastic agents. The Ehrlich ascites carcinoma is sensitive to tubercidin and nebularine, whereas the leukemias L1210 and P388 are much less sensitive to these agents. (2) Drs. Jakobs and Fu have shown in our laboratory that administration of high, potentially lethal doses of nebularine together with host-protecting doses of dilazep were effective in treatment of mice implanted with the Ehrlich ascites carcinoma in that long-term survivors resulted. However, dilazep and tubercidin at dosages tolerated as single agents became toxic when coadministered.

III. ADENOSINE AND REGULATION OF IMMUNE FUNCTION

14. MECHANISMS OF TOXICITY OF ADENOSINE, 2′-DEOXYADENOSINE, AND INHIBITORS OF ADENOSINE DEAMINASE

J. FRANK HENDERSON

The mechanisms of toxicity of adenosine, deoxyadenosine, and adenosine deaminase inhibitors such as coformycin, deoxycoformycin, and EHNA (erythro-9-(2-hydroxy-3-nonyl)adenine) clearly are related in some manner, but exactly how? To what extent are these identical, to what extent similar, and to what extent quite unrelated? All have a relationship to the enzyme adenosine deaminase, but exactly what is the role of adenosine deaminase and its inhibition in the toxicity of each compound? This chapter reviews the present status of these and related questions and attempts to raise critical questions regarding both published and future work on this topic. Although the toxicities of adenosine, deoxyadenosine, and the several inhibitors of adenosine deaminase have been studied extensively [1–4], much still remains to be learned. (Because of space limitations, only selected references can be given.)

TOXIC EFFECTS

It is important, first, to recall that these agents have a very wide variety of toxicities, depending on the biological system employed, the parameter measured, drug dose and treatment schedule, and other experimental conditions. Thus, in experimental animals *in vivo*, adenosine or deoxyadenosine

The preparation of this chapter was supported by the National Cancer Institute of Canada, and the original work reported was described by the National Cancer Institute and Medical Research Council of Canada.

R.M. Berne, T.W. Rall, and R. Rubio (eds.), Regulatory Function of Adenosine, *pp. 223–234.*

treatment, or both, have been reported to produce liver cell dysfunction and necrosis, reduced clotting time, hemolysis, leukopenia, whole-body immobility and lethargy, weight loss, and death [5]. Treatment with one or another adenosine deaminase inhibitor, even without added adenosine or deoxyadenosine, may produce hemolysis, leukopenia, lymphoid depletion of the spleen, weight loss, growth retardation, increased spleen and lung weight, inhibition of tumor growth, and immunosuppression [6–11]; almost all these conditions have *not* been observed at other doses or schedules. Not only chemotherapeutic responses (usually of relatively short duration) have been obtained in human cancer patients treated with deoxycoformycin [12], but also central nervous system toxicity (lethargy, somnolence, seizures), renal insufficiency, liver enzyme abnormalities, conjunctival inflammation, and other toxic effects [13, 14].

When studied *in vitro*, adenosine or deoxyadenosine, or both, can inhibit growth and produce cell death in a very wide range of cultured cell systems [1, 2]; when lymphoblast lines of T and B cell origin are compared, the former have usually been found to be more sensitive. Other toxic effects that have been reported involve inhibition of blastogenesis by peripheral and spleen lymphocytes and by thymocytes, inhibition of lymphocyte-mediated cytolysis, inhibition of monocyte maturation, death of nonproliferating peripheral blood lymphocytes, immunoglobulin production, and prevention of the expression of a particular cell surface receptor in a subpopulation of circulating T lymphocytes [1, 2, 15, 16].

Treatment of one or another biological system *in vitro* by EHNA or deoxycoformycin, or both, has been reported to inhibit blastogenesis by peripheral blood lymphocytes, to inhibit monocyte activation, to inhibit thymocyte differentiation, to prevent differentiation of bone marrow precursor T cells, to delay the appearance of specific lymphocyte surface receptors, to inhibit T lymphocyte colony formation, to prevent spindle elongation in mitotic cells and microtubule-based intracellular motility, and to inhibit the replication of herpes simplex virus [17–25].

The very wide variety of biological effects produced by adenosine, deoxyadenosine, and adenosine deaminase inhibitors leads to the question of whether they are all due to one common mechanism or to many, whether even a single one of these compounds acts by a single mechanism in all of these systems or by several. It seems useful to divide the aforementioned observations of toxicity into the following classes (although this may be entirely too simple): (1) effects on rapidly dividing cells *in vitro* or *in vivo*, (2) effects on nondividing lymphoid cells, and (3) effects on other types of cells. The possibility or probability that different mechanisms may be involved in each has to be considered.

PHARMACOLOGICALLY ACTIVE FORMS

With this great biological complexity in mind, it is now possible to move on to a more biochemical question, namely, whether adenosine, deoxyadenosine, and the adenosine deaminase inhibitors are active of themselves or act through metabolites or other molecules.

In addition to considering the question of the active form of EHNA, coformycin, and deoxycoformycin, it is also necessary to ask whether these compounds might also inhibit enzymes other than adenosine deaminase. Here, in fact, the situation is growing increasingly complicated, and in various systems one or another adenosine deaminase inhibitor has been shown to inhibit the uptake of adenosine, the phosphorylation of this nucleoside, the relative rate of conversion of inosinate to adenine and guanine nucleotides, and purine nucleotide synthesis *de novo* and from adenine, guanine, and the like [26, 27]. In addition, the triphosphate of coformycin is an inhibitor of RNA polymerase [28]. Furthermore, a number of reports have shown that adenosine deaminase inhibitors can also inhibit adenylate deaminase [26, 29–31]. Although earlier publications found that the latter enzyme was considerably less sensitive to inhibition than was the former, a recent study has reported that in some cell types just the opposite is true: adenylate deaminase is more sensitive than adenosine deaminase [31].

Clearly, these additional effects of the adenosine deaminase inhibitors will be dependent on the cell type under investigation, as well as on the drug dose and treatment schedule employed. In many *in vitro* systems employing intact cells, it seems possible to use drug concentrations that will preferentially inhibit adenosine deaminase, although it appears that it is no longer possible to take this for granted [31]. In studies *in vivo,* however, the situation is very much more complicated, because of the relatively low permeability of many cells to at least some adenosine deaminase inhibitors, relatively rapid clearance and excretion, and so forth. Thus, relatively high drug doses sometimes have to be given to inhibit adenosine deaminase in animal tissues *in vivo* [8, 32], and at these doses adenylate deaminase (and possibly other enzymes as well) may also be at least partially inhibited. In mouse liver, for example, adenylate deaminase is very substantially inhibited at deoxycoformycin doses that are required to totally inhibit liver adenosine deaminase [D. Siepen and J.F. Henderson, unpublished]. It may be pointed out that doses of adenosine deaminase inhibitors that have been used in various studies both *in vivo* and *in vitro* have varied enormously; in some cases one wonders how much inhibition of adenosine deaminase actually was attained, and at high doses one wonders what other enzymes were also affected. Adequate controls have not always been carried out.

Another complication—one that especially affects studies *in vivo*—concerns the duration of inhibition of adenosine deaminase following treatment with inhibitors. This factor varies, first, depending on whether a very tight binding inhibitor (e.g., deoxycoformycin) or a less tight binding inhibitor (e.g., EHNA) was used [33]. In addition, however, it has been shown that duration of inhibition also varies with the tissue, as new, uninhibited adenosine deaminase is synthesized at different rates in different cell types [8, 34]. As might be expected, inhibition is very prolonged in erythrocytes.

The final complication, which will be mentioned only briefly, is that EHNA and the coformycins appear in some biological systems to have qualitatively different effects; the basis of such observations has never been investigated.

If these considerable complications are now put aside and the simplest case assumed, the following question still may be asked: Do adenosine deaminase inhibitors exert their toxicity in themselves or through other metabolites, or both? There have been, and to some extent still are, two different views regarding this question. An early view associated high adenosine deaminase activity with sensitivity to inhibitors of this enzyme. Implicit in this approach is the idea that inhibition of adenosine deaminase in a target cell such as rapidly dividing thymocytes is, by itself and in the absence of exogenous metabolites, an event that leads directly to growth inhibition and cell death. This view appears to be held at least tacitly by many investigators who measure adenosine deaminase activities in extracts of different types of malignancies and implicitly or explicitly propose that drug sensitivity is not only associated with but is also based upon high activities of this enzyme [35, 36]. In this case, therefore, deoxycoformycin (or EHNA) is thought to be toxic in and of itself.

The other view is that adenosine deaminase inhibitors act only in a passive or permissive way to allow either exogenous or endogenous adenosine, deoxyadenosine, or both, to exert their own toxic effects [32]. In this case, adenosine deaminase is viewed as having a protective role, and its inhibition merely makes the cell susceptible to the effects of adenosine or deoxyadenosine, effects that, in themselves, are quite independent of adenosine deaminase. In this view, then, it is adenosine or deoxyadenosine that is the real toxic agent.

There is much evidence to support the key role of adenosine or deoxyadenosine in the toxicity of adenosine deaminase inhibitors, and in at least the many cases in which these inhibitors have been found to be active in the absence of added adenosine or deoxyadenosine, it seems likely that one or another nucleoside was already present in the medium used. In addition, one could make the case that, from a pharmacological point of view, it would be easier for a given dose of drug to inhibit a small amount of target enzyme (i.e., adenosine deaminase) than a large amount.

Yet the belief prevails that high adenosine deaminase activity is associated with sensitivity to inhibitors of this enzyme and to adenosine and deoxyadenosine. One must ask, in the first place, how rigorously this hypothesis has actually been tested. For example, although cultured T lymphoblasts generally have higher adenosine deaminase activities and greater sensitivities to EHNA or deoxycoformycin than B lymphoblast lines, this is not always the case [37, 38]; furthermore, differences in the adenosine deaminase activities of peripheral blood T and B lymphocytes are small [39, 40]. Immunohistological assays of adenosine deaminase show considerable heterogeneity in activity, even among populations of thymocytes and of peripheral blood lymphocytes [41]. Furthermore, when adenosine deaminase activities are measured using intact rather than disrupted cells, nonsensitive peripheral blood granulocytes have higher activities than do lymphocytes [J.F. Henderson, unpublished]. Adenosine deaminase activities also vary widely among mammalian species [42, 43], yet blastogenesis is inhibited by adenosine even in horse lymphocytes [44], which have very low levels of this enzyme.

Whether the relationship in some cell types between high adenosine deaminase activity and high sensitivity to inhibition by adenosine, deoxyadenosine, and adenosine deaminase inhibitors is fortuitous or pertains to certain species (e.g., humans) remains to be studied further. Another possibility is that high adenosine deaminase is a marker for a particular class of cell (especially lymphatic cells) that for completely independent reasons has a high sensitivity to the toxic actions of these compounds.

An additional question concerns the relative importance of adenosine and of deoxyadenosine in the toxicity of adenosine deaminase inhibitors. Both nucleosides are toxic in terms of growth inhibition and cell death in cultured cell systems, and the current focus on deoxyadenosine is based on observations (1) that this compound accumulates and is excreted in patients who have an inherited deficiency of adenosine deaminase and (2) that deoxyadenosine and dATP accumulate in experimental animals and in human patients treated with adenosine deaminase inhibitors. The fate and possible effects of adenosine in such circumstances really have not been adequately studied. It may be noted, however, that plasma levels of both adenosine and deoxyadenosine are elevated in both situations, although urinary excretion only of deoxyadenosine is increased [45, 46].

Finally, the source or sources of adenosine and deoxyadenosine in animals treated with adenosine deaminase inhibitors needs to be considered. One question is whether the toxic adenosine or deoxyadenosine comes from within the target cell or from within the tissue or organ in which toxicity is exerted. Alternatively, does the toxic nucleoside come from outside the target cell or organ? This point remains to be studied with respect to adenosine; based on the numerous physiological roles of adenosine in individual tissues, however, one may suppose that much, if not all, of the adenosine that produces toxicity is formed within target organs.

This is less likely to be the case for deoxyadenosine, however, as DNA may be presumed to be substantially stable in target cells, deoxyribonucleotide pools are very small, and the regulation of ribonucleotide reductase would ensure that adenine deoxyribonucleotides are not produced in excess. Alternative sources would seem to include (1) the diet, (2) dead and dying cells, and (3) the nuclei that are lost during erythropoiesis. Recent studies using mice have shown that although dietary DNA does not seem to be an important source of deoxyadenosine in the body, both dying cells and erythroblast nuclei do contribute deoxyadenosine that is accumulated by other cells [32, 47]. It may be significant that the thymus is said to contain a subpopulation of cells that normally turn over (i.e., die); hence, this tissue would contain its own source of deoxyadenosine. However, dATP concentrations are no higher in this tissue than in liver or peripheral lymphocytes [C.M. Smith and J.F. Henderson, unpublished].

The problem of the active form of adenosine may be subdivided into the following specific questions: (1) Does adenosine have to enter cells to exert toxicity? (2) Is adenosine itself capable of exerting toxicity? (3) Does adenosine have to be converted to nucleotide metabolites before producing toxicity? (4) Exactly which nucleotide metabolites increase in concentration as a result of adenosine treatment?

It has been demonstrated clearly that in at least one system in which toxicity is manifested by growth inhibition and cell death, adenosine actually does have to enter cells [1]. However, in view of the presence of cell surface adenosine receptors in many cell types, and the rapidity with which some types of toxicity are expressed, this point really remains open and needs to be established in each system studied.

The need for nucleotide formation from added adenosine has been established in a number of experimental systems in which adenosine inhibits cell growth, usually through comparison of parent cells and sublines that lack adenosine kinase [1]. However, a few studies have shown nucleotide formation not to be a requirement [1].

Three candidate nucleotides that sometimes are thought to be involved in adenosine-caused growth inhibition are ATP, cyclic AMP, and dATP. A detailed review of measurements of ATP concentrations in cells treated with adenosine shows some cases in which these are elevated and in which this elevation could be related to inhibition of cell growth. In other cases, however, ATP concentrations have been elevated equally by toxic and nontoxic concentrations of ATP, or under other conditions when there was no growth inhibition. In addition, some of the increases in ATP concentrations that have been observed were quite small, and their significance is not at all certain [1].

Strong evidence in favor of cyclic AMP as the important nucleotide in adenosine toxicity has been presented in at least one biological system, that of lymphocyte mediated cytolysis, but equally good evidence for the lack of importance of this cyclic nucleotide has been found in other systems [1].

Finally, there are several observations that dATP concentrations are elevated in cells treated with adenosine, although the possible role of this adenosine metabolite has not been studied in most systems [1].

The same questions may be asked about the metabolism of deoxyadenosine as about adenosine. Does it have to enter cells, and does it have to be converted to dATP in order to exert toxic effects? Consideration of nucleotide synthesis from deoxyadenosine is very much complicated by the existence of a variety of kinases, depending on the biological system; thus, adenosine kinase, deoxycytidine-deoxyguanosine-deoxyadenosine kinase, deoxyguanosine-deoxyadenosine kinase, and very specific deoxyadenosine kinases all may be involved [2]. In any case, the general picture is that deoxyadenosine enters cells and is converted to nucleotides, and the concentrations of dATP are elevated.

There are problems, however, when one tries to relate elevations in dATP concentration to growth inhibition. Although in some cases there are good correlations among nucleoside dose, dATP concentrations, time, and extent of growth inhibition, in other cases the correlation is not good, and dATP may even be elevated without growth inhibition [2]. In many tissues *in vivo,* dATP concentration may be elevated without any toxicity being exerted [32]. Thus, dATP formation may be required for toxicity in some systems, but this certainly is not sufficient.

BIOCHEMICAL EFFECTS AND MECHANISMS OF ACTION

The final points to be considered are the mechanisms of action of adenosine and deoxyadenosine. This is not unrelated to the preceding question, of course, inasmuch as some putative target enzymes are inhibited by nucleotide derivatives, while others are inhibited by the free nucleosides.

Two potential target enzymes that have received the most attention are S-adenosylhomocysteine hydrolase and ribonucleotide reductase [1, 2]. The main evidence in favor of the involvement of ribonucleotide reductase is (1) increased concentrations of a potential endproduct inhibitor, dATP; (2) concomitant decreases in the concentrations of a distal product, dCTP; and (3) complete or partial protection against deoxyadenosine toxicity by deoxycytidine. In some systems, this evidence indeed looks convincing, although in no case has it been determined by direct measurement, using intact cells, that ribonucleotide reductase actually is inhibited. Questions can be raised in other cases, however, either because deoxycytidine does not protect or because dCTP concentrations can be restored to normal without actual prevention of toxicity, or because the concentrations of other deoxyribonucleoside triphosphates do not change in the expected manner.

Other questions regarding the rate of elevated concentrations of dATP and of inhibition of ribonucleotide reductase have recently been raised by studies of deoxyguanosine toxicity. Treatment of cultured cells with this nucleoside also leads to increased concentrations of an end-product inhibitor of the reductase, dGTP, and decreases in those of dCTP; partial protection and reversal are effected by deoxycytidine. In this case, however, growth inhibition is not characterized by accumulation of cells at the G_1-S boundary or the early S phase, as would be expected, but instead by accumulation in the middle of S phase [C.M. Smith and J.F. Henderson, unpublished]. Such observations raise many new questions, which have not yet been addressed.

Finally, observations that deoxyadenosine or deoxycoformycin still are toxic in nondividing lymphocyte or thymocyte populations [48, 49], or before DNA synthesis begins in mitogen-stimulated lymphocytes and thymocytes [50, 51], also are inconsistent with ribonucleotide reductase as the sole site of action.

Another important potential target enzyme, especially in nondividing cell populations, is S-adenosylhomocysteine hydrolase; evidence relating to this was presented earlier in this symposium. Inhibition or inactivation of this enzyme, or both, is believed to affect methylation reactions involving S-adenosylmethionine. However, it has been shown that adenosine, 2′-deoxyadenosine, and cordycepin (3′-deoxyadenosine) all can have different effects on methylation [52]; hence, it seems inappropriate to make too many generalizations at this stage. What seems necessary, therefore, is, first, a careful and thorough analysis and description of the effects of various hydrolase inhibitors on the many methylated compounds in cells and, second, a critical evaluation of the consequences of each case of undermethylation for toxicity. This evaluation needs to keep in mind almost certain differences between dividing and nondividing cell populations with respect to the consequences of undermethylation.

Finally, other biochemical effects of treatment with adenosine, deoxyadenosine, or adenosine deaminase inhibitors have also been described, and their contributions to the question of mechanisms of toxicity remain to be evaluated. Thus, ATP concentrations can be very much reduced in erythrocytes under certain conditions of deoxycoformycin or deoxyadenosine treatment; this has also been observed in some but not all lymphoid cell populations [6, 32]. This can lead to changes in cell membrane structure, to increased susceptibility to attack by extracellular phospholipases [53], and, at least potentially, to accelerated removal of cells from blood by the reticuloendothelial system. A wide variety of other effects on purine and pyrimidine metabolism and on RNA and protein metabolism have also been described in cells treated with one or another of these toxic agents, and their role in toxicity remains to be evaluated.

REFERENCES

1. Henderson JF, Scott FW: Inhibition of animal and invertebrate cell growth by naturally occurring purine bases and ribonucleosides. *Pharmacol Ther* 8:539–571, 1980.
2. Henderson JF, Scott FW, Lowe JK: Toxicity of naturally occurring purine deoxynucleosides. *Pharmacol Ther* 8:573–604, 1980.
3. Henderson JF: Inhibition of microbial growth by naturally occurring purine bases and ribonucleosides. *Pharmacol Ther* 8:605–627, 1980.
4. Glazer RI: Adenosine deaminase inhibitors: Their role in chemotherapy and immunosuppression. *Cancer Chemother Pharmacol* 4:227–235, 1980.
5. Paine RM, Weston BJ, Clink HM, Kohn J, Neville AM, McGhee KG, Harrap KR: Toxicity and immunosuppressive activity of binary combinations of 2′-deoxycoformycin and 2′-deoxyadenosine. *Cancer Treat Rep* 65:259–266, 1981.
6. Smith CM, Belch A, Henderson JF. Hemolysis in mice treated with deoxycoformycin, an inhibitor of adenosine deaminase. *Biochem Pharmacol* 29:1209–1210, 1980.
7. Smyth JF, Young RC, Young DM: In vivo toxicity to lymphoid tissue by 2′-deoxycoformycin. *Cancer Chemother Pharmacol* 1:49–51, 1978.
8. Agarwal RP: Deoxycoformycin toxicity in mice after long-term treatment. *Cancer Chemother Pharmacol* 5:83–87, 1980.
9. Hall JG, Gyure L, Peppard J, Orlans E: Levels of adenosine deaminase in some experimental animal tumours and the possible therapeutic effect of the ADA inhibitor 2-deoxy-coformycin. *Br J Cancer* 40:750–754, 1979.
10. Lum CT, Sutherland DER, Payne WD, Gorecki P, Matar AJ, Najarian JS: Prolongation of mouse and rat pancreatic islet cell allografts by adenosine deaminase inhibitors and adenine arabinoside. *J Surg Res* 28:44–48, 1980.
11. Tedde A, Balis ME, Ikehara S, Pahwa R, Good RA, Trotta PP: Animal model for immune

dysfunction associated with adenosine deaminase deficiency. *Proc Natl Acad Sci USA* 77:4899–4903, 1980.
12. Kanofsky JR, Roth DG, Smyth JF, Baron JM, Sweet DL, Ultmann JE: Treatment of lymphoid malignancies with 2′-deoxycoformycin. *Am J Clin Oncol* 5:179–183, 1982.
13. Major PP, Agarwal RP, Kufe DW: Deoxycoformycin: Neurological toxicity. *Cancer Chemother Pharmacol* 5:193–196, 1981.
14. Grever MR, Siaw MFE, Jacob WF, Neidhart JA, Miser JS, Coleman MS, Hutton JJ, Balcerzak SP: The biochemical and clinical consequences of 2′-deoxycoformycin in refractory lymphoproliferative malignancy. *Blood* 57:406–417, 1981.
15. Bessler H, Djaldetti M, Kupfer B, Moroz C: Two T lymphocyte subpopulations isolated from human peripheral blood following "in vitro" treatment with adenosine. *Biomedicine* 32:66–71, 1980.
16. Tsuchiga S, Nakae S, Konno T, Tada K: S-adenosylhomocysteine hydrolase activity in a lymphoblastoid cell line from a patient with adenosine deaminase deficiency disease. *J Inher Metab Dis* 4:197–201, 1981.
17. Yagawa K, Okamura J: Role of adenosine deaminase in activation of macrophages. *Infect Immun* 32:394–397, 1981.
18. Thuillier L, Garresi F, Cartier P: Inability of immunocompetent thymocytes to produce T-cell growth factor under adenosine deaminase deficiency conditions. *Cell Immunol* 63:81–90, 1981.
19. Ballet J-J, Insel R, Merler E, Rosen FS: Inhibition of maturation of human precursor lymphocytes by coformycin, an inhibitor of the enzyme adenosine deaminase. *J Exp Med* 143:1271–1276, 1976.
20. Shore A, Dosch HM, Gelfand EW: Role of adenosine deaminase in the early stages of precursor T cell maturation. *Clin Exp Immunol* 44:152–155, 1981.
21. Zuckerman SH, Kay NE, Douglas SD: Effect of adenosine deaminase inhibitors on Fcμ receptor expression in human T-cell cultures. *Cell Immunol* 56:112–119, 1980.
22. Lum CT, Schmidtke JR, Sutherland DER, Najarian JS: Inhibition of human T-cell rosette formation by the adenosine deaminase inhibitor erythro-9-(2-hydroxy-3-nonyl)adenine hydrochloride (EHNA). *Clin Immunol Immunopath* 10:258–261, 1978.
23. Aye MT, Dunne JV: Effect of 2′-deoxycoformycin on erythroid, granulocytic, and T-lymphocyte colony growth. *Blood* 58:1043–1046, 1981.
24. Cande WZ: Inhibition of spindle elongation in permeabilized mitotic cells by erythro-9-[3-(2-hydroxynonyl)]adenine. *Nature* 295:700–701, 1982.
25. North TW, Cohen SS: *Erythro*-9-(2-hydroxy-3-nonyl)adenine as a specific inhibitor of herpes simplex virus replication in the presence and absence of adenosine analogues. *Proc Natl Acad Sci USA* 75:4684–4688, 1978.
26. Henderson JF, Brox L, Zombor G, Hunting D, Lomax CA: Specificity of adenosine deaminase inhibitors. *Biochem Pharmacol* 26:1967–1972, 1977.
27. Plagemann PGW, Wohlhueter RM: 2′-deoxycoformycin inhibition of intracellular phosphorylation of adenosine in Novikoff rat hepatoma cells. *Biochem Pharmacol* 30:417–426, 1981.
28. Schröder HC, Schuster DK, Zahn RK, Müller WEG: A novel metabolic effect of the adenosine deaminase inhibitor coformycin, a potentiator of antiviral adenosine analogues. *Antiviral Res* 1:383–391, 1981.
29. Agarwal RP, Parks RE Jr: Potent inhibition of muscle 5′-AMP deaminase by the nucleoside antibiotics coformycin and deoxycoformymin. *Biochem Pharmacol* 26:663–666, 1977.
30. Debatisse M, Berry M, Buttin G: The potentiation of adenine toxicity to Chinese hamster cells by coformycin: Suppression in mutants with altered regulation of purine biosynthesis or increased adenylate deaminase activity. *J Cell Physiol* 106:1–11, 1981.
31. Fishbein WN, Davis JI, Winkert JW, Strong DM: Levels of adenosine deaminase, AMP deaminase, and adenylate kinase in cultured human lymphoblast lines: Exquisite sensitivity of AMP deaminase to adenosine deaminase inhibitors. *Biochem Med* 26:377–386, 1981.
32. Smith CM, Henderson JF: Deoxyadenosine triphosphate accumulation in erythrocytes of deoxycoformycin-treated mice. *Biochem Pharmacol,* 1982 (in press).
33. Plunkett W, Alexander L, Chubb S, Loo TL: Comparison of the activity of 2′-deoxycoformycin and erythro-9-(2-hydroxy-3-nonyl)adenine in vivo. *Biochem Pharmacol* 28:201–206, 1979.
34. Agarwal RP: Recovery of 2′-deoxyadenosine-inhibited adenosine deaminase of mouse erythrocytes and leukemia L1210 in vitro. *Cancer Res* 39:1425–1427, 1979.

35. Storch H, Krüger W, Ratzsch W: Adenosine deaminase activity in plasma and blood cells of patients with haematological and autoimmune diseases. *Acta Haematol (Basel)* 65:183–188, 1981.
36. Koya M, Kanoh T, Sawada H, Uchino H, Ueda K: Adenosine deaminase and ecto-5′-nucleotidase activities in various leukemias with special reference to blast crisis: Significance of ecto-5′-nucleotidase in lymphoid blast crisis of chronic myeloid leukemia. *Blood* 58:1107–1111, 1981.
37. Tritsch GL, Minowada J: Differences in purine metabolizing enzyme activities in human leukemia T-cell, B-cell and null-cell lines. *J Natl Cancer Inst* 60:1301–1304, 1978.
38. Mitchell BS, Mejias E, Daddona PE, Kelley WN: Purinogenic immunodeficiency diseases: Selective toxicity of deoxyribonucleosides for T cells. *Proc Natl Acad Sci USA* 76:5011–5014, 1978.
39. Nishida Y, Okudaira K, Tanimoto K, Akoata I: The differences in purine metabolism between T and B lymphocytes. *Exp Hematol* 8:593–598, 1980.
40. MacDermott RP, Tritsch GL, Formeister JF: Adenosine deaminase and nucleoside phosphorylase activities in normal human blood mononuclear cell subpopulations. *Clin Exp Immunol* 42:303–307, 1980.
41. Chechik BE, Schrader WP, Minowada J: An immunomorphic study of adenosine deaminase distribution in human thymus tissue, normal lymphocytes, and hematopoietic cell lines. *J Immunol* 126:1003–1007, 1981.
42. Carson DA, Kaye J, Wasson DB: Differences in deoxyadenosine metabolism in human and mouse lymphocytes. *J Immunol* 124:8–12, 1980.
43. Peters GJ, Oosterhof A, Veerkamp JH: Adenosine and deoxyadenosine metabolism in mammalian lymphocytes. *Int J Biochem* 13:445–455, 1981.
44. Magnuson NS, Perryman LE: In vitro effects of adenosine on lymphocytes and erythrocytes from horses with combined immunodeficiency. *J Clin Invest* 64:89–101, 1979.
45. Hirschhorn R, Roegner V, Rubinstein A, Papageorgiou P: Plasma deoxyadenosine, adenosine, and erythrocyte deoxy ATP are elevated at birth in an adenosine deaminase-deficient child. *J Clin Invest* 65:768–771, 1980.
46. Venner PM, Glazer RI, Blatt J, Sallan S, Rivera G, Holcenberg JS, Lipton J, Murphy SB, Poplack DG: Levels of 2′-deoxycoformycin, adenosine, and deoxyadenosine in patients with acute lymphoblastic leukemia. *Cancer Res* 41:4508–4511, 1981.
47. Chan T-S: Purine excretion by mouse peritoneal macrophages lacking adenosine deaminase activity. *Proc Natl Acad Sci USA* 76:925–929, 1979.
48. Kefford RF, Fox RM: Purine deoxynucleoside toxicity in nondividing human lymphoid cells. *Cancer Res* 42:324–330, 1982.
49. Kefford RF, Fox RM: Deoxycoformycin-induced response in chronic lymphocytic leukaemia: Deoxyadenosine toxicity in non-replicating lymphocytes. *Br J Haematol* 50:627–636, 1982.
50. Uberti J, Lightbody JJ, Johnson RM: The effect of nucleosides and deoxycoformycin on adenosine and deoxyadenosine inhibition of human lymphocyte activation. *J Immunol* 123:189–193, 1979.
51. Thuillier L, Garreau F, Cartier PF: Consequences of adenosine deaminase deficiency on thymocyte metabolism. *Eur J Immunol* 11:788–794, 1981.
52. Glazer RI, Hartman KD: Evidence that the inhibitory effect of adenosine, but not cordycepin, on the methylation of nuclear RNA is mediated by S-adenosylhomocysteine hydrolase. *Mol Pharmacol* 18:483–490, 1980.
53. Higgins TJC, Bailey PJ, Allsopp D: The influence of ATP depletion on the action of phospholipase C on cardiac myocyte membrane phospholipids. *J Mol Cell Cardiol* 13:1027–1030, 1981.

DISCUSSION

DR. H.J. LEIGHTON: (1) In those tissues which are relatively resistant to adenosine deaminase, do you view this as a case for different forms of enzyme or a case for rapid turnover of the enzyme? (2) Do you have any data on the relative adenosine deaminase activity in mast cells, basophils and lymphocytes?

DR. HENDERSON: (1) There is evidence that both could be the case, but it is also possible that drug distribution also affects the extent of inhibition in different tissues. (2) No.

DR. T. SPECTOR: (1) We have isolated adenosine deaminase from jejunum (rat, dog, and human) and have found it to be sensitive to inhibition by EHNA and deoxycoformycin (dCF). (2) We have confluent nondividing serum-starved cells that have very low basal levels of ribonucleotide reductase and still are killed by hydroxy urea and thiosemicarbazones, which are specific reductase inhibitors. In view of this, perhaps the importance of reductase inhibition in nondividing peripheral lymphocytes may not be discounted as a possible mechanism of dCF toxicity.

DR. HENDERSON: (1) Thank you. It is possible, then, that the observed difficulty in completely inhibiting interstitial adenosine deaminase *in vivo* may be due either to problems in drug distribution or simply to the large amount of enzyme that is present in intestine, especially in rodents. (2) This is a very important observation. Further study of this system should be very helpful.

DR. A.C. NEWBY: There is evidence from several sources that deoxycoformycin penetrates different cells at different rates. What is known about the mechanism of penetration of dCf into cells? Can this account for the observed differences in rates of entry?

DR. HENDERSON: Perhaps Dr. Plagemann can help you on this point.

DR. P.G.W. PLAGEMANN: Deoxycoformycin seems to be a substrate for the nucleoside transporter of cultured animal cells, but a relatively inefficient one. The Michaelis-Menten constant for the transport has not been determined, but indirect results indicate that it falls in the millimolar range. Deoxycoformycin enters cells relatively slowly compared to the velocity of transport for adenosine. Whether these data also pertain to tissue cells in the body is not known at present.

DR. T.P. ZIMMERMAN: Concerning nonadenosine deaminase-related effects of EHNA, we have found that this compound is a hyperbolic mixed-type inhibitor of the high-affinity cAMP phosphodiesterase assayed in lymphocyte homogenates. The K_i value is below 100 μM. It is therefore important to limit the concentration of EHNA used in biological test systems to $\leq 10\ \mu M$ in order to avoid this type of nonspecific side effect.

DR. HENDERSON: Thank you. This may help explain why EHNA sometimes has effects that deoxycoformycin does not have.

DR. P.M. UELAND: Dr. Helland and I have recently done some preliminary experiments involving continuous administration of 2-deoxycoformycin (dCf) into mice from Alget microosmotic pumps. We found inactivation of S-adenosylhomocysteine (AdoHcy) hydrolase to 10% to 20% of the activity in nontreated animals in most tissues, but a persistent, residual activity seems sufficient to prevent accumulation of

AdoHcy, even in spleen, where the increase in AdoHcy content is slight and probably not significant. There seems to be a relation between 2′-deoxyadenosine accumulation and AdoHcy hydrolase inactivation. (1) What is your opinion about the role of AdoHcy in the biological effects of dCf? (2) Is it possible that AdoHcy hydrolase inactivation has implications not readily recognized—that is, not related to AdoHcy content in tissues?

DR. HENDERSON: It would be interesting to see if methylation reactions are inhibited under these conditions, even if AdoHcy does not appear to accumulate.

DR. J.S. FEDAN: Could the mechanisms of toxicity you described be occurring during experiments (*in vivo*) on preparations comprised of nondividing cells, e.g., smooth muscle? Are they factors to be considered when evaluating mechanisms of response to adenosine occurring over a period of a few hours?

DR. HENDERSON: There may not be too much toxicity associated with responses to adenosine involving physiological mechanisms. It really has not been well-characterized.

DR. J.P.R.M. VON LAARHOVEN: You have told us about the inhibitory effects of adenosine and deoxyadenosine on lymphocyte function. We have stimulated human peripheral blood lymphocytes with T cell-dependent B cell mitogen and studied the [^{3}H]leucine and [^{3}H]thymidine incorporation as well as immunoglobulin secretion. We have observed that in the presence of EHNA, deoxyadenosine at low concentrations ($\leq 1\ \mu M$) increases both protein synthesis and Ig secretion, whereas [^{3}H]thymidine in corporation is significantly decreased. Adenosine does not show significant effects under these conditions. Could you speculate on the biochemical implications of these findings?

DR. HENDERSON: The growth and development of some cells are stimulated by low concentrations of purines, even though high purine concentrations are inhibitory. Since deoxyadenosine can be cleaved to adenine even in the presence of adenosine deaminase inhibitors, I wonder if adenine is such a stimulatory purine under these conditions.

DR. V. CHAGOYA DE SANCHEZ: You mentioned that adenosine deaminase activity is low in horse lymphocytes. Do you know if these cells are less sensitive to adenosine toxicity or blastogenesis inhibition or if they accumulate adenosine?

DR. HENDERSON: Horse lymphocytes still are sensitive to inhibition for blastogenesis by adenosine, but I do not remember whether the dose-response relationship is the same as in humans. They also take up and metabolize adenosine.

15. CLINICAL EXPRESSION OF BLOCKS OF PURINE NUCLEOTIDE DEGRADATION

IRVING H. FOX

Blocks of purine nucleotide degradation may lead to an accumulation of the intermediates or their metabolites in human body fluids or cells. The accumulation of such intermediates may be toxic to certain cells and may lead to clinical disorders such as immunodeficiency.

In 1972 two patients with severe combined immunodeficiency disease were found to have an absence of adenosine deaminase [1]. This observation had major impact in that it focused on a new association between the purine pathways and the immune system. A second disorder, purine nucleoside phosphorylase deficiency associated with T cell immunodeficiency, was observed in 1975 [2]. These enzyme deficiencies, which represent blocks in the pathway of purine nucleotide degradation (Fig. 15.1), appear to be etiologically related to the immune dysfunction. The observation of these inborn errors of metabolism has therapeutic relevance in directing investigators to innovative approaches to pharmacologic immunosuppression. This has already occurred on an experimental basis.

In this discussion we will examine the clinical expression of both inherited and acquired blocks of purine nucleotide degradation.

The work reported here was supported by USPHS grants AM19674 and 5 M01 RR00042 and a grant from the Michigan Heart Association and Warner/Lambert–Parke/Davis. The author wishes to thank Jean Wolter for excellent typing of the manuscript.

R.M. Berne, T.W. Rall, and R. Rubio (eds.), Regulatory Function of Adenosine, *pp. 235–247.*

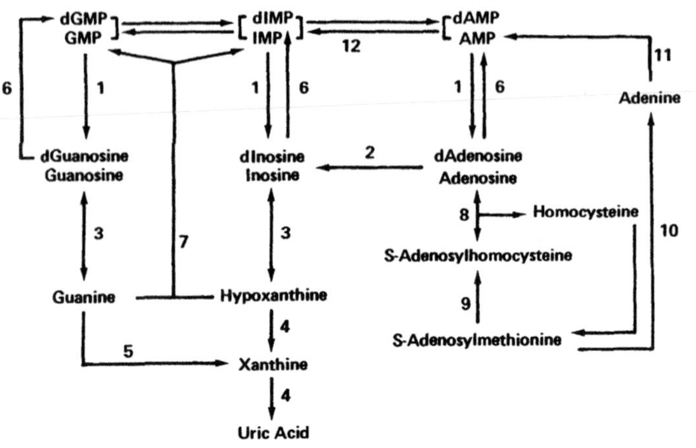

Figure 15.1. Pathway of purine nucleotide degradation. Purine nucleoside monophosphate derivatives are degraded to uric acid in humans by a final common pathway. Nucleotides are dephosphorylated to form the nucleoside derivatives. Specific 5′-phosphomonoesterase (EC: 3.1.3.5) and nonspecific phosphatases (EC: 3.1.3.2) hydrolyze AMP, IMP, GMP, or their deoxynucleoside monophosphates to the deoxyinosine, deoxyguanosine, and guanosine are converted to purine bases and ribose-1-phosphate or deoxyribose-1-phosphate by purine nucleoside phosphorylase (EC: 2.4.2.1, reaction 3). Adenosine and deoxyadenosine are deaminated to inosine and deoxyinosine by adenosine deaminase (EC: 3.5.4.6, reaction 2). Guanine is deaminated to xanthine by guanine deaminase (EC: 3.5.4.3, reaction 5). Hypoxanthine and xanthine are oxidized to uric acid by xanthine oxidase (EC: 1.2.3.2, reaction 4) located primarily in the liver and jejunum of mammals. These pathways are interrupted by reactions that allow the resynthesis of nucleotides. Nucleoside kinases catalyze the conversion of adenosine to AMP, deoxyadenosine to dAMP, deoxyinosine to dIMP, and deoxyguanosine to dGMP (reaction 6). Hypoxanthine-guanine phosphoribosyltransferase (EC: 2.4.2.8, reaction 7) converts hypoxanthine or guanine to IMP or GMP. Adenine phosphoribosyltransferase (EC: 2.4.2.7, reaction 11) catalyzes the formation of AMP from adenine. The precursor substrates of the final common pathway of purine nucleotide degradation are formed from the digestion of dietary nucleoprotein, the degradation of nucleic acids to nucleoside monophosphates, and the formation of nucleotides by de novo purine synthesis and the purine salvage pathways. In addition, S-adenosylhomocysteine is degraded to adenosine and homocysteine by S-adenosylhomocysteine hydrolase (EC: 3.3.1.1, reaction 8) in an essential step of S-adenosylmethionine-mediated methylation reactions (reaction 9). S-adenosylmethionine may also be decarboxylated to enter polyamine metabolism (reaction 10). Degradation of deoxyribonucleotides follows a similar pattern, as indicated. However, dAMP may be converted to dIMP at only 6% of the V_{max} of AMP (reaction 12), whereas dIMP may be converted to dGMP or dAMP at virtually the same V_{max} as IMP. dGMP may be converted back to dIMP at about 1% of the V_{max} of GMP. The dephosphorylation of deoxynucleotides, the deamination of deoxyadenosine, and the phosphorolysis of deoxyinosine or deoxyguanosine are all similar to the reactions of the ribosyl derivatives. GMP, guanosine 5′-monophosphate; dGMP, deoxyguanosine 5′-monophosphate; IMP, inosine 5′-monophosphate; dIMP, deoxyinosine 5′-monophosphate; AMP, adenosine 5′-monophosphate; dA,P, deoxyadenosine 5′-monophosphate.

ADENOSINE DEAMINASE DEFICIENCY

Inborn error

Clinical features

The hallmark of this disorder is severe combined immunodeficiency disease [1, 3, 4]. This presentation occurs in 85% to 90% of adenosine deaminase-deficient patients within the few first months of life. Usually by three months of age, the patients are subject to recurrent chronic infections of viral, fungal, protozoal, and bacterial organisms, which frequently are lethal. These types of infections in the respiratory, dermal, and gastrointestinal systems are characteristic of impairment of both cell-mediated and antibody-dependent immunity. Tonsils, adenoids, and thymus become atrophic during the first few months of life. Patients have diarrhea, malabsorption, and candidiasis. Affected children are lymphopenic, with a markedly reduced number of T cells. *In vitro* tests of lymphocyte function are depressed. These include lymphocyte-proliferative responses to mitogen, antigen, and the mixed lymphocyte response. Clinical evidence of delayed hypersensitivity impairment includes inability to reject a foreign skin graft and inability to mount a delayed hypersensitivity skin response to antigens or to alter sensitization and challenge by dinitrochlorobenzene. Agammaglobulinemia usually occurs, and specific antibody responses are absent.

Approximately 10% to 15% of patients suffer from a less severe disease. The disorder may become evident later than three to six months, and immunoglobulin retention may occur. These children have normal or near normal amounts of immunoglobulins, with normal to increased percentages of B cells.

Other abnormalities noted in these patients include X-ray evidence of osteoporosis, small or absent thymus gland, chondro-osseous dysplasia at the costochondral junction, apophysis of iliac bones, and vertebral bodies. Renal anomalies have occurred in 10% to 15% of the patients including transient renal tubular acidosis, ureteropelvic anomalies, and obliteration of renal glomerular capillaries. Fine or sparse hair has been noted in about 10% of patients. Neurological abnormalities, including spasticity, tremor, nystagmus choreoathetoid movements, head lag, and developmental delay, have occurred in 10% to 15% of patients [5]. There is also a disorder of decreased platelet aggregation [6, 7].

Biochemical features

$$\begin{matrix}\text{Adenosine} \\ \text{Deoxyadenosine} + H_2O\end{matrix} \rightarrow \begin{matrix}\text{Inosine} \\ \text{Deoxyinosine} + NH_3\end{matrix}$$

Severe combined immunodeficiency disease associated with adenosine deaminase deficiency may be diagnosed by assay of peripheral circulating

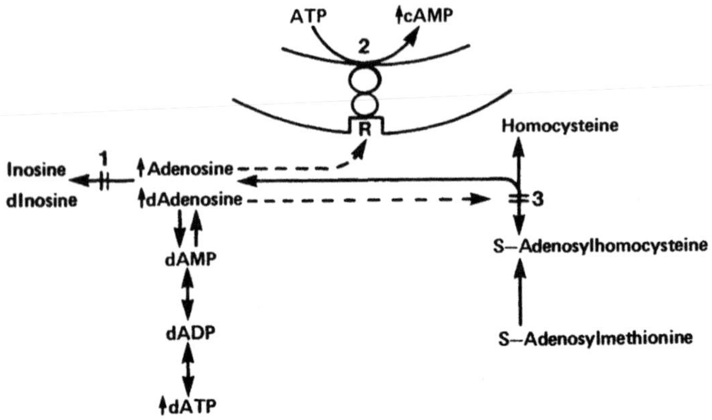

Figure 15.2. Biochemical changes associated with adenosine deaminase deficiency. When adenosine deaminase is deficient (reaction 1), adenosine may accumulate and modify the activity of the adenosine receptor and, subsequently, intracellular cAMP (reaction 2). Deoxyadenosine accumulation leads to increased intracellular dATP levels and inactivation of S-adenosylhomocysteine hydrolase (reaction 3).

erythrocytes. Assay of adenosine deaminase generally shows little or no activity. The diagnosis may also be confirmed by showing substantially decreased enzyme activities in cultured diploid fibroblasts or cultured lymphoblasts.

Biochemical abnormalities appear to account for the clinical syndrome. Elevated adenosine and deoxyadenosine concentrations in the urine, plasma, and erythrocytes result from the adenosine deaminase deficiency (Fig. 15.2). Increased concentrations of dATP and dADP in erythrocytes, lymphocytes, and bone marrow cells and, in some instances, decreased concentration of ATP in erythrocytes from these patients results from the deoxyadenosine accumulation and may have toxic properties toward the immune system [8, 9]. Increased lymphocyte and platelet cyclic AMP concentrations may also contribute to the dysfunction of these cells [7, 10]. A secondary severe deficiency of erythrocyte S-adenosylhomocysteine hydrolase has been observed [11, 12]. This may be related to the deoxyadenosine accumulation, since deoxyadenosine has been shown to inactivate S-adenosylhomocysteine hydrolase.

Adenosine deaminase is structurally altered in tissues from patients with severe combined immunodeficiency disease (reviewed in [13]). The properties of the mutant enzymes provide support for heterogeneity of the mutations involving the structural gene coding for adenosine deaminase. Structural alterations of adenosine deaminase also occur in patients with the enzyme deficiency and normal immune function. In these patients, white cell adenosine deaminase is less severely deficient; this amount is adequate for normal function of the immune system.

Genetics

In general, severe combined immunodeficiency diseases are inherited as X-linked recessive or autosomal recessive diseases. Adenosine deaminase is coded on chromosome 20, and the deficiency accounts for about half the patients with the autosomal recessive form of inheritance. In most instances, the enzyme deficiency seems to be related to a structural gene mutation, based on the studies of mutant enzyme described earlier.

Unequivocal diagnosis of two affected children has been made prenatally by amniocentesis [4]. Ideally, prenatal diagnoses should be done with prior knowledge of the quantitative and qualitative characteristics of adenosine deaminase activity in cells of the affected child from the particular family at risk.

Heterozygote detection for this disorder is possible for peripheral erythrocytes. However, 10% of obligate heterozygotes have erythrocyte adenosine deaminase activity within two standard deviations of the normal mean [4].

Treatment

Experimental therapies attempt to restore immune function in patients with severe combined immunodeficiency and adenosine deaminase deficiency. The results of these trials support an etiological relationship between the enzyme deficiency and the immune disorder. About 50% of the patients with severe combined immunodeficiency disease and adenosine deaminase deficiency respond to enzyme replacement therapy with packed irradiated erythrocytes [14]. The response includes an increase in the absolute lymphocyte count and improved responsiveness to the mixed lymphocyte culture and to phytohemagglutinin stimulation. Also, there is a diminution in the tissue deoxyadenosine triphosphate concentrations, a decrease in the excretion of adenosine and deoxyadenosine, and an elevation of erythrocyte S-adenosylhomocysteine hydrolase [15]. Bone marrow transplantation corrects most of the immunological and many biochemical abnormalities and appears to be the treatment of choice [16, 17]. It provides lymphoid stem cells and partially clears abnormally increased metabolites from nonlymphoid body compartments [17]. Erythrocyte dATP levels were markedly decreased from pretreatment but remained significantly elevated. Plasma adenosine was unchanged, whereas plasma deoxyadenosine was indistinguishable from normal. Urinary deoxyadenosine excretion was markedly decreased but remained abnormal, whereas adenosine excretion remained in the normal range. Erythrocyte S-adenosylhomocysteine hydrolase remained abnormal.

Acquired enzyme deficiency

Adenosine deaminase may be inhibited in patients with the administration of deoxycoformycin, an analogue of deoxyadenosine. Deoxycoformycin is a unique tight binding transition state inhibitor with an ID_{50} of 0.01 to 1 nM. The K_i for deoxycoformycin was found to be 2.5 pM [18]. This compound has now been administered to patients with lymphoproliferative malignancies or solid tumors [19–25] in doses ranging from 0.1 to 1.0 mg/kg/day and is capable of virtual complete inhibition of both erythrocyte and leukocyte adenosine deaminase.

The administration of deoxycoformycin to patients causes biochemical alterations similar to the changes seen in patients with inherited adenosine deaminase deficiency. Plasma levels of adenosine and deoxyadenosine are elevated and dATP levels are increased in the erythrocytes and leukocytes. Adenine deoxynucleotides accumulate in other tissues such as spleen, liver, and kidney [23]. An important observation appears to be the decrease in ATP levels in circulating erythrocytes during therapy.

The effects of deoxycoformycin administration on the patients is in many instances dramatic. Profound lymphocytotoxicity is observed in the majority of patients with doses as low as 0.1 mg/kg, showing over 50% reduction in patients within 48 hours to 88% to 97% reduction of lymphocyte count. In patients who are being treated for lymphoblastic malignancies, the lymphoblast count in many instances is reduced in the peripheral blood as well. No changes have been observed in the immunoglobulin levels determined five days after the start of treatment [21]. The two major and dose-limiting toxicities reported by investigators involve the central nervous system and the kidney. In the central nervous system, the most common abnormality observed is somnolence; lethargy, and occasionally coma and seizures have been observed. In some instances, it is unclear whether this is related to the drug or to the nature of the disease process. In many patients, deterioration of renal function has occurred with evidence of renal insufficiency, as measured by a rise in creatinine. Renal failure has been reported occasionally, due to acute tubular necrosis. Other documented reactions include nausea and vomiting, which are very frequent, diarrhea, liver function abnormalities, conjunctivitis, and pulmonary edema.

It is not clear from these studies what the basis is for the alterations observed in patients. However, they ultimately result from the accumulation of adenosine and deoxyadenosine. The possibilities include dATP accumulation, ATP depletion, S-adenosylhomocysteine hydrolase inactivation with inhibition of methylation, and direct effects of adenosine accumulation.

PURINE NUCLEOSIDE PHOSPHORYLASE DEFICIENCY

Inborn error

Clinical features

The hallmark of this disease is a disorder of cellular immunity [2, 26]. The patients have relatively intact immune function early in life, with progressive deterioration of function over time. Death has occurred anywhere from 18 months up to 12 years from fatal infections that develop from progressive disease from small pox immunization or exposure to viral infections such as varicella or graft versus host reaction subsequent to blood transfusion. From an immunological point of view, the patients have lymphopenia, markedly decreased T cells in the periphery as measured by rosette-forming cells, negative skin tests or delayed hypersensitivity, and a decreased proliferative response to mitogens and allogeneic cells. The thymus is severely hypocellular.

Humoral immune function appears intact or enhanced, as shown by normal or elevated immunoglobulin concentrations, normal numbers of peripheral B lymphocytes, and normal antibody response to antigens. The occurrence of monoclonal gammopathy, rheumatoid factor, antinuclear antibody, and Coombs positive hemolytic anemia indicate excessive antibody production. Features resembling systemic lupus erythematosus are evident in some of these patients, suggesting that the enzyme deficiency may provide a model for this autoimmune disorder.

Three of the ten patients described have had evidence for a central nervous system disorder consisting of ataxia, tremor, spasticity, and tetraplegia [26]. However, it is unclear at present whether the neurological abnormalities are related to purine nucleoside phosphorylase deficiency or are secondary to a central nervous system viral infection.

Biochemical features

Guanosine Inosine		Guanine		Ribose-1-P
	+ P_i ↔		+	
Deoxyinosine Deoxyguanosine		Hypoxanthine		Deoxyribose-1-P

The observation of a patient with a disorder of cellular immunity and hypouricemia should suggest the existence of this abnormality. A definitive diagnosis can be made from the assay of peripheral erythrocytes, which show a marked deficiency or absence of purine nucleoside phosphorylase. Substantially decreased activities may be observed in fibroblast or lymphoblast cell lines derived from these patients.

Patients with this disorder have evidence of a block of purine nucleotide degradation. There is decreased synthesis of uric acid with hypouricemia, hypouricosuria, and an excessive urinary excretion of inosine, guanosine, deoxyinosine, and deoxyguanosine. These patients are overproducers of purines despite the hypouricemia. The overproduction of purines is accompanied by an elevation in the concentration of phosphoribosylpyrophosphate in erythrocytes to a level similar to that observed in patients with a deficiency of hypoxanthine-guanine phosphoribosyltransferase. This is related to a loss of active reutilization of purine bases in purine nucleoside phosphorylase

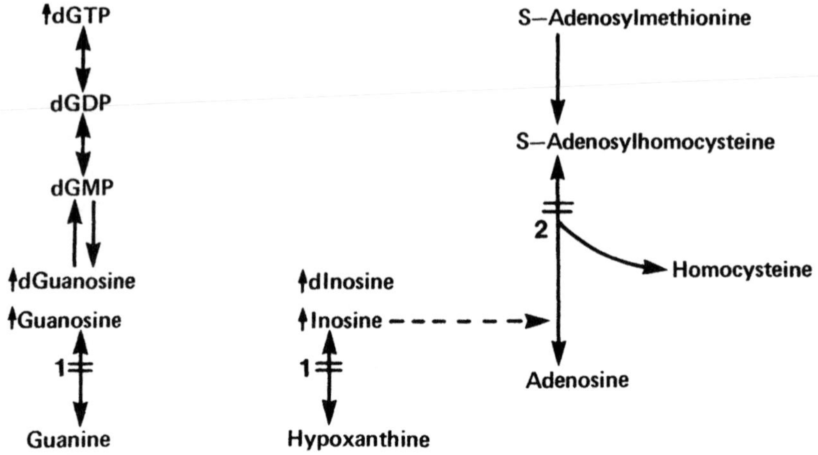

Figure 15.3. Biochemical changes associated with purine nucleoside phosphorylase deficiency. When purine nucleoside phosphorylase is deficient (reaction 1), deoxyguanosine may accumulate and lead to increased dGTP levels. Inosine, which builds up during the enzyme deficiency, inactivates S-adenosylhomocysteine hydrolase (reaction 2).

deficiency from the lack of formation of hypoxanthine or guanine, substrates for hypoxanthine-guanine phosphoribosyltransferase. Deoxyguanosine triphosphate concentrations are elevated in erythrocytes [27, 28]. This is believed to result from the accumulation of deoxyguanosine and its subsequent phosphorylation (Fig. 15.3). Increased concentrations of dGTP may be the toxic factor to the immune system. The degree of abnormality in uric acid levels, nucleoside excretion, and dGTP concentrations reflects the severity of the enzyme deficiency. These patients also have an 80% decrease in erythrocyte S-adenosylhomocysteine hydrolase activity, but the significance of this observation remains unclear [12, 29].

Structural alterations of the decreased purine nucleoside phosphorylase provide evidence for structural gene mutations and genetic heterogeneity in this disorder (reviewed in [13]).

Genetics

The inheritance of this disorder is autosomal recessive. The gene that codes for purine nucleoside phosphorylase has been localized to chromosome 14. Studies of structural alterations of the mutant enzyme provide evidence for structural gene mutations and genetic heterogeneity in this disorder. Heterozygotes may be detected by assay of erythrocyte purine nucleoside phosphorylase, which is about half the normal value and greater than two standard deviations below the mean. Amniocentesis has been attempted in a pregnancy in a family that had previously borne a child with an enzyme deficiency. In this case, the enzyme activity was normal and a normal fetus was indeed delivered.

Treatment

Enzyme replacement therapy with blood transfusions has only partial success in patients with purine nucleoside phosphorylase deficiency [26, 30, 31]. Two of four patients had modest improvement. There was an increase in the percentage of peripheral E-rosette-forming cells without an increase in the lymphocyte count. The lymphocytes stimulated with phytohemagglutinin. A slight reaction to skin tests became evident in one patient. Transfusion therapy in these patients increased the serum urate concentration and the urine uric acid excretion and decreased urine nucleoside excretion. Partial and transient responses to thymosin administration or thymus epithelial transplants have been reported [26].

Acquired deficiency

No drug that blocks purine nucleoside phosphorylase has been administered to humans to date. One drug, 8-aminoguanosine, has promising features with potent specific inhibitory properties in enzyme kinetic studies and in a cell culture system [32].

SUMMARY

Blocks of purine nucleotide degradation lead to an accumulation of intermediates in human body fluids and cells, which lead to immunodeficiency. Adenosine deaminase deficiency is an autosomal recessive disease. It presents by age 3 months, with recurrent chronic infections of viral, fungal, protozoal, and bacterial organisms to which patients are likely to succumb. There is impairment of both cell-mediated and antibody-dependent immunity. Affected children are lymphopenic, with a markedly reduced number of T cells. *In vitro* tests of lymphocyte function are depressed. Agammaglobulinemia usually occurs, and specific antibody responses are absent. Tonsils, adenoids, and thymus become atrophic during the first few months of life. Other abnormalities noted in these patients include diarrhea, candidiasis, bone abnormalities, renal anomalies, fine or sparse hair, neurological abnormalities, and decreased platelet aggregation. Biochemical abnormalities appear to account for the clinical syndrome and include elevated adenosine and deoxyadenosine concentrations in the urine, plasma, and erythrocytes; increased concentrations of dATP in erythrocytes, lymphocytes, and bone marrow cells; increased lymphocyte and platelet cyclic AMP concentrations; and a secondary severe deficiency of erythrocyte S-adenosylhomocysteine hydrolase.

Acquired adenosine deaminase deficiency has been induced clinically by intravenous deoxycoformycin as therapy for patients with lymphoblastic leukemias. Lysis of tumor cells, toxicity to the patient, and similar biochemical changes to the inherited deficiency characterize this enzyme deficiency state.

Purine nucleoside phosphorylase deficiency with a disorder of cellular immunity is an autosomal recessive disease. The patients have relatively intact immune function early in life, with progressive deterioration of function over time. Death occurs from 18 months up to 12 years from fatal infections after small pox immunization or exposure to viral infections. The patients have lymphopenia, markedly decreased T cells, and reduced *in vitro* T cell function. Humoral immune function appears intact or enhanced, as shown by normal or elevated immunoglobulin concentrations, normal numbers of peripheral B lymphocytes, and normal antibody response to antigens. Features resembling systemic lupus erythematosus are evident in some patients. Some patients have a central nervous system disorder. Patients have a block of purine nucleotide degradation, with hypouricemia, hypouricosuria, and an excessive urinary excretion of inosine, guanosine, deoxyinosine, and deoxyguanosine. In erythrocytes, deoxyguanosine triphosphate concentrations are elevated and S-adenosylhomocysteine hydrolase is decreased by 80%.

These blocks of purine nucleotide degradation provide models for how purines regulate the immune system and for the development of innovative therapeutic approaches to the immune system.

REFERENCES

1. Giblett ER, Anderson JE, Cohen F, Pollara B, Meuwissen HG: Adenosine-deaminase deficiency in two patients with severely impaired cellular immunity. *Lancet* 2:1067–1069, 1972.
2. Giblett, ER, Ammann AJ, Wara DW, Diamond LK: Nucleoside-phosphorylase deficiency in a child with severely defective T-cell immunity and normal B-cell immunity. *Lancet* 1:1010–1013, 1975.
3. Meuwissen HJ, Pickering RJ, Pollara B, Porter IH: *Combined Immunodeficiency and Adenosine Deaminase Deficiency: A Molecular Defect.* New York, Academic Press, 1975.
4. Hirschhorn R: Clinical delineation of adenosine deaminase deficiency in *Enzyme Defects and Immune Dysfunction,* Ciba Foundation Symposium No. 68. Amsterdam, Elsevier, 1979, pp 35–49.
5. Hirschhorn R, Papageorgiou PS, Kesarwala HĤ, Taft LT: Amelioration of neurologic abnormalities after enzyme replacement in adenosine deaminase deficiency. *N Engl J Med* 303:377–380, 1980.
6. Schwartz AL, Polmar SH, Stern RC: Abnormal platelet aggregation in severe combined immunodeficiency disease with adenosine deaminase deficiency. *Br J Haematol* 39:189–194, 1978.
7. Lee CH, Evans SP, Rozenberg MC, Bagnara AS, Ziegler JB, Van der Weyden MB: *In vitro* platelet abnormality in adenosine deaminase deficiency and severe combined immunodeficiency. *Blood* 53:464–471, 1979.
8. Coleman MS, Donofrio J, Hutton JJ, Hahn L, Daoud A, Lampkin B, Dyminski J: Identification and quantitation of adenine deoxynucleotides in erythrocytes of a patient with adenosine-deaminase deficiency and severe combined immunodeficiency disease. *J Biol Chem* 253:1619–1626, 1978.
9. Cohen AR, Hirschhorn R, Horowitz SD, Rubinstein A, Polmar SH, Hong R, Martin DW Jr: Deoxyadenosine triphosphate as a potentially toxic metabolite in adenosine deaminase deficiency. *Proc Natl Acad Sci USA* 75:472–476, 1978.
10. Schmalstieg FC, Nelson JA, Mills GC, Monahan T, Goldman AS, Goldblum RM: Increased purine nucleotides in adenosine deaminase deficiency lymphocytes. *J Pediatr* 91:48–51, 1977.
11. Hershfield MS, Kredich NM, Ownby DR, Ownby H, Buckley R: *In vivo* inactivation of erythrocyte S-adenosylhomocysteine hydrolase by 2′-deoxyadenosine in adenosine deaminase-deficient patients. *J Clin Invest* 63:807–811, 1979.
12. Kaminska J, Fox IH: Decreased S-adenosylhomocysteine hydrolase in inborn errors of purine metabolism. *J Lab Clin Med* 96:141–147, 1980.

13. Fox IH: Metabolic bases for disorders of purine nucleotide degradation. *Metabolism* 30:616–634, 1981.
14. Polmar SH: Enzyme replacement and other biochemical approaches to therapy of adenosine deaminase deficiency, in *Enzyme Defects and Immune Dysfunction,* Ciba Foundation Symposium No. 68. Amsterdam, Elsevier, 1979, pp 213–223.
15. Hirschhorn R, Roegner V, Rubinstein A: Plasma deoxyadenosine, adenosine and erythrocyte dATP are elevated at birth in an adenosine deaminase-deficient child. *J Clin Invest* 65:768–771, 1980.
16. Chen S, Ochs HD, Scott CR: Adenosine deaminase deficiency disappearance of adenosine deoxynucleotides from a patient's erythrocytes after successful marrow transplantation. *J Clin Invest* 62:1386–1389, 1978.
17. Hirschhorn R, Roegner-Maniscalco V, Kuritsky L, Rose FS: Bone marrow transplantation only partially restores purine metabolites to normal in adenosine deaminase-deficient patients. *J Clin Invest* 68:1387–1393, 1981.
18. Glazer RI: Adenosine deaminase inhibitors: Their role in chemotherapy and immunosuppression. *Cancer Chemother Pharmacol* 4:227–235, 1980.
19. Mitchell BS, Koller CA, Heyn R: Inhibition of adenosine deaminase activity results in cytotoxicity to T-lymphoblasts *in vivo. Blood* 56:556–559, 1980.
20. Siaw MFE, Mitchell BS, Koller CA, Coleman MS, Hutton JJ: ATP depletion as a consequence of adenosine deaminase inhibition in man. *Proc Natl Acad Sci USA* 77:6157–6160, 1980.
21. Smyth JF, Paine RM, Jackman AL, Harrap KR, Chassin MM, Adamson RH, Johns DG: The clinical pharmacology of the adenosine deaminase inhibitor 2′-deoxycoformycin. *Cancer Chemother Pharmacol* 5:93–101, 1980.
22. Poplack DG, Sallan SE, Rivera G, Holcenberg J, Murphy SB, Blatt J, Lipton JM, Venner P, Glaubiger DL, Ungerleider R, Johns D: Phase 1 study of 2′-deoxycoformycin in acute lymphoblastic leukemia. *Cancer Res* 41:3343–3346, 1981.
23. Grever MR, Siaw MFE, Jacob WF, Neidhart JA, Miser JS, Coleman MS, Hutton JJ, Balcerzak SP: The biochemical and clinical consequences of 2′-deoxycoformycin in refractory lymphoproliferative malignancy. *Blood* 57:406–417, 1981.
24. Major PP, Agarwal RP, Kufe DW: Clinical pharmacology of deoxycoformycin. *Blood* 58:91–96, 1981.
25. Russell NH, Prentice HG, Lee N, Piga A, Ganeshaguru K, Smyth JF, Hoffbrand AV: Studies on the biochemical sequelae of therapy in Thy-acute lymphoblastic leukemia with adenosine deaminase inhibitor 2′-deoxycoformycin. *Br J Haematol* 49:1–9, 1981.
26. Ammann AJ: Immunological aberrations in purine nucleotide phosphorylase deficiencies, in *Enzyme Defects and Immune Dysfunction,* Ciba Foundation Symposium No. 68. New York, Exerpta Medicus, 1979, pp 55–69.
27. Cohen A, Guda LJ, Ammann AJ, Staal GEJ, Martin DW Jr: Deoxyguanosine triphosphate as a possible toxic metabolite associated with purine nucleoside phosphorylase deficiency. *J Clin Invest* 61:1405–1409, 1978.
28. Wortmann RL, Andres CM, Kaminska J, Mejias E, Gelfand E, Arnold W, Rich W, Fox IH: Purine nucleoside phosphorylase deficiency: Biochemical properties and heterogeneity in two families. *Arthritis Rheum* 22:524–531, 1979.
29. Hershfield MS: Proposed explanation for S-adenosylhomocysteine hydrolase deficiency in purine nucleoside phosphorylase and hypoxanthine-guanine phosphoribosyltransferase deficient patients. *J Clin Invest* 67:696–701, 1981.
30. Zegers BJM, Stoop JW, Staal GEJ, Wadman SK: An approach to the restoration of T-cell function in a purine nucleoside phosphorylase deficient patient, in *Enzyme Defects and Immune Dysfunction,* Ciba Foundation Symposium No. 68. Amsterdam, Elsevier, 1979, pp 231–247.
31. Rich KC, Mejias E, Fox IH: Nucleoside phosphorylase deficiency: Improved immune function with enzyme replacement therapy. *N Engl J Med* 303:973–977, 1980.
32. Kazmers IS, Mitchell BS, Dadonna PE, Wotring LL, Townsend LB, Kelly WN: Inhibition of purine nucleoside phosphorylase by 8-aminoguanosine selective toxicity for T-lymphoblasts. *Science* 214:1137–1139, 1981.

DISCUSSION

DR. D. GERMAN: (1) Can you comment on homocysteine levels in the plasma of these immunodeficient patients? (2) If homocysteine levels are decreased, and increases in S-adenosylhomocysteine (SAH) inhibit transmethylation, there are two ways to inhibit methylation—that is, decreased homocysteine for remethylation and known inhibition of transmethylation by S-adenosylmethionine (SAM) due to increased SAH. Transcobolamin II deficiency is another immune deficiency state that prevents remethylation of homocysteine to methionine and could lend more support to the idea that methylation is important here. Please comment. (3) Have patients with adenosine deaminase deficiency or purine nucleoside phosphorylase deficiency been treated with pharmacologic doses of vitamin B_{12}?

DR. FOX: (1) No. Drs. Hershfield and Kredich of Duke University would be most likely to have this information. (2) Investigators have focused on the agammaglobulinemia of this disorder and hypogammaglobulinemia of acquired B_{12} deficiency. They have proposed that perhaps the disordered methylation hypothesis could account for abnormal B cell function. This may be naive. Your comment may have validity but will require further study.

DR. S.H. POLMAR: B_{12} has been given to a patient with purine nucleoside phosphorylase deficiency with no effect.

DR. J.E. SEEGMILLER: It seems to me we have a most curious anomaly to explain. From our presentation and discussion of the past two days we might predict that patients with adenosine deaminase deficiency who have plasma concentrations of adenosine from 10- to 100-fold increases above normal should have at least some evidence of chronic vasodilation or cardiovascular abnormality; yet only the immune system shows pathological function. How can we reconcile this disparity?

DR. R.M. BERNE: Patients with adenosine deaminase deficiency have long-standing elevation of their plasma adenosine levels and it has been demonstrated by Dr. Hester that prolonged exposure to vasodilator doses of adenosine results in a gradual return of blood flow to control levels over a period of several hours, despite continuous administration of adenosine. The mechanism for this abolition of a vascular response to adenosine is unknown.

DR. SEEGMILLER: You argue, then, that chronic exposure to elevated plasma concentrations of adenosine in these children leads to a compensating readjustment of the threshold for a vasodilator response.

DR. BERNE: Yes.

DR. POLMAR: Adenosine deaminase deficiency may be quite heterogeneous in its clinical and immunologic severity. Is there an enzymatic or metabolic basis for this heterogeneity?

DR. FOX: This has not been clearly demonstrated. In purine nucleoside phosphorylase deficiency, there has been reported a family with 0.5% of normal enzyme activity with less severe immunologic and metabolic alterations.

DR. J.P. MANFREDI: Why do you exclude 5′-nucleotidase deficiency as a possible cause of immune dysfunction?

DR. FOX: It seems that 5′-nucleotidase deficiency (ecto-5′-nucleotidase) is a result of abnormal lymphocytes rather than leading to lymphocyte dysfunction. During lymphocyte development 5′-nucleotidase increases. Therefore, immature lymphocytes or incompletely differentiated cells will have lower enzyme values. In addition, 5′-nucleotidase deficiency has been observed in other diseases, such as adenosine deaminase deficiency. Finally, inhibiting 5′-nucleotidase does not change *in vitro* measurements of lymphocyte function.

DR. J.P.R.M. VAN LAARHOVEN: Recently I had a discussion with Dr. Prentice from the group of Drs. Janossy and Hoffbrant from London. They were treating T-ALL patients with deoxycoformycin. I was told that they were forced to stop the administration of the drug because of severe renal toxicity, which might be due to this treatment. What is your comment on this?

DR. FOX: There are two ways by which deoxycoformycin therapy could cause nephrotoxicity: (1) acute uric acid nephropathy or xanthine nephropathy from the massive lysis of cells and degradation of purine to their end products and (2) direct toxic properties related to deoxycoformycin by mechanisms that have been discussed by Dr. Henderson.

DR. C.I. THOMPSON: Part of the renal dysfunction with deoxycoformycin treatment may be due to a direct effect of high plasma adenosine on renal function (i.e., GFR), or on the renin-angiotensin system. Why does adenosine deaminase deficiency present such an increased level of adenosine (and deoxyadenosine), since these cells presumably have sufficient levels of adenosine kinase, which should rephosphorylate adenosine or deoxyadenosine readily?

DR. FOX: The answer is not clear. It is possible that there may be inhibition of adenosine phosphorylation by deoxyadenosine, which is a substrate for adenosine kinase, or by S-adenosylhomocysteine, which is an inhibitor of adenosine kinase.

DR. SEEGMILLER: However, less severe deficiency of purine nucleoside phosphorylase does result in an altered clinical presentation. Would you care to describe for us your experiences with this type of disorder?

DR. FOX: That is correct. Together with Dr. E. Gelfand's group in Toronto, we evaluated two brothers with purine nucleoside phosphorylase deficiency with 0.5% of enzyme activity. These patients appeared to have a milder disease from an immunologic point of view. Metabolically, they had evidence for a lesser block of purine nucleoside phosphorylase than other patients. Their serum urate concentrations were normal and their urine uric acid excretion was within the normal range.

16. ADENOSINE AND CYCLIC NUCLEOTIDES IN MODULATION OF IMMUNE RESPONSES

STEPHEN H. POLMAR, RUTH E. BIRCH, RAYA MANDLER,
STEPHEN A. RUDOLPH

The discovery of an immunodeficiency disease associated with the deficiency of adenosine deaminase focused attention on the relationship between adenosine metabolism and immune function [1]. Adenosine receptors similar to those described by Sattin and Rall [2] in brain have been identified on mouse [3] and human lymphocytes [4, 5]. Adenosine activates lymphocyte adenylate cyclase and increases cAMP levels by its interaction with adenosine receptors [4, 5] with similar properties to R sites on other tissues [6]. Since β-adrenergic catecholamines, prostaglandins of the E series (PGE), histamine, and other pharmacologic agents that increase cAMP synthesis influence a wide variety of immune responses [7], it is not surprising that adenosine would have effects on immune function as well. In most instances, agents that increase cAMP levels inhibit immune functions such as lymphocyte proliferation [8], cell-mediated cytolysis [3, 9] and IgE-mediated release of histamine from mast cells [10]. However, in some instances adenosine acts differently from other agents that also increase cAMP levels. For example, adenosine potentiates mast cell histamine release, whereas isoproterenol inhibits it [11]. In this chapter, the immunoregulatory properties of adenosine will be discussed and, as in the

This work was supported by a grant from the USPHS National Institutes of Health (AI-17570). The authors gratefully acknowledge the technical assistance of Mrs. Jill Pugh and the assistance of Mrs. Nancy Ginsburg in the preparation of the manuscript.

R.M. Berne, T.W. Rall, and R. Rubio (eds.), Regulatory Function of Adenosine, *pp. 249–259.*

foregoing example, adenosine's immunologic actions are markedly different from those of other adenylate cyclase activating agents.

EFFECTS OF ADENOSINE ON THE EXPRESSION OF LYMPHOCYTE CELL SURFACE RECEPTORS AND ANTIGENS

In our initial studies, we examined the effects of a variety of pharmacologic agents on the expression of receptors for the Fc portion of IgG (RFcγ) on human lymphocytes, particularly human T lymphocytes. Moretta and coworkers [12] reported that T lymphocytes bearing RFcγ functioned as suppressors of B cell differentiation *in vitro*. Moreover, the binding of IgG containing immune complexes to the RFcγ-bearing T cells enhanced their immunosuppressive activity [13]. Therefore, pharmacologic agents capable of modulating the expression of RFcγ on T cells might also be expected to have interesting immunoregulatory effects.

Human peripheral blood lymphocytes were isolated from heparinized venous blood by Ficoll-Hypaque density centrifugation and were depleted of monocytes by adherence to plastic. T-cell-enriched and non-T-cell (i.e., B-cell-enriched) preparations were obtained by rosetting with sheep erythrocytes (E) followed by Ficoll-Hypaque centrifugation, as previously described [14]. T lymphocyte preparations were further fractionated according to the theophylline stability of their E receptors by the method of Limatibul et al. [15]. Theophylline-resistant (T_R) lymphocytes function as helper/inducer cells for T-cell-dependent B cell differentiation, whereas theophylline-sensitive (T_S) lymphocytes suppress B cell differentiation. T_R lymphocyte preparations are depleted of RFcγ-bearing cells, whereas T_S lymphocyte preparations are enriched for RFcγ-bearing cells [14, 16]. RFcγ were determined using bovine erythrocytes coated with rabbit IgG antibody, as previously described [14].

Incubation of peripheral blood lymphocytes with adenosine at 37°C resulted in a 2- to 3-fold increase in the percentage of cells expressing receptors for the Fc portion of IgG (RFcγ). Increased RFcγ expression was detectable within 15 min of exposure to adenosine, was maximal at 30 min, and decreased to baseline after 60 min. If adenosine was removed at 30 min by washing the cells with fresh medium, the newly expressed RFcγ were stable for at least 24 hr in culture. Optimal RFcγ induction was observed at 10 μM adenosine [14]. RFcγ could not be induced if cells were incubated with adenosine at 4°C or in the presence of sodium azide.

The lymphocyte population and subpopulation specificity of adenosine-induced RFcγ enhancement is shown in Figure 16.1. Adenosine causes an increase in RFcγ expression in peripheral blood lymphocytes and in T lymphocyte and T_R lymphocyte preparations but not in the non-T cell (B-cell-enriched) or T_S lymphocyte preparations. The increased percentage of lymphocytes expressing RFcγ is due to increased RFcγ expression in the T_R lymphocyte subpopulation [14]. The effect of adenosine on the expression of other lymphocyte surface markers is shown in Table 16.1. E receptors, surface

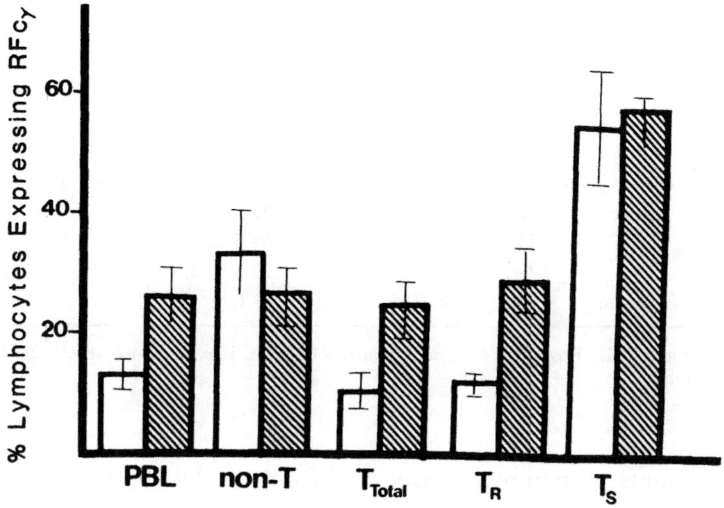

Figure 16.1. Effect of adenosine upon RFcγ expression in lymphocyte subpopulations. The cross-hatched bars indicate lymphocyte preparations incubated with 10^{-5} *M* adenosine for 30 min at 37°C. The open bars indicate lymphocyte preparations incubated without adenosine.

Table 16.1. The effect of adenosine (Ado) on surface marker expression of T helper/inducer (T_R) lymphocytes

Marker	% Lymphocytes expressing marker (± SEM) Control	Ado ($10^{-5}M$)	*p*
RFcγ	7.2 ± 1.3	16.2 ± 2.0	<.01
E_S	88.0 ± 2.2	90.4 ± 1.6	n.s.
sIg	0	0	–
Clq	3.6 ± 1.1	2.0 ± 0.6	n.s.
Ia	5.0 ± 3.0	2.9 ± 0.8	n.s.
β2μ	40.5 ± 4.5	60.7 ± 5.1	<.025

Source: From Birch, R.E., Rosenthal, A.K., and Polmar, S.H.: *Clin. Exp. Immunol.* 48:231–237, 1982.
n.s. = not significant

immunoglobulin (sIg), membrane Clq, and Ia antigen expression on T_R (helper/inducer) lymphocytes (Table 16.1) and unfractionated T lymphocytes (data not shown) are unaffected by adenosine under conditions that maximally enhance RFcγ expression [17]. Clq was studied because this protein can bind immune complexes; Ia antigen is associated with activation of T-lymphocytes. However, there was a significant increase in the percentage of T lymphocytes with readily detectable membrane-associated β_2 microglobulin when assayed by indirect immunofluorescent antibody technique. The appearance of in-

Table 16.2. The effect of adenosine (Ado) on the expression of T lymphocyte subset antigens detected by monoclonal antibodies on T helper/inducer (T_R) lymphocytes

Monoclonal antibody	% Lymphocytes expressing antigen (± SEM)		
	Control	Ado (10^{-5} *M*)	*p*
OKT3	92.9 ± 1.1	93.6 ± 0.9	n.s.
OKT4	71.8 ± 4.7	58.3 ± 3.0	<.005
OKT5	20.2 ± 2.3	20.5 ± 2.8	n.s.
OKT8	16.5 ± 3.8	33.0 ± 4.3	<.0025

Source: From Birch, R.E., Rosenthal, A.K. and Polmar, S.H.: Clin. Exp. Immunol. 48:231–237, 1982.
n.s. = not significant

creased amounts of membrane-associated β_2 microglobulin may be related to increased RFcγ expression, since Birch and co-workers [18] have shown that incubation of T lymphocytes with β_2 microglobulin results in an increase in RFcγ+ T cells.

Monoclonal antibodies have been developed to surface membrane antigens that identify functionally distinct human T lymphocyte subsets [19]. The antigen defined by the monoclonal antibody OKT3 is present on all mature T lymphocytes. The OKT4 antigen is present primarily on T-helper/inducer cells, whereas the OKT5 and OKT8 antigens are present on T-suppressor and T-cytotoxic lymphocytes. As shown in Table 16.2, T_R lymphocyte preparations are enriched for OKT4+ cells. Incubation of T_R cells with adenosine under conditions that enhance RFcγ expression results in a significant decrease in the percentage of OKT4+ cells and an equivalent increase in OKT8+ cells; expression of OKT3 and OKT5 antigens is unaffected by adenosine [18]. Changes in OKT4 and OKT8 expression occur rapidly; the decrease in OKT4 antigen expression is detectable within 5 min of incubation with adenosine and is maximal in 15 min. The increase in OKT8 expression is detectable within 15 min and is maximal at 30 min.

EFFECTS OF ADENOSINE ON THE IMMUNOREGULATORY FUNCTION OF T LYMPHOCYTES

Adenosine causes an increase in T lymphocyte receptors and antigens associated with suppressor activity (i.e., RFcγ and OKT8) while causing a decrease in the expression of antigens associated with T-helper/inducer function (i.e., OKT4). The effects of adenosine on the immunoregulatory activity of T lymphocyte subsets were studied using an *in vitro* B lymphocyte differentiation assay [12]. In this assay, pokeweed mitogen (PWM) induces differentiation of B lymphocytes into immunoglobulin-secreting plasma cells; T-helper/inducer cells (T_R) facilitate differentiation, whereas T suppressor cells (T_S) inhibit the appearance of plasma cells in culture. As shown in Figure 16.2,

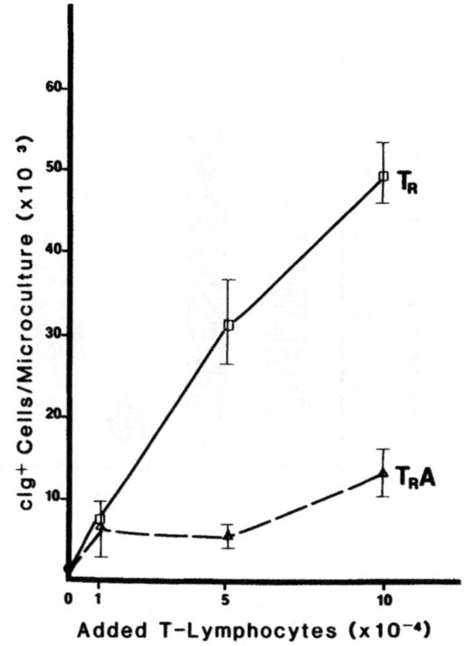

Figure 16.2. Effect of adenosine preincubation on the ability of T helper/inducer (T_R) lymphocytes to enhance B cell differentiation *in vitro*. T_RA designates T_R cells that have been preincubated with 10^{-5} *M* adenosine for 30 minutes at 37°C, then washed with fresh medium to remove adenosine, prior to use in the B cell differentiation assay.

addition of increasing numbers of T_R cells (1–10 × 10^4/culture) to a constant number of non-T cells (B cell source, 5 × 10^4/culture) results in the appearance of significantly increased numbers of plasma cells in PWM-stimulated cultures after 6 days. However, if the T_R cells are preincubated with adenosine $10^{-5}M$ at 37°C for 30 min and then washed to remove the adenosine, the adenosine-treated T_R cells (T_RA) lose their ability to facilitate B cell differentiation. As shown in Figure 16.3, the T_RA cells exhibit suppressor activity by inhibiting the development of plasma cells in cultures containing non-T cells and untreated T_R cells [20]. The suppressor activity that develops in the adenosine-treated T_R cell preparations appears to be due to the 10% to 20% of the cells that express RFcγ after exposure to adenosine. Removal of RFcγ+ cells from T_RA preparations restores the ability of the T_RA cells, now RFcγ-depleted, to express helper activity, whereas the T_RA RFcγ+-enriched preparations continue to function as potent suppressor cells (Fig. 16.4). Thus, adenosine-induced suppression of B cell differentiation is intimately associated with the appearance of RFcγ+ cells. The T_RA suppressor cells differ from the theophylline-sensitive (T_S) suppressor cells in that they retain a theophylline-resistant E receptor, have stable RFcγ in culture, are radioresistant, and need not proliferate in order to express their suppressor activity [20].

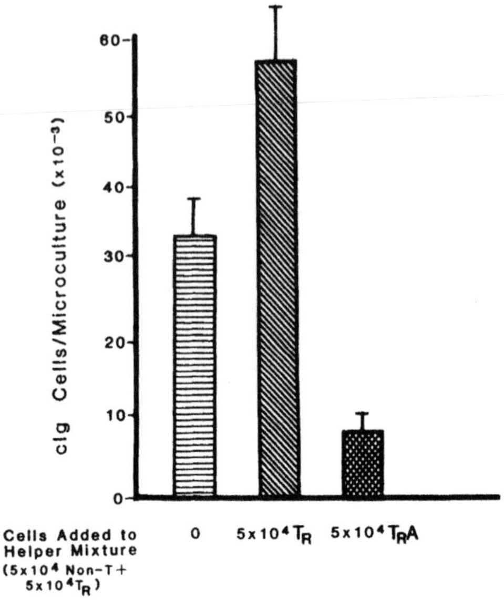

Figure 16.3. Suppression of B cell differentiation by adenosine-treated T helper/inducer lymphocytes (T_Ra).

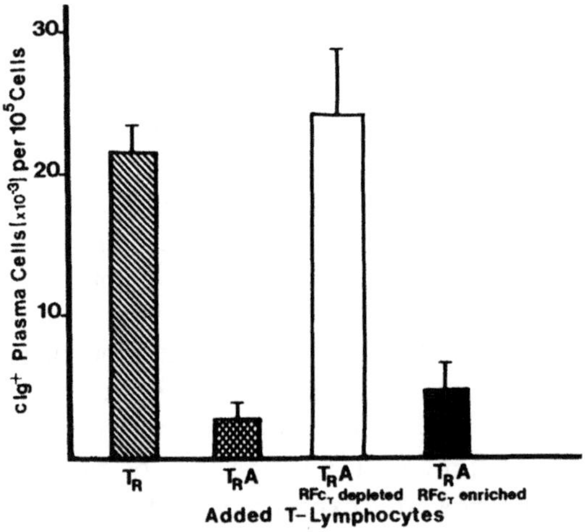

Figure 16.4. Suppressor activity of adenosine-treated T helper/inducer (T_RA) cells is due to those cells bearing newly expressed RFcγ after incubation with adenosine.

Table 16.3. Effects of adenosine and adenosine analogues on RFcγ expression of T helper/inducer (T_R) lymphocytes

	RFcγ+ cells, % of untreated control
Adenosine (10^{-5} *M*)	188 ± 4.6
IBMX (10^{-4} *M*)	94 ± 27
Adenosine (10^{-5} *M*) + IBMX (10^{-4} *M*)	123 ± 24
Adenine (10^{-5} *M*)	82 ± 17
3'-deoxyadenosine (10^{-5} *M*)	100 ± 3
2',5'-dideoxyadenosine (10^{-4} *M*)	80 ± 10

POSSIBLE MECHANISMS OF ADENOSINE MEDIATED IMMUNOSUPPRESSION: RELATIONSHIP TO ADENOSINE RECEPTORS AND CYCLIC AMP

The appearance of RFcγ on the T-helper/inducer (T_R) subpopulation occurs within 30 minutes of their exposure to adenosine and the cells which bear these newly expressed RFcγ are responsible for the immunosuppressive activity. Therefore, the appearance of RFcγ provides a convenient way to examine the early pharmacologic events associated with adenosine-induced immunosuppression. Adenosine-induced RFcγ expression appears to require interaction of adenosine with an R-type adenosine receptor. The response is maximal at 10 μ*M* adenosine, the same concentration found to be optimal for activation of lymphocyte adenylate cyclase [5]. The response is inhibited by methylxanthines (e.g., theophylline and isobutyl methylxanthine IBMX) and requires the presence of an unmodified ribose ring (Table 16.3).

Adenosine induces a rapid but transient rise in T_R lymphocyte cyclic AMP levels in the absence of a phosphodiesterase inhibitor. The peak cAMP level is seen within 1–5 min and returns to baseline by 10 min (Fig. 16.5). Inhibition of the cAMP rise by preincubation of T_R cells with 2'5'-dideoxyadenosine prior to the addition of adenosine inhibits increased RFcγ expression. Thus, the initial cAMP rise appears to be necessary. However, other agents that increase T_R lymphocyte cAMP to levels comparable to those achieved with 10 μ*M* adenosine, such as isoproterenol and PGE_2, do not significantly increase RFcγ expression (Table 16.4). These observations might be interpreted as indicating that adenosine enhancement of RFcγ expression either is a highly compartmentalized response or is unrelated to cAMP. However, the cAMP peaks produced by isoproterenol and PGE_2 are more prolonged than the adenosine cAMP peak. Both an initial rapid rise and fall in cAMP appear to be necessary for release of histamine from mast cells [21]. Adenosine has been found to potentiate both the cyclic AMP rise and fall and to enhance histamine release, whereas isoproterenol produces more sustained cAMP levels and inhibits histamine release [11]. It is possible that both the rapid rise and fall in cAMP seen

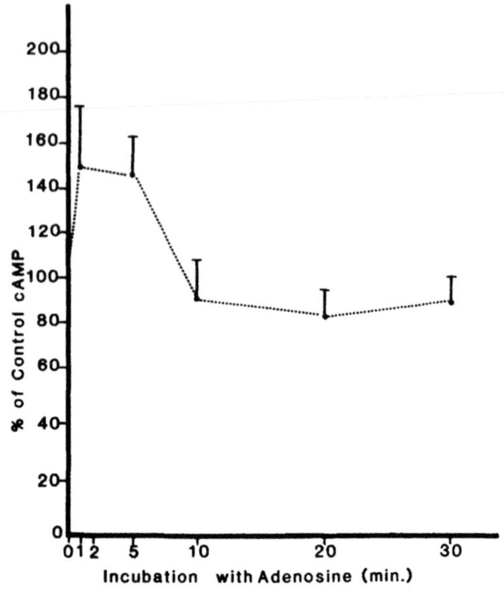

Figure 16.5. Changes in T helper/inducer (T_R) cAMP levels over time after incubation with 10^{-5} *M* adenosine at 37°C.

Table 16.4. Effects of agents that modulate cAMP levels on RFcγ expression of T helper/inducer (T_R) lymphocytes

	RFcγ+ cells, % of untreated control
Adenosine (10^{-5} *M*)	188 ± 4
2′,5′-dideoxyadenosine	80 ± 10
Adenosine (10^{-5} *M*) + 2′,5′-dideoxyadenosine (10^{-4} *M*)	83 ± 17
Prostaglandin E_2 (10^{-6} *M*)	123 ± 23
Isoproterenol (10^{-6} *M*)	117 ± 47
2-chloroadenosine (10^{-5} *M*)	100 ± 4
8-Br-cAMP (10^{-5} *M*)	197 ± 24

with adenosine stimulation are important for the development of RFcγ and immunosuppressive activity. The rapid termination of the adenosine cAMP peak may result, in part, from inhibition of adenylate cyclase activity by adenosine via a P site adenosine receptor [6]. This hypothesis is consistent with the finding that a 2-chloroadenosine does not increase RFcγ expression (Table 16.4). This analogue activates cAMP via an R adenosine receptor but is very slowly taken up and metabolized by cells [22] and does not appear to stimulate P sites [M.E. Maguire, personal communication]. Moreover, incubation of T_R

lymphocytes with 8-Br-cAMP for 5 min, followed by rapid removal of the agent by washing with fresh medium, results in enhanced RFcγ expression comparable to that seen with adenosine (Table 16.4) [23]. However, if 8-Br-cAMP is left in the incubation mixture for longer periods of time, RFcγ are not expressed. Whereas the adenosine effect on RFcγ is antagonized by IBMX, the 8-Br-cAMP effect is not inhibited by IBMX, indicating that the 8-Br-cAMP enhancement of RFcγ is not mediated by an R site adenosine receptor. Exogeneous cAMP is unable to increase RFcγ expression, whereas 8-Br-cAMP as well as 8-N_3-cAMP can. The two cAMP analogues are rapidly taken up by lymphocytes, but cAMP itself is not. These findings suggest that RFcγ enhancement and immunosuppression require changes in the intracellular concentrations of cAMP.

Adenosine also increases the occupancy of intracellular cAMP binding sites, thought to be the regulatory subunits of protein kinase, and increases protein phosphorylation in T and T_R lymphocytes within 10 min [23, 24]. Increased phospholipid methylation and arachidonic acid release are observed within 30 min of incubation of T and T_R lymphocytes with adenosine [24]. The relationship among these later events, RFcγ expression and the development of immunosuppressive activity, remains to be elucidated.

CONCLUDING REMARKS

Distinct immunoregulatory T lymphocyte subpopulations exist in humans. These subpopulations can be identified by the presence of specific antigens and receptors on the lymphocyte surface, and these cell surface characteristics are closely correlated with cellular function [19]. We have shown that adenosine can modulate the expression of some subset-specific antigens and receptors as well as modify the immunoregulatory function of some T lymphocytes. Since a cell's response to stimuli in its environment often depends on the presence of specific receptors (e.g., hormone receptors) on the cell membrane, agents that modify cell surface receptor display might also be expected to modify cellular behavior. In this regard, adenosine, through its action on adenosine receptors, is capable of modulating the expression of immunologic receptors. This heterologous receptor regulation appears to require the interaction of adenosine with an R site adenosine receptor, followed by adenylate cyclase activation and increased cAMP levels. The other biochemical events initiated by this adenosine-cAMP signal that lead to the development of immunosuppressive activity are currently under investigation.

REFERENCES

1. Giblett EL, Anderson JE, Cohen F, Pollara B, Meuwissen HJ: Adenosine deaminase deficiency in two patients with severely impaired cellular immunity. *Lancet* 2:1067–1069, 1972.
2. Sattin A, Rall TW: The effect of adenosine and adenine nucleotides on the cyclic adenosine 3′,5′-phosphate content of guinea pig cerebral cortex slices. *Mol Pharmacol* 6:13–23, 1970.
3. Wolberg G, Zimmerman TP, Hiemstra K, Winston M, Chu L-C: Adenosine inhibition of

lymphocyte-mediated cytolysis: Possible role of cyclic adenosine monophosphate. *Science* 187:957–959, 1975.

4. Schwartz AL, Stern RC, Polmar SH: Demonstration of an adenosine receptor on human lymphocytes *in vitro* and its possible role in the adenosine deaminase-deficient form of severe combined immunodeficiency. *Clin Immunol Immunopathol* 9:499–505, 1978.
5. Marone G, Plaut M, Lichtenstein LM: Characterization of a specific adenosine receptor on human lymphocytes. *J Immunol* 121:2153–2159, 1978.
6. Londos C, Wolff L: Two distinct adenosine-sensitive sites on adenylate cyclase. *Proc Natl Acad Sci USA* 74:5482–5486, 1977.
7. Bourne HR, Lichtenstein LM, Melmon KL, Henney CS, Weinstein Y, Shearer GM: Modulation of inflammation and immunity by cyclic AMP. *Science* 184:19–28, 1974.
8. DeRubertis FR, Zenser TV, Adler WH, Hudson T: Role of cyclic adenosine 3′,5′-monophosphate in lymphocyte mitogenesis. *J Immunol* 113:151–161, 1974.
9. Henney CS, Lichtenstein LM: The role of cyclic AMP in the cytolytic activity of lymphocytes. *J Immunol* 197:610–612, 1971.
10. Lichtenstein LM, Margolis S: Histamine release *in vitro:* Inhibition by catecholamines and methylxanthines. *Science* 161:902–903, 1968.
11. Marquardt DL, Parker CW, Sullivan TJ: Potentiation of mast cell mediator release by adenosine. *J Immunol* 120:871–878, 1978.
12. Moretta L, Webb JR, Grossi CE, Lydyard PM, Cooper MD: Functional analysis of two human T-cell populations: Help and suppression of B-cell responses by T-cells bearing receptors for IgM and IgG. *J Exp Med* 146:184–200, 1977.
13. Moretta L, Mingari MC, Moretta A, Cooper MD: Human T-lymphocyte subpopulations: Studies of the mechanism by which T-cells bearing Fc receptors for IgG suppress T-dependent B-cell differentiation induced by pokeweed mitogen. *J Immunol* 122:984–990, 1979.
14. Birch RE, Polmar SH: Induction of Fcγ receptors on a subpopulation of human T-lymphocytes by adenosine and impromidine, an H_2-histamine agonist. *Cell Immunol* 57:455–467, 1981.
15. Limatibul S, Shore A, Dosch HM, Gelfand EW: Theophylline modulation of E-rosette formation: An indicator of T-cell maturation. *Clin Exp Immunol* 33:503–513, 1978.
16. Polmar SH, Birch RE: Regulation of Fc receptor expression on T-lymphocyte subsets mediated by adenosine and histamine receptors, in Krakauer RS, Cathcart MK (eds): *Immunoregulation and Autoimmunity*. Amsterdam, Elsevier North-Holland, 1980, pp 11–19.
17. Birch RE, Rosenthal AK, Polmar SH: Pharmacological modification of immunoregulatory T-lymphocytes: II. Modulation of T-lymphocyte cell surface characteristics. *Clin Exp Immunol* 48:231–238, 1982.
18. Birch RE, Fanger MW, Bernier GM: β_2 microglobulin enhances human lymphocytes surface receptor expression for IgG. *J Immunol* 122:997–1001, 1979.
19. Reinherz EL, Schlossman SF: The differentiation and function of human T-lymphocytes. *Cell* 19:821–827, 1980.
20. Birch RE, Polmar SH: Pharmacological modification of immunoregulatory T-lymphocytes: I. Effect of adenosine, H_1 and H_2 histamine agonists upon T lymphocyte regulation of B lymphocyte differentiation *in vitro*. *Clin Exp Immunol* 48:218–230, 1982.
21. Sullivan TJ, Parker KL, Kulczycki A, Parker CW: Modulation of cyclic AMP in purified rat mast cells: III. Studies on the effects of concanavalin A and anti-IgE on cyclic AMP concentrations during histamine release. *J Immunol* 117:713–716, 1976.
22. Sturgill TW, Schrier BK, Gilman AG: Stimulation of cyclic AMP accumulation by 2-chloradenosine: Lack of incorporation of nucleoside by cyclic nucleotides. *J Cyclic Nucleotide Res* 1:21–30, 1975.
23. Mandler R, Birch RE, Polmar SH, Kammer GM, Rudolph SA: Abnormal adenosine-induced immunosuppression and cyclic AMP metabolism in T-lymphocytes of patients with systemic lupus erythematosus. *Proc Natl Acad Sci USA* 79:7542–7546.
24. Birch RE, Mandle R, Rudolph SA, Polmar SH: Immunologic and pharmacologic events associated with adenosine induced immunosuppression. *Fed Proc* 41:799, 1982 (abst).

DISCUSSION

DR. T.W. RALL: I have two questions relating to theophylline. First, have any observations been made on the effects of theophylline in patients with adenosine deaminase deficiency? Second, could you expand on the use of theophylline in identifying and/or separating different subpopulations of lymphocytes?

DR. POLMAR: (1) I am only aware of studies *in vitro* on adenosine deaminase-deficient lymphocytes from an ADA-SCID patient. Cells were obtained after the patient was treated with enzyme replacement therapy. The lymphocytes from this patient were more sensitive to proliferation inhibition by the theophylline than were normal cells. (2) T lymphocytes can be divided into two subpopulations, based on the sensitivity of their sheep erythrocyte (E) receptor to theophylline ($10^{-3}M$, 37°C, 60 min). Approximately 20% of the T lymphocytes will lose the E-rosette-forming capacity; these are theophylline-sensitive (T_S) cells and function as suppressor cells. The E rosette receptor will reappear in 1 hr at 37° but will not reappear if the cells are kept at 4°C. The remaining 80% of the cells have theophylline-resistant E receptors and function as helper cells. T cells that form rosettes (T_R) are separate from the non-rosette-forming (T_S) cells by E rosette technique followed by separation by Ficoll-Hypaque gradient centrifugation.

DR. J.P.R.M. VAN LAARHOVEN: Do you have any information about the effects of deoxyadenosine in your PWM system on the helper/suppressor ratio? The reason I ask this is that, as I said earlier, we have observed an increased Ig secretion in PWM-stimulated human peripheral blood lymphocytes at low concentrations of deoxyadenosine ($\leq 1\ \mu M$) in the presence of EHNA. [^{3}H]thymidine incorporation was inhibited under these conditions. In our opinion, this points to an impairment of suppressor activity.

DR. POLMAR: In our system, 2′-deoxyadenosine does not change expression of lymphocyte surface markers. The conditions used were an incubation time of 30 min at 37°. These conditions differ significantly from those that you describe. I think your findings are very interesting and your interpretation—namely, that low concentrations of 2′-deoxyadenosine inhibit suppressor function—seems reasonable. Most suppressor cells must proliferate in order to suppress, and thus inhibition of proliferation by 2′-deoxyadenosine would be expected to inhibit suppression. In contrast, the adenosine-induced suppressor (T_RA) need not proliferate in order to suppress.

DR. B.B. FREDHOLM: Is the effect of adenosine in your system affected by an uptake inhibitor, or, in other words, is it intra- or extracellular?

DR. POLMAR: We have used dipyridamole to inhibit adenosine transport. We have not used any of the newer nucleoside transport inhibitors. Dipyridamole blocked the appearance RFcγ on adenosine-treated lymphocytes. This finding is consistent with our observation that 2′-chloroadenosine, which enters cells slowly, does not induce RFcγ.

17. MECHANISMS OF ADENOSINE AND 2′-DEOXYADENOSINE INHIBITION OF IMMUNE FUNCTION IN MOUSE LYMPHOCYTES

THOMAS P. ZIMMERMAN, GERALD WOLBERG, GAIL S. DUNCAN

Adenosine (Ado) has been recognized as having an inhibitory influence on the immune system since 1970, when it was reported that this purine nucleoside inhibits the mitogenic stimulation of human peripheral blood lymphocytes by phytohemagglutinin [1]. Widespread interest in the possible immune modulatory role of Ado was sparked by the 1972 report of Giblett and her colleagues [2] describing the genetic deficiency of the enzyme Ado deaminase (ADA) in two children with severe combined immunodeficiency. This inherited disorder is characterized by a marked reduction of functional T and B lymphocytes. With the discovery of additional immunodeficient patients lacking ADA, an extensive research effort was begun to elucidate the biochemical mechanisms by which ADA deficiency results in the rather selective demise of the immune system. From 1972 until 1978, it was widely assumed that this immunodeficiency disease results from the buildup of high concentrations of Ado and/or adenine ribonucleotides in ADA-deficient patients. Indeed, in 1976 Mills et al. [3] reported that plasma Ado and adenine, and erythrocyte ATP, were elevated in an ADA-deficient child. Accordingly, on the basis of results obtained with a number of *in vitro* model systems, several hypotheses were introduced to explain the role of Ado in the pathophysiology of ADA-deficient immunodeficiency, the more prominent of which were Ado-induced

We gratefully acknowledge the excellent technical assistance of Robert L. Veasey and Mrs. Marvin S. Winston and express our appreciation to Dr. Gertrude B. Elion for her helpful comments during the preparation of the manuscript and to Beverly H. Nobles for her preparation of the graphics.

R.M. Berne, T.W. Rall, and R. Rubio (eds.), Regulatory Function of Adenosine, *pp. 261–273.*

pyrimidine starvation, Ado-stimulated increases in cAMP, and Ado-mediated accumulation of S-adenosylhomocysteine (AdoHcy) and consequent inhibition of methylation reactions (for reviews see [4–7]). Subsequently, in 1978, it was reported that 2′-deoxyadenosine (dAdo) was elevated in the plasma and urine of ADA-deficient children and that dATP was elevated in their erythrocytes and lymphocytes. With the further important observation that dAdo is more toxic than Ado to proliferating lymphoid cells, the emphasis in biochemical studies concerning the causal relationship between ADA deficiency and severe combined immunodeficiency shifted away from Ado toward dAdo. The historical development of this interesting field has been the subject of several recent reviews [4–7].

An *in vitro* model employed in this laboratory for the study of cellular immunity is the lysis of tumor cells by specifically sensitized mouse T lymphocytes. This phenomenon, termed lymphocyte-mediated cytolysis (LMC), may play an important role in tumor and graft rejection (for review see [8]). Although the biochemical mechanism of this immunological process is poorly understood, LMC provides a convenient and rapid *in vitro* assay of a lymphocyte function, based on the release of an appropriate radioisotope from labeled target cells. Adenosine [9] and a large number of Ado analogues [10–12], including dAdo, have been found to inhibit LMC. This chapter summarizes our current knowledge of the biochemical effects of Ado on the cytolytic T lymphocyte and of how these various effects relate to the inhibition of lymphocyte function caused by Ado. For comparative purposes, parallel results obtained with dAdo are included wherever possible.

MATERIALS AND METHODS

Periodate-oxidized Ado was synthesized by Lowrie M. Beacham (Wellcome Research Laboratories) according to a published procedure [13]. 5-iodotubercidin was a gift from Leroy B. Townsend (University of Michigan). dAdo was passed through a small column of boric acid gel, under alkaline conditions, to eliminate possible contamination by Ado [14]. Other materials were from sources identified elsewhere [9–12, 14, 15].

C57BL leukemia EL4 cells were maintained, harvested, and labeled with $Na_2{}^{51}CrO_4$, as previously reported [10]. Cytolytic lymphocytes were obtained from CD-1 mice 10 or 11 days after intraperitoneal injection of 3×10^7 EL4 cells [10]. Normal spleen lymphocytes were isolated from CD-1 mice [15]. Dulbecco's phosphate-buffered saline supplemented with 5% fetal calf serum (heat-inactivated) was the medium for all cell incubations, except for the use of Gey's solution (Gey's balanced salt solution supplemented with 2% bovalbumin and 10 m*M* Hepes and adjusted to pH 7.2) in the measurement of protein carboxymethylation. Unless specified otherwise, 8 μ*M* *erythro*-9-(2-hydroxy-3-nonyl)adenine was included in all cellular incubations.

The *in vitro* assay of LMC has been described in detail [16]. Briefly, this assay monitors the release of ^{51}Cr during a 70 min incubation at 37° of ^{51}Cr-

labeled EL4 cells and specifically sensitized cytolytic lymphocytes. With a 1:1 ratio of cytolytic lymphocytes to EL4 target cells, approximately 10% tumor cell lysis occurs in the control assays.

The effect of various agents on the relative levels of [^{3}H]AdoHcy in cytolytic lymphocytes was assessed by reversed-phase high-performance liquid chromatography (HPLC) after prelabeling cells for 60 min with L-[2-^{3}H] methionine (100 μCi per 2–4 × 10^8 cells in 10 ml of medium) and then incubating the washed cells (1–3 × 10^7 cells/10 ml of medium) for 30 min with saline or test compound [12, 14]. The effect of agents on the absolute level of AdoHcy in normal spleen lymphocytes was determined by reversed-phase HPLC after 60 min incubations of cells (2.5 × 10^8 cells /100 ml of medium) with saline or test compound.

Protein carboxymethylation (PCM) was assayed in cytolytic lymphocytes, essentially as previously described [17].

Acid-soluble extracts of cells were analyzed for cAMP by radioimmunoassay and for ribonucleoside 5'-triphosphates by anion-exchange HPLC [10].

RESULTS

Effects of Ado and dAdo on lymphocyte function

Ado by itself was mildly inhibitory to LMC, exhibiting a 50% inhibitory concentration of >10 μ*M* (Fig. 17.1). The simultaneous addition of 8 μ*M* *erythro*-9-(2-hydroxy-3-nonyl)adenine (EHNA), an inhibitor of ADA [18], to the LMC assay resulted in a substantial decrease in the 50% inhibitory concentration of Ado to approximately 3 μ*M* (Fig. 17.1). Both 4-(3-butoxy-4-methoxybenzyl)-2-imidazolidinone (Ro 20-1724), an inhibitor of cAMP phosphodiesterase [19], and L-homocysteine (Hcy), a cosubstrate for AdoHcy hydrolase [20], enhanced the inhibitory activity of Ado in the LMC assay. By contrast, theophylline (Theo), an antagonist of adenylate cyclase-coupled Ado receptors [21], largely prevented the inhibition of LMC caused by low concentrations of Ado (Fig. 17.1).

Uridine (Urd), an inhibitor of Ado transport [22], augmented slightly the inhibition of LMC caused by Ado, both in the absence and presence of Ro 20-1724 (Table 17.1). However, Urd partially antagonized the inhibitory effect of the Ado/Hcy combination in the LMC assay. 5-iodotubercidin (ITu), an inhibitor of Ado kinase [23], resembled Urd in its slight enhancement of the LMC-inhibitory activity of Ado ± Ro 20-1724 (Table 17.1). Unlike Urd, however, ITu further increased the inhibition of LMC caused by the Ado/Hcy combination. Theo was more effective in antagonizing the inhibition of LMC caused by Ado ± Ro 20-1724 than of that caused by the Ado/Hcy combination.

In the absence of added Ado, EHNA, Ro 20-1724, Hcy, Theo, Urd, and ITu were not inhibitory to LMC at the concentrations employed in the present experiments. Of considerable interest, periodate-oxidized Ado (Adox), a po-

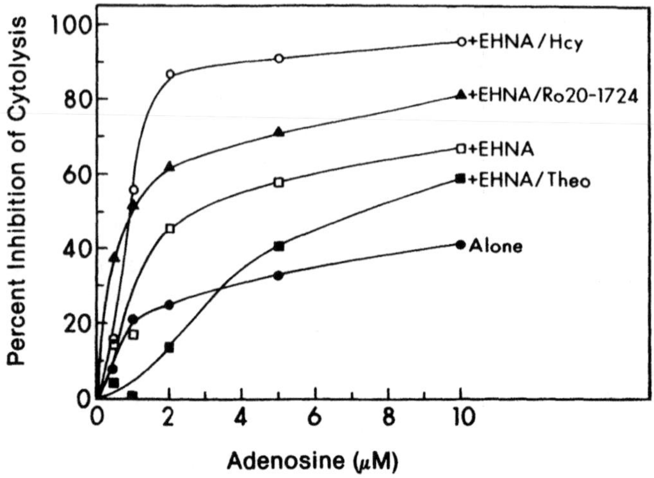

Figure 17.1. Inhibition of LMC by Ado alone or in the presence of various agents. The concentrations of EHNA, theophylline (Theo), Ro 20-1724, and L-homocysteine thiolactone (Hcy) were 8, 75, 25, and 200 μ*M*, respectively. Each point represents the mean of duplicate assays.

Table 17.1. Effects of theophylline, uridine, and 5-iodotubercidin on the inhibition of lymphocyte-mediated cytolysis (LMC) by adenosine in the absence or presence of Ro 20-1724 or L-homocysteine

	Other additive (% inhibition of LMC)			
Additive	None	Theo	Urd	ITu
Ado	32	9	49	47
Ado + Ro 20-1724	60	35	65	63
Ado + Hcy	74	66	45	97

Each assay was performed in duplicate in the presence of 8 μ*M* EHNA. At the concentrations employed, Ro 20-1724 (25 μ*M*), L-homocysteine thiolactone (Hcy, 200 μ*M*), theophylline (Theo, 75 μ*M*), uridine (Urd, 20 m*M*) and 5-iodotubercidin (ITu, 0.1 μ*M*), either alone or in the specified combinations, were not inhibitory to LMC. The concentration of adenosine (Ado) was 2.0 μ*M*.

tent inhibitor of AdoHcy hydrolase [13], was also not inhibitory to LMC at concentrations up to 100 μ*M*.

The inhibition of LMC caused by Ado ± Ro 20-1724 was rapidly reversible upon transfer of the cytolytic lymphocytes to drug-free medium. However, the inhibition of LMC resulting from a 60 min pretreatment of cytolytic lymphocytes with 10 μ*M* Ado plus 200 μ*M* Hcy required >60 min of lymphocyte incubation in drug-free medium for full restoration of cytolytic activity.

In the presence of EHNA, dAdo inhibited LMC with a 50% inhibitory concentration of 92 μ*M*. In the absence of EHNA, dAdo was completely inactive in the LMC assay, even at 150 μ*M*. Unlike the case with Ado, the

LMC-inhibitory activity of dAdo (75–150 μ*M*) was unaffected by Ro 20-1724 (25 μ*M*), Hcy (200 μ*M*), Theo (75 μ*M*) and ITu (0.1 μ*M*). However, Urd (20 m*M*) completely prevented this biological activity of dAdo. The inhibition of LMC caused by dAdo was not increased by a 60 min pretreatment of the cytolytic lymphocytes with this agent and was reversed rapidly following transfer of the lymphocytes to drug-free medium.

Effects of Ado and dAdo on lymphocyte biochemistry

Nucleotide metabolism

A number of biochemical parameters have been monitored within mouse lymphocytes in an attempt to elucidate the mechanisms whereby Ado and dAdo inhibit the cytolytic function of these cells. The lymphocyte pools of UTP, ATP, and GTP exhibited minimal changes after 60 min incubation of the cells with either 18.8 μ*M* Ado or 150 μ*M* dAdo; the largest effect observed was a 23% increase in cellular ATP caused by Ado [11]. Although our methodology has been shown capable of determining dATP in cell extracts, no dATP formation was detected in the foregoing experiment or in subsequent experiments utilizing [^{3}H]dAdo [11].

Urd (20 m*M*) and ITu (0.1 μ*M*) inhibited the uptake and conversion of 2 μ*M* [^{3}H]Ado to [^{3}H]ATP in cytolytic lymphocytes by 97% and 75%, respectively, during a 30 min incubation, whereas Theo (75 μ*M*), Ro 20-1724 (25 μ*M*), and Adox (50 μ*M*) had no effect on this metabolic pathway. Hcy (200 μ*M*) enhanced this [^{3}H]Ado→[^{3}H]ATP conversion by 29%.

cAMP levels

Ado alone caused a dose-dependent elevation of cAMP in the cytolytic lymphocytes, and this effect was enhanced significantly by EHNA (Fig. 17.2). Both Ro 20-1724 and Hcy considerably augmented this Ado-stimulated increase in lymphocyte cAMP (Fig. 17.2). Theo caused a lowering of basal cAMP and antagonized the ability of Ado ± Ro 20-1724 or Hcy to increase lymphocyte cAMP (Table 17.2). Urd and ITu had no apparent effect on cAMP levels when both cells *and* medium were acid-extracted for cAMP determinations (Experiment I in Table 17.2) but slightly enhanced the Ado-stimulated cAMP elevation when cells were harvested by centrifugation prior to their acid extraction (Experiment II in Table 17.2). These lymphocytes release cAMP into the medium, and, typically, 40% to 60% of the total cAMP of a cell suspension is found in the medium. In Experiment II in Table 17.2, ITu enhanced the lymphocyte cAMP elevation stimulated by the Ado/Hcy combination.

In the presence of EHNA, dAdo (150 μ*M*) caused a 60% increase in lymphocyte cAMP after a 30 min incubation, and this effect was not potentiated by Ro 20-1724 [11]. Adox (50 μ*M*) caused a 49% elevation in lymphocyte cAMP.

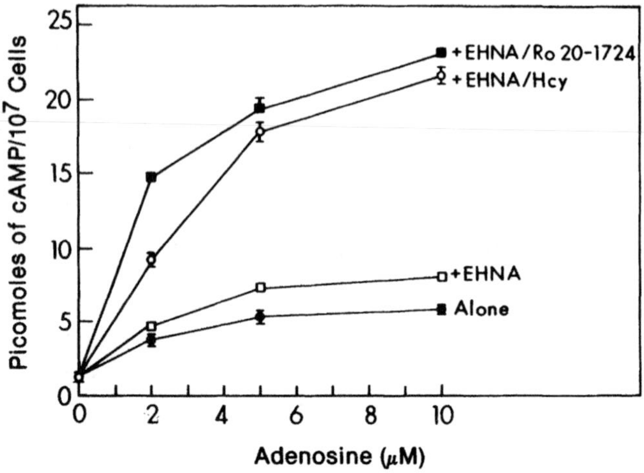

Figure 17.2. Effect of Ado alone or in combination with various agents on lymphocyte cAMP levels. Cells (7.2×10^6/5.0 ml of medium) were incubated for 30 min at 37° with the specified agents. Entire cell suspensions (cells plus medium) were then acid-extracted for cAMP determinations. All incubations were performed in duplicate, and each column-purified extract was radioimmunoassayed in duplicate for cAMP. Each point represents the mean ± SEM for four determinations. The concentrations of EHNA, Hcy, and Ro 20-1724 were 8, 200, and 25 μ*M*, respectively.

Table 17.2. Effects of theophylline, uridine, and 5-iodotubercidin on the elevation of cyclic AMP caused by adenosine in the absence or presence of Ro 20-1724 or L-homocysteine

	Other additive (pmol of cAMP/10^7 cells)			
Additive	None	Theo	Urd	ITu
Experiment I				
None (saline)	2.38 ± 0.06	1.86 ± 0.06	1.87 ± 0.09	2.07 ± 0.13
Ro 20-1724	3.47 ± 0.04	2.78 ± 0.08	3.08 ± 0.05	3.58 ± 0.14
Ado	9.12 ± 0.35	3.01 ± 0.17	8.54 ± 0.31	8.52 ± 0.14
Ado + Ro 20-1724	23.44 ± 0.75	4.79 ± 0.20	22.83 ± 0.81	26.57 ± 1.33
Experiment II				
None (saline)	1.27 ± 0.05	0.34 ± 0.01	1.24 ± 0.07	1.21 ± 0.15
Hcy	0.75 ± 0.10	0.26 ± 0.01	0.83 ± 0.05	0.34 ± 0.08
Ado	1.97 ± 0.15	0.47 ± 0.04	2.62 ± 0.13	2.80 ± 0.47
Ado + Hcy	3.37 ± 0.18	0.93 ± 0.02	2.79 ± 0.16	4.74 ± 0.28

Cytolytic lymphocytes (7.0 or 8.2×10^6 cells/5.0 ml of medium) were preincubated for 10 min at 37°, supplemented with 8 μ*M* EHNA plus the specified additives and incubated for 30 min at 37°. In Experiment I, the entire cell suspensions (i.e., cells plus medium) were acid-extracted. In Experiment II, cells were harvested by centrifugation (730 x*g* for 5 min in the cold) and only the cells were acid-extracted. Concentrations of reagents were the same as in Table 17.1. All cellular incubations and cAMP radioimmunoassays were performed in duplicate. Each value represents the mean ± SEM for four determinations.

AdoHcy metabolism

Mammalian cells treated with Ado exhibit an accumulation of AdoHcy due either to simple inhibition of AdoHcy hydrolase or to condensation of this Ado with endogenous Hcy via this same enzyme [20]. Because of both the small pool of AdoHcy present in cytolytic lymphocytes and the paucity of this particular tissue, it has not proved feasible to measure directly the AdoHcy present in extracts of these cells except when the cells have been treated with Ado in the presence of Hcy [15, 24]. However, Ado has been shown to cause a dose-dependent buildup of [^{3}H]AdoHcy in cytolytic lymphocytes prelabeled with L-[2-^{3}H]methionine [24]. Within one such experiment, 2 μM Ado was observed to cause a 4-fold elevation in lymphocyte [^{3}H]AdoHcy after 30 min; this effect of Ado was not altered by Theo (75 μM) but was inhibited 89% by Urd (20 mM) and was enhanced 1.5- and 2.8-fold by Ro 20-1724 (25 μM) and ITu (0.1 μM), respectively. Adox (50 μM) was 1.8 times more effective than 10 μM Ado in elevating lymphocyte [^{3}H]AdoHcy.

dAdo also causes an increase in [^{3}H]AdoHcy in L-[2-^{3}H]methionine-labeled lymphocytes [14]. However, dAdo (150 μM) is considerably less effective than 10 μM Ado or 50 μM Adox in this respect. Also, dAdo, unlike many other purine nucleosides, does not appear to undergo metabolism to its corresponding 5′-deoxy-5′-S-homocysteinyl derivative in these cells [14].

As reported elsewhere [15], incubation of cytolytic lymphocytes for 60 min with 10 μM Ado in the presence of Hcy results in the intracellular accumulation of AdoHcy in excess of 2 mM. This massive increase in lymphocyte AdoHcy caused by the Ado/Hcy combination is over 100 times greater than that caused by Ado in the absence of Hcy. More recently, it has been found that ITu (0.1 μM) increases 3-fold the lymphocyte buildup of AdoHcy from 2 μM Ado plus 200 μM Hcy during a 60 min incubation, that Urd (20 mM) inhibits this metabolic event by >95%, and that Theo (75 μM) has no effect under these same conditions.

The efficacy of Ado, dAdo, Hcy, and Adox in increasing the absolute level of cellular AdoHcy was assessed with normal spleen lymphocytes, a tissue more abundant than immune cytolytic lymphocytes. The basal level of AdoHcy found with spleen lymphocytes was $\leq$4 pmol/10^7 cells (approximately 2 nmol/ml of packed cells). Sixty-minute treatment of these cells with 10 μM Ado, 150 μM dAdo, 200 μM Hcy, or 50 μM Adox resulted in 7.5-, 4.8-, 5.7- and 13.5-fold increases, respectively, in lymphocyte AdoHcy. These results both confirm the findings with [^{3}H]methionine-labeled cells, showing that Ado, dAdo, and Adox cause substantial elevations in lymphocyte AdoHcy, and reveal that Hcy alone also causes a large increase in AdoHcy in these cells.

Protein carboxymethylation (PCM)

These various agents were also investigated for their ability to inhibit a representative methylation reaction (i.e., PCM) in the cytolytic lymphocytes. In the absence of Hcy, Ado at 5 and 10 μ*M* inhibited PCM by 24% and 52%, respectively. By comparison, Hcy (200 μ*M*) and Adox (50 μ*M*), two compounds not inhibitory to LMC, inhibited PCM by 46% and 70%, respectively. Although Hcy potentiates the inhibition of LMC caused by Ado (Fig. 17.1, Table 17.1), this amino acid did not potentiate the inhibition of PCM caused by Ado. In fact, the inhibition of PCM caused by the Ado (5 or 10 μ*M*)/Hcy combinations was less than the sum of the inhibitory effects of Ado and Hcy alone. The inhibition (65%) of PCM by the combination of 10 μ*M* Ado plus Hcy was slightly less than the effect of Adox on this reaction. dAdo (150 μ*M*) inhibited PCM by 69%.

cAMP metabolism

In the course of these studies concerning Ado, AdoHcy, and cAMP, it has come to light that cells allowed to accumulate high levels of AdoHcy exhibit a greatly enhanced cAMP response to a variety of agents (e.g., prostaglandin E_1, isoproterenol, and cholera toxin) that stimulate adenylate cyclase [15]. This effect of the Ado/Hcy combination on lymphocyte cAMP metabolism was shown to be due both to a small (approximately 2-fold) enhancement in adenylate cyclase in Ado/Hcy-treated cells and to competitive inhibition (K_i = 1.7 m*M*) of the high-affinity cAMP phosphodiesterase by metabolically formed AdoHcy [15].

COMMENTARY

In the presence of an ADA inhibitor, Ado is inhibitory to the cytolytic function of mouse T lymphocytes at micromolar concentrations. This inhibitory effect of Ado is enhanced by Urd, an inhibitor of Ado transport [22], and by ITu, an inhibitor of Ado kinase [23]. These results indicate that Ado does not require cellular uptake or phosphorylation in order to inhibit this lymphocyte function. Moreover, the LMC-inhibitory activity of Ado is potentiated by Ro 20-1724, an inhibitor of cAMP phosphodiesterase [19], and is prevented partially by Theo, an antagonist of Ado receptors [21]. These pharmacological interactions suggest strongly that Ado is affecting lymphocyte function by reason of its ability to bind to, and to activate, cell surface receptors coupled to adenylate cyclase [25]. Direct measurement of cAMP levels in lymphocytes treated with these various agents support this conclusion. It is well established that lymphocytes from human [26, 27], mouse [28], guinea pig [29], and rat [30] possess Ado R_a/A_2 receptors coupled to adenylate cyclase. The enhancement by Urd and ITu of the LMC-inhibitory activity of Ado can be attributed to the maintenance of higher extracellular Ado concentrations, due to inhibition of Ado uptake and/or phosphorylation, and to a consequently prolonged occupancy of these cell surface Ado receptors.

The effects of the Ado/Hcy combination on LMC and on cAMP metabolism appear to be somewhat more complex than the effects of Ado ± Ro 20-1724. Cells incubated with Ado in the presence of exogenous Hcy accumulate high concentrations of AdoHcy as a result of the condensation of these two substrates by AdoHcy hydrolase [14, 15, 20, 24]. The AdoHcy formed intracellularly causes a small (approximately 2-fold) activation of adenylate cyclase and acts as a competitive inhibitor (K_i = 1.7 m*M*) of the high-affinity cAMP phosphodiesterase [15]. For these reasons, metabolically formed AdoHcy augments the elevation of lymphocyte cAMP stimulated by residual extracellular Ado. The inhibition of Ado uptake by Urd prevents this intracellular condensation of Hcy with Ado and thereby partially antagonizes the LMC-inhibitory effect of the Ado/Hcy combination. By contrast, the enhancement of the lymphocyte buildup of AdoHcy caused by ITu results in a further augmentation of the inhibition of LMC by the Ado/Hcy combination.

This cAMP-dependent mechanism for the LMC-inhibitory activity of Ado is supported by the observation that a number of other compounds that cause an elevation of lymphocyte cAMP also inhibit LMC. These chemically diverse agents include histamine, isoproterenol, prostaglandins E_1 and E_2, cholera enterotoxin [31], and the diterpene forskolin [G. Wolberg and G.S. Duncan, unpublished observation].

Two types of evidence indicate that the inhibition of LMC caused by Ado (in the absence of exogenous Hcy) is not due to an elevation of lymphocyte AdoHcy and to a consequent inhibition of one or more critical methyltransferases. First, Urd, an inhibitor of Ado transport, prevents the adenosine-induced increase in lymphocyte AdoHcy while, at the same time, it augments the LMC-inhibitory activity of Ado. Second, Adox is more effective than Ado both in increasing lymphocyte AdoHcy and in inhibiting lymphocyte PCM, yet it is not inhibitory to LMC.

In the presence of EHNA, dAdo is also an inhibitor of LMC but is considerably less potent than Ado. Current evidence indicates that dAdo inhibits LMC by a mechanism independent of cAMP metabolism. Antagonism by Urd of the LMC-inhibitory activity of dAdo suggests that dAdo requires transport into the cytolytic lymphocytes in order to affect this cell function. Comparative biochemical studies with dAdo and Adox indicate that dAdo is not acting in these cells via effects on AdoHcy metabolism or on methylation reactions. The lack of increase in inhibitory potency of dAdo with increasing time of lymphocyte preincubation, as well as the rapid reversal of its inhibition, implies that dAdo itself, rather than a metabolite, is the active agent within the cytolytic lymphocyte. Further work is necessary to define the mechanism of action of dAdo in the LMC system.

Assuming that cAMP is the "second messenger" that mediates the inhibition of LMC by Ado, it is of great importance to identify the molecular process that cAMP (in concert with cAMP-dependent protein kinase) regulates within the cytolytic lymphocytes. Since the molecular details of LMC are poorly understood [8], it is not readily apparent which proteins are likely candidates to modulate this cytolytic function according to their state of phosphorylation. However, it is known that LMC is a receptor-directed event requiring extensive cell-to-cell contact between the cytolytic lymphocyte and its target cell [8]. It has also been reported that brief treatment of mouse cytolytic lymphocytes with Ado impairs their ability to form conjugates with target cells [32]. It is possible that the actomyosin contractile system is an essential participant in these membrane-associated events during LMC and that myosin light chain kinase [33] is the critical substrate of cAMP-dependent protein kinase, whose phosphorylation state regulates this lymphocyte function. This hypothesis is supported by the recent evidence that actin is concentrated in the cytolytic lymphocytes at the area of attachment to their target cells and that lymphocytes exhibit a high motility in this contact area [34].

The rapid effect of Ado on lymphocyte surface markers described by Dr. S.H. Polmar and his colleagues (see Chapter 16) may be important to the mechanism of LMC-inhibition by Ado.

It is not presently known whether Ado receptor activation in lymphoid tissue contributes to the symptomatology of severe combined immunodeficiency associated with ADA deficiency. It seems likely that elevated tissue levels of dAdo and the selective trapping of this deoxyribonucleoside in lymphoid cells as 5′-nucleotides is the principal cause of the lymphopenia observed in these patients [4–7]. However, it has been found that the residual lymphocytes of an ADA-deficient patient responded poorly to mitogens and that this response could be increased substantially by the addition of ADA to the cell cultures [35]. The complementary finding of elevated cAMP levels in lymphocytes from an ADA-deficient child [36] suggests that the impairment of function observed with such cells could be due to Ado activation of adenylate cyclase. Of related interest, platelets from ADA-deficient children are defective in their aggregation response [37, 38], contain elevated levels of cAMP, and can be restored to normal function by treatment with exogenous ADA [38].

REFERENCES

1. Hirschhorn R, Grossman J, Weissmann G: Effect of cyclic 3′ ,5′-adenosine monophosphate and theophylline on lymphocyte transformation. *Proc Soc Exp Biol Med* 133:1361–1365, 1970.
2. Giblett ER, Anderson JE, Cohen F, Pollara B, Meuwissen HJ: Adenosine-deaminase deficiency in two patients with severely impaired cellular immunity. *Lancet* 2:1067–1069, 1972.
3. Mills GC, Schmalstieg FC, Trimmer KB, Goldman AS, Goldblum RM: Purine metabolism in adenosine deaminase deficiency. *Proc Natl Acad Sci USA* 73:2867–2871, 1976.
4. Fox IH, Kelley WN: The role of adenosine and 2′-deoxyadenosine in mammalian cells. *Annu Rev Biochem* 47:655–686, 1978.
5. Polmar SH: Metabolic aspects of immunodeficiency disease. *Semin Hematol* 17:30–43, 1980.
6. Thompson LF, Seegmiller JE: Adenosine deaminase deficiency and severe combined immunodeficiency disease. *Adv Enzymol* 51:167–210, 1980.
7. Martin DW Jr, Gelfand EW: Biochemistry of diseases of immunodevelopment. *Annu Rev Biochem* 50:845–877, 1981.
8. Berke G: Interaction of cytotoxic T lymphocytes and target cells. *Prog Allergy* 27:69–133, 1980.
9. Wolberg G, Zimmerman TP, Hiemstra K, Winston M, Chu L-C: Adenosine inhibition of lymphocyte-mediated cytolysis: possible role of cyclic adenosine monophosphate. *Science* 187:957–959, 1975.

10. Zimmerman TP, Rideout JL, Wolberg G, Duncan GS, Elion GB: 2-fluoroadenosine 3′ :5′-monophosphate, a metabolite of 2-fluoroadenosine in mouse cytotoxic lymphocytes. *J Biol Chem* 251:6757–6766, 1976.
11. Wolberg G, Zimmerman TP, Duncan GS, Singer KH, Elion GB: Inhibition of lymphocyte-mediated cytolysis by adenosine analogs: Biochemical studies concerning mechanism of action. *Biochem Pharmacol* 27:1487–1495, 1978.
12. Zimmerman TP, Wolberg G, Duncan GS: Inhibition of lymphocyte-mediated cytolysis by 3-deazaadenosine: evidence for a methylation reaction essential to cytolysis. *Proc Natl Acad Sci USA* 75:6220–6224, 1978.
13. Hoffman JL: The rate of transmethylation in mouse liver as measured by trapping S-adenosylhomocysteine. *Arch Biochem Biophys* 205:132–135, 1980.
14. Zimmerman TP, Wolberg G, Duncan GS, Elion GB: Adenosine analogues as substrates and inhibitors of S-adenosylhomocysteine hydrolase in intact lymphocytes. *Biochemistry* 19:2252–2259, 1980.
15. Zimmerman TP, Schmitges CJ, Wolberg G, Deeprose RD, Duncan GS, Cuatrecasas P, Elion GB: Modulation of cyclic AMP metabolism by S-adenosylhomocysteine and S-3-deazaadenosylhomocysteine in mouse lymphocytes. *Proc Natl Acad Sci USA* 77:5639–5643, 1980.
16. Wolberg G, Hiemstra K, Burge JJ, Singler RC: Reversible inhibition of lymphocyte-mediated cytolysis by dimethyl sulfoxide (DMSO). *J Immunol* 111:1435–1443, 1973.
17. Pike MC, Kredich NM, Snyderman R: Requirement of S-adenosyl-L-methionine-mediated methylation for human monocyte chemotaxis. *Proc Natl Acad Sci USA* 75:3928–3932, 1978.
18. Schaeffer HJ, Schwender CF: Enzyme inhibitors: 26. Bridging hydrophobic and hydrophilic regions on adenosine deaminase with some 9-(2-hydroxy-3-alkyl)adenines. *J Med Chem* 17:6–8, 1974.
19. Sheppard H, Wiggan G: Analogues of 4-(3,4-dimethoxybenzyl)-2-imidazolidinone as potent inhibitors of rat erythrocyte adenosine cyclic 3′,5′-phosphate phosphodiesterase. *Mol Pharmacol* 7:111–115, 1971.
20. Kredich NM, Martin DW Jr: Role of S-adenosylhomocysteine in adenosine-mediated toxicity in cultured mouse T lymphoma cells. *Cell* 12:931–938, 1977.
21. Bruns RF: Adenosine antagonism by purines, pteridines and benzopteridines in human fibroblasts. *Biochem Pharmacol* 30:325–333, 1981.
22. Cohen A, Ullman B, Martin DW Jr: Characterization of a mutant mouse lymphoma cell with deficient transport of purine and pyrimidine nucleosides. *J Biol Chem* 254:112–116, 1979.
23. Henderson JF, Paterson ARP, Caldwell IC, Paul B, Chan MC, Lau KF: Inhibitors of nucleoside and nucleotide metabolism. *Cancer Chemother Rep* 3 (Part 2):71–85, 1972.
24. Wolberg G, Zimmerman TP, Duncan GS, Deeprose RD: Inhibition of lymphocyte-mediated cytolysis by adenosine: mechanistic studies, in Pollara B, Pickering RJ, Meuwissen HJ, Porter IH (eds): *Inborn Errors of Specific Immunity*. New York, Academic Press, 1979, pp 315–325.
25. Daly JW: Adenosine receptors: Targets for future drugs. *J Med Chem* 25:197–207, 1982.
26. Schwartz AL, Stern RC, Polmar SH: Demonstration of an adenosine receptor on human lymphocytes *in vitro* and its possible role in the adenosine deaminase-deficient form of severe combined immunodeficiency. *Clin Immunol Immunopathol* 9:499–505, 1978.
27. Marone G, Plaut M, Lichtenstein LM: Characterization of a specific adenosine receptor on human lymphocytes. *J Immunol* 121:2153–2159, 1978.
28. Bonnafous J-C, Dornand J, Mani J-C: Hormone-like action of adenosine in mouse thymocytes and splenocytes. *FEBS Lett* 107:95–99, 1979.
29. Fredholm BB, Sandberg G, Ernström U: Cyclic AMP in freshly prepared thymocyte suspensions: Evidence for stimulation by endogenous adenosine. *Biochem Pharmacol* 27:2675–2682, 1978.
30. Nordeen SK, Young DA: Separation of effects of adenosine on energy metabolism from those on cyclic AMP in rat thymic lymphocytes. *J Biol Chem* 252:5324–5331, 1977.
31. Bourne HR, Lichtenstein LM, Melmon KL, Henney CS, Weinstein Y, Shearer GM: Modulation of inflammation and immunity by cyclic AMP. *Science* 184:19–28, 1974.
32. Bradley TP, Anisman D, Bonavida B: Molecular interactions in cell-mediated cytotoxicity: Role of adenosine in the binding of cytotoxic T lymphocytes to target cells. *Fed Proc* 39:921, 1980 (abst).
33. Adelstein RS, Pato MD, Conti MA: The role of phosphorylation in regulating contractile proteins. *Adv Cyclic Nucleotide Res* 14:361–373, 1981.

34. Ryser J-E, Rungger-Brändle E, Chaponnier C, Gabbiani G, Vassalli P: The area of attachment of cytotoxic T lymphocytes to their target cells shows high motility and polarization of actin, but not myosin. *J Immunol* 128:1159–1162, 1982.
35. Polmar SH, Wetzler EM, Stern RC, Hirschhorn R: Restoration of *in vitro* lymphocyte responses with exogenous adenosine deaminase in a patient with severe combined immunodeficiency. *Lancet* 2:743–746, 1975.
36. Schmalstieg FC, Nelson JA, Mills GC, Monahan TM, Goldman AS, Goldblum RM: Increased purine nucleotides in adenosine deaminase-deficient lymphocytes. *J Pediatr* 91:48–51, 1977.
37. Schwartz AL, Polmar SH, Stern RC, Cowan DH: Abnormal platelet aggregation in severe combined immunodeficiency disease with adenosine deaminase deficiency. *Br J Haematol* 39:189–194, 1978.
38. Lee CH, Evans SP, Rozenberg MC, Bagnara AS, Ziegler JB, Van der Weyden MB: *In vitro* platelet abnormality in adenosine deaminase deficiency and severe combined immunodeficiency. *Blood* 53:465–471, 1979.

DISCUSSION

DR. S.H. POLMAR: The marked enhancement in cAMP production seen by the addition of homocysteine (Hcy) with PGE is similar to that seen with colchicine. The colchicine effect is thought to be due to increased mobility of receptors with catalytic units caused by inhibition of microtubules. Could a similar mechanism be responsible for Hcy enhancement of cAMP synthesis?

DR. ZIMMERMAN: That is a good suggestion. We do not have any information at the moment concerning the possible participation of microtubules in this AdoHcy effect on cAMP metabolism.

DR. H. MCILWAIN: Another similarity to CNS systems: responses parallel to those reported in lymphocytes Ado + Hcy + prostaglandin E were found with neocortex using Ado + Hcy + norepinephrine. Colchicine in such tissues is also an inhibitor of adenosine uptake.

DR. ZIMMERMAN: Thank you. We have also found that colchicine augments the cAMP response of these mouse cytolytic lymphocytes to PGE_1.

DR. B.B. FREDHOLM: Do you believe that the effect of adenosine is due to inhibition of binding or to inhibition of the programmed killing? The reason I ask is that, in studies with Dr. Jondal of the Department of Tumor Biology, Karolinska Institutet, we obtained some data suggesting that procedures that elevate cyclic AMP levels inhibit binding rather than the lytic step, whereas other agents may have the opposite effect.

DR. ZIMMERMAN: Bradley, Anisman, and Bonavida have reported in an abstract [*Fed. Proc.* 39:921, 1980] that treatment of cytolytic lymphocytes with adenosine results in inhibition of their ability to form conjugates with target cells. However, similar experiments in our laboratory have failed to detect an effect of adenosine on conjugate formation. This apparent discrepancy in results may be due to the technical difficulties associated with the determination of conjugate formation.

IV. ADENOSINE AND PLATELET FUNCTION

18. RECEPTORS FOR NUCLEOSIDES AND NUCLEOTIDES ON BLOOD PLATELETS

D.C.B. MILLS, D.E. MACFARLANE, B.W.G. LEMMEX, R.J. HASLAM

The specific interaction of platelets with ADP was discovered by Hellem and his co-workers at the University of Oslo [1] when they identified ADP as the compound present in an acidic extract of red blood cells that caused platelets to become trapped in a column of glass beads. The effect was specific for ADP, as neither AMP nor a range of naturally occurring nucleoside polyphosphates were active. Born [2] showed that aggregation of platelets could be studied quantitatively by a nephelometric technique and found that adenosine and 2-chloroadenosine [3] inhibit aggregation induced by ADP in a competitive manner. The initial conclusion that these nucleosides were competing with ADP for a specific receptor was revised in the light of several observations. The inhibitory effect of the nucleosides increases with time, and AMP is less active than adenosine. Furthermore, adenosine inhibits platelet aggregation induced by agents other than ADP (Fig. 18.1) [4], and in the case of nucleated thrombocytes of the frog, which aggregate when stimulated by thrombin but not with ADP, adenosine was still an inhibitor [5]. These discrepancies were resolved by the demonstration that adenosine increases the cyclic AMP content of platelets [6] and that it stimulates the adenylate cyclase activity of platelet membrane fragments [7]. Figure 18.2 shows that in platelets treated with different concentrations of adenosine either alone or in the presence of the cyclic AMP phosphodiesterase inhibitor, papaverine, there is a close correspondence between the increase in cyclic AMP level achieved and the degree of inhibition of aggregation observed. The inhibitory effect of adenosine is re-

R.M. Berne, T.W. Rall, and R. Rubio (eds.), Regulatory Function of Adenosine, *pp. 277–289.*

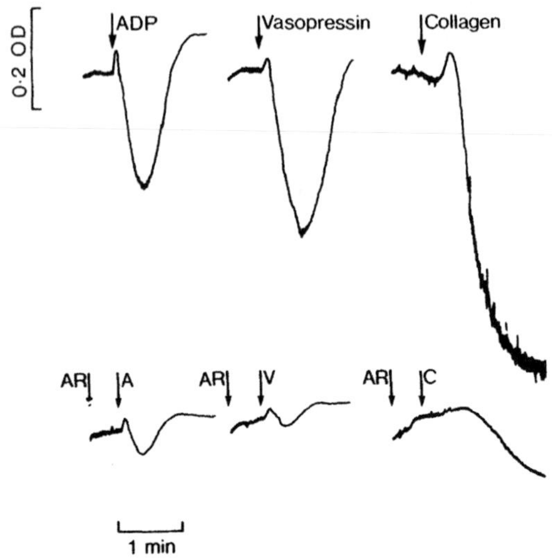

Figure 18.1. Inhibition of platelet aggregation by adenosine. Aggregation in human citrated platelet-rich plasma was induced with (1) ADP (1.0 μ*M*); (2) arginine vasopressin (100 mU/ml); and (3) bovine tendon collagen (5 μg/ml). A higher concentration of adenosine (40 μ*M*) was required to inhibit the response to collagen than the concentration (2 μ*M*) used with ADP or with vasopressin.

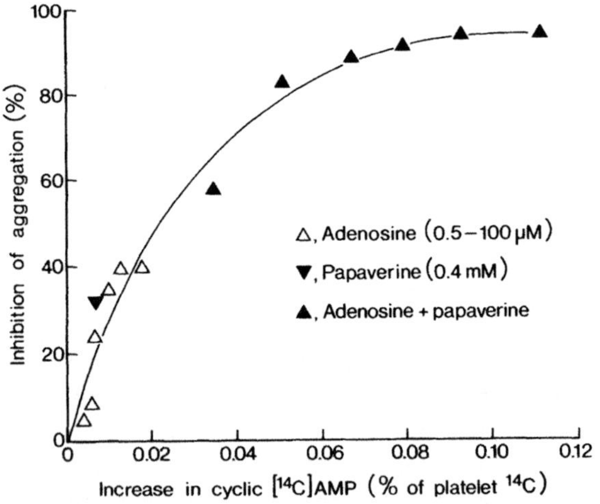

Figure 18.2. Relationship between the inhibitory effect of adenosine on platelet aggregation and platelet cyclic AMP content. Aggregation was measured as the change in light absorbance occurring 30 sec after the addition of 0.5 μ*M* ADP, with adenosine and papaverine as indicated. Cyclic [^{14}C]AMP was measured at the same time in identically treated samples of platelet-rich plasma that had previously been allowed to incorporate [^{14}C]adenine. For details, see [8].

duced by theophylline and caffeine, which also prevent the increase in cyclic AMP [6, 8]. ADP also reduces the increase in platelet cyclic AMP caused by adenosine but, in contrast to the methylxanthines, acts noncompetitively [8]. The effect of adenosine does not depend on its penetration into the platelets, as it is potentiated by drugs of the pyrimido-pyrimidine type, which, besides inhibiting cyclic AMP phosphodiesterase, powerfully antagonize the adenosine uptake mechanism [9]. From these observations, it became clear that the effects of ADP and adenosine on platelets are mediated by two different specific receptors on the cell surface.

PLATELET ADENOSINE RECEPTORS

The stimulation of cyclic AMP accumulation in platelets exposed to adenosine increases with dose to a maximum at about 50 μM, after which it declines. This decline does not occur when transport of adenosine into the platelet is blocked by p-nitrophenylthioguanosine, which does not inhibit cyclic AMP phosphodiesterase [8]. This indicates that adenosine can cause inhibition of cyclic AMP accumulation in platelets through an intracellular mechanism. In platelet lysates, increasing concentrations of adenosine first stimulate and then inhibit adenylate cyclase activity, but in the presence of PGE_1, a powerful stimulator of platelet adenylate cyclase, only the inhibitory effect is seen [7]. Some adenosine analogues, such as N^6-phenyladenosine, stimulate platelet adenylate cyclase but have no inhibitory activity, either alone or in the presence of PGE_1 [10, 11]. These effects of adenosine on platelets can be interpreted to be the result of simultaneous stimulation of adenylate cyclase through an extracellular receptor and inhibition via an intracellular site [8]. Londos and Wolff [12] later designated these sites as R and P, respectively, on the basis of a requirement for intact ribose and purine moieties of adenosine, respectively. The true receptors or R sites have now been classified into those that stimulate adenylate cyclase—R_a (also referred to as A_2)—and those that in certain other tissues inhibit the enzyme—R_i (also referred to as A_1) [13, 14]. The platelet receptor clearly belongs to the A_2 type, and the recent demonstration that 5′-N-ethylcarboxamidoadenosine (NECA) is more active than adenosine as an inhibitor of platelet aggregation and as a stimulator of cyclic AMP accumulation [15] indicates that the structural specificity of the platelet A_2 receptor is analogous to that of similar receptors on other cell types [12, 13]. Of the many adenosine analogues that have been tested as inhibitors of aggregation, only NECA and compounds substituted in the 2 and N^6 positions have activities approaching or exceeding that of adenosine [11].

The powerful P site adenylate cyclase inhibitor, 2′,5′-dideoxyadenosine (DDA), has been a useful tool for demonstrating the role of cyclic AMP in the inhibitory effect of agents such as PGE_1 on platelet aggregation [17]. Inhibition of aggregation by PGE_1 is accompanied by an increase in cyclic AMP. In the presence of DDA, both the increase in cyclic AMP and the inhibition of aggregation are blocked. DDA itself can lower the basal adenylate cyclase

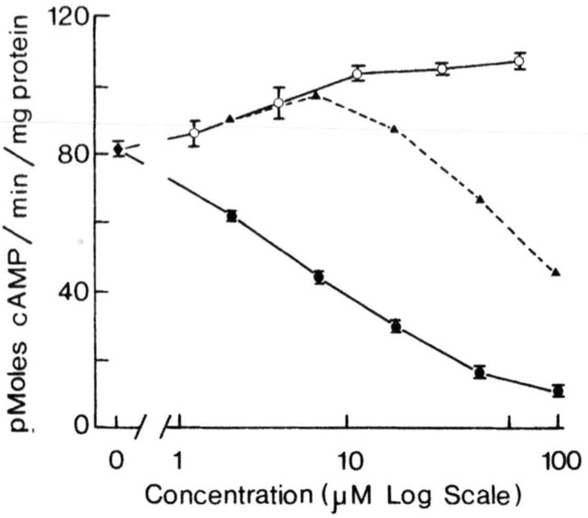

Figure 18.3. Stimulation of platelet adenylate cyclase activity by adenosine (▲,△) and by 2-azidoadenosine (●,○) assayed at two concentrations of Mg^{2+}. Adenylate cyclase was measured in 10 min incubations of a particulate fraction from frozen and thawed rabbit platelets at either 0.8 m*M* Mg^{2+} (open symbols) or 5 m*M* Mg^{2+} (filled symbols). Other details of the assay were as in [17]. Results are means ± SEM from triplicate assays.

activity of platelets, although it does not induce platelet aggregation nor does it potentiate the aggregation induced by ADP or by vasopressin [17]. This indicates that cyclic AMP is a unidirectional mediator involved only in inhibition of aggregation, not in its induction.

The contrasting effects of adenosine, DDA, and NECA on the adenylate cyclase activity of rabbit platelet homogenate are shown in Figure 18.3; under these assay conditions, the observed stimulations of adenylate cyclase were quite small. Incubation of a particulate fraction of platelet lysate with adenosine deaminase causes approximately a 20% reduction of the observed adenylate cyclase activity, indicating the presence of endogenous adenosine. However, the inclusion of either indomethacin to inhibit the formation of endogenous prostaglandins or propranolol to block effects mediated by β-adrenergic receptors had little or no effect. 2-azidoadenosine, first described by Cusack and Born [18], has been investigated as a possible photoaffinity label for the platelet adenosine (A_2) receptors. This compound, which is not a substrate of adenosine deaminase [18], stimulated platelet adenylate cyclase much more effectively in the presence of this enzyme (Table 18.1). Figure 18.4 shows that the effects of adenosine and 2-azidoadenosine on the adenylate cyclase activity of a particulate fraction from frozen and thawed rabbit platelets are also enhanced in percentage terms when a low (e.g., 0.8 m*M*) Mg^{2+} concentration was used in the assays. An optimal stimulation of approximately 200% was obtained when the action of 2-azidoadenosine on rabbit platelet

Table 18.1. Effects of adenosine deaminase, propranolol, and indomethacin on adenylate cyclase activity of a particulate fraction from frozen and thawed rabbit platelets assayed in the presence of 5 μM Mg^{2+}

Additive	Adenylate cyclase activity (mean ± SEM) (pmol/min/mg protein)		
	Control	+ 10 μ*M* 2-azidoadenosine	Increase in activity (%)
None	27.5 ± 0.6	44.1 ± 1.3	60.4
Adenosine deaminase (AD) 8.5 U/ml	23.9 ± 0.7	42.3 ± 0.6	77.0
Propranolol (Pro) 10 μ*M*	28.6 ± 0.3	43.6 ± 1.4	52.4
Indomethacin (Ind) 20 μ*M*	28.2 ± 0.7	43.3 ± 0.9	53.5
AD + Pro	22.3 ± 0.5	42.9 ± 0.9	92.4
AD + Ind	21.2 ± 0.6	44.6 ± 1.2	110.4
Pro + Ind	26.1 ± 1.1	45.5 ± 1.3	74.4
AD + Pro + Ind	21.2 ± 0.1	43.2 ± 0.9	103.8

adenylate cyclase was studied in the presence of adenosine deaminase and 0.8 m*M* $MgCl_2$. 2-azidoadenosine was slightly more potent than adenosine as a stimulator of adenylate cyclase, but, like the N^6-substituted analogues, it lacked the ability to cause inhibition at high concentrations [17]. Radioactivity from 2-azido[8-^{3}H]adenosine was incorporated into platelet proteins when the compound was exposed to ultraviolet light in the presence of a platelet particulate fraction, but this also occurred if a particulate fraction was added to the compound after irradiation [19]. Neither adenosine nor N^6-cyclohexyladenosine inhibited the labeling. Thus, the adenosine receptor was not identified using this technique.

PLATELET ADP RECEPTORS

In addition to causing a functional response of platelets, with a change in shape followed by aggregation, ADP also inhibits the accumulation of cyclic AMP in platelets exposed to activators of adenylate cyclase, including prostaglandins and adenosine [11]. This inhibition is noncompetitive [20] and appears to be due to a receptor-mediated inhibition of adenylate cyclase [21–23]. This receptor has many properties in common with the receptor involved in shape change and aggregation. Both aggregation and the effect of ADP on cyclic AMP accumulation are inhibited by ATP but not by AMP or other nucleoside diphosphates [24]. As with the shape change, inhibition of cyclic AMP accumulation occurs in the presence of calcium chelators such as EDTA [25]. The effect of ADP on cyclic AMP accumulation is very rapid, as is the reversal of this effect on addition of excess ATP. When ADP is added to platelets 10 sec after PGE_1, the intracellular cyclic AMP level begins to fall within 1 sec, and when an excess of ATP is added to platelets incubated for 15 sec with both

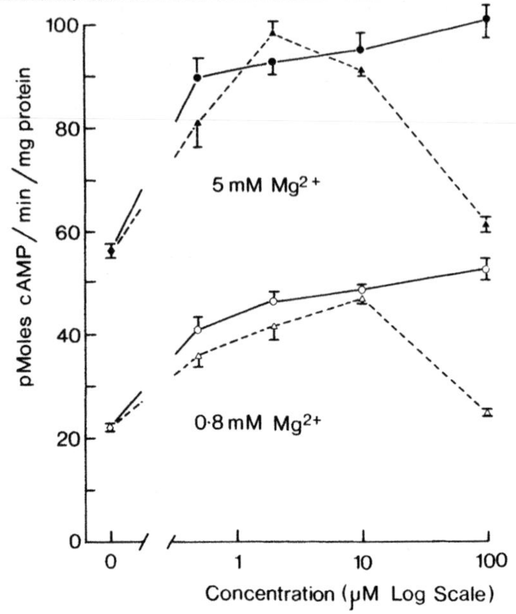

Figure 18.4. Effect of adenosine (▲—▲), 5′-N-ethylcarboxamidoadenosine (NECA) (○—○), and 2′,5′-dideoxyadenosine (●—●) on the adenylate cyclase activity of frozen and thawed rabbit platelets assayed in the presence of 5 m*M* Mg^{2+}, 100 m*M* Na^{+}, 1 m*M* cyclic AMP, and no phosphodiesterase inhibitor. Results are plotted as means of triplicate or duplicate determinations. Where appropriate, the error bars indicate ±SEM.

PGI_2 and ADP, the intracellular cyclic AMP begins to rise within 2 sec [25].

Several attempts have been made to measure the binding of ADP to its presumed receptor, either on intact platelets or on membrane fragments. These experiments have been criticized [11]. In an early study, we found that the amount of ADP bound to human platelets and measured after their centrifugation through silicone oil, was equivalent to less than 1000 molecules per platelet [unpublished results]. The small number of ADP receptors and their comparatively low affinity for ADP indicated that an analogue of ADP with higher affinity for the receptor was required. Among the derivatives of ADP tested as aggregating agents, the only modifications that were not associated with a reduced affinity were at the 2 position in the purine ring. In fact, 2-chloroADP and 2-methylthioADP were shown to be 10- and 30-fold more potent than ADP as aggregating agents in the sheep [26, 27], and 2-azidoADP, a photolysable analogue and therefore potentially a photoaffinity probe, was 5-fold more active than ADP with human platelets [28]. 2-azidoADP (2-N_3ADP) was prepared from 2-chloroadenosine via 2-hydrazinoadenosine, 2-azidoadenosine, and 2-azidoAMP. The final phosphorylation was performed with ^{32}P-labeled P_i and carbonyl diimidazole [29], giving a product labeled in the β-phosphate with a specific activity in the range 2–10 Ci/mmole. 2-N_3ADP proved to be both an aggregating agent (Fig. 18.5) and an inhibitor of PGE_1-stimulated accumulation of cyclic AMP (Fig. 18.6). In each case, after

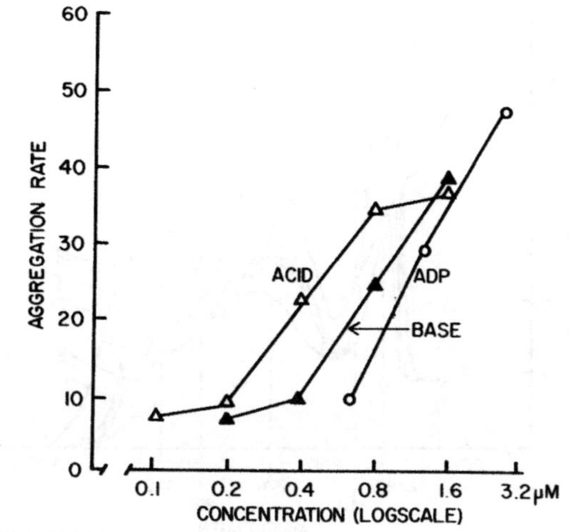

Figure 18.5. Aggregation of human platelets in citrated platelet-rich plasma induced by ADP (O—O) and by 2-azidoADP, measured as the maximum rate of change of absorbance in arbitrary units. 2-azidoADP was stored for 2 hr at 4° before use either in 0.1 *M* NaOH (▲—▲) or in 0.1 *M* HCl (△—△).

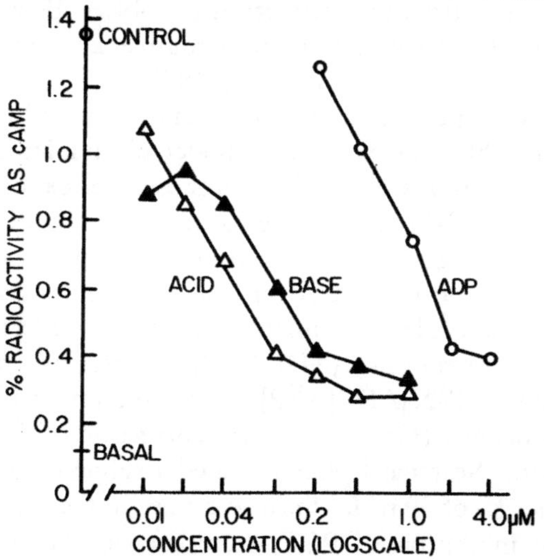

Figure 18.6. Inhibition of the PGE_1-induced accumulation of cyclic [^{14}C]AMP in platelets preincubated with [^{14}C]adenine caused by ADP (O—O) or by the acidic (△—△) or basic (▲—▲) forms of 2-azidoADP. Cyclic AMP accumulation was measured as the percentage of intracellular [^{14}C] converted to cyclic [^{14}C]AMP 1 min after addition of 2 μ*M* PGE_1 with the phosphodiesterase inhibitor RA 233 (0.2 m*M*) and ADP or 2-azidoADP.

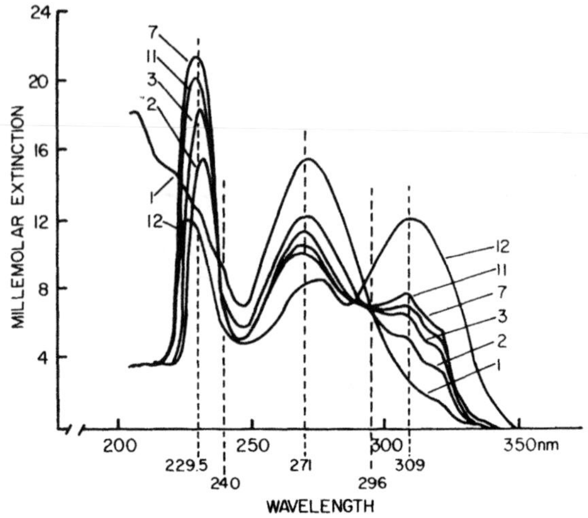

Figure 18.7. Ultraviolet absorption spectra of 2-azidoAMP recorded at different values of pH from 1 to 12. The solutions were kept at the indicated pH for 2 hr before scanning.

storage in dilute NaOH, the activity was about 2-fold lower than after storage in dilute acid. The spectroscopic properties of 2-N_3AMP were studied in detail. The acidic form had an absorption maximum at 271 nm, whereas the basic form had a peak absorption at 309 nm (Fig. 18.7). The basic form was slightly fluorescent, with absorption and emission maxima at 312 and 460 nm and a quantum yield of 0.002, and was stable under ultraviolet light, whereas the acidic form was not fluorescent and decomposed on exposure to ultraviolet light. The spectrum of 2-N_3AMP stored at neutral pH indicated the presence of both forms. By analogy with other purine azides, we assume that the basic form represents one or both of the two possible tetrazolo-derivatives, whereas the acidic form is the azide. Presumably, similar considerations apply to 2-N_3ADP. The acidic form of 2-N_3ADP[β^{32}P] was used as a radioligand for the platelet ADP receptor. 2-N_3ADP[β^{32}P] rapidly and reversibly bound to about 700 sites on each platelet (Fig. 18.8). After correction of the pellet counts for trapped supernatant, Scatchard plots revealed a single class of noninteracting binding sites with a K_d of 30 nM, which was close to the concentration required for half maximal inhibition of PGE_1-stimulated cyclic AMP accumulation (Fig. 18.6). Binding of 2-N_3ADP was competitively inhibited by ADP and by ATP, with half-maximal displacement at 8.3 and 10 μM, respectively. AMP and other purine nucleoside diphosphates were about 100-fold less active. Binding and dissociation of 2-N_3ADP were both rapid, being essentially complete within 30 sec.

When 2-N_3ADP[β^{32}P] was exposed to ultraviolet light in the presence of washed human platelets, several components of the membrane incorporated radioactivity covalently. On analysis by SDS-polyacrylamide gel elec-

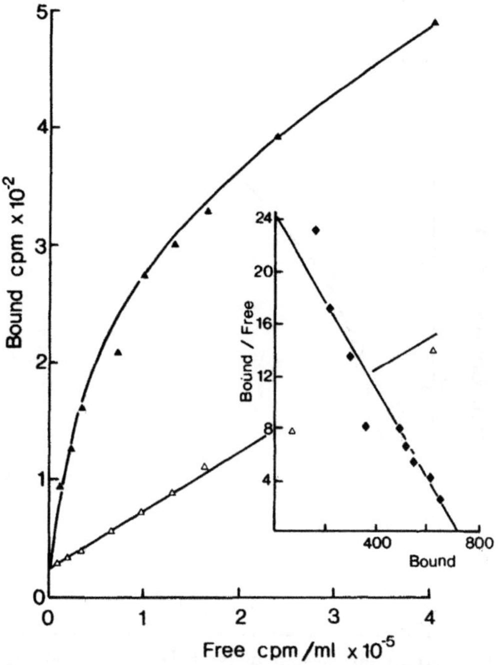

Figure 18.8. Equilibrium binding of 2-azidoADP to human platelets measured by the technique of Born and Feinberg [32]. Samples of citrated platelet-rich plasma at 37° were mixed with 2-azidoADP [β^{32}P] (7.5 n*M* to 1.5 μ*M*) in the presence of 4.5 m*M* EDTA added to prevent aggregation, and the platelets were centrifuged through a layer of silicone oil. Radioactivity in the pellet is plotted against radioactivity in a sample of the supernatant fraction. The open triangles represent binding in the presence of 1 m*M* ADP and correspond to trapped plasma (2.3 fL/platelet) were used to obtain specific binding by difference. Inset: A Scatchard plot of the corrected data.

trophoresis, four major bands in the range 80,000–150,000 daltons were found to be labeled, although no radioactivity was associated with the major intracellular proteins, actin and myosin. No difference in the pattern of labeling was detected when excess ATP or ADP were present during photolysis, indicating that none of the labeled bands corresponds to the site identified by equilibrium binding. Since ADP analogues with comparatively bulky substituents in the 2 position retain activity both as aggregating agents [11] and as inhibitors of cyclic AMP accumulation [29], the failure of the nitrene generated by photolysis of 2-N_3ADP to label the ADP binding site covalently may be due to orientation of the 2 position of the purine away from the receptor. The same difficulty may account for the failure, so far, of attempts to photolabel platelet adenosine receptors with 2-azidoadenosine.

The presence of about 700 binding sites for ADP on the platelet surface has been confirmed with 2-methylthioADP, a light-stable ADP analogue that is both an aggregating agent [27] and a powerful inhibitor of cyclic AMP accumulation [30]. 2-methylthioADP[β^{32}P] binds rapidly and reversibly to

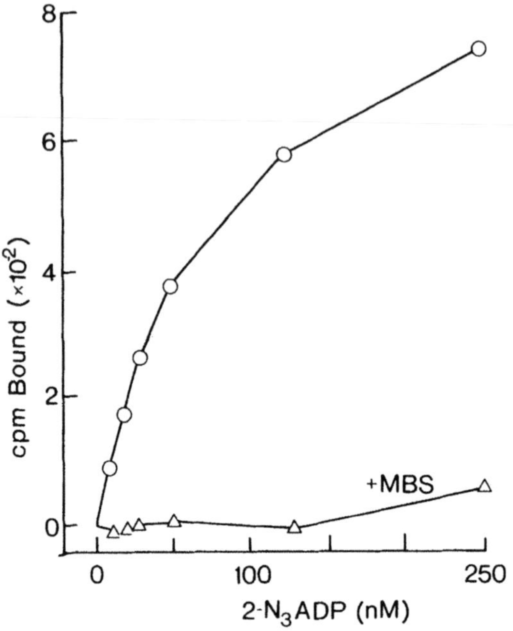

Figure 18.9. Effect of p-mercuribenzenesulphonate (MBS) on the binding of 2-azidoADP to platelets. Samples of citrated platelet-rich plasma were incubated with 4.5 m*M* EDTA without (O—O) and with (Δ—Δ) 0.6 mM MBS for 3 min before adding the appropriate concentrations of 2-azidoADP [$\beta^{32}P$]. Bound radioactivity is shown after subtraction of radioactivity bound in the presence of excess (1 m*M*) ADP.

platelets with a dissociation constant (K_d) of 5–15 × 10^{-9} *M*. The relative activities of 2-N_3ADP and 2-methylthioADP as inhibitors of cyclic AMP accumulation and as ligands for the ADP binding site strongly suggest that this site is the receptor through which ADP inhibits PGE_1-stimulated adenylate cyclase. Binding of 2-N_3ADP to platelets was blocked by the mercurial thiol reagent, p-chloromercuribenzenesulphonate (MBS) (Fig. 18.9). This compound also blocks the effect of ADP and its analogues on cyclic AMP accumulation. The shape change of platelets that precedes aggregation induced by ADP is unaffected by MBS, suggesting either that shape change and aggregation are mediated by a different receptor from that responsible for adenylate cyclase inhibition or that far fewer receptors need to be occupied to induce shape change than are required for inhibition of cyclic AMP accumulation. Evidence that aggregation and inhibition of cyclic AMP accumulation by ADP are mediated by the same receptor has recently been reported by Cusack and Hourani [31], using competitive inhibitors of the effects of ADP with widely differing affinities. A correlation was found between the ability of these compounds to inhibit ADP-induced aggregation and to inhibit the effects of ADP on cyclic AMP accumulation. Whereas the discovery of a single compound that competitively inhibits the two effects of ADP with widely differing potencies would serve to refute this correlation, these data favor the view that a single receptor is involved in both effects.

The platelet ADP receptor is, in our present state of ignorance, unique to this tissue. It differs from the P_2 purinergic receptor present in various smooth muscles and neurons in its absolute specificity for the diphosphate. Similar ADP receptors may also be present on tissues less accessible to investigation than platelets. The use of high-affinity analogues of ADP should make a search for such receptors a feasible undertaking.

REFERENCES

1. Gaarder A, Jonsen J, Laland S, Hellem A, Owren PA: Adenosine diphosphate in red cells as a factor in the adhesiveness of human platelets. *Nature* 192:531–532, 1961.
2. Born GVR: Aggregation of blood platelets by adenosine diphosphate and its reversal. *Nature* 194:927–929, 1962.
3. Born GVR: Strong inhibition by 2-chloroadenosine of the aggregation of blood platelets by adenosine diphosphate. *Nature* 202:95–96, 1964.
4. Haslam RJ, Rosson GM: Aggregation of human blood platelets by vasopressin. *Am J Physiol* 223:958–967, 1972.
5. Kien M, Belamarich FA, Shepro D: Effect of adenosine and related compounds on thrombocyte and platelet aggregation. *Am J Physiol* 220:604–608, 1971.
6. Mills DCB, Smith JB: The influence on platelet aggregation of drugs that affect the accumulation of adenosine 3′,5′-cyclic monophosphate in platelets. *Biochem J* 121:185–196, 1971.
7. Haslam RJ, Lynham JA: Activation and inhibition of blood platelet adenylate cyclase by adenosine or by 2-chloroadenosine. *Life Sci* 2(Part II):1143–1154, 1972.
8. Haslam RJ, Rosson GM: Effects of adenosine on levels of adenosine cyclic 3′,5′-monophosphate in human blood platelets in relation to adenosine incorporation and platelet aggregation. *Mol Pharmacol* 11:528–544, 1975.
9. Mills DCB, Smith JB, Born GVR: Pharmacology of platelet aggregation inhibition. *Thromb Diath Haemorrh* (Suppl) 42:176–184, 1972.
10. Haslam RJ, Davidson MML, Desjardins JV, Fox JEB, Lynham JA: Factors affecting the formation and actions of cyclic AMP in blood platelets. *Adv Pharmacol Ther* 4:75–85, 1978.
11. Haslam RJ, Cusack NJ: Blood platelet receptors for ADP and for adenosine, in Burnstock G (ed): Purinergic Receptors. London, Chapman and Hall, 1981, pp 222–285.
12. Londos C, Wolff J: Two distinct adenosine sensitive sites on adenylate cyclase. *Proc Natl Acad Sci USA* 74:5482–5486, 1977.
13. Londos C, Cooper DMF, Wolff J: Subclasses of external adenosine receptors. *Proc Natl Acad Sci USA* 77:2551–2554, 1980.
14. Van Calker D, Müller M, Hamprecht B: Adenosine inhibits the accumulation of cyclic AMP in cultured brain cells. *Nature* 276:839–841, 1978.
15. Cusack NJ, Hourani SMO: 5′-N-ethylcarboxamidoadenosine: A potent inhibitor of human platelet aggregation. *Br J Pharmacol* 72:443–447, 1981.
16. Bruns RF: Adenosine receptor activation in human fibroblasts: Nucleoside agonists and antagonists. *Can J Physiol Pharmacol* 58:673–691, 1980.
17. Haslam RJ, Davidson MML, Desjardins JV: Inhibition of adenylate cyclase by adenosine analogues in preparations of broken and intact human platelets: Evidence for the unidirectional control of platelet function by cyclic AMP. *Biochem J* 176:83–95, 1978.
18. Cusack NJ, Born GVR: Inhibition of adenosine deaminase and of platelet aggregation by 2-azidoadenosine, a photolysable analogue of adenosine. *Proc R Soc Lond [Biol]* 193:307–311, 1976.
19. Lemmex BWG: Activation of platelet adenylate cyclase by 2-azidoadenosine: A potential photoaffinity label. MSc thesis, McMaster University, 1980.
20. Haslam RJ: Interactions of the pharmacological receptors of blood platelets with adenylate cyclase. *Ser Haematol* 6:333–350, 1973.

21. Cooper DMF, Rodbell M: ADP is a powerful inhibitor of human platelet plasma membrane adenylate cyclase. *Nature* 282:517–518, 1979.
22. Mellwig KP, Jakobs KH: Inhibition of platelet adenylate cyclase by ADP. *Thromb Res* 18:7–17, 1980.
23. Mills DCB, Macfarlane DE: Platelet receptors, in Gordon JL (ed): *Platelets in Biology and Pathology*. Amsterdam, Elsevier North-Holland, 1976, pp 159–202.
24. Macfarlane DE, Mills DCB: The effects of ATP on platelets: Evidence against the central role of released ADP in primary aggregation. *Blood* 46:309–320, 1975.
25. Macfarlane DE, Mills DCB: Inhibition by ADP of prostaglandin-induced accumulation of cyclic AMP in intact human platelets. *J Cyclic Nucleotide Res* 7:1–11, 1981.
26. Maguire MH, Michel F: Powerful new aggregator of blood platelets—2-chloroadenosine-5′-diphosphate. *Nature* 217:571–573, 1968.
27. Gough GR, Maguire MH, Penglis F: Analogues of adenosine 5′-disphosphate—new platelet aggregators: Influence of purine ring and phosphate substitutions on the platelet-aggregating potency of adenosine 5′-diphosphate. *Mol Pharmacol* 8:170–177, 1972.
28. Cusack NJ, Born GVR: Effects of photolysable 2-azido analogues of adenosine, AMP and ADP on human platelets. *Proc R Soc Lond [Biol]* 197:515–520, 1977.
29. Macfarlane DE, Mills DCB, Srivastava PC: Binding of 2-azidoadenosine [^{32}P] diphosphate to the receptor on intact human blood platelets which inhibits adenylate cyclase. *Biochemistry* 21:544–549, 1982.
30. Macfarlane DE, Srivastava PC, Mills DCB: 2-methylthioadenosine [β^{32}P]diphosphate. An agonist and radio ligand for the receptor that inhibits the accummulation of cyclic AMP in intact blood platelets. J Clin Invest 71, 1983 (in press).
31. Cusack NJ, Hourani SMO: Adenosine 5′-diphosphate antagonists and human platelets: No evidence that aggregation and inhibition of stimulated adenylate cyclase are mediated by different receptors. *Br J Pharmacol* 76:221–227, 1982.
32. Born GVR, Feinberg H: Binding of adenosine diphosphate to intact human platelets. *J Physiol (Lond)* 251:803–816, 1977.

DISCUSSION

DR. H.P. BAER: Do platelets contain a number of phosphorolytic ecto-enzymes?

DR. MILLS: There is evidence for several ecto-nucleotidases on the platelet membrane, including an ATP phosphohydrolase, ADP phosphohydrolase, and 5′-nucleotidase. Nucleoside diphosphate kinase activity has also been demonstrated, but as this enzyme is also found in soluble form in supernatant fractions from platelet suspensions, its classification as an ecto-enzyme is insecure.

DR. M.H. MAGUIRE: (1) Would you care to comment on the number of adenine nucleotide and adenosine sites on platelets? (2) Could you comment on the claims by other workers, e.g., Handen and Colman, to have isolated the ADP aggregation receptor site?

DR. MILLS: Aside from intracellular P site, the platelet has an adenosine receptor and an ADP receptor on its membrane and an adenosine uptake mechanism. There is no evidence associating the ADP-binding proteins identified by Nachman and Handin in isolated platelet membranes with either of the physiological effects of ADP. They may well represent binding sites present on the inner surface of the membrane. Fluorosulfonylbenzoyl adenosine labels a single 100,000 dalton peptide in intact platelets. It does not correspond to the binding site for 2 MeSADP.

V. VASCULAR EFFECTS OF ADENOSINE

19. BLOOD FLOW REGULATION BY ADENOSINE IN HEART, BRAIN, AND SKELETAL MUSCLE

ROBERT M. BERNE, H. RICHARD WINN, ROBERT M. KNABB,
STEPHEN W. ELY, RAFAEL RUBIO

Although it has long been recognized that adenosine is a potent vasodilator, only during the past 20 years has serious consideration been given to its possible role in the local regulation of blood flow. Prior to that time other potential mediators of regional blood flow regulation, such as reduced Po_2, increased Pco_2, or decreased pH were often suggested, but definitive experimental proof for a physiological role for these agents was lacking. In more recent years, more evidence has appeared in support of a reduced pH, and additional substances such as potassium ions, inorganic phosphate, and increased osmolarity have been suggested with varying degrees of acceptance. In all likelihood, several factors, including a myogenic response, are involved in local blood flow regulation, but adenosine appears to be the primary factor. Since most of the studies on adenosine have been focused on its mediation of the control of coronary, cerebral, and skeletal muscle blood flow, in that order, discussion will be limited to these three tissues. This in no way means that adenosine does not act similarly in several other tissues. In the kidney, however, adenosine induces vasoconstriction, a subject considered in detail by Dr. Osswald in Chapter 25.

This work was supported in large part by USPHS grants HL 10384 and NS 7477 and a grant from the American Heart Association. H.R. Winn is recipient of a Teacher Investigator Award, K07-500404.

R.M. Berne, T.W. Rall, and R. Rubio (eds.), Regulatory Function of Adenosine, *pp. 293–317.*

HEART

Reduced oxygen supply

The earliest attempts to demonstrate adenosine release from hypoxic or ischemic hearts met with failure, because only the degradative products of adenosine, namely, inosine and hypoxanthine, were recovered in the effluents of isolated perfused hearts or in blood-perfused hearts of the open-chest dog [1]. Nevertheless, it was demonstrated by Gerlach et al. [2] and by Imai et al. [3] that adenosine was formed in extirpated unperfused cardiac tissue incubated in moist chambers at 37°C. Adenosine was first found in the perfusate of isolated perfused hearts in 1964 [4], with the use of an uncoupler of oxidative phosphorylation, and in 1966 [5], with the aid of an adenosine deaminase inhibitor. Since that time, several studies have appeared showing that adenosine is released following brief periods of ischemia or hypoxia in anesthetized animals, or in isolated perfused hearts, and is also present in the normal heart [6]. Furthermore, the magnitude of the increase in adenosine release and the increment in the coronary conductance showed a striking parallelism [7, 8]. Finally, it was demonstrated that, within a single cardiac cycle, a significant increase in adenosine occurred during systole when compared to diastole [9]. This observation emphasizes the rapidity of the adenosine response as well as its close association with the contractile state of the myocardium.

With respect to the regulation of coronary blood flow by adenosine, the key problem is whether adenosine release is coupled to myocardial metabolism, since the cardiac oxygen demand is the parameter that is altered under physiological conditions, not the oxygen supply. Nevertheless, there is obviously a balance between oxygen supply and oxygen demand of the heart and, as depicted in Figure 19.1, adenosine may play a key role in adjusting blood flow (oxygen supply) to myocardial needs. How this balance can be disturbed is illustrated in Figure 19.2. The line to the right, which originates near the abscissa, represents the increase in coronary flow obtained with reduction in the oxygen content of the perfusion fluid in the isolated perfused guinea pig heart. The line to the left, which also originates at the same point near the abscissa, represents the increase in coronary flow when the coronary vasodilator nitroglycerin is added to the perfusion fluid. In the latter instance, flow, and hence oxygen supply, exceeds the needs of the myocardium and there is a progressive decrease in the adenosine content of the heart as flow increases at a high oxygen content in the perfusion fluid. However, in physiological states, we do not have excess perfusion; blood flow usually closely parallels changes in oxygen demand.

Increased oxygen demand

When myocardial metabolic activity is increased, even in the presence of a large oxygen supply, there is an associated increment in adenosine formation

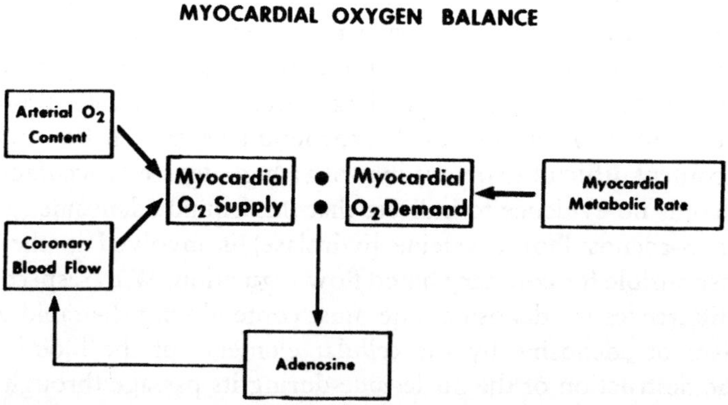

Figure 19.1. Oxygen balance in the heart can be maintained by regulation of coronary blood flow by adenosine.

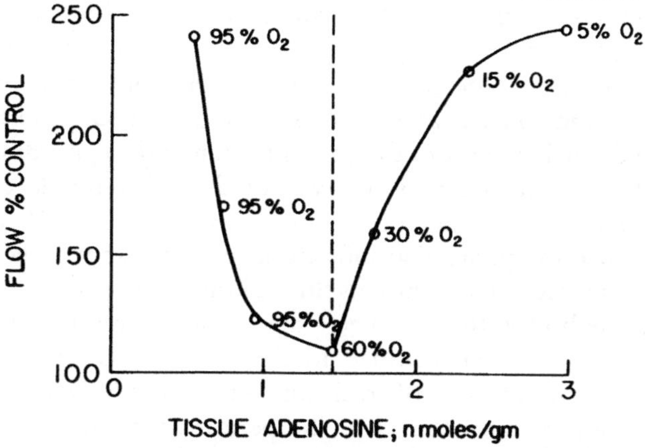

Figure 19.2. Effect of hypoxia and an extrinsic vasodilator on coronary flow and on the adenosine content of the isolated perfused guinea pig heart. The curve to the right was produced by reducing the oxygen content (% O_2) to the heart and resulted in an increase in tissue adenosine levels. The curve to the left represents the effect of nitroglycerin and shows that excess flow at 95% O_2 in the perfusion fluid results in a decrease in tissue adenosine levels. Source: Reproduced by permission from *Prog. Cardiovasc. Dis.* [33].

in the heart. For example, constriction of the aorta just distal to the origin of the innominate artery in the open-chest rat with a normal P_{AO_2} results in large increases in the adenosine content of the heart, despite a high coronary blood flow secondary to the elevated coronary perfusion pressure [10]. Similar increases in cardiac adenosine levels were observed in the open-chest dog with stellate ganglia stimulation [11] and with other interventions [12].

In the stellate ganglia-stimulation experiments, a new technique was used to estimate the amount of adenosine released by the heart. Previously, either samples of myocardium were frozen *in situ* and then analyzed for adenosine or arteriovenous differences (coronary sinus minus arterial concentration) of adenosine were determined. Each of these methods has serious drawbacks. In the case of myocardial sampling the large pool of intracellular protein-bound adenosine results in a considerable background level that makes small changes in tissue content difficult to determine with a high degree of accuracy. Furthermore, there is no evidence to indicate that this bound adenosine (presumably bound to S-adenosylhomocysteine hydrolase) is involved in the adenosine release responsible for coronary blood flow regulation. With respect to arteriovenous differences in adenosine, one must contend with the rapid uptake and metabolism of adenosine by the cellular elements of the blood as well as enzymatic destruction of the nucleoside during its passage through the vessel wall.

To circumvent some of the problems associated with tissue sampling and arteriovenous blood differences, we injected a small volume (25–40 ml) of Krebs-Henseleit solution into the pericardial sac during control and experimental conditions and then withdrew the solution after 4.5 min and analyzed it for adenosine, inosine, and hypoxanthine. The amount of adenosine and its degradative products in the pericardial infusate did not reach equilibrium with the interstitial fluid concentration of these metabolites but nevertheless served as a useful and reliable index of the concentration in the epicardium [11]. Since a concentration gradient across the wall of the left ventricle could only be demonstrated when coronary blood flow was restricted [13], the epicardial levels, and hence the pericardial infusate levels, probably reflect the changes that occur with various interventions throughout the normal heart. One drawback of this method is that it reflects the adenosine release from the whole heart, not from any given chamber. Ideally, one should directly measure the interstitial fluid concentration of adenosine, but methods for doing so have not yet been developed; until they are, the pericardial infusate method, in our view, is probably the least objectionable one available.

Studies have been carried out in the anesthetized and the conscious dog utilizing the pericardial infusate method to estimate cardiac adenosine release, and in some experiments this was coupled with arteriovenous adenosine concentration differences across the heart. Figure 19.3 illustrates the effect of several interventions on left ventricular oxygen consumption, left circumflex coronary blood flow, and the adenosine concentration in the pericardial infusate [14]. It is evident that, regardless of the type of intervention used, correlation among the three variables was observed. In other words, enhanced myocardial oxygen consumption was paralleled by an increment in adenosine release and coronary blood flow. This tight coupling of myocardial metabolic activity, adenosine production, and coronary blood flow was also observed in the conscious dog [15, 16]. In our experiments [16], the animals were surgically instrumented under sterile conditions for measurement of arterial pressure and left circumflex coronary blood flow, for sampling of arterial and

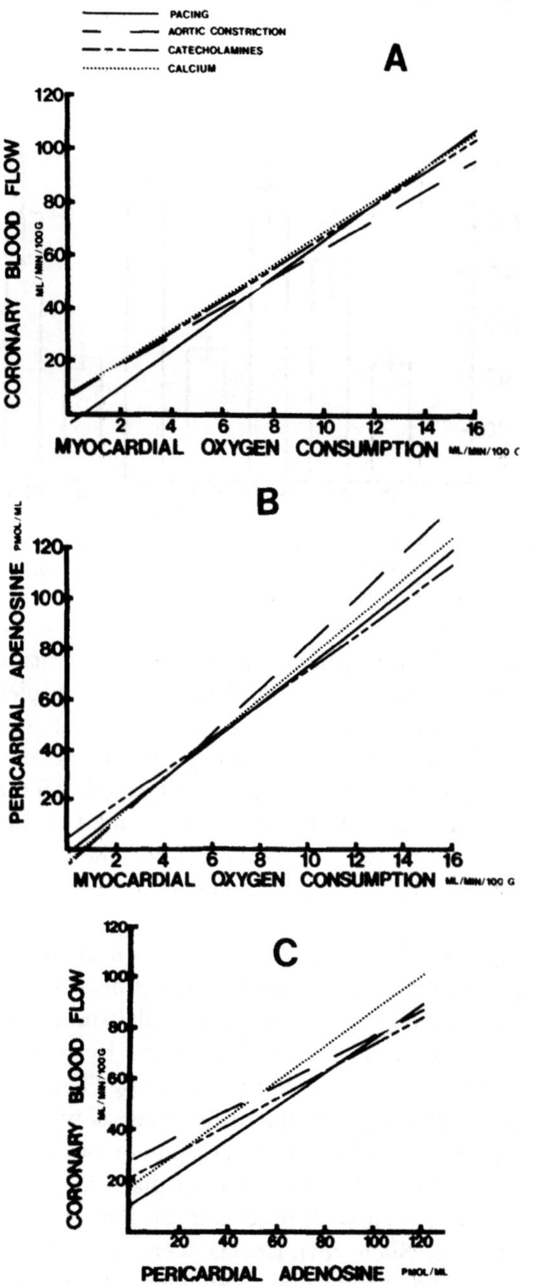

Figure 19.3. Effect of four interventions in the open-chest dog heart on the relationship between myocardial oxygen consumption and coronary flow (A), between oxygen consumption and pericardial infusate concentration of adenosine (B), and between pericardial adenosine concentration and coronary blood flow (C). Correlation coefficients of the different parameters were significant ($p < .05$).

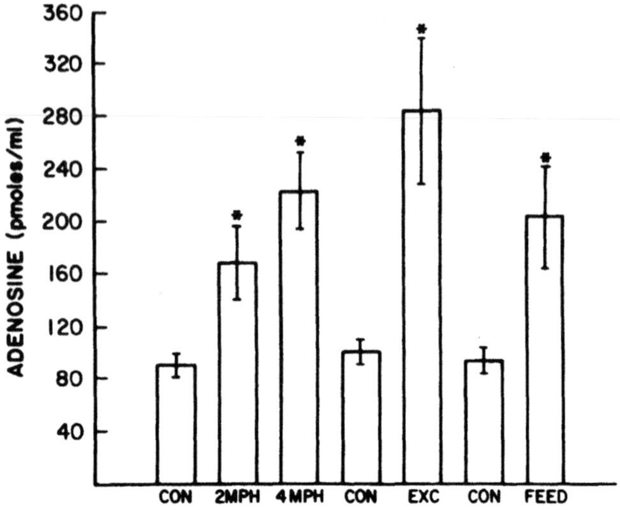

Figure 19.4. Effect of treadmill exercise at 2 and 4 mph, at a 10% grade, excitement (exc) produced by loud noise, and feeding (feed) in the trained unanesthetized dog. Con = control (rest). Source: Reproduced by permission from *Am. J. Physiol.* [16].

coronary sinus blood, for infusion and withdrawal of pericardial infusate, and for recording of electrical activity in the heart. After complete recovery from surgery (1–2 weeks), studies were initiated. Figure 19.4 shows the effects of treadmill exercise at 2 and 4 mph at a 10% grade, of excitement produced by noise, and of feeding, with controls (rest) interspersed. All three perturbations produced significant increases in the adenosine content of the pericardial infusate, and, as in the case of the anesthetized dogs, the conscious animals also showed proportionate increments in myocardial oxygen consumption and coronary blood flow (Fig. 19.5).

With respect to the effect of cardiac pacing on adenosine production, a controversy exists. Manfredi and Sparks [17] found that atrial pacing in the anesthetized open-chest dog increased myocardial oxygen consumption but failed to increase adenosine release, as determined by coronary sinus–arterial adenosine differences. In a similar study on pacing, Knabb et al. [18], using both coronary sinus–arterial adenosine differences as well as pericardial infusate concentrations of adenosine for measurement of adenosine release, observed the same correlation between myocardial oxygen consumption and adenosine release as was seen with other stimuli that produced enhanced cardiac metabolic activity. Such correlations were also observed during atrial pacing in the resting conscious dog [19] when the resting heart rate was significantly less than that of the anesthetized dog (Fig. 19.6). The reason for these opposite results from two different laboratories has not yet been determined.

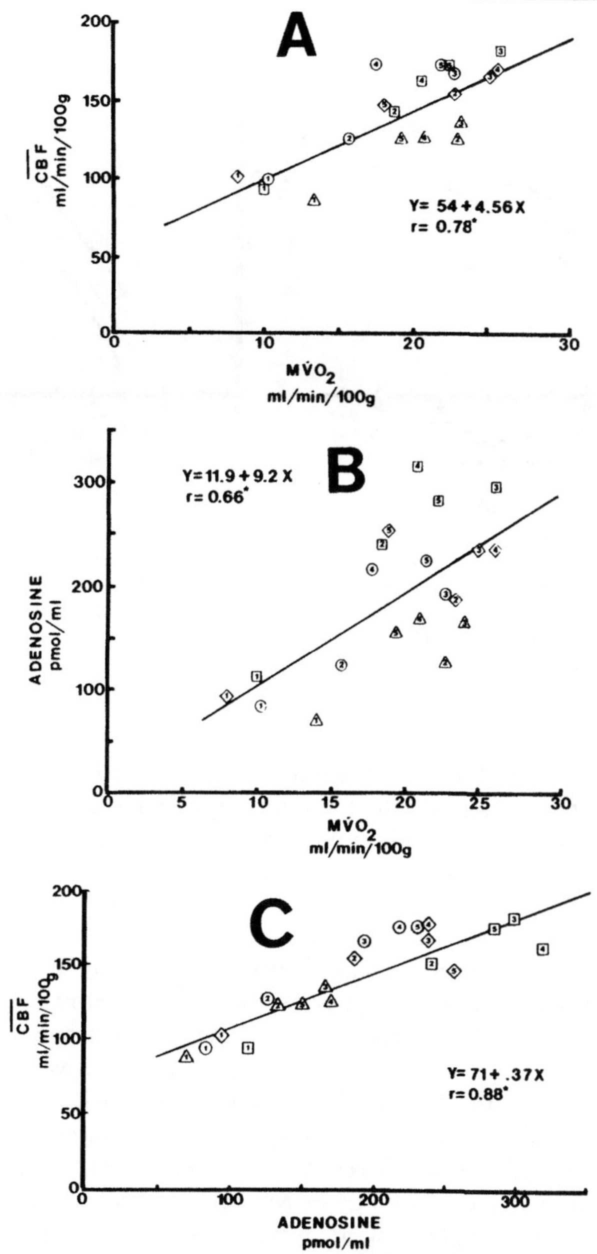

Figure 19.5. Relationship of myocardial oxygen consumption ($M\dot{V}O_2$) and coronary blood flow (CBF) (A), $M\dot{V}O_2$ and the adenosine concentration in the pericardial infusate (B), and adenosine concentration in the pericardial infusate and CBF (C) in the conscious dog. Each symbol (○, △, □, ◇) represents an individual dog and the numbers indicate: 1, control; 2, treadmill exercise at 2 mph; 3, exercise at 4 mph; 4, excitement; 5, feeding. Source: Reproduced by permission from the *Am. J. Physiol.* [16].

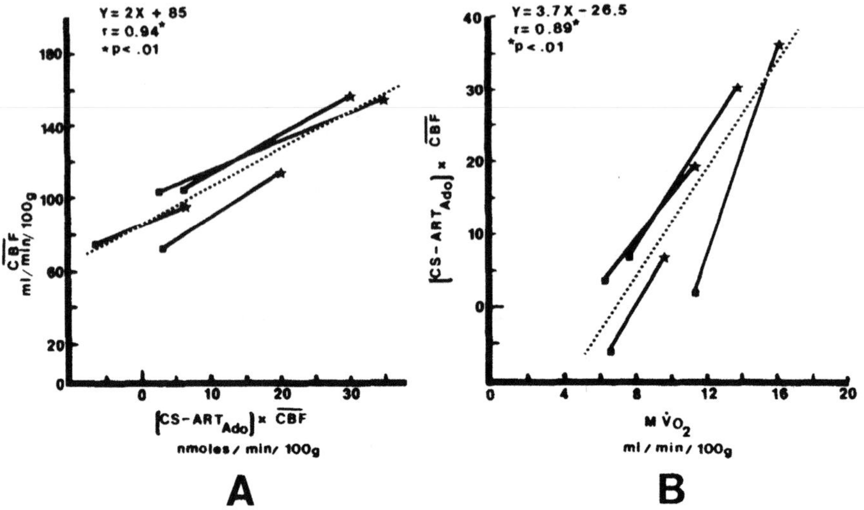

Figure 19.6. Effect of atrial pacing on adenosine release from the heart, as measured by coronary sinus minus arterial adenosine levels times coronary blood flow ([CS − Art Ado] × $\overline{CBF}$), CBF, and $M\dot{V}O_2$ in conscious dogs. Dotted lines in A and B indicate the mean values. Source: Reproduced by permission of *Fed. Proc.* [19].

Dipyridamole

In 1970, Kübler et al. [20] made the interesting observation that in the isolated perfused dog heart subjected to hypoxia, dipyridamole administration produced an increase in tissue adenosine levels but a decrease in the release of adenosine into the perfusate. The interpretation given these findings was that adenosine was trapped in the myocardial cells and could not reach the vascular smooth muscle to induce vasodilation. Hence, the observed vasodilation could not be caused by adenosine release, since dipyridamole apparently blocked adenosine transport out of the myocytes. Similar observations [21] were made on the isolated perfused rat heart and posed a serious challenge to the adenosine hypothesis for the regulation of coronary blood flow. However, the data are also compatible with the concept that dipyridamole blocked adenosine uptake by myocardial cells as well as its transport into the vascular space but was without effect on adenosine release from the myocardium. In agreement with this interpretation are the findings of Mustafa et al. [22], who reported that dipyridamole blocked uptake of adenosine by cultured chick-embryo cardiac cells but had no effect on hypoxia-induced adenosine release by these cultured myocytes. Extrapolation to the intact heart could mean that adenosine release into the interstitial fluid was enhanced by hypoxia but that the dipyridamole blocked reuptake by the cardiac cells as well as movement across the vessel wall. Hence, there would be accumulation of adenosine in the interstitial fluid, where it could act upon the vascular smooth muscle.

To test this hypothesis, a study was carried out on the hearts of open-chest dogs in the presence and absence of dipyridamole during basal and experimental conditions that enhanced myocardial oxygen consumption [23]. In these experiments, pericardial infusate adenosine concentrations were used as an index of interstitial fluid adenosine levels. As shown in Figure 19.7, dipyridamole increased coronary blood flow and pericardial infusate adenosine levels during the basal states as well as during various interventions, but since the drug had essentially no effect on myocardial oxygen consumption, the curves relating myocardial oxygen consumption to coronary blood flow (Fig. 19.7A) or to adenosine release (Fig. 19.7B) were significantly displaced from those obtained in the absence of dipyridamole. However, the relationship between coronary blood flow and adenosine release was unchanged by dipyridamole (Fig. 19.7C). This coupling of adenosine release and coronary blood flow in the absence or presence of dipyridamole, regardless of the level of myocardial oxygen consumption, strongly suggests that the increment in coronary blood flow is in large part attributable to the increase in the interstitial concentration of adenosine. These observations are also in agreement with the hypothesis that dipyridamole produces accumulation of adenosine in the interstitial space by blocking uptake and transvessel movement without influencing adenosine release from the myocardium.

Adenosine deaminase

If adenosine is indeed an important mediator of coronary dilation, one should observe impaired dilation in response to hypoxia in the presence of a substance that inactivates adenosine. Adenosine deaminase converts adenosine to inosine, which lacks significant vasodilator properties. In the open-chest dog, Saito et al. [24] observed a 30% to 39% decrease in repayment of the flow debt in reactive hyperemia when adenosine deaminase was infused into the coronary arterial blood. Similar findings were obtained in the isolated perfused guinea pig heart when active subunits (M.W. *ca.* 4,000–25,000) of adenosine deaminase (produced by pronase treatment of commercially obtained adenosine deaminase) were added to the perfusion fluid (Fig. 19.8). In these experiments the hypoxia-induced increase in coronary flow was reduced by 48%, whereas during normoxia a small but insignificant flow reduction was observed with adenosine deaminase [25]. Associated with these changes in coronary flow were significant reductions in tissue adenosine levels by adenosine deaminase and significant increases in tissue inosine and hypoxanthine levels under normoxic and (to a greater extent) hypoxic conditions (Fig. 19.8). Since we do not know the concentration or the distribution of the deaminase in the interstitial fluid because of its relatively large molecular size, it is quite possible that all the available adenosine is not inactivated before it reaches the vascular smooth muscle. Therefore, the attenuation of the hypoxia-induced dilation probably represents the minimal effect that can be observed under the conditions of these experiments.

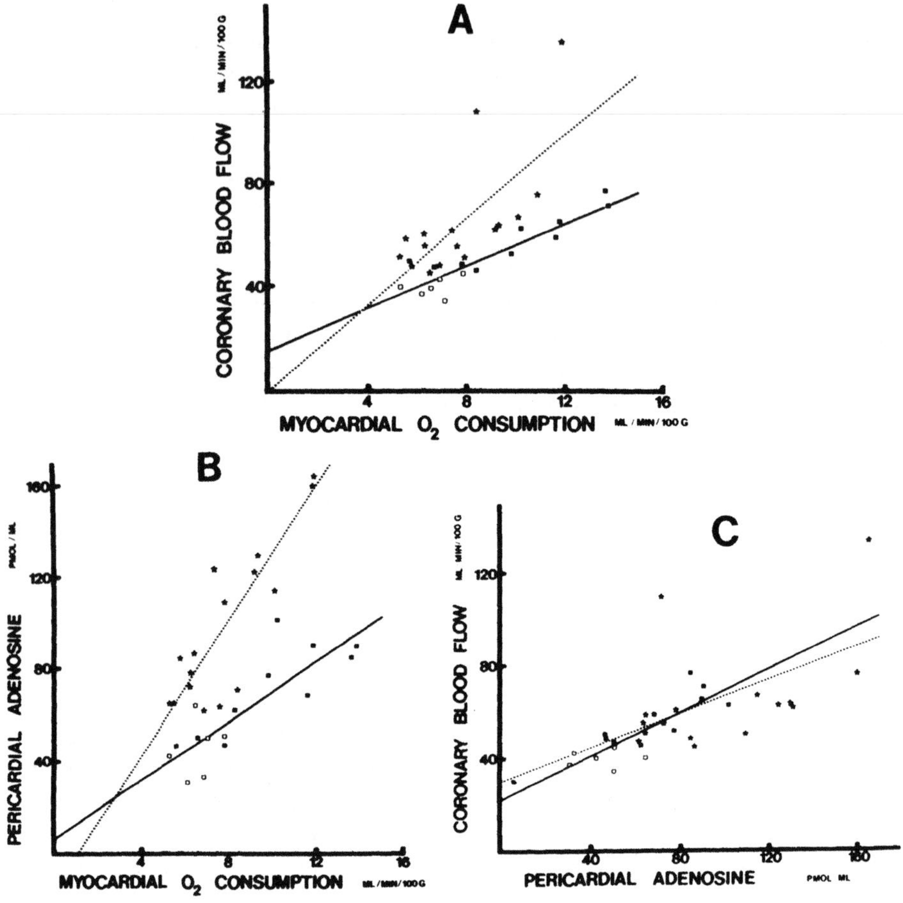

Figure 19.7. Effect of dipyridamole on the interrelationships between myocardial oxygen consumption and coronary blood flow (A), myocardial oxygen consumption and pericardial infusate concentration of adenosine (B), and pericardial infusate concentration of adenosine and coronary blood flow (C) in the anesthetized open-chest untreated dog (solid lines) during basal conditions (open squares) and interventions (open stars), such as aortic constriction, calcium infusion, norepinephrine infusion, and atrial pacing. The closed squares and stars represent the control and experimental procedures, respectively, when carried out in the presence of dipyridamole (dotted lines). Note that in A and B, dipyridamole shifts the curves to the left (greater flow for the same oxygen consumption). In C, the close parallelism between adenosine release and coronary blood flow is unaffected by dipyridamole (see text).

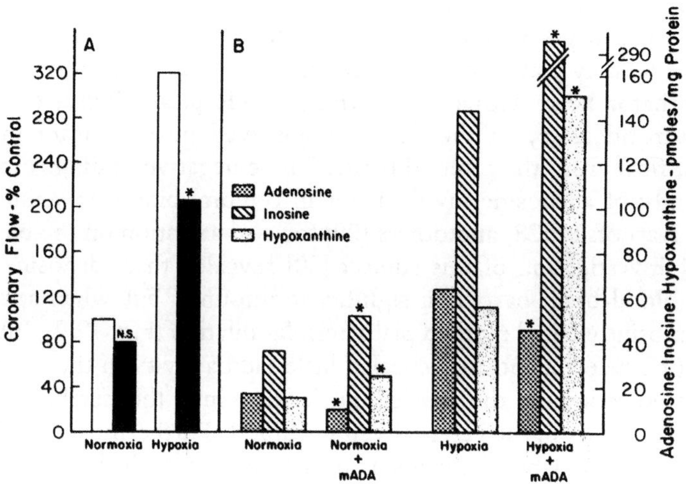

Figure 19.8. Effect of small molecular adenosine deaminase (mADA) on coronary flow and the concentration of adenosine, inosine, and hypoxanthine in the hearts of isolated perfused guinea pigs during normoxia and hypoxia (substitution of oxygen by nitrogen in the perfusion fluid). In A, the open bars represent the absence of mADA and the solid bars the presence of mADA. The star in A represents a significant ($p < .05$) difference between normoxia plus mADA and normoxia alone and also between hypoxia plus mADA and hypoxia alone. The stars in B represent significant differences ($p < .05$) between adenosine, inosine, and hypoxanthine with administration of mADA when compared with controls (absence of mADA).

Conclusions

From the indirect studies summarized here, and the data obtained with dipyridamole and adenosine deaminase, it can be concluded that adenosine definitely plays a role in the regulation of coronary blood flow and hence is really no longer a hypothesis. What remains to be determined is how much of a role adenosine plays in the mediation of coronary dilation, since other factors, such as pH, potassium, and yet unrecognized factors, are undoubtedly also involved in the regulation of coronary blood flow. One of the key factors that remains to be accurately determined is the interstitial fluid concentration of adenosine under a variety of physiological and pathophysiological states. Precise measurement of adenosine and other vasoactive substances to which the resistance vessels are exposed will go a long way in clarifying the relative importance of these agents in the control of the coronary circulation.

BRAIN

Reduced oxygen supply

Ever since Deuticke and Gerlach [26] showed that the ischemic brain, like the ischemic heart, displayed increased adenosine levels, the idea has haunted us that adenosine may play a role in the regulation of vascular resistance in brain similar to that in heart. However, Buyniski and Rapela [27] infused adenosine into the carotid artery of the dog and observed no significant increase in venous outflow from the cerebral tissue. These negative findings discouraged further study of adenosine involvement in cerebral blood flow until experiments by Karnofsky [28] and others [29] focused attention on the blood-brain barrier. Reinvestigation of this subject [30] revealed that adenosine failed to cross the blood-brain barrier in significant amounts, but when applied topically, adenosine elicited marked pial arteriolar dilation (Fig. 19.9). Intracarotid injection of labeled adenosine revealed little radioactivity in the cerebrospinal fluid, and injection of the labeled nucleoside into the cerebrospinal fluid showed few counts in the cerebral venous blood. Furthermore, a bolus intravenous injection of labeled adenosine in the rat showed little uptake by brain compared to other tissues, such as the heart [30].

After a series of experiments on the brains of rats subjected to a variety of stimuli (hypoxia, hypercapnia, hypotension, and electrical stimulation), in which tissue samples were obtained by freeze-clamping, showed changes in cerebral adenosine levels [31], the technique of Veech et al. [32] was adopted. With this technique, the supratentorial part of the rat brain could be blown out of the cranium and between aluminum blocks precooled to the temperature of liquid nitrogen within a few milliseconds. The results of these experiments are illustrated in Figure 19.10 [33]. Within 5 sec of the onset of ischemia, the brain showed a significant increase in adenosine levels [34]. Hypoxia and hypotension also resulted in highly significant increments in cerebral adenosine levels [35, 36]. Furthermore, increases in cerebral blood flow, as measured in the retroglenoid vein, were significantly attenuated in the presence of theophylline, a competitive antagonist of adenosine (Fig. 19.10D) [37]. In the dog, Emerson et al. [38] also observed attenuation of the hypoxia-induced cerebral blood flow increase with theophylline.

Increased oxygen demand

From the studies of Ingvar and Lassen [39] it is well established that regional increases in cortical blood flow can be achieved with sensorimotor activation and that global flow increments are observed with the induction of seizures produced by bicuculline injection [40]. During seizures induced by the injection of bicuculline in rats, in which the brain tissue was obtained by the freeze-blow technique, a large increase in cerebral adenosine levels was produced, as shown in Figure 19.11 [41]. The enhanced adenosine levels were not affected by a 2-fold increase in P_{AO_2}, but when a limited oxygen supply existed (P_{AO_2}

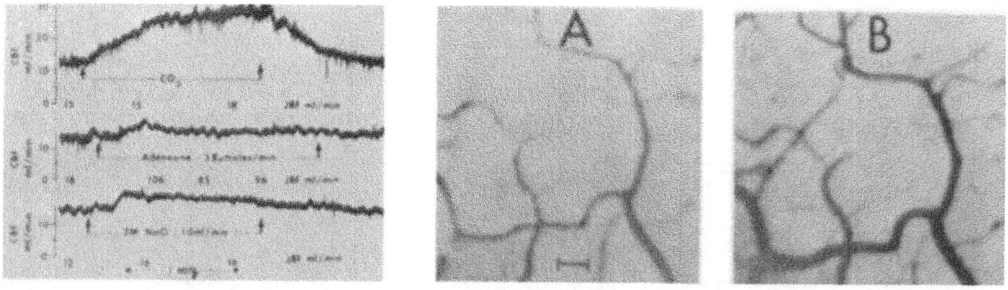

Figure 19.9. Effect of adenosine on pial arterioles. On the left, 7% CO_2 inhalation increased cerebral blood flow (CBF) in the dog, as measured by venous outflow. JBF refers to jugular blood flow and represents cephalic but not cerebral blood flow. Intraarterial adenosine had essentially no effect on cerebral blood flow but greatly increased cephalic, noncerebral blood flow. On the right, A illustrates the diameter of cat pial vessels before topical administration of adenosine and B the effect of adenosine (10^{-5} *M*) on the same vascular bed. Source: Reproduced by permission from the American Heart Association [30].

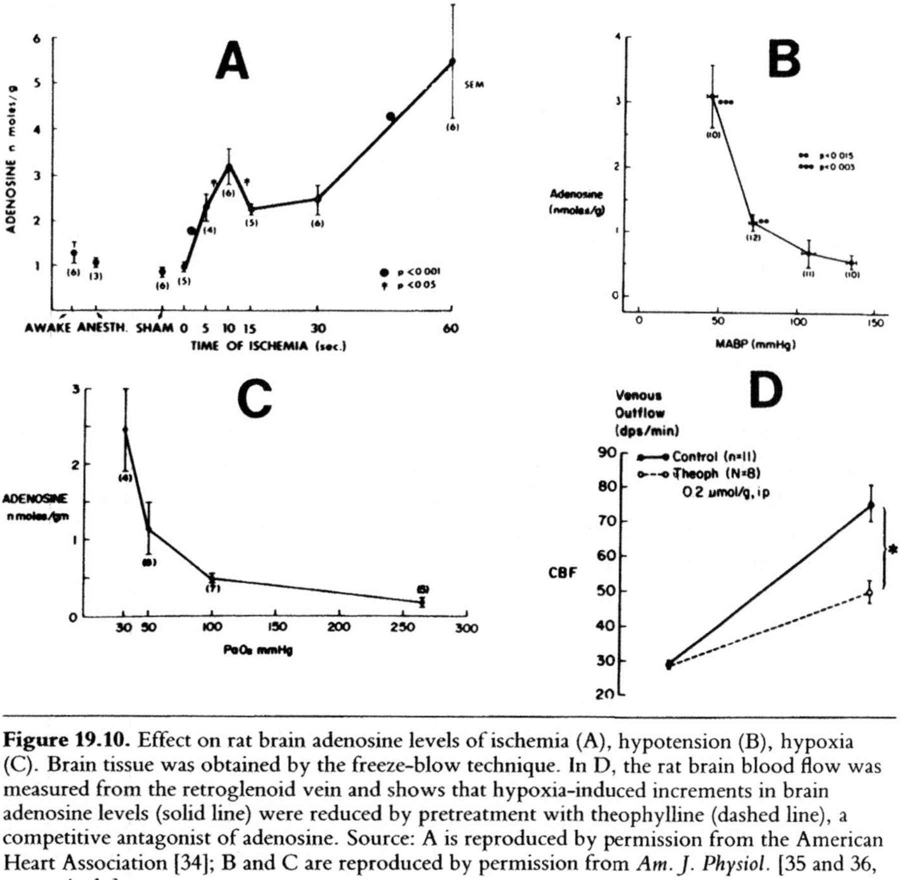

Figure 19.10. Effect on rat brain adenosine levels of ischemia (A), hypotension (B), hypoxia (C). Brain tissue was obtained by the freeze-blow technique. In D, the rat brain blood flow was measured from the retroglenoid vein and shows that hypoxia-induced increments in brain adenosine levels (solid line) were reduced by pretreatment with theophylline (dashed line), a competitive antagonist of adenosine. Source: A is reproduced by permission from the American Heart Association [34]; B and C are reproduced by permission from *Am. J. Physiol.* [35 and 36, respectively].

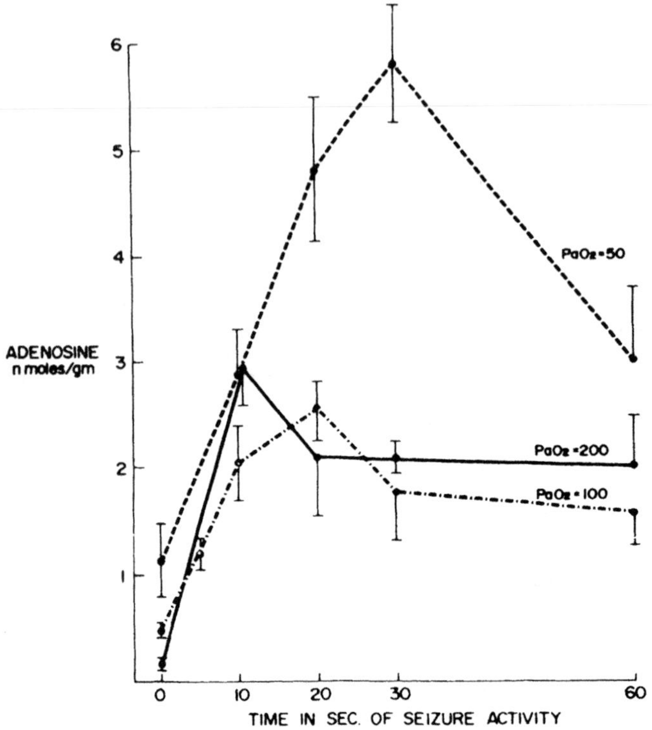

Figure 19.11. Effect of seizures produced by bicuculline injections in rats on brain adenosine levels at different arterial blood oxygen tensions. Brain tissue was obtained by the freeze-blow technique. Doubling of the PA_{O_2} had no effect on brain adenosine levels produced by seizures, but halving the P_{O_2} resulted in a greater production of adenosine in the brain (increased metabolism plus decreased oxygen supply). Source: Reproduced by permission from the American Heart Association [41].

of 50 mm Hg), the increment in adenosine rose above that seen at a normal PA_{O_2}. Hypoglycemia also increased cerebral adenosine levels 2- to 3-fold (1.38 ± 0.10 nmol/gm to 3.31 ± 0.40 nmol/gm) at a blood glucose of 27 mg/dl and increased brain adenosine content 18-fold (to 25 ± 3.65 nmol/gm) at a blood glucose level of 16 mg/dl [42].

Adenosine deaminase

As is true for the heart, adenosine deaminase attenuates the response of the resistance vessels to hypoxia. In cats with cranial windows with ports for superfusion of the visible pial surface of the brain, Wei and Kontos [43] observed that the addition of adenosine deaminase to the mock cerebrospinal fluid produced a marked reduction in hypoxia-induced (PA_{O_2}, 24 mm Hg) dilation of small (51 μm) and large (160 μm) pial arterioles. This effect was greatest on the smaller vessels (Fig. 19.12). Similar attenuation of pial arteriolar dilation was produced by adenosine deaminase in hypotension-induced

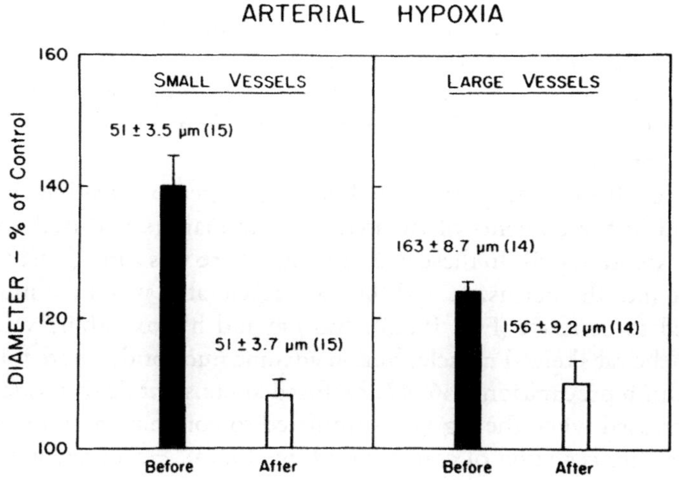

Figure 19.12. Effect of adenosine deaminase on hypoxia-induced vasodilation of pial vessels of the cat, as viewed through a cranial window before (solid bars) and after (open bars) the addition of the enzyme to the superfusion solution. Source: Reproduced by permission from *J. Cerebral Blood Flow Metab.* [43].

dilation of these vessels [Kontos, personal communication]. During normoxia, the addition of adenosine to the mock cerebrospinal fluid produced a dose-dependent dilation of the pial arterioles that was prevented by prior administration of adenosine deaminase [43].

Conclusions

The rapid appearance of adenosine in brain tissue and cerebrospinal fluid when cerebral oxygen supply is inadequate for cerebral oxygen needs, and the attenuating effects of adenosine deaminase and theophylline on hypoxia-induced changes in pial arterioles and cerebral blood flow strongly support a role for adenosine in the regulation of cerebral blood flow. Whether the location of 5′-nucleotidase in the glial foot plates that practically abut on the microvessels is responsible for the strategic release of adenosine in the brain remains to be determined [44]. Also, it is not known to what extent neuronal and glial cells contribute to adenosine production during metabolic regulation of the cerebral circulation.

SKELETAL MUSCLE

Ischemia and restricted blood flow

The original analysis of skeletal muscle by simple paper chromatography revealed the presence of only the degradation products of adenosine (inosine and hypoxanthine) [26]. Subsequent studies [3] using chromatographic methods and enzymatic analysis also failed to show the existence of adenosine in skeletal muscle, even after periods of ischemia that resulted in significant reductions in creatine phosphate and adenine nucleotides. Since skeletal muscle is rich in adenylic acid deaminase but poor in adenosine deaminase, it was felt that any adenylic acid formed was deaminated to inosinic acid prior to dephosphorylation, so that the inosine and hypoxanthine arose from inosinic acid without the formation of adenosine.

With the advent of improved methodology (greater sensitivity), adenosine was found in the effluents of isolated rat hindquarters perfused with Krebs-Henseleit solution [45]. In these experiments, there was a progressive release of adenosine into the perfusate, and this was accelerated when the muscles were stimulated to contract (Fig. 19.13). Inosine and hypoxanthine were also released by the rat skeletal muscle, but no adenine nucleotides were found. In the dog hindlimb preparations [46, 47] perfused at constant flow, tissue adenosine levels increased when the leg was stimulated to contract and returned to control levels after cessation of contraction (Fig. 19.13). However, with free-flow perfusion, tissue adenosine content did not increase with exercise [48]. The conclusion reached by these investigations [48] was that adenosine was found in skeletal muscle only when the oxygen supply was limited by the restricted arterial inflow.

Contraction during free-flow perfusion

In the blood-perfused gracilis muscle of the dog under conditions of free flow, it was recently demonstrated that tissue adenosine content increased progressively with increments in the frequency of stimulation and in parallel with an increase in vascular conductance (Table 19.1) [49]. Hence, adenosine can be formed in skeletal muscle under physiological conditions, and the small amount produced may be because the enzyme 5′-nucleotidase, which forms adenosine from adenylic acid, is restricted to the vicinity of the blood vessels and is not widely distributed in the cell membranes, as in the case of the heart [50]; consequently, the nucleoside is diluted by the adenosine-free tissue included with the adenosine-rich tissue during tissue sampling.

Adenosine deaminase

Adenosine release was also studied in the transilluminated cremaster muscle of the hamster during resting conditions and during stimulation of the muscle [51]. In this preparation, the addition of adenosine deaminase to the suffusion solution adjacent to the arterioles significantly reduced the exercise-induced dilation of the muscle arterioles, regardless of the PO_2 of the suffusion solution (Fig. 19.14). These data definitely incriminate adenosine in vasodilation observed in the metabolically active muscle, but they account for only a minor fraction of the dilation observed. Conceivably, the large molecular weight enzyme could not reach and inactivate all the adenosine released before the nucleoside could reach the vascular smooth muscle of the arterioles; active small molecular subunits of the adenosine deaminase might have produced a

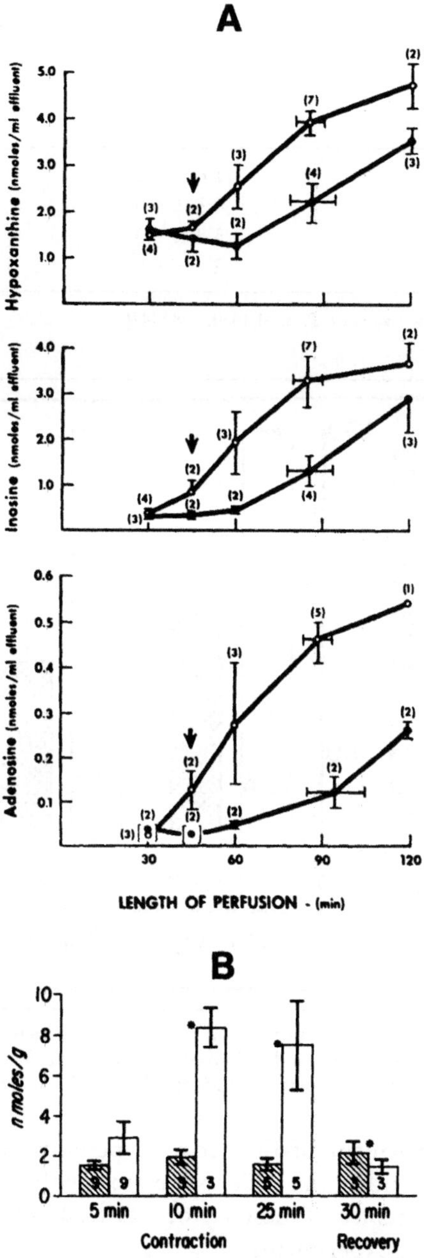

Figure 19.13. Effect of hypoxia and exercise on adenosine formation in skeletal muscle. In A, the rat hindquarters were perfused with Krebs-Henseleit solution. The lower lines indicate the release of adenosine, inosine, and hypoxanthine with perfusion alone. The upper lines (starting at the arrows) represent the release of these compounds during stimulation of the nerves to the muscles of the legs. B represents the tissue concentration of adenosine (ordinate) after various intervals of stimulation of the dog hindlimb muscles during constant-flow perfusion with the animals' own blood. The star indicates significant increases in adenosine formation ($p < .05$). Source: A is reproduced by permission from *Pflueger's Arch.* [45]; B is reproduced by permission from *Am. J. Physiol.* [46].

Table 19.1. Effect of stimulation of the dog gracilis muscle on vascular resistance and tissue adenosine levels during free-flow blood perfusion of the muscles

	Vascular resistance (pru)	Tissue adenosine (nmol/gm)
Control	11.6 ± 1.1	0.57 ± 0.08
Exercise 0.5 Hz	6.3 ± 0.9	0.77 ± 0.28
Exercise 1.0 Hz	5.3 ± 0.3	1.13 ± 0.28
Exercise 2.0 Hz	2.7 ± 0.2	2.08 ± 0.59

Source: Data from Bockman et al., *Fed. Proc.* 41:1680, 1982 [47].

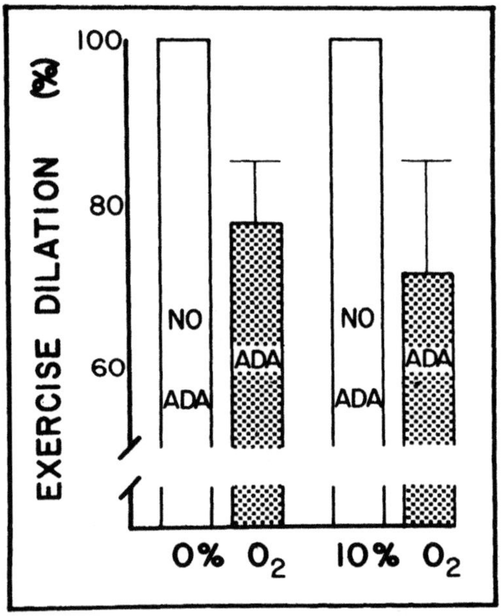

Figure 19.14. Effect of adenosine deaminase on exercise-induced vasodilation of the arterioles of the hamster cremaster muscle at 0% and 10% oxygen in the superfusion fluid. ADA, adenosine deaminase. Source: Reproduced by permission from *Am. J. Physiol.* [50].

more dramatic impairment of the active hyperemia. Of course, it is quite probable that factors other than adenosine are involved in active hyperemia, but the relative importance of these factors, as well as that of adenosine, must await more definitive experiments.

CONCLUDING REMARKS

Adenosine undoubtedly is a mediator of metabolic vasodilation in heart, brain, and skeletal muscle and possibly also in several other tissues. Furthermore, in the heart there is tight coupling between adenosine production and coronary blood flow under all conditions examined. The role of adenosine and other modulators of tissue blood flow cannot be elucidated further until true interstitial fluid concentrations can be accurately determined. It is also important to uncover the source of adenosine that is responsible for blood flow regulations. Does it come from an intracellular source, such as S-adenosylhomocysteine or intracellular 5′-nucleotidase? If the source is intracellular, then why is there such a high concentration of 5′-nucleotidase in the external membranes of cardiac and glial cells? Finally, what is the role of the endothelial cells with respect to adenosine production, transport, and uptake? Most exogenously injected adenosine is probably trapped in the endothelial cells because of the large endothelial uptake of adenosine [52], and hence it would not reach the vascular smooth muscle, whereas endogenous adenosine reaches the vascular smooth muscle before it reaches the endothelium. Therefore, one cannot determine the effective concentration of adenosine on the resistance vessels *in vivo* from arterial injections or from a tissue source. This problem further emphasizes the need to obtain precise values for interstitial fluid adenosine concentrations.

REFERENCES

1. Berne RM: Cardiac nucleotides in hypoxia: Possible role in regulation of coronary blood flow. *Am J Physiol* 204:317–322, 1963.
2. Gerlach E, Deuticke B, Dreisbach RH: Der Nucleotid Abbau im Herzmuskel bein Sauerstoffmangel und seine Mögliche Bedentung für die Coronardurchblutung. *Naturwissenschaften* 50:228–229, 1963.
3. Imai S, Riley AL, Berne RM: Effect of ischemia on adenine nucleotides in cardiac and skeletal muscle. *Circ Res* 15:443–450, 1964.
4. Richman HG, Wyborny L: Adenine nucleotide degradation in the rabbit heart. *Am J Physiol* 207:1139–1145, 1964.
5. Katori M, Berne RM: Release of adenosine from anoxic hearts: Relationship to coronary flow. *Circ Res* 19:420–425, 1966.
6. Berne RM: The role of adenosine in the regulation of coronary blood flow. *Circ Res* 47:807–813, 1980.
7. Rubio R, Wiedmeier VT, Berne RM: Relationship between coronary flow and adenosine production and release. *J Mol Cell Cardiol* 6:561–566, 1974.
8. Olsson RA, Snow JA, Gentry MK: Adenosine metabolism in canine myocardial reactive hyperemia. *Circ Res* 42:358–362, 1978.
9. Thompson CI, Rubio R, Berne RM: Changes in adenosine and glycogen phosphorylase activity during the cardiac cycle. *Am J Physiol* 238:H389–H398, 1980.
10. Foley DH, Herlihy JT, Thompson CI, Rubio R, Berne RM: Increased adenosine formation by rat myocardium with acute aortic constriction. *J Mol Cell Cardiol* 10:293–300, 1978.
11. Miller WL, Belardinelli L, Bacchus A, Foley DH, Rubio R, Berne RM: Canine myocardial adenosine and lactate production, oxygen consumption and coronary blood flow during stellate ganglia stimulation. *Circ Res* 45:708–718, 1979.
12. Saito D, Nixon DG, Vomacka RB, Olsson RA: Relationship of cardiac oxygen usage, adenosine content, and coronary resistance in dogs. *Circ Res* 47:875–882, 1980.
13. Foley DH, Miller WL, Rubio R, Berne RM: Transmural distribution of myocardial adenosine content during coronary constriction. *Am J Physiol* 226:H833–H838, 1979.
14. Knabb RM, Bacchus AN, Rubio R, Berne RM: Parallel changes in myocardial oxygen consumption and adenosine production. *Physiologist* 23:5, 1980 (abst).
15. McKenzie JE, Steffen RP, Haddy FJ: Relationship between adenosine and coronary resistance in conscious exercising dogs. *Am J Physiol* 242:H24–H29, 1982.

16. Bacchus AN, Ely SW, Knabb RM, Rubio R, Berne RM: Adenosine and coronary blood flow in conscious dogs during normal physiologic stimuli. *Am J Physiol* 243:628–633, 1982.
17. Manfredi JP, Sparks HV Jr: Lack of evidence for involvement of adenosine in the hyperemia induced by pacing. *Physiologist* 23:5, 1980 (abst).
18. Knabb RM, Ely SW, Bacchus A, Rubio R, Berne RM: Changes in myocardial oxygen consumption and adenosine production with alterations in heart rate. *Physiologist* 24:454, 1981 (abst).
19. Berne RM, Knabb RM, Ely SW, Rubio R: Adenosine in the local regulation of blood flow: A brief overview. *Fed Proc*, 1982 (in press).
20. Kübler W, Spieckermann PG, Bretschneider HJ: Influence of dipyridamole (Persantin) on myocardial adenosine metabolism. *J Mol Cell Cardiol* 1:23–38, 1970.
21. Degenring FH, Curnish RR, Rubio R, Berne RM: Effect of dipyridamole on myocardial adenosine metabolism and coronary flow in hypoxia and reactive hyperemia in the isolated perfused guinea pig heart. *J Mol Cell Cardiol* 8:877–888, 1976.
22. Mustafa SJ, Berne RM, Rubio R: Adenosine metabolism in cultured chick-embryo heart cells. *Am J Physiol* 228:1474–1478, 1975.
23. Knabb RM, Ely SW, Rubio R, Berne RM: Effects of dipyridamole on relationships among myocardial oxygen consumption, coronary blood flow, and pericardial infusate adenosine, in Berne RM, Rall T, Rubio R (eds): *Regulatory Function of Adenosine*. The Hague, Martinus Nijhoff Medical Division, 1983 (abst).
24. Saito D, Steinhart CR, Nixon DG, Olsson RA: Intracoronary adenosine deaminase reduces canine myocardial reactive hyperemia. *Circ Res* 49:1262–1267, 1981.
25. Rubio R, Knabb RM, Ely SW, Berne RM: 50% decrease in the coronary flow responses to hypoxia caused by "micro" adenosine deaminases. *Fed Proc* 41:1599, 1982 (abst).
26. Deuticke B, Gerlach E: Abbau freier Nucleotide in Herz, Skeletmuskel, Gehir, und Leben der Ratte bei Sauerstoffmangel. *Pfluegers Arch* 292:239–254, 1966.
27. Buyniski JP, Rapela CE: Cerebral and renal vascular smooth muscle responses to adenosine. *Am J Physiol* 217:1660–1664, 1969.
28. Karnovsky MJ: The ultrastructural basis of transcapillary exchanges. *J Gen Physiol* 52:64–95, 1968.
29. Bodenheimer TS, Brightman MW: A blood-brain barrier to peroxidase in capillaries surrounded by perivascular spaces. *Am J Anat* 122:249–267, 1968.
30. Berne RM, Rubio R, Curnish RR: Release of adenosine from ischemic brain: Effect on cerebral vascular resistance and incorporation into cerebral adenine nucleotides. *Circ Res* 35:262–271, 1974.
31. Rubio R, Berne RM, Bockman EL, Curnish RR: Relationship between adenosine concentration and oxygen supply in rat brain. *Am J Physiol* 228:1896–1902, 1975.
32. Veech RL, Harris RL, Veloso D, Veech EH: Freeze-blowing: A new technique for the study of brain *in vivo*. *J Neurochem* 20:183–188, 1973.
33. Berne RM, Winn HR, Rubio R: The local regulation of cerebral blood flow. *Prog Cardiovasc Dis* 24:243–260, 1981.
34. Winn HR, Rubio R, Berne RM: Brain adenosine production in the rat during 60 seconds of ischemia. *Circ Res* 45:486–492, 1979.
35. Winn HR, Rubio R, Berne RM: Brain adenosine concentration during hypoxia in rats. *Am J Physiol* 241:H235–H242, 1981.
36. Winn HR, Welsh JE, Rubio R, Berne RM: Brain adenosine production in rat during sustained alteration in systemic blood pressure. *Am J Physiol* 239:H636–H641, 1980.
37. Winn HR, Morii S, Ngai AC, Berne RM: The effects of theophylline on cerebral blood flow. *Physiologist*, 25:256, 1982 (abst).
38. Emerson TE Jr, Raymond RM: Involvement of adenosine in cerebral hypoxic hyperemia in the dog. *Am J Physiol* 241:H134–H138, 1981.
39. Ingvar DH, Lassen NA: Regional blood flow of the cerebral cortex determined by krypton-85. *Acta Physiol Scand* 54:325–338, 1962.
40. Meldrum BS, Nilsson B: Cerebral blood flow and metabolic rate early and late in prolonged epileptic seizures induced in rats by bicuculline. *Brain* 99:523–542, 1976.
41. Winn HR, Welsh JE, Rubio R, Berne RM: Changes in brain adenosine during bicuculline-induced seizures in rats. *Circ Res* 47:568–577, 1980.
42. Winn HR, Weaver DD, Ngai AC, Berne RM: Changes in brain adenosine concentrations during hypoglycemia in the rat. *Stroke* 13:123, 1982 (abst).

43. Wei EP, Kontos HA: Role of adenosine in cerebral arteriolar dilation from arterial hypoxia. *J Cerebral Blood Flow Metabol* 1:S395–S396, 1981.
44. Kreutzberg GW, Barron KD, Schubert P: Cytochemical localization of 5′-nucleotidase in glial plasma membranes. *Brain Res* 158:247–257, 1978.
45. Bockman EL, Berne RM, Rubio R: Release of adenosine and lack of release of ATP from contracting skeletal muscle. *Pfluegers Arch* 355:229–241, 1975.
46. Bockman EL, Berne RM, Rubio R: Adenosine and active hyperemia in dog skeletal muscle. *Am J Physiol* 230:1531–1537, 1976.
47. Belloni FL, Phair RD, Sparks HV: The role of adenosine in prolonged vasodilation following flow-restricted exercise of canine skeletal muscle. *Circ Res* 44:759–766, 1979.
48. Phair RD, Sparks HV: Adenosine content of skeletal muscle during active hyperemia and ischemic contraction. *Am J Physiol* 237:H1–H9, 1979.
49. Bockman EL, Steffen RP, McKenzie JE, Yachnis AT, Haddy FJ: Adenosine and active hyperemia in dog gracilis muscle. *Fed Proc* 41:1680, 1982 (abst).
50. Rubio R, Berne RM, Dobson JG Jr: Sites of adenosine production in cardiac and skeletal muscle. *Am J Physiol* 225:938–953, 1973.
51. Proctor KG, Duling, BR: Adenosine and free-flow functional hyperemia in striated muscle. *Am J Physiol* 242:H688–H697, 1982.
52. Pearson JD, Carleton JS, Hutchings A, Gordon JL: Uptake and metabolism of adenosine by pig aortic endothelial and smooth-muscle cells in culture. *Biochem J* 170:265–271, 1978.

DISCUSSION

DR. R. VAN BORSTEL: Have you tried a combination of adenosine deaminase and dipyridamole in hypoxic hearts? This might be interesting in view of Dr. Schrader's observation that another nucleoside transport inhibitor, NBMPR, prevents hypoxic adenosine release from the heart while it (NBMPR) is still acting as a coronary vasodilator. I am wondering about the possible vasodilatory action of adenosine transport inhibitors in hypoxia in the absence of extracellular adenosine.

DR. BERNE: No, we have not administered adenosine deaminase along with dipyridamole in hypoxic hearts. On the basis of Dr. Mustafa's experiments on cultured embryonic chick heart cells in which he found that dipyridamole blocked the uptake but *not* the release of labeled adenosine, I would expect that the results would not differ from the experiments done with adenosine deaminase alone. However, I agree that such an experiment should be performed, particularly with other adenosine transport blockers, such as NBMPR, that appear to block adenosine release from the myocardium.

DR. J.P. MANFREDI: Studies on adenosine's role in coronary vasoregulation have passed from a phase in which adenosine was implicated in flow regulation to a phase in which the magnitude of adenosine's contribution to flow regulation is being examined. We are particularly interested in the results obtained with atrial pacing because this may be a situation in which other vasoactive substances are involved, perhaps in addition to adenosine. In view of the information that may be provided with this atrial pacing maneuver, I encourage other labs to investigate the importance of adenosine in the vasodilation accompanying atrial pacing. However, I also warn these investigators about the difficulties in obtaining accurate measurements of adenosine. I suspect that the discrepant results obtained in the Charlottesville and East Lansing labs will eventually be resolved on the basis of adenosine measurements.

DR. BERNE: I am sure that the discrepancy in the results on the effect of cardiac pacing on adenosine release from the myocardium obtained in Charlottesville and East Lansing will be resolved. To do so, it is essential that we repeat the experiments and closely follow each other's protocol. This can be done by performing the experiments in each other's laboratory, as was done a number of years ago preceding a symposium on the autoregulation of blood flow.

DR. H.V. SPARKS: (1) Problems with antagonists: Would you comment on the lack of effects of methylxanthine and adenosine "autoblockade" on metabolic vasodilation in heart and skeletal muscle? (2) Facilitated transport across endothelial cells: With reference to the effect of dipyridamole to increase tissue content and decrease release, we have preliminary evidence that indicates that dipyridamole reduces capillary transport of adenosine relative to that of a nontransported analogue, arahypoxanthine. This means that adenosine could be restricted to the interstitial space by dipyridamole.

DR. BERNE: (1) With respect to the methylxanthines, it appears that they are effective in partially blocking hypoxia-induced vasodilation in brain [Kontos; Winn] and in some studies in heart [Curnish]. We have not studied methylxanthines in skeletal muscle. Dramatic effects have been obtained with these compounds on the dromotropic effects of hypoxia or adenosine in heart [Belardinelli] and on a number of tissue preparations. Nevertheless, I am not at all comfortable with the use of the methylxanthines as a tool in intact preparations because of possible interfering side effects. The mechanism of adenosine autoblockade, produced by a large bolus injection of adenosine, is unclear. It may saturate or in some manner interfere with the sensitivity of the adenosine receptors since prolonged (4–6 hr) exposure to high concentrations of adenosine results in a return of blood flow to control levels despite continued adenosine administration and a normal hyperemic response to brief periods of ischemia in skeletal muscle [Hester]. Probably factors other than adenosine regulation of blood flow come into play. (2) I am pleased to know of your results of dipyridamole on adenosine transport across endothelial cells since they are in agreement with the observations made by Dr. R.M. Knabb in our laboratory that dipyridamole increases pericardial infusate adenosine levels (and presumably interstitial fluid concentration) under a variety of experimental interventions. These observations, taken in conjunction with those of Mustafa, lend credence to our hypothesis that, in hypoxia, adenosine release from myocardial cells is unaffected but transport back into the myocytes or through the endothelial cells is blocked. Hence, less adenosine leaves the heart in the cardiac effluent and builds up in the interstitial fluid, where it elicits relaxation of the coronary vascular smooth muscle.

DR. J.E. MCKENZIE: There are problems in all methods of measuring adenosine, whether it be tissue, blood, or pericardial perfusates. It must be remembered that pericardial perfusate levels are subepicardial adenosine and that there are transmural changes in adenosine under certain conditions.

DR. BERNE: As yet, there is no ideal way of determining the interstitial fluid concentration of adenosine in the heart, and I agree that the three methods you mention all have their drawbacks. We feel that the pericardial infusate method is the least objectionable and hope to prove this to everyone's satisfaction. With respect to a transmural difference in the adenosine concentration across the free wall of the left ventricle, neither you nor we found any difference between epicardial and endocardial layers unless we restricted coronary inflow. Under the latter condition the endo/epi ratio decreased.

DR. A.C. NEWBY: Have you ever tried using methylxanthines in the pericardial infusate? Your talk has emphasized the importance of interstitial adenosine. Perhaps the concentration of methylxanthines in this compartment is also critical.

DR. BERNE: To date, we have not added methylxanthines to the pericardial infusate. I have no idea what this might do or, if we observed a change in the adenosine concentration of the infusate, what it would mean. It might be interesting to administer the methylxanthines intravenously and see if this procedure altered the pericardial infusate concentration of adenosine and/or coronary blood flow. However, as I mentioned earlier, I am a bit wary of the use of methylxanthines in studies of coronary blood flow because of the controversial results obtained by ourselves and others, and because of possible interfering side effects of these drugs.

DR. J. LAMMERANT: If adenosine is to be considered as a link between cardiac metabolism and O_2 consumption on the one hand, and coronary circulation on the other hand, it might be anticipated that adenosine by itself should not have any effect on cardiac oxygen consumption, at least in the absence of β-adrenergic stimulation. Is there sufficient evidence that it is so?

DR. BERNE: We have found no effect of adenosine on cardiac oxygen consumption in the absence or presence of a β-adrenergic blocker.

DR. H.P. BAER: I am confused regarding the effects of adenosine transport blockers. The data by Kübler, Spieckermann, and Bretschneider you referred to suggest that adenosine efflux from dog hearts is not affected; other studies have shown that this drug does block adenosine uptake. Earlier, Dr. Schrader reported studies in which one of the nucleoside-type transport blockers (N^6-thiobenzyl derivative) did block adenosine efflux in guinea pig heart. I wonder, are there species differences or differences in drug specificity, and what is the significance of a separate and distinct pathway for adenosine generation and extrusion (i.e., distinct from facilitated diffusion) described in one of the posters? Maybe this is an area in need of further study?

DR. BERNE: Kübler et al. and later Degenring et al. in our laboratory observed that in hypoxia of the isolated perfused heart, dipyridamole resulted in a decrease in adenosine release by the heart but an increase in tissue content. This was interpreted as a block of adenosine release from the myocytes and hence an intracellular buildup of adenosine. Therefore, less adenosine reached the resistance vessels and could not account for the increase in coronary flow. However, in studies on cultured cardiac cells and cultured vascular smooth muscle cells, adenosine blocked uptake but *not* release, and we postulated that the adenosine was trapped in the interstitial fluid space and not in the myocytes. Our studies [Knabb] with pericardial infusates support this concept. Other adenosine transport blockers, such as NBMPR, may act differently and therefore our experiments should be repeated with the other transport blockers and a more precise method for measuring interstitial fluid adenosine levels must be found in order to verify our results with dipyridamole.

DR. J. LAIRD: As a simple plumber, I have been enlightened and a bit bewildered by all I have heard in the last days. If I understand the body of literature and your presentation, one can conclude that the production rate and uptake/degradation rate are of the same magnitude. This would imply that the mass balance determinants of interstitial adenosine are not directly sensitive to flow itself or washout. Therefore, is flow really controlled under normoxic conditions? You cannot *control* something if you do not measure it, and the washout term is the only "flowmeter" in the system. Perhaps adenosine is a "modulator," not a controller. In passing, I would also point out one intervention in which there is a negative relationship between oxygen consumption and vascular conductance. When you increase perfusion pressure, at constant heart rate and L.V. pressure, oxygen consumption goes up by 15% and conductance goes down by a factor of two or three.

DR. BERNE: What you say regarding the release and uptake of adenosine by the myocardium is correct. Very little leaves via the circulation (washout). The critical factor is the interstitial fluid adenosine concentration—that in contact with the vascular smooth muscle. With hypoxia, more adenosine is released and, consequently, more is taken up by the myocardium and more reaches the vascular lumen. When this occurs, the concentration of adenosine in the interstitial fluid increases and serves to dilate the arterioles and to match coronary flow to the metabolic needs of the heart. Even though the washout is small, it nevertheless reflects the interstitial fluid concentration, under normoxic conditions or with various interventions. Of course, it is true that you can manipulate the conditions so as to dissociate oxygen consumption and conductance. In your example you are "overperfusing" the heart and the heart is trying to oppose the high perfusion pressure (autoregulation). Were you to measure adenosine release under those conditions, you would probably find that it decreased. Perhaps you recall the slide in which nitroglycerin was infused to increase flow above that required by the myocardium. The result was a decrease in adenosine release. Also, with dipyridamole we dissociated myocardial oxygen consumption and coronary blood flow, but the adenosine levels in the pericardial infusate still correlated well with the increase in flow produced by the dipyridamole. The increase in oxygen consumption with elevated perfusion pressure could be due to the opening of previously closed capillaries and a greater tissue oxygen supply. The decrease in conductance could be due to autoregulation, which in turn could be the result of a reduced interstitial fluid adenosine concentration.

20. MECHANISMS OF ACTION OF ADENOSINE ON VASCULAR SMOOTH MUSCLE AND CARDIAC CELLS

RAFAEL RUBIO, MAUREEN T. KNABB, TOKUMASA TSUKADA,
ROBERT M. BERNE

Although it has been suspected for some time that precise coupling exists between blood flow and parenchymal cell metabolism, it has only recently been clearly demonstrated by the availability of more sophisticated modern techniques. The findings clearly indicate that metabolism and blood flow are regulated, at least in part, by common feedback signals. One such signal is adenosine. The release of adenosine into the surrounding interstitial fluid is associated with parenchymal cell ATP-dependent functions such as transmembrane electrical regenerative processes, muscle contraction, and cell secretion (transmitters, hormones, ions). These effects of adenosine are most likely exerted at the level of the cell membrane through changes in adenylate cyclase activity or ionic permeability to Ca^{2+} or K^+ or Cl^-. This chapter will be restricted to the effects of adenosine on ionic membrane permeation.

EFFECTS OF ADENOSINE ON Ca^{2+} INFLUX IN ATRIA AND VASCULAR SMOOTH MUSCLE

Studies in mammalian atria

In K^+-depolarized guinea pig atria treated with isoproterenol, electrical stimulation results in slow-rising action potentials caused by an influx of Ca^{2+} [1, 2]. Addition of $10^{-6}M$ adenosine results in a decrease of the rate of rise and amplitude, and ultimately complete suppression of these action potentials (Fig. 20.1, upper traces). To demonstrate that this effect of adenosine on the action potential was exerted on the outside of the membrane, Schrader et al. [3]

R.M. Berne, T.W. Rall, and R. Rubio (eds.), Regulatory Function of Adenosine, *pp. 319–332.*

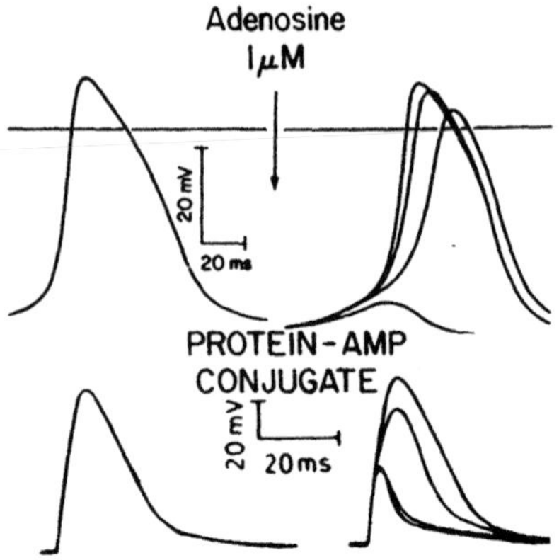

Figure 20.1. Blockade of atrial Ca^{2+}-dependent slow action potentials by adenosine (upper traces) and by AMP-protein conjugate (lower traces). The upper and lower traces on the left represent control traces. Note that adenosine acts on successive action potentials by depressing amplitude and rate of rise. However, the AMP-protein conjugate depresses amplitude by enhancing the rate of repolarization without affecting the rate of depolarization. Source: Reproduced by permission from *J. Mol. Cell. Cardiol.* and *Pfluegers Arch.* [1, 3].

covalently bound 5′-AMP to carbonic anhydrase and found that this complex also depressed the Ca^{2+}-dependent atria action potentials by enhancing the rate of repolarization without affecting the depolarization (upstroke) (Fig. 20.1, lower traces). Thus, it appears that adenosine acts in a different manner than do large conjugates of adenosine. To determine whether this differential effect is associated with molecular size, we investigated the effects of various molecular weight adenosine polymers on the rat atrial slow action potential. These polymers were [Adenylyl-(3′-5′)]$_n$-adenosine, where n can vary from 0 to 9. The higher molecular weight polymers tested consisted of 10 and 6 adenosine residues and depressed the action potentials by decreasing their duration without affecting their upstroke velocity (Fig. 20.2, lower traces). The intermediate molecular weight polymers, consisting of 3 and 4 adenosine residues, depressed the action potentials by decreasing both the upstroke velocity and duration (Fig. 20.2, right top traces); however, the smallest polymers (2 residues) and adenosine acted only on the upstroke velocity (Figs. 20.1 and 20.2). These data suggest that adenosine decreases action potential upstroke velocity by decreasing Ca^{2+} influx and decreases action potential duration by increasing K^+ influx [4]. Since these two effects can be separated when high and low molecular weight adenosine polymers are used, two different adenosine receptors with different degrees of accessibility may be involved. One site appears to be superficially located and is acted upon by small and large size ligands, whereas the other site is less accessible and responds only to adenosine and to small polymers. In accordance with this concept, NBMPR (a blocker of adenosine uptake) [5] in the presence of adenosine produces an effect that mimics that of large polymers. This observation suggests that blocking

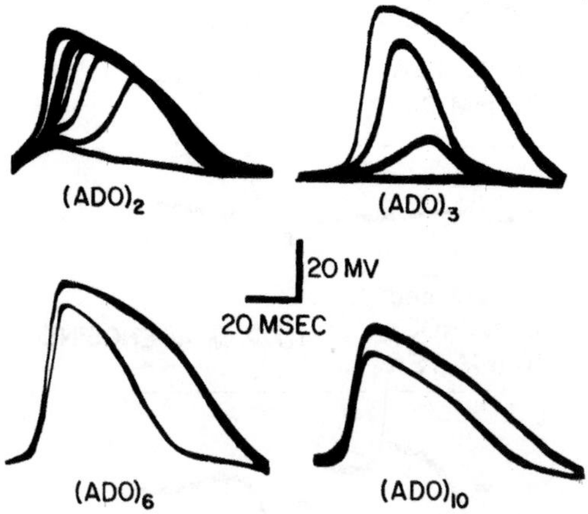

Figure 20.2. Differential effects of various molecular size adenosine polymers (ADO)$_N$ on the rat atrial slow action potential. Upper traces: effects of 5×10^{-5} *M*, 2-adenosine residues (left traces) and 10^{-5} *M*, 3-adenosine residues (right traces). Note that following the administration of the polymers, the successive action potentials, in the case of 2-adenosine residues, have a lower amplitude and upstroke velocity, whereas in the case of 3-adenosine residues, there is depression of upstroke velocity and duration. Lower traces: effects of 5×10^{-6} *M* and 8×10^{-7} *M*, respectively, of 6- and 10-adenosine residues. Note that the effect is essentially on the duration of the action potential and that in the case of 6-adenosine residues, there is a small effect on the rising phase.

the movement of adenosine across the membrane elicits a response that is identical to that obtained with a large adenosine polymer or a protein conjugate of adenosine. The stimulatory effects of adenosine on K^+ membrane conductance is also suggested by the hyperpolarization observed in the frog sinus venosus [6]. Perhaps this particular cell system possesses only the external adenosine receptor.

Vascular smooth muscle

In addition, effects of adenosine on Ca^{2+} permeability have been demonstrated in vascular smooth muscle. In small coronary arteries in which the large K^+ conductance is depressed with tetraethylammonium (TEA), electrical stimulation results in the generation of transmembrane action potentials that are Ca^{2+}-dependent (Fig. 20.3, upper traces) [7] and are also blocked by adenosine (Fig. 20.3, lower traces). Furthermore, adenosine also depresses the uptake of $^{45}Ca^{2+}$ caused by K^+ depolarization in swine carotid arteries (Fig. 20.4) and in cultured vascular smooth muscle cells [8].

ADENOSINE DEAMINASE

An important question is whether endogenous adenosine modulates ionic membrane permeabilities. To study this problem in rat atria, we inactivated released adenosine with a mixture of small molecular polypeptides that had

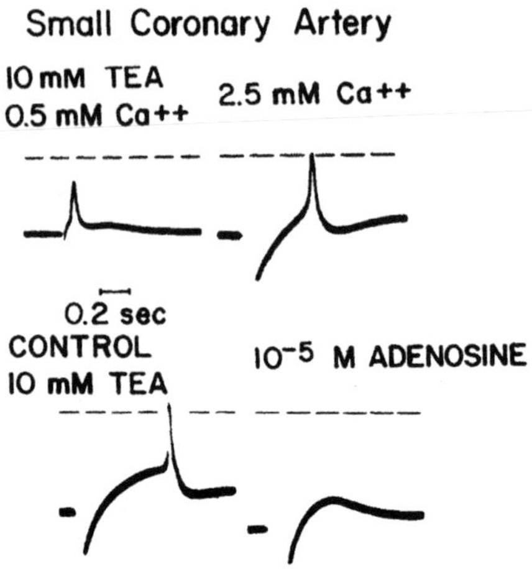

Figure 20.3. Ca^{2+} dependency and blockade by adenosine of TEA-induced smooth muscle action potentials in small coronary vessels. Upper traces: amplitude of action potential is augmented by elevating extracellular Ca^{2+} concentration. Lower traces: blockade of action potential by adenosine. Source: Reproduced by permission from the American Heart Association [7].

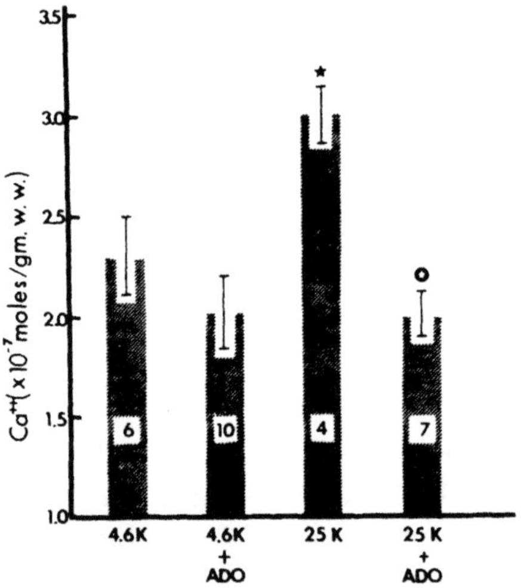

Figure 20.4. Effect of adenosine on $^{45}Ca^{2+}$ uptake by normally polarized (left two columns) and potassium depolarized (right two columns) media strips from swine carotid arteries. K^+ depolarization-enhanced Ca^{2+} uptake was prevented by 10^{-6} *M* adenosine. Source: Reproduced by permission from *Am. J. Physiol.* [8].

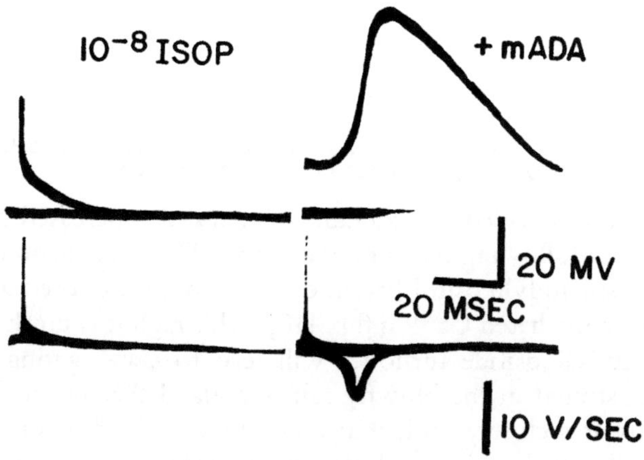

Figure 20.5. Potentiation of the responses to isoproterenol by selective destruction of endogenous adenosine by "micro" adenosine deaminase (mADA). Rat atria depolarized with 22 m*M* K^+. Upper traces: transmembrane potentials as recorded with glass micropipettes filled with 3*M* KCl. Lower traces: first derivative of transmembrane potential. Left traces: lack of response to electrical stimulation in the presence of 10^{-8} *M* isoproterenol. Right traces: after equilibration of the tissue with "microdeaminase" isoproterenol (10^{-8} *M*) induced a slow action potential when electrically stimulated.

adenosine deaminase activity. This mixture of polypeptides was obtained by digesting adenosine deaminase from calf intestinal mucosa with pronase. The resulting active polypeptides had molecular weights between 4,000 and 25,000 daltons. These smaller polypeptides were used because they could penetrate more readily into the interstitial space and, hence, deaminate endogenous adenosine. It is well known that the amplitude and rate of rise of atrial slow action potentials are dependent upon isoproterenol in a dose-related fashion. Upon addition of exogenous adenosine, the dose-response curve for isoproterenol is displaced to the right as a function of the concentration of adenosine, indicating antagonism between adenosine and isoproterenol. We reasoned that if endogenous adenosine had appreciable effects, then treatment with "micro" adenosine deaminase should increase the sensitivity of the response to isoproterenol. This proved to be the case, as demonstrated in Figure 20.5. Prior to equilibration of the rat atria with "micro" deaminase, 10^{-8}*M* isoproterenol failed to induce slow action potentials (Fig. 20.5, left upper trace); however, when the "micro" deaminase was added, administration of 10^{-8}*M* isoproterenol induced a large response (Fig. 20.5, right upper trace). Further, upon addition of the enzyme, the sigmoidal dose-response curve was displaced to the left by approximately one order of magnitude [9], indicating that isoproterenol becomes more potent because it is no longer opposed by the endogenously released adenosine. Similar results were obtained using theophylline, a competitive antagonist for the adenosine receptor site. Therefore, these experiments demonstrate that endogenous adenosine modulates

Ca^{2+} ionic permeability of K^+-depolarized isoproterenol-treated atria. However, it has not been demonstrated that this is true for other ions and for other cell systems.

POSSIBLE ROLE OF MEMBRANE PHOSPHOLIPIDS AND PHOSPHOLIPASE D IN Ca^{2+} GATING AND THEIR CONTROL BY ADENOSINE

Recent work has provided considerable evidence for increased phosphatidylinositol turnover following receptor activation. The phosphatidic acid formed following phosphatidylinositol breakdown has been proposed as a necessary step in receptor-mediated Ca^{2+} influx [10]. Although it is not known exactly how phosphatidylinositide turnover will lead to Ca^{2+} gating, it has been clearly demonstrated in the blowfly salivary gland that phosphatidylinositol turnover correlates closely with transmembrane Ca^{2+} fluxes [11], and it has been suggested that the generated phosphatidic acid acts as an ionophore [12, 13]. This has been demonstrated in a Pressman chamber [12]. Submicromolar concentrations of phosphatidate produce contractions in isolated smooth muscle cells [14] as well as an increase in Ca^{2+} accumulation in synaptosomes [13]. Phosphatidylinositol turnover increases when smooth muscle is treated with high K^+ (smooth muscle has potential-sensitive Ca^{2+} gates), but its turnover is not increased when the pancreas is depolarized with K^+ (pancreas does not have potential-sensitive Ca^{2+} gates) [10]. This suggests that phosphatidylinositol turnover and/or phosphatidic acid play a physiological role in Ca^{2+} influx.

We investigated the effects of phosphatidic acid on rat atrium partially depolarized with 22 mM K^+. Phosphatidic acid (10 μg/ml) induced slow action potentials upon electrical stimulation (Fig. 20.6), with a mean amplitude of 49.7 mV and a maximum rate of rise of 13.2 V/sec. We tested the Ca^{2+} sensitivity of the slow response by changing extracellular Ca^{2+} and Na^+ concentrations. Increase in $[Ca^{2+}]_o$ from 0.5 to 5.0 mM caused the amplitude of the slow action potential to rise linearly with the logarithm of the $[Ca^{2+}]_o$. The calculated slope of this line was 21.1 mV for a 10-fold change in $[Ca^{2+}]_o$. Decrease in $[Na^+]_o$ from 101 to 60 mM caused no change in action potential amplitude. Ca^{2+} antagonists, Mn^{2+}, verapamil, and adenosine (Fig. 20.6) blocked this slow response.

Enzymes such as phospholipase C and phospholipase D, which break down phospholipids into 1,2-diacylglycerol and phosphatidic acid, respectively, were examined for their ability to generate slow action potentials in rat atria. Phospholipase D, but not phospholipase C, was able to induce a slow response, possibly through phosphatidic acid formation (Fig. 20.7). Adenosine also blocked phospholipase D-induced action potentials, and aminophylline reversed the effects of adenosine.

The observation that phosphatidate and phospholipase D generate Ca^{2+}-dependent slow action potentials in depolarized rat atrium further supports a role for phosphatidates in mediating Ca^{2+} influx through control of phos-

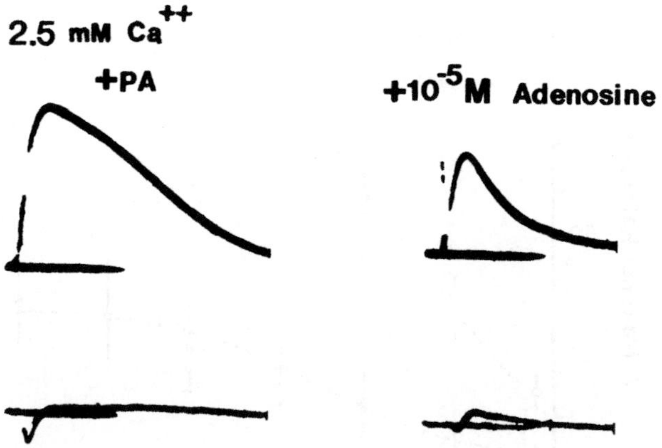

Figure 20.6. Induction slow Ca^{2+}-dependent action potentials by phosphatidic acid (PA) in rat atria and their blockade by adenosine. Transmembrane potential recordings (upper traces) and their first derivative (lower traces). In rat atria in 22 mM K^+, electrical stimulation failed to induce a response. However, upon addition of 10 μg/ml of phosphatidic acid, electrical stimulation generated a Ca^{2+}-dependent action potential (left trace). Adenosine depressed this response (right trace).

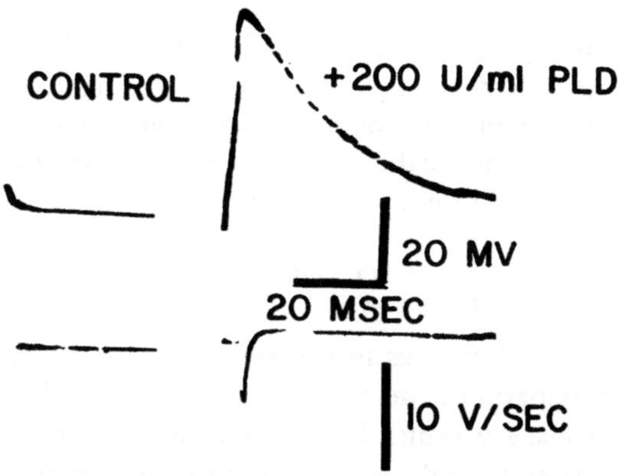

Figure 20.7. Induction of slow action potentials by phospholipase D (PLD) in rat atria depolarized by 22 mM K^+. Left trace: before addition of PLD. Right trace: after PLD treatment. Adenosine depressed this response (not shown). Phospholipase C failed to induce slow action potentials, indicating that generation of phosphatidic acid may be a step in the induction of Ca^{2+} slow action potentials.

pholipid turnover or interference with phosphatidate-induced channel formation. If Ca^{2+} gating depends exclusively on membrane phosphatidylinositide turnover, then the adenosine receptor for Ca^{2+} influx has to be an element in this biochemical pathway.

However, there is an observation indicating that the inotropic effects of adenosine may be partially independent of Ca^{2+} influx. Studies by Verhaeghe

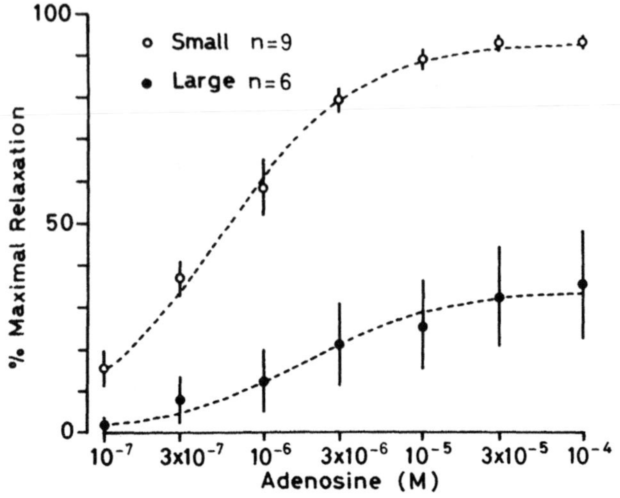

Figure 20.8. Comparison of the relaxing effects induced by adenosine in large and small dog coronary arteries. On the ordinates: 100% = relaxation obtained by adding EGTA. Coronary vessels were contracted with 20 m*M* K^+. At all adenosine concentrations, the level of relaxation was much greater for the small coronary vessels (0.5 mm O.D.) than for the large vessels.

[15] indicate that in dog saphenous vein strips contracted by norepinephrine in the presence of EGTA or in the presence of verapamil (a Ca^{2+} blocker), adenosine was still capable of inducing relaxation. This isolated observation needs to be confirmed, and further work is necessary to elucidate the mechanism of action of adenosine under these experimental conditions.

DIFFERENCES IN SENSITIVITY TO ADENOSINE AND NITROGLYCERIN BETWEEN LARGE AND SMALL CORONARY VESSELS

Important segmental differences in the smooth muscle properties have been described at various parts of the vascular tree and have been clearly demonstrated for the coronary vascular tree. One example is provided by the large difference in sensitivities to adenosine and to nitroglycerin that exists between large and small coronary vessels [16]. The large coronary arteries are almost unresponsive to adenosine, whereas the small coronary arteries are very sensitive (Fig. 20.8); the opposite is true for nitroglycerin. In large coronary arteries, the action potentials induced by TEA are not blocked by adenosine but are abolished by nitroglycerin (Fig. 20.9) [7]. Similarly, strips of small coronary vessels show predominance of β-adrenergic receptors, whereas strips of large coronary arteries show a predominance of α-adrenergic receptors [17, 18]. The mechanism responsible for the difference in responsiveness of the vascular smooth muscle of various segments of the same vascular tree is not known. It is possible that these differences between small and large coronary arteries are related to the distribution of nerve endings in the vessel wall or to

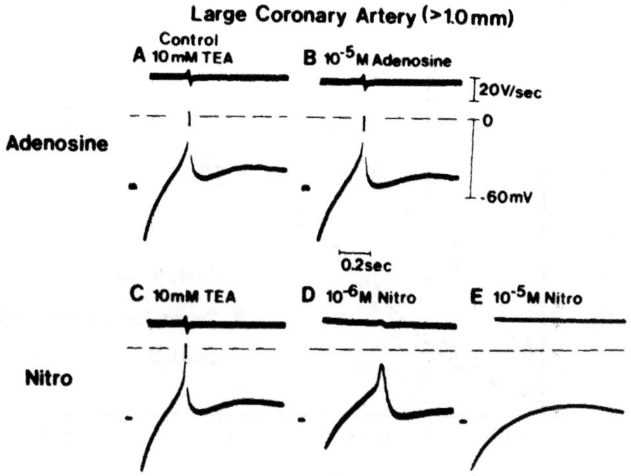

Figure 20.9. Lack of effects of adenosine (upper trace) and blockade by nitroglycerin of TEA-induced Ca^{2+}-dependent action potentials in the smooth muscle of large coronary arteries. Source: Reproduced by permission from the American Heart Association [7].

the distance between the vascular smooth muscle and the parenchymal cells. The nerve endings in the coronary vessels tend to remain in the adventitia-media junction; hence, only the peripheral smooth muscle cells are innervated. The greater the coronary vessel diameter (media of five to six smooth muscle cells thick), the lesser the fraction of vascular smooth muscle fibers that are innervated, whereas the opposite is true for smaller diameter vessels in which the media thickness approaches one vascular smooth muscle cell. All the smooth muscle cells of small diameter vessels may be affected by secretions of the parenchymal cells, but with vessels of greater diameter, only the outer smooth muscle cells may be affected. A difference in the properties between the inner and outer smooth muscle layers of the sheep carotid artery has been described [19, 20]. The inner layer of smooth muscle is about 100-fold more sensitive to norepinephrine than the outer layer [19]. Furthermore, the inner cells have a stable resting potential of about −60 mV and when depolarized to about −40 mV show extensive delayed rectification (attributed to increased potassium permeability). The smooth muscle cells of the outer layer have a low resting potential of about −30 mV and show irregular spike discharges that sometimes are superimposed on slow oscillations [20]. The relative electrical stability of the inner muscle has been attributed to the powerful delayed rectification [20]. It has been suggested that the differences between the inner and outer smooth layers are probably imposed by the nerves on the outer muscle, or by norepinephrine released from them [19]. In other vessels, a trophic influence exerted by the adrenergic nerves on the smooth muscle has also been suggested.

To test whether the differences between large and small coronary arteries to

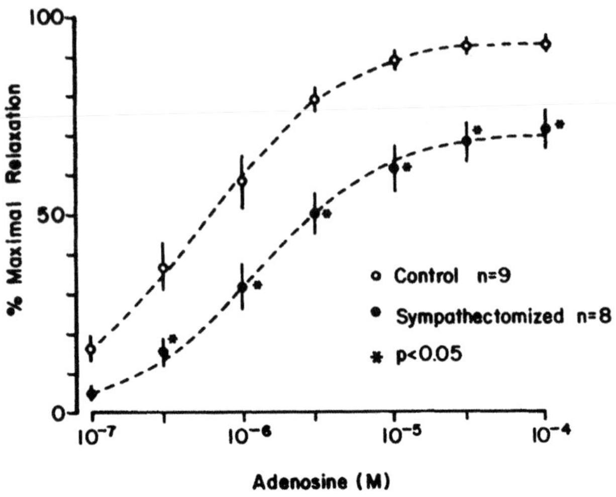

Figure 20.10. Diminished responsiveness to adenosine of small coronary vessels following chronic cardiac denervation. On the ordinates: 100% = relaxation obtained by adding EGTA. Coronary vessels contracted with 20 mM K^+. In experimental dogs, cardiac denervation was produced 6 weeks prior to experiment. At all adenosine concentrations, the level of relaxation obtained was less in vessels of chronically denervated hearts than in those from normally innervated hearts.

catecholamine stimulation [17] and their differential responsiveness to nitroglycerin and adenosine are related to innervation, we compared the dose-response curve to adenosine of small coronary vessels from normal and chronically sympathectomized dog hearts [18]. The results show that chronic cardiac denervation reduces the relaxation responses induced by adenosine at all concentrations (Fig. 20.10). These studies indicate that innervation partially determines responsiveness of vascular smooth muscle to adenosine, and they further suggest that adenosine receptor site density and affinity are factors subjected to modulation. Thus, regional environmental influences may be responsible for the heterogeneity of the responses of vascular smooth muscle observed throughout the vascular tree. Modulation of receptor density/affinity by environmental chemical influences and innervation has been clearly shown in various biological systems, e.g., β-adrenergic receptors [21], acetylcholine receptors [22, 23], and glutamate receptors [24].

EFFECTS OF ADENOSINE ON CL^- MEMBRANE PERMEABILITY

The effects of adenosine on Cl^- permeability can be clearly demonstrated in some cell epithelia [26, 27], such as the isolated saline-perfused dogfish rectal gland (Fig. 20.11). Fluid secretion by this gland is the result of Cl^- excretion that results from a selective increase in Cl^- permeability of the cell membrane facing the secretory side. Circumstantial evidence indicates that this effect of adenosine may be coupled to regulation of adenylate cyclase [26, 27].

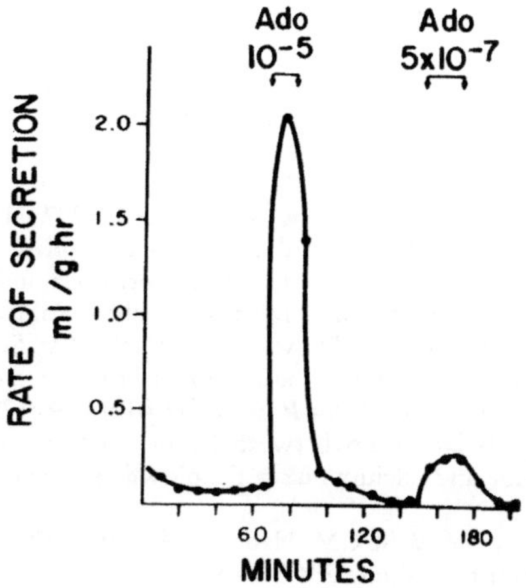

Figure 20.11. Effects of adenosine on the rate of fluid secretion of the isolated perfused dogfish rectal gland. Fluid secretion is proportional to Cl^- excretion and responds promptly to infusion of adenosine. Source: Reproduced by permission from *Bull. Mount Desert Island Biol. Lab.* [26].

CONCLUSIONS

There is considerable evidence that exogenous and endogenous adenosine may modulate membrane permeabilities to Ca^{2+}, K^+, and Cl^- ions. In some cardiac tissues and in vascular smooth muscle, inhibition of Ca^{2+} influx by adenosine results in relaxation. Adenosine apparently acts through two receptor sites with different degrees of accessibility. The most external site may control K^+ efflux and the deeper and external sites together may control Ca^{2+} influx. Innervation modulates sensitivity to adenosine of coronary vascular smooth muscle. Phosphatidylinositides and, in particular, phosphatidic acid may be involved in the generation of Ca^{2+} action potentials in atrial muscle, and regulation of phospholipid turnover may be the mechanism whereby adenosine controls Ca^{2+} gating. Finally, adenosine may cause relaxation of smooth muscle by a process that is independent of Ca^{2+} influx.

REFERENCES

1. Schrader J, Rubio R, Berne RM: Inhibition of slow action potential of guinea pig atrial muscle by adenosine: A possible effect on Ca^{++} influx. *J Mol Cell Cardiol* 7:427–433, 1975.
2. Belardinelli L, Rubio R, Berne RM: Blockade of Ca^{++}-dependent rat atrial slow action potentials by adenosine and lanthanum. *Pfluegers Arch* 380:19–27, 1979.
3. Schrader J, Nees S, Gerlach E: Evidence for a cell surface adenosine receptor on coronary myocytes and atrial muscle cells: Studies with an adenosine derivative of high molecular weight. *Pfluegers Arch* 369:251–257, 1977.
4. Knabb MT, Rubio R, Berne RM: Differential effects of adenosine and high molecular weight adenosine polymers on the rat atrial slow action potential. *Fed Proc* 41:1384, 1982 (abst).
5. Paterson ARP: Adenosine transport, in Baer HP, Drummond GI (eds): *Physiological and Regulatory Functions of Adenosine and Adenine Nucleotides.* New York, Raven Press, 1979, pp 305–313.

6. Hartzell HC: Adenosine receptors in frog sinus venosus: Slow inhibitory potentials produced by adenine compounds and acetylcholine. *J Physiol* 293:23–49, 1979.
7. Harder DR, Belardinelli L, Sperelakis N, Rubio R, Berne RM: Differential effects of adenosine and nitroglycerin on the action potentials of large and small coronary arteries. *Circ Res* 44:176–182, 1979.
8. Fenton RA, Bruttig SP, Rubio R, Berne RM: Effect of adenosine on calcium uptake by intact and cultured vascular smooth muscle. *Am J Physiol* 242:H797–H304, 1982.
9. Knabb MT, Rubio R, Berne RM: "Micro" adenosine deaminase and theophylline potentiate isoproterenol-induced slow action potentials during normoxia, in Berne RM, Rall TW, Rubio R (eds): *Regulatory Function of Adenosine*. Martinus Nijhoff Publishers, The Hague, 1983.
10. Mitchell RA, Kirk CJ, Jones LM, Downes CP, Creba JA: Stimulation of inositol lipid metabolism that accompanies calcium mobilization in stimulated cells: defined characteristics and unanswered questions. *Philos Trans R Soc Lond [Biol]* 296:123–137, 1981.
11. Fain JN, Berridge MJ: Relationship between hormonal activation of phosphatidylinositol hydrolysis, fluid secretion and calcium flux in the blowfly salivary gland. *Biochem J* 178:45–58, 1979.
12. Putney JW, Weiss SJ, Van de Walle CM, Haddas RA: Is phosphatidic acid a calcium inophore under neurohumoral control? *Nature* 284:345–347, 1980.
13. Harris RA, Schmidt J, Hitzemann BA, Hitzemann RJ: Phosphatidate as a molecular link between depolarization and neurotransmitter release in the brain. *Science* 212:1290–1291, 1981.
14. Salmon DM, Honeyman TW: Proposed mechanism of cholinergic action in smooth muscle. *Nature* 284:344–345, 1980.
15. Verhaeghe RH: Action of adenosine and adenine nucleotides on dog's isolated veins. *Am J Physiol* 233:H114–H121, 1977.
16. Schnaar RC, Sparks HV: Response of large and small coronary arteries to nitroglycerin, $NaNO_2$, and adenosine. *Am J Physiol* 223:223–228, 1972.
17. Zuberbuhler RC, Bohr DF: Response of coronary smooth muscle to catecholamines. *Circ Res* 16:431–440, 1965.
18. Tsukada T, Rubio R, Berne RM: Effects of chronic cardiac sympathectomy on the responsiveness of large and small coronary arteries of dog to various chemical agents. *Fed Proc* 41:1528, 1982 (abst).
19. Keatinge WR, Graham JM: Influence of papaverine and nitrite on the responses to norepinephrine of inner and outer smooth muscle of sheep carotid arteries. *Blood Vessels* 11:229–240, 1974.
20. Mekata F, Keatinge WR: Electrical behavior of inner and outer smooth muscle of sheep carotid artery. *Nature* 258:534–535, 1975.
21. Levitzki A: Catecholamine receptors, in Cuatrecasas P, Greaves MF (eds): *Receptors and Recognition* 2A. London, Chapman and Hall, 1976, pp 200–229.
22. Miledi R: The acetylcholine sensitivity of frog muscle fibers after complete or partial denervation. *J Physiol* 151:1–23, 1960.
23. Katz B, Miledi R: The development of acetylcholine sensitivity in nerve-free segments of skeletal muscle. *J Physiol* 170:389–396, 1964.
24. Cull-Candy SG: Effects of denervation and local damage on extrajunctional L-glutamate receptors in locust muscle. *Nature* 258:530–531, 1975.
25. Spinowitz BC, Zadunaisky JA: Action of adenosine on chloride active transport of isolated frog cornea. *Am J Physiol* 237:F121–F127, 1979.
26. Erlij D, Silva P, Rubio R: Effects of adenosine analogues on secretion by the isolated rectal gland of the dogfish squalus acanthias. *Bull Mount Desert Island Biol Lab* 20:145–146, 1980.
27. Forrest JN, Rieck D, Murdaugh A: Evidence for a ribose specific adenosine receptor (Ra) mediating stimulation of chloride secretion in the rectal gland of Squalus Acanthias. *Bull Mount Desert Island Biol Lab* 20:152–155, 1980.

DISCUSSION

DR. A.C. NEWBY: Is it possible that the adenosine effect on Ca^{2+} channels may be mediated through intracellular adenosine?

DR. RUBIO: It is possible. However, although it is known that there is bound intracellular adenosine, it is not known if there is free intracellular adenosine.

DR. H.H. ROTMENSCH: I agree with Dr. Rubio that there is evidence suggesting that adenosine (Ado) may inhibit Ca^{2+} influx as well as enhance K^+ efflux and that Ado causes inhibition of force of contraction of atrial tissue. Although physiologically it would make sense to link those two processes, one may question whether those two phenomena are indeed causally related or whether other mechanisms may be operative, at least in some species. The results of our studies on the spontaneously beating carp atrium show that adenosine-induced negative inotropy surprisingly coincided with a >50% increase in tissue calcium content from 2.2 to 3.4 nmol per gm wet weight tissue. We feel that adenosine may cause sequestration of Ca^{2+} or, in other words, may inhibit release of Ca^{2+} from stores resulting in lack of Ca^{2+} availability at the contractile machinery leading to relaxation. A similar mechanism may be operative in the smooth muscle preparation you have mentioned, where relaxation occurred in a Ca^{2+}-free solution.

DR. RUBIO: This question can best be answered by Dr. Fenton.

DR. R.A. FENTON: Dr. Bruttig and I have recently reported data obtained from experiments utilizing cultured vascular smooth muscle cells (VSM) that suggest adenosine-enhanced cellular sequestration of calcium [Fenton et al., *Am. J. Physiol.* 242:H797–H804, 1982]. Additional experiments were conducted in an attempt to quantitate the exchangeable calcium sequestered. VSM cultures were equilibrated with PSS enriched with ^{45}Ca, in the presence or absence of 10^{-7} *M* 2-chloroadenosine. The cells were then exposed to 20 consecutive 1 min washes of 20 m*M* Ca-EGTA. After 10 min of washout (after ECS Ca^{2+} had been shown previously to be depleted) $10^{-4}M$ dinitrophenol and $10^{-3}M$ iodoacetate were added to the wash solution to deplete cellular energy stores. Cellular calcium lost from the cells during the washout was integrated. The amount of additional Ca^{2+} released was 93.4 ± 16.5 (S.E.) pM $Ca^{2+}/10^6$ cells when adenosine was present during the ^{45}Ca equilibration period. This would support further the concept of adenosine-enhanced cellular calcium sequestration. The site of adenosine's action cannot be ascertained at this time.

DR. J. WOLFF: An alternate pathway for phosphatidylinositol involvement is the C-protein kinase of Nishizuka (in Kobe, Japan) in which a phospholipase C, specific for PI, produces diacylglyceride (with an unsaturated β fatty acid), which in turn sensitizes the kinase to Ca^{2+} by about two orders of magnitude.

DR. RUBIO: It is certainly another possibility to be considered.

DR. C.I. THOMPSON: Can one block the effect of adenosine on slow channels by the use of methylxanthines? In particular, is the small molecule accessible site equally sensitive to methylxanthines?

DR. RUBIO: The depression of slow action potentials by large and small size adenosine polymers is reversed by methylxanthines.

DR. H.J. LEIGHTON: Regarding your observation with large, intermediate, and small molecular weight adenosine derivatives, are your experimental observations made at equilibrium? One might think that with prolonged incubation to establish equilibrium conditions, the large molecular weight polymer would eventually affect the K^+ and Ca^{2+} conductance.

DR. RUBIO: Although I showed the transient effects, the results at equilibrium were the same. At equilibrium, large polymers enhance the rate of repolarization, possibly due to an effect on g_K, whereas small polymers depress the rate of rise, possibly due to an effect on $g_{Ca^{2+}}$.

DR. S.J. MUSTAFA: (1) We reported two years ago [*Fed. Proc.* 1980], using small coronary arteries from dogs, that adenosine has no significant effects on the influx of ^{45}Ca. However, adenosine significantly increased the efflux of Ca^{2+}, and this efflux was blocked by theophylline. You reported using rat aortic vascular smooth muscle cells in which inhibition of ^{45}Ca influx was caused by adenosine. Would you comment on this discrepancy? (2) Did you run a dose-response curve for adenosine in large coronary arteries in sympathectomized animals? If so, how does it compare to small coronary vessels?

DR. RUBIO: (1) We could only demonstrate ^{45}Ca uptake depression with adenosine in K^+ depolarized cells but not in normally polarized cells. I really have no further comments that could explain this discrepancy. (2) Yes. Dose-response curves to adenosine of large coronaries are not affected by denervation.

DR. R. BÜNGER: The "micro" ADA experiment is important regarding the question of the effects of endogenously formed adenosine. Would a "micro" NP experiment be a proper control?

DR. RUBIO: I don't know what would be the proper control or, particularly, why "micro" nucleoside phosphorylase would be a more appropriate control.

DR. D.G. SATCHELL: Since adenosine has a greater effect on the smaller coronary vessels, which contain a higher proportion of innervated smooth muscle cells, has the possibility been considered that adenosine might be exerting its action via release of one of the new putative neurotransmitters, e.g., peptides?

DR. RUBIO: It is possible. I am sure many other substances besides transmitters are released by the nerve.

21. ADENOSINE UPTAKE AND ADENINE NUCLEOTIDE METABOLISM BY VASCULAR ENDOTHELIUM

JEREMY D. PEARSON, PAUL G. HELLEWELL, JOHN L. GORDON

THE PHYSIOLOGY OF VASCULAR ENDOTHELIUM

Control of vascular tone

Recent research has emphasized that extracellular purines (particularly adenosine and ATP) may be important regulators of vessel tone, commonly producing vasodilation, through their binding to receptors currently classified as P_1 (adenosine) and P_2 (ATP) [1]. The direct control of vascular tone has usually been thought of in terms of the action of mediators at receptors on vascular smooth muscle cells, causing contraction or relaxation. However, within the last two years it has been recognized that endothelial cells are necessary modulators of the vasodilator response to several agents, including ATP [2–4]; before the publication of these results, such a role for endothelium in the control of vascular tone had been largely unconsidered. Indeed, until fairly recently, the concept that endothelial cell functions could contribute in any way to the regulation of blood flow, for example, by the metabolism of circulating vasoactive compounds, was not widely recognized.

The lack of attention paid to endothelial cell function in the past seems to have stemmed from a tendency to view large blood vessels in cross section, where the thin monolayer of endothelial cells lining the lumen, though perhaps appearing to be a physical barrier to the exudation of molecules and cells from the bloodstream, forms only a very small proportion of the tissue. If interest is focused on the microcirculation, however, then the potential impor-

R.M. Berne, T.W. Rall, and R. Rubio (eds.), Regulatory Function of Adenosine, *pp. 333–345.*

tance of endothelium as an active modulator of vascular homeostasis becomes more readily apparent. Here the surface area of endothelial cells exposed to unit blood volume increases more than 100-fold compared to the major vessels (e.g., in a typical capillary bed such as in the heart there is a surface area of 500 cm^2 of endothelium per gram of tissue [5]), and blood flow decreases by a similar factor, dramatically increasing the probability of functional interactions between blood components and endothelium. In addition, the likelihood that such interactions will be of physiological consequence for the tissue concerned becomes obvious when it is realized that the proportion of cells that are endothelial within an organ is considerable; for example, there are about 10^8 endothelial cells per gram of heart tissue, whereas in the lung, with its extensive capillary network, this figure is about 10^9. Endothelial cells thus represent approximately 5% and 40% of the total cell population in each case [5, 6].

Metabolism of circulating vasoactive agents

The development of techniques to isolate and culture endothelial cells has been responsible for rapid growth in the knowledge of endothelial cell physiology and for the realization that endothelium is multifunctional, with active roles in many aspects of vascular homeostasis. Examples of specific endothelial cell functions discovered or explored in detail by such *in vitro* studies include the inactivation of circulating thrombin [7, 8]; the secretion of pro- and anticoagulant factors [9, 10]; and interactions with circulating leukocytes [11].

The first demonstrations that endothelial cells may be important regulators of the levels of circulating vasoactive mediators were those in which perfused organs, notably, the lung, were shown to take up avidly and to metabolize certain amines and prostaglandins [12, 13]; several types of evidence suggest that capillary endothelium is responsible [14, 15]. It now seems likely that endothelial cells are the main site for the metabolism of circulating vasoactive compounds; this metabolism can be subdivided into three categories:

1. Inactivation by uptake and intracellular metabolism, as noted earlier for amines and prostaglandins
2. Stimulated synthesis and secretion, e.g., of prostacyclin, the potent vasodilator and inhibitor of platelet aggregation [16, 17]
3. Metabolism by ectoenzymes; this may form either inactive or more active products—for example, angiotensin-converting enzyme [18], which both activates angiotensin I to angiotensin II and inactivates bradykinin.

In the following sections, we describe how endothelial cells metabolize circulating adenosine and adenine nucleotides.

ADENOSINE UPTAKE AND METABOLISM BY ENDOTHELIUM

Initially, it was supposed that circulating adenosine was inactivated by deamination in plasma or blood cells [20]; subsequently, it was briefly reported that the pulmonary bed removed circulating adenosine, that inosine was found in the venous effluent, and that the process was inhibited by dipyridamole [21, 22].

Partly because of our interest in the fate of circulating adenosine, and partly because of the continuing debate as to the sources of circulating adenosine and adenine nucleotides (see later discussion), we began an investigation of the uptake and release of adenosine by endothelium. Our initial studies were with cells cultured from large vessels (usually aorta) in the pig [19]. We also compared our results with those obtained from smooth muscle cells cultured from the same source. Additionally, since capillary endothelium may have different properties from aortic endothelium, and since procedures for obtaining pulmonary capillary endothelium in culture do not yet exist (although the recent development of techniques for culture of capillary endothelium from a few other sites [23, 24] makes it likely that it will become possible in future), we have studied the fate of adenosine in the pulmonary bed of pig isolated perfused lungs.

Pig aortic endothelial cells *in vitro* rapidly take up adenosine, and in experiments with incubation times of a few minutes, two saturable processes can be discerned. This first is of high affinity (apparent K_m 3 μM) and is selectively and potently inhibited by dipyridamole; $10^{-7}M$ dipyridamole inhibits uptake of $10^{-5}M$ adenosine by >90%. It has been known for many years that very low doses of dipyridamole block nucleoside uptake in many cell types and that this is due to direct interaction with the nucleoside carrier [25, 26]. Despite this, the clinical efficiency of the drug as an antithrombotic agent is usually ascribed to its inhibition of phosphodiesterase in the platelet, which only occurs at much higher concentrations (>10^{-5} M) [27].

Adenosine uptake by aortic endothelium is linear for several minutes, and adenosine is incorporated into intracellular nucleotides (predominantly ATP) with little accumulation of intracellular adenosine or release of inosine or hypoxanthine; from these data, we originally suggested that the K_m value may be that of the adenosine carrier, although we recognized that the cell/medium adenosine concentration ratio reaches unity in a matter of seconds and therefore that we were not measuring initial uptake rates [19]. For a detailed discussion of the problems involved in measuring kinetic parameters of nucleoside transport see Plagemann and Wohlhueter [26] and Chapter 12 in this volume. The difficulties are most pronounced with adenosine, which can be metabolized intracellularly by at least two pathways—via adenosine kinase to give AMP and via adenosine deaminase to give inosine—and perhaps by a third, via adenosine phosphorylase to yield adenine. It has recently been shown that adenosine transport parameters in other cells can be measured accurately only by using extremely short incubations (a few seconds) and by blocking intracellular metabolism or using cells genetically lacking the enzymes concerned [28]. In view of these results, the K_m value given earlier may represent that of the highest affinity enzyme, adenosine kinase, inside endothelial cells, rather than that of the adenosine carrier.

Kinetic analysis of endothelial adenosine uptake revealed a second process of lower affinity (K_m approximately 250 μM), which could be distinguished from the high-affinity process by its relative insensitivity to dipyridamole and its selective inhibition by adenine. This K_m value may be that of adenosine deaminase. Although it is not immediately obvious why this enzyme should be inhibited by adenine, this purine base is itself rapidly transported into endothelial cells [29]. The reported high affinity of this process ($K_m \sim 2\ \mu M$) is likely, for the same reasons given earlier concerning adenosine uptake, to be that of the first step in the intracellular metabolism of adenine (via adenine phosphoribosyltransferase to yield adenosine; see [26] and Chapter 12, this volume); thus, adenine conversion to intracellular adenosine will inhibit the uptake and metabolism of adenosine itself.

The nucleoside carrier on other cell types is nonselective, i.e., it transports adenosine and other nucleosides [26, 28, 30], although the affinities of each nucleoside for the carrier vary. In practice, endothelial cell uptake of 10 μM adenosine was not inhibited in the presence of 100 μM inosine [19]. Whether the endothelial nucleoside carrier is particularly selective in favor of adenosine has yet to be determined, but this result is of practical importance for the clearance of circulating adenosine, as inosine is present at higher plasma concentrations than is adenosine [31].

We have also studied adenosine uptake and metabolism in the pig isolated perfused lung, using the single-circulation paired-tracer dilution technique. An example of the dilution curves obtained following rapid bolus injection of small amounts of [^{3}H]adenosine with [^{14}C]sucrose (an extracellular reference marker) is shown in Figure 21.1A. Adenosine was efficiently (>40%) removed during one passage through the lung and, as in cultured aortic endothelium, there was no subsequent release of radioactivity; ^{3}H was almost exclusively recovered in nucleotides in lung tissue. For comparison, the dilution curve profile following injection of [^{3}H]prostaglandin $F_{2\alpha}$ is shown in Figure 21.1B. In this case there is uptake, catabolism, and subsequent release of the tritiated products, so the ^{3}H curve has a long tail above the reference curve.

Varying the concentration of unlabeled adenosine in the perfusing medium showed that an uptake process of similar high affinity to that in aortic endothelium was present in the lung. Chromatographic analysis of the venous effluent revealed, however, that a small proportion (approximately 10%) of the adenosine was converted to inosine. The shape of the dilution curves suggested that this deamination took place without adenosine leaving the circulation—in other words, either an ecto-enzyme was responsible or intracellular deaminase activity was released to the perfusate. This was confirmed by demonstrating that dipyridamole blocked uptake without affecting the percentage of inosine in the perfusate. Control recirculation experiments showed that the adenosine deaminase activity was not attributable to enzymes released into the circulation, and it therefore appears that endothelium in the pulmonary bed possesses an ecto-enzyme that deaminates adenosine.

We have no evidence that this enzyme is present on cultured aortic endothelium. This may represent a specific functional difference between aortic

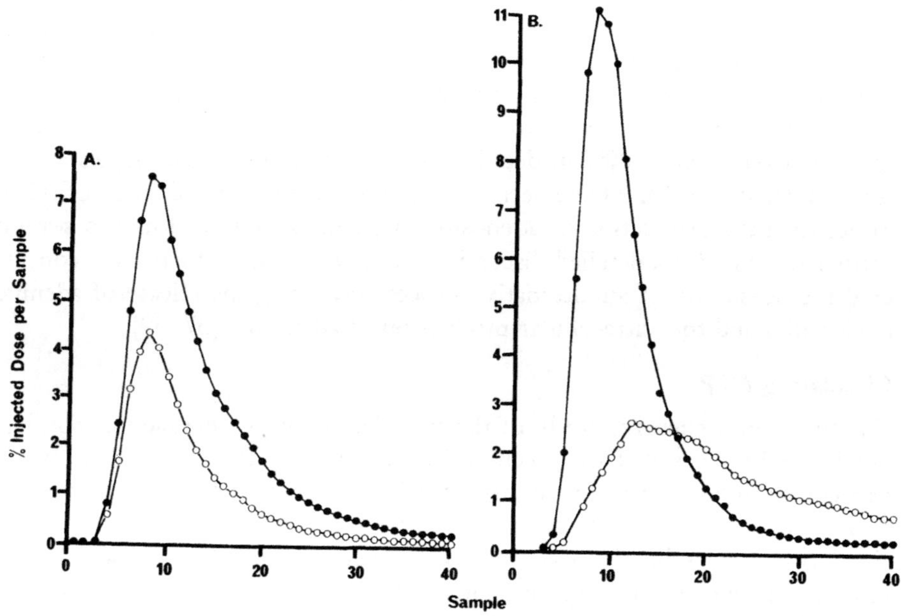

Figure 21.1. A: Appearance of radioactivity in venous effluent from pig lung, perfused at 10 ml/min with Krebs solution + 4.5% (w/v) Ficoll 70, following bolus injection of 0.1 ml of a mixture of [^{3}H]adenosine (0.15 nmol) and [^{14}C]sucrose. Overall uptake of adenosine in this example was 50%. B: Appearance of radioactivity following bolus injection of [^{3}H]prostaglandin $F_{2\alpha}$ (3 pmol) and [^{14}C]sucrose. For details see [56]. In both cases, 40 samples were collected over approximately 60 sec; closed symbols [^{14}C], open symbols [^{3}H].

and pulmonary capillary endothelium, as seems to be true in other instances—e.g., the pulmonary endothelial prostaglandin and serotonin uptake systems are apparently absent in aortic endothelium [32, 33]. Alternatively, the ecto-enzyme may be lost during culture; this has yet to be investigated but is a possibility in view of the known loss of another ecto-enzyme, angiotensin-converting enzyme, from porcine endothelium during conventional monolayer culture [34, 35]. Other ecto-enzymes involved in nucleotide catabolism (see later discussion) are retained, although the activity of one of these has been reported to decline in culture [36].

Aortic smooth muscle cells also take up adenosine, but much less rapidly than does endothelium, and the uptake has no high-affinity component, being almost linearly related to adenosine concentrations up to 1 m*M* [19]. A carrier-mediated process is presumably involved because uptake is inhibited by various agents, including dipyridamole (though at much higher concentrations than are required to block endothelial adenosine transport).

PATHWAYS OF EXTRACELLULAR ADENOSINE PRODUCTION

The regulatory role of circulating adenosine in the production of reactive hyperemia following ischemia and the mechanism whereby adenosine acts on vascular smooth muscle have been discussed in the previous two chapters. Although it seems clear from the evidence presented that the action of adenosine is limited by endothelial uptake, it has been less easy to understand how adenosine is produced (e.g., in ischemic myocardium) and then released to the circulation. The intracellular pathways of nucleotide catabolism during hypoxia have been studied in detail in polymorphonuclear leukocytes and in isolated hepatocytes and have been shown to involve the deamination of AMP rather than the production of adenosine [37, 38]. Because of this and several further reasons (for a detailed discussion, see [29]) we and others have considered the possibility of an alternative process involving the release of adenine nucleotides and the extracellular production of adenosine [37–39].

Circulating ATP

The first comprehensive study of the vascular action of adenine nucleotides, notably ATP, was that of Green and Stoner [40] in the 1940s. They were particularly interested in the ability of ATP to induce shock, and they were also the first to show that adenosine and probably ATP are released into the circulation after ischemia or trauma. They were intrigued by the fact that "a substance of such high energy potential as ATP, when in an extracellular site, is capable of producing such a profound lowering of the energy output of the body" [40, p. 199], and they documented the dilation of peripheral vascular beds and the constriction of the pulmonary bed induced by ATP.

Since then, the release of ATP into the circulation after relatively mild ischemia has been confirmed [41, 42], and an extensive literature characterizing the direct vasoaction of ATP via receptors (P_2) that can be pharmacologically distinguished from those for adenosine (P_1) now exists (for reviews, see [42, 43]). Nonetheless, three major questions still need to be answered: first, what the cellular sources and mechanisms of release of circulating adenine nucleotides and adenosine may be; second, how the action of extracellular nucleotides is terminated; and third, how the responses to extracellular nucleotides are mediated (i.e., by what cells and through what mechanisms).

ATP release from endothelium

ATP is released locally at high concentrations into the bloodstream following cell damage or during intravascular platelet aggregation [44]. Born and colleagues [45], investigating the concept that ADP, formed from ATP, may be the initiator of platelet activation during hemostasis, have recently been able to measure free ATP in blood immediately after cutting small blood vessels. They have suggested that red cells were the cellular source of this ATP. Our interest in the likelihood that extracellular secretion of ATP from cells of the vessel wall itself could be an important source of vasoactive or platelet active purines was promoted by our discovery that ATP can be selectively released in biologically significant quantities from cultured endothelial (or smooth muscle) cells by stimuli such as thrombin or extremely mild physical trauma (e.g.,

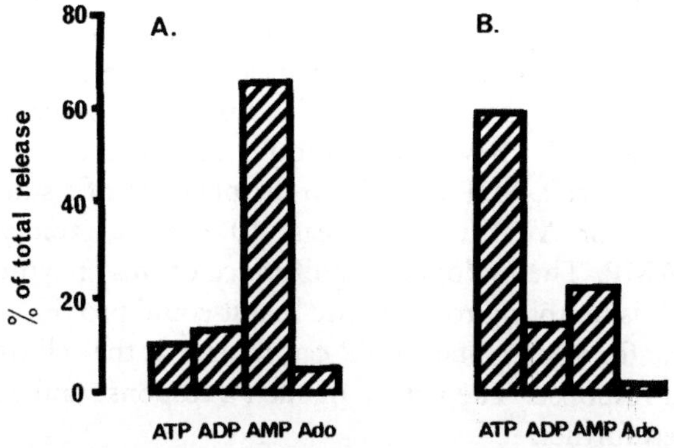

Figure 21.2. Release of nucleotides from cultured aortic smooth muscle cells and the effect of removing Mg^{2+}. Cells were preincubated with [^{3}H]adenosine for 140 min, rinsed four times with phosphate-buffered saline (PBS), and then incubated for 5 min in PBS with 0.7 m*M* EGTA plus 0.8 m*M* Mg^{2+} (calcium alone chelated; frame A), or in PBS with 0.7 m*M* EDTA (calcium and magnesium chelated; frame B). Although [^{3}H] is mainly found in AMP in the presence of Mg^{2+}, it is mainly in ATP in the absence of Mg^{2+}. Total release in 5 min amounted to 16.2% (A) and 15.1% (B) of the [^{3}H]incorporated into the cells.

merely changing the culture medium above the cells) [46]. This release, which appears to involve a reversible change in membrane permeability to ATP, has been confirmed by other workers [47] and does not lead to measurable leakage of macromolecules (e.g., lactate dehydrogenase).

Extracellular metabolism of ATP by endothelium

In the foregoing experiments [46], we noted that after intracellular ATP was secreted, it was metabolized by ecto-enzymes of the vascular cell membrane to yield adenosine. In further studies [48], we have characterized the degradative pathway. Both endothelial and smooth muscle cells possess three separate ecto-enzymes (nucleoside triphosphatase, nucleoside diphosphatase, 5′-nucleotidase), distinguished by selective inhibition with nucleotide analogues, that convert ATP→ADP→AMP→Adenosine. The enzymes are each Mg^{2+}-stimulated, so that in the absence of Mg^{2+}, the ATP released from the cells by a mild stimulus is broken down much more slowly than in the presence of Mg^{2+}. An example of this effect is shown for smooth muscle cells in Figure 21.2. Similar results are obtained with endothelial cells. The rate-limiting enzyme of the sequence in subcultured pig aortic endothelium is 5′-nucleotidase, well known as an ecto-enzyme on other cell types and previously described in endothelium, where the activity in freshly isolated cells was higher than in subcultured cells [36]. The rate of adenosine formation from AMP, even in subcultured cells, is faster than adenosine can be transported back into the cells, so that adenosine accumulates transiently.

We found no nonspecific phosphatase, pyrophosphatase, AMP deaminase, or (as noted earlier) adenosine deaminase as ecto-enzymes on the cultured vascular cells. (For a review of ecto-enzymes capable of metabolizing nucleotides on other cells, see [49]). When trying to examine whether nucleoside triphosphates inhibited the ectodiphosphatase, however, we obtained a surprising result indicating the presence of a further ecto-enzyme, nucleoside diphosphate kinase. This enzyme on endothelium has a K_m of about 100 μM for ADP and can recycle ADP to ATP in the presence of a suitable phosphate donor (e.g., GTP or ATP itself) at least 10× more rapidly than ADP is catabolized to AMP. The biological significance of this enzyme remains to be elucidated, but it is tempting to speculate that it could provide a mechanism to prolong the time for which intact ATP can remain at the cell surface to stimulate a biological response. The nature of such a response and its consequences will be discussed further.

PHYSIOLOGICAL ROLE OF ENDOTHELIAL ECTONUCLEOTIDASES

Based on the results described earlier, we proposed, first, that the selective release of a small proportion of cellular ATP (the intracellular ATP concentration is normally several millimolars) followed by its extracellular degradation to adenosine could be at least partly responsible for the production of circulating adenosine in reactive hyperemia. Second, the endothelial ecto-enzymes together with the adenosine carrier form a system that can regulate intravascular purine metabolism and hence the levels of vascular or platelet responses to circulating purines (from whatever source).

The first of these proposals has yet to be proved to occur *in vivo;* recent evidence in favor of intracellular production and release of adenosine following ischemia in the perfused heart has been presented by Schütz, Schrader, and Gerlach [39], although conditions in which adenosine can be produced extracellularly have now been described by Schrader (Chapter 9). Thus, it is probable that both intracellular and extracellular pathways, depending on the stimuli involved, can operate to form adenosine.

It seems certain that the ecto-enzyme system that we have characterized on endothelium *in vitro* also operates *in vivo*. The inactivation of circulating ATP on passage through the lung was first demonstrated in 1950 [50]; since then, the Ryans and others [51, 52] have studied the metabolism of infused adenine nucleotides by the lung and have shown directly that the inactivation takes place at the surface of pulmonary endothelial cells.

ENDOTHELIAL SITE OF ACTION FOR EXTRACELLULAR ATP

Until recently, most vasodilators were assumed to act directly on receptors on the smooth muscle cell membrane, and in some cases this concept has been verified. However, the surprising discovery that the action of certain vasodilators (including ATP but not adenosine) requires the presence of endothelium [2–4] suggests that an ATP receptor controlling vasodilation exists on endothelial cells, which then modulate the vasodilation—either directly via electrical coupling of endothelial cells with subjacent smooth muscle cells or

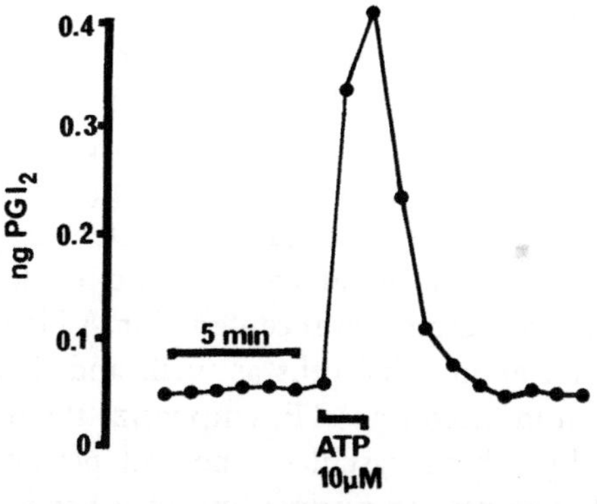

Figure 21.3. Release of prostacyclin from endothelial cells stimulated by ATP. Approximately 10^6 cells, grown on Biosilon microcarrier beads (about 0.5 ml packed volume) were perfused with HEPES-buffered Dulbecco's medium at 0.25 ml/min. 10 μM ATP was present in the perfusate for 2 min, as shown.

by generating a second chemical messenger. We had previously studied two cellular responses to vasoactive mediators in smooth muscle and endothelium, namely, changes in transmembrane potassium flux and stimulation of prostaglandin secretion [53, 54], and we have now shown that ATP stimulates both endothelial potassium flux and prostacyclin synthesis [55; Slakey, unpublished observations]. Because the prostaglandin response to ATP is transient, it cannot be seen easily in monolayer cultures, where incubation times of many minutes are required during which there is a background level of release and, more seriously, where the act of replacing the culture medium itself induces prostacyclin production; changing the medium above endothelial cells in a multiwell culture plate induces more prostacyclin formation than a further hour of undisturbed incubation. We have therefore used endothelial cells grown on solid microcarrier beads, packed in columns and continuously perfused; this avoids artefacts associated with medium replacement; and because the cell number exposed to unit volume of medium is at least 100 times greater than in conventional cultures, samples of the perfusate can be taken every minute or less for analysis, so that transient responses are easily measured. An example of the response to ATP is shown in Figure 21.3; we are currently using the microcarrier system to characterize the types of purinergic receptors expressed by vascular cells and the biochemical responses of the cells to receptor occupation.

CONCLUSION

Endothelial cells possess a system of ectonucleotidases and efficiently take up extracellular adenosine by a carrier-mediated process; together, these form an important physiological regulator of the levels of vasoactive and platelet-active purines in the bloodstream. For example, the circulating adenosine that appears in response to ischemia might be formed intracellularly, released, and its action terminated by endothelial uptake. Alternatively, cellular ATP may be secreted and catabolized extracellularly via the nucleotidases to yield adenosine. ATP secretion from endothelium has been demonstrated *in vitro* in response to specific stimuli; a tiny fractional depletion of endothelial ATP levels by this mechanism *in vivo* would be sufficient to provide biologically active concentrations of extracellular adenine nucleotides and adenosine. Finally, the discoveries that there are receptors for ATP not only on vascular smooth muscle but on endothelial cells as well, and that endothelium modulates the vasodilation induced by ATP, emphasize the need for further study of the mechanisms by which vascular tone and permeability are regulated through endothelial responses to purines and to other vasoactive stimuli.

REFERENCES

1. Burnstock G (ed): *Purinergic Receptors,* Receptors and Recognition Series B, volume 12. London, Chapman and Hall, 1981.
2. Furchgott RF, Zawadzki JV: The obligatory role of endothelial cells in the relaxation of arterial smooth muscle by acetylcholine. *Nature* 288:373–376, 1980.
3. De Mey JD, Vanhoutte PM: Role of the intima in cholinergic and purinergic relaxation of isolated canine femoral arteries. *J Physiol* 316:347–355, 1981.
4. Cherry PD, Furchgott RF, Zawadzki JV, Jothianandan D: Role of endothelial cells in relaxation of isolated arteries by bradykinin. *Proc Natl Acad Sci USA* 79:2106–2110, 1982.
5. Bassingthwaite JP, Yipintsoi T, Harvey RB: Microvasculature of the dog left ventricular myocardium. *Microvasc Res* 7:229–249, 1974.
6. Weibel ER, Gehr P, Haies D, Gil J, Bachofen M: The cell population of the normal lung, in Bouhuys A (ed): *Lung Cells in Disease.* Amsterdam, Elsevier North-Holland, 1976, pp 3–16.
7. Lollar P, Owen WG: Clearance of thrombin from circulation in rabbits by high-affinity binding sites on endothelium. *J Clin Invest* 66:1222–1230, 1980.
8. Chan TK, Chan V: Antithrombin III, the major modulator of intravascular coagulation, is synthesized by human endothelial cells. *Thromb Haemost* 46:504–506, 1981.
9. Loskutoff DJ, Edgington TS: Synthesis of a fibrinolytic activator and inhibitor by endothelial cells. *Proc Natl Acad Sci USA* 74:3903–3907, 1977.
10. Jaffe EA, Hoyer LE, Nachman RL: Synthesis of Von Willebrand factor by cultured human endothelial cells. *Proc Natl Acad Sci USA* 71:1906–1909, 1974.
11. Pearson JD, Gordon JL: Granulocyte interaction with endothelium, in Dingle JT, Gordon JL (eds): *Cellular Interactions.* Amsterdam, Elsevier North-Holland, 1981, pp 107–118.
12. Gaddum JH, Hebb CO, Silver A, Swan AAB: 5-hydroxytryptamine: Pharmacological action and destruction in perfused lungs. *Q J Exp Physiol* 38:255–262, 1953.
13. Ferreira SH, Vane JR: Prostaglandins: Their disappearance from and release into the circulation. *Nature* 216:868–873, 1967.
14. Gillis CN, Roth JA: Pulmonary disposition of circulating vasoactive hormones. *Biochem Pharmacol* 25:2547–2553, 1976.
15. Nicholas TE, Strum, JM, Angelo LS, Junod AF: Site and mechanism of uptake of ^{3}H-1-norepinephrine by isolated perfused rat lungs. *Circ Res* 35:670–680, 1974.
16. MacIntyre DE, Pearson JD, Gordon JL: Localisation and stimulation of prostacyclin production in vascular cells. *Nature* 271:549–551, 1978.
17. Weksler BB, Ley CW, Jaffe EA: Stimulation of endothelial cell prostacyclin production by thrombin, trypsin, and the ionophore A23187. *J Clin Invest* 62:923–930, 1978.
18. Ryan US, Ryan JW, Whitaker C, Chiu A: Localisation of angiotensin converting enzyme. II Immunocytochemistry and immunofluorescence. *Tissue Cell* 8:125–145, 1976.
19. Pearson JD, Carleton JS, Hutchings A, Gordon JL: Uptake and metabolism of adenosine by pig aortic endothelial and smooth muscle cells in culture. *Biochem J* 170:265–271, 1978.
20. Clarke DA, Davoli J, Philips FS, Brown GB: Enzymatic deamination and vasodepressor effects of adenosine analogues. *J Pharmacol Exp Ther* 106:291–302, 1952.

21. Pfleger K, Schondorf H: Zur Aufklarung des Wirkungsmechanismus von Dipyridamol. *Arzneim Forsch* 19:97–98, 1969.
22. Das DK, Das S, Steinberg H: Uptake and metabolism of adenosine by the isolated rat lung. *Fed Proc* 37:222, 1978 (abst).
23. Folkman J, Haudenschild C: Induction of capillary growth *in vitro*, in Dingle JT, Gordon JL (eds): *Cellular Interactions*. Amsterdam, Elsevier North-Holland, 1981, pp 119–133.
24. De Bault LE, Henriquez E, Hart MN, Cancilla PA: Cerebral microvessels and derived cells in tissue culture: II. Establishment, identification and preliminary characterization of an endothelial cell line. *In Vitro* 17:480–494, 1981.
25. Kübler W, Bretschneider HJ: Die Permeation von Adenosin durch die Erythrocyten Membran des Hundes. *Pfluegers Arch* 277:141–149, 1963.
26. Plagemann PGW, Wohlhueter RM: Permeation of nucleosides and nucleic acid bases and nucleotides in animal cells. *Curr Top Membr Transp* 14:275–330, 1980.
27. Mills DCB, Smith JB: The influence on platelet aggregation of drugs that affect the accumulation of adenosine 3':5'-cyclic monophosphate in platelets. *Biochem J* 121:185–196, 1971.
28. Lum CT, Marz R, Plagemann PGW, Wohlhueter RM: Adenosine transport and metabolism in mouse leukemia cells and in canine thymocytes and peripheral blood leukocytes. *J Cell Physiol* 101:173–200, 1979.
29. Pearson JD, Gordon JL: Metabolism of serotonin and adenosine, in Jaffe EA (ed): *The Biology of the Endothelial Cell*. The Hague, Martinus Nijhoff, 1983 (forthcoming).
30. Cass CE, Paterson ARP: Mediated transport of nucleosides by human erythrocytes: Specificity towards purine nucleosides as permeants. *Biochim Biophys Acta* 291:734–746, 1973.
31. Rubio VR, Weidmeier T, Berne RM: Nucleoside phosphorylase: Localization and role in the myocardial distribution of purines. *Am J Physiol* 222:550–555, 1972.
32. Trevethick MA, Olverman HJ, Pearson JD, Gordon JL, Lyles GA, Callingham BA: Monoamine oxidase activities of porcine vascular endothelial and smooth muscle cells. *Biochem Pharmacol* 30:2209–2216, 1981.
33. Ody C, Dieterle Y, Wand I, Stalder H, Junod AF: PGA_1 and $PGF_{2\alpha}$ metabolism by pig pulmonary endothelium, smooth muscle and fibroblasts. *J Appl Physiol* 46:211–216, 1979.
34. Hayes LW, Goguen CA, Ching S-F, Slakey LL: Angiotensin-converting enzyme: Accumulation in medium from cultured endothelial cells. *Biochem Biophys Res Commun* 82:1147–1153, 1979.
35. Gordon JL, Pearson JD: Responses of endothelial cells to injury, in Nossel H, Vogel A (eds): *The Pathobiology of Endothelium*. New York, Academic Press, 1982, pp 433–454.
36. Hayes LW, Goguen CA, Stevens AL, Magargal WW, Slakey LL: Enzyme activities in endothelial cells and smooth muscle cells from swine aorta. *Proc Natl Acad Sci USA* 76:2532–2535, 1979.
37. Newby AC: Role of adenosine deaminase, ecto-(5'-nucleotidase) and ecto-(non-specific phosphatase) in cyanide-induced adenosine monophosphate catabolism in rat polymorphonuclear leukocytes. *Biochem J* 186:907–918, 1980.
38. Vincent M-F, Van den Berghe G, Hers H-G: The pathway of adenine nucleotide catabolism and its control in isolated rat hepatocytes subjected to anoxia. *Biochem J* 202:117–123, 1982.
39. Schütz W, Schrader J, Gerlach E: Different sites of adenosine formation in the heart. *Am J Physiol* 240:H963–H970, 1981.
40. Green HN, Stoner HB: *Biological Actions of the Adenine Nucleotides*. London, Lewis, 1950.
41. Paddle BM, Burnstock G: Release of ATP from perfused heart during coronary vasodilation. *Blood Vessels* 11:110–119, 1974.
42. Forrester T: Adenosine or adenosine triphosphate? in Vanhoutte PM, Leusen I (eds): *Vasodilation*. New York, Raven Press, 1981, pp 205–229.
43. Su C: Purinergic receptors in blood vessels, in Burnstock G (ed): *Purinergic Receptors*, Receptors and Recognition Series B, volume 12. London, Chapman and Hall, 1981, pp 93–117.
44. Mills DCB, Robb IA, Roberts GCK: The release of nucleotides, 5-hydroxytryptamine and enzymes from human blood platelets during aggregation. *J Physiol* 195:715–729, 1968.
45. Born GVR, Görög P, Kratzer MAA: Aggregation of platelets in damaged vessels. *Philos Trans R Soc Lond [Biol]* 294:241–250, 1981.
46. Pearson JD, Gordon JL: Vascular endothelial and smooth muscle cells in culture selectively release adenine nucleotides. *Nature* 281:384–386, 1979.
47. Pearson JD, Carleton JS, Hutchings A, Gordon JL: Metabolism of adenine nucleotides by

ectoenzymes of vascular endothelial and smooth muscle cells in culture. *Biochem J* 190:421–429, 1980.
48. Lollar P, Owen WG: Active-site-dependent, thrombin-induced release of nucleotides from cultured human endothelial cells. *Ann NY Acad Sci* 370:51–56, 1981.
49. Manery JF, Dryden EE: Ectoenzymes concerned with nucleotide metabolism, in Baer HP, Drummond GI (eds): *Physiological and Regulatory Functions of Adenosine and Adenine Nucleotides.* New York, Raven Press, 1979, pp 323–339.
50. Binet L, Burstein M: Poumon et action vasculaire de l'adenosine-triphosphate. *Presse Med* 58:1201–1203, 1950.
51. Ryan JW, Smith U: Metabolism of adenosine 5'-monophosphate during circulation through the lungs. *Trans Assoc Am Physicians* 84:297–306, 1971.
52. Crutchley DJ, Eling TE, Anderson MW: ADPase activity of isolated perfused rat lung. *Life Sci* 22:1413–1420, 1978.
53. Martin W, Gordon JL: Spontaneous and agonist-induced ^{86}Rb efflux from rabbit aortic smooth muscle cells in culture: a comparison with fresh tissue. *J Cell Physiol* 115: 1983 (in press).
54. Ager A, Gordon JL, Moncada S, Pearson JD, Salmon JA, Trevethick MA: Effects of isolation and culture on prostaglandin synthesis by porcine aortic endothelial and smooth muscle cells. *J Cell Physiol* 110:9–16, 1982.
55. Gordon JL, Martin W: Endothelium-dependent relaxation of the pig aorta: relationship to stimulation of ^{86}Rb efflux from isolated endothelial cells. *Br J Pharmacol*, 1983 (in press).
56. Hellewell PG, Pearson JD: Effect of sulphasalazine on pulmonary metabolism of prostaglandin $F_{2\alpha}$ in the pig. *Br J Pharmacol* 76:319–326, 1982.

DISCUSSION

DR. S.N. JARVIS: (1) Do you know how the ATP released from cultured endothelial cells is transported across the cell membrane? (2) You have mentioned that porcine lung endothelial cells may be a major cell type in regulating circulating adenosine. However, if you do similar calculations with respect to erythrocytes, it would also appear that for some species, e.g., human, rabbit, and pig erythrocytes, these cells would be equally important in removing adenosine from the circulation, especially in those organs with a large capillary bed, such as the lung.

DR. PEARSON: (1) We do not know if the ATP is transported, although this is a possibility [see Plagemann and Wohlhueter in *Curr. Top. Membr. Transp.* 14:275, 1980]. The release is not accompanied by macromolecular release but may reflect a transient increase in membrane permeability to small molecules. (2) This is *a priori* clearly possible. Infusion of adenosine into the bloodstream (or perfusion of isolated tissues with blood + adenosine), however, leads to efficient extraction of adenosine by the lung or other tissue concerned, and not to accumulation in erythrocytes.

DR. J.P. MANFREDI: Might the low-affinity uptake mechanism reflect adenosine deaminase activity rather than adenosine transport?

DR. PEARSON: This is a possibility, but we have other data that would agree with Dr. Nees's suggestion that adenosine deaminase activity is rather low in endothelium, and more work needs to be done to determine the levels of adenosine-metabolizing enzymes in endothelium.

DR. H.P. BAER: Comparing the capacity of endothelial and smooth muscle cells with respect to metabolism and uptake of adenosine and ATP on the basis of cell number could be misleading. How different are these cells in size, and what are their surface appearances?

DR. PEARSON: In culture, the size and surface of these cells are similar—certainly not different enough to account for uptake differences. *In vivo,* endothelial cells are undoubtedly smaller than smooth muscle cells.

22. ADENINE NUCLEOTIDE AND ADENOSINE METABOLISM IN CULTURED CORONARY ENDOTHELIAL CELLS: FORMATION AND RELEASE OF ADENINE COMPOUNDS AND POSSIBLE FUNCTIONAL IMPLICATIONS

S. NEES AND E. GERLACH

It is well recognized that in heart tissue, adenine nucleotides and their degradation products are compartmentalized [1–4]. Some years ago, it was demonstrated in our laboratory [5] that under normoxic conditions, isolated perfused guinea pig hearts prelabeled with ^{14}C-adenosine or ^{14}C-adenine continuously released various dephosphorylated adenine nucleotide degradation products (adenosine, inosine, hypoxanthine), the specific activities of which exceeded by far the specific activities of the precursor nucleotides. However, after a brief period of ischemia, the specific activities of the released compounds became considerably reduced and approached the activities of the myocardial precursor substances. These and further observations were interpreted to indicate compartmentation, particularly of adenine nucleotides and of adenosine. The question remained unanswered, however, whether compartmentation of these substances is a subcellular phenomenon pertaining to cardiomyocytes or whether different cellular compartments are involved.

With respect to a possible cellular compartmentation, cardiomyocytes and the endothelial cells of the coronary system deserve particular attention. Cardiomyocytes, which constitute about 85% of the heart weight [6], possess a surface area of about 250-500 cm^2/gm heart, as was estimated from cell counts [7, 8] and the mean surface of cardiomyocytes in guinea pig hearts. However, the total population of endothelial cells in the heart amounted to only 2% to 4% of the heart weight [6, 9] and displayed an exceptionally high plasmalemmal surface area of about 1000 cm^2/gm heart [8, 10]. Furthermore,

R.M. Berne, T.W. Rall, and R. Rubio (eds.), Regulatory Function of Adenosine, *pp. 347–360.*

the endothelial cells that form a continuous lining of the vascular space occupy an important strategic position, since the transfer of substances to and from the cardiomyocytes necessarily involves a passage through the endothelium. In addition, during the past few years, increasing evidence has been obtained concerning a generally high metabolic activity of vascular endothelial cells in various organs [11].

All these observations prompted us to study some metabolic features of coronary endothelial cells, especially with respect to the problem of compartmentation of myocardial adenine metabolites. Because of the methodological difficulties associated with studying biochemical features of endothelial cells *in situ,* we performed most of our experiments on cultured coronary endothelial cells (CEC).

ISOLATION AND CULTIVATION OF CORONARY ENDOTHELIAL CELLS

Hearts from ether-anesthetized female guinea pigs (250–300 gm) were aseptically isolated and perfused by means of a peristaltic pump for 5 min with phosphate buffered saline (PBS), according to the Langendorff technique. While still connected to the peristaltic pump, the hearts were completely immersed into tissue culture medium 199, containing 20% sucrose (Fig. 22.1). Dispersion of the endothelial lining of the coronary vessels was achieved within 15 min at 40°C by intermittent perfusions with proteolytic enzymes (collagenase and trypsin) dissolved in PBS. The effluent perfusate containing the detached endothelial cells formed a layer on top of the more dense sucrose immersion medium and could be collected easily. The cells were sedimented, washed with culture medium, and finally purified by means of Percoll density gradient centrifugation [13].

Pure CEC were cultivated in culture dishes under standard conditions (38°C, water-saturated air with 3% CO_2) and identified by application of various morphological, cell-biological, and biochemical methods. From primary cultures grown to confluency within 2 to 4 weeks, secondary cultures were derived that were used as confluent monolayers for all our studies.

BIOCHEMICAL FEATURES OF ADENINE NUCLEOTIDE METABOLISM IN CORONARY ENDOTHELIAL CELLS

By use of highly sensitive HPLC techniques, tissue contents of the major adenine metabolites were determined, both in cultured CEC and in ventricular myocardium obtained from isolated perfused guinea pig hearts. Quantitative figures are summarized in Table 22.1. Most interesting is the finding that CEC contain adenine nucleotides, particularly ATP, in extraordinarily high amounts. The sum of ATP, ADP, and AMP is almost three times higher in endothelial cells than in cardiac tissue [14]. To our knowledge, such high levels of adenine nucleotides have not been reported for any other mammalian tissue except for platelets, in which, however, a large fraction of the adenine nucleotides is stored in metabolically inactive pools [15]. As we recently demon-

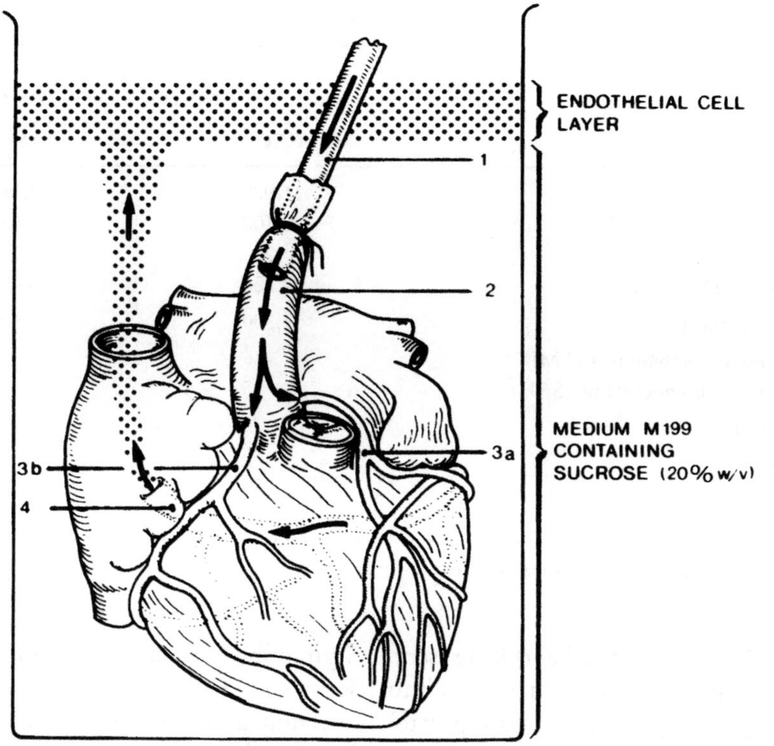

Figure 22.1. Schematic representation of steps involved in the perfusion of the guinea pig heart and in the collection of dissociated coronary endothelial cells. The heart is completely immersed into culture medium containing 20% sucrose. The coronary endothelial cells are enzymatically detached by perfusion of the coronary system with isotonic collagenase-trypsin solution (aortic valves are closed). The perfusate containing the endothelial cells overlayers the sucrose solution, thus preventing contact of the heart surface with the enzymes. 1, cannula; 2, aorta; 3a, left, 3b, right coronary artery; 4, coronary sinus; arrows, direction of perfusion. Source: From [12] with permission.

strated, creatine phosphate is also present in CEC in remarkable quantities. Thus, in CEC, high-energy phosphate compounds (ATP, ADP, creatine phosphate) exceed by far those in the myocardium.

Another interesting feature of endothelial cells concerns their high contents of SAM and SAH, which indicates that CEC are generally capable of performing methylating reactions.

The common adenine nucleotide degradation products, such as adenosine, inosine, hypoxanthine, and adenine, are present in CEC in 40 to 60 times higher concentrations than in the myocardium. Among these compounds, adenosine deserves particular attention. Under the assumption that the adenosine content of cultured CEC is in the same range as that in the coronary endothelium *in situ,* the following rough estimate seems of interest: 1 mg endothelial cells contains 65–75 pmol adenosine. The adenosine content of the

Table 22.1. Adenine metabolites and creatine phosphate in cultured coronary endothelial cells (CEC) and in the myocardium

	Content (nmol/gm tissue wet weight)	
	CEC	Myocardium
ATP	11.900	4.280
ADP	2.100	1.050
AMP	230	160
Σ ATP, ADP, AMP	14.230	5.490
Creatine phosphate	14.300	9.600
S-adenosyl-methionine (SAM)	136	9
S-adenosyl-homocysteine (SAH)	5	0.5
Adenosine	72	1.9
Inosine	85	1.3
Hypoxanthine	50	0.8
Adenine	60	1.2

Mean values from five to seven analyses each.

20–30 mg of endothelial cells present in 1 gm of heart tissue amounts to 1.400–2.200 pmol. These values are almost identical to the adenosine content of 1.800–2.200 pmol that was separately determined in 1 gm of normoxic heart tissue. This estimate favors the assumption that the adenosine measured in the normoxic myocardium is mainly present in the endothelial compartment.

A further characterization of CEC was possible by determinations of activities of various enzymes involved in adenine nucleotide degradation and synthesis. As is evident from the quantitative figures in Table 22.2, most enzyme activities are considerably higher in coronary endothelial cells than in the ventricular myocardium. A complete cascade of Mg^{2+}-dependent ectonucleotidases that catalyzes a rapid extracellular dephosphorylation of adenine nucleotides from ATP via ADP and AMP to adenosine appears not to be a specific feature of CEC. Endothelial cells from other vascular beds were shown to display a similar nucleotidase pattern [16, 17], although quantitative differences exist. CEC are also equipped with a cascade of endonucleotidases of appreciable activities; thus, these cells can also dephosphorylate adenine nucleotides intracellularly to adenosine.

It is interesting to note that adenosine deaminase is the only enzyme whose specific activity is strikingly lower in endothelial cells than in the myocardium. Most likely, this low activity is partly responsible for the rather high adenosine content of CEC.

All our data concerning content of adenine metabolites and enzyme patterns suggest that CEC must exhibit a very active adenine nucleotide metabolism and are capable of producing adenosine. In order to maintain the extraordinary

Table 22.2. Enzymes involved in adenine nucleotide metabolism of cultured coronary endothelial cells (CEC) and myocardium

		Specific activity (nmol/min/mg tissue)	
		CEC	Myocardium
ATPase (EC: 3.6.1.15)	ecto	6.9	3.5
	endo	6.8	
ADPase (EC: 3.6.1.6)	ecto	12.3	3.1
	endo	5.1	
5′-nucleotidase (EC: 3.1.3.5)	ecto	18.0	1.2
	endo	2.4	
Adenylate kinase (EC: 2.7.4.3)		1.5	0.7
Adenosine deaminase (EC: 3.5.4.4)		0.17	1.6
Purine nucleoside phosphorylase (EC: 2.4.2.5)		280.0	2.7
Adenosine kinase (EC: 2.7.1.20)		0.53	0.10

Mean values from four to six assays each.

levels of high-energy phosphate compounds, these cells seem to be mainly dependent on energy production via aerobic glycolysis. This assumption is derived from special experiments in which we determined rates of glycolytic and oxidative glucose degradation in CEC by use of U-^{14}C-glucose and 6-^{14}C-glucose in the absence of other substrates.

DYNAMIC ASPECTS OF ADENINE NUCLEOTIDE AND ADENOSINE METABOLISM IN CORONARY ENDOTHELIAL CELLS

The velocity of the uptake of adenosine and its incorporation into intracellular metabolites was determined after 30–60 min incubation of CEC with 8-^{14}C-adenosine. In accordance with observations on pig aortic endothelial cells [18], we could demonstrate also that CEC rapidly take up adenosine. Lineweaver-Burk plots revealed a high-affinity transport system with a K_m value of $1 \times 10^{-6} M$. The rate of adenosine uptake was determined to be 170 pmol/min/mg CEC at an adenosine concentration of 1 μM.

Incorporation of adenosine into the different intracellular adenine nucleotides proved to be fast. After an incubation period of 60 min, the specific activities of ATP, ADP, AMP, and adenosine were almost identical. These findings prove that adenine nucleotides in CEC, contrary to platelets [15], are not stored in metabolically inactive pools.

CEC prelabeled with 8-^{14}C-adenosine and incubated in fresh medium under strictly controlled conditions continuously released labeled adenine metabo-

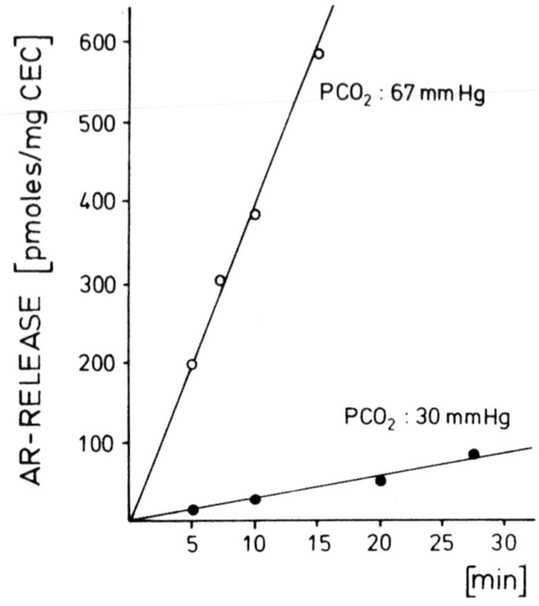

Figure 22.2. Release of adenosine (AR) from prelabeled coronary endothelial cells at different P_{CO_2} values. Mean values from four experiments.

lites in measurable quantities into the culture medium. Among these compounds, cAMP, AMP, adenosine, as well as inosine and hypoxanthine could be identified. Whether all these substances are released as such or whether some of them are derived from a primary precursor remains to be clarified.

Intact CEC from guinea pig hearts—in contrast to endothelial cells of pig aorta [19]—are apparently not capable of releasing ATP or ADP. Utilizing the extremely sensitive firefly assay for ATP and ADP directly applied in the incubation medium overlying the cell cultures, and using a specially developed luminescence photometer, we could conclude that these nucleotides appear extracellularly.

Among the adenine metabolites that are directly or indirectly released from CEC, adenosine is of particular functional significance. Therefore, the kinetics of adenosine release were studied more in detail. As shown in Figure 22.2, adenosine is constantly liberated from prelabeled CEC superfused with Krebs-Henseleit solution under standard conditions (P_{CO_2} = 30 mm Hg, P_{O_2} = 150 mm Hg, 38°C). The rate of basal adenosine release amounted to about 3 pmol/min/mg CEC. Increasing the P_{CO_2} from 30 to 67 mm Hg resulted—most likely through a decrease of pH—in a 12- to 14-fold enhancement of adenosine liberation. The adenosine release proved to be sensitive not only to changes of P_{CO_2}. Diminution of P_{O_2} or addition of α- or β-adrenergic catecholamines, of histamine, or of thrombin to the superfusate resulted in a more or less pronounced release of adenosine. However, these changes were generally smaller than those observed after elevation of PCO_2.

From a physiological point of view, it is of interest to relate the basal adenosine release measured in cultures of CEC to the adenosine release into the coronary effluent of isolated perfused hearts. Provided that CEC in culture and *in situ* release adenosine at the same rate, one can calculate that the coronary endothelium of 1 gm heart tissue is capable of liberating 60–90 pmol adenosine per minute. Adenosine released from normoxic perfused isolated guinea pig hearts was indeed shown to be in this range [20]. It thus seems possible that adenosine present in and released from the isolated guinea pig heart under normoxic conditions can in fact originate almost entirely from the endothelial compartment. Furthermore, an enhanced release of adenosine from CEC, caused, for instance, by an elevation of PCO_2 or by increases in catecholamine levels, could significantly contribute to the rise of the adenosine concentration and thus to the metabolically induced adaptation of coronary flow.

From all our findings, it appears that, by virtue of its very active adenine nucleotide and adenosine metabolism, the coronary endothelium must actually be considered to be of quantitative significance with respect to the compartmentation of adenine metabolites in the myocardium. Therefore, the aforementioned great differences between the specific activity of adenosine released into the perfusate and the specific activities of myocardial tissue precursor nucleotides could well be due to a preferential labeling of the endothelium.

A more direct support of this assumption was obtained in special experiments on isolated guinea pig hearts, which were perfused for 30 min with different concentrations of adenosine in the presence of 8-^{14}C-adenosine. The total uptake of adenosine per gram ventricular tissue was determined as the difference of the radioactivities in the inflow and effluent perfusate. CEC of these hearts were enzymatically detached and isolated by centrifugation. The radioactivity present in these cells was related to the total endothelial cell count of 1 gm ventricular heart tissue, reported to be 2×10^8 cells [7, 8].

It is obvious from the data summarized in Figure 22.3 that after perfusion of hearts with labeled adenosine in concentrations of $10^{-8}M$, $10^{-7}M$, and $5 \times 10^{-7}M$, more than 80% of the radioactivity taken up was recovered from the endothelial cells. Eighteen percent of the radioactivity was released during isolation and was determined in the supernatant fraction of the endothelial harvest. Most likely, this radioactivity was also liberated from CEC, since we could demonstrate that collagenase and trypsin treatment of prelabeled coronary endothelial cells in culture resulted in a quantitatively similar release of radioactivity. After perfusion of isolated hearts with $10^{-6}M$ adenosine, endothelial cells contained only 56% of the total radioactivity, whereas the percentage of the released radioactivity was again in the 20% range. The difference between the total radioactivity taken up by the heart and the radioactivity recovered in these two fractions amounted to more than 20% and can most

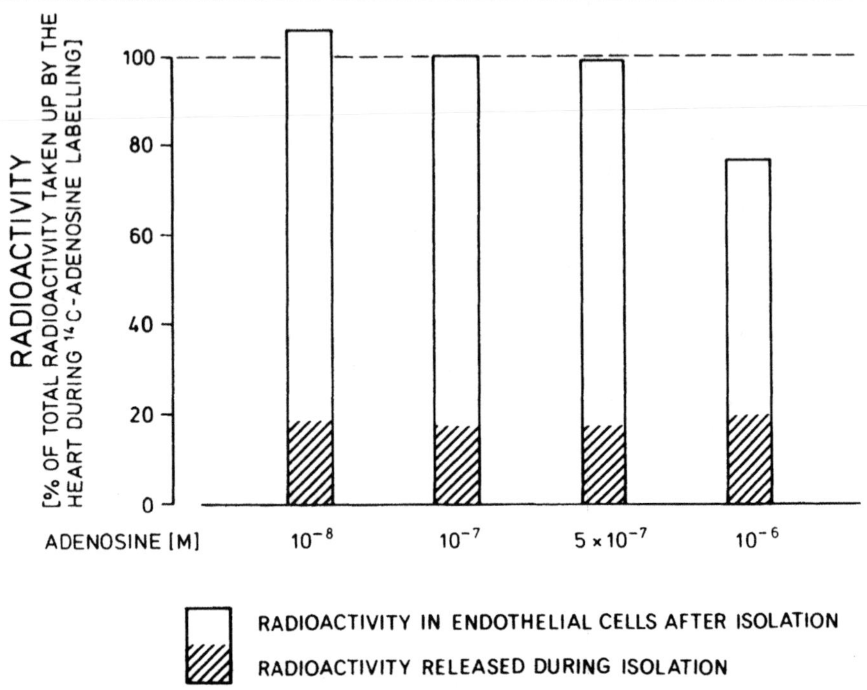

Figure 22.3. Fractional distribution of radioactivity after 30 min prelabeling isolated perfused guinea pig hearts with 8-^{14}C-adenosine and subsequent isolation of coronary endothelial cells. Mean values from three to four experiments.

likely be attributed to a corresponding incorporation of adenosine into cardiomyocytes.

These tracer studies demonstrate that perfusion of isolated hearts with adenosine in concentrations $<10^{-6}M$ results in a preferential adenosine uptake into the endothelial compartment. At perfusate concentrations of adenosine $>10^{-6}M$, the endothelial transport system becomes saturated and adenosine is then also taken up by cardiomyocytes.

CONCLUSIONS AND PHYSIOLOGICAL IMPLICATIONS

From our investigations on cultured CEC, it is evident that the coronary endothelium is characterized by an extraordinarily high content and a very active metabolism of adenine nucleotides. Therefore, it must in fact be considered as a significant compartment of adenine nucleotides and of adenosine in the heart.

In the normoxic myocardium, adenosine seems to originate mainly in the endothelium. Under certain conditions, the coronary endothelium may contribute to the metabolic adaptation of coronary flow by an accelerated formation and release of adenosine.

It is suggested that vascular endothelial cells, through their high capacity for release and reuptake of adenosine, are generally involved in the maintenance of constant plasma levels of this nucleoside. Furthermore, it appears very likely that adenosine of endothelial origin, because of its antiaggregatory properties, is one essential factor contributing to the well-known antithrombogenicity of the luminal surface of the vascular system.

REFERENCES

1. Gudbjarnason S, Mathes P, Ravens KG: Functional compartmentation of ATP and creatine phosphate in the heart muscle. *J Mol Cell Cardiol* 1:325–339, 1970.
2. Liu MS, Feinberg H: Incorporation of adenosine 8-^{14}C and inosine 8-^{14}C into rabbit heart adenine nucleotides. *Am J Physiol* 220:1242–1248, 1971.
3. Wiedmeier VT, Rubio R, Berne RM: Incorporation and turnover of adenosine-U-^{14}C in perfused guinea pig myocardium. *Am J Physiol* 223:51–54, 1972.
4. Yagi K, Mase R: Possible compartmentation of adenine nucleotides in a coupled reaction system composed of F-actomyosin-adenosine triphosphatase and creatine kinase, in Ebashi S, Oosawa F, Sekine T, Tonomura Y (eds): *Molecular Biology of Muscular Contraction*. New York, Elsevier, 1965, pp 109–123.
5. Schrader J, Gerlach E: Compartmentation of cardiac adenine nucleotides and formation of adenosine. *Pfluegers Arch* 367:129–135, 1976.
6. Mall G, Mattfeldt T, Rieger P, Volk B, Frolov VA: Morphometric analysis of the rabbit myocardium after chronic ethanol feeding: Early capillary changes. *Basic Res Cardiol* 77:57–67, 1982.
7. Sasaki R, Watanabe Y, Morishita T, Yamagata S: Estimation of the cell number of heart muscles in normal rats. *Tohoku J Exp Med* 95:177–184, 1968.
8. Simionescu M, Simionescu N: Isolation and characterization of endothelial cells from the heart microvasculature. *Microvasc Res* 16:426–452, 1978.
9. Anversa P, Olivetti G, Melissari M, Loud AV: Stereological measurement of cellular and subcellular hypertrophy and hyperplasia in the papillary muscle of adult rat. *J Mol Cell Cardiol* 12:781–795, 1980.
10. Bassingthwaite JB, Yipintsoi T, Harvey RB: Microvasculature of the dog left ventricular myocardium. *Microvasc Res* 7:229–249, 1974.
11. Ryan JW, Ryan US: Endothelial metabolism, in Effros RM, Schmid-Schöngein H, Ditzel J (eds): *Microcirculation: Current Physiologic, Medical, and Surgical Concepts*. New York, Academic Press, 1981, pp 147–169.
12. Nees S, Gerbes AL, Gerlach E: Isolation, identification, and continuous culture of coronary endothelial cells from guinea pig hearts. *Eur J Cell Biol* 24:287–297, 1981.
13. Pertoft H, Laurent TC: Isopycnic separation of cells and cell organelles by centrifugation in modified colloidal silica gradients, in Catsimpoalas N (ed): *Methods of Cell Separation*, Vol. 1. New York, Plenum Press, 1977, pp 25–65.
14. Nees S, Willershausen-Zönnchen B, Gerbes AL, Gerlach E: Studies on cultured coronary endothelial cells. *Folia Angiol* 38:64–68, 1980.
15. Holmsen H, Day HJ, Storm E: Adenine nucleotide metabolism of blood platelets: VI. Subcellular localization of nucleotide pools with different functions in the platelet release reaction. *Biochim Biophys Acta* 186:254–266, 1969.
16. Pearson J, Carleton S, Gordon JL: Metabolism of adenine nucleotides by ectoenzymes of vascular endothelial and smooth-muscle cells in culture. *Biochem J* 190:421–429, 1980.
17. Crutchley DJ, Ryan US, Ryan JW: Effects of aspirin and dipyridamole on the degradation of adenosine diphosphate by cultured cells derived from bovine pulmonary artery. *J Clin Invest* 66:29–35, 1980.
18. Pearson JD, Carleton JS, Hutchings A, Gordon JL: Uptake and metabolism of adenosine by pig aortic endothelial and smooth-muscle cells in culture. *Biochem J* 170:265–271, 1978.
19. Pearson JD, Gordon JL: Vascular endothelial and smooth muscle cells in culture selectively release adenine nucleotides. *Nature* 281:384–386, 1979.
20. Schrader J, Haddy FJ, Gerlach E: Release of adenosine, inosine and hypoxanthine from the isolated guinea pig heart during hypoxia, flow-autoregulation and reactive hyperemia. *Pfluegers Arch* 369:1–6, 1977.

DISCUSSION

DR. R.M. BERNE: (1) In your experiments in which you enzymatically removed the endothelial cells from isolated perfused guinea pig hearts, did you then measure the adenosine and adenine nucleotide content of the myocardium after removal of the endothelium? The reason I ask this question is that you claim that all of the adenosine can be accounted for by the endothelial cells. Therefore, there should be none in the myocardium after removal of the endothelial cells. (2) I was impressed with your electron micrographs after proteolytic enzyme treatment. They showed total absence of endothelial cells in large vessels and capillaries. Was this representative of the entire heart, or were certain sections of the hearts selected for examination?

DR. NEES: The yield of enzymatically detached endothelial cells was only about 0.1% to 1% of the total population of coronary endothelial cells (CEC). Therefore, of course, most of the radioactivity taken up by the hearts was still present in the myocardium. In order to refer radioactivity in the isolated CEC to the total amount of CEC present in 1 gm heart, we assume 2×10^8 endothelial cells to be present in 1 gm heart, as was independently reported by several investigators. Referring to your question concerning the vascular origin of the isolated CEC, I must confess that we really do not know from which part of the coronary circulation our endothelial cell cultures were derived. However, due to the exceptionally high relative percentage of capillary cells in the coronary system, we assume that we were mainly, if not exclusively, dealing with capillary endothelial cells.

DR. R.A. OLSSON: (1) Method of weighing cells: The estimated levels of adenine nucleotides and some of the purine-metabolizing enzymes are higher than in other cell types. Since these data are referred to the wet weight of the cells, the accuracy of your nucleotide and enzyme estimates depends directly on the accuracy of the weighing. How did you weigh those cells? (2) Percent weight difference plate I cells: Since your estimate consists of weighing a culture dish before and after the endothelial cells were removed, what was the *percentage* difference in weight?

DR. NEES: (1) The culture medium of confluent CEC cultures in standard petri dishes (Falcon, 35 mm) was sucked off. The cell monolayers were then washed with fresh medium containing [^{3}H] inulin or ^{14}C-sucrose, respectively (10^6 cpm/ml). After 1 min of incubation at 20°C, the medium was carefully sucked off as completely as possible. Then the culture dish containing the wet monolayer of cells was immediately weighed (this weight was corrected for constant evaporation of water from the dish within the time interval between medium removal and weight determination). Thereafter, the cells were extracted by cold 0.4N $HClO_4$ and scraped off the dishes by means of a rubber policeman. This extract was analyzed for radioactivity and DNA content. The remaining empty dish was washed with water, lyophilized, and finally weighed. The wet weight of the cells could then be calculated and corrected for the extracellular space. Inulin and sucrose spaces proved to be identical. (2) The weight difference between dishes with confluent wet monolayers and empty dishes was about 5% and proved to be well reproducible.

DR. T.D. WHITE: I wonder if you would care to comment on why you were unable to detect ATP release, whereas Dr. Pearson observed a release of ATP with thrombin.

DR. NEES: We could exclude ATP concentrations higher than 10^{-13} M/ml medium when we were dealing with intact monolayers of coronary endothelial cells (CEC) grown in standard culture media.

DR. J.D. PEARSON: (1) We have not done exactly the same experiments as you; however, I would agree that hypoxia released little or no ATP from cultured EC. We are sure that thrombin and trypsin do, and this has been confirmed by others. The reason for any differences in our results probably relates to experimental protocol but may also depend on the cells; you are using coronary cells and we use aortical cells. (2) Because some people here have expressed concern at the use of wet weight to compare cells with tissue, would it be useful to tell us what DNA and protein content (per wet weight) are in each case?

DR. NEES: Thank you for your statement. Confluent CEC subjected to hypoxia, an increased PCO_2, catecholamines, and low thrombin concentrations certainly do not release appreciable amounts of ATP and ADP. But I agree that prolonged treatment of the cells with proteases such as trypsin—and also with thrombin at higher concentrations (>10 NIH units/ml—did in fact result in the liberation of nucleotides. In our opinion, this release of nucleotides is not a specific reaction of the cells but rather a nonspecific increase in permeability of the plasma-membrane most likely caused by partial digestion of its protein constituents. Referring to your second question: 1 mg wet endothelial tissue is equivalent to 2.5×10^6 cells that contain 16–17 μg DNA (1.6% to 0.7% of wet weight). These values are in good agreement with weight estimates using morphometrically determined volumes of endothelial cells and their known density of only 1.04 gm/cm^3 [S. Nees et al., *Eur. J. Cell Biol.*, 1981].

DR. H. OSSWALD: (1) What is the time course of the tissue levels of ATP, adenosine, SAM, SAH, etc.? Is there a change following isolation after 1, 5, 10 days of culture? (2) What is the wet weight to dry weight relation in your CEC compared to that of the isolated perfused heart? A difference in this relation might explain the different tissue levels. (3) Could it be that the high adenosine, SAM, and SAH levels are the result of the isolation procedure? Are these 1% of CEC representative? Are there repair mechanisms including membrane methylation? Do you get the same tissue levels using a different isolation procedure?

DR. NEES: Tissue contents of adenosine metabolites in proliferating endothelial cells are different from those obtained in confluent monolayers. From a physiological point of view, analyses of adenine metabolites in confluent monolayers are of special interest, since the individual cells within endothelial sheets are in the same phase of growth (G_o) as they are in the endothelium of the normal vessels. Therefore, for all our metabolic studies, we used microscopically controlled confluent CEC about 3 to 4 weeks after seeding. Concerning your second question, I would like to point out that the relation between wet weight of cultured coronary endothelial cells and their dry weight primarily depended on the duration of culture. This was most likely due to extracellular protein and polysaccharide synthesis of the cell layer. Determinations of the dry weight are therefore not well suited to characterize the cell mass of a given culture. With respect to your third question, I would like to refer you to my answer concerning your first question.

DR. C.E. STERLING: I have a comment regarding Dr. Berne's question. We have injected adenosine in the left ventricle of hamster hearts. Autoradiography, 30 min later, showed that nearly all of the label derived from [^{3}H]adenosine is in the endothelial cells. Very recent quantitative estimates indicate that >80% of the label is in the endothelial cells.

DR. NEES: Thank you. Your results support our observations.

DR. S.P. BRUTTIG: The stimulation of adenosine production by endothelial cells in response to elevated P_{CO_2} is similar to our results [Bruttig et al., *Physiologist,* 1981] for vascular smooth muscle. In that study, we attempted to dissociate the effects of CO_2 from those of pH. Would you comment on whether the results you obtained result from changes in P_{CO_2} *per se* or from changes in pH?

DR. NEES: During all our experiments we used bicarbonate buffered media, the pH of which depended upon the P_{CO_2} applied; increased P_{CO_2} resulted in a decrease in pH. The accompanying increase in purine release observed under these conditions could be exactly imitated in experiments in which we induced similar pH changes by addition of acids (HCl, lactic acid) to the medium. Therefore, we tend to assume that the increase in H^+ ions was mainly responsible for the observed acceleration of basal endothelial release. The control experiments—increased P_{CO_2} at constant pH in the medium—have not been performed as yet.

DR. G. RABERGER: I would like to know whether you are measuring a time-dependent release or perhaps a decreased reuptake when you measure an increase in P_{CO_2}. I am asking this because we could show markedly enhanced activity of exogenously applied adenosine by raising P_{CO_2} or lowering pH [Raberger et al., *Naunyn Schmiedebergs Arch. Pharmacol.* 271:301, 1971, and *CEPP* 2:373, 1975).

DR. NEES: I am sure that we were really dealing with adenosine release from the cultured cells. By means of specially designed culture dish and perfusion devices, we could rapidly separate released purines from the cell layers prior to the quantitative analyses of these substances in the perfusate.

DR. G. MISTRY: (1) What percentage of your isolated endothelial cells are viable and grow in the culture? (2) For how many passages do endothelial cells in culture retain their morphological and biochemical characteristics? (3) Have you looked at the adenylate cyclase of these cultured coronary endothelial cells?

DR. NEES: About 30% of the freshly isolated cells were viable, as was estimated by phase contrast microscopy one day after seeding the cells. Concerning your second question, we used only primary or secondary cultures, which revealed identical morphological and biochemical features. Cells in the third passage revealed reduced proliferation rates. To your third question, CEC do contain adenylate cyclase activity, but we have not characterized the enzyme as yet.

DR. MISTRY: I share your and Dr. Pearson's views on the importance of endothelial cells in blood flow and endothelial permeability. We have isolated capillaries from rabbit heart and have looked at the regulation of cAMP by various agents. Our studies show β_2, α, $A_2(R_a)$, and prostaglandin receptors linked to adenylate cyclase and also the present P site.

DR. R. BÜNGER: What is the rate of oxygen uptake of the cultured endothelial cells? Is aerobic glycolysis the main energy-yielding process in the monolayer cells?

DR. NEES: As I have already pointed out, aerobic glycolysis is the main energy-yielding process in confluent endothelial cells. According to the relatively small number of mitochondria present in endothelial cells, oxidative phosphorylation under normoxic conditions is not very active. For technical reasons, it is therefore not possible to measure oxygen consumption of the 2 mg of endothelial tissue present in a standard culture dish. Recent application of the microbead cultivation technique, however, resulted in CEC densities of about 0.5–1 gm cells/100 ml culture medium. These cultures should therefore be well suited to measure oxygen uptake directly.

DR. P.M. UELAND: A minor point: I was somewhat puzzled by the low concentration of S-adenosylmethionine (SAM) in heart tissue. If I am not wrong, this is far below the value reported in heart by others and is below the values reported for other tissues, including skeletal muscle. Have you any explanation for this finding? Have you compared SAM synthesis activity in endothelial cells and heart tissues?

DR. NEES: In order to determine SAM free of contaminating UV-absorbing substances, we used a two-step HPLC method. SAM was first prefractionated on a reverse-phase column (C_{18}), using acid potassium phosphate buffers. The SAM-containing fraction was heated in the presence of ammonia. Adenine, quantitatively hydrolyzed from SAM, was then analyzed by a second C_{18} reverse-phase chromatography, using a potassium phosphate buffer/methanol gradient for its elution. It may be, therefore, that the relatively low SAM values reported are due to this special experimental protocol. We have not yet performed assays of enzymes involved in SAM and SAH synthesis. Recently, we succeeded in the isolation of functionally stable adult cardiomyocytes from guinea pig hearts that do not show the calcium paradox. We intend to compare the relative distribution of SAM synthase and SAH hydrolase in cardiomyocytes as well as in CEC.

DR. S.J. MUSTAFA: Do you see any free intracellular pools of adenosine in these endothelial cells, or is it mostly bound?

DR. NEES: For biochemical reasons, most of the adenosine determined in cultured CEC cannot be present as free nucleoside; rather, it must be bound to more or less specific binding sites. One possible candidate for a specific adenosine binding site certainly is SAH-hydrolase. As was recently reported by Karnovsky et al., endothelial cells are capable of heparin and heparan synthesis. These highly acidic polysulfonated polysaccharide structures may also bind additional amounts of adenosine in endothelial tissues.

DR. R. RUBIO: Does PCO_2 affect the adenosine cell levels?

DR. NEES: According to results of preliminary experiments, adenosine levels do not appear to be profoundly altered.

DR. C.I. THOMPSON: In view of the report by Furchgott et al. suggesting a role for endothelial cells in mediation of acetylcholine (Ach)-induced vasodilation, have you measured the release of purines during administration of Ach?

DR. NEES: Acetylcholine, even in concentrations of up to $10^{-3}M$, did not increase the basal rate of purine release from cultured coronary endothelial cells.

DR. C. SU: Acetylcholine, as well as thrombin and KCl, causes release of [^{3}H]purine in the perfused rabbit pulmonary artery. However, this purine release is too small in quantity to account for the vasodilator effect of acetylcholine.

DR. S. NEES: Acetylcholine never increased basal purine release from cultured coronary endothelial cells, even when it was applied in concentrations as high as $10^{-3}M$.

DR. E. GERLACH: The situation regarding the vasodilator effect of adenosine (AR) is getting more and more complex. In my opinion, we have now clearly been able to distinguish between two AR compartments: (1) the vascular system containing AR exogenously administered or AR continuously released from the endothelium and (2) the interstitial space in conjunction with the cardiomyocyte compartment containing AR derived from contracting myocardial cells. It appears very likely that AR—like AMP, ADP, and ATP—present in the coronary vascular system can exert a vasodilator effect by means of a still unidentified mediator substance released from the intact endothelium. The endothelium itself apparently functions as a kind of trap for AR when its concentration in the vascular space is in the range of 10^{-8} to $10^{-6}M$. This exogenously applied AR in low concentrations does not seem to act per se on smooth muscle cells. Only when AR concentration is elevated above $10^{-6}M$ may AR also reach the smooth muscle cells, the interstitial space, and the cardiomyocyte, because then the endothelial compartment overflows. However, AR produced in the cardiomyocytes under well-defined conditions (e.g., inotropic stimulation, hypoxia) can reach the smooth muscle cells via the interstitial space and directly exert a relaxing effect. Most likely, this direct effect can be facilitated or amplified by AR in the vascular system, which becomes effective via the aforementioned mediator substance. Based on the concept outlined, several unexplained findings regarding the effects of exogenously applied and of endogenously formed AR may be better understood. In addition, the suggestion of two AR compartments separated from each other by the endothelium does not at all interfere with the principal concept of AR being the important factor in the metabolic regulation of coronary blood flow.

VI. INFLUENCE OF ADENOSINE ON CARDIAC AND RENAL FUNCTION

23. ANTIADRENERGIC EFFECTS OF ADENOSINE IN THE HEART

JAMES G. DOBSON, JR., RICHARD A. FENTON

Adenosine has been implicated as a modulator of catecholamine-induced responses in the myocardium. This action of adenosine is in addition to its well-known vasodilator effect on the coronary circulation. A main interest of this laboratory has been to elucidate the properties and the mechanisms responsible for the modulatory role of the nucleoside. This chapter will be restricted to salient developments that fostered the concept that adenosine may serve as an important negative-feedback modulator of catecholamine-elicited contractile and glycogenolytic responses in the heart.

Adenosine has only recently been reported to reduce the catecholamine-induced elevations of myocardial cyclic adenosine 3,5′-monophosphate (cyclic AMP) [1, 2] and glycogen phosphorylase activity [1]. The nucleoside has also been reported to attenuate catecholamine-induced increases in the inotropic state of rat atria [3], rabbit papillary muscles [4], isolated perfused hearts [2, 5], and the *in situ* dog heart [6]. This effect of adenosine has not been found to be mimicked by inosine, adenine, adenosine 5′-monophosphate, or guanosine [3]. The antiadrenergic action of adenosine is observed when low concentrations (1–10 μM) [1, 3] of the nucleoside are employed. These concentrations are similar to those reported to exist in the normal oxygenated ventricular

We are indebted to E. George, L. Griffin-Shea, and J.M. Praderio for their excellent technical assistance. This work was supported by P.H.S. Grants 22828 and 18280 from the National Heart, Lung and Blood Institute. J.G. Dobson, Jr., is a recipient of Research Career Development Award HL-00602.

R.M. Berne, T.W. Rall, and R. Rubio (eds.), Regulatory Function of Adenosine, *pp. 363–376.*

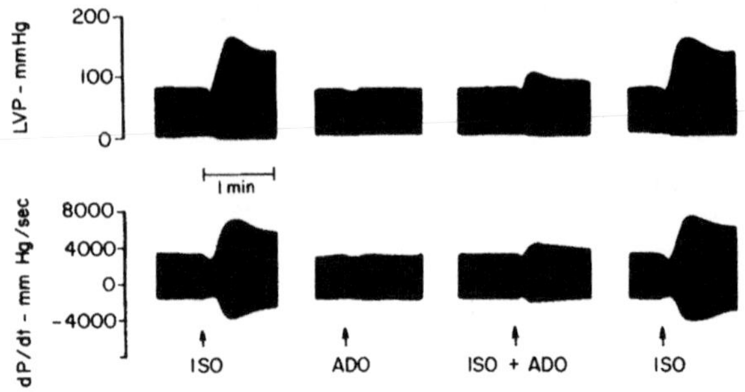

Figure 23.1. Effect of adenosine on the isoproterenol-induced increase in left ventricular pressure (LVP) and the maximum rate of left ventricular pressure development (dP/dt) in the isolated rat heart. These contractile parameters were recorded from a balloon cannula placed in the lumen of the left ventricle. The heart was paced at 300 Hz and perfused at a constant flow of 10 ml/min with oxygenated physiological saline solution (PSS) maintained at 37°C. Isoproterenol (ISO) at 10^{-7} *M*, adenosine (ADO) at 10^{-5} *M*, or both (ISO + ADO) were infused for 1 min just proximal to the aorta. Between infusion periods, the contractile parameters were allowed to return to control values (5–10 min).

[7, 8] and atrial [9] myocardium. The ability of adenosine to attenuate catecholamine-induced responses appears to be different [3] from the direct negative inotropic action of the nucleoside on the atrial myocardium [10, 11], which is caused by higher concentrations (0.1–1 m*M*) of the nucleoside. High concentrations of adenosine have no appreciable influence on the ventricular myocardium [12, 13].

ADENOSINE ATTENUATION OF THE CONTRACTILE RESPONSES ELICITED BY CATECHOLAMINES

Adenosine reduces the increases in left ventricular pressure and the rate of left ventricular pressure development produced by isoproterenol stimulation of perfused rat hearts, as illustrated in Figure 23.1. Adenosine alone at 10^{-5} *M* had virtually no effect on the two contractile parameters shown. When infused simultaneously with isoproterenol, the nucleoside markedly reduced the increases in left ventricular pressure and rate of ventricular pressure development compared to the observed responses obtained with the catecholamine alone. Since isoproterenol and adenosine were administered simultaneously, the action of adenosine to attenuate the isoproterenol elicited responses was very rapid. The effect of adenosine was found to be reversible, because 5 min after removing both agents from the perfusion fluid, infusion of isoproterenol alone produced normal catecholamine contractile responses. Not only does adenosine attenuate the positive inotropic responses of catecholamines, but the nucleoside has been reported to decrease the positive chronotropic response resulting from catecholamine stimulation [6, 14, 15] as well as to decrease atrioventricular conduction velocity [16].

ADENOSINE ATTENUATION OF THE INCREASES IN CYCLIC AMP, PROTEIN KINASE, AND PHOSPHORYLASE ELICITED BY CATECHOLAMINES

An important mechanism whereby catecholamines elicit enhancement of the inotropic state and metabolism is thought to involve an elevation of myocardial cyclic AMP [17, 18]. This cyclic nucleotide fosters dissociation of the regulatory and catalytic subunits of cyclic AMP-dependent protein kinase, resulting in the subsequent phosphorylation and activation of phosphorylase kinase [18, 19]. The protein kinase and phosphorylase kinase may then, in turn, catalyze the phosphorylation of a number of myocardial proteins [20, 21] that are thought to be responsible for the augmentation of contractility and metabolism, especially with regard to glycogen phosphorylase activity. Thus, it was of interest to determine whether adenosine would reduce the aforementioned cellular processes thought to be responsible for the observed contractile and glycogenolytic manifestations of catecholamines.

In perfused rat hearts, adenosine reduced the isoproterenol-induced increases in myocardial cyclic AMP concentration and the activation of both cyclic AMP-dependent protein kinase and phosphorylase, as shown in Figure 23.2. The relatively low concentration (10^{-5} *M*) of adenosine employed had no effect alone on cyclic AMP, protein kinase, or phosphorylase. Theophylline, a known antagonist of adenosine [2, 22], was used to assess the specificity of adenosine's action. When the methylxanthine was infused simultaneously with adenosine at a concentration that did not inhibit the phosphodiesterase activity, as judged by no detectable increase in myocardial cyclic AMP levels, the attenuating effects of the nucleoside were completely prevented.

Similar results have been observed in a variety of preparations. Adenosine has been reported to reduce catecholamine-induced increases in myocardial cyclic AMP in ventricular slices of rat heart [1] and in perfused hearts of the rat [5] and guinea pig [2]. The attenuation of catecholamine-induced activation of phosphorylase has been observed in rat heart ventricular slices [1]. In perfused hearts, the nucleoside has been found to reduce the catecholamine-induced activation of cyclic AMP-dependent protein kinase and phosphorylase [5] and the elevation of myocardial glucose-1-phosphate formation [2].

ADENOSINE ATTENUATION OF CATECHOLAMINE-SENSITIVE ADENYLATE CYCLASE ACTIVITY

The β-adrenergic agonists are thought to cause an elevation of cytoplasmic cyclic AMP levels by the activation of adenylate cyclase, the enzyme responsible for synthesis of the cyclic nucleotide. It was of interest to consider this enzyme as a pivotal point in the modulation by adenosine of catecholamine-induced contractile and metabolic activities. Thus, the ability of adenosine to influence adenylate cyclase activity was ascertained using a crude myocardial membrane preparation from perfused rat hearts.

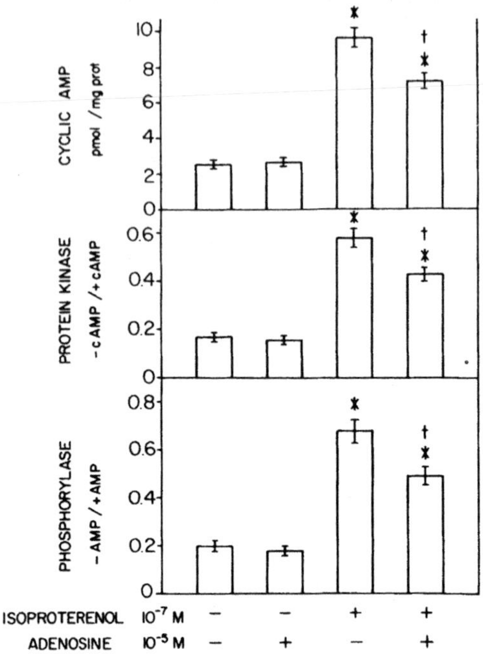

Figure 23.2. Effect of adenosine on the isoproterenol-induced increase in myocardial cyclic adenosine 3′,5′-monophosphate content (CYCLIC AMP) and activity ratios of cyclic AMP-dependent protein kinase (PROTEIN KINASE) and glycogen phosphorylase (PHOSPHORYLASE) in isolated rat hearts. Hearts were perfused as described in the legend of Figure 23.1. Isoproterenol, adenosine, or both were infused into the perfusing PSS for 2 min and the hearts frozen. An increase in the activity ratios of protein kinase and phosphorylase reflect activation of the enzymes. Each value represents the mean ± 1 SE of six experiments. Asterisks denote a significant difference ($p < 0.05$) from the corresponding value in the absence of isoproterenol. Daggers denote the significant difference from the isoproterenol value in the absence of adenosine.

Adenosine at 10^{-5} *M* significantly decreased the enhancement of adenylate cyclase activity caused by isoproterenol in the presence of 15 m*M* Mg^{2+}, as illustrated in Figure 23.3. However, the nucleoside did not attenuate the isoproterenol-induced increase in activity of the enzyme when assayed at 1 m*M* Mg^{2+}. Under both assay conditions, adenosine did not have an inherent effect on the basal enzyme activity. Magnesium at the concentration of 1 m*M* probably best reflects the actual intracellular activity of the cation to which the cytoplasmic facing membrane enzyme is exposed [23]. Others have reported that adenosine attenuates catecholamine-sensitive myocardial adenylate cyclase activity when the assay Mg^{2+} concentration is 15 m*M* [2, 24]. Yet at 2 m*M* Mg^{2+}, the nucleoside has been found to have little, if any, effect on basal adenylate cyclase activity from canine heart [25]. Thus, evidence to date strongly suggests that adenosine may mediate its negative-feedback action by

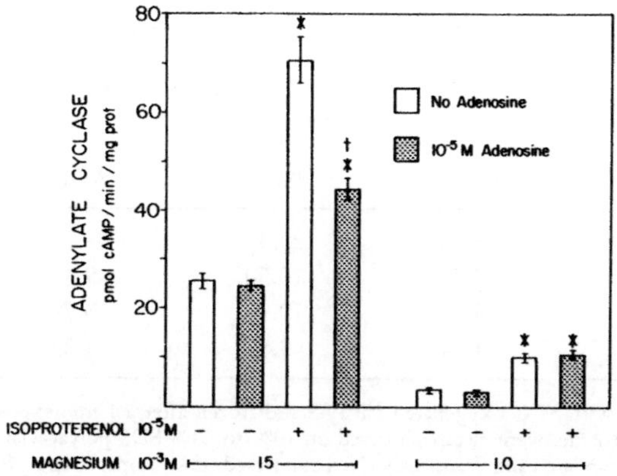

Figure 23.3. Effect of adenosine on the isoproterenol-induced elevation of adenylate cyclase activity in a crude myocardial membrane preparation. Adenylate cyclase activity was assessed in the presence of 15 or 1 m*M* Mg^{2+} by determining the rate of formation of ^{32}P-cAMP from [γ-^{32}P]ATP. Each value represents the mean ± 1 SE of six experiments. Asterisks denote a significant difference ($p < 0.05$) from the corresponding values in the absence of isoproterenol. Daggers denote a significant difference from the isoproterenol value in the absence of adenosine.

attenuating the catecholamine activation of adenylate cyclase. However, this point requires further verification.

We have found that adenosine does not exert its antiadrenergic actions in heart via the release of endogenous acetylcholine or prostaglandins, since the use of atropine, a muscarinic antagonist, and indomethacin, a prostaglandin synthesis inhibitor, do not modify the attenuating effects of adenosine. It has been reported that adenosine at 0.1 m*M* does not appear to influence myocardial phosphodiesterase activity [2].

ADENOSINE ATTENUATION OF CATECHOLAMINE-INDUCED ^{32}P INCORPORATION INTO MYOCARDIAL PROTEINS

The catecholamine-elicited activation of protein kinases (e.g., cyclic AMP-dependent protein kinase and phosphorylase kinase) is thought to produce covalent modification via phosphorylation [20] of myocardial proteins that may play an important role in the catecholamine augmentation of contractile and metabolic states [18, 21]. To ascertain whether adenosine would influence ^{32}P incorporation into myocardial proteins, isolated rat hearts were perfused with oxygenated physiological saline containing ^{32}P. Infusion of 10^{-7} *M* isoproterenol for 1 min enhanced the ^{32}P incorporation into four proteins having approximate molecular weights of 155,000, 92,000, 29,000, and 20,000 daltons, as shown in Table 23.1. These proteins are thought to represent the C-protein of myosin, phosphorylase, troponin I, and myosin P-light chain, respectively. The simultaneous infusion of 10^{-5} *M* adenosine reduced the isoproterenol-enhanced incorporation of ^{32}P into these proteins. The nu-

Table 23.1. Effect of adenosine on the isoproterenol-induced ^{32}phosphate incorporation into myocardial proteins of isolated perfused hearts.

	cpm/gm protein			
Molecular weight (daltons)	Cont	ISO	ADO	ISO + ADO
155,000	15.03	26.40	14.32	23.20
92,000	10.42	18.50	9.40	15.22
29,000	5.13	19.96	3.79	10.08
20,000	4.54	9.87	4.33	6.31

Hearts were perfused with 37°C oxygenated ^{32}P PSS and frozen after a 1 min exposure to the above agents. Whole homogenate proteins were electrophoresed on 10% to 20% SDS polyacrylamide slab gels. Molecular weights are based on standards simultaneously electrophoresed. ISO, isoproterenol (10^{-7} *M*); ADO, adenosine (10^{-5} *M*); Cont, control. Each value is normalized for specific activity of [γ-^{32}P] ATP and represents the mean of two hearts.

cleoside alone had no appreciable effect on the basal levels of ^{32}P incorporation. Enhanced ^{32}P incorporation with adrenergic agents has been reported previously in studies performed with ventricular slices [20] and perfused hearts [21, 26].

RELATIONSHIP BETWEEN MYOCARDIAL TISSUE AND PERFUSATE ADENOSINE AND THE ATTENUATION OF CATECHOLAMINE-INDUCED MECHANICAL AND METABOLIC RESPONSES

To determine the modulatory efficacy of endogenously produced tissue adenosine as compared to that which is exogenously administered, perfused hearts were rendered hypoxic for 10 min. Hypoxia was associated with an increase in both tissue and perfusate adenosine concentrations, concomitant with a marked reduction in the ability of isoproterenol to enhance the rate of ventricular pressure development as illustrated in Figure 23.4. However, hypoxia did not significantly reduce the isoproterenol-induced activation of phosphorylase. Thus, hypoxia substantially reduced only the isoproterenol-induced increase in contractile state when compared to oxygenated hearts. Hypoxia alone activated the myocardial phosphorylase activity, as has been reported previously, and appears to occur independent of a catecholamine–cyclic AMP mechanism [27]. The reduction of the catecholamine-induced change in contractility correlated with the elevation in tissue and coronary perfusate adenosine levels. The increase in tissue and perfusate adenosine observed with hypoxia has been reported previously by other investigators [28, 29].

The importance of endogenous adenosine was investigated further. Isolated rat atria were either oxygenated or rendered hypoxic in muscle baths. Adenosine deaminase, an enzyme known to render adenosine inactive by deamination, was employed to minimize the level of the extracellular nucleoside. The geometry of the atrial preparation optimized access of the deaminase to extracellular endogenous adenosine. The administered adenosine

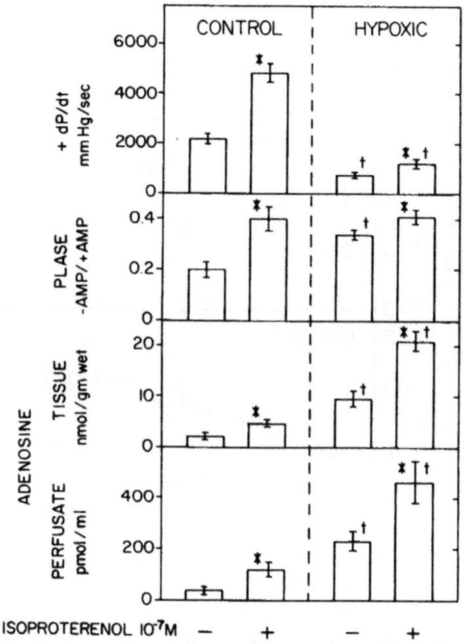

Figure 23.4. Effect of hypoxia on isoproterenol-induced increase in the maximum rate of ventricular pressure development (+dP/dt), glycogen phosphorylase activity ratio (PLASE), and adenosine content of tissue and coronary perfusate from isolated perfused rat hearts. Control hearts were perfused with oxygenated (95% O_2 – 5% CO_2) physiological saline solution (PSS), whereas hypoxic hearts were perfused with PSS gassed with 95% N_2 – 5% CO_2 for 10 min. Isoproterenol was administered to the PSS during the last 2 min of either perfusion condition, and the hearts were frozen. Values for +dP/dt were determined as described in the legend in Figure 23.1. Adenosine concentrations were ascertained enzymatically. Each value represents the mean ± 1 SE of five experiments. Asterisks denote a significant difference ($p < 0.05$) from the appropriate value in the absence of isoproterenol. Daggers denote a significant difference from the corresponding control value.

deaminase significantly potentiated the isoproterenol-induced increase in the rate of force development in oxygenated atria, as shown in Figure 23.5. The enzyme also reduced the fade (progressive decline) of the augmented contractile activity in response to a 6 min exposure of the atria to the catecholamine. When boiled (denatured) adenosine deaminase was added to the muscle bath, the enhanced contractile response to isoproterenol was absent. Since catecholamine stimulation of cardiac tissue has been reported to enhance adenosine formation [28, 30], the present results suggest that endogenous adenosine not only prevents maximal expression of catecholamine-induced contractile responses but facilitates the fade of these responses with prolonged catecholamine stimulation.

In atria rendered hypoxic for 12 min, the rate of force development exhibited a steady decline, as shown in Figure 23.5. Isoproterenol at a concentration of 10^{-7} M tended to offset this loss of force development. Adenosine deami-

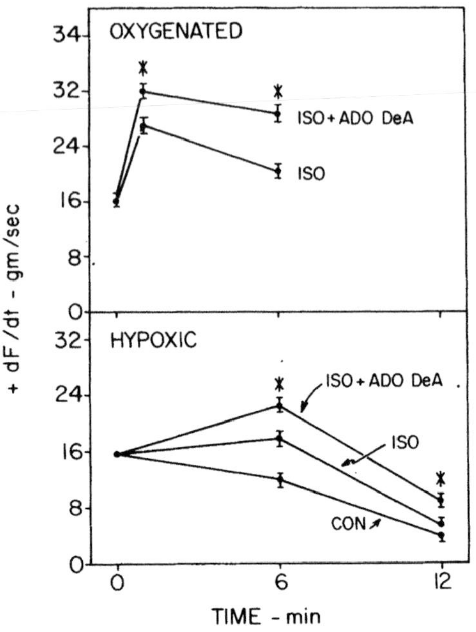

Figure 23.5. Effect of adenosine deaminase on the isoproterenol-induced enhancement of the maximum rate of force development (+dF/dt) in oxygenated and hypoxic isolated rat atria stimulated to contract at 2 Hz. Isoproterenol (ISO) at $10^{-7}M$ or isoproterenol plus 4 U/ml adenosine deaminase (ISO + ADO DeA) was administered to the oxygenated (95% O_2 − 5% CO_2) PSS bathing-designated atrial preparations at zero time. Other atria were rendered hypoxic beginning at zero time by gassing the PSS with 95% N_2 − 5% CO_2. Isoproterenol was administered to hypoxic atria either in the absence (ISO) or presence (ISO + ADO DeA), of adenosine deaminase for 1 min at 5 and 11 min. Determinations of +dF/dt were also obtained from hypoxic atria not subjected to any additions (CON). When adenosine deaminase was used, it was present in the PSS throughout the 12 min of hypoxia. Each value represents the mean ± 1 SE of seven experiments. Asterisks denote a significant difference ($p < 0.05$) from the isoproterenol (ISO) value at the same time point.

nase significantly potentiated the isoproterenol response at 6 and 12 min of hypoxia. It is well known that the mechanical function of hypoxic cardiac tissue is depressed (27, 31), presumably due largely to the decreased availability of ATP and creatine phosphate [7, 32] and reduced intracellular pH [33]. To a large extent, this may explain the depressed catecholamine contractile response observed in hypoxic cardiac muscle. However, the present findings suggest that endogenous adenosine may, in part, be responsible for attenuating the positive inotropic response produced by catecholamine stimulation of the hypoxic myocardium.

CONCLUSIONS

The findings reported here, as well as those reported previously, indicate that adenosine may serve an important negative-feedback function to regulate cate-

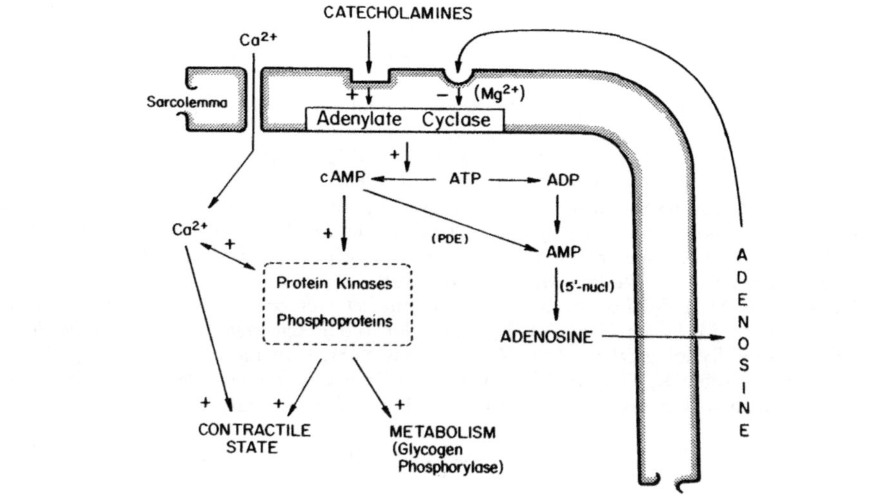

Figure 23.6. A schematic diagram of a myocardial cell, illustrating the negative-feedback role of adenosine that is proposed to modulate catecholamine-induced elevations in cardiac contractile state and metabolism. The abbreviations PDE and 5′-nucl represent the enzymes phosphodiesterase and 5′-nucleotidase, respectively.

cholamine-elicited responses in the heart. This regulatory role of adenosine, diagrammatically illustrated in Figure 23.6, represents an action of the nucleoside that is different from its well-known function as an important regulator of vascular smooth muscle tone. The impact of catecholamines in enhancing contractile state and phosphorylase activity in the heart appears to be modulated by adenosine at the level of adenylate cyclase in the presence of an optimal magnesium activity. This, in turn, influences the formation of cyclic AMP, activation of protein kinases, protein phosphorylation, and calcium availability, all of which play an important role in the biochemical regulation of the myocardial contractile state and metabolism. The importance of the antiadrenergic action of adenosine can be visualized in two ways. First, since adenosine formation via 5′-nucleotidase activity is known to be elevated by catecholamines in the oxygenated heart, the nucleoside may modulate the elicited responses so that they are commensurate with the existing level of metabolism. This would tend to protect the heart from overresponding to excessive or prolonged catecholamine stimulation. Second, when adenosine formation is markedly increased, as is the case with myocardial hypoxia or ischemia, the nucleoside may markedly attenuate the catecholamine-induced mechanical and metabolic responses, thereby protecting this tissue against excessive energy loss. The onset of necrosis in ischemic tissue might then be delayed by curtailing the rate of energy utilization.

REFERENCES

1. Dobson JG Jr: Reduction by adenosine of the isoproterenol-induced increase in cyclic adenosine 3′,5′-monophosphate formation and glycogen phosphorylase activity in rat heart muscle. *Circ Res* 43:785–792, 1978.
2. Schrader J, Baumann G, Gerlach E: Adenosine as inhibitor of myocardial effects of catecholamines. *Pfluegers Arch* 372:29–35, 1977.
3. Rockoff JB, Dobson JG Jr: Inhibition by adenosine of catecholamine-induced increase in rat atrial contractility. *Am J Physiol* 239:H365–H370, 1980.
4. Endoh M, Yamashita S: Adenosine antagonizes the positive inotropic action mediated via β- but not α-adrenoceptors in the rabbit papillary muscle. *Eur J Pharmacol* 65:445–448, 1980.
5. Dobson JG Jr: Adenosine inhibition of catecholamine-elicited contractile and glycogenolytic responses in rat heart. *Fed Proc* 39:1173, 1980 (abst).
6. Juhasz-Nagy A: Modulation by adenosine of sympathetic responses in the in situ canine heart, in Szentivanyi M, Juhasz-Nagy A (eds): *Factors Influencing Adrenergic Mechanisms in the Heart,* Advances in Physiological Sciences, Vol. 27. New York, Pergamon Press, 1981, pp 85–97.
7. Berne RM, Rubio R, Dobson JG Jr, Curnish RR: Adenosine and adenine nucleotides as possible mediators of cardiac and skeletal muscle blood flow regulation. *Circ Res* 28(Suppl. 1):115–119, 1971.
8. Rubio R, Berne RM, Dobson JG Jr: Sites of adenosine production in cardiac and skeletal muscle. *Am J Physiol* 225:938–953, 1973.
9. Thomas RA, Rubio R, Berne RM: Comparison of the adenine nucleotide metabolism of dog atrial and ventricular myocardium. *J Mol Cell Cardiol* 7:115–123, 1975.
10. Hollander PB, Webb JL: Cellular membrane potentials and contractility of normal rat atrium and the effects of temperature, tension and stimulus frequency. *Circ Res* 3:604–612, 1955.
11. deGubareff T, Sleator WJ: Effects of caffeine on mammalian atrial muscle and its interaction with adenosine and calcium. *J Pharmacol Exp Ther* 148:202–214, 1965.
12. Buckley NM: Cardiac effects of nucleosides after propranolol treatment of isolated hearts. *Am J Physiol* 218:1399–1405, 1970.
13. Shah A, Kechejian SJ, Kavaler F, Fisher VJ: Effect of adenine nucleotides on contractility of normal and postischemic myocardium. *Am Heart J* 87:740–749, 1974.
14. Sadavongvivad C, Satayavivad J, Prachayasittikul V, Mo-Suwun L: Modification of relative potencies of beta-adrenoceptor agonists by negative chronotropic agents in rat atrium. *Gen Pharmacol* 8:263–267, 1977.
15. Samet MK, Rutledge CO: Interaction of adenosine derivatives and catecholamines in isolated rat atria. *Fed Proc* 41:1055, 1982 (abst).
16. Belardinelli L, Mattos EC, Berne RM: Evidence for adenosine mediation of atrioventricular block in the ischemic canine myocardium. *J Clin Invest* 68:195–205, 1981.
17. Dobson JG Jr, Ross J Jr, Mayer SE: The role of cyclic adenosine 3′,5′-monophosphate and calcium in the regulation of contractility and glycogen phosphorylase activity in guinea pig papillary muscle. *Circ Res* 39:388–395, 1976.
18. Dobson JG Jr: Protein kinase regulation of cardiac phosphorylase activity and contractility. *Am J Physiol* 234:H638–H645, 1978.
19. Brown DF, Honeyman TW, Dobson JG Jr: Properties of epinephrine-induced activation of cardiac adenosine 3′,5′-monophosphate-dependent protein kinase. *Biochim Biophys Acta* 544:462–473, 1978.
20. Dobson JG Jr: Catecholamine-induced phosphorylation of cardiac muscle proteins. *Biochim Biophys Acta* 675:123–131, 1981.
21. Dobson JG Jr: Cyclic AMP-dependent activation of protein kinases in the myocardium, in Deluis W, Gerlach E, Grobecker H, Kubler W (eds): *Catecholamines in the Heart: Recent Advances in Experimental and Clinical Research.* Berlin, Springer-Verlag, 1981, pp 128–140.
22. Bunger R, Haddy FJ, Gerlach E: Coronary responses to dilating substances and competitive inhibition by theophylline in the isolated perfused guinea pig heart. *Pfluegers Arch* 358:213–224, 1975.
23. Polimeni PI, Page E: Magnesium in heart muscle. *Circ Res* 33:367–374, 1973.
24. Baumann G, Schrader J, Gerlach E: Inhibitory action of adenosine on histamine- and dopamine-stimulated cardiac contractility and adenylate cyclase in guinea pigs. *Circ Res* 48:259–266, 1981.

25. Webster S, Olsson RA: Adenosine regulation of canine cardiac adenylate cyclase. *Biochem Pharmacol* 30:369–373, 1981.
26. Cogoli JM, Dobson JG Jr: An easy and rapid method for the measurement of [γ-^{32}P] ATP specific radioactivity in tissue extracts obtained from *in vitro* rat heart preparations labeled with ^{32}Pi. *Anal Biochem* 110:331–337, 1981.
27. Dobson JG Jr, Mayer SE: Mechanisms of activation of cardiac glycogen phosphorylase in ischemia and anoxia. *Circ Res* 33:412–420, 1973.
28. Katori M, Berne RM: Release of adenosine from anoxic hearts: Relationship to coronary flow. *Circ Res* 19:420–425, 1966.
29. Schutz W, Schrader J, Gerlach E: Different sites of adenosine formation in heart. *Am J Physiol* 240:H963–H970, 1981.
30. Wiedmeier VT, Spell LH: Effects of catecholamines, histamine and nitroglycerin on flow, oxygen utilization, and adenosine production in the perfused guinea pig heart. *Circ Res* 41:503–508, 1977.
31. Bing, OHL, Brooks WW, Messer JV: Effect of isoproterenol on heart muscle performance during myocardial hypoxia. *J Mol Cell Cardiol* 4:319–328, 1972.
32. Imai S, Riley AL, Berne RM: Effect of ischemia on adenine nucleotides in cardiac and skeletal muscle. *Circ Res* 15:433–450, 1964.
33. Bers DM, Ellis D: Intracellular calcium and sodium activity in sheep heart purkinje fibres: Effect of changes of external sodium and intracellular pH. *Pfluegers Arch* 393:171–178, 1982.

DISCUSSION

DR. H.V. SPARKS: The mechanical effects of adenosine occurred at 10^{-7} *M*, but the effects on adenylate cyclase occurred at 10^{-6} *M* to 10^{-5} *M*. This raises the possibility that adenosine exerts its action by a mechanism other than adenylate cyclase.

DR. DOBSON: Yes, with the data that we have obtained thus far, adenosine at a given concentration appears to cause a greater attenuation of the catecholamine-induced increase in contractility compared to the decrease in hormone-sensitive adenylate cyclase. The action of adenosine on adenylate cyclase, we believe, is an important focal point for modulation, but we suspect that adenosine might also exert an action independent of adenylate cyclase.

DR. M.D. DEVOUS: When considering adenosine for a role in modulation of catecholamine stimulation, I have recently shown that inosine exerts a potentiating effect of isoproterenol-induced increases in inotropy and chronotropy. Under conditions of stress, much more inosine is produced than adenosine and could counteract any adrenergic reactions of adenosine. Have you examined inosine in any of your preparations, and, if so, what effects have you observed?

DR. DOBSON: We have never observed any potentiating effects of inosine on catecholamine-induced increases in contractile state at concentrations as high as 100 μ*M* on either isolated atria or perfused hearts. In addition, the compound seems to have no effect on contractile state when administered alone. Of course, our experiments are much different from your experiments in the anesthetized dog, where you infuse inosine intravenously.

DR. DEVOUS: You use adenosine deaminase to remove the effects of endogenous adenosine on isoproterenol-induced myocardial responses. I would like to point out that this procedure not only decreases adenosine but increases inosine concentrations. Since inosine potentiates isoproterenol-induced increases in heart rate and dP/dt, this could explain your observation. In addition, inosine is itself a positive inotrope, which could synergistically lead to the observed prolongation of the apparent isoproterenol effect. Please comment on this alternative interpretation.

DR. DOBSON: From the data that we have obtained thus far, your alternative interpretation seems unlikely. I say this because using inosine at concentrations as high as 100 μ*M* in isolated atria, we see no synergistic actions in the presence of isoproterenol. In fact, inosine at these concentrations neither attenuates the isoproterenol-induced increase in contractile state nor has any effect on mechanical function by itself.

DR. M.G. COLLIS: What are the relative roles of adenosine's pre- and postfunctional inhibitory effects in the heart? Do you feel that prejunctional inhibition also has a major effect on cardiac responses to adrenergic drive?

DR. DOBSON: I think you have asked a very important question. As you know, we have only studied the postjunctional inhibitory effects of adenosine in the heart. Thus, I am afraid I cannot give you an answer as to the relative importance of the pre- and postjunctional effects of the nucleoside. From the literature, it would appear that the prejunctional inhibitory effects of adenosine in the heart might be most important.

DR. R. RUBIO: (1) Do you think ^{32}P incorporation in your electrophoretic patterns reflects phosphorylation or turnover of ^{32}P? (2) You saw a greater change in ^{32}P incorporation in the 29,000 and 20,000 dalton bands. Do you think there is a special significance to these findings?

DR. DOBSON: (1) In response to your first question, the data that we have just presented merely reflect an increase in ^{32}P incorporation into the 155,000, 92,000, 29,000, and 22,000 dalton proteins, and adenosine attenuated this incorporation. However, last year we reported [*Biochim. Biophys. Acta*] that isoproterenol elicited phosphorylation in acid-susceptible protein obtained from hearts labeled with ^{32}P. (2) Regarding your second question, we have not to date attached a special significance to the ability of adenosine to decrease the ^{32}P incorporation into the 29,000 and 20,000 dalton proteins, but I must admit that we are excited about the findings.

DR. C. LONDOS: The data showing adenosine inhibition of adenylate cyclase in isolated membranes appear to be P site-mediated rather than receptor-mediated. One difficulty with demonstrating adenosine receptor-mediated inhibition of hormone-stimulated cyclase with the use of adenosine *per se* is that the stimulated enzyme is far more sensitive to adenosine inhibition at the P site than is the basal enzyme. For selective detection of receptor-mediated inhibition, would it not be better to eliminate adenosine from the assay medium with adenosine deaminase and then test for inhibition with receptor-selective adenosine analogues?

DR. DOBSON: Yes, I agree with you on this point. We have used adenosine deaminase and one receptor-selective adenosine analogue, PIA, in a few recent experiments. The results are preliminary, but I think we are seeing the appropriate responses that indicate that we have receptor-mediated effects.

DR. J.P. MANFREDI: You showed that the isoproterenol-induced increase in contractility waned with continued stimulation by isoproterenol. Does adenosine release decrease in parallel with this decline in contractility?

DR. DOBSON: The fade or progressive decrease of isoproterenol-induced increase in cardiac contractility with continued catecholamine stimulation has been observed by a number of investigators, including ourselves. Mainly, we see an increase in adenosine release with catecholamine stimulation in isolated perfused hearts. However, the enhanced release seems to decrease slightly with time. I think Dr. Schrader has reported similar findings in perfused guinea pig hearts.

DR. O. KRAUPP: In intact animals (dogs), the vasodilator effects of adenosine, of adenosine uptake blockers, and of long-acting N_6- or 5′-substituted adenosine derivatives are accompanied by tachycardia and even by a positive inotropic effect. This increase in sympathetic activity is secondary to a decrease in blood pressure. The observed positive chronotropic effects are dose-dependent and occur in a dose range at which coronary dilation is already maximal. In humans, tachycardia is also the most serious side effect of these drugs and prevents the clinical evaluation of adenosine-like substances. Only in a far higher dose range do negative chrono- and inotropic effects become apparent and overcome the reflex sympathetic influence.

DR. DOBSON: If I understand your comment correctly, I believe I must agree. It is over the physiological ranges of adenosine content that we believe the antiadrenergic actions of adenosine are observed to occur. We think of this action as being an indirect effect of the nucleoside. At very high levels of adenosine, not only would there be a very strong antiadrenergic action but now the direct effect of the nucleoside would be observed. An example of the direct action would be the negative inotropic effects of adenosine alone on atrial contractility. Perhaps we might consider that as the endogenous adenosine content of the myocardium increases, one first sees the indirect action of the nucleoside. Then, as the content of adenosine increases further, the direct actions are apparent. Thus, both actions of adenosine could counteract sympathetic activity in the heart. However, we think that the indirect actions of adenosine are the most important because they are observed to occur at lower (physiological) concentrations of the nucleoside.

24. CHRONOTROPIC AND DROMOTROPIC EFFECTS OF ADENOSINE

LUIZ BELARDINELLI, ALEXANDER WEST, RICHARD CRAMPTON,
ROBERT M. BERNE

Since 1929, adenosine has been known to slow sinus rate and cause atrioventricular (A-V) block [1]. Although the coronary vasodilator action of adenosine has received considerable attention in the last two decades, reports on the direct effects of adenosine on the myocardium and specialized conduction system have been relatively few. It is of more than historic interest that the A-V block observed in the guinea pig heart first drew the attention of Drury and Szent-Györgyi to the physiologial activity of the nucleic acid derivatives. They used the induction of A-V block in guinea pigs as a bioassay for the presence of a substance derived from heart muscle extracts, later found to be adenylic acid. Adenosine obtained from yeast nucleic acid was subsequently shown to have a similar action [1].

This chapter will indicate that in addition to its coronary vasodilator effects, adenosine may also modulate two other important cardiac functions, impulse formation in the sinoatrial node and His bundle and impulse propagation through the atrioventricular node. Evidence suggests that these myocardial actions of adenosine may have greater physiological importance than previously recognized. Furthermore, the fact that both the slowing of sinus rate and the A-V block caused by adenosine can be demonstrated in humans makes the

This work was supported in part by Grant-in-Aid 81-911 from the American Heart Association, AHA-Virginia Affiliate grant HL-10384, and grant BRS-5S07RR05431-19 from the University of Virginia School of Medicine. Dr. Belardinelli is the recipient of NIH Research Career Development Award 1K04-HL00969-01.

R.M. Berne, T.W. Rall, and R. Rubio (eds.), Regulatory Function of Adenosine, *pp. 377–398.*

animal laboratory findings more relevant and quite likely to have important clinical implications.

CHRONOTROPIC ACTION OF ADENOSINE

Depression of Sinus Node by Adenosine

The depressant effects of adenyl compounds on cardiac rhythmicity, notably, on the sinoatrial (S-A) node, have been observed in several different species and experimental models [1–9]. The negative chronotropic action of adenosine has been demonstrated in intact animals, by intravenous infusion of adenosine [1, 10] or by direct perfusion of the sinus node artery with solutions containing adenosine [3]; in isolated perfused dog atria [10] and rat and guinea pig hearts [4, 9, 11]; in isolated sinus pacemakers [6, 7]; and in Purkinje fibers [7, 8]. Slowing of the heart rate produced by bolus infusions of adenosine into the sinus node artery of the dog was the first demonstration of a direct action of adenosine on the sinus pacemaker [3]. A similar finding has been reported for the isolated blood-perfused dog atria [12].

Figure 24.1 illustrates the sinus bradycardia caused by a bolus infusion of adenosine into the S-A node artery of the dog. During sinus slowing, the P wave nearly disappears, indicating a possible shift in pacemaker site. This response to adenosine resembles the effect of bolus infusions of acetylcholine (Ach) into the sinus node of the dog [13]. However, in this canine experimental model, the effect of adenosine on the S-A node cannot be easily interpreted because, concomitant with near disappearance of the P wave, an A-V junctional rhythm appeared. Since subsidiary pacemakers may have different sensitivities to adenosine, the full effect of adenosine on cardiac automaticity can only be appreciated in isolated pacemaker preparations. Use of small, isolated S-A node preparations [14] precludes pacemaker shifts and therefore permits determination of the true sensitivity of the S-A node to various agents. In Figure 24.2, the effects of adenosine on a spontaneously beating rabbit S-A node preparation (dimension 300 μm × 300 μm) is shown. Adenosine slows the sinus node by decreasing the slope of phase 4 depolarization, and, upon washout, the rate returns to control. In guinea pig, Szentmiklosi has shown that automaticity of the S-A node is decreased by adenosine in a dose-dependent manner [7]. The concentration required for inducing half-maximal inhibition (IC_{50}) of the sinus pacemaker activity was 105 ± 6.2 μM. The study of Szentmiklosi was conducted in isolated, spontaneously beating right atria, and therefore shift of the pacemaker cannot be excluded.

With regard to the mechanism of adenosine-induced slowing of the sinus rate, little is known, especially in the mammalian heart. In the frog sinus venosus, adenosine and adenine nucleotides cause a long-lasting hyperpolarization of the membrane [6]. This Ach-type hyperpolarization is likely to be mediated by an increase in K^+ permeability, since adenosine, like Ach, pro-

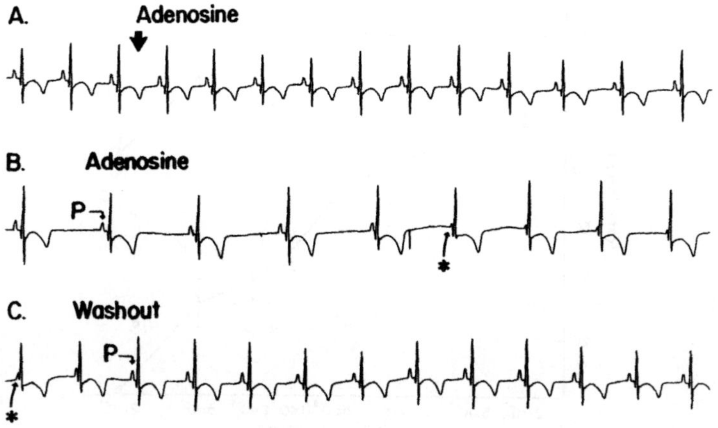

Figure 24.1. Negative chronotropic effects of adenosine injected into the sinoatrial node artery (SANA) of the *in situ* dog heart. A: A 500 μl bolus of normal saline containing adenosine 5 μgm was injected (arrow) into the SANA. B: Marked slowing with near disappearance of the P wave into QRS. C: Washout of adenosine with reappearance of P wave and restoration of sinus rate. ECG lead II (0.5 mV/cm) recorded at 25 mm/sec. See text for discussion.

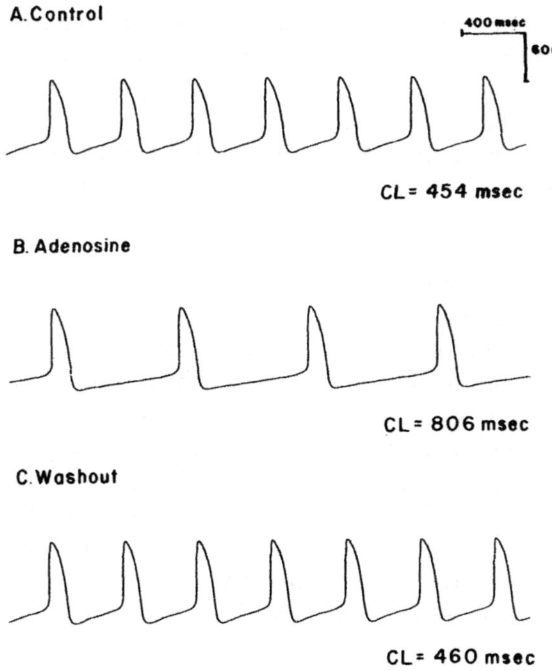

Figure 24.2. Effect of adenosine on firing rate of rabbit isolated S-A node pacemaker cells. Adenosine slows rate and depresses phase 4 depolarization. Preparation is approximately 300 by 300 μm. A: Control, cycle length = 454 msec. B: Adenosine, 4×10^{-5} *M*, cycle length = 806 msec. C: Washout (5 min), cycle length = 460 msec. Temp. = 34°C.

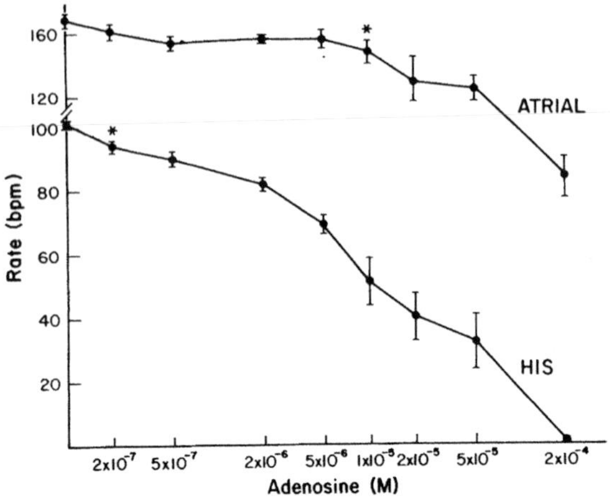

Figure 24.3. Deceleration of atrial and His bundle rhythms by adenosine of isolated guinea pig heart with complete A-V block. Adenosine causes a greater deceleration of His bundle than atrial rhythm. The asterisk denotes first significant change in rate.

duces an increase in efflux of ^{86}Rb and ^{42}K in the tortoise heart sinus venosus [15]. The magnitude of the increase in the efflux of ^{86}Rb and ^{42}K varied for different parts of the atria and is consistent with the observation of different sensitivities of sinus pacemakers to adenosine. In guinea pig atria, adenosine also increases the efflux of ^{42}K and potassium outward current [16]. However, efflux studies have limitations that should be taken into consideration. Potassium efflux depends on the membrane potential and on the rate of stimulation [17]. In addition, precise determination of which cells are responsible for K efflux is difficult, especially in S-A node preparations. Moreover, the underlying K^+ currents responsible for the measured K efflux can only be determined under voltage clamp conditions [17].

Effects of adenosine on ventricular automaticity have only recently been investigated. The automaticity of Purkinje fibers from guinea pigs [7] and dogs [8] is also depressed by adenosine in a dose-dependent manner. The reported IC_{50} for guinea pig Purkinje fibers is $3.5 \pm 0.4\ \mu M$, indicating greater sensitivity to adenosine than the S-A node of the same species [7]. In keeping with this finding, we recently showed that a His bundle rhythm in the isolated guinea pig heart with surgically induced heart block is two-to-three orders of magnitude more sensitive to adenosine than is the atrial rhythm (Fig. 24.3). When these data are considered along with another report [7], it appears that, in the guinea pig heart, a hierarchy in the sensitivity of pacemakers to adenosine may exist (Fig. 24.4). However, a similar hierarchy in sensitivity to adenosine may not be true for other species.

Adenosine also depresses A-V junctional rhythms induced in dog hearts [5, 18]. Since adenosine antagonizes the inotropic and some of the metabolic

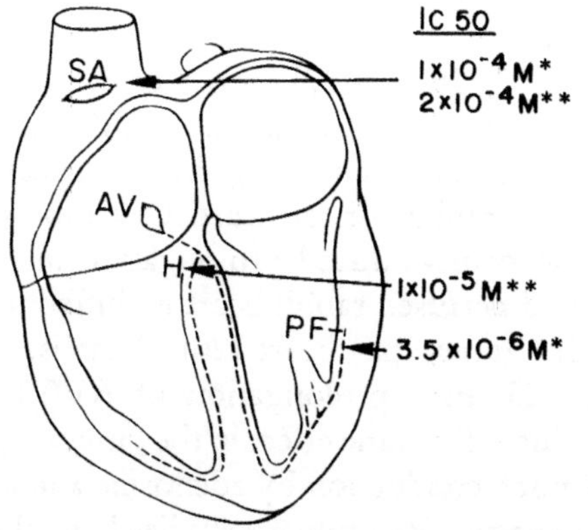

Figure 24.4. Hierarchy of adenosine sensitivity in cardiac pacemakers of isolated guinea pig heart. H = His bundle; PF = Purkinje fiber. Source: Data from [7]* and [9]**.

effects of β-adrenergic agonists [19], it is anticipated that adenosine would also attenuate the positive chronotropic effect of catecholamines. A clear demonstration of such antagonism in pacemaker tissues is rather difficult to achieve, since adenosine also has a direct negative chronotropic action that is independent of β-adrenergic stimulation. Therefore, a direct and indirect, i.e., anti-β-adrenergic, action of adenosine cannot be clearly distinguished.

Sinus Node Depression Induced by Ischemia

With regard to cardiac automaticity, it has been postulated that release of adenine compounds from ischemic cells could be responsible for sinus arrest or arrhythmias that sometimes occur in myocardial infarction [6, 20]. Occlusion of the arteries supplying the sinus node of the dog results in slowing of sinus rate, which has been attributed to metabolites released by the ischemic cells [20]. In isolated mammalian hearts, hypoxia and ischemia often result in A-V block, slight slowing of sinus rate, and very slow ventricular escape rates [21, 22]. In this context, it is interesting to note that adenosine-induced A-V block is accompanied by very slow ventricular escape rates and, occasionally, ventricular standstill, and by relatively less slowing of the sinus rate (Fig. 24.3 and 24.6). A low ventricular escape rate has also been noted during A-V block caused by injection of adenosine into the atrioventricular node artery of the dog [23]. Thus, it is possible that endogenously released adenosine from hypoxic and ischemic myocardial cells may influence cardiac rhythmicity. In addition, agents known to attenuate or potentiate the effects of adenosine may therefore have profound influence on cardiac automaticity.

DROMOTROPIC ACTIONS OF ADENOSINE

Drury and Szent-Györgyi first showed that adenosine and related compounds can cause heart block [1]. Stafford [24] reported that adenosine produced a dose-dependent atrioventricular conduction block (A-V block) in guinea pig hearts. There are differences in the responsiveness of the mammalian A-V node to adenosine [5, 9, 18, 24–27], some of which are probably due to differences in species or experimental models. Although no detailed study of these differences exists, the guinea pig heart appears more sensitive to adenosine than the dog, rabbit, or rat heart. One can estimate that the dose-response curve of adenosine-induced prolongation of A-V conduction time (AVCT) in the isolated perfused rabbit heart is shifted to the right about 100-fold compared to the guinea pig heart [26]. That is, 10^{-4} *M* adenosine is required to produce 25 msec prolongation of AVCT in the rabbit heart, whereas 10^{-6} *M* produces the same effect in the guinea pig heart. Nevertheless, the slowing of A-V node conduction by adenosine and adenine compounds is reproducible among species. Preliminary studies have shown that bolus injections of adenosine cause A-V block when injected into the right coronary artery of humans [27].

In comparing differences in sensitivity of the A-V node to adenosine, one should take into consideration spontaneous or paced heart rates, temperature, and catecholamine levels. First, the AVCT is a function of heart rate, and hence, the degree of its prolongation by adenosine will vary with heart rate. Second, at lower temperatures, the adenosine-induced prolongation of AVCT and the production of A-V block are enhanced [1, 26, 28]. The increased sensitivity to adenosine by lowering temperature can be partially explained by reduced inhibition of adenosine uptake at lower temperatures [28]. Third, catecholamines and adenosine are antagonists [29]. A-V transmission is highly dependent on sympathetic tone and endogenous catecholamine levels. In fact, we have recently observed [unpublished data] that hearts from reserpinized guinea pigs become markedly sensitized to adenosine. Moreover, the basal level of catecholamines in these hearts is approximately 10^{-8} *M*, a concentration that could partially antagonize the effects of adenosine [19].

Our recent studies have confirmed previous observations that adenosine causes A-V node conduction delay and block [26, 28]. In addition, it was established that the effects of adenosine on A-V conduction are dose-dependent and are due to increased conduction delay between the atria and His bundle, i.e., an increase in the A-H interval of the His bundle electrogram (Figs. 24.5). In contrast, His bundle to ventricle (H-V interval) conduction is not altered by adenosine, even at high concentrations (10^{-4} *M*). Thus, the selective action of adenosine on the A-H interval resembles other Ca^{2+} channel blockers, such as verapamil and $MnCl_2$ [30, 31]. Moreover, the A-V block caused by adenosine is accompanied by inhibition or blockade of the "slow" Ca^{2+}-Na^{+}-mediated action potentials in the A-V node cells (32), a finding consistent with adenosine depression of Ca^{2+}-mediated action potentials in mammalian atria [33, 34], and with the fact that the proximal A-V junction is more sensitive to drugs that interfere with slow Ca^{2+}-Na^{+} inward current than is the distal portion. However, the study of Rubio et al. [32] was

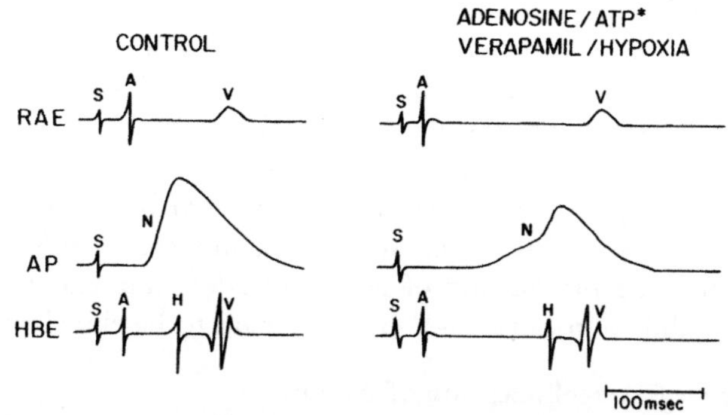

Figure 24.5. Prolongation of A-V conduction time by various interventions in isolated perfused guinea pig heart. Note that A-H interval increased while H-V interval remained constant. S, stimulus artifact; A, atrial; H, His bundle; V, ventricular; N, nodal potential. Upper tracing: RAE, right atrial electrogram. Middle tracing: AP, transmembrane action potential recorded from A-V nodal cell. Lower tracing: HBE, His bundle electrogram. ★ATP is probably hydrolyzed to adenosine (see text for discussion).

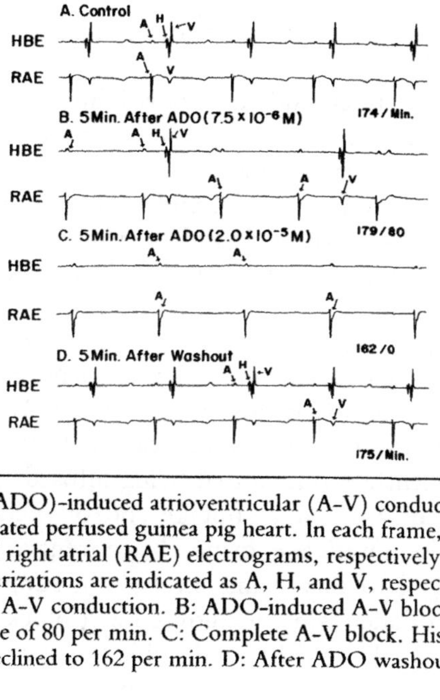

Figure 24.6. Adenosine (ADO)-induced atrioventricular (A-V) conduction block in the normal spontaneously beating isolated perfused guinea pig heart. In each frame, upper and lower traces are His bundle (HBE) and right atrial (RAE) electrograms, respectively. Right atrial, His bundle, and ventricular depolarizations are indicated as A, H, and V, respectively. A: Control normal sinoatrial rhythm and A-V conduction. B: ADO-induced A-V block with atrial rate of 179 per min and His escape rate of 80 per min. C: Complete A-V block. His escape rhythm was abolished and atrial rate declined to 162 per min. D: After ADO washout.

done with a single microelectrode and no mapping of the A-V node, and hence, the site of adenosine action cannot be determined. For example, it is possible that adenosine, like Ach, blocks conduction at the atrial/A-V node border, the AN region. A shortening of the action potential in the AN region could result in failure of impulse propagation and thus result in A-V block. Like Ach, the shortening of the guinea pig atrial action potential by adenosine is accompanied by an increase in ^{42}K efflux, suggesting an increase in K^+ conductance [16]. Therefore, the precise site of action of adenosine and the mechanism underlying its ability to cause A-V block remain to be elucidated.

Evidence for an Extracellular Site of Action

Several of the postulated physiological actions of adenosine have been assumed to be mediated by specific receptor sites on the external surface of membranes [35–37]. The hypothesis that adenosine must first bind to a specific receptor is based on two lines of evidence: (1) the actions of adenosine persist or are potentiated by nucleoside transport inhibitors (e.g., dipyridamole and nitrobenzylthioinosine); and (2) impermeant analogues of adenosine cause equal or greater responses than does free adenosine. More direct evidence for specific binding of adenosine in brain, fat, and heart cell membranes has been obtained only recently [37–39].

The rationale of using nucleoside transport inhibitors to determine whether adenosine action is mediated via an extracellular site is illustrated in Figure 24.7. By inhibition of adenosine uptake, the concentration of the nucleoside in the extracellular space should increase, and hence, more will be available for activation of the receptors. All nucleoside transport blockers listed in Figure 24.7 have been shown to inhibit adenosine uptake [11, 28, 40–43]. The various nucleoside transport inhibitors reduce the uptake of ^{14}C-adenosine in a dose-dependent manner. Furthermore, it was shown that the uptake inhibition of exogenous adenosine by dipyridamole is accompanied by an increase in the extracellular concentration of adenosine [28]. Table 24.1 shows the levels of adenosine in the effluent of isolated guinea pig hearts. When adenosine is infused, a lower concentration than that being infused appears in the effluent, indicating a significant amount of uptake and/or metabolism. When dipyridamole is infused along with adenosine, the concentration of adenosine in the effluent is nearly equal to that being infused. In support of these findings, many of the actions of adenosine, including heart block and negative chronotropism, appear to be enhanced by nucleoside transport inhibitors [6, 10, 11, 24, 28]. Figures 24.8 and 24.9 illustrate that dipyridamole markedly augments both adenosine-induced deceleration of His bundle rhythm A-H prolongation. Dipyridamole ($5 \times 10^{-8}M$–$5 \times 10^{-6}M$) alone did not cause any significant change in either A-V conduction or His bundle rate. Irrespective of the nucleoside transport inhibitor used or its concentration, the lesser the uptake of adenosine by the heart, the greater the effect of adenosine (Fig. 24.9).

The finding that, in the rat, $5 \times 10^{-8}M$ dipyridamole neither inhibited

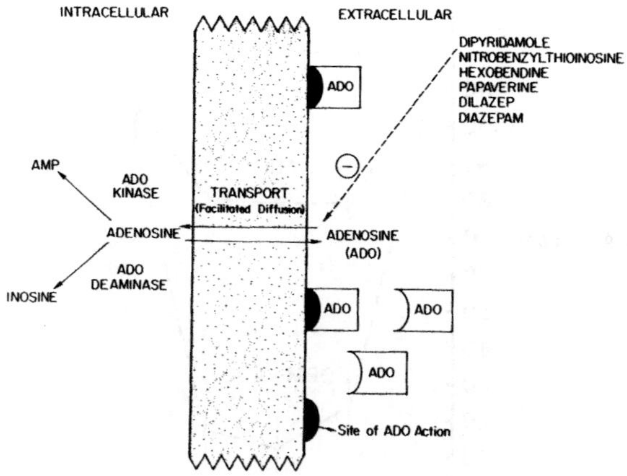

Figure 24.7. Diagram of rationale of using nucleoside transport inhibitors to determine the site of adenosine (ADO) action. See text for discussion.

Table 24.1. Adenosine concentration in the effluent of isolated guinea pig heart

Intervention	Adenosine in effluent (± SEM)
Pre-ADO control	$7.47 \times 10^{-9} \pm 0.32\ M$
ADO	$1.76 \times 10^{-6} \pm 0.01\ M$*
Post-ADO control	$7.94 \times 10^{-9} \pm 0.43\ M$
DPM + ADO	$5.01 \times 10^{-6} \pm 0.02\ M$*

ADO = adenosine in perfusion fluid ($5 \times 10^{-6}\ M$); DPM = dipyridamole in perfusion fluid ($5 \times 10^{-6}\ M$). * = statistically different ($P < 0.05$) from pre-ADO control.

adenosine uptake nor enhanced adenosine-induced A-H prolongation further supports the correlation between adenosine uptake and its action on the A-V node [28]. This species difference regarding the effects of dipyridamole on adenosine transport and the sensitivity of the A-V node to adenosine are in partial agreement with previous reports [11, 24].

An alternative explanation for the ability of nucleoside transport blockers (e.g., dipyridamole) to potentiate the actions of adenosine is the possibility that they may also alter the activity of other enzymes, such as adenosine deaminase [44, 45] and adenosine kinase [46, 47]. These effects are unlikely to explain our previous results [28], since they require concentrations of dipyridamole of the order of $10^{-4}M$ and thus far exceed the concentration necessary to significantly inhibit the uptake of adenosine. Modification of the binding characteristics of adenosine by nucleoside transport inhibitors may be involved, although only tentative evidence at present suggests such a role [48].

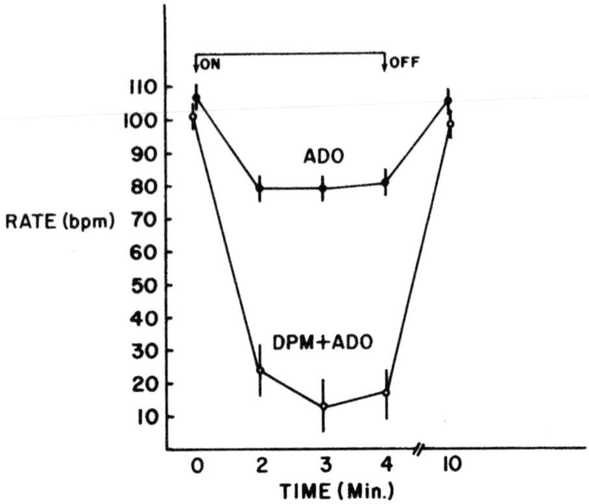

Figure 24.8. Potentiation of effect of adenosine (ADO) on His bundle rate by dipyridamole (DPM). ADO and DPM concentrations were 7.3×10^{-6} *M* and 5×10^{-8} *M*, respectively. Top line indicates start (on) and stop (off) of infusion of ADO.

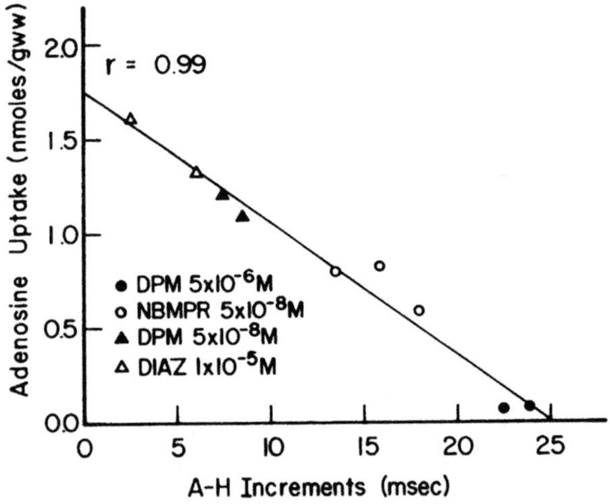

Figure 24.9. Correlation between degree of adenosine uptake and the observed change in the A-H interval in isolated guinea pig hearts. A-H increments obtained by subtracting the A-H interval values (msec) obtained with adenosine plus nucleoside transport blockers from those obtained with adenosine alone. Correlation is significant irrespective of the nucleoside blocker used or its concentration. DPM, dipyridamole; NBMPR, nitrobenzylthioinosine; DIAZ, diazepam. Source: Reproduced from [28] with permission.

Moreover, since the same electrophysiological effect can be demonstrated with different nucleoside transport blockers, it is unlikely that any diverse characteristic could be responsible for such results, other than a common action on adenosine transport.

The second line of evidence to suggest an external receptor for the actions of adenosine is based on the effects of impermeant analogues of adenosine. For example, the oligonucleotide adenyl (3′-5′)-adenosine $[AP]_9$ has a molecular weight of about 3,300 and is too large to enter the intact cell. This analogue has been shown to be about 2.6-fold more potent (per mole) than free adenosine in producing hyperpolarization in frog sinus venosus cells [6]. In the A-V node, the polymer $[AP]_9$-adenosine was approximately 1.8 times more potent per mole than adenosine in increasing A-V conduction time [28]. Additional evidence for an extracellular site of the action of adenosine in the A-V node is the lack of effect of the P site analogue 2′-deoxyadenosine [28]. The R site analogue N^6-methyladenosine was equipotent to adenosine in impairing A-V transmission, and its effects were also enhanced by dipyridamole. With respect to the direct effects of adenosine on the S-A and A-V nodes, present evidence strongly favors an extracellular receptor site. However, indirect effects, such as the antiadrenergic action of adenosine on the S-A and A-V nodes, could conceivably be mediated by an intracellular receptor site.

Antagonism Between Adenosine and Methylxanthines

Sattin and Rall [49] first reported that methylxanthines in micromolar concentrations inhibit the actions of adenosine in brain, and subsequent reports have confirmed a similar antagonism in other tissues [50–53]. However, methylxanthines are also known to inhibit phosphodiesterase, thereby enhancing the intracellular accumulation of cyclic AMP (cAMP) and release of catecholamines from the adrenal medulla and nerve terminals [51]. Thus, the pharmacological actions of methylxanthines can be attributed to several different mechanisms.

In heart, methylxanthines antagonize the vasodilation and the negative chronotropic and dromotropic actions of adenosine (Fig. 24.10) [6, 26, 28, 53–55]. In the A-V node, the antagonism of the action of adenosine by aminophylline seems to be selective, since the effects of Ach or $MnCl_2$ cannot be reversed by aminophylline [26]. Furthermore, the antagonism is competitive [28]. This finding is consistent with the observation that aminophylline competitively antagonizes the hyperpolarization of frog sinus venosus by ATP [6]. The adenosine-induced slowing of pacemaker activity of the His bundle is also antagonized in a dose-dependent manner by aminophylline (not shown). Caffeine [53] and aminophylline [23] attenuate the slowing of the sinus rate and the A-V block caused by injections of adenosine directly into the S-A and A-V node arteries, respectively. The ability of aminophylline to reverse the actions of adenosine in isolated guinea pig heart is not related to inhibition of phosphodiesterase and/or release of catecholamines [28]. In propranolol-

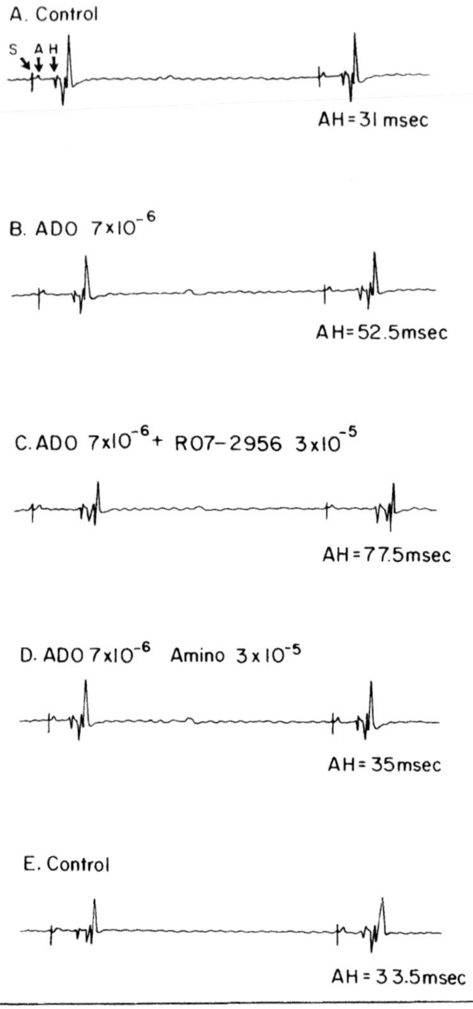

Figure 24.10. Adenosine-induced A-H prolongation in isolated guinea pig heart antagonized by aminophylline but not by a specific phosphodiesterase (PDE) inhibitor. A: Control, A-H = 31 msec. B: Adenosine, 7×10^{-6} *M;* A-H = 52.5 msec. C: Adenosine, 7×10^{-6} *M,* plus PDE inhibitor RO7-2956, 3×10^{-5} *M;* A-H = 77.2 msec. D: Adenosine, 7×10^{-6} *M,* plus aminophylline, 3×10^{-5} *M;* A-H = 35 msec. E: Control washout, A-H = 33.5 msec. Pacing frequency = 2.5 Hz.

treated hearts or reserpinized guinea pigs, aminophylline still antagonizes the effects of adenosine on A-V conduction [28]. Aminophylline in concentrations as high as $5 \times 10^{-5}M$ causes no release of norepinephrine. Furthermore, the compound RO7-2956 (Hoffman-La Roche) in concentrations known to inhibit phosphodiesterase and thus increase cAMP does not reverse the actions of adenosine on A-V conduction (Fig. 24.10) [54]. Aminophylline, 1.8×10^{-3} *M,* and theophylline, $2 \times 10^{-3}M,$ are required for half-maximal inhibition of

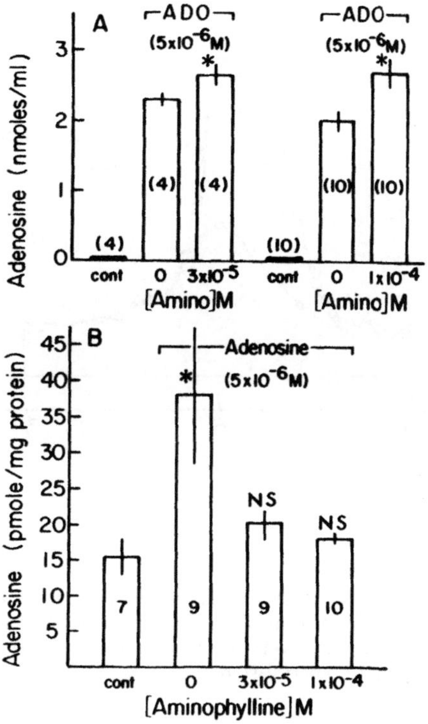

Figure 24.11. Effects of aminophylline (Amino) on effluent and tissue levels of adenosine (ADO) in isolated perfused guinea pig hearts. Each bar shows mean ± SEM for number of hearts indicated. A: ADO concentration in effluent during control (cont) and during infusion of 5×10^{-6} *M* ADO alone or combined with Amino. Statistical comparisons were based on paired data. B: Myocardial ADO concentration in control (cont) hearts and in hearts perfused with 5×10^{-6} *M* ADO and Amino. Statistical comparisons based on unpaired data. Note that Amino increased ADO concentration in the effluent, whereas it decreased myocardial ADO levels. $^{*}p < 0.05$.

phosphodiesterase in guinea pig hearts [55], a finding consonant with our recent observation that concentrations of aminophylline as high as $10^{-4}M$ did not cause any increase in cAMP, whereas $10^{-2}M$ aminophylline caused a significant accumulation of cAMP [28]. Thus, the concentrations of aminophylline required to produce an increase in myocardial cAMP far exceed the concentrations necessary for antagonizing the effects of adenosine. The coronary vascular effects of adenosine and impermeant theophylline covalently linked to oxidized stachyose have also been shown to be antagonistic and suggest that the antagonism between adenosine and theophylline occurs at an extracellular site. Consistent with these findings are the observations that reversal of the adenosine-induced increase in A-V conduction time is accompanied by an increased concentration of adenosine in the cardiac venous effluent and with a concomitant reduction in myocardial adenosine levels (Fig. 24.11). Since aminophylline does not affect adenosine uptake or release from

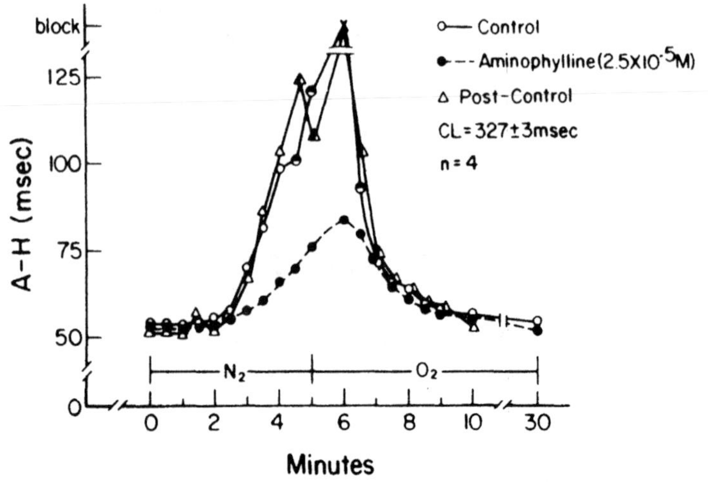

Figure 24.12. Time course of A-H interval during hypoxic perfusion and recovery. Control, aminophylline, and postcontrol runs were performed in order, with no intervening time. Aminophylline was infused from 3 min before the second hypoxic period until 5 min of the subsequent recovery period. Half-filled symbols signify that one or more preparations had developed second-degree heart block by that time. X signifies that all hearts had developed second-degree block. Measurements were made every 30 sec. n, number of preparations; CL, cycle length. Source: Reproduced from [26] with permission.

cardiac cells [56], it is reasonable to suggest that adenosine can be readily displaced from receptor sites by aminophylline.

Adenosine in Hypoxia- and Ischemia-Induced Disturbances of A-V Conduction

We recently presented evidence that adenosine may play a role as a biochemical mediator of the A-V conduction disturbances observed when myocardial oxygen availability is limited [23, 26]. In 1931, it was recognized by Bennet and Drury [57] that myocardial injury caused heart block when the perfusate effluent of the injured heart was utilized to perfuse a second heart in series. Myocardial injury, caused by burning the first heart with electric current, resulted in heart block in the second heart. Moreover, it was postulated that adenosine-like substances released from the injured heart were responsible for the A-V block. More recently, ATP released from ischemic cells was proposed as a candidate for the metabolic mediation of the early ischemic A-V block [58]. This hypothesis is based on the observation that ATP (1–10 mg/ml) infused into the A-V node artery or the anterior septal artery produces complete A-V block. The A-V block caused by ATP has been observed previously by others [59]. However, in the isolated guinea pig heart, present data indicate that ATP-induced prolongation of the A-H interval and the A-V block is a result of ATP degradation to adenosine [54]. This conclusion is based on three independent observations: (1) the less hydrolyzable ATP analogue, β-γ-methylene ATP, has essentially no effect on A-V conduction; (2) nucleoside transport blockers markedly potentiate the effects of ATP; and (3) aminophylline antagonizes the ATP-induced A-H prolongation.

Any proposed candidate for the biochemical mediation of the A-V conduction delay or block associated with hypoxia or ischemia should meet the following criteria: (1) the substance should be produced during hypoxia and ischemia and should attain a concentration sufficient to account for its effects; (2) the substance should increase in concentration at the site of action with a sufficiently fast time course to account for the effects of hypoxia and ischemia; (3) the substance should mimic the effects of hypoxia and ischemia; and (4) agents known to attenuate or potentiate the action of the substance should reduce or enhance the effects of hypoxia and ischemia, respectively. The evidence for the first two criteria is indirect and, hence, more speculative. The concentration of adenosine in the myocardium and interstitium is difficult to measure accurately, and thus a wide range of values has been reported [60–62]. However, when oxygen availability becomes limited, the adenosine levels rise sharply. From the values reported, we would anticipate that after 5 min of hypoxia, the myocardial adenosine levels would be high enough to yield an interstitial fluid concentration of about $10^{-5}M$ to $10^{-4}M$ [61, 62]. Adenosine at $10^{-5}M$ could account for the A-V conduction disturbances seen after 5 min of hypoxia in guinea pig heart. Whether or not adenosine is produced by the A-V nodal cells is not known. However, ventricular as well as atrial cells do produce adenosine in response to hypoxia and ischemia. With regard to the remaining two criteria, more conclusive evidence exists. As shown in Figures 24.5 and 24.12, ischemia, hypoxia, and asphyxia are known to impair A-V conduction [23, 26, 63–66]. In addition, A-V node action potentials are depressed by hypoxia, and, concomitantly, the A-H interval is markedly prolonged [65]. Thus, with regard to A-V conduction, adenosine mimics the effects of hypoxia and ischemia. Aminophylline, a competitive antagonist of the actions of adenosine, also attenuates the hypoxia- and ischemia-induced A-V conduction delay and block (Figures 24.10, 24.12). Conversely, dipyridamole enhances the nucleoside actions and potentiates the ischemia-induced A-V block [23].

CLINICAL IMPLICATIONS

Intravenous administration of adenosine in humans causes a transient sinus bradycardia and second-degree A-V block (Fig. 24.13) [67]. Recently, it was shown that when adenosine is injected directly into the human right coronary artery, from which the A-V node artery originates in 85% to 90% of hearts, it causes transient A-H prolongation and A-V block [27]. Complete A-V block in these patients was accompanied by transient ventricular standstill. Similar findings have been reported with intravenous injection of ATP, which is very effective in terminating supraventricular tachycardia (SVT) due to reentry in

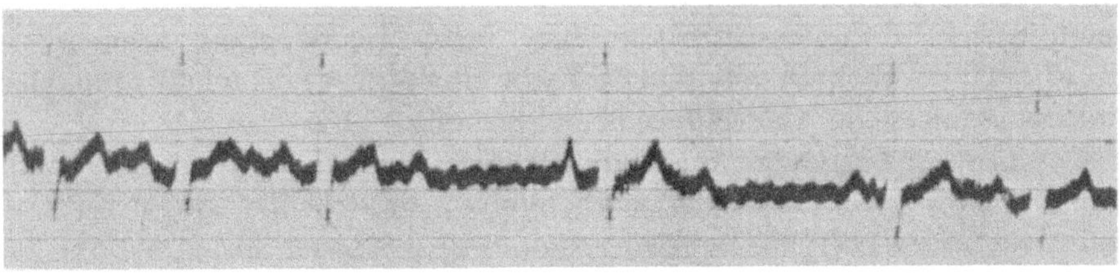

Figure 24.13. Adenosine-induced transient 2:1 A-V block in human heart. Record obtained after administration of 50 mg adenosine intravenously. Source: Reproduced from [67] with permission.

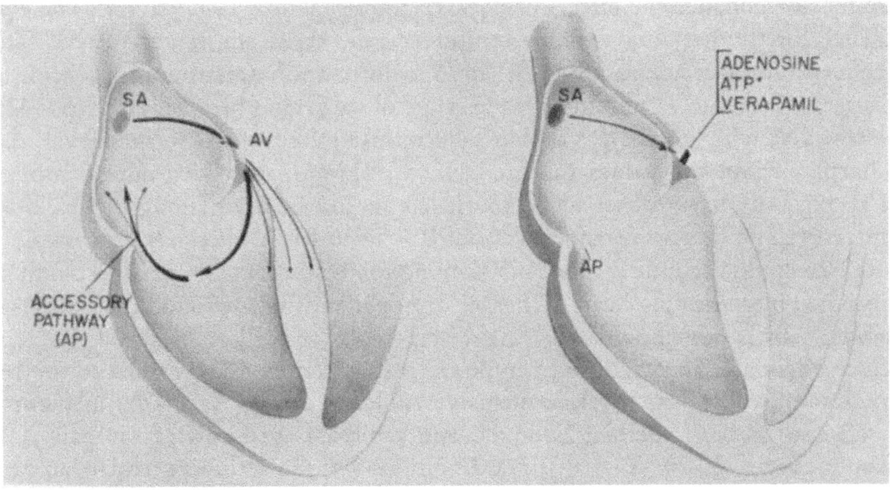

Figure 24.14. Possible mechanism of adenosine termination of supraventricular tachycardia when reentrant pathway includes the A-V node. *ATP is likely to be hydrolyzed to adenosine. See text for discussion.

the A-V node [68–70]. As discussed earlier, the effects of ATP are probably due to its degradation to adenosine. The mechanism whereby ATP and adenosine terminate SVT can be explained by their ability to increase A-V conduction time (A-H prolongation) like verapamil (Fig. 24.14). As advocated for ATP [70], adenosine could also serve as a diagnostic tool for determination of whether or not one of the limbs of the reentrant pathway during SVT includes the A-V node. In cardiologic practice, intravenous injection of ATP is now used to diagnose SVT of unclear type. During ATP-induced transient A-V block, atrial flutter or fibrillation can be unmasked on the body surface ECG recording and appropriate drug or electrotherapy can be undertaken [68, 70].

Since the human S-A and A-V nodes are sensitive to and respond to adenosine in a fashion similar to observations in animal experiments, it is conceivable that adenosine may mediate some of the rhythm and conduction (A-V) disturbances during myocardial infarction. For example, bradycardia, hypotension, and A-V block frequently occur in patients with acute inferior myocardial infarction. Reversal of these bradyarrhythmias by atropine has led investigators to attribute them to augmentation of vagal tone [71]. However, a subset of patients who develop A-V block 1 to 2 days after onset of symptoms does not respond to atropine [71]. In these patients, atropine may accelerate the sinus rate while simultaneously decelerating the A-V transmission rate to the ventricle. Finally, atropine may both increase sinus rate and accelerate A-V conduction, whether or not inferior myocardial infarction is present. Hence, the ability of atropine to reverse the bradyarrhythmias in the case of inferior myocardial infarction does not permit definitive conclusions regarding the mechanisms responsible for these disturbances. The release of negative dromotropic substances and transient hypoxia have also been considered to play a role [72, 73]. Thus, one should seriously consider the possibility that the pathophysiology underlying bradyarrhythmias and A-V block associated with inferior myocardial infarction may consist of two mechanisms: first, an increase in vagal activity, and second, release by the ischemic cells of a negative dromotropic metabolite. In human diaphragmatic infarction complicated by complete A-V block, the escape rhythm is characterized by a narrow QRS complex preceded by a His bundle spike [74]. This observation is quite similar to those made during adenosine- and ischemia-induced A-V block in isolated guinea pig hearts (Fig. 24.6).

In brief, regardless of the possible physiological role of adenosine in normal oxygenated hearts, it seems likely that endogenously released adenosine from hypoxic and ischemic cells may be responsible for some of the rhythm and A-V conduction disturbances accompanying myocardial infarction. Any agent capable of attenuating or potentiating the actions of adenosine may therefore lessen or exacerbate, respectively, the effects of myocardial oxygen deprivation. For example, if adenosine specifically causes acute ischemic A-V block, it might be prudent to consider a low dose infusion of a methylxanthine, an adenosine antagonist, to reverse the A-V block. Low doses of methylxanthines, or of an impermeant theophylline analogue, would possibly avoid undesirable ventricular arrhythmias that have been associated with elevated levels of cAMP [75]. Furthermore, the risks of introducing a pervenous cardiac pacing electrode would be avoided [76].

REFERENCES

1. Drury AN, Szent-Györgi A: The physiological activity of adenine compounds with special reference to their action upon the mammalian heart. *J Physiol (Lond)* 68:213–237, 1929.
2. Green HN, Stoner HB: The effect of purine derivatives on the cardiovascular system, in Green HN, Stoner HB (eds): *Biological Actions of the Adenine Nucleotides*. London, H.K. Lewis and Co, 1950, pp 65–103.

3. James TN: The chronotropic action of ATP and related compounds studied by direct perfusion of the sinus node. *J Pharmacol Exp Ther* 149:233–247, 1965.
4. Versprille A: The chronotropic effect of adenine and hypoxanthine derivatives on isolated rat hearts before and after removing the sinoauricular node. *Pfluegers Arch* 291:267, 1966.
5. Chiba S, Hashimoto K: Differences in chronotropic and dromotropic responses of the SA and AV nodes to adenosine and acetylcholine. *Jpn J Pharmacol* 22:273–274, 1972.
6. Hartzell HC: Adenosine receptors in frog sinus venosus: Slow inhibitory potential produced by adenosine compounds and acetylcholine. *J Physiol (Lond)* 293:23–49, 1979.
7. Szentmiklosi AJ, Nemeth M, Szegi J, Papp JG, Szekeres L: Effect of adenosine in sinoatrial and ventricular automaticity of the guinea pig. *Naunyn Schmiedebergs Arch Pharmacol* 311:147–149, 1980.
8. Rosen MR, Weis RM, Danilo P: The actions of adenosine on normal and abnormal cardiac impulse initiation. *Circulation* 64:1X-50, 1981 (abst).
9. Belardinelli L, West A, Crampton RS, Berne RM: Deceleration of atrial and His-bundle rhythmicity by adenosine in guinea pig heart. *Circulation* 64:IV-327, 1981 (abst).
10. Chiba S: Potentiation of the negative chronotropic and inotropic effects of adenosine by dipyridamole. *Tohoku J Exp Med* 114:45–48, 1974.
11. Kolassa N, Pfleger K, Tram M: Species differences in action and elimination of adenosine after dipyridamole and hexobendine. *Eur J Pharmacol* 13:320–325, 1971.
12. Furukawa Y, Chiba S: Inotropic and chronotropic responses to inosine in isolated and blood perfused dog atria. *Eur J Pharmacol* 67:339–345, 1980.
13. James TN, Nadeau RA: Direct perfusion of the sinus node: An experimental model for pharmacologic and electrophysiologic studies of the heart. *Henry Ford Hosp Med Bull* 10:21–25, 1962.
14. Noma A, Irisawa H: Membrane currents in the rabbit sinoatrial node cell as studied by double microelectrode method. *Pfluegers Arch* 364:45–52, 1976.
15. Hutter OF, Rankin AC: The effect of adenosine on potassium permeability of sinus venosus of heart studied using tracer flux measurement, in Berne RM, Rall TW, Rubio R (eds): *Regulatory Function of Adenosine*. The Hague, Martinus Nijhoff, 1983 (abst).
16. Jochem G, Nawrath H: Adenosine activates a potassium conductance in atrial heart muscle, in Berne RM, Rall TW, Rubio R (eds): *Regulatory Function of Adenosine*. The Hague, Martinus Nijhoff, 1983 (abst).
17. Vereecke J, Isenberg G, Carmeliet E: K efflux through inward rectifying K channels in voltage clamped Purkinje fibers. *Pfluegers Arch* 384:207–217, 1980.
18. Urthaler F, James TN: Effects of adenosine and ATP on AV conduction and on AV junctional rhythm. *J Lab Clin Med* 79:96–105, 1972.
19. Schrader J, Bauman G, Gerlach E: Adenosine as inhibitor of myocardial effects of catecholamines. *Pfluegers Arch* 372:29–35, 1977.
20. Billette J, Elharrar V, Porlier G, Nadeau RA: Sinus slowing produced by experimental ischemia of the sinus node in dogs. *Am J Cardiol* 31:331–335, 1973.
21. Baba N, Leighton RG, Weissler AM: Experimental cardiac ischemia. Observation of the sinoatrial and atrioventricular nodes. *Lab Invest* 23:168–178, 1970.
22. Carbonin P, DiGennaro M, Valle R, Weisz AM: Inhibitory effect of anoxia on reperfusion- and digitalis-induced ventricular tachyarrhythmias. *Am J Physiol* 240:H730–H737, 1981.
23. Belardinelli L, Mattos EC, Berne RM: Evidence for adenosine mediation of atrioventricular block in the ischemic canine myocardium. *J Clin Invest* 68:195–205, 1981.
24. Stafford A: Potentiation of adenosine and the adenine nucleotides by dipyridamole. *Br J Pharmacol Chemother* 28:218–277, 1966.
25. Drury AN: The physiological activity of nucleic acid and its derivatives. *Physiol Rev* 16:292–325, 1936.
26. Belardinelli L, Belloni FL, Rubio R, Berne RM: Atrioventricular conduction disturbances during hypoxia: Possible role of adenosine in rabbit and guinea pig heart. *Circ Res* 47:684–691, 1980.
27. Favale S, DiBiase M, Rizzo U, Belardinelli L, Rizzon P: Effeti dell'ATP i adenosina sulla conduzioni soprahissiana nell'uomo (in press).
28. Belardinelli L, Fenton R, West A, Linden J, Althaus J, Berne RM: Extracellular action of adenosine and the antagonism by aminophylline on the atrioventricular conduction in isolated perfused guinea pig and rat hearts. *Circ Res* 51:569–579, 1982.

29. Bauman G, Schrader J, Gerlach E: Inhibitory action of adenosine on histamine- and dopamine-stimulated cardiac contractility and adenylate cyclase in guinea pigs. *Circ Res* 48:259–266, 1981.
30. Benitez D, Mascher D, Alanis J: The electrical activity of the bundle of His. The fast and slow inward currents. *Pfluegers Arch* 345:61–72, 1973.
31. Zipes DP, Mendez C: Action of manganese ions and tetrodotoxin on atrioventricular nodal transmembrane potentials in isolated rabbit hearts. *Circ Res* 32:447–454, 1973.
32. Rubio R, Belardinelli L, Thompson CI, Berne RM: Cardiac adenosine: Electrophysiological effects, possible significance in cell function, and mechanisms controlling its release, in Baer HP, Drummond GI (eds): *Physiological and Regulatory Functions of Adenosine and Adenine Nucleotides*. New York, Raven Press, 1979, pp 167–182.
33. Schrader J, Rubio R, Berne RM: Inhibition of slow action potentials of guinea pig atrial muscle by adenosine: A possible effect on Ca^{2+} influx. *J Mol Cell Cardiol* 7:427–433, 1975.
34. Belardinelli L, Rubio R, Berne RM: Blockade of CA^{2+} dependent rat atrial slow action potentials by adenosine and lanthanum. *Pfluegers Arch* 380:19–27, 1979.
35. Olsson RA, Davis CJ, Khouri EM, Patterson RE: Evidence for an adenosine receptor on the surface of dog coronary myocytes. *Circ Res* 39:93–98, 1976.
36. Schrader J, Nees S, Gerlach E: Evidence for a cell surface adenosine receptor on coronary myocytes and atrial muscle cells. *Pfluegers Arch* 369:251–257, 1977.
37. Schwabe U, Trost T: Characterization of adenosine receptors in rat brain by $(-)$-[^{3}H]N^6-phenylisopropyladenosine. *Naunyn Schmiedebergs Arch Pharmacol* 313:179–187, 1980.
38. Trost T, Schwabe U: Adenosine receptors in fat cells: Identification by $(-)$-N^6-[^{3}H]phenylisopropyladenosine binding. *Mol Pharmacol* 19:228–235, 1981.
39. Dutta P, Mustafa SJ: Saturable binding of adenosine to the dog heart microsomal fraction: Competitive inhibition by aminophylline. *J Pharmacol Exp Ther* 211:496–501, 1979.
40. Kolassa N, Pfleger K, Rumel W: Specificity of adenosine uptake into the heart and inhibition of dipyridamole. *Eur J. Pharm* 9:265–268, 1970.
41. Paterson ARP: Adenosine transport, in Baer HP, Drummond GI (eds): *Physiological and Regulatory Functions of Adenosine and Adenine Nucleotides*. New York, Raven Press, 1979, pp 305–313.
42. Schütz W, Schrader J, Gerlach E: Different sites of adenosine formation in the heart. *Am J Physiol* 240:H963–H970, 1981.
43. Clanachan AS, Marshall RJ: Therapeutic concentrations of diazepam potentiate the effects of adenosine on isolated cardiac and smooth muscle. *Br J Pharmacol* 68:148P-149P, 1980 (abst).
44. Deuticke B, Gerlach E: Kompetitive Hemmung der adenosin-desaminase als mögliche ursache der coronardilatierenden wirkung einer pyrimidopyrimidin-verbindung. *Naunyn Schmiedebergs Arch Pharmacol* 255:107–199, 1966.
45. Kübler W, Spieckermann PG, Bretschneider HJ (1970): Influence of dipyridamole (Persantin) on myocardial adenosine metabolism. *J Mol Cell Cardiol* 1:23–28.
46. Olsson RA, Snow JA, Gentry MK, Frick GP: Adenosine uptake by canine heart. *Circ Res* 31:767–778, 1972.
47. de Jong JW, Kiejzer E: Dipyridamole affects myocardial adenosine kinase activity. *Arch Int Physiol Biochim* 87:525–532, 1979.
48. Cohen S, Rotmensch H, Rubinstein R, Lass Y: Lack of uptake or degradation of adenosine in the termination of its action in the beating carp atrium. *Biochem Pharmacol* 30:890–893, 1981.
49. Sattin A, Rall TW: The effect of adenosine and adenine nucleotides on the cyclic adenosine 3′,5′-phosphate content of guinea pig cerebral cortex slices. *Mol Pharmacol* 6:13–23, 1970.
50. Fredholm BB: Are methylxanthine effects due to antagonism of endogenous adenosine? *Trends Pharmacol Sci* 1:129–132, 1980.
51. Rall TW: Central nervous system stimulants, in Gilman AG, Goodman LS, Gilman A (eds): *The Xanthines in the Pharmacological Basis of Therapeutics*. New York, Macmillan, 1981, pp 592–607.
52. Bunger R, Haddy FJ, Gerlach E: Coronary responses to dilating substances and competitive inhibition by theophylline in the isolated perfused guinea pig heart. *Pfluegers Arch* 358:213–214, 1975.
53. Chiba S, Kubota K, Hashimoto K: Inhibition of the negative chronotropic action of adenosine by caffeine in the dog. *Eur J Pharmacol* 21:281–285, 1973.
54. West A, Belardinelli L, Berne RM: Effects of purines on the AV node of isolated guinea pig

hearts, in Berne RM, Rall TW, Rubio R (eds): *Regulatory Function of Adenosine*. The Hague, Martinus Nijhoff Publishers, 1983 (abst).
55. Kukovetz WR, Poch G, Wurm A: Quantitative relations between cyclic AMP and contraction as affected by stimulators of adenylate cyclase and inhibitors of phosphodiesterase. *Adv Cyclic Nucleotide Res* 5:395–414, 1975.
56. Mustafa SJ, Berne RM, Rubio R: Adenosine metabolism in cultured chick embryonic heart cells. *Am J Physiol* 228:1474–1478, 1975.
57. Bennet DW, Drury AN: Further observations relating to the physiological activity of adenine compounds. *J Physiol* 72:288–320, 1931.
58. James TN, Bear ES, Frink RJ, Urthaler F: Pharmacologic production of atrioventricular block with and without initial bundle branch block. *J Pharmacol Exp Ther* 179:338–346, 1971.
59. Gillespie JH: The biological significance of the linkages in adenosine triphosphoric acid. *J Physiol (Lond)* 80:345–359, 1934.
60. Arch JRS, Newsholme EA: The control of the metabolism and the hormonal role of adenosine. *Essays Biochem* 14:82–123, 1978.
61. Thomas RA, Rubio R, Berne RM: Comparison of the adenine nucleotide metabolism of dog atrial and ventricular myocardium. *J Mol Cell Cardiol* 7:115–123, 1975.
62. Rubio R, Wiedmeier VT, Berne RM: Relationship between coronary flow and adenosine production and release. *J Mol Cell Cardiol* 6:517–566, 1974.
63. Zumino AZP, Parisi IM, DeCeretti ERP: Effect of ischemia and low-sodium medium on atrioventricular conduction. *Am J Physiol* 218:1489–1494, 1970.
64. Bagdonas AA, Stuckey JH, Plera J, Amer NS, Hoffman BF: Effects of ischemia and hypoxia on the specialized conducting system of the canine heart. *Am Heart J* 61:206–218, 1961.
65. Senges J, Mizutani T, Pelzer D, Brachmann J, Sonnhof U, Kübler W: Effect of hypoxia on the sinoatrial node, atrium and atrioventricular node in the rabbit heart. *Circ Res* 44:856–863, 1979.
66. Alanis J, Lopez E, Mandoki JJ, Pilar G: Propagation of impulses through the atrioventricular node. *Am J Physiol* 196:1171–1174, 1959.
67. Honey RM, Ritchie WT, Thomson WAR: The action of adenosine upon the human heart. *Q J Med* 23:485–489, 1930.
68. Motté G, Waynberger M, Lebars A, et Bouvrain, Y: L'adénosine triphosphorique dans les tachycardies paroxystiques. Intérêt diagnostique et thérapeutique. *Nouv Presse Med* 1:3057–3061, 1972.
69. Leclerq JF, Coumel P: Les effets de l'adenosine triphosphate (ATP) sur le noeud sinusal et le noeud auriculo-ventriculaire chez l'homme: Variations selon le lieu d'injection. *Coeur Med Interne* 17:541–546, 1978.
70. Perrot B, Faivre G: Effects of adenosine triphosphate on the accessory pathways. *Circulation* 64:IV-544, 1981 (abst).
71. Adgey AAJ, Geddes JS, Mulholland HC, Keegan DAJ, Pantridge JF: Incidence, significance, and management of early bradyarrhythmia complicating acute myocardial infarction. *Lancet* 2:1097–1101, 1968.
72. James TN: Evidence for adrenergic alpha receptor depressant activity in the heart. *Am J Physiol* 215:1366–1375, 1968.
73. Sutton R, Davies M: The conduction system in acute myocardial infarction complicated by heart block. *Circulation* 38:987–992, 1968.
74. Rosen KM, Loeb HS, Chuquima R, Sinno MZ, Rahimtoola SH, Gunnar RM: Site of heart block in acute myocardial infarction. *Circulation* 42:925–933, 1970.
75. Opie LH, Nathan D, Lubbe WF: Biochemical aspects of arrhythmogenesis and ventricular fibrillation. *Am J Cardiol* 43:131–148, 1979.
76. Austin JL, Preis LK, Crampton RS, Beller GA, Martin RP: Analysis of pacemaker malfunction and complications of temporary pacing in the coronary care unit. *Am J Cardiol* 49:301–306, 1981.

DISCUSSION

DR. G. MISTRY: Dr. Belardinelli's data showed the effect of catecholamines on Ca^{2+} current. Has Dr. Dobson thought of looking at phosphorylation or dephosphorylation of sarcolemma proteins that may be responsible for Ca^{2+} currents? Adenosine may be inhibiting catecholamine inotropic effects by acting on ion fluxes through the sarcolemma.

DR. J.G. DOBSON: Yes, we have considered the possibility and are currently working on it. We also plan to see if adenosine influences the catecholamine-induced phosphorylation of membrane proteins.

DR. D. GERMAN: (1) For the treatment of Wolfe-Parkinson-White syndrome, are you using surgical treatment such as His bundle or A-V node ablation? (2) Have you measured adenosine levels in the effluent from the perfused heart of anesthetized dogs during arrhythmias induced by pacing?

DR. BELARDINELLI: (1) No. (2) No, not yet.

DR. H.J. LEIGHTON: Does adenosine have a role in critical care for arrhythmias?

DR. BELARDINELLI: It is too premature to predict whether adenosine or one of its analogues will have any role in critical care for arrhythmias.

DR. LEIGHTON: Which aspects of adenosine are similar to calcium channel antagonists and which are different?

DR. BELARDINELLI: The effects of adenosine and Ca^{2+} channel blockers in the A-V node are very similar, but this does not mean that the mechanism is the same. In the ventricular myocardium, adenosine appears to only antagonize the effects of β-adrenergic agents, whereas Ca^{2+} blockers depress contractility in the absence or presence of isoproterenol. In the atrial muscle, according to Dr. Nawrath, adenosine increases potassium permeability.

DR. R. VON BORSTEL: Doses of theophylline as low as 6 mg/kg in intact rats can produce tachycardia, which can be prevented by pretreatment with reserpine, hexamethonium, and propranolol. (This dose of theophylline results in blood levels of about $10^{-9}M$). This may represent (or might not) an antagonism of adenosine either in the CNS, sympathetic ganglia, or at active sympathetic terminals. I just want to point out that methylxanthines (and possibly adenosine) can act on the heart by fairly indirect mechanisms, which should, however, be taken into account in assessments of physiological significance. Also, I have found that it takes arterial adenosine levels of about 10 μM (elevated by intravenous infusion and monitored by sampling arterial blood for adenosine assay) to induce significant bradycardia in conscious, unrestrained, catheterized rats.

DR. BELARDINELLI: I do agree that it is also possible that the site of the antagonism between adenosine and theophylline could be the CNS, sympathetic ganglia, or sympathetic terminals. However, I am unaware of any clear evidence for such sites of action. In regard to your second comment, it is difficult to compare doses of adenosine necessary to produce bradycardia or A-V block in isolated perfused hearts with conscious, unrestrained, catheterized rats.

DR. G. RABERGER: (1) Could you please comment on the effects of ATP and adenosine given to the patients? (2) Is the bradycardia increased after the hypotension phase of adenosine by an increased sympathetic tone?

DR. BELARDINELLI: (1) In brief, the A-V block caused by adenosine is very transient, i.e., it lasts 25 to 50 sec. In general, the A-V block is accompanied by transient (80 sec) ventricular arrest. These effects are dose-dependent. (2) It is very likely that the tachycardia after ATP injection is caused by increased sympathetic tone due to the hypotension. That is the reason why we injected 20 μg or less adenosine directly into the right coronary artery. With such small amounts of the nucleoside, hypotension is not seen, and thus a late tachycardia does not occur.

DR. DOBSON: How low a concentration of adenosine have you used to detect a decrease in the inward current elicited by catecholamine stimulation of your voltage-clamped isolated cell experiments?

DR. BELARDINELLI: The lowest concentration of adenosine used was $2 \times 10^{-5}M$.

DR. F.J. HADDY: Compare the threshold concentration for a dromotropic effect with the threshold concentration for coronary vasodilation.

DR. BELARDINELLI: The threshold concentration for a dromotropic effect of adenosine is about 10 times greater than that for coronary vasodilation. However, if one lowers the basal catecholamine levels (e.g., with reserpine) the sensitivity of the A-V node is greatly increased. Thus, the presence of catecholamines may lead to underestimation of the true threshold concentration.

DR. H. ROTMENSCH: With regard to ATP on the A-V node, the limited experience I have had with patients in Israel is that intravenous ATP has a short-lasting (20–30 sec) effect on A-V block. ATP has been used in the diagnosis as well as in the treatment, i.e., conversion of supraventricular tachycardia (SVT). The patient usually experiences severe transient reduction of blood pressure with syncopal features. A certain percentage will convert; in others, the diagnosis of SVT or VT can be ascertained.

DR. BELARDINELLI: Thank you for your comment. Your experience is similar to that of others who are using ATP or adenosine to terminate SVT.

25. ADENOSINE AND RENAL FUNCTION

H. OSSWALD

This review on the role of adenosine in renal function is divided into three parts. In the first part, the action of exogenous adenosine on renal blood flow, glomerular filtration rate, and renin secretion is considered. In the second part, evidence that renal tissue levels of adenosine correlate inversely with those of ATP is presented. The third part describes adenosine as the metabolic mediator in the kidney that regulates glomerular filtration rate and renal renin secretion in response to the sodium chloride load of the thick ascending limb of Henle's loop.

EXOGENOUS ADENOSINE

The first observation of a renal vasoconstriction induced by adenosine was reported in 1929 by Drury and Szent-Györgyi [1]. This finding was confirmed by Thurau in 1964 [2] and later by other investigators [3–7]. Following intrarenal injection of adenosine, the renal vasculature responds immediately with marked constriction, as shown in Figure 25.1. However, during continuous intraarterial infusion of adenosine, renal blood flow decreases initially and then returns to preinfusion levels within 2–5 min. Following cessation of adenosine infusion, there is a short-lasting increase of renal blood flow (see also Fig. 25.4). Although renal vascular resistance is unchanged or even reduced during adenosine infusion, glomerular filtration rate is decreased by 10% to

The author's research mentioned in this chapter was supported by the Deutsche Forschungsgemeinschaft, Os 42/1–6. The secretarial assistance of Ms. E. Abrahams is gratefully acknowledged.

R.M. Berne, T.W. Rall, and R. Rubio (eds.), Regulatory Function of Adenosine, *pp. 399–415.*

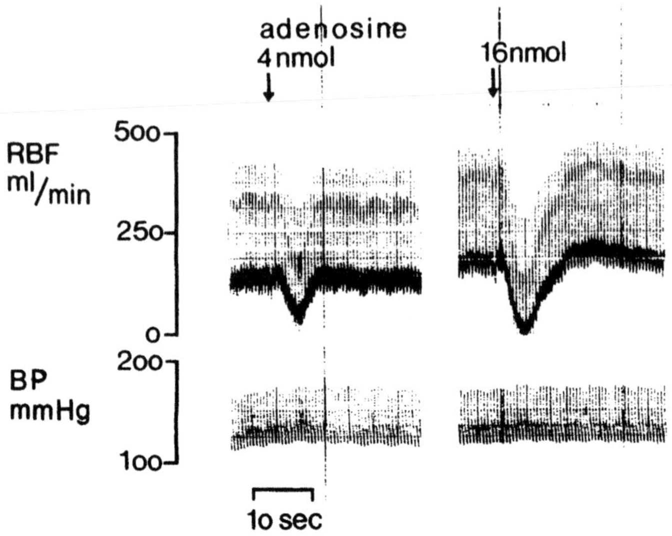

Figure 25.1. Original tracing of renal blood flow (RBF) and arterial blood pressure in a dog. Injections of 8 or 16 nmol of adenosine into the renal artery result in an immediate and short-lasting fall of RBF. Note the complete stop of flow at diastolic pressures when 16 nmol of adenosine were injected. Arterial pressure is not affected by adenosine [5].

20% [5, 8]. Micropuncture studies in dogs [8] and rats [9] demonstrated that adenosine induces sustained afferent arteriolar vasoconstriction of the outer cortex (Fig. 25.2). Since whole kidney vascular resistance did not change, it was concluded that the vasculature of the deep cortex dilates in response to intrarenal adenosine infusion. In fact, deep cortical vasodilation due to adenosine was demonstrated by Spielman et al. [10], using radioactive labeled microspheres in dogs.

In this situation, the question arises of what the physiological response of the renal vasculature is to adenosine. To answer this question, one should keep in mind that single injections of adenosine unequivocally induce vasoconstriction. Also, endogenous adenosine, when transiently released by the kidney, induces renal vasoconstriction (see later discussion). At present, we do not know whether deep cortical vasodilation due to exogenous adenosine can also be elicited by endogenous adenosine. Under certain conditions, exogenous adenosine-induced vasoconstriction does not wane but persists during the entire infusion time in dogs. One such condition is elevation of ureteral pressure to 65 mm Hg [Spielman and Osswald, 1981, unpublished observation]; the other is furosemide-induced diuresis [11; unpublished observation].

Therefore, we can state that the vascular response of the kidney to intraarterial infusion of adenosine is characterized by outer cortical vasoconstriction and deep cortical vasodilation, and the deep cortical dilation is variable and probably controlled by other unknown factors.

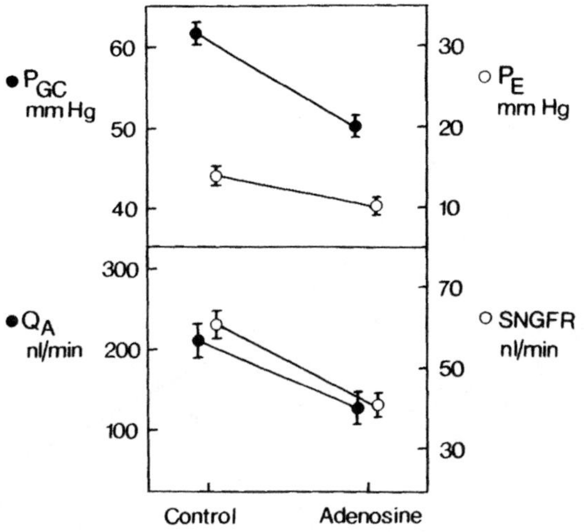

Figure 25.2. Upper half: Hydrostatic pressures in glomerular capillaries (P_{GC}) and in the star vessels of the peritubular circulation (P_E) of anesthetized dogs under control conditions and during continuous infusion of 0.3 μmol/min adenosine into the renal artery. Lower half: Glomerular plasma flow (Q_A) and single nephron glomerular filtration rate (SNGFR) before and during adenosine infusion in the same dogs in which the pressures were obtained. The calculated resistances of the afferent arterioles increased from 1.38 ± 0.25 to $3.04 \pm 0.43 \cdot 10^{10}$ dynes · sec · cm^{-5}, $p < 0.003$ [8].

The hemodynamic effect of adenosine on the kidney is dependent on the dietary status of the animals; sodium chloride-loaded and therefore renin-depleted rats and dogs do not respond with a decrease in renal blood flow or glomerular filtration rate, whereas sodium chloride-restricted animals with high plasma renin activity show an enhanced response to adenosine [12, 13]. Figure 25.3 shows the individual data of the adenosine-induced fall in glomerular filtration rate obtained in 19 dogs. A similar correlation was found in rats [12]. These results suggest that the vasoconstrictive action of adenosine is indirect and somehow related to the plasma renin activity. It is possible that adenosine constricts the resistance vessel in the kidney via angiotensin II (AII). Therefore, we carried out experiments in sodium-depleted dogs using a specific antagonist of AII to inhibit adenosine-induced renal vasoconstriction. It was found that the AII antagonist Sar^1-Ile^8-angiotensin II in a dose of 3 μg/min intrarenally—sufficient to block the vasoconstrictive action of high concentrations of circulating AII—did not affect adenosine-induced vasoconstriction. When, however, the dose of the AII antagonist was increased to 50 μg/min, adenosine-induced vasoconstriction was attenuated by a factor of 20 [14]. Figure 25.4 shows an original tracing of the renal blood flow response to adenosine before and during intrarenal infusion of the AII antagonist Sar^1-Ile^8-angiotensin II. Since the AII antagonist doses necessary to block circulating AII are distinctly different from those required to block adenosine-induced vaso-

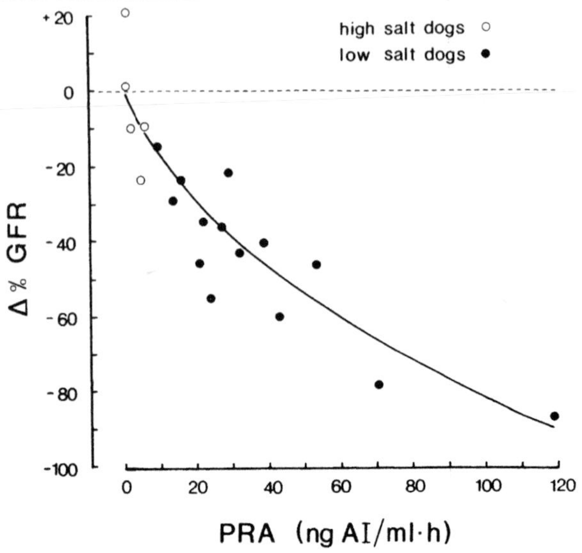

Figure 25.3. Correlation between plasma renin activity (PRA) and the decrease in glomerular filtration rate (GFR) expressed as percentage change compared to the preinfusion GFR (Δ%) produced by adenosine (0.3 μmol/min intrarenally) in 14 sodium chloride-depleted and 5 sodium chloride-loaded dogs [13].

constriction, it was concluded that adenosine acts to constrict the kidney by an intrarenal angiotensin mechanism [14]. The nature of this mechanism, however, is not known at present.

Besides its effects on renal hemodynamics, adenosine inhibits renin secretion. Tagawa and Vander [3] were the first to report that adenosine compounds lower plasma renin activity in dogs. In sodium-depleted rats, infusion of adenosine into the aorta above the renal arteries reduced net renin secretion rate by more than 60% [12]. This effect of adenosine on the kidney appears to occur at much lower concentrations than the hemodynamic effects [15]. In addition, it has been shown that adenosine inhibits release of norepinephrine from intrarenal nerve terminals [16]. Whether this effect is related to the inhibitory action of adenosine on renin release remains to be determined.

The effects of adenosine on renal vasculature, renal renin secretion, and norepinephrine release can be antagonized by theophylline [5, 15, 16]. The half-maximal inhibition of the adenosine effects can be seen at theophylline concentrations of about 2–5 μmol/l. The type of antagonism seems to be competitive in nature [5]. Therefore, theophylline may be used as a tool to evaluate the actions of endogenously generated adenosine.

ENDOGENOUS ADENOSINE

The metabolic breakdown of ATP might proceed via AMP, IMP, and inosine to uric acid. It could also be that AMP is dephosphorylated to adenosine,

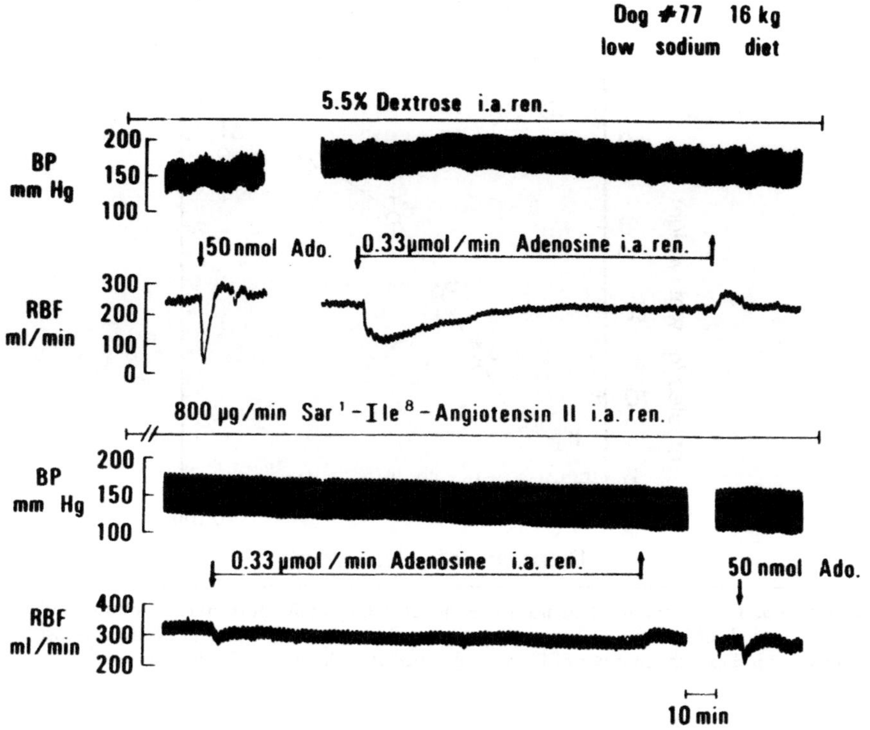

Figure 25.4. Original tracing of renal blood flow (RBF) and arterial blood pressure in a sodium chloride-depleted dog. Note the nearly complete shutdown of RBF following 50 nmol of adenosine injected into the renal artery (upper panel). When 800 µg/min of the AII antagonist were infused into the renal artery (lower panel), adenosine-induced vasoconstriction was nearly abolished [13].

which then is deaminated to inosine. Application of sensitive methods demonstrated that adenosine is present in normoxic rat kidneys with a tissue content of 5 nmol per gm of wet weight [17]. Kidneys were clamp-frozen in liquid nitrogen and acid-extracted. Adenosine, inosine, and hypoxanthine were separated by charcoal adsorption, and thin-layer chromatography. The nucleoside was measured enzymatically in a spectrometer [17]. As shown in Figure 25.5, the adenosine tissue levels increased 3-fold within 30 sec of ischemia and reached a value of 38.1 ± 6.3 nmol per gm of wet weight after 10 min of ischemia. Tissue levels of inosine and hypoxanthine were also elevated by ischemia in the rat kidney [17]. Adenosine is present under normoxic conditions and also accumulates following ischemia in kidneys of dogs and cats [18] as well as in humans [unpublished observation]. These results demonstrate that ATP hydrolysis in the kidney can proceed via AMP dephosphorylation to generate adenosine.

A method for studying the relation between metabolism and hemodynamics in an organ is the interruption of arterial blood supply and subsequent observa-

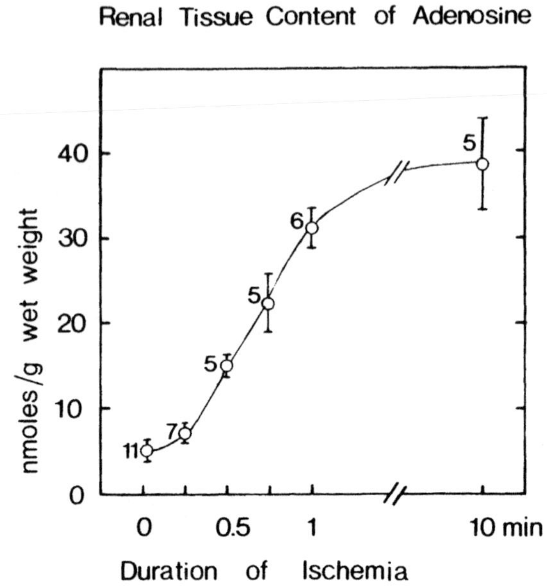

Figure 25.5. Tissue content of adenosine in the rat kidney under normoxic conditions and during brief periods of ischemia. The data are expressed as mean values ± SE. Numbers close to the mean value indicate the number of kidneys clamped at each period of ischemia.

tion of the postocclusive blood flow. When this method is applied to the kidney, a postocclusive vasoconstriction can be observed. This renal response to ischemia is unique, since all other organs so far studied exhibit only a postocclusive hyperemia. The first observation of a "no-reflow" phenomenon of the kidney was made by Bier in 1898 [19]. Later, a postocclusive reduction of renal blood flow was found by several investigators [17, 20–23]. However, this response was variable, and postocclusive hyperemia was also observed [24–26]. Spielman and Osswald [7] found a reproducible postocclusive ischemia in cats when renal prostaglandin synthesis was inhibited by meclofenamate.

What is the mechanism whereby the kidney vasculature responds with constriction following brief periods of ischemia? Before trying to answer this question, it seems necessary to remember the basic requirements of renal blood flow regulation in relation to renal metabolism. The metabolic theory of local blood flow regulation assumes that a vasoactive metabolite (or more than one) generated from the energy metabolism of the organ acts on the vasculature to adapt the oxygen and substrate supply to the specific functional demands of the organ. However, direct metabolic control of renal blood flow and glomerular filtration rate has been considered unlikely to exist, as stated in the *Handbook of Physiology* by Renkin and Gilmore [27]. In the kidney, most of the oxygen supply is consumed for tubular sodium reabsorption. If an increase

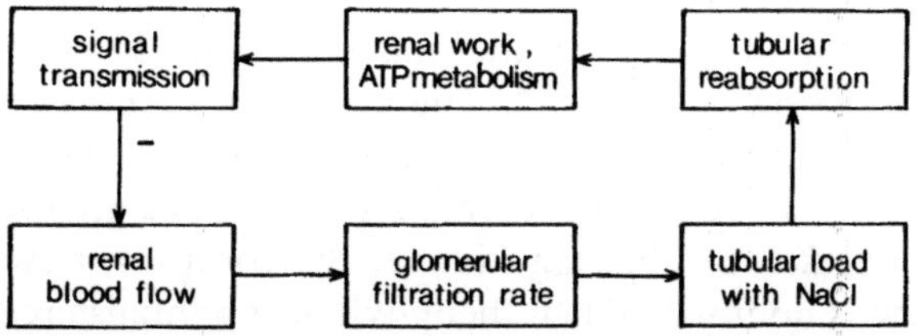

Figure 25.6. Schematic diagram of the interrelationship between renal blood flow and tubular sodium chloride reabsorption. Note that if renal metabolism would produce a vasodilating metabolite, a positive feedback would be initiated.

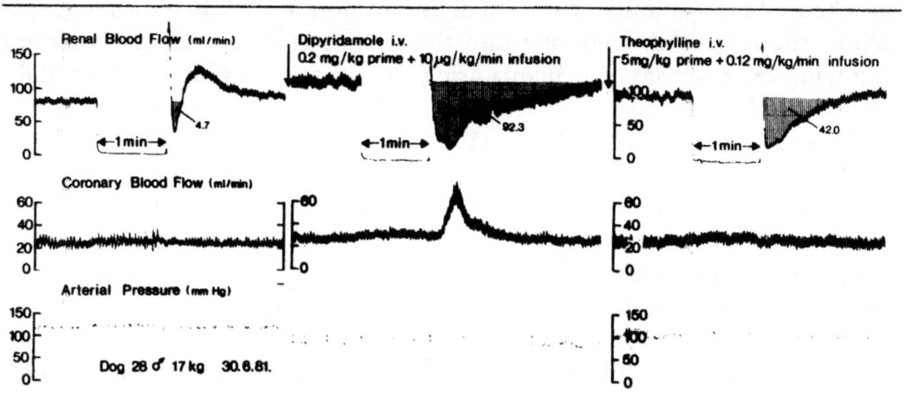

Figure 25.7. Original tracing of renal blood flow, coronary blood flow (ramus anterior descendens) and arterial blood pressure of a dog on a normal diet. On the left of the figure, one can see the renal response to 1 min renal artery occlusion. No response of coronary blood flow or arterial blood pressure can be observed. In the middle of the figure, the renal response to 1 min renal artery occlusion is shown during continuous infusion of dipyridamole. Note the increase of flow debt of the kidney (shaded area) and the steep increase in coronary blood flow when renal blood flow is minimal. On the right side of the figure, one can see the inhibition of the postocclusive ischemia of the kidney and the complete blockade of coronary dilation by theophylline.

of sodium reabsorption would lead to vasodilation, a positive feedback would be initiated: increased renal blood flow → increased glomerular filtration → increased tubular load with NaCl → increased tubular sodium reabsorption → increased metabolism (Fig. 25.6). Such a positive feedback would in no way contribute to the known constancy of renal blood flow or glomerular filtration. If we introduce, however, the idea of a metabolic vasoconstriction and the hypothesis that adenosine is the mediator of such a vasoconstriction controlled by the metabolic rate of the kidney, we might understand the postocclusive ischemia.

The experiment shown in Figure 25.7 provides evidence that adenosine mediates the postocclusive vasoconstriction in the kidney. A rather typical mixed response to 1 min renal artery occlusion is shown on the left (Fig. 25.7). Only a short period of vasoconstriction can be seen. The following hyperemic phase is less pronounced when the renal artery is occluded for only 30 sec. When the uptake of adenosine by the cells was inhibited by dipyridamole in this dog, arterial blood pressure fell. The vascular response of the kidney to 1 min arterial occlusion shows a dramatic enhancement of the vasoconstriction phase. No hyperemia could be observed. As a result of a prolonged half-life of the adenosine released by the dog kidney, the coronary vessels dilate when renal blood flow is minimal. When theophylline is administered in addition to the continuous infusion of dipyridamole, arterial blood pressure increases, the postocclusive ischemia in the kidney is reduced by half, and the coronary dilation is completely blocked. Such vascular responses of the kidney and the heart as shown in Figure 25.7 can be easily reproduced in pentobarbital-anesthetized mechanically ventilated dogs on a normal diet. In the same preparation, the release of adenosine into the venous effluent was demonstrated following brief periods of ischemia using an adenosine radioligand assay [28]. The postocclusive reduction of renal blood flow was also observed in rats and could be blocked by theophylline [17]. In cats, two inhibitors modified the renal response to brief periods of ischemia and to intrarenal adenosine injection in the same direction: meclofenamate potentiated the postocclusive ischemia and the adenosine-induced vasoconstriction as well [7]. However, a specific angiotensin II antagonist blocked the adenosine-induced fall of renal blood flow as well as the postocclusive vasoconstriction [14].

Taken together, the available data provide evidence that intrarenal-generated adenosine mediates the postocclusive ischemia in the kidney. This conclusion supports the concept of a metabolic vasoconstriction in the kidney.

To further test such a hypothesis, we carried out experiments designed to increase the reabsorptive work of the kidney [29]. Hypertonic sodium chloride (1.0 mol/l) solution was infused into the thoracic aorta at a rate of 0.275 ml/min to elevate sodium concentration in the plasma flowing into the kidney to 160–180 mmol/l. This maneuver increased the tubular sodium load and thus the reabsorptive work of the kidney. The results are shown in Figure 25.8. Under control conditions, tubular sodium reabsorption was 64.5 ± 5 μmol/min · 100 gm of body weight. The corresponding adenine nucleotide tissue content in μmoles per gm of wet weight was (mean ± S.E.) ATP, 2.55 ± 0.05; ADP, 0.60 ± 0.09; and AMP, 0.25 ± 0.06, respectively. The adenosine tissue content was 5.69 ± 1.2 μmol per gm of wet weight. Following hypertonic sodium chloride infusion for 10 min (n = 8), tubular sodium reabsorption increased to 82.8 ± 5.9 μmol/min · 100 gm, whereas ATP tissue levels were lowered to 1.64 ± 0.1 mol per gm of wet weight, those of ADP were 0.47 ± 0.04, and those of AMP 0.22 ± 0.05 μmol per gm of wet weight, respectively. Adenosine tissue content was slightly elevated to 7.69 ± 1 nmol per gm of wet weight. To further increase the ability of the kidney to reabsorb the high tubular load of sodium chloride, we repeated the experiments in rats that were initially sodium-depleted by 50 mg/kg furosemide and then kept on a low sodium diet for 10–15 days (daily sodium intake was approximately 0.7

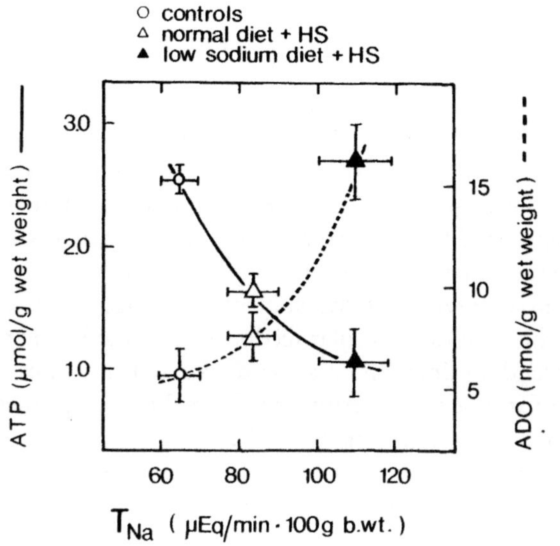

Figure 25.8. Tissue levels of ATP (left ordinate, connected with the solid line) and of adenosine (right ordinate, broken line) in relation to the corresponding tubular sodium reabsorption rate (abscissa) of the left kidney in control rats (n = 6), open circles; in rats infused with hypertonic sodium chloride into the thoracic aorta (n = 8), open triangles; and in rats on a low sodium diet infused with hypertonic sodium chloride (n = 7), solid triangles.

mmol), a condition associated with hyperaldosteronism. In this group (n = 7), tubular sodium transport rose to 109 ± 9 µmol/min · 100 gm because of the hypertonic sodium chloride. ATP tissue content was further lowered to 1.05 ± 0.26 µmol per gm of wet weight (ADP, 0.43 ± 0.05; AMP, 0.25 ± 0.05). Adenosine tissue levels were elevated 3-fold above control values to 16.2 ± 1.3 nmol per gm of wet weight. These data demonstrate an inverse relationship between renal tissue levels of ATP and those of adenosine under conditions of increased tubular sodium reabsorption.

The vascular response of the kidney to hypertonic NaCl was studied in a dog preparation in which systemic effects of the hypertonic solution are reduced. Infusion of hypertonic sodium chloride into the renal artery induced sustained renal vasoconstriction [30]. In addition to the vasoconstriction, hypertonic saline reduced renal renin secretion rate. Spielman and co-workers [15] provided evidence that the renal response to hypertonic saline in dogs is mediated by adenosine. Two observations lead to this conclusion: first, the effects of hypertonic saline (i.e., a fall in glomerular filtration rate and a reduction in renin secretion rate) could be completely blocked by intrarenal infusion of theophylline (5 µmol/min); second, the effects of hypertonic saline could be mimicked by intrarenal infusion of adenosine, which also could be blocked by addition of theophylline [15].

The results of the experiments using hypertonic saline to increase the reabsorptive work of the kidney support the concept of an adenosine-mediated mechanism that couples metabolism to hemodynamics and renin secretion.

ADENOSINE AND TUBULOGLOMERULAR FEEDBACK

The change of single nephron glomerular filtration rate (SNGFR) in response to variations in the loop of Henle flow rate—the tubuloglomerular feedback—was first demonstrated by Schnermann [31]. Thus far, many specific details of this feedback mechanism have been described, and there is a recent excellent and comprehensive review of the subject by Wright and Briggs [32]. The sensing site in the tubuloglomerular feedback seems to be the region at the end of the thick ascending limb, the macula densa, which makes contact with the vascular pole of its own glomerulus to form the juxtaglomerular apparatus. The best method for interfering with the signal transmission of the tubular fluid composition to the glomerular vasculature seems to be the micropuncture technique. Single nephrons can be perfused *in situ* without affecting systemic blood pressure or salt and water balance. The vascular response to an increase of sodium chloride load to the thick ascending limb of Henle's loop can be monitored continuously using micropressure pipettes. A semisolid wax block is injected into the first accessible loop of the proximal tubule. Upstream of this block, the hydrostatic stop-flow pressure (SFP), an equivalent of the hydrostatic pressure in the glomerular capillaries, can be measured by a servo-controlled micropressure device [33]. Another pipette filled with a modified Ringer solution and connected with a nanoliter pump is inserted into the last loop of the proximal tubule. The sodium chloride load to the loop of Henle is varied by setting the perfusion rate from zero to 50 nl/min. Figure 25.9 shows an original tracing of the SFP recorded in the proximal tubule and its response to elevation of the perfusion rate of the loop of Henle. The response is blocked completely 15 min after intravenous administration of theophylline (8 mg/kg) (Fig. 25.9, upper panel). In contrast, dipyridamole (5 mg/kg i.v.) potentiates the SFP response (Fig. 25.9, lower panel). The results are summarized in Table 25.1. The most likely interpretation of this finding is that adenosine is involved in the vascular response of the glomerulus to an increased load of NaCl to the loop of Henle. Experiments with other adenosine synergistic (erythrohydroxynonyladenosine) or antagonistic (adenosine deaminase) substances [34] support the foregoing interpretation.

CONCLUSION

The metabolic work of the kidney is the tubular reabsorption of sodium. Most of the other solutes are reabsorbed secondarily to the primary active sodium translocation achieved by Na-K-ATPase [35]. The activity of this enzyme is the main source for renal ATP hydrolysis. This review puts forward the

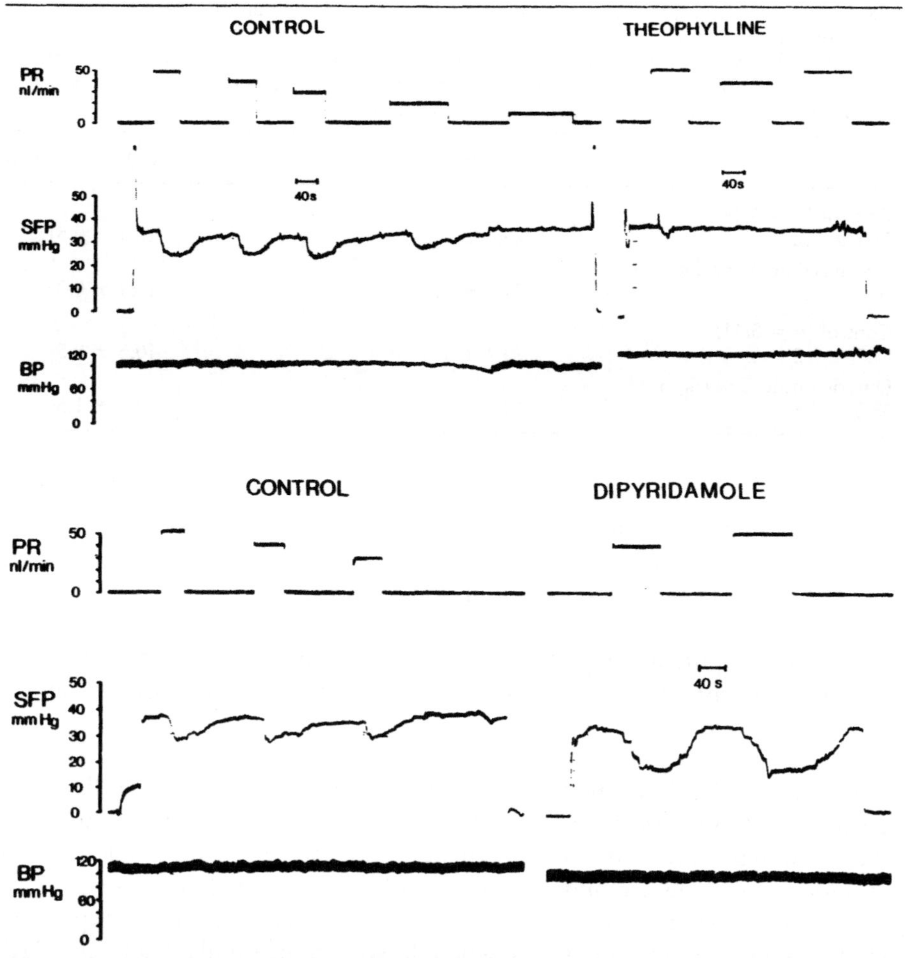

Figure 25.9. Original tracing of endproximal perfusion rate (PR), stop-flow pressure in the early proximal tubule (SFP), and arterial blood pressure (BP). Theophylline (upper panel) was administered intravenously at a dose of 8 mg/kg. Note the complete disappearance of the SFP fall in response to the increased PR from 0 to 50 nl/min. Lower panel: Potentiation of the SFP response 20 min after i.v. administration of 5 mg/kg dipyridamole [29].

hypothesis of a metabolic control of renal function. Adenosine meets the criteria necessary for a substance to function as a metabolic mediator in the kidney:

1. Adenosine has a vasoconstrictive action on renal vasculature. A dilator metabolite is excluded as a candidate because it would induce a positive feedback between metabolism and renal blood flow.
2. Adenosine is present in renal tissue and can be released at an enhanced rate by renal cells when ATP hydrolysis prevails over ATP synthesis.

Table 25.1. Stop-flow pressures (mm Hg) in the early proximal tubule at end-proximal perfusion rates of 0 and 50 nl/min recorded in Sprague-Dawley rats prepared for micropuncture in a standard fashion [9]. Measurements were obtained during control periods and 15 min after drug administration for a period of one hour. Numbers in parentheses indicate the number of animals/tubules studied in each group (mean values ± SE).

0 nl/min	50 nl/min	Δ 0 − 50 nl/min
Control (*n* = 5/14)		
35.7 ± 1.2	25.6 ± 1.2	10.1 ± 0.51
Theophylline (8 mg/kg i.v.)		
36.4 ± 1.2	34.5 ± 1.2	1.89 ± 0.37
Control (*n* = 5/11)		
35.7 ± 1.2	25.4 ± 1.0	10.5 ± 0.51
Dipyridamole (5 mg/kg i.v.)		
38.5 ± 2.5	19.1 ± 1.7	19.4 ± 1.5

3. The hemodynamic action of adenosine is rapid in onset and short in duration. Autoregulation of renal blood flow and tubuloglomerular feedback are phenomena that adjust vascular resistance to new settings within seconds.
4. Adenosine-potentiating substances (e.g., dipyridamole) decrease and adenosine-antagonistic substances (e.g., theophylline) increase glomerular filtration rate.

Adenosine also has a powerful action on the renin-containing cells to decrease the release of the hormone. This action might provide better insight into the intrarenal mechanisms that regulate renin release. It is possible that the renin-secreting cells are under tonic inhibition by intrarenal adenosine. When the tubules receive an increased NaCl load (for instance, due to hypertonic saline), renin secretion is reduced; when, however, the tubular load with NaCl is curtailed by reduction in renal perfusion pressure and therefore in glomerular filtration rate, the tubular reabsorption rate falls and renin secretion is elevated.

It should be mentioned that the concept of a metabolic vasoconstriction in the kidney applies only to the cortex in which blood flow determines metabolism. The circumstances are quite different in the renal medulla. In this region, adenosine might function, as in other organs, as a metabolic vasodilator. We are beginning to understand the renal actions of adenosine, and much experimental work has to be done to elucidate its role in renal function.

REFERENCES

1. Drury AN, Szent-Györgyi A: The physiological activity of adenine compounds with special reference to their action upon the mammalian heart. *J Physiol (Lond)* 68:213–237, 1929.
2. Thurau K: Renal hemodynamics. *Am J Med* 36:689–719, 1964.

3. Tagawa H, Vander AG: Effects of adenosine compounds on renal function and renin secretion in dogs. *Circ Res* 26:327–338, 1970
4. Hashimoto K, Kumakura S: The pharmacological features of the coronary, renal, mesenteric and femoral arteries. *Jpn J Physiol* 15:540–551, 1965.
5. Osswald H: Renal effects of adenosine and their inhibition by theophylline. *Naunyn Schmiedebergs Arch Pharmacol* 288:79–86, 1975.
6. Osswald H, Schmitz H-J, Heidenreich O: Adenosine response of the rat kidney after saline loading, sodium restriction and hemorrhagia. *Pfluegers Arch* 357:323–333, 1975.
7. Spielman WS, Osswald H: Characterization of the postocclusive response of renal blood flow in the cat. *Am J Physiol* 235:F286–F290, 1978.
8. Osswald H, Spielman WS, Knox FC: Mechanism of adenosine mediated decrease in glomerular filtration rate. *Circ Res* 43:465–469, 1978.
9. Haas JA, Osswald H: Adenosine induced fall in glomerular capillary pressure: Effect of ureteral obstruction and aortic constriction in the Munich-Wistar rat kidney. *Naunyn Schmiedebergs Arch Pharmacol* 317:86–89, 1981.
10. Spielman WS, Britton SL, Fiksen-Olsen MJ: Effect of adenosine on the distribution of renal blood flow in dogs. *Circ Res* 46:449–456, 1980.
11. Osswald H, Sacher R, Leser KH: Conversion of adenosine into a pure renal vasoconstrictor by furosemide. 8th Int. Congr. Nephrology, Vol. of Abstracts: *Hemodynamics,* HE-038, Athens, 1981.
12. Osswald H, Schmitz H-J, Kemper R: Renal action of adenosine: Effect on renin secretion in the rat. *Naunyn Schmiedebergs Arch Pharmacol* 303:95–99, 1978.
13. Osswald H, Hermes HH, Nabokowski G: The role of adenosine in signal transmission of tubuloglomerular feedback. *Kidney International,* Vol. 22, Suppl. 12:S136–S142, 1982.
14. Spielman WS, Osswald H: Blockade of the postocclusive renal vasoconstriction by an angiotensin II antagonist: Evidence for an angiotensin-adenosine interaction. *Am J Physiol* 237:F463–F467, 1979.
15. Spielman WS, Knox FG, Dousa TP: Evidence that intrarenal adenosine mediates the decrease in renin release during hypertonic saline infusion. *Fed Proc* 39(3):2949, 1980 (abst).
16. Hedqvist P, Fredholm BB, Ölundh S: Antagonistic effects of theophylline and adenosine on adrenergic neuroeffector transmission in the rabbit kidney. *Circ Res* 43:592–598, 1978.
17. Osswald H, Schmitz H-J, Kemper R: Tissue content of adenosine, inosine and hypoxanthine in the rat kidney after ischemia and postischemic recirculation. *Pfluegers Arch* 371:45–49, 1977.
18. Miller WL, Thomas RA, Berne RM, Rubio R: Adenosine production in the ischemic kidney. *Circ Res* 43:390–397, 1978.
19. Bier A: Die Entstehung des Kollateralkreislaufs. Teil II. Der Rückfluß des Blutes aus ischämischen Körperteilen. *Virchows Arch* [*Pathol Anat*] 153:435–465, 1898.
20. Stierlen G: Untersuchungen über die Nierengefässreaktion bei Mangeldurchblutung. *Pfluegers Arch* 727–737, 1937.
21. Hashimoto K, Yasuda K, Satoh S: Renal vascular effect of dipyridamole: Potentiation of norepinephrine and adenosine. *Eur J Pharmacol* 10:385–388, 1970.
22. Hinshaw IB, Page BB, Brake CM, Emerson TE: Mechanisms of intrarenal hemodynamic changes following acute arterial occlusion. *Am J Physiol* 205:1033–1041, 1963.
23. Sakai K, Yasuda K, Hashimoto K: Role of catecholamine and adenosine in the ischemic response following release of renal artery occlusion. *Jpn J Physiol* 18:673–685, 1968.
24. Honda N, Morikawa A, Nicei H, Aizawa C, Yashitoshi Y: Postocclusive vascular responses in isolated perfused kidney of rabbits. *Am J Physiol* 222:1581–1586, 1972.
25. Herbaczynska-Cedro K, Vane JR: Contribution of intrarenal generation of prostaglandin to autoregulation of renal blood flow in the dog. *Circ Res* 33:428–436, 1973.
26. Aizawa C, Honda N: Failure to abolish reactive hyperemia by indomethacin in denervated kidneys of rabbits. *Am J Physiol* 233:F89–F93, 1977.
27. Renkin EM, Gilmore JP: Glomerular filtration, Chapter 9 in Orloff J, Berliner RW (eds): *Handbook of Physiology,* Section 8: Renal Physiology. Washington, D.C., American Physiological Society, 1973, p 220.
28. Sacher R, Osswald H: Measurement of adenosine in picomole amounts in tissue samples by means of an adenosine binding protein from heart and kidney. *Naunyn Schmiedebergs Arch Pharmacol* 311(Suppl):R49, 1980 (abst).

29. Osswald H, Nabakowski G, Hermes H: Adenosine as a possible mediator of metabolic control of glomerular filtration rate. *Int J Biochem* 12:263–267, 1980.
30. Gerber JG, Branch RA, Nies AS, Hollifield JW, Gerkens JF: Influence of hypertonic saline on canine renal blood flow and renin release. *Am J Physiol* 237:F441–F446, 1979.
31. Schnermann J, Wright FS, Davis JM, von Stackelberg W, Grill G: Regulation of superficial nephron filtration rate by tubuloglomerular feedback. *Pfluegers Arch* 318:147–175, 1970.
32. Wright FS, Briggs JP: Feedback control of glomerular blood flow, pressure and filtration rate. *Physiol Rev* 59:958–1006, 1979.
33. Wiederhielm CA, Woodburg JW, Kirk S, Rushmer RF: Pulsatile pressures in the microcirculation of frog's mesentery. *Am J Physiol* 207:173–176, 1964.
34. Osswald H, Hermes H: Potentiation and inhibition of tubulo-glomerular feedback by adenosine synergistic and antagonistic substances. *Fed Proc* 39:1084, 1980 (abst).
35. Ullrich KJ, Frömter E: Active and passive transtubular transport in the proximal convolution. *Proc. 7th. Int. Congr. Nephrology*. Basel, Karger, 1978, pp 147–154.

DISCUSSION

DR. J. SCHRADER: What is the minimal concentration of adenosine in plasma required to induce autoblockade (Hashimoto block) in the kidney?

DR. OSSWALD: About 100 μM.

DR. C.I. THOMPSON: Histological studies suggest the production of adenosine within the proximal tubules; however, stop-flow distal perfusion experiments that bypass the proximal tubule elicit the proposed metabolic response. Where do you believe the sites of adenosine production and action are?

DR. OSSWALD: The site of production and the site of action of adenosine seem to be in close proximity. Adenosine ($10^{-6}M$) infused into the tubule from the end of the proximal tubule into the loop of Henle does not affect stop-flow pressure of the same nephron. Adenosine production by the epithelium of the proximal convoluted tubule presumably does not have access to the vasa afferentia but is taken up by the peritubular capillaries. The sites of adenosine production when tubuloglomerular feedback is initiated have to be the tubular cells in close proximity to the macula densa cells, the macula densa cells by themselves, and possibly the Goormaghtigh cells of the juxtaglomerular apparatus. However, a definite answer cannot be given at present.

DR. F.L. BELLONI: (1) Why does it require a higher dose of the AII antagonist to block the intrarenal AII than it takes to block exogenous AII? (2) Do intrarenal AII and exogenous AII act at the same site?

DR. OSSWALD: (1) The concentration of intrarenally formed AII might be much higher in the juxtaglomerular apparatus than in the renal lymph or in the plasma. The AII antagonist may not reach the site of intrarenally formed AII at the same rate as it reaches the sites of circulating AII action. (2) Circulating AII seems to act more at the efferent arterioles than at the afferent arterioles.

DR. E. GERLACH: I have two brief questions: (1) Do you have any idea in which population of renal cells adenosine is being formed during ischemia? Are these tubular cells or vascular endothelial cells? (2) In your scheme, adenosine production is directly coupled to Na transport. Do you have quantitative data in support of your hypothesis?

DR. OSSWALD: (1) The most likely candidates of cells producing adenosine during ischemia are the tubular cells, especially those that exhibit high transport rates. The tubular cells represent more than 90% of the kidney wet weight. Ischemia of shorter periods (30 min) induces damage mainly to tubular cells. The endothelium does not show such damage. In addition, the renal response to hypertonic saline in rats was an increased tubular sodium transport. The vascular cells are not involved in active reabsorption of the filtered load. Therefore, it seems to be reasonable to assume that the major fraction of kidney cells that respond to ischemia or to hypertonic saline are the tubular cells. (2) Yes, we have quantitative data that show an enhanced rate of adenosine release when active sodium transport is stimulated. These experiments were done during the summer of 1981, together with Dr. John Forrest at the Mount Desert Island Biological Laboratories in Maine, in the rectal gland of the spiny dogfish (squalus acanthius). This gland increases the rate of adenosine release from 4 to 16 pmol/min in the rectal fluid when the NaCl transport rate was stimulated by vasoactive polypeptide 3-fold. Dr. Rubio has measured decreased ATP tissue levels upon stimulation of electrolyte transport in this tissue. The point is that the transepithelial transport of electrolytes requires work and that the cells hydrolyze ATP for this transport.

DR. GERLACH: Your answer to my first question does not fully convince me. In my opinion, adenosine in the kidney could well originate, at least partially, in the renal vascular endothelium. This opinion is derived from many observations according to which endothelial cells possess a very high metabolic activity, particularly with regard to adenine nucleotides and adenosine metabolism. Tomorrow we shall present data demonstrating that coronary endothelial cells on which we performed our studies contain extraordinarily high amounts of adenine nucleotides and also of adenosine, which is continuously released in substantial quantities. Provided that renal vascular endothelium exhibits similar features, which seems to be very likely, adenosine of endothelial origin also has to be considered to be of special importance in the kidney.

DR. H.V. SPARKS: If adenosine causes vasoconstriction by increasing angiotensin II action on afferent arterioles, how do you explain the observed decrease in renin release? Are they related?

DR. OSSWALD: It could be that AII by itself acts to inhibit renin release. It could also be that the renin-containing cells are under a constant stimulus by locally released catecholamines. It was shown by Hedqvist and co-workers that adenosine inhibits norepinephrine release. Therefore, the inhibitory action of adenosine on renin release could be partially mediated by such a presynaptic mechanism. At present, we do not know how the vasoconstrictor action of adenosine is linked to its action to decrease the renin secretion rate. Both actions can be antagonized by theophylline.

DR. R. BÜNGER: What is "metabolically" regulated—renal blood flow and thus substrate and oxygen supply to the transport epithelia, or GFR and thus the Na^+ (or work) load on the tubules?

DR. OSSWALD: Let me first remind you of this basic fact: most of the renal blood flow (>95%) goes through the glomeruli. There are two regulations. The acute response to an increase in tubular load (and therefore sodium reabsorption) is vasoconstriction, resulting in a fall of the glomerular filtration rate. The chronic response to an increase of a NaCl load to the tubules is a reduction in renin secretion rate. This results in a decrease of aldosterone plasma levels and transforms the kidney into a sodium-wasting state. Finally, there might be a neurally mediated response via renal afferent nerves that have access to the vasomotor center in the central nervous system to regulate sodium excretion via a pressure diuresis.

DR. T.W. RALL: I think I understand the general idea of negative feedback regulation mediated by adenosine. But your evidence shows that the vasoconstrictor effects of adenosine depend on the presence (and action) of AII *and* that adenosine produces a decline in plasma renin activity. My question is: What do you visualize as the initial action of adenosine? Does adenosine initially cause a release of renin and is the resultant AII responsible for the vasoconstriction?

DR. OSSWALD: The action of adenosine on the resistance vessels and the renin-containing cells is sensitive to antagonism by theophylline. In that respect, a similar receptor can be assumed. The renin-containing cells are located in the walls of the afferent arterioles of the glomeruli. In regard to the action of adenosine on these cell types (renin and smooth muscle cells), they are interdependent. What comes first in the very early phase of adenosine action (AII release or renin release inhibition) we clearly do not know. As early as 5 min after the onset of adenosine infusion into the renal artery, renal vein renin concentration is already decreased and remains at the same reduced concentration for the following 50 min of adenosine infusion. I think it is necessary to separate the release of renin into the systemic circulation from intrarenal AII-dependent vasoconstriction. We do not know how many cells mediate renal adenosine action on the juxtaglomerular apparatus (JGA).

DR. B.B. FREDHOLM: One possible explanation could be that adenosine acts not by releasing AII but by enhancing the vasoconstrictor action of preformed AII, much the same way as adenosine sometimes enhances α-adrenoceptor Ach muscarinic receptor-mediated actions. In other words, the AII would be there all the time but would be "seen" only when Ado is released.

DR. OSSWALD: I agree with you that the vasoconstrictor action of adenosine could be due to a potentiating mechanism of constrictors already present in the JGA. However, neither α-adrenergic nor β-adrenergic antagonists inhibit the vasoconstrictive action of adenosine. The synergistic effect between exogenous norepinephrine or other vasoconstrictors, such as vasopressin, can be explained on the basis of an increase in vascular resistance. At high renal perfusion pressures, adenosine is a powerful constrictor, whereas at a renal perfusion of 50 mm Hg, adenosine is not constrictive at all.

DR. V. ALABASTER: (1) Have you looked at the renal effects of selective adenosine agonists in an attempt to characterize the kidney receptor, or do you think that the atypical response to adenosine in the kidney vasculature indicates a completely different type of adenosine receptor? (2) Have you used a converting enzyme inhibitor, such as captopril, and do the data support your proposed hypothesis?

DR. OSSWALD: (1) I have not investigated the renal effects of other adenosine-like agonists. However, theophylline inhibits adenosine action in a fashion compatible with the assumption of a competitive antagonism. I believe this indicates an adenosine R receptor type also in the kidney. (2) I have infused SQ14225 into the renal artery of two dogs and found inhibition of adenosine-induced vasoconstriction and blockade of post-occlusive ischemia similar to that observed with the angiotensin II antagonist. These data support the hypothesis that adenosine-induced vasoconstriction in the kidney is mediated by an angiotension II-dependent mechanism.

VII. ACTIONS OF ADENOSINE IN THE NERVOUS SYSTEM

26. THE ROLE OF ADENOSINE IN CENTRAL NEUROMODULATION

JOHN W. PHILLIS, PETER H. WU

The last decade has witnessed an enormous expansion of the research commitment into the actions of adenosine and its nucleotides on nerve and muscle cells. Interest in adenosine is not, however, a new phenomenon, and over fifty years have elapsed since the first observations of the actions of adenosine and adenosine 5′-monophosphate (AMP) on smooth and cardiac muscles [1]. The intervening years have seen a continuing interest in the peripheral actions of adenosine and its nucleotides, with the publication of an important monograph, *Biological Actions of the Adenosine Nucleotides* [2] in 1950. This text described the actions of adenine nucleotides and related purines on the cardiovascular and respiratory systems and clearly enunciated some of the structural requirements necessary for activation of purinergic receptors. The modern era of research on adenosine was initiated by the hypothesis, proposed by Burnstock and his colleagues [3], that adenosine 5′-triphosphate (ATP) is the transmitter released from so-called purinergic nerves, which form a third division of the autonomic nervous system. The hypothesis itself and the supporting data were elegantly outlined in a subsequent review [4]. It is curious that, although the status of the purinergic hypothesis remains somewhat uncertain, at least in the form in which it was originally presented, there has been widespread acceptance of the concept of adenosinergic modulation of transmitter release in both the central and peripheral nervous systems.

In recognition of the existence of a separate adenosine receptor in peripheral tissues, Burnstock [5] proposed the P_1/P_2 nomenclature for purinergic recep-

R.M. Berne, T.W. Rall, and R. Rubio (eds.), Regulatory Function of Adenosine, *pp. 419–437.*

tors. Adenosine interacts with the P_1 receptor, where its action is antagonized by the methylxanthines, such as caffeine and theophylline. P_2 receptors prefer ATP and adenosine 5′-diphosphate (ADP) as agonists, and adenosine or AMP have little action at this site. Quinidine, imidazolines, apamine, and 2-2′ pyridylisatogen are rather weak and nonspecific antagonists of the P_2 receptor. Arylazidoaminopropionyl ATP has recently been proposed as a specific P_2 purine receptor antagonist [6]. Activation of the P_2 receptor leads to the synthesis of prostaglandins, which may mediate some of the pharmacological actions of ATP [7]. Other purine and pyrimidine nucleotides are able to activate the P_2 receptor in peripheral tissues.

Adenosine is also a regulator of cyclic AMP formation in brain slices and homogenates, acting via membrane-localized receptor mechanisms. There are two extracellularly located adenylate cyclase-coupled adenosine receptors. A high-affinity (A_1) receptor mediates an inhibitory action of adenosine on the isoproterenol-stimulated enzyme, and a somewhat lower affinity (A_2) receptor site has a stimulatory action on the enzyme [8, 9]. Adenosine inhibits adenylate cyclase at concentrations in the range of 10^{-8}–10^{-6} *M*. L-N^6-phenylisopropyl adenosine (L-PIA) and N^6-cyclohexyladenosine (CHA) are potent agonists at the A_1 site. 5′-N-ethylcarboxamide adenosine (NECA) is less potent than adenosine. D-PIA is 70 to 100 times less potent than the L-isomer. NECA and adenosine 5′-cyclopropylcarboxamide (NCPCA) are potent agonists at the A_2 receptor, being about 10-fold more active than adenosine, L-PIA, and 2-chloroadenosine. L-PIA is only 5 to 10 times more potent that the D-isomer at A_2 sites. Activation by adenosine of both the A_1 and A_2 sites is antagonized by the methylxanthines.

The relationship of the P_1 site to either of the cyclase-associated sites remains questionable. It has been claimed that the P_1 site is identical with the A_1 site [10, 11] or, alternatively, with the A_2 receptor [12]. It is also possible that the P_1 sites represent separate, noncyclase-related receptors [13] and that the pharmacological actions of adenosine are not necessarily mediated by cyclic AMP. This concept will be explored in greater detail in a later section of this chapter.

The evidence supporting a neuromodulatory role of adenosine in the central nervous system is multifaceted; a discussion of the relevant findings will form the substance of the remaining sections of this chapter.

REGIONAL DISTRIBUTION OF ADENOSINE AND ATP IN BRAIN

The distribution of ATP has been studied in rat brains fixed by microwave irradiation [14, 15] to minimize ATP hydrolysis. The ATP content of rat cerebral cortex (2.098 mmol per kg wet weight) obtained by this method is comparable to those observed in brains prepared by the rapid freezing techniques, and the microwave irradiation method offers the marked advantage that regional distribution can be measured in microdissected segments of brain slices. Higher concentrations of ATP were associated with gray rather than white matter. The sensorimotor cortex, entorhinal cortex, hippocampus, hypothalamus, and thalamus had the highest ATP concentrations of the various brain regions studied, but, overall, the regional variability in ATP distribution was disappointingly minimal. Perhaps, however, this was not an altogether unexpected finding in the light of the major involvement of the adenosine derivatives in the metabolic and genetic functions of individual cells.

The distribution of adenosine in rat brain has recently been described [16]. Unfortunately, the brains used in that study were not rapidly frozen *in situ*, and the adenosine levels are considerably higher than those reported in brains frozen or fixed *in situ* without interruption of the circulation or prior traumatization of the tissues [13]. Regional variability in adenosine levels was not pronounced. Cerebral cortex had the highest levels (309 $\pm$ 13 μmol/kg) of adenosine, followed, in descending order of content, by the striatum, mesencephalon plus thalamus, brain stem, hypothalamus, and cerebellum. Further investigations using microwave fixed or frozen *in situ* brains will be necessary to clearly establish that adenosine levels are relatively uniform in cerebral tissues.

RELEASE OF PURINES FROM BRAIN

There is now a considerable volume of data on the release of adenosine and its nucleotides from nervous tissues. ATP release from peripheral nerve terminals was first reported by Holton [17], and Silinsky [18] has described the release of endogenous ATP from motor nerve terminals. The release of labeled adenosine derivatives from isolated desheathed rabbit vagus nerves is, however, mainly in the form of inosine and hypoxanthine, with adenine and adenosine accounting for less than 3% and nucleotides less than 1% of the efflux [19].

Adenosine is released by electrical stimulation or K^+ depolarization from *in vitro* cerebral tissues (slices or synaptosomal preparations) that have been preincubated with labeled adenine or adenosine [20–22]. As described earlier for the isolated vagal nerve, only a small fraction of the material released from such *in vitro* brain preparations is present as nucleotides, even though much of the labeled material had initially been incorporated into intracellular nucleotides. The evidence suggests that release from such *in vitro* preparations is primarily in the form of nucleosides and that these are not formed by the extracellular hydrolysis of released nucleotides [19, 23].

A very different situation apparently pertains to purine release from intact normoxic brain, where much of the released material is in the form of nucleotides. Sulakhe and Phillis [24] measured purine release from the *in situ* cat cerebral cortex after prelabeling with [^{3}H]adenosine. Nucleotides accounted for 33% of the spontaneous efflux and 34.5% of the enhanced release elicited by direct electrical stimulation. Adenosine, inosine, and hypoxanthine accounted for the remainder of the labeled material.

Studies on purine release from the rat cerebral cortex confirm that release from the intact brain is largely in the form of nucleotides. Sixty-six percent of

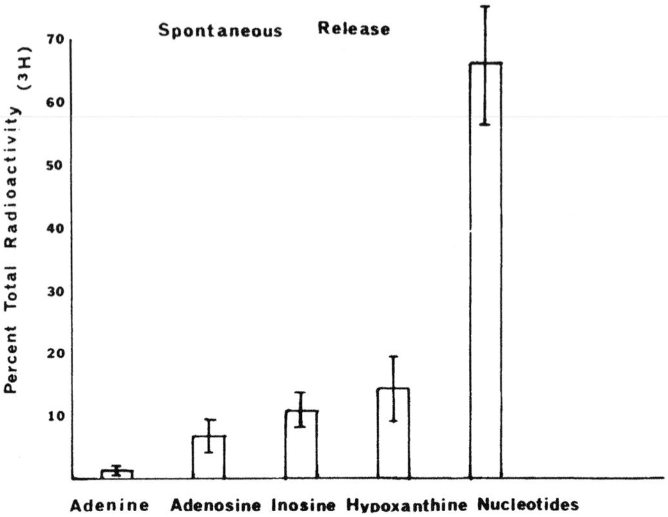

Figure 26.1. Release of [^{3}H]adenine derivatives into a rat cerebral cortical cup. [^{3}H]adenosine was incubated in a rat cortical cup for 40 min. The cup was then washed repeatedly with Ringers solution and filled with 0.4 ml of fresh saline solution. This was collected and replaced at 15 min intervals. The collected samples were pooled, lyophilized, and redissolved in 200 μl of distilled water. 25 μl of this solution was applied to the origin of a TLC silica gel plate. The plate was then developed in a solvent system of n-butanol: ethylacetate: methanol: ammonium hydroxide (7:4:3:4 v/v). The metabolites were located under ultraviolet light. The silica from areas identified as containing adenosine, adenine, inosine, hypoxanthine, and nucleotides was transferred to separate vials and its radioactivity counted.

the labeled material released from unstimulated rat cortex prelabeled with [^{3}H]adenosine was in the form of nucleotides, with adenosine, inosine, and hypoxanthine accounting for most of the rest of the labeled material (Fig. 26.1) [25]. Comparable results were obtained in another study on the rat cerebral cortex [26]. Release was calcium-dependent. Applications of L-glutamate, potassium, or veratridine to the cortex enhanced the efflux of labeled material without much alteration in the relative proportions of nucleosides to nucleotides. The release of endogenous adenine derivatives from hypoxic brain is primarily in the form of adenosine [27], a finding that correlates well with the pronounced increase in brain adenosine levels observed during hypoxia, ischemia, or severe hypotension [28–30].

It is apparent from these results that adenine nucleotides account for a much higher proportion of the total release from *in situ* cerebral tissues with an intact vascular system than for release from isolated tissues. Release from metabolically intact nerve terminals may be predominantly in the form of ATP, whereas release from hypoxic or isolated tissues may be predominantly in the form of nucleosides, at the expense of the nucleotides. Therefore, it is likely that the relative proportions of nucleotide and nucleoside released by different preparations will vary, reflecting the metabolic status of the preparation and the nature of the stimulus used to elicit the release.

Information on the actual levels of adenosine in the extracellular fluid in brain has recently become available. The perfusate from a thin dialysis tube inserted into the unanaesthetized rat caudate contained 0.33 μM adenosine and 0.21 μM inosine [31]. Nucleotide levels were not measured. It can be estimated from these values that the actual concentration of adenosine with free access to the extracellular space would be 1–2 μM. This represents an adenosine concentration that would affect synaptic transmission [32].

In the absence of any direct evidence for the existence of adenosinergic or ATP-ergic nerve in the central nervous system, there has been considerable interest in the possibility of nucleotide storage in and release from synaptic vesicles as a cotransmitter. ATP is known to be packaged in cholinergic and noradrenergic vesicles and may be released together with these transmitters [13]. Silinsky [18] has demonstrated a calcium-dependent release of ATP from phrenic nerve preparations, and there is evidence for endogenous ATP release from noradrenergic nerve terminals in the vas deferens [6] and of ^{3}H-labeled purine release from noradrenergic terminals in arterial tissues [33].

ATP may also be released from postsynaptic sites in some tissues. Israel et al. [34] have detected a postsynaptic release of endogenous ATP during individual discharges of the Torpedo electric organ. An ATP release from exercizing muscles, which was very probably postsynaptic in origin, has also been reported [35].

CENTRAL ACTIONS OF ADENOSINE AND ATP

Behavioral effect

Adenosine and its analogues have sedative, hypothermic, and anticonvulsant actions when administered intraperitoneally [13, 36]. It is questionable, however, whether these effects of peripherally administered purines are exerted directly on the central nervous system or occur as a result of actions elsewhere in the body. Adenosine analogues have a powerful hypotensive action, which could account for many of their sedative and anticonvulsant effects. This possibility is supported by the finding that the blood-brain barrier effectively limits the entry of adenosine into the brain [27, 37].

When the blood-brain barrier is circumvented by direct intracerebral administration into the cerebral ventricles, adenosine and its analogues produce ataxia, muscular weakness, and sleepiness [38–40]. Intracisternally administered adenosine and its analogues have an antinocisponsive effect on mice tested for hot plate reaction times [41]. NECA and NCPCA are more potent than 2-chloroadenosine and adenosine. Caffeine and theophylline antagonize the effects of the adenosine analogues.

The methylxanthine adenosine antagonists, caffeine and theophylline, are potent stimulants of the central nervous system [42]. Caffeine enters the brain more rapidly after peripheral administration and for this reason tends to be preferred in behavioral studies. Caffeine enhances wakefulness, stimulates locomotor activity, and increases sensitivity to nociceptive stimulation. Humans ingesting caffeine may experience nervousness or anxiety, restlessness, insomnia, hyperaesthesia, and muscle tremors [13]. At large doses (3–10 gm), focal and generalized seizures may result.

Iontophoretic studies on central neurons

Adenosine and its analogues have a depressant action on the firing of neurons in several regions of the rat brain, including the cerebral and cerebellar cortices, hippocampus, caudate nucleus, thalamus, superior colliculus, and olfactory bulb [43]. A number of purine and pyrimidine nucleotides and nucleosides were tested on the spontaneous firing of cerebral cortical neurons. Consistent depressant activity was observed only with adenosine and adenosine analogues, including its nucleotides. The structure-activity relationships for adenosine analogues suggest that an amino group at the 6 position of the purine ring is essential for full agonist activity. Inosine was only a weak agonist, and purine riboside was inactive. Substitutions at the C2 position of the ring enhanced and prolonged the action of the parent compound (2-chloroadenosine, 2-fluoroadenosine), reflecting the resistance of these compounds to deamination and removal by the adenosine uptake system. Substitutions at the C6 and C8 positions reduced or abolished activity (6-phenyladenosine, 8-bromoadenosine).

Ribose substituents at C-2′ and C-3′ reduced activity (2′-deoxyadenosine, 3′-deoxyadenosine). Substituents at the C-5′ position were generally compatible with activity (5′-deoxyadenosine and adenosine 5′-mononicotinate). Adenosine 5′-sulphate was virtually inactive. Phosphate groups at the C-2′, C-3′, and C-5′ positions did not affect depressant potency, possibly because these compounds are rapidly metabolized to adenosine by ectonucleotidases. When metabolically stable analogues of the adenine nucleotides, which are resistant to hydrolysis to adenosine, were tested, depressant activity was greatly reduced. Thus the α, β-methylene isosteres of ADP and ATP had only weak depressant activity, and the methylene isostere of AMP (ACP) and homoadenosine triphosphate (ACPOPOP) both lacked depressant activity. The loss of depressant activity with methylene substitutions near the ribose ring indicates that nucleotides must be converted to adenosine in order to elicit depressant activity.

Uptake inhibitors (dipyridamole, hexobendine, papaverine, lidoflazine, and 2-hydroxy-5-nitrobenzyl-thioguanosine) potentiated the action of both adenosine and the adenine nucleotides, as did the adenosine deaminase inhibitors, deoxycoformycin and erythro-9-(2-hydroxy-3-nonyl)adenine. Adenosine uptake inhibitors and adenosine deaminase inhibitors powerfully depressed neuronal firing, indicating that sufficient adenosine can be released endogenously to affect neuronal excitability.

Adenosine antagonists (theophylline, caffeine, isobutylmethylxanthine, oxy-

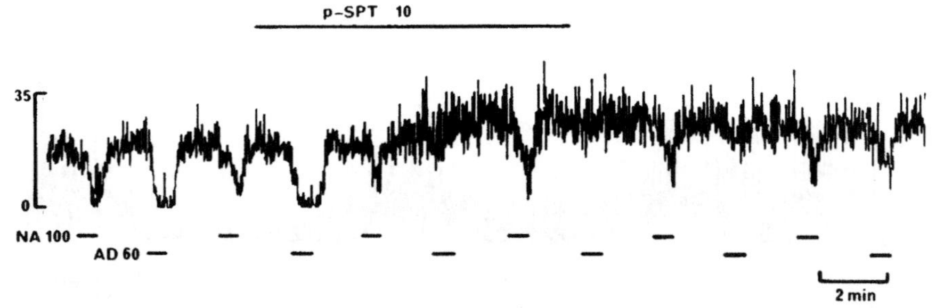

Figure 26.2. Adenosine (AD, 60 nA) and noradrenaline (NA, 100 nA) depressed the spontaneous firing of this rat cerebral cortical neuron. During and after an application of 8-(p-sulphophenyl) theophylline (p-SPT, 10 nA), the depressant effects of adenosine were blocked. Those of noradrenaline were not affected. This is a rate meter recording with the number of action potentials per second on the ordinate. Horizontal bars represent periods of drug application.

phyllin (theophylline isopropanolamine), and 8-(p-sulphophenyl) theophylline increased the rate of firing of cortical neurons and antagonized the depressant actions of adenosine and its nucleotides, but they did not affect norepinephrine-evoked inhibitions (Fig. 26.2).

The antagonism of both adenosine and its nucleotides by theophylline and caffeine and the lack of depressant activity of the metabolically stable methylene isosteres of the nucleotides indicate that a P_1 type of receptor is involved in the depression of neuronal firing. To obtain further information regarding the nature of the P_1 receptor and its relationship to adenylate cyclase, a number of the more selective ligands for A_1 and A_2 receptors have been tested. The adenosine carboxamides, adenosine 5′-methylcarboxamide (NMCA), NECA, and NCPCA, proved to be exceptionally potent depressants of the firing of cerebral cortical neurons [44]. These compounds were even more potent than 2-chloroadenosine in terms of application currents, although they did not elicit the relatively persistent depression observed with the latter.

L-PIA and CHA were less effective than adenosine or AMP in terms of application current magnitude, but both compounds elicited a rather prolonged depression of cortical neuronal firing (Fig. 26.3). The iontophoretic studies therefore indicate that selective A_2 receptor agonists (the adenosine carboxamides) are more potent depressants of neuronal firing than the A_1 receptor agonists. Whether the central P_1-type receptor can be equated with the A_2 adenylate cyclase receptor remains, however, an unanswered question. Studies on the depression of evoked potentials recorded from isolated slice preparations suggest that the receptors mediating the pharmacological actions of adenosine are not identical with those associated with adenylate cyclase [45, 46]. The possibility that the P_1 receptor is distinct from both A_1 and A_2 receptors must therefore be given serious consideration. Even the acceptance of the P_1 nomenclature must be considered as suspect, since Burnstock [5] has identified the P_1 receptor as an adenylate cyclase-associated site.

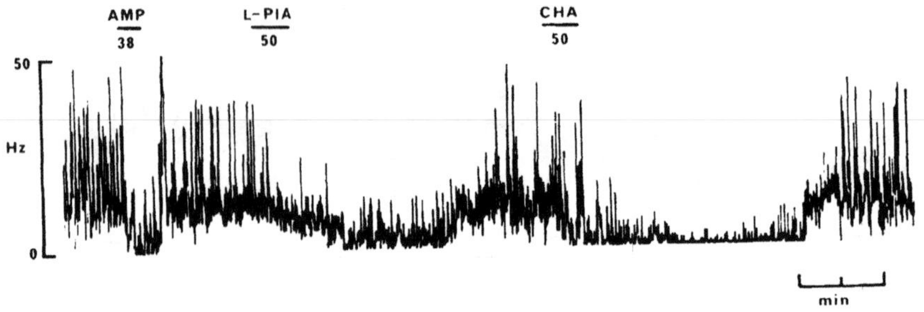

Figure 26.3. Firing frequency record of a rat cerebral cortical neuron. AMP (30 nA), L-PIA (L-phenylisopropyladenosine, 50 nA), and CHA (N^6-cyclohexyladenosine, 50 nA) depressed the spontaneous firing. Note the long duration of depression elicited by L-PIA and CHA. Firing rate in Hz.

Intracellular recordings have been made from cortical neurons in order to study the mechanism of action of adenosine and its nucleotides [37]. The most marked effect of AMP was a hyperpolarization accompanied by a decrease or abolition of spontaneously occurring and evoked synaptic potentials. There was no significant change in input resistance or in the threshold for action potential generation. It was concluded that the most likely explanation for the depression of neuronal firing by adenosine and the adenine nucleotides was an impairment of synaptic transmission as a result of inhibition of presynaptic excitation-transmitter release coupling.

The polyphosphates of all the purines and pyrimidines tested had excitant actions on neurons in the cerebral cortex and elsewhere in the brain, being most pronounced with quiescent or slowly firing neurons. These excitant effects had a short latency of onset and declined rapidly once the application had ceased. Excitation was most pronounced with the triphosphates and especially with homoadenosine triphosphate, which, lacking depressant activity, was a very effective excitant. Excitations were less pronounced with diphosphates and were infrequently observed with the monophosphates.

EFFECTS OF PERIPHERALLY ADMINISTERED PURINES ON BLOOD PRESSURE AND NEURONAL FIRING

Intravenously administered adenosine has a hypotensive action. The threshold dose of adenosine required to elicit a discernible fall in rat blood pressure is 5–10 μg/kg, and at this dose level adenosine did not alter the firing frequency of single cerebral cortical neurons [37]. Larger doses of adenosine (200–1000 μg/kg) caused more pronounced but relatively transient falls in blood pressure, which, if they exceeded 40–50 mm Hg, were frequently associated with a reduction in the firing frequency of cortical neurons (Fig. 26.4). To resolve the issue of whether these depressant effects of intravenously administered adenosine on neuronal firing were direct or were a result of the fall in blood pressure, adenosine was administered by close intraarterial injection into the

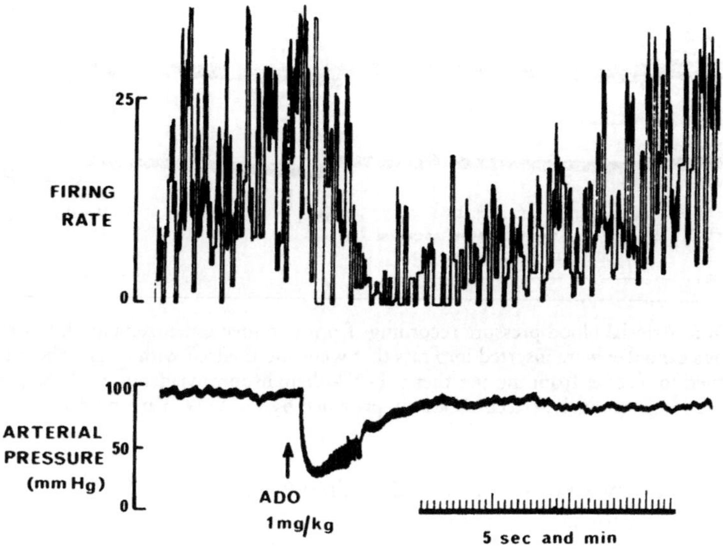

Figure 26.4. Simultaneous recordings of the spontaneous firing of a cerebral cortical neuron and femoral arterial blood pressure in a nitrous oxide-methoxyflurane anesthetized rat. Adenosine (mg/kg) was administered by the femoral vein. Firing rate in Hz.

carotid artery. Again, there was no alteration in the firing of cerebral cortical neurons until sufficient adenosine was administered to cause a large fall in blood pressure. The effects of peripherally administered adenosine on cerebral cortical neuron firing are therefore likely to be a result of a fall in blood pressure.

Autoregulation of cerebral blood flow, as systemic blood pressure is decreased within the physiologic range, is a well-established phenomenon [47–50]. A decrease in blood pressure below about 60 mm Hg is required to surmount the autoregulatory process and induce a decrease in cerebral blood flow and ischemia. Winn et al. [30] have demonstrated that a fall in mean arterial blood pressure (MABP) in rats from 135 mm Hg to 72 mm Hg, which is within the range of cerebral autoregulation, resulted in a doubling of brain adenosine levels; with moderate to severe hypotension (MABP = 45 mm Hg), brain adenosine levels increased almost 6-fold. The decreases in cortical neuronal firing observed during adenosine-evoked hypotension may well have been a result, therefore, of the cerebral ischemia and the resultant increase in brain adenosine levels. This would explain the observation that a 40–50 mm Hg fall in blood pressure was necessary to elicit any decrease in cortical neuronal firing frequency. Smaller reductions in blood pressure would have been compensated for by the autoregulatory process described earlier.

In recent studies, some of the more potent and selective purine ligands have been tested for their effects on blood pressure and cortical neuronal firing.

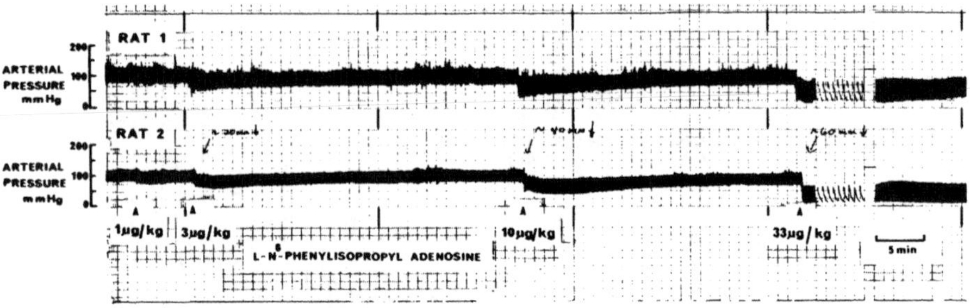

Figure 26.5. Arterial blood pressure recordings from two unanesthetized rats. Intracarotid and intravenous cannulae were inserted into rats that were anesthetized with ether. The rats were then allowed to recover from the anesthetic. L-N^6-phenylisopropyladenosine (1–33 µg/kg) was administered intravenously. (Records kindly provided by Dr. G.G. Yarbrough).

1-methylisoguanosine (1-methyl-2-hydroxyadenosine), L-PIA (Fig. 26.5), CHA, and 2-chloroadenosine were approximately equipotent as hypotensive agents, with an intravenous dose of 0.5–2 µg/kg required to elicit a discernible fall in blood pressure [44]. The threshold dose for intraperitoneal injections was some 10 to 20 times higher. None of these compounds affected the rate of firing of cerebral cortical neurons unless there was a large (>40–50 mm Hg) fall in systemic blood pressure.

The adenosine carboxamides (NMCA, NECA, and NCPCA) were extraordinarily potent hypotensive agents in the rat [44]. Intravenous administration of these agents in doses as low as 10–20 ng/kg was sufficient to elicit a discernible fall in blood pressure. Once again, these compounds failed to affect cortical neuronal firing unless there was a large fall in arterial blood pressure (Fig. 26.6).

Further evidence that the alterations in firing frequency were secondary to alteration in arterial blood pressure was obtained from experiments in which hypotension was elicited nonpharmacologically by hemorrhaging the rats. Removal of blood from a femoral vein caused an immediate controlled fall in blood pressure, which could be reversed by reinjection of the blood. Reductions in the firing rate or cortical neurons were observed only when sufficient blood had been withdrawn to cause large (>40 mm Hg) falls in arterial blood pressure (see Fig. 6 in [36]).

These experiments utilizing combined neuronal and blood pressure recordings illustrate the need for caution in any assessment of the results obtained when purines are administered peripherally. Depression of neuronal firing could be secondary to a fall in cerebral blood flow and an enhanced release of endogenous adenosine. Behavioral effects might have a similar origin.

PRESYNAPTIC MODULATION OF TRANSMITTER RELEASE

Adenosine and the adenine nucleotides reduce transmitter release at a variety of peripheral cholinergic and adrenergic junctions [10, 23, 51]. Centrally,

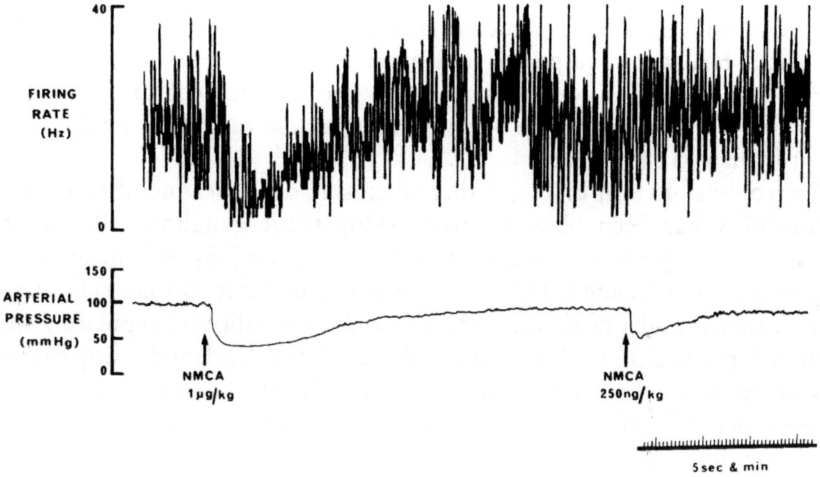

Figure 26.6. Simultaneous recordings of the spontaneous firing of a rat cerebral cortical neuron and femoral arterial blood pressure. NMCA (adenosine 5′-methylcarboxamide, 1 μg/kg and 0.25 μg/kg) was administered via the femoral vein. Firing rate in Hz.

adenosine depresses the release of acetylcholine (Ach) from motor axon collateral terminals on Renshaw cells [52] and the release of Ach by cerebral cortex [53]. Adenosine antagonists enhance cortical Ach release [25, 54]. Norepinephrine release from nerve terminals is inhibited as adenosine reduces the duration of the inhibition of cerebral cortical neurons evoked by stimulation of noradrenergic neurons in the locus ceruleus [55]. Adenosine depresses the release of labeled norepinephrine, Ach, dopamine, serotonin, and γ-aminobutyric acid from brain slices or synaptomes [56–59]. Adenine, inosine, guanosine, uridine, cytidine, and thymidine are inactive. The inhibitory effects of adenosine on transmitter release are antagonized by the methylxanthines, caffeine and theophylline.

IONIC BASIS OF ADENOSINE ACTION

The possibility that adenosine inhibits transmitter release by an action on Ca^{2+} fluxes has received considerable attention. The effects of adenosine are most apparent when extracellular Ca^{2+} is reduced and can be abolished by elevated extracellular Ca^{2+} levels [32, 60]. Adenosine's depressant action on transmitter release may therefore result from interference with the entry of calcium into the nerve terminal [61]. There have also been suggestions that adenosine may reduce the levels of intracellular free calcium [62, 63]. Recent studies in our laboratory have demonstrated that the effects of adenosine on the uptake of Ca^{2+} by K^+-depolarized rat brain cerebral cortical synaptosomes are antagonized by theophylline and enhanced by dipyridamole and indicate that an extracellular adenosine receptor of the P_1 type is involved [64]. The effects of adenosine were shared by a number of analogues, including 2-

chloroadenosine, CHA, L-PIA, D-PIA, NECA, and ATP. 2′, 5′-dideoxyadenosine was inactive at concentrations of up to $10^{-5}M$. 2-chloroadenosine and CHA were equipotent and more active than L-PIA and adenosine. D-PIA was 5-fold less potent than L-PIA. These pharmacological characteristics suggest that the receptor is more comparable to the A_2 cyclase-related receptor than to the A_1 site.

Electrophysiological evidence that adenosine depresses membrane calcium permeability has been obtained from sympathetic ganglion cells, where it specifically antagonizes calcium-dependent potentials [65]. Adenosine also depresses the calcium-mediated slow potentials in heart muscle cells [66]. In some instances, adenosine can influence cell excitability by regulating membrane K^+ permeability. Adenosine and the adenine nucleotides hyperpolarize cells of the sinus venosus of frog and tortoise hearts by increasing potassium permeability [67, 68]. ATP hyperpolarizes intestinal smooth muscle cells by increasing potassium permeability [69, 70]. In this tissue, a P_2-type receptor is involved, which is not blocked by theophylline. The ATP-elicited increase in K^+ permeability in guinea pig taenia caeci appears to be mediated by an initial influx of calcium, which then regulates the opening of the K^+ channels [71]. The extent to which such K^+-dependent actions of adenosine are involved in regulating transmitter release remains obscure at this time.

CONCLUSIONS

Adenosine has a depressant action on the spontaneous and evoked firing of central neurons, which can most readily be explained by postulating a presynaptic site of action, with inhibition of transmitter release. The receptor involved in this action appears to be specific for adenosine, and adenine nucleotides must be converted to adenosine by ecto-enzymes in order to exert this effect. Release studies have demonstrated that adenosine depresses the release of acetylcholine, norepinephrine, dopamine, serotonin, and, to a lesser extent, GABA from brain preparations. The presynaptic actions of adenosine probably result from a reduction in the calcium permeability of the nerve terminal membrane; less calcium enters during action potentials resulting in reduced transmitter release. At some junctions, notably, those between optic tract fibers and cells in the lateral geniculate nucleus and superior colliculus, adenosine does not depress postsynaptic potentials [72]. The transmitter released by this pathway has not been defined.

Adenosine levels in the extracellular fluid of brain tissue (1–2 μM) are sufficient to exert an ongoing modulation of synaptic transmission, and adenosine antagonists indeed excite central neurons, whereas uptake and deaminase inhibitors depress their firing. Adenosine release probably originates from at least two pools. ATP released as a cotransmitter provides one source of adenosine. The release during intense activity or hypoxia probably originates from a nonvesicular pool, derived from the intracellular hydrolysis of ATP.

Adenosine formed extracellularly from released ATP by the action of ectonucleotidases will act presynaptically to limit the further release of transmitter. These presynaptic receptors can be viewed as autoreceptors since they mediate a negative feedback regulation of ATP and cotransmitter release. ATP released into the synaptic cleft may act postsynaptically at receptors that control membrane conductance, or it may enhance the responses to its cotransmitter. The overall effect of this activation of pre- and postsynaptic receptor mechanisms will be to enhance the fidelity of transmission across central synapses.

Our studies with benzodiazepines, phenothiazines, and analgesic drugs suggest that release of adenosine or block of its reuptake may be mechanisms of action common to several groups of sedative, anxiolytic, anticonvulsant, and analgesic agents [13, 36, 73].

Adenosine release may also constitute an important factor in protecting central neurons from excessive damage during profound hypoxia or anoxia. Under such conditions, cerebral activity decreases rapidly, and EEG activity may disappear within 30 sec in the anoxic brain [74]. Adenosine release by the anoxic brain can account for many of the initial effects of oxygen deprivation on central neurons, including the hyperpolarization and the disappearance of synaptic potentials [75]. The adenosine uptake inhibitor lidoflazine and a structurally related analogue, flunarizine, markedly enhance cerebral reperfusion blood flows and reduce the degree of neurologic deficit in dogs following cardiac arrest [76, and unpublished observations by the same group]. These beneficial effects of lidoflazine and flunarizine may be a consequence of adenosine uptake inhibition and an adenosine-elicited relaxation of cerebral blood vessels. Adenosine would assist in the prevention of Ca^{2+} entry into the ischemic brain cells, an event that has been equated with the onset of neuronal death. Improved neurologic recovery has also been noted in dogs treated postischemia with indomethacin [77], a potent inhibitor of adenosine uptake [78].

REFERENCES

1. Drury AN, Szent-Györgyi A: The physiological activity of adenine compounds with especial reference to their action upon the mammalian heart. *J Physiol (Lond)* 68:213–237, 1928.
2. Green HN, Stoner HB: *Biological Actions of the Adenine Nucleotides*. London, H.K. Lewis and Co., 1950.
3. Burnstock G, Campbell G, Satchell D, Smythe D: Evidence that adenosine triphosphate or a related nucleotide is the transmitter substance released by non-adrenergic inhibitory nerves in the gut. *Br J Pharmacol* 40:668–688, 1970.
4. Burnstock G: Purinergic nerves. *Pharmacol Rev* 24:509–581, 1972.
5. Burnstock G: A basis for distinguishing two types of purinergic receptor, in Bolis L, Straub RW (eds): *Cell Membrane Receptors for Drugs and Hormones: A Multidisciplinary Approach*. New York, Raven Press, 1978, pp 107–118.
6. Fedan JS, Hogaboom GK, O'Donnell JP, Colby J, Westfall DP: Contribution by purines to the neurogenic response of the vas deferens of the guinea pig. *Eur J Pharmacol* 69:41–53, 1981.
7. Brown CM, Burnstock G: The structural confirmation of the polyphosphate chain of the

ATP molecule is critical for its promotion of prostaglandin biosynthesis. *Eur J Pharmacol* 69:81–86, 1981.
8. Van Calker D, Müller M, Hamprecht B: Adenosine regulates via two different types of receptors, the accumulation of cyclic AMP in cultured brain cells. *J Neurochem* 33:999–1005, 1979.
9. Cooper DMF, Londos C, Rodbell M: Adenosine receptor-mediated inhibition of rat cerebral cortical adenylate cyclase by a GTP-dependent process. *Mol Pharmacol* 18:598–601, 1980.
10. Paton DM: Structure-activity relations for presynaptic inhibition of noradrenergic and cholinergic transmission by adenosine: evidence for action on A_1 receptors. *J Auton Pharmacol* 1:287–290, 1981.
11. Reddington M, Lee KS, Schubert P: An A_1-adenosine receptor, characterized by [^{3}H] cyclohexyl adenosine binding, mediates the depression on evoked potentials in a rat hippocampal slice preparation. *Neurosci Lett* 28:275–279, 1982.
12. Brown CM, Collis MG: Evidence for an Ra purine receptor in the guinea pig trachealis muscle. *Br J Pharmacol* 75:157P, 1982 (abst).
13. Phillis JW, Wu PH: The role of adenosine and its nucleotides in central synaptic transmission. *Prog Neurobiol* 16:187–239, 1981.
14. Wu, PH, Phillis JW: Distribution and release of adenosine triphosphate in rat brain. *Neurochem Res* 3:563–571, 1978.
15. Wu PH, Moore KC, Phillis JW: Topographical distribution of ATP in rat brain. *Experientia* 35:881–883, 1979.
16. Gharib A, Sarda N, Chabannes B, Cronenberger L, Pacheco H: The regional concentrations of S-adenosyl-L-methionine, S-adenosyl-L-homocysteine, and adenosine in rat brain. *J Neurochem* 38:810–815, 1982.
17. Holton P: The liberation of adenosine triphosphate on antidromic stimulation of sensory nerves. *J Physiol (Lond)* 145:494–504, 1959.
18. Silinsky EM: On the association between transmitter secretion and the release of adenine nucleotides from mammalian motor nerve terminals. *J Physiol (Lond)* 247:145–162, 1982.
19. Maire JC, Medilanski J, Straub RW: Uptake of adenosine and release of adenine derivatives in mammalian non-myelinated nerve fibres at rest and during activity. *J Physiol* 323:589–602, 1982.
20. Pull I, McIlwain H: Output of [^{14}C] adenine nucleotides and their derivatives from cerebral tissues. *Biochem J* 136:893–901, 1973.
21. Daval JL, Barberis C, Gayet J: Release of [^{14}C] adenosine derivatives from superfused synaptosome preparations: Effects of depolarizing agents and metabolic inhibitors. *Brain Res* 181:161–174, 1980.
22. Bender AS, Wu PH, Phillis JW: The rapid uptake and release of [^{3}H] adenosine by rat cerebral cortical synaptosomes. *J Neurochem* 36:651–660, 1981.
23. Fredholm BB, Hedqvist P: Modulation of neurotransmission by purine nucleotides and nucleosides. *Biochem Pharmacol* 29:1635–1643, 1980.
24. Sulakhe PV, Phillis JW: The release of [^{3}H] adenosine and its derivatives from cat sensorimotor cortex. *Life Sci* 17:551–556, 1975.
25. Phillis JW, Jiang ZG, Chelack BJ, Wu PH: The effect of morphine on purine and acetylcholine release from rat cerebral cortex: Evidence for a purinergic component in morphine's action. *Pharmacol Biochem Behav* 13:421–427, 1980.
26. Jhamandas K, Dumbrille A: Regional release of [^{3}H] adenosine derivatives from rat brain *in vivo:* Effect of excitatory amino acids, opiate agonists, and benzodiazepines. *Can J Physiol Pharmacol* 58:1262–1278, 1980.
27. Berne RM, Rubio R, Curnish R: Release of adenosine from ischemic brain. *Circ Res* 25:262–272, 1974.
28. Winn HR, Rubio R, Berne RM: Brain adenosine production in the rat during 60 seconds of ischemia. *Circ Res* 45:485–492, 1979.
29. Winn HR, Rubio R, Berne RM: Brain adenosine concentration during hypoxia in rats. *Am J Physiol* 241:H235-H242, 1981.
30. Winn HR, Welsh JE, Rubio R, Berne RM: Brain adenosine production in rat during sustained alteration in systemic blood pressure. *Am J Physiol* 239:H636–H641, 1980.
31. Zetterström T, Vernet L, Ungerstedt U, Tossman U, Jonzon B, Fredholm BB: Purine levels

in the intact rat brain: Studies with an implanted perfused hollow fibre. *Neurosci Lett* 29:111–115, 1982.
32. Kuroda Y: Physiological roles of adenosine derivatives which are released during neurotransmission in mammalian brain. *J Physiol (Paris)* 74:463–470, 1978.
33. Katsuragi T, Su C: Purine release from vascular adrenergic nerves by high potassium and a calcium ionophore, A-23187. *J Pharmacol Exp Ther* 215:685–690, 1980.
34. Israel M, Lesbats B, Meunier FM, Stinnakre J: Postsynaptic release of adenosine triphosphate induced by single impulse transmitter action. *Proc R Soc Lond [Biol]* 193:461–468, 1976.
35. Forrester T, Lind AR: Identification of adenosine triphosphate in human plasma and the concentration in the venous effluent of forearm muscles before, during and after sustained contractions. *J Physiol* 204:347–364, 1969.
36. Phillis JW, Wu PH: Roles of adenosine and adenine nucleotides in the C.N.S., in Daly JW, Kuroda Y, Phillis JW, Shimizu H, Ui M (eds): *Physiology and Pharmacology of Adenosine Derivatives*. New York, Raven Press, 1983, pp. 219–236.
37. Phillis JW, Edstrom JP, Kostopoulos GK, Kirkpatrick JR: Effects of adenosine and adenine nucleotides on synaptic transmission in the cerebral cortex. *Can J Physiol Pharmacol* 57:1289–1312, 1979.
38. Feldberg W, Sherwood SL: Injections of drugs into the lateral ventricle of the cat. *J Physiol (Lond)* 123:148–167, 1954.
39. Buday PV, Carr CJ, Miya TS: A pharmacologic study of some nucleosides and nucleotides. *J Pharm Pharmacol* 13:290–299, 1961.
40. Haulica I, Ababei L, Brănisteanu D, Topoliceanu F: Preliminary data on the possible hypnogenic role of adenosine. *J Neurochem* 21:1019–1020, 1973.
41. Yarbrough GG, McGuffin-Clineschmidt JC: In vivo behavioral assessment of central nervous system purinergic receptors. *Eur J Pharmacol* 76:137–144, 1981.
42. Daly JW, Bruns RF, Snyder SH: Adenosine receptors in the central nervous system: Relationship to the central actions of methylxanthines. *Life Sci* 28:2083–2097, 1981.
43. Kostopoulos GK, Phillis JW: Purinergic depression of neurons in different areas of the rat brain. *Exp Neurol* 55:719–724, 1977.
44. Phillis JW: Evidence for an A_2-like adenosine receptor on cerebral cortical neurons. *J Pharm Pharmacol* 34:453–454, 1982.
45. Smellie FW, Daly JW, Dunwiddie TV, Hoffer BJ: The dextro and levorotatory isomers of N-phenylisopropyladenosine: Stereospecific effects on cyclic AMP-formation and evoked synaptic responses in brain slices. *Life Sci* 25:1739–1748, 1979.
46. Reddington M, Schubert P: Parallel investigations of the effects of adenosine on evoked potentials and cyclic AMP accumulation in hippocampal slices of the rat. *Neurosci Lett* 14:37–42, 1979.
47. Finnerty, FA, Witkin L, Fazekas JF: Cerebral haemodynamics during cerebral ischemia induced by acute hypotension. *J Clin Invest* 33:1227–1232, 1954.
48. Lassen NA: Cerebral blood flow and oxygen consumption in man. *Physiol Rev* 39:183–238, 1959.
49. Harper AM: Autoregulation of cerebral blood flow: Influence of the arterial blood pressure on the blood flow through the cerebral cortex. *J Neurol Neurosurg Psychiatry* 29:398–403, 1966.
50. Fitch W, Ferguson GG, Sengupta D, Garisi J, Harper AM: Autoregulation of cerebral blood flow during controlled hypotension in baboons. *J Neurol Neurosurg Psychiatry* 39:1014–1022, 1976.
51. Vizi ES: Presynaptic modulation of neurochemical transmission. *Prog Neurobiol* 12:181–290, 1979.
52. Lekić D: Presynaptic depression of synaptic response of Renshaw cells by adenosine 5′-monophosphate. *Can J Physiol Pharmacol* 55:1391–1393, 1977.
53. Jhamandas K, Sawynok J: Methylxanthine antagonism of opiate and purine effects on the release of acetylcholine, in Kosterlitz HW (ed): *Opiates and Endogenous Opioid Peptides*. Amsterdam, Elsevier North-Holland, 1976, pp 161–168.
54. Phillis JW, Siemens RK, Wu PH: Effects of diazepam on adenosine and acetylcholine release from rat cerebral cortex: Further evidence for a purinergic mechanism in action of diazepam. *Br J Pharmacol* 70:341–348, 1980.
55. Taylor DA, Stone TW: The action of adenosine on noradrenergic neuronal inhibition induced by stimulation of locus coeruleus. *Brain Res* 183:367–376, 1980.

56. Harms HH, Wardeh G, Mulder AH: Adenosine modulates depolarization-induced release of ^{3}H-noradrenaline from slices of rat brain neocortex. *Eur J Pharmacol* 49:305–308, 1978.
57. Harms HH, Wardeh G, Mulder AH: Effects of adenosine on depolarization-induced release of various radiolabelled neurotransmitters from slices of rat corpus striatum. *Neuropharmacology* 18:577–580, 1979.
58. Michaelis ML, Michaelis EK, Myers SL: Adenosine modulation of synaptosomal dopamine release. *Life Sci* 24:2083–2092, 1979.
59. Hollins C, Stone TW: Adenosine inhibition of γ-aminobutyric acid release from slices of rat cerebral cortex. *Br J Pharmacol* 69:107–112, 1980.
60. Ribeiro JA: The decrease of neuromuscular transmission by adenosine depends on previous neuromuscular depression. *Arch Int Pharmacol* 255:59–67, 1982.
61. Ribeiro JA, Sá-Almeida AM, Namorado JM: Adenosine and adenosine triphosphate decrease ^{45}Ca uptake by synaptosomes stimulated by potassium. *Biochem Pharmacol* 28:1297–1300, 1979.
62. Ginsborg BL, Hirst GDS: The effect of adenosine on the release of transmitter from the phrenic nerve of the rat. *J Physiol (Lond)* 224:629–645, 1972.
63. Branisteanu DD, Haulica ID, Proca B, Nitue BG: Adenosine effects upon transmitter release parameters in the Mg^{2+}-paralyzed neuromuscular junction of frog. *Naunyn Schmiedebergs Arch Pharmacol* 308:273–279, 1979.
64. Wu PH, Phillis JW, Thierry DL: Adenosine receptor agonists inhibit K^+-evoked Ca^{2+} uptake by rat brain cortical synaptosomes. *J Neurochem* 39:700–708, 1982.
65. Henon BK, Turner DK and McAfee DA: Adenosine receptors: Electrophysiological actions at pre- and postsynaptic sites on mammalian neurons. *Soc Neurosci Abstr* 6:257, 1980.
66. Schrader J, Rubio R, Berne RM: Inhibition of slow action potentials of guinea pig atrial muscle by adenosine: A possible effect on Ca^{++} influx. *J Mol Cell Cardiol* 7:427–433, 1975.
67. Hartzell HC: Adenosine receptors in frog sinus venosus: Slow inhibitory potentials produced by adenine compounds and acetylcholine. *J Physiol (Lond)* 293:23–49, 1979.
68. Hutter OF, Rankin AC: Increase by adenosine and adenine nucleotides in potassium permeability of sinus venosus of tortoise heart. *J Physiol,* 1982 (in press).
69. Jager LP: The effect of catecholamines and ATP on the smooth muscle cell membrane of the guinea pig taenia coli. *Eur J Pharmacol* 25:372–382, 1974.
70. Maas AJJ, Den Hertog A, Ras R, Van Der Akker J: The action of apamin on guinea pig taenia caeci. *Europ J Pharmacol* 67:265–274, 1980.
71. Den Hertog, A: Calcium and the action of adrenaline, adenosine triphosphate and carbachol on guinea pig taenia caeci. *J Physiol* 325:423–439, 1982.
72. Okada Y, Saito M: Inhibitory action of adenosine, 5-HT (serotonin) and GABA (γ-aminobutyric acid) on the postsynaptic potential (PSP) of slices from olfactory cortex and superior colliculus in correlation to the level of cyclic AMP. *Brain Res* 160:368–371, 1979.
73. Phillis JW, Wu PH: The effect of various centrally active drugs on adenosine uptake by the central nervous system. *Comp Biochem Physiol [C]* 72C:179–187, 1982.
74. Sugar O, Gerard R: Anoxia and brain potentials. *J Neurophysiol* 1:558–572, 1938.
75. Grossman RG, Williams VF: Electrical activity and ultrastructure of cortical neurons and synapses in ischemia, in Brierley JB, Meldrum BA (eds): *Brain Hypoxia*. London, W. Heinemann, 1971, pp 61–75.
76. White BC, Gadzinski DS, Hoehner PJ, Krome C, Hoehner T, White JD, Trombley JH: Effect of flunarizine on canine cerebral cortical blood flow and vascular resistance post cardiac arrest. *Ann Emerg Med* 11:119–126, 1982.
77. Hallenbeck JM, Furlow TW: Prostaglandin I_2 and indomethacin prevent impairment of post-ischemic brain reperfusion in the dog. *Stroke* 10:629–637, 1979.
78. Phillis JW, Wu PH: Indomethacin, ibuprofen and meclofenamate inhibit adenosine uptake by rat brain synaptosomes. *Eur J Pharmacol* 72:139–140, 1981.

DISCUSSION

DR. M. REDDINGTON: The difference in our interpretation of the effects of various derivatives in terms of A_1 or A_2 receptors might well be due to differences in experimental design. In the hippocampal slice preparation, we apply the drugs until a new steady response is obtained, i.e., to equilibrium. This would seem to allow better comparison with ligand binding studies where equilibrium conditions are used. In iontophoretic experiments, you might be measuring an initial receptor activation rate rather than a true affinity.

DR. T.D. WHITE: (1) Cotransmission of ATP in the brain is not easy to demonstrate. Neither botulinum toxin, which inhibits acetylcholine release, nor 6-hydroxydopamine pretreatment, which destroys both noradrenergic and dopaminergic nerve terminals, diminishes K^+- or veratridine-evoked release of ATP from rat brain synaptosomes. In contrast, 6-hydroxydopamine pretreatment diminishes ATP release from isolated peripheral varicosities from the myenteric plexus. (2) What stimulation parameters did you use to evoke ATP release into cortical cups? Was release blocked by tetrodotoxin?

DR. PHILLIS: The stimulation parameters were square wave pulses 1–5 mA, 0.2 msec duration, 20 pulses per sec, applied to the cortical surface with a bipolar electrode. Similar parameters have been used by other investigators to elicit GABA release from the cortex. Tetrodotoxin blocks glutamate and veratridine (but not K^+) induced release of labeled adenosine and adenine nucleotides from the rat cerebral cortex [Jhamandas and Dumbrille 1980].

DR. Y. KURODA: I also observed inhibitory action of adenosine in olfactory cortex slices [Kuroda, 1975, 1976] in which we controlled the concentration of agent in the extracellular medium. With iontophoretic application of agents, we do not know the concentration applied. CHA, less than 1 μM concentration of adenosine, had a strong inhibitory action in brain slices. CHA is very lipid-soluble so possibly it is trapped by membranes around the tip of the electrode. Is there any possibility that CHA did not reach the site of action?

DR. PHILLIS: Sequestration of iontophoretically applied CHA by membrane lipids could account for its slow onset of action. The point that I wanted to make was that its concentration in the lipids around the receptor could account for the very high level of activity observed when brain slices are exposed to CHA for periods of several minutes.

DR. KURODA: CHA (1 μM) did inhibit the postsynaptic potential within 5 sec, so that accumulation might not have occurred.

DR. M. WILLIAMS: In view of your comments on an endogenous purinergic tone in CNS, what physiological role do you feel autoreceptors for specific putative neurotransmitters play in regard to determining negative-feedback inhibition mechanisms?

DR. PHILLIS: Autoregulation by norepinephrine or acetylcholine occurs in the central nervous system, but these systems have not been thoroughly characterized to date. Adenosine appears to be a very potent modulating substance, with an ability to affect a variety of differing types of nerve terminals. Its action may be more generalized than those of some of the other neurotransmitters.

DR. H. MCILWAIN: It is very encouraging that you find long-term effects on cell firing by small additions of adenosine and its derivatives. One mechanism for this persistence, after the disappearance of the initial substance, is already available in the long-term persistence of cyclic nucleotide changes. This brings me to ask what the concentrations were of the substances that were added and how the concentrations compare with those involved with already characterized receptor systems?

DR. PHILLIS: A limitation of the iontophoretic technique is that one does not have the ability to monitor drug concentrations at the receptor site. Studies on isolated preparations suggest that adenosine concentrations of between 10^{-7} to $10^{-6}M$ are required to exert a threshold effect. In the presence of uptake and deaminase inhibitors, adenosine is even more potent. Its potency therefore compares favorably with those of other better characterized neurotransmitters.

DR. D.G. SATCHELL: We have found in intestinal smooth muscle preparations, as you have found in brain, that certain analogues of ATP and adenosine that may be quite potent have very long times of onset of action. We have no explanation of why some compounds are most rapid in onset of action and others slow in onset.

DR. A. SATTIN: Faden and Holoday at Walter Reed have shown massive peripheral release of opiate peptides following acute hemorrhage. It is possible that there might be a central nervous system counterpart of this resulting in release of adenosine, as proposed by Trevor Stone.

DR. PHILLIS: This is an interesting suggestion, and one would have to look at the time course of the opiate-like peptide release during hemorrhage. Our data confirm that morphine enhances adenosine release from the rat cerebral cortex. This appears to represent a true increase in release rather than an inhibition of adenosine uptake, with a subsequent enhancement of efflux from the cortex.

DR. T.W. RALL: In view of the powerful effects of adenosine on accumulation of cyclic AMP in brain tissue and in view of the other indications that cyclic AMP may be associated with enhanced release of transmitter, I am led to conclude that the powerful (and prevalent) presynaptic inhibitory effects of adenosine are not mediated by cyclic AMP. I would appreciate comments from Dr. Phillis or from later speakers on this issue or on the possibility that reduction of cyclic AMP is involved in this response.

DR. PHILLIS: This is a complicated issue. There are good indications that enhanced levels of cAMP are associated with enhanced transmitter release, and Dr. Kuroda may wish to comment on this. The problem with the converse situation is that extracellular adenosine levels are normally in the range that stimulate cAMP formation (1–2 μM). One would anticipate that the inhibitory receptor must exist in a state of near-maximal activation and that small changes in adenosine levels would have little additional effect at this receptor. Hence, the A_1 receptor may not be involved in the depression of transmitter release.

DR. KURODA: Adenosine-induced increase of cyclic AMP seems to have facilitory function in the synapses of the olfactory cortex. When we incubate the olfactory cortex slices with adenosine to increase the levels of tissue cyclic AMP 4-fold and then wash it out quickly, we find that the evoked postsynaptic potentials are facilitated. The time course and the magnitude of the facilitation is very similar to postsynaptic potentiation observed in the slices. Adenosine increased the cyclic AMP in presynaptic terminals and cyclic AMP analogues caused the facilitation of postsynaptic potentials and an increase in the high K^+-dependent Ca^{2+} influx to nerve terminals in the olfactory cortex. These results suggest that adenosine-induced increases of cyclic AMP in presynaptic terminals can mediate the facilitation (probably posttetanic potentiation) that appears only when adenosine derivatives in the synapse are removed and no *direct* inhibitory action is present. Hence, there would be two adenosine receptors in presynaptic terminals, one for the indirect inhibition (ADO-1) and another, also indirect, via cyclic AMP facilitation (ADO-2). Both control transmitter release with different time courses.

DR. G. BURNSTOCK: May I make a comment on an issue raised both in your talk and at other times in this conference? Namely, how can one know whether an action of ATP is on its own (P_2) receptor or is being degraded to AMP and adenosine before acting on a P_1 purinoceptor? Four tests can be carried out, which would indicate that ATP is broken down to adenosine before acting:

1. if ATP is equipotent or less potent than adenosine;
2. if the responses to ATP are blocked by methylxanthines and potentiated by dipyridamole;
3. if the slowly degradable analogues, α,β-methylene ATP and α,β-methylene ATP, are ineffective (bearing in mind that the α,β analogue is broken down relatively quickly and that the α,β analogue sometimes has other actions);
4. if the combination of AMP deaminase and adenosine deaminase blocks the responses to ATP as well as AMP and adenosine.

DR. H.P. BAER: Commenting on Dr. Burnstock's remarks, I would like to say that, although his suggestion for distinguishing effects of ATP and adenosine is basically correct and logical, one must be cautious. There are reports that ATP and adenosine act on the same receptor (thus, potency comparisons would not be conclusive), and the α,β-methylene phosphonate analogues of ATP or ADP have been shown in at least one system to cause adenosine-like or ATP-like effects via a receptor or pathway distinct from those for adenosine or ATP. May I also ask whether forskolin has been applied to cortical neurons to probe for cyclic AMP involvement?

DR. PHILLIS: We have not tested forskolin on cerebral cortical neurons, but I believe that Dr. Dunwiddie has tested this compound on his slice preparations. Perhaps he would like to comment on his results.

DR. T. DUNWIDDIE: In the rat hippocampus maintained *in vitro*, forskolin did not appear to have any significant effect on synaptically evoked responses; however, we did not determine whether forskolin interacted in any way with adenosine responses.

DR. M.H. MAGUIRE: Do the hypotensive responses to the phenylisopropyl and cyclohexyl derivatives reflect the slow onset of action of the compounds in the brain? Many adenosine analogues have long-lasting hypotensive effects but rapid onset of hypotensive responses similar to that of adenosine.

DR. PHILLIS: The slow chart speed at which we recorded the blood pressure effects of these adenosine derivatives precluded an accurate comparison of the rate of onset of their effects. However, there did not appear to be any gross differences in their actions.

27. SYNAPTIC MODULATION BY ADENOSINE: ELECTROPHYSIOLOGICAL AND BIOCHEMICAL CHARACTERISTICS

PETER SCHUBERT, KEVIN LEE, MARTIN REDDINGTON, GEORG KREUTZBERG

The concept of purinergic modulation of CNS activity by adenine compounds has recently attracted substantial experimental attention. This notion was originally formulated on the basis of observations that certain purine compounds are capable of depressing neuronal activity and are also potent stimulants of cAMP synthesis in the brain [1, 2]. More recently, the identification of multiple adenosine receptors and the availability of radiolabeled ligands with high specific activity has broadened the conceptual and technical approaches within this field. A correlative approach to these different sets of data was one of the primary goals of our current experiments. On the basis of evidence gathered from these technically disparate avenues of approach, a better understanding of purine modulation of CNS function has begun to emerge.

While embracing a variety of technical approaches, the present investigations will focus on a single central structure, the hippocampus. This region of the brain provides a particularly useful test system, owing to the simplicity of its organization, in terms of both intrinsic cellular structure and afferent distribution. Furthermore, this structure is readily adapted to an *in vitro* test system (i.e., the hippocampal slice), which has been extensively characterized.

Utilizing extra- and intracellular recordings, we have examined the depressive effect of adenosine and adenine nucleotides on synaptic transmission in both excitatory and inhibitory neuronal circuits. The latter is of particular functional interest, since a modulation of inhibitory circuits would change the delicate balance between excitation and inhibition, resulting in a qualitative

R.M. Berne, T.W. Rall, and R. Rubio (eds.), Regulatory Function of Adenosine, *pp. 439–454.*

change in the evoked activity pattern. To elucidate the possible factors by which purinergic modulation in the CNS may be specified, emphasis was given to the identification and localization of the adenosine receptor mediating this effect. Additional correlative electrophysiological and biochemical experiments were undertaken to test whether the strength of adenosine action is determined to some extent by the regional distribution and local density of adenosine receptors. An additional manner in which the strength and time course of purinergic modulation is thought to be determined is via the release and subsequent uptake of adenine compounds from the extracellular space. Attention was therefore directed to the identification of the sites of release and uptake and how these processes might be controlled.

ELECTROPHYSIOLOGICAL EVALUATION OF THE DEPRESSIVE EFFECT OF ADENOSINE ON SYNAPTIC TRANSMISSION AT EXCITATORY AND INHIBITORY NEURONAL CONNECTIONS

When tested in a hippocampal slice preparation, micromolar concentrations of adenosine were found to depress, in a dose-dependent manner, extracellularly recorded, evoked potentials from CA_1 neurons generated by stimulation of afferent fibers in the stratum radiatum [3, 4] (Fig. 27.1). The observed reduction in the size of the population spike, recorded from the soma layer of CA_1 neurons, indicates that adenosine reduces the number of neurons that generate action potentials in response to a given amount of afferent activation. The depression of the population spike is accompanied by a reduction in the slope and amplitude of the synaptic potentials recorded in the dendritic region, where the activated afferent fibers terminate. This finding suggests that adenosine is acting at synaptic sites, a conclusion that is supported by the data obtained from a current source-density analysis [3]. This technique provides more precise information about the size and distribution of current fluxes during neuronal activation. In the presence of adenosine, the synaptic currents (i.e., the inward currents generated in the area of afferent fiber termination) are attenuated (Fig. 27.2). Further support for a synaptic site of adenosine action was obtained from intracellular recordings using the same experimental preparation [5] (Fig. 27.2). The postsynaptic excitatory potentials that were elicited by mild afferent stimulation (below the threshold for generating action potentials) were found to be reduced in size if adenosine was present. In low adenosine concentrations (up to 10 μM), such a depression was seen without any noticeable change in membrane polarization and input resistance, indicating that the electrical properties of the postsynaptic neurons apparently remain unchanged. Although the possibility of postsynaptic adenosine effects at remote dendritic sites cannot be ruled out, the observed reduction in the size of the evoked postsynaptic potentials is consistent with a presynaptic site of adenosine action [6].

The postsynaptic effects of adenosine as reported in the literature [7], i.e., a hyperpolarization of the membrane potential in the postsynaptic neurons,

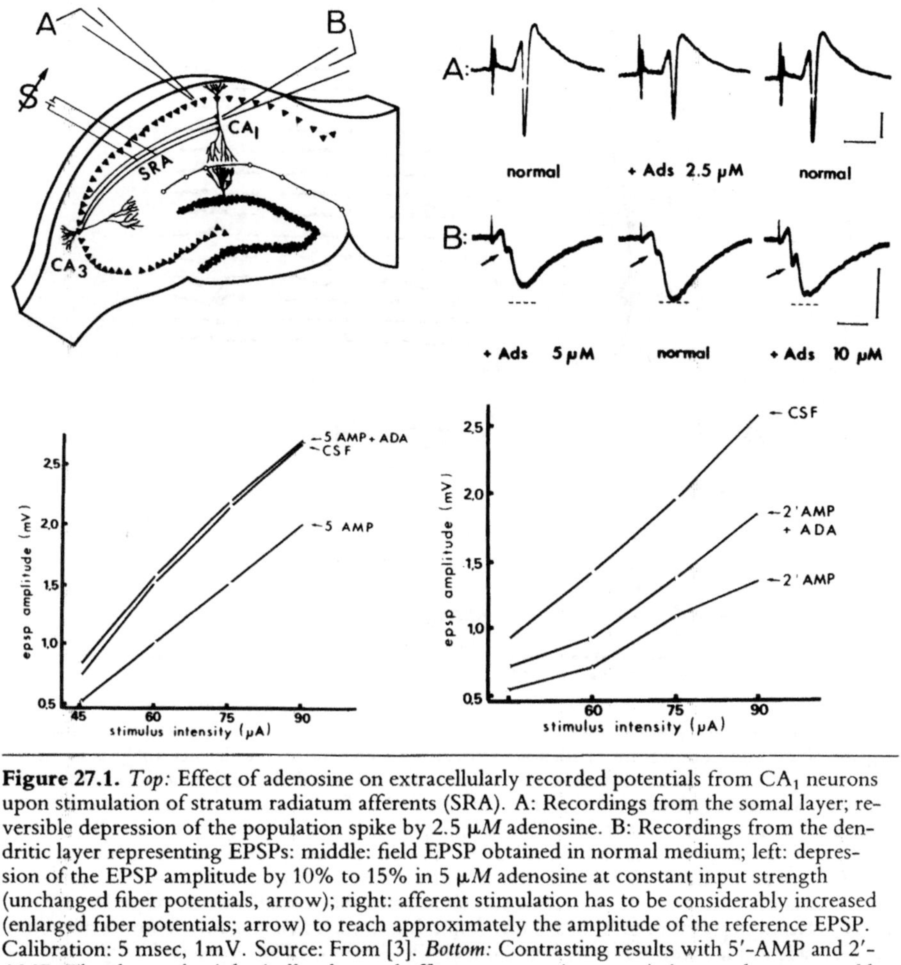

Figure 27.1. *Top:* Effect of adenosine on extracellularly recorded potentials from CA_1 neurons upon stimulation of stratum radiatum afferents (SRA). A: Recordings from the somal layer; reversible depression of the population spike by 2.5 μ*M* adenosine. B: Recordings from the dendritic layer representing EPSPs: middle: field EPSP obtained in normal medium; left: depression of the EPSP amplitude by 10% to 15% in 5 μ*M* adenosine at constant input strength (unchanged fiber potentials, arrow); right: afferent stimulation has to be considerably increased (enlarged fiber potentials; arrow) to reach approximately the amplitude of the reference EPSP. Calibration: 5 msec, 1mV. Source: From [3]. *Bottom:* Contrasting results with 5′-AMP and 2′-AMP. The electrophysiologically observed effects on synaptic transmission are demonstrated by input-output graphs in which the sizes of the evoked extracellular EPSPs (abscissa) are plotted against the stimulus intensities used (ordinate). The graphs have been originated in the presence and absence of the compounds indicated and reveal a depressive effect with both nucleotides. The depression with 5′-AMP (left) is completely antagonized by adenosine deaminase (ADA); the depression with 2′-AMP is only partially antagonized (right). Source: From [20].

were only observed if higher adenosine concentrations were used (exceeding 10 μ*M*). At these concentrations, the observed adenosine effects resembled those seen in the presence of N^6-monobutyryl cyclic AMP, a cyclic AMP derivative that is believed to be membrane-permeable and is relatively resistant to hydrolysis. Therefore, it seems possible that high adenosine concentrations lead to a further depression of the efficiency of synaptic transmission by producing a membrane hyperpolarization, perhaps mediated by stimulation of the synthesis of cyclic AMP. However, to reiterate, the depressive effect of

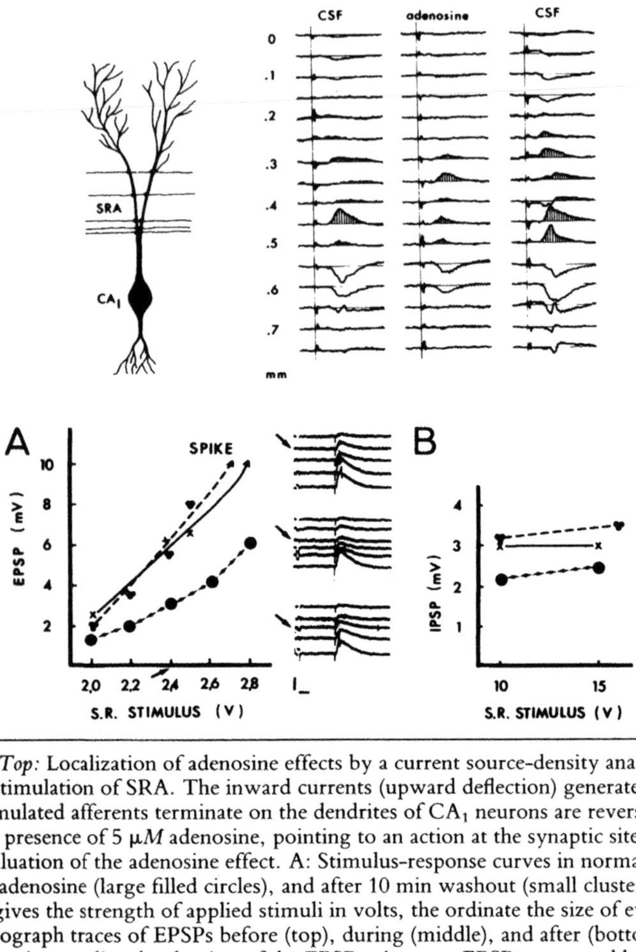

Figure 27.2. *Top:* Localization of adenosine effects by a current source-density analysis in the CA_1 area on stimulation of SRA. The inward currents (upward deflection) generated in the area where the stimulated afferents terminate on the dendrites of CA_1 neurons are reversibly depressed in the presence of 5 μM adenosine, pointing to an action at the synaptic sites. *Bottom:* Intracellular evaluation of the adenosine effect. A: Stimulus-response curves in normal medium (X), in 5 μM adenosine (large filled circles), and after 10 min washout (small clustered circles). The abscissa gives the strength of applied stimuli in volts, the ordinate the size of evoked EPSPs in mV. Oscillograph traces of EPSPs before (top), during (middle), and after (bottom) adenosine show the adenosine-mediated reduction of the EPSPs. Arrows, EPSPs generated by identical 2.4 V stimuli. B: Stimulus-response curve demonstrating the adenosine-induced depression of IPSPs. Calibration: 10 mV, 20 msec. Source: From [5].

adenosine that was observed at low concentrations seems unrelated to stimulation of cyclic AMP formation.

It is interesting that a depressive effect on intracellularly recorded evoked synaptic potentials was seen not only on excitatory but also on inhibitory postsynaptic potentials [5]. The possibility of a modulation of inhibitory synaptic events, as previously suggested [8], was directly tested on a well-defined inhibitory circuit [9]. In the hippocampus, collaterals branching off from CA_1 pyramidal cell axons innervate inhibitory interneurons, which in turn feed back to the pyramidal neurons (see Fig. 27.3). This inhibitory loop is activated by firing of the pyramidal cell neurons, thus reducing their own excitability via feedback inhibition. The efficiency of this inhibitory circuit can be tested in

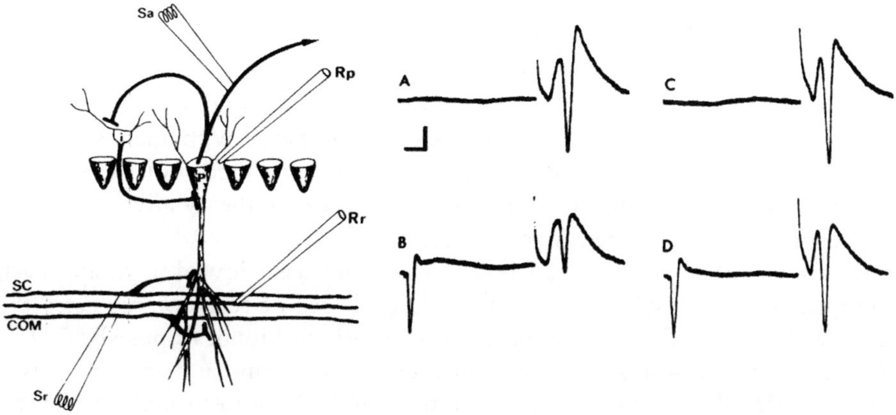

Figure 27.3. Effect of adenosine on paired-pulse inhibition. *Left:* Illustration of the circuitry under study in CA_1. Stimulation in the alveus (Sa) antidromically activates the pyramidal cells (P) and additionally the pyramidal cell axon collaterals, which innervate inhibitory interneurones (i). In these experiments, stimulation was presented to (1) Sr alone (unpaired) and (2) Sa first and then to Sr 35 msec later (paired). *Right:* A: Unpaired stimulation (Sr) in control medium. B: Paired stimulation (Sa–Sr) in control medium. An obvious inhibition of the paired response is seen. C: Unpaired stimulation (Sr) in the presence of 10 μM adenosine. D: Paired stimulation (Sa–Sr) in the presence of 10 μM adenosine. In C and D, the intensity of Sr stimulation was increased so as to evoke an unpaired population spike of similar size to that seen in control medium. By comparing the effect of Sr stimulation in B and D, the pronounced attenuation of paired pulse inhibition by adenosine can be seen. Source: From [9].

a slice preparation utilizing a paired-pulse stimulation paradigm. Following activation of this circuit by an antidromic conditioning stimulus, the number of pyramidal cells that are activated in response to subsequent afferent stimulation will be reduced. This effect can be assessed by the reduced size of the extracellularly recorded population spike. Paired-pulse inhibition is greatly attenuated if micromolar concentrations of adenosine are present, and this effect is completely reversible upon reinfusion of control medium (Fig. 27.3). Interfering with the regulatory influence of interneurons on the excitability of principal neurons is one of the most subtle ways to modulate signal processing in the brain and, as indicated by the findings, may be achieved by using adenine compounds as modulatory signals.

IDENTIFICATION OF THE RECEPTOR MEDIATING THE DEPRESSIVE EFFECT OF ADENOSINE ON SYNAPTIC TRANSMISSION

In recent years, biochemical studies on the interactions of adenosine and various nucleoside derivatives with adenylate cyclase have led to the classification of adenosine receptors into two types. In several cells, including those of CNS origin, adenosine has been shown to attenuate cyclic AMP synthesis [10, 11]. The receptor mediating this response has been designated A_1. The second receptor, designated A_2, however, is associated with stimulation of adenylate cyclase. These two receptor types can be distinguished by their different affinities for adenosine and various adenosine analogues. Thus the A_1 receptor is characterized by affinities in the nanomolar range, by a high stereoselectivity for the L- and D-enantiomers of phenylisopropyladenosine (PIA), and by the greater efficacy of N^6-substituted derivatives compared with adenosine. The A_2 receptor has a lower affinity for nucleosides (in the micromolar range) and shows little stereospecificity and little difference in the relative effects of adenosine and several adenosine derivatives.

The availability of several adenosine analogues has allowed us to approach the question as to which of the foregoing classes of receptor mediates the depressive effect of adenosine on evoked EPSPs in hippocampus slices [12]. Studies on the dose-response relationships for adenosine and several derivatives revealed that nanomolar concentrations of the N^6-substituted derivatives, L-PIA and cyclohexyl-adenosine (CHA), were sufficient to obtain 50% inhibition of the population EPSPs in hippocampal CA_1 neurons. L-PIA and D-PIA showed a marked difference (about 50-fold) in apparent affinities, demonstrating stereospecificity of the receptor mediating the depressive effect of nucleosides on synaptic transmission. Furthermore, adenosine was found to be considerably less effective than L-PIA and CHA. This observed order of potencies is consistent with the properties of an A_1 adenosine receptor. This finding supports previous suggestions that an increase in cyclic AMP synthesis, which would be associated with an A_2 receptor, is not involved in producing the modulatory effect of adenosine on synaptic transmission [13].

A direct biochemical approach to adenosine receptor characterization has been made possible by the recent availability of [^{3}H]adenosine analogues of high specific radioactivity [14–17]. Using [^{3}H]CHA, electrophysiologically the most potent adenosine derivative tested in the hippocampus slice preparation, we have shown that the high-affinity [^{3}H]CHA binding site of rat brain membranes, like that of guinea pig brain [14], has the properties of an A_1 receptor [12]. Dose-response curves based on the ability of various adenosine analogues to inhibit [^{3}H]CHA binding reveal the same general potency relationships seen in the electrophysiological studies (Fig. 27.4). It would therefore appear that the [^{3}H]CHA binding site does indeed represent the receptor mediating the depressive effect of adenosine on synaptic transmission in the hippocampus.

The similarity of the [^{3}H]CHA binding site to a physiological receptor suggests that [^{3}H]CHA binding can be employed as a marker for the presence of A_1 adenosine receptors. One such application is the use of autoradiographic methods to study the distribution of such receptors, knowledge of which is clearly a prerequisite for our ultimate understanding of the physiological significance of purinergic modulation. Autoradiograms show that A_1 receptors are selectively distributed in the brain (see Fig. 27.4). In the dorsal aspect of the rat hippocampus, for example, [^{3}H]CHA binding was particularly high in the dendrites of CA_1 neurons; little activity was found in association with the cell

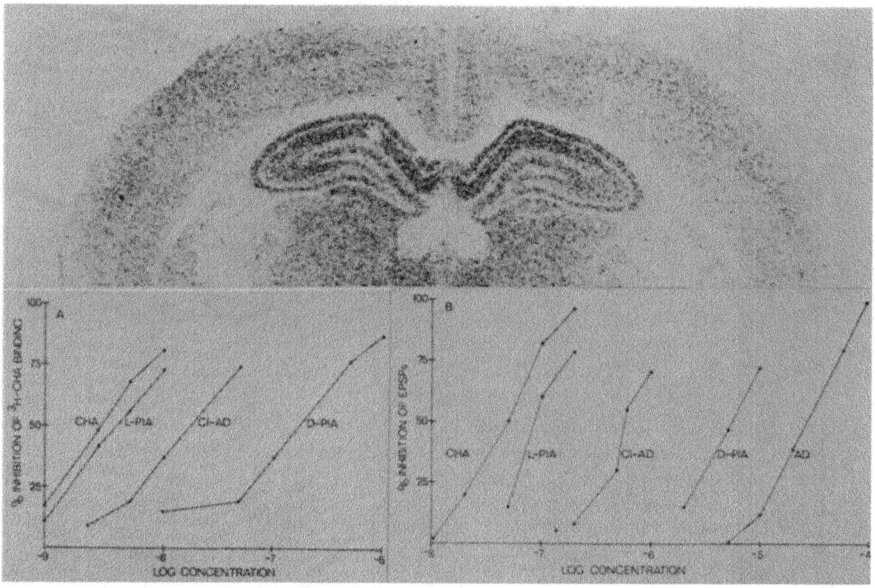

Figure 27.4. Autoradiography of [^{3}H]CHA binding and dose-response curves for EPSP depression and inhibition of CHA binding. *Top:* A light microscopic autoradiograph of the rat brain, in which the dark areas represent specific binding of [^{3}H]CHA. *Bottom left:* Dose-response curves for several adenosine analogues, which indicate their capacity to displace [^{3}H]CHA binding to membranes prepared from rat forebrain. *Bottom right:* Dose-response curves for the capacity of the same analogues to depress extracellularly recorded population EPSPs in the stratum radiatum of CA_1. The relative potencies of these analogues for the displacement of binding and the depression of EPSPs are identical, suggesting that the binding site in the displacement studies is indeed that which mediates the depression of the EPSPs. CHA, cyclohexyladenosine; PIA, phenylisopropyladenosine; C1-AD, 2-chloroadenosine; AD, adenosine. Source: From [12].

body layer. Such a highly selective receptor distribution presumably reflects a topographical specificity of purinergic modulation. It might be expected that the strength of purine action would be determined not only by the local extracellular adenosine concentration but also by the receptor density. This assumption proved to be approachable in the rat hippocampus, where we found differences in [^{3}H]CHA binding between the dorsal aspect and the ventral part. A more detailed analysis of the data showed that the higher binding observed in the dorsal hippocampus reflected a greater concentration of binding sites, rather than differences in their affinities for [^{3}H]CHA. This dorsal/ventral difference also held true in the case of the CA_1 subregion (Fig. 27.5). In parallel electrophysiological experiments, the efficacy of adenosine in reducing evoked EPSPs was tested in slices cut from the dorsal and ventral hippocampus. At all concentrations tested (1–100 μM), a more pronounced depression of evoked potentials was found dorsally than ventrally [unpublished observations] (Fig. 27.5). This correlation of the strength of adenosine action and the

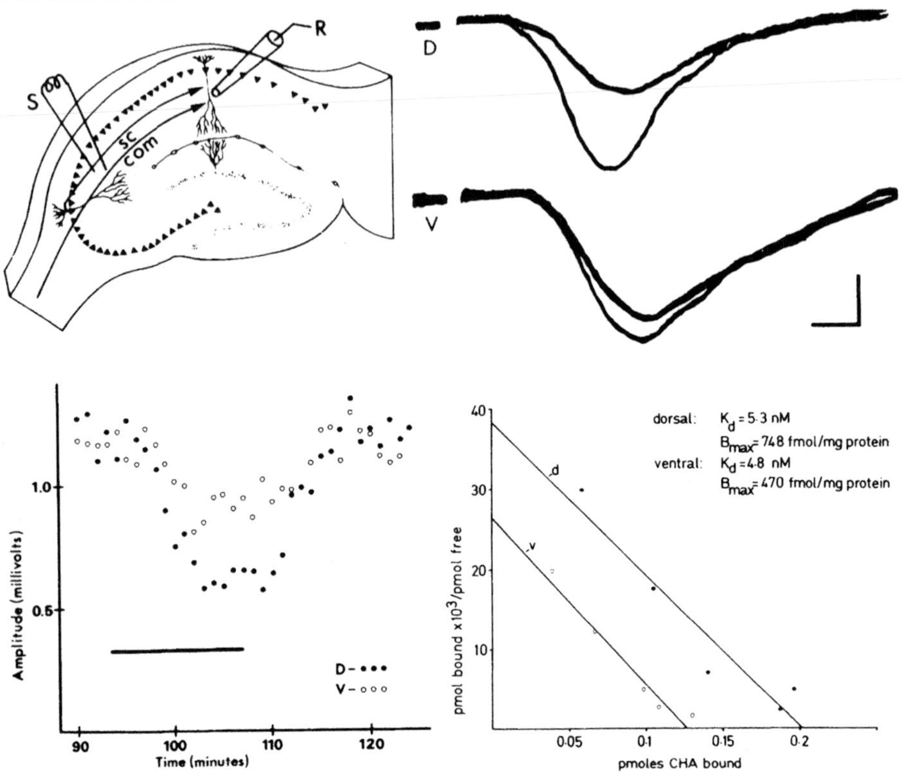

Figure 27.5. Differential depression of evoked potentials in the dorsal and ventral hippocampus. The combined Schaffer collateral (sc) and commissural (com) inputs were tested by placing stimulating (S) and recording (R) electrodes as depicted. *Top right:* Extracellularly recorded population EPSPs (simultaneously evoked) taken from slices of the dorsal (D) and ventral (V) aspects of the same hippocampus. In both cases, the larger negativity represents the response taken in control medium, whereas the smaller responses were elicited in the presence of 25 μ*M* adenosine. The depression caused by adenosine is clearly greater in the dorsal slice (calibration bars: 2 msec and 0.5 mV). *Bottom left:* The time course of another experiment in which the amplitudes of evoked potentials were measured with a MINC 11 computer. The dorsal slice (filled circles) was depressed to a greater extent than was its ventral counterpart (open circles). The solid bar indicates the time during which 25 μ*M* adenosine was superfused. The *y*-axis of the graph is the amplitude of the evoked responses in millivolts and the *x*-axis is the age of the preparation in minutes. *Bottom right:* A Scatchard plot of the binding to membranes from dorsal and ventral aspects of the hippocampus. The number of [^{3}H]CHA binding sites is greater in the dorsal aspect, whereas the apparent affinities are similar [unpublished observations].

number of [^{3}H]CHA binding sites strongly suggests that the local concentration of A_1 receptors is indeed an important determinant of the extent of purinergic modulation of synaptic transmission.

FURTHER FACTORS SPECIFYING THE MODULATORY ACTION OF ADENOSINE ON SYNAPTIC TRANSMISSION

In addition to the local concentration of adenosine receptors, it might be expected that factors affecting local adenosine concentrations also play a role in

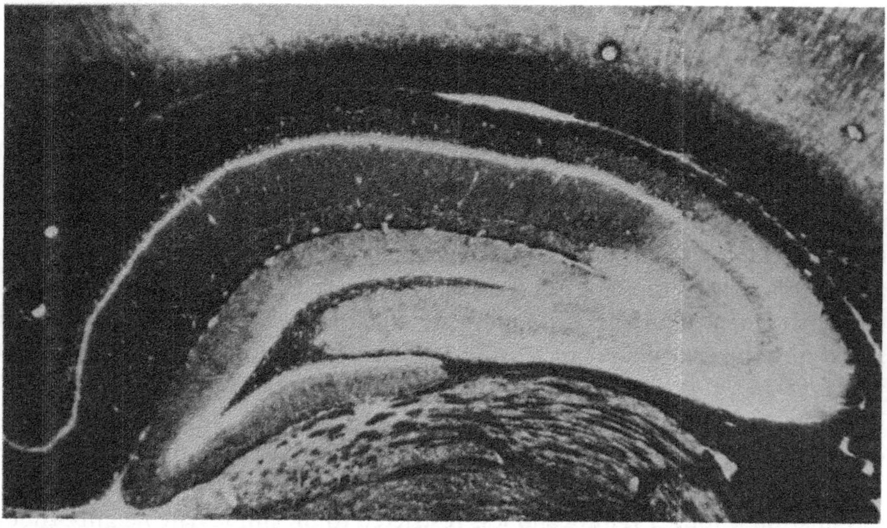

Figure 27.6. Enzyme histochemical demonstration of the 5′-nucleotidase activity in the hippocampus, revealing a highly selective distribution.

determining the overall characteristics of purinergic modulation. One such factor is the amount of the enzyme, 5′-nucleotidase, which hydrolyses 5′-AMP to produce adenosine. This enzyme shows a highly selective distribution within the brain, as demonstrated by light microscopical histochemical techniques (Fig. 27.6). Cytochemical studies at the electron microscopical level have shown 5′-nucleotidase activity often associated with the plasma membrane of glial cells, and astrocytic processes carrying 5′-nucleotidase are frequently found to engulf synaptic complexes [18]. The enzyme would therefore appear to be in a suitable position to form adenosine from nucleotides, probably released from synaptic terminals (see, e.g., [19]).

In order to elucidate the presumed function of 5′-nucleotidase in relation to purinergic modulation, we tested whether nucleotides by themselves are able to produce the depressive effect or whether the formation of adenosine is a prerequisite [20]. The addition of 5′-AMP to the superfused medium of the hippocampal slice preparation produces a marked depression of evoked potentials in the CA_1 region, an area rich in 5′-nucleotidase activity. In this case, the effect is actually elicited by adenosine after hydrolysis of 5′-AMP. Thus, adenosine deaminase, which converts adenosine into the electrophysiologically rather ineffective inosine, completely abolishes the depression of evoked potentials by 5′-AMP (Fig. 27.1). However, 2′-AMP, which is only poorly hydrolyzed by brain membrane preparations, also revealed a depression of evoked potentials, which in this case was not antagonized by adenosine deaminase (Fig. 27.1). It thus appears that nucleotides by themselves might be capable of producing a modulatory effect. This is in accordance with the biochemical observation that 5′-AMP inhibits [^{3}H]CHA binding to rat brain membranes if tested in the presence of a 5′-nucleotidase inhibitor. Thus 5′-AMP can interact with the adenosine receptor mediating the electrophysiological effect. The relative potency of nucleotides compared to adenosine in producing purinergic modulation is as yet unknown.

Although the formation of adenosine would seem not to be an absolute prerequisite for eliciting purinergic modulation, 5′-nucleotidase may still represent an important controlling factor. By hydrolyzing 5′-AMP to adenosine, it brings the membrane-impermeable nucleotide into a permeable form, thus allowing a rapid removal from the extracellular space by reuptake into cellular compartments. It is therefore possible that 5′-nucleotidase is involved in specifying the time course of adenosine action by regulating the rate of purine inactivation. It is interesting that the selective distribution of this enzyme resembles in some ways the distribution of [^{3}H]CHA binding sites. Some exceptions are found, however, where high binding is found in areas of low enzyme activity. It is interesting to speculate that, in these areas, purinergic modulation is longer lasting and less restricted to the immediate neighborhood of the site of release than in an area rich in 5′-nucleotidase.

DOES PURINERGIC MODULATION OCCUR IN THE NORMALLY FUNCTIONING BRAIN?

Although pharmacological experiments of the type described clearly demonstrate that adenine compounds are capable of producing a depression of synaptic transmission, they do not show whether purinergic modulation does in fact occur in the normally functioning brain. In particular, the question remains to be answered as to whether adenine compounds, artificially added in the experimental situation, are physiologically present in sufficient extracellular concentrations to exert a modulatory influence on synaptic transmission. In conjunction with this idea, we have to consider which sites are the endogenous sources that sustain and regulate the local extracellular concentration of adenosine in the brain. Furthermore, the conditions controlling the delivery from these sources should be clarified.

From experiments in which the effect of adenosine deaminase on evoked potentials was examined, the physiologically present concentration of adenosine was shown to be sufficient to elicit a slight, but clear, tonic effect on synaptic transmission. In the presence of adenosine deaminase, which converts adenosine into the electrophysiologically ineffective inosine, the amplitudes of the evoked potentials were found to be slightly enhanced [20, 21]. Since the amount of depression produced by a given concentration of adenosine is experimentally known, the physiologically effective extracellular concentration of adenosine can be estimated. The amount of observed "disinhibition" with this enzyme was found to correspond with the depression elicited by low micromolar concentrations of exogenously applied adenosine. This value is consistent with the reported values derived from direct measurements of the extracellular adenosine concentration in brain tissue [22].

Since the brain appears to be a closed system for adenine nucleotides, well shielded from foreign influences by an effective blood brain barrier [23], the extracellular concentration of these compounds within the brain should be determined to a great extent by the local release from intracellular compartments. Specific information about a possible release from glial cells is lacking so far, but there is evidence for a controlled release occurring from nerve cells. By an effective intracellular transport system, operative in dendrites as well as in axons, nucleotides may be shifted within neurons over long distances before being released. This process might well produce local variations of adenosine concentration. Release of adenosine derivatives was found to occur from a postsynaptic site (i.e., from the dendritic branches of the neuron), as well as from the axon and axon terminals [24–27]. The latter has been shown for several pathways in the central nervous system and in particular for major afferent projections to the hippocampus. In the hippocampus, a presynaptic release of axonally transported adenosine derivatives was autoradiographically demonstrated to occur from fibers originating in (1) the entorhinal cortex [28] (Fig. 27.7), (2) the contralateral hippocampus, and (3) the septum [29]. The release from septal fibers (which have been shown to be cholinergic) indicates that the release of adenine nucleotides is a general phenomenon and may occur in addition to the release of the principal transmitter. With respect to the presumed purinergic modulation of interneuron function (see earlier discussion), it is interesting to note that the release of adenosine derivatives from septal fibers appears to be directed to individual neurons that resemble interneurons [29]. This suggests that such a modulation might be regulated by the selective release from an afferent fiber system.

In this context, it is important that the release of adenine compounds from axon terminals has been shown to be related to nerve cell activity. This has been shown for the entorhino-hippocampus system by *in vivo* experiments [30] and was recently confirmed by a combined *in vivo/in vitro* approach [31]. In the latter case, a defined input system to the hippocampus, i.e., the entorhinal fibers, was first loaded with radioactive adenine compounds, using axonal transport as a vehicle. Slices were then prepared from the hippocampus, allowing the stimulation of precisely defined afferent tracts. Only stimulation of the labeled entorhinal afferents resulted in a significant increase of radioactive efflux. Thus, nerve cell activity seems to be another important factor whereby the local extracellular level of adenosine can be regulated.

On the basis of these data, it was intriguing to test whether the endogenous release of adenine compounds from axon terminals, initiated by afferent stimulation, leads to a sufficient elevation of the extracellular adenosine concentration to exert a modulatory effect on synaptic transmission. If a single test pulse to the stratum radiatum afferents is preceded (20–250 msec) by conditioning stimulation of the same pathway, then the paired "test" response is significantly greater than an unpaired "test" response. This phenomenon, called paired-pulse facilitation, is presumably the result of increased presynaptic calcium concentrations as a consequence of the conditioning stimulation.

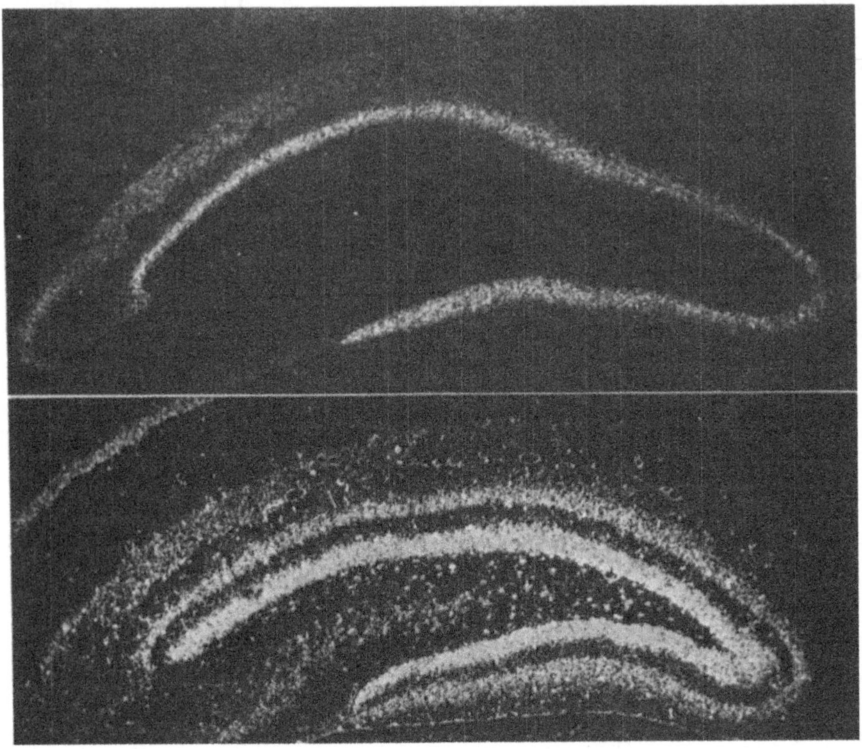

Figure 27.7. Axonal transport of leucine and transneuronal transfer of adenosine. Dark field autoradiographs of the transport of leucine (above) and adenosine (below) to the dentate gyrus, 2 days following injections into the entorhinal cortex. Both cases show labeling of the terminal zone of the entorhinal cortical afferents (i.e., the molecular layer of the dentate gyrus). In the case of adenosine, however, the cell body region of the postsynaptic cells (i.e., the granule cell layer) is additionally labeled, indicating a transneuronal transfer from the pre- to postsynaptic element.

If adenosine is released by the conditioning pulse, then the efficiency of the test pulse should also be influenced by purinergic modulation. Thus, in the presence of adenosine deaminase, such an influence would presumably be abolished and an even greater paired-pulse facilitation should occur. In the present experiments, three conditioning stimuli were delivered at 3 msec intervals, followed by a single test pulse 100 msec later. The amount of paired-pulse facilitation under these stimulation conditions was significantly greater in the presence of adenosine deaminase. It therefore appears that a sufficient amount of adenosine is released by the three conditioning pulses to exert an effect on subsequent afferent activity. Furthermore, the action of the released material is relatively rapid, exerting an effect by at least 100 msec after its release.

CONCLUDING REMARKS

Taken together, these recent data have resulted in a clearer picture of the manner in which adenosine modulates synaptic transmission. Although the depressive effect on neuronal activity has generally been confirmed, the finding that inhibitory circuitry is also affected by adenosine provides another potent mechanism whereby purines can refine neuronal activity. The identification of specific receptors that mediate the physiological phenomenon renders an ubiquitous compound, such as adenosine, capable of restricted and well-defined action. Furthermore, the differential distribution of the receptors and resultant differential physiological responses appear to be a means whereby both the sites and the strength of adenosine modulation are specified. Activity-related release of endogenous stores of adenosine and/or related compounds is capable of modifying neuronal activity, and associated enzymes (e.g., 5′-nucleotidase) appear to be in a position to regulate the time course of their action. Therefore, the emerging concept of purinergic modulation of synaptic transmission appears to be one of a dynamic system capable of significantly contributing to signal processing.

REFERENCES

1. Sattin A, Rall TW: Cyclic AMP content of guinea pig cerebral cortex slices. *Mol Pharmacol* 6:13–23, 1970.
2. Phillis JW, Kostopoulos GK, Limacher JJ: Depression of corticospinal cells by various purines and pyrimidines. *Can J Physiol Pharmacol* 52:1226–1229, 1974.
3. Schubert P, Mitzdorf U: Analysis and quantitative evaluation of the depressive effect of adenosine on evoked potentials in hippocampal slices. *Brain Res* 172:186–190, 1979.
4. Dunwiddie TV, Hoffer BJ: Adenine nucleotides and synaptic transmission in the in vitro rat hippocampus. *Br J Pharmacol* 69:59–68, 1980.
5. Siggins G, Schubert P: Adenosine depression of hippocampal neurons in vitro: An intracellular study of dose-dependent actions on synaptic and membrane potentials. *Neurosci Lett* 23:55–60, 1981.
6. Fredholm BB, Hedqvist P: Modulation of neurotransmission by purine nucleotides and nucleosides. *Biochem Pharmacol* 29:1635–1643, 1980.
7. Phillis JW, Edstrom JP, Kostopoulos GK, Kirkpatrick JR: Effects of adenosine and adenine nucleotides on synaptic transmission in the cerebral cortex. *Can J Physiol Pharmacol* 57:1289–1312, 1979.
8. Stone TW, Taylor DA: Effects of adenosine and related compounds on an inhibitory process in rat cerebral cortex. *Exp Neurol* 70:556–566, 1980.
9. Lee K, Schubert P: Modulation of an inhibitory circuit by adenosine and AMP in the hippocampus. *Brain Res* 246:311–314, 1982.
10. Londos C, Cooper D, Wolff J: Subclasses of external adenosine receptors. *Proc Natl Acad Sci USA* 77:2551–2554, 1980.
11. van Calker D, Müller M, Hamprecht B: Adenosine regulates via two different types of receptors the accumulation of cyclic AMP in cultured brain cells. *J Neurochem* 33:999–1005, 1979.
12. Reddington M, Lee K, Schubert P: Depression of evoked potentials in a hippocampal slice preparation by adenosine: mediation via an A_1 adenosine receptor, characterized by ^{3}H-cyclohexyladenosine binding. *Neurosci Lett* 28:275–279, 1982.
13. Reddington M, Schubert P: Parallel investigations of the effects of adenosine on evoked potentials and cyclic AMP accumulation in hippocampus slices of the rat. *Neurosci Lett* 14:37–42, 1979.
14. Bruns RF, Daly JW, Snyder SH: Adenosine receptors in brain membranes: Binding of N^6-

cyclohexyl [^{3}H]adenosine and 1,3-diethyl-8-[^{3}H]phenylxanthine. *Proc Natl Acad Sci USA* 77:5547–5551, 1980.
15. Schwabe U, Trost T: Characterization of adenosine receptors in rat brain by (−)[^{3}H]N^6-phenylisopropyladenosine. *Naunyn Schmiedebergs Arch Pharmacol* 313:179–187, 1980.
16. Williams M, Risley EA: Biochemical characterization of putative purinergic receptors by using 2-chloro [^{3}H]adenosine, a stable analog of adenosine. *Proc Natl Acad Sci USA* 77:6892–6896, 1980.
17. Wu PH, Phillis JW, Balls K, Rinaldi B: Specific binding of 2-[^{3}H]chloroadenosine to rat brain cortical membranes. *Can J Physiol Pharmacol* 58:576–579, 1980.
18. Kreutzberg GW, Barron KD, Schubert P: Cytochemical localization of 5′-nucleotidase in glial plasma membranes. *Brain Res* 158:247–257, 1978.
19. Barberis C, McIlwain H: 5-adenine mononucleotides in synaptosomal preparations from guinea pig neocortex: Their change on incubation, superfusion and stimulation. *J Neurochem* 26:1021–1025, 1976.
20. Lee KS, Schubert P, Emmert H, Kreutzberg GW: Effect of adenosine versus adenine nucleotides on evoked potentials in a rat hippocampal slice preparation. *Neurosci Lett* 23:309–314, 1981a.
21. Dunwiddie TV: Endogenously released adenosine regulates excitability in the in vitro hippocampus. *Epilepsia* 21:541–548, 1980.
22. Phillis JW, and Wu PH: The role of adenosine and its nucleotides in central synaptic transmission. *Prog Neurobiol* 16:187–239, 1981.
23. Berne RM, Rubio R, Curnish RR: Release of adenosine from ischaemic brain. *Circ Res* 35:262–272, 1974.
24. Schubert P, Kreutzberg G: Axonal transport of adenosine and uridine derivatives and transfer to postsynaptic neurones. *Brain Res* 76:526–530, 1974.
25. Schubert P, Kreutzberg GW: Dendritic and axonal transport of nucleoside derivatives in single motoneurons and release from dendrites. *Brain Res* 90:319–323, 1975.
26. Hunt SP, Künzle H: Bidirectional movement of label and transneuronal transport phenomena after injection of [^{3}H] adenosine in the central nervous system. *Brain Res* 112:127–132, 1976.
27. Wise SP, Jones EG, Berman N: Direction and specificity of the axonal and transcellular transport of nucleosides. *Brain Res* 139:197–217, 1978.
28. Schubert P, Rose G, Lee K, Lynch G, Kreutzberg G: Axonal release and transfer of nucleoside derivates in the entorhinal-hippocampal system: An autoradiographic study. *Brain Res* 134:347–352, 1977.
29. Rose G, Schubert P: Release and transfer of [^{3}H]adenosine derivatives in the cholinergic septal system. *Brain Res* 121:353–357, 1977.
30. Schubert P, Lee K, West M, Deadwyler S, Lynch G: Stimulation-dependent release of [^{3}H]adenosine derivatives from central axon terminals to target neurones. *Nature* 260:541–542, 1976.
31. Lee KS, Schubert P, Gribkoff V, Sherman B, Lynch G: A combined in vivo/in vitro study of the presynaptic release of adenosine derivatives in the hippocampus. *J Neurochem* 38:80–83, 1981.

DISCUSSION

DR. J.W. PHILLIS: Your example of autoinhibition along a given pathway by adenosine is very interesting and reminds me of Ribeiro's experiments, which show that Wedensky inhibition in skeletal muscle may be due to adenosine autoinhibition. My question pertains to your observations of an adenosine-elicited block of recurrent inhibition in the hippocampus. Do you know whether this action of adenosine was exerted at the excitatory synapse on the inhibitory interneuron or at the inhibitory synapse (or both)?

DR. SCHUBERT: We don't know. Because of the well-established effect of adenosine on excitatory synapses at which glutamate is presumably the transmitter, I think that certainly the afferent synaptic connection between the axon collateral of the pyramidal cell and the interneuron is modulated by adenosine. So far, we don't know whether or not the inhibitory GABAergic connection between the axon of the interneuron and the pyramidal cell soma is affected in addition; this seems less likely. Thus, a reduction of GABA release from brain tissue was reported to occur only in the presence of very high, presumably unphysiological GABA concentrations in the millimolar range. Furthermore, in preliminary experiments testing the effect of pressure-injected, locally applied adenosine on synaptic transmission, we did not see any effect upon injection of adenosine into the soma region of the pyramidal cells. This suggests that the inhibitory synapse of the interneuron is less or not affected by adenosine. More detailed experiments along this line, as well as intracellular recordings from the interneurons, should give more precise information.

DR. R. BÜNGER: Dr. L. Sokoloff at NIH has shown very tight correlations between local cerebral blood flow and local glucose utilization (as measured by the deoxyglucose method). Do you think there could be a relationship between the release of vasoactive cerebral adenosine and the rate of glucose utilization?

DR. SCHUBERT: We know that the vasoactive adenosine (presumably in the form of nucleotides) is released from brain tissue in a stimulation-dependent manner. It seems not unlikely, therefore, that adenosine release serves under physiological conditions to link nerve cell activity and regulation of local blood flow. There is no experimental evidence suggesting that the observed concomitant changes in glucose utilization may be directly related to and caused by the adenosine release.

DR. PHILLIS: What is your evidence that adenine nucleotides activate the P_1 receptor directly without prior degradation to adenosine?

DR. SCHUBERT: If the effect of 5′-AMP on synaptic transmission of pyramidal cells is tested in the CA_1 region, a marked depression is seen, which is reversible by adenosine deaminase (ADA). This indicates that the effect is exerted by adenosine, which is rapidly formed in this area rich in 5′-nucleotidase. But if nucleotide analogues are also tested (e.g., 2′-AMP, AOPCP, AIP) that are rather resistant to hydrolysis, a similar depression of evoked responses was observed, and this depression was not abolished by ADA. This suggests that nucleotides on their own are capable of depressing synaptic transmission, and this might be of physiological relevance in areas where enzymatic hydrolysis of AMP is rather poor.

DR. T.W. RALL: I do not understand your response to Dr. Phillis's question. Did you say that 2-chloradenosine's effects were not very susceptible to adenosine deaminase and that this indicated action of nucleotides?

DR. SCHUBERT: No, the depressive effect of 2′-AMP and other adenine nucleotide analogues on synaptic transmission were tested and found to be not abolished in the presence of adenosine deaminase. This rules out a possible action of nucleotides via adenosine, since adenosine deaminase would have abolished this action by converting adenosine into the electrophysiologically rather ineffective inosine.

DR. L.E. GUSTAFSSON: (1) Did you try other low molecular weight compounds, such as other nucleosides or glucose, in your autoradiography transport experiments? (2) Where in the molecule was your label, and did you check whether or not this label was still in the purine backbone after transport?

DR. SCHUBERT: We also tested uridine, cytidine, and guanosine. Uridine and cytidine derivatives were effectively transported in the axon and released to postsynaptic neurons, but not the guanosine derivatives. The relative proportion of observed transneuronal transfer and the time course of postsynaptic accumulation of radioactive material were quite different for the different nucleosides. We used tritium-labeled adenosine, labeled at position 2 and/or 8 of the purine ring. After one day to allow for transport from the entorhinal cortex to the hippocampus, 94% of radioactivity recovered from the hippocampal target area was found to be in the TCA-soluble fraction. If these adenosine compounds transported to the hippocampus were further analyzed by thin-layer chromatography, the majority of radioactivity was found associated with adenosine, inosine hypoxanthine, some also with cyclic AMP, and a minor proportion was still present in the form of nucleotides.

DR. A. SATTIN: Which portions of hippocampus were labeled following injection of [^{3}H]adenosine into medial septal nucleus?

DR. SCHUBERT: After injection of a higher dose and with three days allowed for transport in septal fibers, labeling was found in the granule cell layer and scattered in the CA_4 layer of the dentate gyrus. Labeling was also associated with the pyramidal cell somal layers of regions superior and inferior to the hippocampus. Labeling of CA_3 neurons was more pronounced than that of CA_1 neurons and coincided with the distribution of ACHE activity, thereby marking the areas of cholinergic input. It is interesting that, after injection of a minute amount and with only one day allowed for transport, a highly selective labeling was seen in individual neurons that resembled hippocampal interneurons in size and location. The possibility that they were labeled by retrograde transport seems unlikely, since they were always accompanied by labeled glial cells, indicating local release of radioactive adenosine compounds. Thus, these neurons seem to receive a particular dense projection from septal fibers, markedly exceeding the strength of septal input to other hippocampal neurons.

28. MODULATION OF CALCIUM CURRENTS BY ADENOSINE RECEPTORS ON MAMMALIAN SYMPATHETIC NEURONS

BARBARA K. HENON, DONALD A. McAFEE

The modulation of voltage-dependent Ca currents is a powerful mechanism for the regulation of neuronal activity. Several neurotransmitters have now been shown to have this capability. GABA, noradrenalin, and serotonin inhibit neuronal Ca conductances in the dorsal root ganglion [1]. Noradrenalin also inhibits Ca currents in the postganglionic neurons of the rat superior cervical ganglion [2]. Although it is not clear that adenosine qualifies as a neurotransmitter, we have now shown that adenosine inhibits voltage-dependent Ca currents by activation of adenosine receptors on the somata of these postganglionic neurons [3].

Ca channels are located on the somata as well as the dendrites of most neurons, but Ca channels on the presynaptic axon terminals are strategically located to play a key role in the regulation of neurotransmitter release. In the superior cervical ganglion, adenosine appears to have presynaptic actions in addition to its postsynaptic effect on Ca currents. Adenosine analogues reduce the amplitude of EPSPs in peripheral and central neurons [4–8]. However, we have found that adenosine does not act simply to inhibit neuronal activity, but may actually facilitate synaptic transmission during repetitive firing. These effects of adenosine are also dependent on extracellular Ca^{2+} concentrations, and we will present evidence that the presynaptic actions of adenosine may include inhibition of the Ca currents on the presynaptic axon terminals. In these studies, we have used the rat superior cervical sympathetic ganglion, as detailed in Figure 28.1.

R.M. Berne, T.W. Rall, and R. Rubio (eds.), Regulatory Function of Adenosine, *pp. 455–466.*

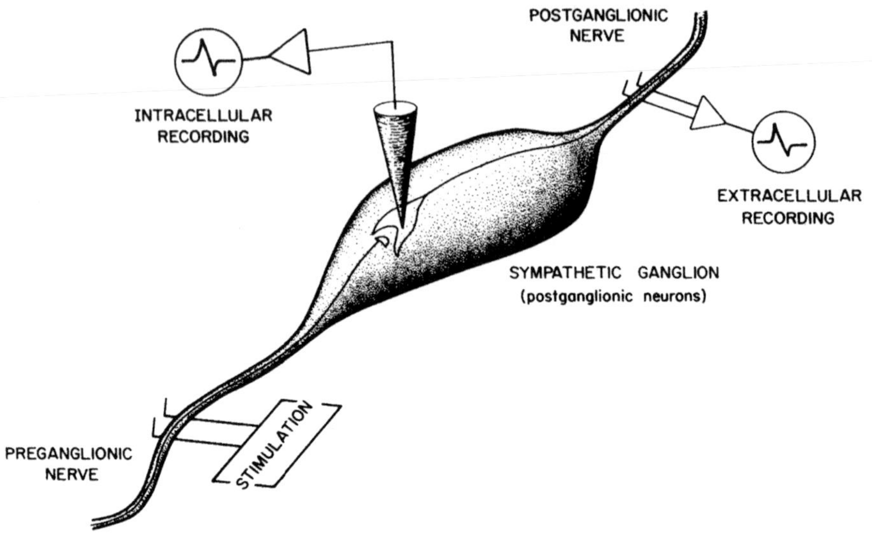

Figure 28.1. *The experimental set-up*. Our studies utilize the superior cervical ganglion of rats as a model system for the investigation of adenosine effects on the voltage-dependent Ca currents and their role in the modulation of synaptic transmission. The SCG has a more simple and homogeneous morphology than brain, and the synaptic membrane physiology and pharmacology are well characterized. It may be maintained *in vitro* for hours with continuous superfusion of Krebs solution. The superfusate also serves as a means of delivering various pharmacologic agents to the ganglion, as described in the text. Postganglionic neurons (50 μm) are visualized by the use of an aus Jena microscope-micromanipulator equipped with Hoffman optics for microelectrode penetration. For synaptic studies, electrodes on the cervical sympathetic nerve are used for preganglionic simulation. Postsynaptic responses are monitored by microelectrodes in the single postganglionic neurons or recorded extracellularly from electrodes on the internal carotid (postganglionic) nerve.

ADENOSINE ACTIONS ON THE SOMA OF POSTGANGLIONIC NEURONS

When present in the bathing solution, adenosine and adenosine analogues (1–200 μ*M*) produce a small transient hyperpolarization in postganglionic neurons of 1–4 mV that lasts approximately 4 min (Figure 28.2A). No measurable change in input resistance is associated with this potential. The adenosine-induced hyperpolarization is similar to that seen in the presence of catecholamines [2], in that the effects are not dose-dependent and were inconsistent in different trials on the same cell. For this reason, our studies have focused on the effects of adenosine on active membrane processes.

Adenosine inhibits Ca-dependent potentials

The Ca-dependent potentials in postganglionic neurons have been well characterized and constitute the hyperpolarizing afterpotential (HAP), a Ca spike, and the shoulder on the normal action potential. The HAP is generated by two different K^+ conductance changes. One is simply voltage-dependent and inac-

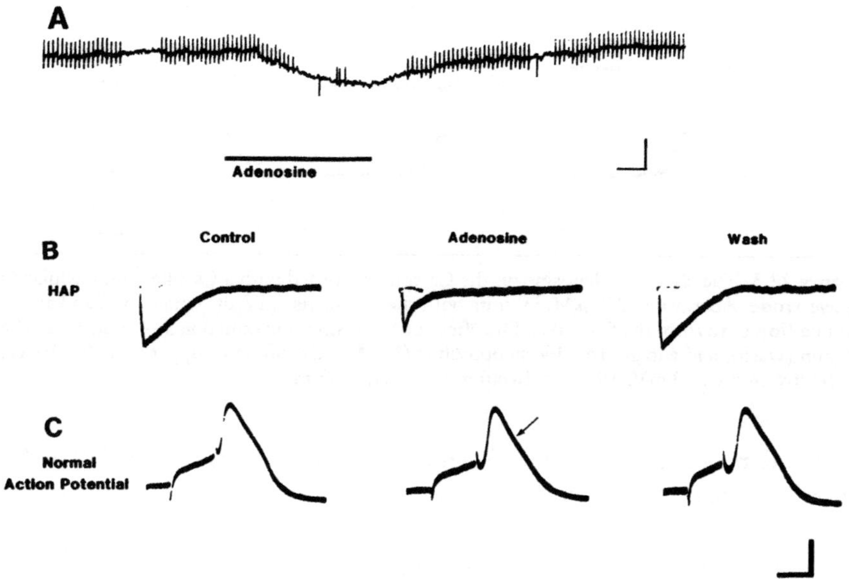

Figure 28.2. The effect of adenosine on resting membrane potentials (A) and on Ca^{2+}-dependent portions of action potentials recorded with K-filled microelectrodes in normal and adenosine-containing Krebs solution. A: Hyperpolarization of a postganglionic neuron in response to bath applied 10 μM adenosine for 5 min (bar). The response shown was larger than usual. Vertical lines are evoked action potentials attenuated by the low-frequency response of the chart recorder. Calibrations are 5 mV, 1 min. Resting potential was −59 mV, and the input impedance (−60 MΩ) did not change in the presence of adenosine. B: The normal (left) hyperpolarizing afterpotential, which follows the spike (offscreen) in postganglionic neurons, is inhibited by 250 μM adenosine (center) for 3 min but recovers completely when adenosine is washed out for 30 min (right). C: The normal (left) action potential in the same neuron has a Ca-sensitive shoulder on the falling phase of the spike. The shoulder (arrow), but not the rising phase of the spike, is reduced by 250 μM adenosine (center) and returns after washout (right). Calibrations: 5 mV, 100 msec in B; 20 mV, 2 msec. in C. Resting potential, −52 mV.

tivates shortly after the action potential returns to the baseline. The other is activated by Ca^{2+} influx during the action potential and inactivates only after the Ca^{2+} is sequestered, a process requiring about 400 msec. Consequently, the amplitude and duration of the HAP are proportional to the extracellular Ca^{2+} concentration [9]. The Ca spike can be studied by intracellular recordings in Krebs solution containing TTX and TEA to block the Na spike and to inhibit the voltage-dependent K conductance [9]. In normal Krebs solution, the Ca component of the action potential is expressed as a shoulder on the falling phase of this Na spike.

Although adenosine affects neither the amplitude nor the rate of rise of the Na spike, it inhibits the HAP and the action potential shoulder in a dose-dependent and reversible fashion [Figure 28.2B, 28.2C]. Adenosine inhibits the peak amplitude and greatly reduces the duration of the HAP. The maximal percentage inhibition of the HAP was 30%, which represents a loss of most of

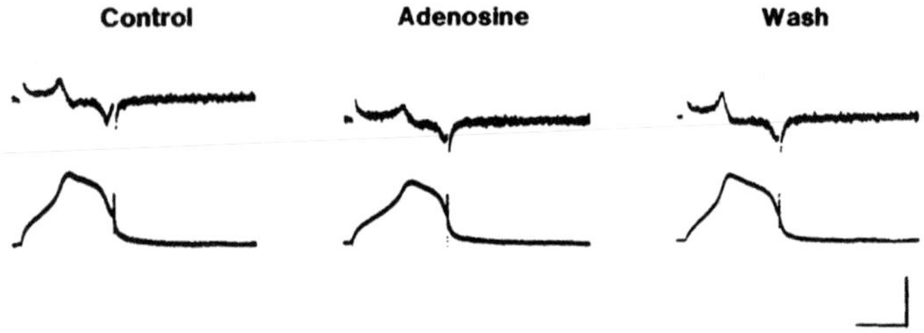

Figure 28.3. The effect of adenosine on the Ca spike recorded with a Cs-filled intracellular microelectrode. Adenosine (250 μ*M*) inhibits both the rate of rise (dV/dt, upper row) and amplitude (lower row) of the Ca spike. The effect of adenosine is maximal in 3–4 min, reversible in 15 min (wash), and antagonized by theophylline (1 m*M*). Calibrations: upper row, 0.4 V/sec, 10 msec; lower row, 20 mV, 10 msec. Resting potential, −45 mV.

the Ca-dependent component. Adenosine concentrations of 100–250 μ*M* produced maximal effects, but higher adenosine concentrations (1 m*M*) were less efficacious. Peak effects of adenosine on the HAP were observed within about 3 min, and the HAP amplitude generally returned to control levels within 10–25 min of washing in normal Krebs solution.

Adenosine markedly increased the threshold and greatly reduced the rate of rise and amplitude of the Ca spike (Figure 28.3). Peak inhibition of the Ca spike amplitude (26% in 250 μ*M* adenosine) occurred within 3–4 min and was reversed by washing 10–25 min in normal Krebs solution. The Ca spike is a more direct indication of the amount of Ca^{2+} influx than the HAP. Presumably, the reduction of Ca^{2+} influx during the Ca spike is responsible for the substantial reduction of the HAP by adenosine.

The most obvious interpretation of our results is that adenosine acts directly to antagonize the calcium conductance mechanism. This interpretation is consistent with observations in cardiac muscle, where adenosine inhibits a Ca-dependent component of the action potential [10], and in the hippocampus, where adenosine inhibits the generation of synaptic currents [11]. However, voltage-dependent Ca currents can also be inhibited indirectly by receptor-mediated increases in K conductance that short-circuit and reduce the Ca current during the action potential [12]. Indeed, adenosine has been shown to evoke a receptor-mediated increase in the resting K conductance in cardiac muscle [13]. Several lines of evidence indicate that the adenosine inhibition of the voltage-dependent Ca current in postganglionic neurons is direct and not mediated by a change in K conductance. If adenosine increased K conductance during the action potential, it could reduce the action potential shoulder, but the peak of the HAP would be increased, not decreased as it was by adenosine. The extent of reduction of the HAP suggests that Ca^{2+} influx throughout the action potential was substantially reduced, even though the total amount of depolarization (voltage-time integral) applied to the Ca conductance mechanism was only slightly reduced.

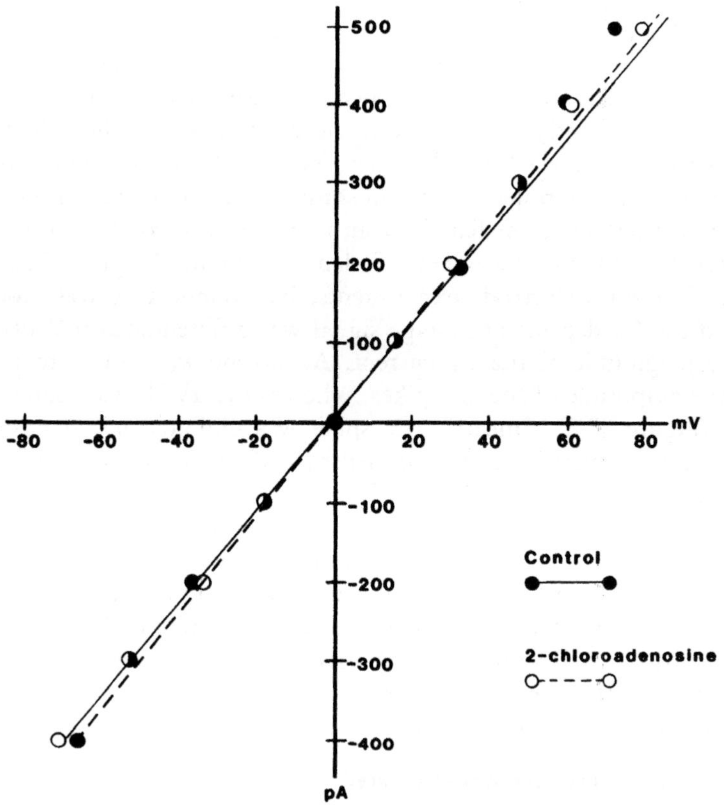

Figure 28.4. Current-voltage characteristics of a postganglionic neuron in Krebs solution containing 1 μ*M* TTX, 10 m*M* TEA, 0.5 m*M* Ca^{2+}, 8 m*M* Mg^{2+}, and 5 m*M* Co^{2+}. Voltage deflections in response to 100 pA current steps from −400 to +500 pA were measured at the end of a 50 msec pulse. Regression lines (plotted) were calculated from the responses to 100 pA current steps between −400 and + 200 pA. There was no significant difference in the slope of the regression line in this and five other experiments in the presence or absence of 10 μ*M* 2-chloroadenosine. Note the difference in the 500 pA steps. The control pulse amplitude shows a sag at the end of the 50 ms pulse. This sag is presumably due to an increase in K conductance following an incomplete block of the Ca current. The response in 2-chloroadenosine approximates the predicted value for a passive membrane. The calculated input resistance was 170 MΩ; the resting potential was −48 mV.

The only way an increased outward current could account for adenosine effects would be if the outward current had exactly the same time course as the inward Ca current. We made I/V measurements under conditions designed to maximally block the Ca conductance. In the presence of low external Ca^{2+} (1.0 m*M*), high Mg^{2+} (8 m*M*), TTX (1 μ*M*), TEA (10 m*M*), and Co^{2+} (5 m*M*) we found no change in the slope resistance of postganglionic neurons in the presence of adenosine (Figure 28.4). Thus, adenosine does not appear to evoke nor potentiate voltage-induced conductances. Furthermore, in these experiments there was a sag in the charging curve at high levels of depolarization (over 400 pA), presumably due to an increase in outward or K current. This sag was reduced, not potentiated, in the presence of adenosine.

Another line of evidence was provided by experiments using Cs-filled electrodes. Cs^+ apparently diffuses from the electrodes into the cell, where it substantially antagonizes K conductance. This results in a large increase in the duration of the action potential and, in some cases, elimination of the HAP. If adenosine actions were mediated by an increase in K conductance, then Cs electrodes should reduce the effects of adenosine on the Ca spike. This was not the case. In the Cs electrode experiments, Na conductance was blocked by TTX and the Ca-dependent voltage signal was differentiated (dV/dt) to estimate the magnitude of the Ca current. Adenosine was shown to markedly reduce the amplitude of the Ca spikes, whereas the dV/dt was reduced by up to 70% (Figure 28.3). Since the Ca spike was inhibited by adenosine even when K currents were severely opposed by Cs^+, it seems most unlikely that adenosine effects could be explained by an increase in K conductance.

Finally, we have made preliminary voltage clamp measurements from postganglionic neurons. Using Cs-filled electrodes in the presence of 1 μ*M* TTX and 10 m*M* TEA, a net inward current of about 750 pA could be induced by depolarizing steps. This current was markedly inhibited by 10 μ*M* 2-chloroadenosine, with only a slight change in the leakage current (Figure 28.5). This result strongly supports our conclusion that adenosine acts to directly inhibit the inward Ca current.

Adenosine effects are receptor-mediated

The adenosine effects on Ca-dependent potentials appear to be mediated by adenosine receptors on the somal surface of postganglionic neurons. Several adenosine analogues were shown to have dose-dependent effects similar to those of adenosine. The order of analogue potency for inhibition of the HAP was as follows: N^6-[L-phenylisopropyl]adenosine (EC_{50} 0.5 μ*M*) > 2-chloroadenosine > adenosine > cyclic AMP = 5′-AMP. Adenosine effects were potentiated by the adenosine uptake blocker dipyridamole (10 μ*M*), suggesting that adenosine acts at an extracellular site. The presence of dipyridamole shifted the dose-response curve for adenosine to the left, making the EC_{50} for adenosine the same as that of 2-chloroadenosine, an analogue with little affinity for adenosine uptake or degradation systems. Additional support for an adenosine receptor was seen in the action of theophylline. This antagonist of adenosine receptors blocked the effects of adenosine and 2-chloroadenosine on the Ca spike. Concentrations of theophylline (50 μ*M*) having no effect on Ca-dependent potentials nor on phosphodiesterase activity were sufficient to antagonize the inhibitory effects of 2-chloroadenosine.

Although cyclic AMP produced similar effects to those of adenosine, this action was completely eliminated by the presence of adenosine deaminase (10 μg/ml). It appears likely that cyclic AMP is metabolized to adenosine in the

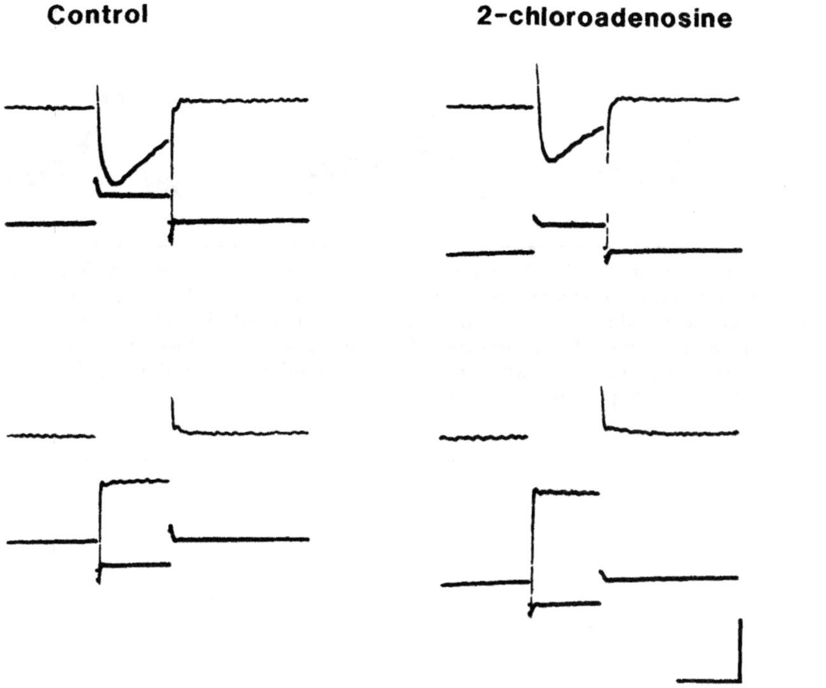

Figure 28.5. Single electrode voltage clamp records from a postganglionic neuron before and after 10 min in 10 μM 2-chloroadenosine. The Krebs solution also contained 1 μM TTX and 10 mM TEA. The electrode was filled with Cs_2SO_4. A depolarizing step (40 mV, 50 msec) induced a 750 pA net inward current (upper trace, upper left frame). The leak current (upper trace, bottom two frames) is estimated from hyperpolarizing steps. The adenosine analogue greatly reduced the peak inward current (upper right frame) and slightly increased the leak current. The voltage is poorly controlled during the first 3 msec of the voltage step but shows no sag during the remainder of the pulse. Outward current tails can be seen following return of the holding potential (−42 mV) in all frames. Those following the depolarizing pulses are more likely to be K and C currents, and those following the hyperpolarizing steps could be inactivating A currents. Calibrations: 500 pA, 100 mV, 50 ms.

extracellular fluid, where it then activates cell surface receptors. Noradrenalin produces effects similar to those of adenosine on the Ca-dependent potentials, by activation of α_2-adrenergic receptors. These responses are blocked by 10 μM phentolamine, whereas adenosine responses are not. We conclude that the α_2 and adenosine receptors act independently to inhibit the voltage-dependent Ca current.

ADENOSINE EFFECTS ON SYNAPTIC TRANSMISSION

Adenosine has a variety of actions in the CNS that seem to be inhibitory or depressant in nature. It reduces the spontaneous firing of central neurons [8], reduces the amplitude of evoked EPSPs and population spikes in the hippocampus [14], and activates receptors on presynaptic axon terminals to in-

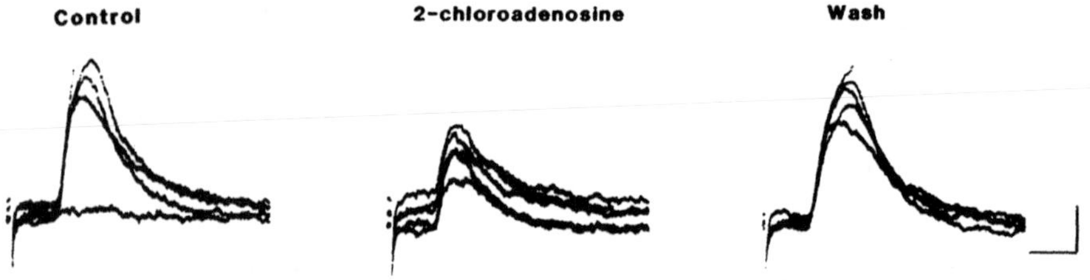

Figure 28.6. Inhibition of EPSP amplitude by 5 μ*M* 2-chloroadenosine. *Left:* Five superimposed intracellularly recorded responses to single-stimulus pulses delivered to the preganglionic nerve. Pulses were delivered at 20 sec intervals, with the ganglion bathed in normal Krebs solution. *Center:* Five superimposed responses after 10 min in 5 μ*M* 2-chloroadenosine. *Right:* Recovery of EPSP amplitude after 20 min wash in normal Krebs solution. Calibrations: vertical, 4 mV; horizontal, 20 ms. Resting potential, −59 mV.

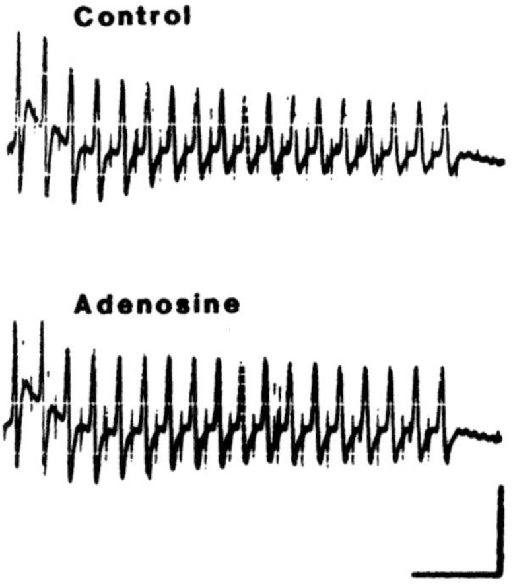

Figure 28.7. Adenosine (100 μ*M*) reduces synaptic depression. The compound action potential recorded extracellularly from the internal carotid nerve decreases in amplitude during repetitive (20 Hz, 1 sec) preganglionic stimulation. Adenosine reduces the decrease in amplitude during the train. Compare first and last spikes. Calibrations: 1 mV; 200 ms.

hibit the release of several different types of neurotransmitters [15]. However, we have found that adenosine does not always act as an inhibitor of neuronal activity. Although 2-chloroadenosine does reduce the amplitude of single EPSPs, it can actually facilitate synaptic transmission during repetitive activity.

Adenosine inhibits EPSPs

When single stimulus pulses are delivered once every 20 sec to the preganglionic nerve, 2-chloroadenosine (10 μ*M*) reduces the EPSP amplitude by 42% ± 9% of control within 10 min (Figure 28.6). Inhibition of EPSP amplitude

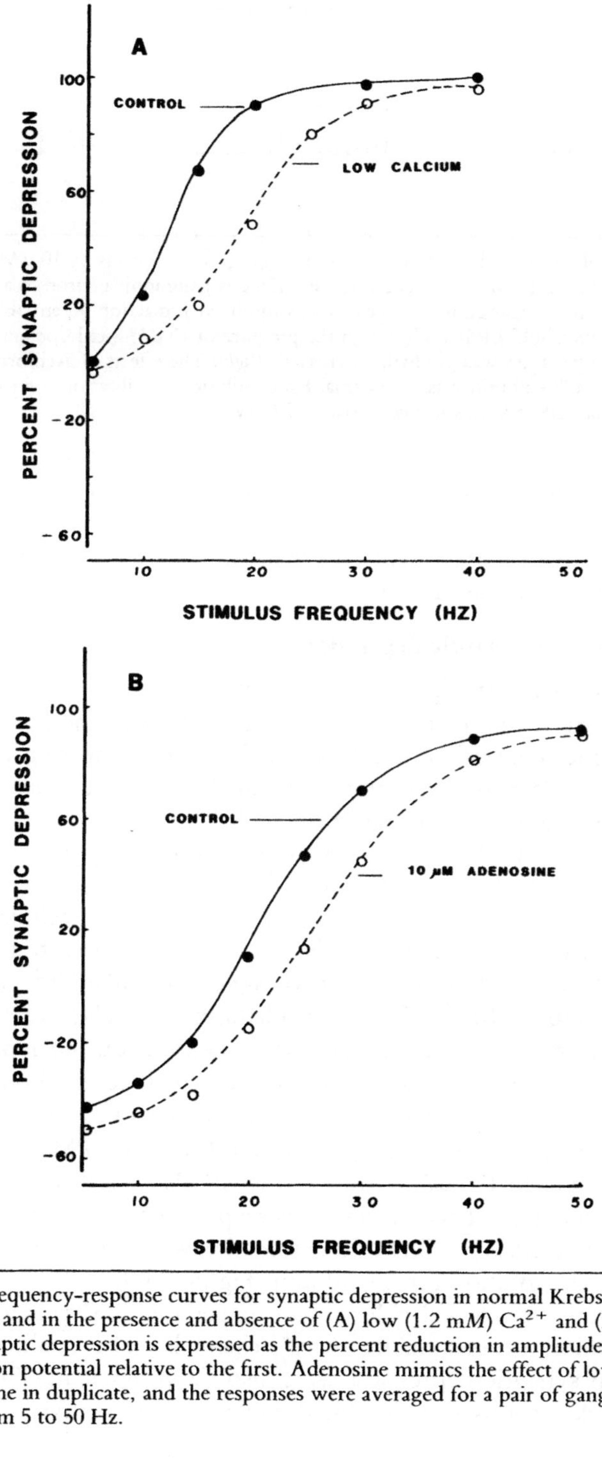

Figure 28.8. Frequency-response curves for synaptic depression in normal Krebs solution (2.2 m*M* Ca^{2+}) and in the presence and absence of (A) low (1.2 m*M*) Ca^{2+} and (B) 10 μ*M* adenosine. Synaptic depression is expressed as the percent reduction in amplitude of the last compound action potential relative to the first. Adenosine mimics the effect of low Ca^{2+}. Experiments were done in duplicate, and the responses were averaged for a pair of ganglia at frequencies ranging from 5 to 50 Hz.

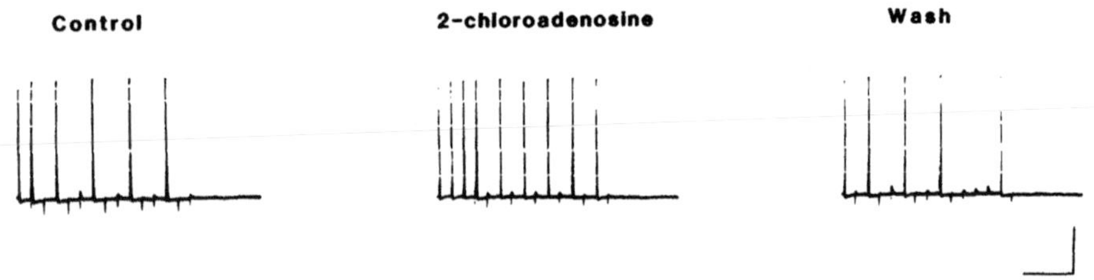

Figure 28.9. Facilitation of firing frequency in postganglionic neurons by 10 μM 2-chloroadenosine. *Left:* Intracellularly recorded response of a postganglionic neuron to a 20 Hz stimulus train delivered to the preganglionic nerve. Some stimuli evoked action potentials, whereas others elicited only subthreshold EPSPs. *Center:* In the presence of 10 μM 2-chloroadenosine, the number of spikes per train was gradually increased. *Right:* The effects of 2-chloroadenosine were reversed by a 20–30 min wash in normal Krebs solution. Calibrations: vertical, 40 mV; horizontal, 200 ms. Resting potential, −79 mV.

occurred without a change in the input impedance of postganglionic neurons or in the sensitivity to bath-applied carbachol (0.5 μM). This result indicates that 2-chloroadenosine is acting presynaptically to reduce the amount of acetylcholine released per impulse.

Adenosine reduces synaptic depression

The facilitory effects of adenosine on repetitive activity were demonstrated in studies of synaptic depression. Extracellular records of the compound action potential during trains of stimuli (5–40 Hz for 1 sec) to the preganglionic nerve show a gradual decrease in amplitude from the first to the last spike (Figure 28.7). This phenomenon is thought to result from depletion of readily releasable transmitter from the nerve terminals [16] or possibly from inactivation of Ca^{2+} channels, which would reduce Ca^{2+} influx during firing [12] and thus diminish transmitter release. Reducing the amount of extracellular Ca^{2+} in the medium reduces the amount of depression during the train. Adenosine (10–100 μM) has the same effect as lowering extracellular Ca^{2+} (Figure 28.8).

Similar facilitory effects of 2-chloroadenosine (10 μM) were observed by intracellular recording from postganglionic neurons during repetitive firing. The response to preganglionic stimulus trains (20 Hz for 0.75 sec) delivered at 20 sec intervals shows evidence of synaptic depression, in that a few stimuli elicit spikes while the remainder generate subthreshold EPSPs. In the presence of 2-chloroadenosine (10 μM), more of the EPSPs exceeded threshold and the number of action potentials increased over a period of 10 min to 181% ± 12% of control. The facilitory effect of 2-chloroadenosine was reversed by washing in normal Krebs solution for 20–30 min (Figure 28.9).

The inhibitory effects of 2-chloroadenosine on the EPSPs could be explained by an inhibition of presynaptic Ca^{2+} channels. The facilitory effects of adenosine on synaptic depression were also Ca-dependent. The mechanism whereby single responses are inhibited while repetitive activity is facilitated is not presently understood, but the overall effect on synaptic transmission would be to reduce the sensitivity to low-frequency events and enhance the transmission of high-frequency impulses, much like a "high-pass filter." This effect could have important consequences for explaining central actions of adenosine.

REFERENCES

1. Dunlap K, Fischback GD: Neurotransmitters decrease the calcium conductance activated by depolarization of embryonic chick sensory neurones. *J Physiol* 317:519–535, 1981.
2. Horn JP, McAfee DA: Alpha-adrenergic inhibition of calcium-dependent potentials in rat sympathetic neurones. *J Physiol* 301:191–204, 1980.
3. Henon BK, McAfee DA: The ionic basis of adenosine receptor actions on postganglionic neurones. *J Physiol,* 1983 (in press).
4. Henon BK, McAfee DA: Adenosine modulation of synaptic efficacy in the superior cervical ganglion of the rat. *Soc Neurosci Abstr* 8:555, 1982.
5. Kuroda Y: Regulation of neurotransmission by adenosine and cyclic AMP in mammalian central nervous system, in Rodnight R, Bachelard HS, Stahl WL (eds): *Chemistry of the Brain.* New York, Churchill Livingstone, 1981, pp 108–118.
6. Dunwiddie TV, Hoffer BJ: Adenine nucleotides and synaptic transmission in the *in vitro* rat hippocampus. *Br J Pharmacol* 69:59–68, 1980.
7. Schubert P, Mitzdorf U: Analysis and quantitative evaluation of the depressive effect of adenosine on evoked potentials in hippocampal slices. *Brain Res* 172:186–190, 1979.
8. Phillis JW, Wu PH: The role of adenosine and its nucleotides in central synaptic transmission. *Prog Neurobiol* 16:187–239, 1981.
9. McAfee DA, Yarowsky PJ: Calcium dependent potentials in the mammalian sympathetic neurone. *J Physiol* 290:507–523, 1979.
10. Schrader J, Rubio R, Berne RM: Inhibition of slow action potentials of guinea pig atrial muscle by adenosine: A possible effect on Ca^{2+} influx. *J Mol Cell Cardiol* 7:427–433, 1975.
11. Schubert P, Mitzdorf U: Analysis and quantitative evaluation of the depressive effect of adenosine on evoked potentials in hippocampal slices. *Brain Res* 172:186–190, 1979.
12. Klein M, Shapiro E, Kandel ER: Synaptic plasticity and the modulation of the Ca^{2+} current, in Kravitz EA, Treherne JE (eds): *Neurotransmission, Neurotransmitters and Neuromodulators.* Cambridge, England, Cambridge University Press, 1980, pp 117–157.
13. Hartzell CH: Adenosine receptors in frog sinus venosus: Slow inhibitory potentials produced by adenine compounds and acetylcholine. *J Physiol* 293:23–49, 1979.
14. Siggins GR, Schubert P: Adenosine depression of hippocampal neurons in vitro: an intracellular study of dose-dependent actions on synaptic and membrane potentials. *Neurosci Lett* 23:55–60, 1981.
15. Fredholm BB, Hedqvist P: Modulation of neurotransmission by purine nucleotides and nucleosides. *Biochem Pharmacol* 29:1635–1643, 1980.
16. Thies RE: Neuromuscular depression and the apparent depletion of transmitter in mammalian muscle. *J Neurophysiol* 28:427–442, 1965.

DISCUSSION

DR. Y. KURODA: (1) We observed facilitation in olfactory cortex slices after wash out. Kuba et al. also observed long-lasting facilitation after the washing out of epinephrine in the ganglion. How did you wash out adenosine, and did you find a change of EPSP after the washing? (2) Did you find any effects of cyclic AMP analogues on the EPSPs and Ca^{2+} and K^+ currents?

DR. MCAFEE: (1) We have not yet studied the behavior of our preparations during the adenosine washout periods. (2) We have not yet studied those cyclic nucleotide analogues which would not be metabolized to active adenosine analogues.

DR. J.W. PHILLIS: What do you perceive to be the physiological functions of the after-hyperpolarizing potentials, and what advantages would adenosine depression of this potential confer on the cell?

DR. MCAFEE: Afterpotentials have always been subjects of study wherever found, whether in brain or peripheral tissues, in mammals or in mollusks. It seems likely that the hyperpolarizing afterpotential that is mediated by a Ca-dependent K^+ conductance would tend to limit the rate of discharge or minimum interspike interval. Obviously, adenosine, by inhibiting this potential, should increase the ability of the tissue to support high-frequency activity. This may be one mechanism for the effect of adenosine on synaptic depression. Dr. Henon and I would like to think of adenosine as a "high-pass filter."

DR. T. DUNWIDDIE: (1) We have been able to demonstrate in the *in vitro* hippocampus that calcium spikes are inhibited by adenosine, suggesting that this is a general feature of neurons that exhibit calcium spikes. (2) We have been struck in our work by how effective adenosine is in antagonizing synaptic transmission. In hippocampus, we observe an approximate 95% decrease in postsynaptic potentials at high adenosine concentrations. However, we have never been able to observe effects of repetitive stimulation (that can be attributed to adenosine) of anywhere near this magnitude. Does this not perhaps suggest that adenosine is not a particularly significant modulator of synaptic transmission, but that it may function as a safety valve, a way of shutting off synaptic transmission during a severe metabolic challenge (e.g., during hypoxia)?

DR. MCAFEE: I am not familiar with your studies on repetitive stimulation, but I hope your lab and others will look into synaptic depression and the effects of adenosine on short (1 sec) trains. In sympathetic ganglia, adenosine may actually *facilitate* bursts of activity and inhibit only low-frequency discharge. Inhibitors are usually tested by their effect on single responses. It is encouraging to see more of us using repetitive activity in evoked responses as targets of physiological control.

DR. KURODA: In olfactory cortex, adenosine did not inhibit the EPSP completely. About 30% of amplitude still remained. Therefore, even if adenosine worked in a negative-feedback manner, it would not stop the synaptic transmission. I think adenosine might have a physiological role only when the input is stimulated tetanically. Adenosine seems to carry information that the impulses pass at high frequency at the synapse when the release and formation of adenosine derivatives overcome their inactivation and adenosine derivatives accumulate in the synapse. Adenosine's modulatory actions, especially facilitory ones, are important factors for the regulation of synaptic plasticity, including the mechanisms of memory and learning.

29. INTERACTIONS OF ADENOSINE WITH OTHER AGENTS

TREVOR W. STONE

The concept that purines might modify the responses of cells to other agonists can be traced back to the original observations of Rall and his colleagues [1, 2]. It was found that adenosine not only would elevate the levels of cyclic AMP in slices of guinea pig cerebral cortex but also would potentiate the increase of cyclic AMP produced by norepinephrine [1–3]. A large amount of evidence has since accumulated suggesting that endogenous purines have a variety of effects that may reflect physiological roles [4, 5], and consequently it is necessary to consider that an important function of purines may be to regulate the sensitivity of receptors for other endogenous agents or exogenously administered drugs.

INTERACTIONS WITH CATECHOLAMINES

Besides the discovery of a potentiative interaction of adenosine and norepinephrine on cyclic AMP levels, which led to the concept of adenosine-dependent and adenosine-independent α receptors [2], it was found that adenosine would prevent the development of "refractoriness" to norepinephrine, that is, the loss of adenylate cyclase sensitivity to norepinephrine that occurred when the amine was applied a second or subsequent time to the same preparation [3]. Similar phenomena can be demonstrated in quite different systems.

Holck and Marks [6], for example, reported that adenosine would enhance responses of the guinea pig vas deferens to exogenous norepinephrine.

R.M. Berne, T.W. Rall, and R. Rubio (eds.), Regulatory Function of Adenosine, *pp. 467–477.*

Adenosine was more effective than AMP in this respect, with ADP having no action and ATP causing depression of norepinephrine responses. If the vas was desensitized to norepinephrine, purines were found to increase the rate of resensitization in the order AMP, adenosine, ADP, with ATP inactive. The suggestion arising from these studies was that a purine receptor might be intimately associated with the α adrenoceptor and might mediate some measure of control over α receptor sensitivity. It is remarkable to recall that, after juggling with molecular models of receptors, Smythies [7] had postulated several years earlier that a purine molecule might form part of the α adrenoceptor complex. It is interesting to speculate that desensitization of adrenoceptors might be associated with loss of the purine component, and resensitization with its replacement. Certainly, Fredholm [8] has noted the loss of purines from adrenergically stimulated tissues and the prevention of that loss by α receptor blockade.

Other examples are known in which activation of α adrenoceptors postjunctionally is enhanced by adenosine, as in some blood vessels [9], although it has recently been reported that adenosine reduces α adrenergic sensitivity postjunctionally in the cat nictitating membrane [10].

Catecholamine effects on β receptors, however, are usually reduced by adenosine. This applies to studies on adenylate cyclase activation [11, 12] as well as to studies on tension in isolated cardiac muscle [13]. In the case of inhibition of adenylate cyclase activity, the action of adenosine appears to result from the fact that adenosine is a weaker agonist than most β receptor agonists and that the adenosine receptor is continuously combined with membranal adenylate cyclase [12], whereas the β receptor is not, collision being required between an activated β receptor and cyclase molecule.

A number of questions are unanswered in this field. First, there are simple discrepancies between the observations of different groups. Thus, whereas Holck and Marks [6] reported that ATP reduced norepinephrine responses of the guinea pig vas, Kazić and Milosavljević [14] observed a potentiation.

Second, it remains to be established that some of the interactions described are indeed direct effects at the postjunctional surface. It is well known that purines can enhance the uptake into and reduce the release of amines from nerve terminals [4, 5] and that such actions might reduce the extracellular amine concentrations sufficiently to increase the rate of resensitization without necessarily affecting the α receptor directly. A single injection of reserpine of 2.5 mg/kg 48 hours before testing [6] may not be adequate to remove this possibility.

Finally, it is well known that Ca^{2+} can modify the extent and rate of receptor desensitization and resensitization [15]. As purines have been shown to modify calcium-binding sites [16] or calcium currents [4], the question arises of whether purine effects on receptors are mediated by changes of calcium activity. Certainly the removal of calcium inhibits norepinephrine-induced accumulation of cyclic AMP in brain slices in the presence of adenosine [18].

INTERACTIONS WITH ACETYLCHOLINE

More detailed electrophysiological analyses of purine effects on acetylcholine receptors have been carried out. Buchthal and Kahlson [17] were among the first to report an increased acetylcholine sensitivity of muscle during the application of ATP. Using intracellular recording at the neuromuscular junction, Ewald [19] later noted that ATP increased responses to iontophoretically applied acetylcholine and decreased the rate of desensitization. Similar observations have been made [20] using voltage clamp methods, showing that although ATP increased the end-plate current produced by acetylcholine, neither the equilibrium potential, the lifetime of the end-plate current, nor the affinity of acetylcholine for its receptor was altered. The findings led to the conclusion that ATP increased acetylcholine sensitivity by increasing either the conductance of unit channels or the total number of available channels.

Again, complicating factors have not been entirely eliminated. For example, ATP was used in these studies at heroic concentrations of 1–10 m*M* [19, 20] for several minutes. As ATP is known to alter potassium and calcium currents [4], the intracellular and membrane concentrations of these ions could have been changed sufficiently to alter the properties of the postjunctional membrane, and the question of a pharmacological action rather than a physiologically meaningful one is raised.

INTERACTIONS WITH BENZODIAZEPINES

Adenosine has an anticonvulsant action [21] but little significance was attached to this until Skolnick and his colleagues extracted from brain compounds that displaced the anticonvulsant diazepam from brain membranes. These compounds proved to be inosine and hypoxanthine [22], and inosine itself has since been found to be anticonvulsant.

At present, the view that endogenous purines may be ligands for the benzodiazepine binding site is not greatly favored, mainly because the concentrations of purines needed to displace benzodiazepines are relatively high (IC_{50} about 1 m*M*). However, this skepticism is difficult to understand, since both adenosine and inosine can almost certainly reach concentrations of this order in brain [23]. It has also been shown that inosine has the same effects as diazepam on cultured spinal neurones and shows cross desensitization [24].

Several other findings suggest that purines of some kind may be relevant to benzodiazepine pharmacology. First, Skolnick et al. have described an adenosine derivative which is able to enhance diazepam binding [24]. Second, two deoxy derivatives (2′-deoxyinosine and 2′-deoxyguanosine) were found to have IC_{50} values of about 300 μM, so that a related endogenous compound could have important actions at the benzodiazepine binding site. Third, Davies et al. [25] have isolated from a marine sponge a purine derivative that has an IC_{50} of only 19 μM against diazepam binding. This compound is 1-methylisoguanosine (or 1-methyl, 2-oxo-adenosine) and it has been shown to have muscle relaxant properties similar to the benzodiazepines after oral administration.

An interesting observation made by Davies et al. [25] was that dipyridamole, well known as an inhibitor of adenosine uptake, would displace diazepam with an even lower IC_{50} of about 300 n*M*. Independently, Phillis and his colleagues [26] showed that benzodiazepines would inhibit the uptake of adenosine into brain preparations at concentrations likely to occur during their clinical administration and that their relative inhibitory potencies correlated with their therapeutic efficacy. These findings raised the possibility that benzodiazepines might act partly by inhibiting the uptake of endogenously released adenosine, an idea consistent with the observation that benzodiazepines would enhance the neuronal depressant effects of adenosine. A clear correlation was subsequently found between the ability of adenosine uptake inhibitors to inhibit uptake and to displace diazepam from brain synaptosomes [27].

The adenosine uptake site therefore warrants further study as a potential site of action of benzodiazepines. However, two cautionary points should be made. First, the potential relevance of the uptake site rests largely on the potency correlations just described, but there is a substantial discrepancy between the IC_{50} values for diazepam displacement obtained by Davies et al. [25] (both papaverine and hexobendine > 1 m*M*) and those reported by Wu et al. [27] (papaverine 40 μM and hexobendine 5 μM). Miraculously, the IC_{50} for dipyridamole was found to be the same, 300 n*M*, by both groups. Second, dipyridamole can have direct actions on the adenosine receptor [28]. Therefore, direct interactions between dipyridamole, adenosine, and benzodiazepines on fragments of postjunctional membrane need to be excluded before interactions at a putative uptake site can be considered to be proved.

INTERACTIONS WITH OPIATES

Morphine is a potent inhibitor of transmitter release from both peripheral and central nervous system neurones, and the reduction of this action by elevated extracellular calcium levels or high-frequency stimulation has suggested to many that a decrease in the availability of calcium could mediate the inhibition. As it is also known that methylxanthines can release and mobilize calcium from membrane and intracellular storage sites, it should in retrospect have come as little surprise that theophylline could reduce the analgesic effects of morphine [29]. However, Sawynok and Jhamandas [30] found that theophylline would block the depressant effects of both adenosine and morphine on electrically evoked contractions of guinea pig ileum. These workers hypothesized that their results "introduce the possibility that morphine might be releasing an adenosine-like substance." Theophylline was also reported to prevent the inhibitory effects of adenosine and morphine on acetylcholine release in the ileum [36] and in the cerebral cortex [31].

The publication of the adenosine mediation hypothesis was followed by reports that morphine would increase the veratridine-induced release of purines from brain synaptosomes [32] and enhance purine release from rat cerebral cortex *in vivo* [33]. We have recently shown increased release of purines induced by ouabain and electrical stimulation from slices of cerebral cortex [34]. An electrophysiological study of neurones in the globus pallidus of anesthetized rats showed that aminophylline, applied by microiontophoresis, would reverse the depression of neuronal firing produced by either adenosine or morphine at doses that did not affect responses to GABA [35]. However, despite the implication of this work that the release of adenosine may mediate the depressant effects of morphine on the nervous system, there are observations that cast doubt on this conclusion. In their original report, Sawynok and Jhamandas [30] noted that the synaptic depressant effects of morphine were not as readily antagonized by theophylline as were the effects of adenosine, even though quite high doses of theophylline were being used. More recent studies in other laboratories have failed to replicate the original observations. Gallant and Clement [37] showed that doses of theophylline or 8-phenyltheophylline can be used that selectively block the purine responses of ileum without affecting morphine. Vizi et al. [38] were also unable to antagonize opiate inhibition of evoked contractions or acetylcholine release in the ileum. Morphine's powerful inhibitory action on the mouse vas deferens is not affected by theophylline [39], and morphine does not release adenosine in this preparation [40].

In electrophysiological experiments, Duggan and Griersmith [41] were unable to block opiate effects with isobutylmethylxanthine. However, this study was performed on neurones of the dorsal horn of the spinal cord, and the morphine effect studied was the suppression of noxious input, not the depression of spontaneous firing; isobutylmethylxanthine (chosen because the authors were concerned with the role of the cyclic AMP system) is notoriously difficult to apply by microiontophoresis. Therefore, a direct comparison with the study of pallidal neurones [35] is not appropriate.

A different line of research that supports the concept of a relationship between opiates and purines is that of the quasimorphine abstinence syndrome (QMAS) [42]. The QMAS refers to the cluster of behaviors produced by methylxanthine administration followed by naloxone in opiate-naive rats. These behaviors closely resemble the symptoms seen when opiate-tolerant and opiate-dependent animals are withdrawn from the opiate or when naloxone is administered. Collier has attributed the methylxanthine QMAS to inhibition of phosphodiesterase on the basis of a correlation between the intensity of QMAS produced by a series of xanthine and nonxanthine compounds and the inhibition *in vitro* of phosphodiesterase extracted from rat brain [42]. However, it may be remarked that a correlation will only reflect a rate-limiting step, such as lipid solubility. It is perhaps more relevant to note that doses of methylxanthines higher than those used to elicit the QMAS have been found *not* to affect phosphodiesterase activity in the brain when the brains of treated animals are analyzed [43, 44]. Thus, the QMAS may result from the blockade of endogenous adenosine or from the mobilization of calcium ions. A novel contribution to this area has recently been made by Yarbrough et al. [45], who have shown that purines administered directly into the CNS are capable of producing analgesia. It would seem to be a logical inference that the methylxanthines induce the QMAS by opposing the actions of endogenous purines.

Is there any way to reconcile any of this sometimes conflicting and usually confusing data? Jurna [46] has examined the effects of aminophylline on the depressant action of morphine on noxious input into the spinal cord. The author concluded that aminophylline was not able to block the action of morphine on the noxious activation of lamina V neurones, where morphine's action is probably on primary afferent terminals to suppress transmitter release. This conclusion is therefore consistent with findings mentioned earlier [41]. However, the depression by morphine of the flexion reflex *was* blocked by aminophylline. In seeking to explain this distinctive action of aminophylline, the author recalls that morphine has an apparently direct inhibitory effect on the spinal motoneurones and suggests that the antagonistic effect of aminophylline might be exerted at this site. This would be consistent with our own observations on spontaneously firing pallidal neurones [35].

It should also be emphasized that the depressant effects of morphine are observed most readily on the cells in the pallidum that show high spontaneous activity. Presumably, these cells tend to release purines into the extracellular space during depolarization, and this high background release may be more readily modulated by opiates than the small release from slowly firing units.

INTERACTIONS WITH ANTIDEPRESSANTS

Twenty years ago, Lewis and his colleagues [47] reported the effects of psychotropic drugs on the levels of adenine compounds in the brain. They found that reserpine and related mood depressants decreased the ratio of ATP to ADP in rat brain, whereas antidepressants increased this ratio. An explanation for these changes may be found in the demonstration that desipramine would increase the conversion of ATP to cyclic AMP [48]. Testing a series of antidepressants on cyclic AMP levels, Sattin et al. [49] subsequently found that they all increased cyclic AMP concentrations in chopped guinea pig cortex; furthermore, the microiontophoretic application of the same antidepressants potentiated the depressant effects of adenosine on neuronal firing [49, 50]. The order of potency of six unrelated antidepressants was almost identical in the two experimental situations.

As the elevation of cyclic AMP levels by antidepressants was found to be blocked by theophylline [49], it seemed likely that a release of purines by the antidepressants might be involved. We have recently begun to look for such an action using the rat cerebral cortex *in vivo*. Small plastic cups are held in place on the cortical surface of urethane-anesthetized rats and, after placing [^{3}H]adenosine into the cup, in artificial CSF, for 20 to 60 min, the contents of the cup, approximately 200 μl, are removed for scintillation counting and replaced by fresh medium. If the cup fluid includes an adenosine uptake inhibitor (after the initial labeling) and relatively high concentrations of antidepressants (500 μ*M*) are used, similar to those found effective in elevating cyclic AMP [49], then an increased purine release can be demonstrated.

Clearly, this is not a potent effect of antidepressant drugs, and the mechanism remains uncertain, that is, whether the purine release is a specific effect or whether it is a concomitant or consequence of increased amine concentrations. However, it must be emphasized that a range of pharmacologically related antidepressants seem to be effective in these systems [49, 50]; these compounds include those known to affect amine uptake and those that do not. In addition, all the experiments to date have been acute. As it is well known that antidepressants require several days of chronic administration to elicit any clinical effects, it is conceivable that effects on purine systems too small to be detected at lower concentrations in acute experiments may have a cumulative effect over a period of time. It is interesting to speculate that since purines can modify catecholamine receptor sensitivity, as discussed earlier, chronic changes of purine activity may be responsible for some of the changes of receptor sensitivity found to accompany antidepressant treatment [51].

REFERENCES

1. Sattin A, Rall TW: The effect of adenosine and adenine nucleotides on the cyclic AMP content of guinea pig cerebral cortex slices. *Mol Pharmacol* 6:13–23, 1970.
2. Sattin A, Rall TW, Zanella J: Regulation of cyclic AMP levels in guinea-pig cerebral cortex by interaction of α-adrenergic and adenosine receptor activity. *J Pharmacol Exp Ther* 192:22–32, 1975.
3. Daly JW: The nature of receptors regulating the formation of cyclic AMP in brain tissue. *Life Sci* 18:1349–1358, 1976.
4. Stone TW: Physiological roles for adenosine and ATP in the nervous system. *Neuroscience* 6:523–555, 1981.
5. Fredholm BB, Hedqvist P: Modulation of neurotransmission by purine nucleotides and nucleosides. *Biochem Pharmacol* 29:1635–1643, 1980.
6. Holck MI, Marks BH: Purine nucleoside and nucleotide interactions on normal and subsensitive α-adrenoreceptor responsiveness in guinea pig vas deferens. *J Pharmacol Exp Ther* 205:104–117, 1978.
7. Smythies JR: Molecular mechanisms of adrenergic receptors. *J Theor Biol* 35:93–102, 1972.
8. Fredholm BB: Release of adenosine-like material from isolated perfused dog adipose tissue following sympathetic nerve stimulation and its inhibition by adrenergic α-receptor blockade. *Acta Physiol Scand* 96:422–430, 1976.
9. Hedqvist P, Fredholm BB: Effects of adenosine on adrenergic neurotransmission: Prejunctional inhibition and postjunctional enhancement. *Arch Pharmacol* 293:217–224, 1976.
10. Luchelli-Fortis MA, Fredholm BB, Langer SZ: Evidence against the presence of presynaptic inhibitory adenosine receptors in the cat nictitating membrane. *J Pharmacol Exp Ther* 219:235–242, 1981.
11. Fain JN, Malbon CC: Regulation of adenylate cyclase by adenosine. *Mol Cell Biochem* 25:143–169, 1979.
12. Braun S, Levitzki A: The attenuation of epinephrine-dependent adenylate cyclase by adenosine and the characteristics of the adenosine stimulatory and inhibitory sites. *Mol Pharmacol* 16:737–748, 1979.
13. Endoh M, Yamashita S: Adenosine antagonizes the positive inotropic action mediated via β, not α-adrenoceptors in rabbit papillary muscle. *Eur J Pharmacol* 65:445–448, 1980.

14. Kazić T, Milosavljević D: Influence of adenosine, cyclic AMP on responses of the isolated terminal guinea pig ileum to electrical stimulation. *Arch Int Pharmacol* 223:187–195, 1976.
15. Nastuk WL: Cholinergic receptor desensitization, in Cottrell GA, Usherwood PNR (eds): *Synapses*. London, Blackie, 1977, pp 177–201.
16. Abood LG: Interrelationships between phosphates and calcium in bioelectric phenomena. *Int Rev Neurobiol* 9:223–261, 1966.
17. Buchthal F, Kahlson G: Application of ATP and related compounds to mammalian striated and smooth muscle. *Nature* 154:178–179, 1944.
18. Rall TW: Regulation of cyclic AMP accumulation in brain tissue, in Baer HP, Drummond GI (eds): *Physiological and Regulatory Roles for Adenosine and Adenine Nucleotides*. New York, Raven Press, 1979, pp 217–227.
19. Ewald DA: Potentiation of postjunctional cholinergic sensitivity of rat diaphragm muscle by high-energy adenine nucleotides. *J Membr Biol* 29:47–65, 1976.
20. Akasu T, Hirai K, Koketsu K: Increase of acetylcholine receptor sensitivity by ATP. *Br J Pharmacol* 74:505–507, 1981.
21. Maitre M, Ciesielski L, Lehmann A, Kempf E, Mandel P: Protective effect of adenosine and nicotinamide against audiogenic seizure. *Biochem Pharmacol* 23:2807–2816, 1974.
22. Paul SM, Marangos PJ, Goodwin FK, Skolnick P: Brain-specific benzodiazepine receptors and putative endogenous benzodiazepine-like compounds. *Biol Psychiatry* 15:407–428, 1980.
23. Pull I, McIlwain H: Adenine derivatives as neurohumoral agents in the brain. *Biochem J* 130:975–981, 1972.
24. Macdonald JF, Barker JL, Paul SM, Marangos PJ, Skolnick P: Inosine may be an endogenous ligand for benzodiazepine receptors in cultured spinal neurons. *Science* 205:715–717, 1979.
25. Davies LP, Cook AF, Poonian M, Taylor KM: Displacement of [^{3}H]diazepam binding in rat brain by 1-methylisoguanosine. *Life Sci* 26:1089–1098, 1980.
26. Phillis JW, Bender AS, Wu PH: Benzodiazepines inhibit adenosine uptake into rat brain synaptosomes. *Brain Res* 195:494–498, 1980.
27. Wu PH, Phillis JW, Bender AS: Inhibition of [^{3}H]diazepam binding to rat brain cortical synaptosomal membranes by adenosine uptake blockers. *Eur J Pharmacol* 65:459–460, 1980.
28. Dowdle EB, Maske R: The effects of dipyridamole on the guinea pig ileal longitudinal muscle–myenteric plexus preparation. *Br J Pharmacol* 71:235–244, 1980.
29. Ho IK, Loh HH, Way EL: Cyclic AMP antagonism of morphine analgesia. *J Pharmacol Exp Ther* 185:336–340, 1973.
30. Sawynok J, Jhamandas KH: Inhibition of acetylcholine release from cholinergic nerves by adenosine, adenine nucleotides and morphine: Antagonism by theophylline. *J Pharmacol Exp Ther* 197:379–390, 1976.
31. Jhamandas K, Sawynok J, Sutak M: Antagonism of morphine action on brain acetylcholine release by methylxanthines and calcium. *Eur J Pharmacol* 49:309–312, 1978.
32. Fredholm BB, Vernet L: Morphine increases depolarization induced purine release from hypothalamic synaptosomes. *Acta Physiol Scand* 104:502–504, 1978.
33. Phillis JW, Jiang ZG, Chelack BJ, Wu PH: Morphine enhances adenosine release from in vivo rat cerebral cortex. *Eur J Pharmacol* 65:97–100, 1979.
34. Stone TW: The effects of morphine and met-enkephalin on the release of purines from cerebral cortex slices of rats and mice. *Br J Pharmacol* 74:171–176, 1981.
35. Perkins MN, Stone TW: Blockade of striatal neurone responses to morphine by aminophylline: Evidence for adenosine mediation of opiate action. *Br J Pharmacol* 69:131–137, 1980.
36. Sawynok J, Jhamandas KH: Interactions of methylxanthines, nonxanthine phosphodiesterase inhibitors and calcium with morphine in guinea-pig myenteric plexus. *Can J Phys Pharm* 57:853–859, 1979.
37. Gallant CA, Clement JG: Methylxanthines antagonise adenosine but not morphine inhibition in guinea pig ileum. *Can J Physiol Pharmacol* 59:886–889, 1981.
38. Vizi ES, Somogyi GT, Magyar K: Evidence that morphine and opioid peptides do not share a common pathway with adenosine in inhibitory acetylcholine release from isolated intestine. *J Auton Pharmacol* 1:413–419, 1981.
39. Stone TW: Theophylline does not affect morphine inhibition of the isolated vas deferens. *Br J Pharmacol* 73:787–789, 1981.
40. Stone TW: Actions of adenine dinucleotides on the vas deferens, guinea pig taenia caeci and bladder. *Eur J Pharmacol* 75:93–102, 1981.

41. Duggan AW, Griersmith BT: Methylxanthines, cyclic AMP and the spinal transmission of nociceptive information. *Br J Pharmacol* 67:51–58, 1979.
42. Butt NM, Collier HOJ, Cuthbert NJ, Francis DL, Saeed SA: Mechanism of quasi-morphine withdrawal behaviour induced by methylxanthines. *Eur J Pharmacol* 53:375–378, 1979.
43. Sattin A: Increase in the content of cyclic AMP in mouse forebrain during seizures and prevention by methylxanthines. *J Neurochem* 18:1087–1096, 1971.
44. Vernikos-Daniellis J, Harris CG: The effects of in vitro and in vivo caffeine, theophylline and hydrocortisone on the phosphodiesterase activity of the pituitary, heart and cerebral cortex of the rat. *Proc Soc Exp Biol Med* 128:1016–1020, 1968.
45. Yarbrough GG, McGuffin-Clineschmidt JC: In vivo behavioural assessment of CNS purinergic receptors. *Eur J Pharmacol* 76:137–144, 1981.
46. Jurna L: Aminophylline differentiates between the depressant effects of morphine on the spinal nociceptive reflex and on the spinal ascending activity evoked from afferent C fibres. *Eur J Pharmacol* 71:393–400, 1981.
47. Lewis JJ, Van Petten GR: The effect of antidepressive drugs on the levels of adenine nucleotides, inorganic phosphate and phosphocreatine in the rat brain. *Br J Pharmacol* 20:462–470, 1963.
48. Huang M, Daly JW: Adenosine elicited accumulation of cyclic AMP in brain slices: Potentiation by agents which inhibit uptake of adenosine. *Life Sci* 14:489–503, 1974.
49. Sattin A, Stone TW, Taylor DA: Biochemical and electropharmacological studies with tricyclic antidepressants in rat and guinea pig cerebral cortex. *Life Sci* 23:2621–2626, 1978.
50. Stone TW, Taylor DA: Antidepressant drugs potentiate suppression by adenosine of neuronal firing in rat cerebral cortex. *Neurosci Lett* 11:93–97, 1979.
51. Dismukes RK, Daly JW: Adaptive responses of brain cyclic AMP-generating systems to alterations in synaptic input. *J Cyclic Nucleotide Res* 2:321–336, 1976.

DISCUSSION

DR. T.W. RALL: I would like to underscore the interaction between methylxanthines and opioid effects. Nearly 20 years ago, Stroud et al. showed that respiratory depression produced by therapeutic doses of meperidine in normal humans could be antagonized by quite low doses of caffeine. I have often wondered whether this effect would be useful clinically.

DR. STONE: I am sure that would be a possibility worth pursuing. I believe there is also some correlation between opioid peptides and childhood apnea, and theophylline is the preferred treatment for that syndrome.

DR. C. LONDOS: Did you mean to imply that the P site has nothing to do with the guanine nucleotide regulatory components?

DR. STONE: No.

DR. G.W. FRANK: Your statement that theophylline may antagonize the effect of opiates on respiratory depression is interesting. Clinical studies have shown that theophylline given to patients with chronic lung disease improves diaphragmatic function. Do you think it is possible that theophylline in this setting may be counteracting endogenous opiates?

DR. T.D. WHITE: The tricyclic antidepressants are all potent anticholinergic agents. Did you test the effect of anticholinergic drugs on the depressing action of adenosine on neuronal firing in the brain?

DR. STONE: No, we did not, although I would expect anything not working by a purine-enhancing mechanism to potentiate the responses to norepinephrine as well as to adenosine.

DR. H.J. LEIGHTON: Regarding your observation that tricyclic antidepressants increase purine release, I point out that tricyclics are not only anticholinergic but also α_1, α_2, and H_1 antagonists. Blockade of these autonomic and autocoid receptors may decrease feedback inhibition at adrenergic neurons and therefore increase not only catecholamine release but also ATP release.

DR. STONE: Yes, we do not know at the moment where the released purines are coming from, and we are not making any claims about the mechanism of the effect. At present, we can only say that the release occurs and that we need to use a variety of agonists and antagonists to clarify the mechanism.

DR. J.W. PHILLIS: There is a well-known use of adenine nucleotides in Asia to enhance the effects of glutamate as gustatory receptors. We see enhancement of glutamate firing by ATP in some critical neurons. If glutamate is a transmitter in the brain, facilitation of its actions may be important.

DR. STONE: That is very interesting. With respect to gustatory receptors, the potentiation of glutamate has also been reported with guanine nucleotides, so this may be a rather general phenomenon among nucleotides.

DR. J.W. DALY: Adenosine regulates many physiological processes via interaction with specific cell surface receptors. A major question is whether or not the enzyme adenylate cyclase is involved in such adenosine receptor-mediated responses. Recently, a new research tool has been introduced that should be useful in this regard. Forskolin, a diterpene, is a general activator (EC_{50} 5–20 μM) for adenylate cyclase in membranes and intact cells [Seamon and Daly, *J. Cyclic Nucleotide Res.* 7:201, 1981]. Thus, any response that involves activation of adenylate cyclase should also be evoked by forskolin, whereas responses involving inhibition of adenylate cyclase should be opposed by forskolin. Forskolin is a vasodilator, antispasmodic, and cardiac stimulant. These effects appear linked to activation of adenylate cyclase. Forskolin causes lipolysis in fat cells, steroidogenesis in adrenal cells, secretion in salivary cells, and inhibition of aggregation in platelets, all cyclic AMP-dependent processes. Not only does forskolin directly activate adenylate cyclase but, more important, forskolin in very low concentrations markedly potentiates the hormone receptor-mediated activation of adenylate cyclase. Thus, in a system where a receptor agonist (such as adenosine) evokes a response via activation of adenylate cyclase, then low concentrations of forskolin will make the agonist appear much more potent. This has been demonstrated in platelets, where the potency of prostaglandins as inhibitors of aggregation can be increased nearly 10-fold by 0.6 μM forskolin [Siegl et al., *Mol. Pharmacol.* 21:680, 1982].

30. ADENOSINE IN THE REGULATION OF NEUROTRANSMITTER RELEASE IN THE PERIPHERAL NERVOUS SYSTEM

BERTIL B. FREDHOLM, LARS E. GUSTAFSSON, PER HEDQVIST, ALF SOLLEVI

About ten years ago, Ginsborg and Hirst [1] found that adenosine reduced the quantum content and the frequency of miniature end-plate potentials in the rat phrenic nerve–diaphragm preparation. The effect of adenosine was partly calcium-dependent and inhibited by theophylline. These findings clearly suggested that adenosine acted on presynaptic receptors to depress the release of acetylcholine from cholinergic nerves. Subsequently, it was shown that adenine nucleotides acted in a similar manner as adenosine and that these compounds depressed the release of several types of transmitters in many tissues, as reviewed elsewhere [2–4].

METHODOLOGICAL CONSIDERATIONS

Generally, the presynaptic effects of a compound can be determined by assessing the depression of the release of transmitter or by measuring changes in the effector response to nerve stimulation. The first method commonly employs radioactively labeled transmitter, and then the results rest on the assumption that endogenous and radioactively labeled transmitters are released in a qualitatively similar fashion. There is little reason to assume that this is not so, even though the release of endogenous and of radioactively labeled transmitter could differ quantitatively. For example, in many tissues, newly synthetized

The original research reported here was supported by the Swedish Medical Research Council (proj. no. 2553 and 4342), by Magnus Bergvalls stiftelse, by Wibergs stiftelse, and by Karolinska Institutet.

R.M. Berne, T.W. Rall, and R. Rubio (eds.), Regulatory Function of Adenosine, *pp. 479–495.*

transmitter is released in preference to transmitter taken up by the membrane pump [5].

The evaluation of presynaptic actions of drugs from effector organ responses is fraught with many more assumptions. First, it is obviously essential that the transmitter that elicits the response of the end organ be known with certainty. The vas deferens is a good case in point. Many of the conclusions regarding the presynaptic effects of adenosine and adenosine analogues are based on studies of the alterations in the contractile responses of the field-stimulated vas. However, the motor response to single or short trains of stimuli does not behave as if it were elicited only by noradrenergic nerves [6–8]. Hence, conclusions about the release of norepinephrine from data on the alterations in motor responses of the vas may be hazardous. The guinea pig ileum has also been used extensively to study presynaptic effects of adenosine and related compounds. This tissue is of course notorious for its extensive neuronal network with a variety of identified and nonidentified transmitters.

The second major problem with this approach to the study of transmitter release is that the putative regulator may affect not only the release of the transmitter but also its actions. Adenosine is known to relax many types of smooth muscle and, at least at higher concentrations, so does the adenosine antagonist theophylline. At least in some tissues, e.g., the cat nictitating membrane, such a direct effect on smooth muscle contractility is the only important action of adenosine (and theophylline) on the contraction induced by nerve stimulation [9]. In other tissues, more complicated interactions occur. Thus, β adrenoceptor-mediated effects of catecholamines seem to be specifically depressed by adenosine and related compounds in many instances. The vasodilator [10] and lipolytic effect [11, 12] of catecholamines in adipose tissue, the action of isoprenaline in the heart [13], and the renin release induced by nerve stimulation or norepinephrine are examples of this type of interaction.

Of particular concern is the fact that adenosine and related compounds may simultaneously depress transmitter release and enhance transmitter action at the effector level [14]. This seems to occur both with effects mediated via α adrenoceptor and muscarinic receptors. In the vas deferens, the contractile effects of norepinephrine are potentiated by adenosine and adenine nucleotides [14–16]. This effect of adenosine and ATP is enhanced by the uptake-blocking agent dipyridamole, suggesting that intracellular receptors are not responsible. In the central nervous system, α adrenoceptor agonists are able to stimulate cyclic AMP accumulation by a mechanism that is largely adenosine-dependent [17]. In the kidney, the vasoconstrictor effect of norepinephrine is markedly potentiated by adenosine and adenine nucleotides [14] (Fig. 30.1). Finally, in subcutaneous vascular beds of the dog *in situ*, adenosine tends to enhance the vasoconstrictor response to nerve stimulation [14, 18]. Thus, there are many examples of a positive interaction between putative purine receptors and α adrenoceptors. Such interactions may also occur at concentrations of

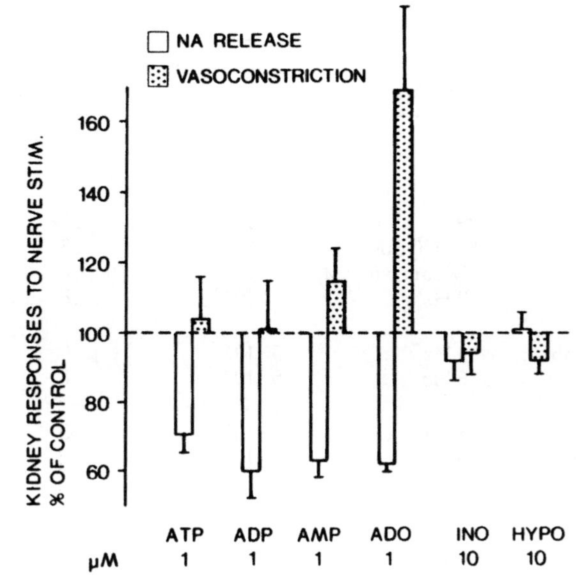

Figure 30.1. Effects of ATP, ADP, AMP, adenosine (ADO), inosine (INO), and hypoxanthine (HYPO) on [^{3}H]noradrenaline release and ensuing vasoconstriction induced by nerve stimulation (5 Hz, 15 sec) in Tyrode's perfused rabbit kidneys. Mean values (± SEM) from 6 to 20 experiments.

adenosine that are at least as low as those required to affect norepinephrine release [14, 15, 18].

In some instances [19, 20], but not in all [21], the effect of acetylcholine is enhanced by adenosine. Since adenosine and related compounds may affect not only the release of the transmitter but also its action, the interpretation of an inhibition of nerve-induced mechanical responses in terms of changes of transmitter release is complicated. The interpretation is much safer if the effects of adenosine on nerve-induced responses are compared to its effects on the responses to exogenous transmitter. It is well known, however, that the response to exogenous transmitter often differs from that to the endogenous one, partly because different sets of receptors may be activated. Even when appropriate controls with exogenous transmitter are used, the possibility exists that an accurate evaluation of the presynaptic effects of adenosine cannot be made. For these reasons, the present overview will deal with actual determination of transmitter release.

REGULATION OF NOREPINEPHRINE RELEASE

Table 30.1 summarizes the published information on the presynaptic effects of adenosine on norepinephrine release. Several conclusions may be drawn from this compilation. First, the effect is observed in several different species. A

Table 30.1. Inhibitory effect of adenosine on the release of norepinephrine from various preparations

	Stimulus	Ado concentration (*M*)	% Inhibition	Reference
In vitro preparation, not perfused				
Dog:				
Saphenous vein	2 Hz	10^{-5}	27	[22]
	K^+ (30 m*M*)	10^{-5}	28	[22]
	Tyramine (6 μ*M*)	10^{-5}	0	[22]
Basilar artery	10 Hz	10^{-6}	28	[23]
Cat:				
Nictitating membrane	4 Hz	$3 \cdot 10^{-4}$	0[a]	[9]
Auricle	5 Hz	$5 \cdot 10^{-4}$	8	[24]
Rabbit:				
Auricle	5 Hz	$5 \cdot 10^{-4}$	53	[24]
Pulmonary artery	3 Hz	$3 \cdot 10^{-4}$	40	[25]
Guinea pig:				
Auricle	5 Hz	$5 \cdot 10^{-4}$	47	[24]
Vas deferens	5 Hz	$5 \cdot 10^{-4}$	57	[24]
Vas deferens	5 Hz	10^{-5}	37[a]	[14]
Rat:				
Vas deferens	5 Hz	$5 \cdot 10^{-4}$	47	[24]
Vas deferens	5 Hz	10^{-5}	60	[26]
Vas deferens	5 Hz	10^{-4}	32	[27]
Salivary gland	5 Hz	$5 \cdot 10^{-4}$	78	[24]
Portal vein	5 Hz	$5 \cdot 10^{-4}$	47	[24]
Portal vein	5 Hz	10^{-4}	45	[28]
Portal vein	5 Hz	10^{-4}	60	[29]
In vitro preparation, perfused				
Cat:				
Spleen	5 Hz	10^{-4}	0[a]	[30]
Rabbit:				
Kidney	6 Hz	10^{-6}	42[a]	[14]
Heart	10 Hz	10^{-5}	60[a,b]	[31]
Rat:				
Heart	K^+ (60 μmol)	$9 \cdot 10^{-6}$	70[a]	[32]
	Tyramine (9 μ*M*)	$9 \cdot 10^{-6}$	20	[32]
Adrenal	1 Hz	$1.8 \cdot 10^{-4}$	40	[33]
In situ preparation.				
Dog:				
Subcutaneous tissue	4 Hz	$3 \cdot 10^{-5}$	55[a]	[14]
Subcutaneous tissue	4 Hz	$5 \cdot 10^{-6}$	57 25[b]	]18]

[a]Dose-response curves presented.
[b]Endogenous norepinephrine.

special case appears to be the cat. In three different cat tissues, no presynaptic inhibitory effect of adenosine was observed, even though very high concentrations were used. Second, within each species there are no major tissue differences. In fact, there is generally good agreement between different preparations in the same laboratory and between laboratories using the same preparation. Third, adenosine appears to affect only depolarization-evoked release (by nerve stimulation or by potassium) but not release evoked by tyramine. Finally, it is of interest that in three saline-perfused tissues the IC_{50} for adenosine is between 1 and 10 μM, whereas in many nonperfused tissues the IC_{50} appears to be above 100 μM. This may represent an artefact caused by differences in the ability of adenosine to reach the presynaptic receptors. The mean diffusion distance through intact tissue capable of inactivating adenosine ought to be considerably less in the case of a perfused compared to a nonperfused tissue. When the tissue is blood-perfused, the rapid inactivation of adenosine in blood is an added complication.

As discussed elsewhere, the presynaptic effect of adenosine is unlikely to be mediated by adrenoceptors, muscarinic receptors, or prostaglandins [2, 3]. The effect is potentiated by drugs that prevent adenosine inactivation, and it is prevented by methylxanthines [2, 3].

REGULATION OF ACETYLCHOLINE RELEASE

As seen in Table 30.2, acetylcholine release is inhibited by adenosine both in tissues where the transmission is nicotinic and where it is muscarinic. The data on nicotinic transmission in ganglia represent inhibition of responses of the postsynaptic cell rather than inhibition of the transmitter. A large part of this effect could be postsynaptic [34–36]. This could perhaps explain the results obtained in cats, since, as noted earlier, norepinephrine release in cats is unaffected by adenosine, indicating a relative absence of presynaptic adenosine receptors in this species. However, a depression of the firing of the sympathetic neuron at the ganglion level could also lead to depression of sympathetic neurotransmission in this species.

The results on nicotinic transmission in skeletal muscle all derive from analyses of end-plate potentials. The potency of adenosine in these studies is quite low, possibly because adenosine is rapidly inactivated in these *in vitro* preparations.

Actual inhibition of acetylcholine release has been determined only in the guinea pig ileum. The potency of adenosine in this system is similar to the potency found for adenosine as an inhibitor of norepinephrine release from different smooth muscle preparations *in vitro* (Table 30.1). The rate of disappearance of adenosine in this preparation is fast [37]. To some extent, this depends on uptake into the tissue but it also depends on inactivation by the bath itself or by microorganisms attached to the tissue chamber [37]. Adenosine accumulates in a well-cleaned bath but not in a bath that has not been extensively cleaned within about six hours of use. This could perhaps

Table 30.2. Effect of adenosine on cholinergic transmission

		Ado concentration (M)	% Inhibition	Reference
Ganglia				
Rat:				
Superior cervical	Single pulse	10^{-5}	50	[36]
Cat:				
Urinary bladder	1–10 Hz	0.05–1 μmol i.a.	Not stated	[34]
Skeletal muscle				
Rat:				
Diaphragm	0.1 Hz	$2.5 \cdot 10^{-4}$	50	[1]
Diaphragm	0.5 Hz	10^{-4}	~50	[39]
Frog:				
Sartorius	0.5 Hz	10^{-3}	~30	[39]
Sartorius	0.5 Hz	10^{-3}	~30	[41]
Cutaneous pectoris		$2.5 \cdot 10^{-4}$	~50	[40]
Torpedo electroplax	5 Hz	10^{-5}	55	[42]
Guinea pig ileum				
	0.1 Hz	$4 \cdot 10^{-5}$	47[a]	[43]
	0.5 Hz	$2 \cdot 10^{-4}$	73[a]	[21]
	3 Hz	$2 \cdot 10^{-4}$	41[a]	[21]
	0.5 Hz	$2 \cdot 10^{-4}$	100	[44]
	0.1 Hz	$3 \cdot 10^{-5}$	76[a]	[45]
	0.1 Hz	$3 \cdot 10^{-5}$	89	[45]
	0.2 Hz	10^{-5}	~50	[38]

[a]Endogenous acetylcholine determined.

explain why neurotransmission apparently works much better in organ baths that are not scrupulously clean, and why it is sometimes difficult to detect a clear-cut potentiation of adenosine action by the adenosine uptake inhibitor dipyridamole.

RECEPTORS AND MECHANISM OF ACTION

Receptors

Several lines of evidence indicate that the presynaptic action of adenosine and adenine nucleotides are receptor-mediated. The effect of adenosine is enhanced by drugs that prevent adenosine uptake. Similarly, several adenosine analogues that are poor substrates for transport are also more potent inhibitors of transmitter release than adenosine. This suggests a site of action on the external surface of the nerve endings. The apparently competitive inhibition of the presynaptic effects of adenosine by xanthine derivatives is also most easily explained by assuming interactions at the receptor level. Finally, there are clear-cut structural specificities for the adenosine analogues that suggest that the response is receptor-mediated.

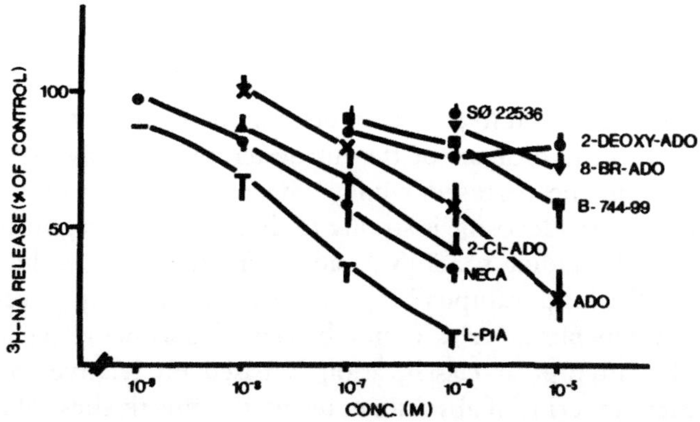

Figure 30.2. The effect of adenosine and some adenosine analogues on [^{3}H] transmitter release from [^{3}H]noradrenaline prelabeled saline-perfused rabbit kidney. After labeling and extensive washing, the sympathetic nerve supply was stimulated for periods of 15 sec at 5 Hz. After periods of control stimulation, adenosine or adenosine analogues were infused to the indicated concentration. In each experiment, a number of concentrations were investigated, followed by a series of control stimulations. In each experiment, the percentage inhibition of transmitter release was calculated. The figures given are means ± SEM of 3 to 14 separate experiments for each compound.

The results shown in Figure 30.1 demonstrate that the natural adenine nucleotides are approximately equipotent with adenosine as antagonists at presynaptic sites regulating norepinephrine release in the rabbit kidney. These results agree rather well with results obtained in other preparations. Thus, most authors find that ATP is essentially equipotent with adenosine in dog blood vessels *in vitro* [22, 23] and *in vivo* [14, 34], in rabbit pulmonary artery [25], and in rat heart [39], adrenal [33], vas deferens [27], and blood vessels [29]. ADP was reported to be more potent than adenosine in rabbit pulmonary artery [25] but less potent in rat heart [32] and blood vessels [29]. AMP is rather less potent than adenosine in blood vessels [29, 33]. The breakdown products of adenosine, i.e., inosine and hypoxanthine, are considerably less potent than the parent compound in rabbit (Fig. 30.1). Even when the inosine concentration was increased to 100 μM, no more than 10% inhibition was found (not shown). Other authors have also found a considerably lower potency for inosine [25, 28, 32].

As discussed at length at this symposium, adenosine receptor subtypes may be classified on the basis of the potency of some stable analogues. The results obtained with a series of such analogues in the rabbit kidney is shown in Figure 30.2. Two of the analogues, SQ22536 and 2′-deoxyadenosine, which are selective for the intracellular P-type receptor [51], had only a marginal effect on the release of transmitter; 8-bromo-adenosine similarly showed a low activity, and it has also been found to possess a low adenosine-like potency in other systems. The other analogues have been shown to possess considerable potency in several biological systems. The most potent was L-N^6-phenylisopropyladenosine, followed by adenosine 5′-N-ethylcarboxamide (NECA) and 2-chloroadenosine. All these analogues were more potent than adenosine. The nitroester of NECA, B744-99, was less potent. This compound has been found to cause a profound hypotensive effect in dogs [52]; however, its activity may depend on hydrolysis of the nitroester, which may not occur at an appreciable rate in the perfused kidney. Since the relative potency of PIA, NECA, and 2-chloroadenosine is similar to that seen for antagonism of lipolysis in fat cells [51], binding to CNS A_1 adenosine receptors, and depression of field EPSPs in the hippocampus [53], the receptor-mediating presynaptic inhibition of norepinephrine release may be classified as being A_1. It should be noted that this attempt at classification is based on relative, not absolute, agonist potency. In terms of absolute potency, the effectiveness of adenosine as inhibitor of transmitter release is not much different from its potency to raise cyclic AMP levels, an archetypal A_2-type response. As already mentioned, adenosine not only depresses norepinephrine release in the kidney but also enhances the effect of the amine. The receptor mediating this response may show slightly different characteristics [unpublished observations].

Concerning cholinergic neurotransmission, the picture is similar. In an electrophysiological study, Silinsky [40] showed that two nonhydrolyzable ATP analogues were devoid of effect on cholinergic neurotransmission. He also showed that 2′-deoxyadenosine, 5′-deoxyadenosine, and 9-β-D-arabinofuranosyladenine, all of which may be classified as P site agonists, did not depress but instead enhanced transmission. However, two R site agonists, 2-chloroadenosine and N^6-methyladenosine, caused a substantial depression. The fact that the latter compound was active could be taken as an indication that presynaptic control of acetylcholine release is influenced by receptors of the A_1 subtype. Other studies attempting to elucidate the receptors involved are unfortunately based on the contractile responses of muscle tissue and consequently are difficult to interpret.

Mechanism behind the presynaptic effect of adenosine

Adenosine receptors are known to regulate cyclic AMP formation in nervous tissue [17]. The receptors regulating transmitter release can apparently be classified as being of the A_1 subtype. These receptors are similar to the R_i receptors of Londos et al. [51], i.e., if anything, they probably decrease the rate of cyclic AMP formation. A depression of the cyclic AMP content of nerve endings could conceivably cause a decrease in transmitter release, since cyclic AMP has been suggested to play a stimulatory role on transmitter release [5]. We do not favor this view, however, since it implies that transmitter release is entirely dependent on cyclic AMP (adenosine may completely inhibit transmitter release). However, the possibility exists that the primary events leading to a decrease in adenylate cyclase activity and to a decrease in transmitter release are similar; for example, they may use the same GTP-binding protein.

Since transmitter release depends on calcium ions, it is obvious that adenosine somehow regulates the handling of calcium in the nerve terminal. The presynaptic effect of adenosine is dependent on the level of extracellular calcium [1, 38, 41]. The effect of adenosine is also frequency-dependent [19, 31, 43]. Such a frequency dependence has, for other presynaptically active substances, been taken as evidence for a calcium-dependent mechanism [5]. The fact that tyramine-induced release is not affected by adenosine is further evidence for this view.

The question of how adenosine modifies the calcium-dependent transmitter release arises. In a recent study, two principally different sites of action were indicated: (1) inhibition of the invasion of the nerve terminals by nerve impulses and (2) inhibition of the stimulus-secretion coupling in the individual varicosity [46]. It was argued by these authors that α-adrenergic inhibition of norepinephrine release was to some extent dependent on inhibition of invasion of the varicosities, because transmitter release induced by nerve stimulation was inhibited to a greater extent than was release induced by potassium depolarization. In the case of adenosine, it appears that both means of stimulating release are inhibited equally well (Table 30.1). We would therefore have to argue that adenosine probably does not act by preventing the invasion of the varicosities. This conclusion would fit with the finding that nerve terminal excitability in the CNS apparently is not affected by adenosine [47]. At least in concentrations above 10 μM, adenosine is able to hyperpolarize pyramidal cells in the rat hippocampus [48–50]. If hyperpolarization also occurs in the nerve terminal, this would be expected to decrease invasion of the terminals by the nerve impulse but not to depress the release of transmitter once the terminal is invaded. Thus, it is interesting that Siggins and Schubert [50] found evidence that the presynaptic effect of adenosine could occur at lower concentrations of adenosine than those required to give hyperpolarization of the pyramidal cell. The possibility exists that the ionic basis of the presynaptic effect on transmitter release is not the same as that analyzed by intracellular recordings in cell bodies. In any event, it is clear that most of the evidence concerning the mechanism whereby adenosine depresses transmitter output is indirect and mostly negative, and that much more information is required.

PHYSIOLOGICAL IMPLICATIONS

Two approaches to determining the physiological role of adenosine as a regulator of norepinephrine release are particularly fruitful: (1) the quantitative comparison of the dose-effect curves with the actually determined levels of adenosine under physiological conditions and (2) the pharmacological evaluation of the effect of drugs that affect either the level or the actions of adenosine on neurotransmission.

Regarding the first of these approaches, it is important that the dose-effect curves for adenosine be done under circumstances such that the nominal concentration agrees with that actually present in the biophase. In organ baths, the potency of adenosine is easily underestimated because of rapid inactivation. In blood-perfused tissue, the elimination of adenosine in the blood stream also seems to underestimate the potency of adenosine as a presynaptic agonist. In saline-perfused tissue or in a preparation pretreated with inhibitors of inactivation, 50% inhibition of transmitter release is obtained with 1–10 μM adenosine. A threshold effect may be obtained with as low as 0.1 μM adenosine [14]. It is difficult to estimate the concentration of adenosine in the biophase; in the case of presynaptic regulation, this refers to the concentration in the extracellular fluid surrounding the nerve terminal. This concentration is not equal to the amount of adenosine in the tissue, since a substantial but variable fraction is sequestered in different tissues. Data obtained in the dog could be taken as evidence that the free extracellular concentration is close to the normal arterial (and venous) plasma concentration [10]. The estimates of this concentration, in turn, varies considerably in the literature—from undetectable (below 0.05 μM) to above 1 μM. However, many investigators, using different techniques, have reported resting values between 0.1 and 0.5 μM in plasma from several species, including humans. If this value is accepted, we must conclude that, under resting conditions, adenosine would be able to modify transmitter release to a very limited extent. Recently, we have used a dialysis technique to determine the extracellular adenosine concentration in the brain [54]. This method estimates the concentration of adenosine to be about 1 μM in both awake and anesthetized rats. At least in the brain, a somewhat larger influence of adenosine as a presynaptic regulator may therefore be envisaged.

Since it employs pharmacological tools, the second approach depends heavily on the specificity of these tools. As antagonists of adenosine action, we rely on xanthine derivatives. Theophylline, caffeine, and isobutylmethylxanthine have many different biological effects. Besides antagonism of adenosine receptors, they exhibit inhibition of phosphodiesterase, mobilization of calcium from intracellular depots, inhibition of carrier-mediated transport, and inhibition of 5′-nucleotidase. It appears that they show a degree of selectivity as adenosine antagonists only over a limited range of concentrations. Thus, the dominant effect of theophylline between 10 and 100 μM may be adenosine antagonism, but appropriate controls are essential. Other analogues, such as 8-phenyltheophylline, may show a greater degree of selectivity and could thus be a superior tool [4]. The inhibitors of adenosine uptake are often phosphodiesterase inhibitors and inhibitors of other transport processes. Of particular concern is the finding that both NBMPR and dipyridamole seem to block not only uptake of adenosine but also outward transport [12]. This means that these two agents cannot be used to potentiate the effects of endogenously released adenosine. In particular, they are not useful as tools to elucidate the importance of adenosine released during nerve activity. Finally, inhibition of adenosine deaminase elevates levels not only of adenosine but also of 2′-deoxyadenosine. The latter compound in particular may influence the levels of S-adenosylhomocysteine and thereby block various transmethylation reactions of potential biological importance.

Using a combination of these two techniques, we have concluded that endogenous adenosine regulates norepinephrine in the perfused rabbit kidney [4, 14, 55]. Thus, endogenous resting levels of adenosine are sufficient to cause some depression of norepinephrine release, and the levels reached following nerve stimulation (especially in the presence of an adenosine deaminase inhibitor) are sufficient to come close to the IC_{50}. Drugs that elevate adenosine levels depress labeled transmitter overflow, whereas alkylxanthines cause an enhanced release that is proportionate to the potency of the drug as an adenosine antagonist. In the guinea pig ileum [3, 4, 19, 37] and rat vas deferens [3, 4, 26, 27], similar results have been obtained. However, studies of subcutaneous tissue in dogs did not provide evidence that endogenous adenosine played an important role in the regulation of transmitter release [2, 4, 18], in contrast to the postjunctional effect, lipolysis [18]. At most, one-fifth of the depression of lipolysis induced by nerve stimulation could be accounted for by presynaptic inhibition.

These findings raise a terminological point of some importance. The word *physiological* is used in at least two senses: to denote the situation in the intact animal and to denote that a mechanism is operative in a given type of preparation. When used in the latter sense, there is little doubt that adenosine is a physiological regulator of transmitter release in a variety of preparations. When used in the former, stricter sense, there is more ambiguity. Whether adenosine truly regulates transmitter release from peripheral nerves in the intact, normal animal or human is not yet clear. When the adenosine level was raised about 100% by a therapeutic dose of dipyridamole, there was no significant change in the concentration of norepinephrine in human plasma [56]. This could mean that adenosine does not cause significant regulation of norepinephrine release in humans under physiological conditions, or it could mean that the resting levels of norepinephrine in plasma are not subject to regulation at all. Theophylline and caffeine are known to elevate plasma catecholamines. This could indicate that there is an influence of endogenous adenosine. However, the increase in plasma adenosine and norepinephrine after addition of these xanthines could also be a reflection of a generalized arousal.

Adenosine and related compounds are released following activation of peripheral cholinergic and noradrenergic pathways [2, 4]. Since the release of the purines was evoked by stimulation of the nerve supply, it is obvious that procedures that inhibit the function of these nerves (TTX, denervation, guanethidine, etc.) depress the release. The purines could originate from nerves (release from purinergic nerves or as cotransmitters) or from the effector tissue. In the tissues we have studied, the release is blocked by receptor-blocking agents, suggesting that release from the effector tissue is quantitatively more important [2, 4]. In particular, we presented evidence that the ATP released together with norepinephrine is unlikely to be quantitatively important [26]. An essentially similar conclusion was drawn by Israël and coworkers [42] from studies of the Torpedo electroplax. Thus, adenosine, or an adenine nucleotide, released by the action of the transmitter could act on the nerve terminal to cause inhibition of transmitter release. To describe this phenomenon, the terms *transsynaptic inhibition* and *retrograde transmission* have been coined [2, 42].

These two terms conjure up a picture in which adenosine or an adenine nucleotide is released at the site of the receptor, located immediately subsynaptically, and moves to the surface of the nerve terminal, where it acts to limit further transmitter release. In other words, the compound may be thought of as being released and acting within the space of the synapse. In this respect, the picture would be quite similar to the commonly drawn representation of how transmitter release is regulated by autoreceptors. However, it may be that the increase in the release of purines is a rather remote consequence of transmitter action. Thus, in adipose tissue, the release of adenosine is caused by the combined influence of α adrenoceptor stimulation in blood vessels leading to vasoconstriction and β adrenoceptor stimulation of fat cells leading to metabolic activation [2, 10]. In many other tissues, such as the heart, the kidney, and smooth muscle organs, the situation may be similar, in that the bulk of the purine compounds is released as a consequence of the organ responding to nerve activation, not as a consequence of initiation (via the receptor) of the response. It is perhaps better, therefore, to picture the role of adenosine as that of a local hormone rather than in terms of a purely synaptic event. With this point of view, it is also easy to envisage a role of other stimuli, such as hypoxia, ischemia, and increased work load influencing the release of transmitter by increasing the tissue concentration of adenosine. There may be no fundamental difference in the role of adenosine as a regulator of transmitter release if it is released by nerve activation or by completely unrelated stimuli.

This way of looking at the role of adenosine also makes teleological sense. The ability of a tissue to respond adequately to a nervous stimulus is a complex function of a variety of factors, including circulation, the supply of metabolic substrates, and intracellular metabolic stores, including the levels of high-energy phosphates, etc. It seems reasonable to think that nature has evolved a mechanism whereby a resultant of these many factors may be signaled to the nervous elements of the tissue, enabling control of local nervous activity depending on the state of the tissue and its ability to respond to nerve activation without compromising its further viability. As discussed elsewhere in this symposium, the role as such a messenger could easily be filled by adenosine.

REFERENCES

1. Ginsborg BL, Hirst GDS: The effect of adenosine on the release of the transmitter from the phrenic nerve of the rat. *J Physiol* (*Lond*) 224:629–645, 1972.
2. Fredholm BB, Hedqvist P: Modulation of neurotransmission by purine nucleotides and nucleosides. *Biochem Pharmacol* 29:1635–1643, 1980.
3. Paton DM: Presynaptic neuromodulation mediated by purinergic receptors, in Burnstock G (ed): *Purinergic Receptors*. London, Chapman and Hall, 1981, pp 199–220.

4. Gustafsson L, Hedqvist P, Fredholm BB: Release and presynaptic effects of adenosine and adenine nucleotides, in Daly JW, Kuroda Y, Phillis JW, Shimizu PH, Ui M (eds): Physiology and Pharmacology of Adenosine Derivatives. New York, Raven Press, 1983, pp 143–153
5. Westfall TC: Local regulation of adrenergic neurotransmission. *Physiol Rev* 57:659–728, 1977.
6. Ambache N, Zar MA: Evidence against adrenergic motor transmission in the guinea pig vas deferens. *J Physiol (Lond)* 216:359–389, 1971.
7. von Euler U, Hedqvist P: Evidence of α- and β_2-adrenoceptor-mediated inhibition of the twitch response in the guinea pig vas deferens by noradrenaline. *Acta Physiol Scand* 93:572–573, 1975.
8. Blakeley AGH, Cunnane TC, Petersen SA: An electropharmacological analysis of the effects of some drugs on neuromuscular transmission in the vas deferens of the guinea pig. *J Auton Pharmacol* 1:367–375, 1981.
9. Luchelli-Fortis MA, Fredholm BB, Langer SZ: Evidence against the presence of presynaptic inhibitory adenosine receptors in the cat nictitating membrane. *J Pharmacol Exp Ther* 219:235–242, 1981.
10. Fredholm BB, Sollevi A: The release of adenosine and inosine from canine subcutaneous adipose tissue by nerve stimulation and noradrenaline. *J Physiol* 313:351–367, 1981.
11. Schwabe U, Ebert R, Erbler HC: Adenosine released from fat cells: Effects on cyclic AMP and hormone actions. *Adv Cyclic Nucleotide Res* 5:569–584, 1975.
12. Sollevi A, Fredholm BB: The antilipolytic effect of endogenous and exogenous adenosine in canine adipose tissue in situ. *Acta Physiol Scand* 113:53–60, 1981.
13. Schrader J: Sites of action and production of adenosine in the heart, in Burnstock G (ed): *Purinergic Receptors*. London, Chapman and Hall, 1981, pp 121–162.
14. Hedqvist P, Fredholm BB: Effects of adenosine on adrenergic neurotransmission: Prejunctional inhibition and postjunctional enhancement. *Naunyn Schmiedebergs Arch Pharmacol* 293:217–223, 1976.
15. Holck MI, Marks BH: Purine nucleoside and nucleotide interactions on normal and subsensitive alpha adrenoceptor responsiveness in guinea pig vas deferens. *J Pharmacol Exp Ther* 205:104–117, 1978.
16. Sakai KK, Hymsom DL, Shapiro R: The effects of adenosine and adenine nucleotides on the guinea pig vas deferens: Evidence for existence of purinergic receptors. Life Sci 24:1299–1308, 1979.
17. Daly JW: Adenosine and cyclic adenosine monophosphate-generating systems in brain tissue, in Baer HP, Drummond GI (eds): *Physiological and Regulatory Functions of Adenosine Nucleotides*. New York, Raven Press, 1979, pp 229–241.
18. Sollevi A, Hjemdahl P, Fredholm BB: Endogenous adenosine inhibits lipolysis induced by nerve stimulation without inhibiting noradrenaline release in canine subcutaneous adipose tissue in vivo. *Naunyn Schmiedebergs Arch Pharmacol* 316:112–119, 1981.
19. Gustafsson L: Influence of adenosine on responses to vagal nerve stimulation in the anaesthetized rabbit. *Acta Physiol Scand* 111:263–268, 1981.
20. Gustafsson L: Adenosine elicits prejunctional inhibition and postjunctional enhancement in rabbit iris sphincter. *Acta Physiol Scand* 114:38A, 1982 (abst).
21. Gustafsson L, Hedqvist P, Fredholm BB, Lundgren G: Inhibition of acetylcholine-release in guinea pig ileum by adenosine. *Acta Physiol Scand* 104:469–478, 1978.
22. Verhaege RH, Vanhoutte PM, Shepherd JT: Inhibition of sympathetic neurotransmission in canine blood vessels by adenosine and adenine nucleotides. *Circ Res* 40:208–215, 1977.
23. Muramatsu I, Fuijiwara M, Miura A, Sakakibara Y: Possible involvement of adenine nucleotides in sympathetic neuroeffector mechanisms of dog basilar artery. *J Pharmacol Exp Ther* 216:401–409, 1981.
24. Wakade AR, Wakade TD: Inhibition of noradrenaline release by adenosine. *J Physiol (Lond)* 282:45–49, 1978.
25. Husted S, Nedergaard OA: Inhibition of adrenergic neuroeffector transmission in rabbit pulmonary artery and aorta by adenosine and adenine nucleotides. *Acta Pharmacol Toxicol* 49:334–353, 1981.
26. Fredholm BB, Fried G, Hedqvist P: Origin of adenosine released from rat vas deferens by nerve stimulation. *Eur J Pharmacol* 79:233–243, 1982.
27. Clanachan AS, Johns A, Paton DM: Presynaptic inhibitory actions of adenine nucleotides and adenosine on neurotransmission in the rat vas deferens. *Neuroscience* 2:597–602, 1977.

28. Moylan RD, Westfall TC: Effect of adenosine on adrenergic neurotransmission in the superfused rat portal vein. *Blood Vessels* 16:302–310, 1979.
29. Enero MA, Saidman BQ: Possible feedback inhibition of noradrenaline release by purine compounds. *Naunyn Schmiedebergs Arch Pharmacol* 297:39–46, 1977.
30. Mueller AL, Moisimann WF, Weiner N: Effects of adenosine on neurally mediated norepinephrine release from the cat spleen. *Eur J Pharmacol* 53:329–333, 1979.
31. Hedqvist P, Fredholm BB: Inhibitory effect of adenosine on adrenergic neuroeffector transmission in the rabbit heart. *Acta Physiol Scand* 105:120–122, 1979.
32. Khan MT, Malik KU: Inhibitory effect of adenosine and adenine nucleotides on potassium-evoked efflux of [^{3}H]noradrenaline from the rat isolated heart: Lack of relationship to prostaglandins. *Br J Pharmacol* 68:551–561, 1980.
33. Wakade AR: Studies on secretion of catecholamines evoked by acetylcholine or transmural stimulation of the rat adrenal gland. *J Physiol (Lond)* 313:463–480, 1981.
34. de Groot WC, Booth AM: Inhibition and facilitation in parasympathetic ganglia of the urinary bladder. *Fed Proc* 39:2990–2996, 1980.
35. Brown DA, Caulfield MP, Kirby PJ: Relation between catecholamine-induced cyclic AMP changes and hyperpolarization in isolated rat sympathetic ganglia. *J Physiol (Lond)* 290:441–451, 1979.
36. McAfee DA, Henon BK, Whiting GJ, Horn JP, Yarowsky PJ, Turner DK: The action of cAMP and catecholamines in mammalian sympathetic ganglia. *Fed Proc* 39:2997–3002, 1980.
37. Gustafsson L, Fredholm BB, Hedqvist P: Theophylline interferes with the modulatory role of endogenous adenosine on cholinergic neurotransmission. *Acta Physiol Scand* 111:269–280, 1981.
38. Dowdle E, Maske R: The effects of calcium concentration on the inhibition of cholinergic neurotransmission in the myenteric plexus of guinea pig ileum by adenine nucleotides. *Br J Pharmacol* 71:245–252, 1980.
39. Ribeiro JA, Walker J: The effects of adenosine triphosphate and adenosine diphosphate on transmission at the rat and frog neuromuscular junctions. *Br J Pharmacol* 54:213–218, 1975.
40. Silinsky EM: Evidence for specific adenosine receptors at cholinergic nerve endings. *Br J Pharmacol* 71:191–194, 1980.
41. Brănisteanu DD, Anulică ID, Proca B, Nhue BG: Adenosine effects upon transmitter release parameters in the Mg^{2+}-paralyzed neuromuscular junction of frog. *Naunyn Schmiedebergs Arch Pharmacol* 308:273–279, 1979.
42. Israël M, Lesbats B, Manarche R, Meunier FM, Frachon P: Retrograde inhibition of transmitter release by ATP. *J Neurochem* 34:923–932, 1980.
43. Vizi ES, Knoll J: The inhibitory effect of adenosine and related nucleotides on the release of acetylcholine. *Neuroscience* 1:391–398, 1976.
44. Wikberg JES, Axelsson KL: A simple and efficient method for studying neurotransmitter release in vitro by a radiotracer technique. *Acta Physiol Scand* 109:123–129, 1980.
45. Hayashi E, Mori M, Yamada S, Kunitomo M: Effects of purine compounds on cholinergic nerves: Specificity of adenosine and related compounds on acetylcholine release in electrically stimulated guinea pig ileum. *Eur J Pharmacol* 48:297–307, 1978.
46. Alberts P, Bartfai T, Stjärne L: Site(s) and ionic basis of α-autoinhibition and facilitation of [^{3}H]noradrenaline secretion in guinea pig vas deferens. *J Physiol* 312:297–334, 1981.
47. Stone TW: Adenosine and related compounds do not affect nerve terminal excitability in the rat CNS. *Brain Res* 182:198–200, 1980.
48. Okada Y, Ozawa S: Inhibitory action of adenosine on synaptic transmission in the hippocampus of the guinea pig in vitro. *Eur J Pharmacol* 68:483–492, 1980.
49. Segal M: Intracellular analysis of a postsynaptic action of adenosine in the rat hippocampus. *Eur J Pharmacol* 79:193–199, 1982.
50. Siggins GR, Schubert P: Adenosine depression of hippocampal neurons in vitro: An intracellular study of dose-dependent actions on synaptic and membrane potentials. *Neurosci Lett* 23:55–60, 1981.
51. Londos C, Wolff J, Cooper DMF: Adenosine as a regulator of adenylate cyclase, in Burnstock G (ed): *Purinergic Receptors*. London, Chapman and Hall, 1981, pp 287–323.
52. Raberger G: Cardiovascular and metabolic actions of adenosine and adenosine analogs, in Baer HP, Drummond GI (eds): *Physiological and Regulatory Functions of Adenosine and Adenine Nucleotides*. New York, Raven Press, 1979, pp 155–166.

53. Dunwiddie TV, Fredholm BB: Adenosine receptors in the rat hippocampus. *Br J Pharmacol* 77:347P (abst), 1982.
54. Zetterström T, Vernet L, Ungerstedt U, Tossman U, Jonzon B, Fredholm BB: Purine levels in the intact rat brain: Studies with an implanted perfused hollow fibre. *Neurosci Lett* 29:111–115, 1982.
55. Hedqvist P, Fredholm BB, Ölund S: Antagonistic effects of theophylline and adenosine on adrenergic neuroeffector transmission in the rabbit kidney. *Circ Res* 43:592–598, 1978.
56. Sollevi A, Östergren J, Hjemdahl P, Fredholm BB, Fagrell B: The effect of dipyridamole on plasma adenosine levels and skin micro-circulation in man. *J Clin Chem Clin Biochem*, 1982 (in press).

DISCUSSION

DR. J.P. MANFREDI: I do not understand the distinction between models of norepinephrine feedback that postulate an influence of local norepinephrine concentrations versus spatially integrated norepinephrine concentrations on norepinephrine release.

DR. FREDHOLM: The distinction is, rather, between receptors located within the synaptic cleft and those located outside it. If the receptors are located outside the cleft, they have the possibility of sensing norepinephrine released not only from one varicosity but from many.

DR. C. SU: (1) Is it correct that, according to the scheme you presented, inhibition of norepinephrine release by adenosine will occur at a site downstream, "distant" or "remote" from the site of release? (2) How do you explain the reports that adrenergic blocking agents do not significantly reduce adenosine release following adrenergic nerve stimulation? (3) Do you have any suggestion as to what differentiates the preparations in which adenosine release is predominantly presynaptic and those in which it is predominantly postsynaptic?

DR. FREDHOLM: (1) I proposed that the relevant adenosine receptors may not be located within the synaptic gap and that the purines released following nerve stimulation may also be formed at a site distant from the subsynaptic cell membrane. (2) It probably means that the purine release is not a consequence of norepinephrine acting on a receptor. It must be mentioned, however, that, in many instances, adrenoceptor blockade inhibits purine release induced by sympathetic nerve stimulation. (3) I am not aware of any preparation in which the purine release is predominantly presynaptic.

DR. L.E. GUSTAFSSON: From experiments in which purine release during nerve stimulation cannot be blocked by receptor-blocking agents, it is often concluded that purine release originates from purines stored in transmitter storage vesicles. The alternative explanation may be that this "prejunctional" purine release reflects the metabolic activation in the neuron in association with action potential conduction, ion transport, etc. This has been contemplated in several early papers dealing with neuroregulation [e.g., Kosterlitz, H.W., & Lees, G.M.: Interrelationships between adrenergic and cholinergic mechanisms. *Handb. Exp. Pharmacol* 33:762–812, 1972].

DR. FREDHOLM: I agree.

DR. D.A. MCAFEE: (1) I would like to simply point out that your model of adenosine controlling propagation into the branches of peripheral sympathetic terminals may not be easily generalized to cholinergic systems that have a different architecture. (2) I would appreciate your expanding your opinions and comments concerning the non-neuronal release of adenosine as an important source of neuromodulation. For instance, do you think this is an important role for glia in the CNS?

DR. FREDHOLM: (1) I agree with you that generalizations from one tissue to another and from one type of neuron to another may be quite hazardous. I would not be surprised if future studies were to show considerable differences in the potency of adenosine between different preparations and that these differences may be related to differences in the architecture of the nerves. (2) In the brain, of course, the effector cells are themselves neurons. Hence, release from the effector cell is not the same as non-neuronal release in the CNS. I agree with you that the role of glia, both as a source of adenosine and as a target for adenosine action, should be thoroughly investigated.

DR. G.W. KREUTZBERG: What do you suppose is the source of adenosine in the peripheral model you sketched on the blackboard?

DR. FREDHOLM: I believe that the source of adenosine varies with the type of stimulus and with the tissue. In the canine subcutaneous tissue and the rabbit kidney, sympathetic nerve stimulation induces purine release that is dependent on vasoconstriction. The purines may then originate from blood vessels (endothelial cells and contracting smooth muscle cells) and from entire regions suffering relative hypoxia. In other tissues, depolarization of pre- or postjunctional elements may lead to release of adenosine and adenine nucleotides. Exocytotic release of nucleotides stored in vesicles may be important in some tissues. The important thing is that the effect of the extracellular adenosine is likely to be much the same irrespective of the source.

DR. KREUTZBERG: During this meeting, we have learned that adenosine may function in the CNS to "sharpen" synaptic function [J. Phillis]. According to your model, adenosine action in the peripheral nervous system lacks this aspect of precision. It is more diffuse.

DR. FREDHOLM: This may be more a semantic issue than a major difference in point of view. I have maintained that the role of the presynaptic inhibitory effect of adenosine may be viewed as a means whereby the synaptic input is "tuned" to the ability of the tissue to respond adequately to the transmitter without compromising its viability. Alternatively, this may be thought of as a means of sharpening the synaptic function by allowing for a dialogue between the pre- and postjunctional elements. Furthermore, a distinction should be made between the pre- and postsynaptic effects of adenosine. Although the presynaptic actions could be considered part of a protective mechanism, the postsynaptic actions may be directly or indirectly involved in signal transfer.

POSTER ABSTRACTS

1. S-ADENOSYLHOMOCYSTEINE FORMATION FROM ENDOGENOUS ADENOSINE IN ISCHEMIC RAT HEART. P.W. Achterberg, P.P. de Tombe, E. Harmsen, J.W. de Jong. Cardiochemical Laboratory, Thoraxcenter, Erasmus University, Rotterdam, The Netherlands.

In heart the coronary vasodilator adenosine (Ado) can be formed either by breakdown of AMP, which is elevated during ischemia, or by synthesis of homocysteine and adenosine from S-adenosylhomocysteine (SAdoHcy). In order to investigate the relative contributions of both pathways under various conditions, we have developed a quick and sensitive HPLC system for the determination of SAdoHyc and Ado. Extracts of freeze-clamped hearts are chromatographed on a Partisil SAX ion-exchange column (buffer: 10 mM KH_2PO_4; pH 2.85; flow 2 ml/min), which retains a.o. high-energy phosphates. Ado and SAdoHcy however are quickly eluted and routed onto a μBondapak C18 (reversed phase) column (buffer KH_2PO_4, 10 g/l; 3% MeOH; pH 4.5; flow: 3 ml/min).

Isolated rat hearts, retrogradely perfused for 30 min with 200 μM homocysteine contained 6-12 nmoles/gwwt of SAdoHcy. When the coronary flow was reduced for 15 min from 11 to 2-2.5 ml/min, SAdoHcy content increased from 6-12 to 85-93 nmoles/g wwt. Adenosine increased from 2-5 to 16-60 nmoles/g wwt. The described method is simple, quick and sensitive (20 pmoles per sample) and seems to be a very useful tool for studying myocardial adenosine metabolism. (Supported by a grant from The Netherlands Organization for the Advancement of Pure Scientific Research FUNGO-ZWO).

2. ADENOSINE-SENSITIVE ADENYLATE CYCLASE IN CULTURED VASCULAR SMOOTH MUSCLE CELLS. Madhu B. Anand-Srivastava, Douglas J. Franks, Marc Cantin and Jacques Genest. Clinical Research Institute of Montreal, 110 Pine Avenue West, Montreal, P.Q. Canada, H2W 1R7.

Adenosine-sensitive adenylate cyclase has been demonstrated in a particulate preparation from cultured vascular smooth muscle cells obtained from rat aorta. Smooth muscle cells were cultured from explants of the intima media layer of rat aortas and cells thus derived were maintained in tissue culture for several months and retained the morphological characteristics of smooth muscle cells as verified by light and electron microscopy. N^6-phenyl-isopropyladenosine (PIA), an analog of adenosine stimulated adenylate cyclase in a concentration dependent manner. The maximal stimulation ($\sim$2.5-fold) was observed at 30 μM PIA and half maximal stimulation was obtained at PIA concentrations as low as 1-5 μM. The stimulation by PIA was demonstrated only in the presence of adenosine deaminase which deaminates endogenous adenosine. The PIA stimulation of adenylate cyclase was dependent on the presence of GTP and Mg^{2+}. The stimulation by PIA was maximal ($\sim$3.5-fold) at low concentrations of Mg^{2+} (1 mM). With increasing concentrations of Mg^{2+} the enzyme activity increased but the degree of stimulation by PIA decreased, suggesting that PIA might interact with the metal binding site of adenylate cyclase. 3-isobutyl-1-methylxanthine (IBMX) inhibited the stimulation by PIA (30 μM) in a concentration dependent manner with complete inhibition at 1 mM IBMX. We have also observed a stimulation of adenylate cyclase by PIA in primary cultures of smooth muscle cells which was also inhibited by IBMX. From these results it can be hypothesized that the relaxant effect of adenosine on smooth muscle may be mediated by activation of adenylate cyclase. (Supported by QHF, MRC, NIH AM 21299, and Juvenile Diabetes Foundation.)

3. ADENOSINE EFFECT ON CYTOSOLIC (NAD^+)/(NADH) RATIO AND TRIACYLGLYCERIDES IN ETHANOL-INTOXICATED LIVER. F. Arias-Mendoza and E. Pina, Depto. de Bioquimica Fac. de Medicina UNAM 04510 Mexico D.F. MEXICO.

It has been shown that exogenous adenosine has been able to diminish the triacylglycerides (TG) accumulation and to increase the (NAD^+)/(NADH) ratio in cytosol of ethanol-intoxicated livers.

This work is devoted to study this effect and the influence of the ratio over the (TG) pool.

Plotting both parameters, a rectangular hyperbola is recorded, pointing out a fine regulation in the (TG) levels by the redox state of such coenzymes.

Considering that lactate and α-glycerophosphate dehydrogenases share the same coenzyme pool, a relationship of their equilibrium constants with a theoretical value of 1.17 is found. The values of 0.74 and 2.36 for the ethanol and ethanol plus adenosine groups respectively, are probably due to competence of such enzymes for the coenzyme altered antagonically by ethanol and adenosine.

4. ADENOSINE RECEPTORS IN SMOOTH MUSCLE. H.P. Baer and M.J. Muller, Department of Pharmacology, University of Alberta, Edmonton, Alberta, Canada, T6G 2H7.

Differences in the structure-activity profiles of adenosine receptors in intestinal smooth muscle and mouse neuroblastoma adenylate cyclase (NBC), serving as a model for cyclase-coupled adenosine receptors, further support the idea that adenosine does not relax smooth muscle via cyclic AMP as a mediator. NBC is effectively stimulated by compounds such as purine ribonucleoside or 6-methylmercaptopurine ribonucleoside, i.e., compounds not bearing an N6-amino function. These compounds are generally devoid of smooth muscle relaxant activity. On the other hand, 2'3'-O-dinitro-5'-N-ethylcarboxamido-adenosine (DNECA) did not stimulate NBC but effectively relaxed several isolated smooth muscle preparations. Such relaxations were partly due to interaction of DNECA with adenosine receptors but also due to the presence of nitro groups. Extensive attempts to demonstrate adenosine effects on adenylate cyclase preparations from different intestinal smooth muscles were unsuccessful. While our data generally support the notion that smooth muscle relaxation by adenosine is not mediated by cyclic AMP, results to the contrary by other authors with beef coronary arteries suggest that differences may exist between adenosine receptors and mechanisms of adenosine action in different smooth muscle systems.

Supported by: Medical Research Council, Alberta Heart Foundation and the Alberta Heritage Foundation for Medical Research.

5. RELATIONSHIP BETWEEN MYOCARDIAL OXYGEN CONSUMPTION (MVO_2), CORONARY FLOW (CF) AND FORMATION OF ADENOSINE (AR). H. Bardenheuer and J. Schrader, Dept. of Physiol., University of Munich, Germany.

The influence of changes in MVO_2 on CF and release of AR into the effluent perfusate was studied in a newly elaborated working heart preparation from isolated guinea pigs. Coronary arteries were perfused according to the Langendorff-technique with a modified Krebs-Henseleit solution, while stroke volume was ejected into an artificial systemic circulation. This made it possible to vary coronary perfusion pressure, afterload and preload independently from each other. MVO_2 was increased either by isoproterenol (ISO), norepinephrine (NE), ouabain, and the cardiotonic agent AR-L-115 or by raising the afterload. AR was determined applying HPLC-techniques.

Results: MVO_2 and CF correlated well under all experimental conditions. ISO ($2\text{-}6 \times 10^{-9}$M), NE ($0.8\text{-}5 \times 10^{-8}$M), ouabain ($4\text{-}9 \times 10^{-7}$M) and AR-L-115 ($3\text{-}5 \times 10^{-4}$M) caused a dose-dependent increase in the release of AR, the concentration of which was within the vasodilatory range of exogenously applied AR. The amount of AR formed with each inotropic stimulus, however, was greatly different, e.g., a 30% increase of MVO_2 induced by AR-L-115, ISO, ouabain, and NE caused the release of AR to increase 2.5-,4-,4-, and 8-fold, respectively. Conversively, the ratio of CF/AR release was different with each intotropic agent. Blockade of coronary alpha-1-adreno-ceptors by prazosin (8×10^{-7}M) significantly diminished the NE-induced release of AR. Similar results were obtained, when oxygen supply to the heart was increased by overperfusion of the coronary arteries.
- Raising the afterload from 60 to 80 mm Hg increased MVO_2 by 37% and CF by 47%, while coronary resistance remained unchanged. Also the release of AR did not increase under these conditions.

Conclusion: 1. AR formed in response to various inotropic stimuli appears to be quantitatively sufficient to be causally involved in the observed changes in CF. However, AR appears not to be the mediator of the afterload-induced changes in CF.
2. The greater increase in the release of AR with ouabain and NE most probably is due to a direct vasoconstrictory action, which leads to a more pronounced imbalance between O_2-supply and demand.
3. Changes in the ratio of O_2-supply/O_2-demand rather than MVO_2 as such are likely to serve as a trigger for the formation of AR by the heart.

6. ADENOSINE-BENZODIAZEPINE INTERACTIONS IN CARDIAC MUSCLE. P.H. Barker and A.S. Clanachan, Department of Pharmacology, University of Alberta, Edmonton, Alberta, Canada, T6G 2H7.

A potentiation of adenosine action by diazepam has been demonstrated in cardiac and smooth muscle *in vitro* and on coronary blood flow *in vivo*. This study compares the effects of several benzodiazepines (BDZs) with nucleoside transport inhibitors on the effects and accumulation of adenosine in guinea pig and rat cardiac muscle *in vitro*. The negative inotropic action of adenosine in guinea pig atria was potentiated by dipyridamole, hydroxynitrobenzylthioguanosine or diazepam. Nucleoside transporter specific accumulation of (^{3}H)adenosine (1 μM) into guinea pig heart was reduced by inhibitors of nucleoside transport (hexobendine > dipyridamole > lidoflazine) and by several BDZs (flunitrazepam = diazepam > clonazepam > lorazepam > flurazepam). The potencies of the BDZs were similar to those estimated for the nucleoside transport system of human erythrocytes through inhibition of the binding of potent transport inhibitor, nitrobenzylthioinosine. Adenosine effects were not potentiated in rat heart. This may be explained by the lower transporter specific accumulation of adenosine (18% of that in guinea pig heart). It is concluded that the potentiation of adenosine in guinea pig heart by several, but not all, centrally active BDZs is possibly due to an inhibition of nucleoside transport. This mechanism may explain the well documented increase in coronary blood flow following diazepam administration in dogs and in man.

Supported by the Alberta Heart Foundation.

7. INTRACELLULAR ADENOSINE POOLS. F.L. Belloni, R. Rubio, and R.M. Berne, University of Virginia, Charlottesville, VA 22908.

Rat liver cells were freshly dispersed using a collagenase perfusion method and were suspended in HEPES-buffered physiological saline. Suspension samples were deproteinated, neutralized, concentrated, passed through PEI-cellulose columns and analyzed for adenosine content using HPLC.

Adenosine levels rose in hypoxic suspensions: this rise was blocked or reversed when adenosine deaminase was added to the suspensions before or during hypoxia, respectively. There was always an adenosine deaminase - resistant (or residual) pool of adenosine, however, averaging 60 pmol/mg protein or about half the adenosine content of whole liver frozen in situ. This residual pool did not exist in the extracellular phase of the suspensions, nor was its magnitude diminished by theophylline or cell disruption. The size of this pool was increased 69% by the combination of dipyrimadole and hypoxia. Preliminary experiments with radioactive tracers indicate that this residual pool may undergo measurable turnover during short time periods (minutes).

Thus, isolated rat liver cells appear to contain a sizable intracellular adenosine pool, probably bound to the adenosine binding protein described by others. In some respects, this pool appears to be labile. The role it plays in vascular regulation is unclear, but measurements of whole tissue adenosine must be interpreted with extreme caution when they are related to adenosine's vasoregulatory function.

Supported by USPHS grant HL 10384.

8. POSSIBLE ROLE OF ADENOSINE IN GLYCEROL-INDUCED ACUTE RENAL FAILURE. Anil K. Bidani and Paul C. Churchill, Dept. Nephrology, Rush-Presbyterian St. Luke's Med. Center, Chicago, Ill., and Dept. Physiol., Wayne State Univ. Sch. Med., Detroit, MI.

Filtration fraction and glomerular filtration rate are severely reduced during the initiation phase of acute renal failure. These effects could be produced, at least in part, by increased concentration of an endogenous substance which constricts afferent and dilates efferent arterioles. On theoretical grounds, adenosine is a likely candidate for the endogenous mediator of these renal hemodynamic effects. The purpose of these experiments was to determine if aminophylline (theophylline, an adenosine-receptor antagonist, compounded with ethylenediamine) altered the severity of acute renal failure in the rat. After measuring baseline serum creatinine, myohemoglobin-uric acute renal failure was induced with glycerol in 2 groups of rats.

Subsequently, 1 group (n=24) received daily saline injections, the other group ((n=24) received daily aminophylline injections. As can be seen below, mean serum creatinines were lower in the aminophylline-treated rats during the 3-day follow up period; moreover, 5 of the saline-injected and only 1 of the aminophylline-injected rats died during the follow up. The lower creatinines and lower mortality rates are indicative of a protective effect of aminophylline in this model of acute renal failure in rats.

	Serum Creatinine		
	saline	p	aminophylline
baseline	0.37±0.30	>0.05	0.40±0.04
1 day	2.83±0.28	<0.04	2.10±0.19
2 days	4.25±0.51	<0.002	2.22±0.36
3 days	4.86±0.71	<0.005	2.37±0.42

9. "ACTIVE HYPEREMIA" AND ADENOSINE IN SUBCUTANEOUS ADIPOSE TISSUE OF DOGS. E.L. Bockman and F.P. McCoy, Department of Physiology, Uniformed Services University, Bethesda, MD 20814.

The relationship between metabolic activity and vascular resistance in adipose tissue as well as the possible role of adenosine as a mediator of the vascular response was studied in the vascularly-isolated, denervated inguinal fat pad of female dogs (N=30). Venous outflow was measured with a drop counter. A wide range of vascular resistances was elicited by intravenous infusion of norepinephrine and/or glucose and by reducing the temperature of the fat pad to $<31^{\circ}$ C. Vascular resistance correlated inversely with glucose uptake (r=-0.71; $P < 0.001$) as well as with lipolysis as indicated by the release of both free fatty acids (r=-0.57; $P < 0.005$) and glycerol (r=-0.55; $P < 0.005$). Vascular resistance also correlated inversely with tissue adenosine content (r= -0.68; $P < 0.001$). In addition, there was a direct relationship between tissue adenosine content and glucose uptake by the adipose tissue (r= 0.64; $P < 0.001$). The relationship between tissue adenosine content and glucose uptake indicates that re-esterification of free fatty acids, perhaps by providing AMP as the result of the activation of free fatty acids, may provide the link between metabolic events and the enhanced production of adenosine. The findings of the present study indicate that there is metabolically-linked local control of blood flow in adipose tissue and that adenosine may have a role in mediating the vascular response.

Supported by USPHS grant HL 26345 and USUHS grant R07639.

10. EVIDENCE FOR AN A_2 ADENOSINE RECEPTOR IN THE GUINEA PIG TRACHEA. C.M. Brown and M.G. Collins. I.C.I. (Pharmaceuticals Division), Alderley Park, Macclesfield, Cheshire SK10 4TG, U.K.

An attempt was made to determine whether the adenosine receptor present in the guinea pig trachea was of the A_1 or A_2 sub-type. Dose response curves to the inhibitory effects of adenosine and its 5' and N^6 substituted analogues were constructed in the guinea pig isolated trachea, contracted with carbachol (0.5 μM). The order of potency was 5'-N'cyclopropylcarboxamideadenosine > 5'-N- ethyl carboxamideadenosine > 2,chloroadenosine > L-N^6-phenylisopropyladenosine (L-PIA) $\geq$

adenosine D-N^6-phenylisopropyladenosine (D-PIA). The difference in potency between D- and L-PIA was at the most five-fold. Responses to low doses of adenosine and its analogues were attenuated after treatment with either theophylline or 8,phenyltheophylline. The results support the existence of an extracellular A_2 sub-type of adenosine receptor in the guinea pig trachea.

11. BEHAVIORAL ACTIVATION PRODUCED BY ADENOSINE ANALOGS. Ronald G. Browne, Pfizer Inc., Groton, CT 06340.

Adenosine and more metabolically stable analogs which bind to brain A1 receptors have been reported to produce pronounced behavioral sedation in rodents (Snyder et al, PNAS, 78: 3260-3264, 1981. We have conducted a series of experiments examining the behavioral effects produced by adenosine analogs on locomotor activity, rearing behavior, and food and water consumption in rats for up to 24 hrs after SC administration. We observed that in addition to sedation, animals also exhibited periods of hyperactivity and marked increases in eating. Frequently, the temporal pattern of activation was observed hours after drug administration, and sometimes interspersed among periods of sedation.

In addition, we examined the ability of various adenosine analogs to displace high affinity 3H-N6-cyclohexyladenosine (N6-CHA) binding to rat brain membranes. In general, there was a good correlation between sedative activity and receptor binding potency. However, several compounds appear divergent in being either more (5'-cyclopropylcarboxamide adenosine) or less (D-N6-phenylisopropyl adenosine) sedating than predicted based on binding data. These results will be discussed in relation to cardiovascular and glucose oxidation effects of these analogs.

12. SUBCELLULAR DISTRIBUTION OF CARDIAC ADENYLATES AS EVALUATED BY A NON-AQUEOUS FRACTIONATION METHOD. Rolf Bünger, Sybille Soboll, Dept. of Physiol., Uniformed Services University, Bethesda, MD (USA), and Dept. of Physiol. Chem. I, University of Düsseldorf (West Germany).

Using density gradient centrifugation of lyophylized myocardial homogenates in heptane/carbon tetrachloride the subcellular distribution of creatine phosphate (CrP) and adenine nucleotides (ATP, ADP) was measured (Hoppe-Seyler's Z. Physiol. Chem. 362, 125-132 (1981)) in a pyruvate perfused working guinea pig heart (Pflügers Arch. 380, 259-276 (1979)). The free cytosolic ADP and AMP concentrations were calculated (creatine kinase mass action ratio (K_{CPK}=2.04 x 10^{-9}), adenylate kinase equilibrium (K_{AK}=0.95)). The cytosolic CrP concentration was approx. 20 mM. CrP was not detectable in the mitochondria. The concentration ratio of free ATP and ADP in the cytosol was 10- to 40-fold higher than in the mitochondria indicating asymmetrical translocation of ADP across the mitochondrial membrane in working heart. The calculated free cytosolic ADP and AMP concentrations were near 100 μM and 1 μM, respectively. Increased O_2 uptake due to norepinephrine infusion (0.08 μM) produced a fall in the cytosolic ATP/ADP x P_i ratio while the extramitochondrial AMP concentration approx. doubled. In contrast, the mitochondrial ATP/ADP ratio did not change during adrenergic stimulation. Thus, in intact working hearts, the extramitochondrial CrP and ATP systems seem to play a decisive role in the control of myocardial respiration and 5-nucleotidase activity.

(light) of the rat. The physiological meaning of this cycle in relation to the sleep-awake cycle as well as the maintenance of the energetic homeostasis of the blood will be commented.

15. SPECIES DIFFERENCES IN NUCLEOSIDE TRANSPORT BY MAMMALIAN ERYTHROCYTES.
A.S. Clanachan, A.R.P. Paterson, J.R. Hammond, and S.M. Jarvis, Department of Pharmacology & Cancer Research Unit (McEachern Laboratory, University of Alberta, Edmonton, Alberta, Canada T6G 2H7.

Although human erythrocytes possess a facilitated diffusion mechanism for purine and pyrimidine nucleosides, its activity is not demonstrable in erythrocytes from most sheep. This report compares nucleoside transport in erythrocytes from several commonly used laboratory mammals. A kinetic study of the inward transport of uridine, a non-metabolized permeant, demonstrated that the V_{max} (mean $\pm$ SEM (n)) of this process differed markedly among species: human, 26$\pm$3(6); rabbit 29$\pm$2(3); mouse 6.8$\pm$0.8(4); rat, 0-7$\pm$0.1(6) and guinea pig, 0.07$\pm$0.01(3) x 10^{-19} mole/cell/sec. K_m values were similar (0.2 mM). Nitrobenzylthioinosine (NBMPR) bound to these erythrocytes with high affinity and saturation values (B_{max}) were proportional to the V_{max} for transport human, 11040$\pm$640(9); rabbit, 9290$\pm$280(3); mouse, 3310$\pm$480(4); rat, 300$\pm$30(4) and guinea pig 27$\pm$4(3) sites/cell. Transport of uridine or adenosine was not detected in dog erythrocytes, nor was saturable, high affinity binding of NBMPR demonstrable. These results indicate that the translocation capacity of each transporter in these cells is similar ($\sim$150 molecules uridine/site/sec at 22°C). Species differences in transport capacity are due to variations in the cell surface content of functional transport sites, rather than to differences in the kinetic properties of these sites.

Supported by the National Cancer Institute of Canada and the Alberta Heart Foundation.

16. EVIDENCE FOR AN ADENOSINE A_1 RECEPTOR IN THE GUINEA PIG ATRIUM. Michael G. Collis, I.C.I. (Pharmaceuticals Division), Alderley Park, Macclesfield, Cheshire SK10 4TG, U.K.

Adenosine (P_1) receptors have been divided into A_1 and A_2 sub-types. We have attempted to sub-classify the adenosine receptor in the guinea pig atrium. Left atria were mounted in organ baths for isometric tension recording. The atria were paced electrically. Adenosine and its 2,N6 and 5' substituted analogues evoked negative inotropic responses with the following order of potency: 5'-N-cyclopropylcarboxamide-adenosine = N^6 cyclohexyladenosine > L-N^6-phenylisopropyladenosine = 2 chloroadenosine >adenosine > D-N^6-phenylisopropyladenosine. The purine transport inhibitor dipyridamole potentiated the effects of adenosine but not those of other agonists. The adenosine receptor antagonist theophylline, attenuated responses to all agonists. These results indicate that the adenosine receptor in the guinea pig atrium is probably of the A_1 sub-type.

13. DISCRIMINATIVE STIMULUS PROPERTIES AND OTHER BEHAVIORAL EFFECTS OF ADENOSINE ANALOGS. J.M. Carney and V.L. Coffin, Univ. of Oklahoma HSC., Okla City, OK 73190.

Characterization of the discriminative stimulus properties (analogous to subjective effects) was obtained by training rats to respond on the appropriate lever when given either saline vehicle or 0.032 mg/kg L-Phenylisopropyladenosine (L-PIA). The rats were trained to make this drug discrimination under a 30 response fixed ratio schedule. On test days, all conditions were the same, except no food (45 mg rodent food pellet, BioServ #T 101) was delivered. Not all rats were able to learn this difficult discrimination. Those that did learn the L-PIA/saline discrimination required more than 40 days of training. L-PIA discriminating rats showed a dose-related generalization to the 0.032 mg/kg training condition when tested with lower doses. 2-Chloroadenosine and N^6-cyclohexyladenosine also generalized (produced CNS effects similar to L-PIA) to the L-PIA cue. In contrast to these analogs, 2-chloroadenine arabinoside did not produce subjective effects that were similar to L-PIA. The discriminative stimulus properties of L-PIA were completely blocked by 3.2 mg/kg caffeine. The order of relative potency (compared to L-PIA) for disruption of FR 30 behavior was N^6-L-Phenyisopropyl-adenosine = N^6-Cyclohexyladenosine > N-ethylcarboxyadenosine cycloheptyl-adenosine > phenyladenosine > D-Phenylisopropyl adenosine > Benzyl-adenosine > Inosine. The disruptive effects of L-PIA could be blocked by caffeine and theophylline, but not by 8 (p-sulfophenyl)-theophylline.

14. CIRCADIAN VARIATIONS OF ADENOSINE IN THE BLOOD, BRAIN AND LIVER OF THE RAT. V. Chagoya de Sanchez, R. Hernandez-Munoz, M. Diaz-Munoz, W. Glender, J. Suarez, S. Virdrio and L. Yanez, Depto. Bioenergetica, Centro. Inv. Fisiologia Cellular, Universidad Nacional Autonoma de Mexico, 04510, Mexico DF. Mexico.

Several physiological actions of adenosine are hormone-related. Whether its action as a metabolic regulator is a direct one or it in turn modulates hormone action is not known, but it seems that its concentration fluctuates along with the physiological changes during the day. This point was explored in adult male rats adapted to a natural dark-light cycle 2 weeks prior to the experiment. The animals were sacrificed either every 2 or 4 hours along a 24-hour period. Blood, brain, and liver were studied, since the liver is the main source of circulating purines; the brain maintains its purine pools through the purine salvage pathway and it should receive a preformed precursor.
Blood. During the first twelve hours (8 A.M. - 8 P.M.) there is a low level of adenosine, which starts to increase at the onset of the dark period, reaching a peak value (6-7 x 10^{-7}M) around 2-7 A.M. The inosine and hypoxanthine profiles are opposed to this one; their concentrations are low at night and start to rise at 5 o'clock at noon respectively reaching an acrophase at 10 P.M. (1.5 x 10^{-6}M). The uric acid changes indicate that the decrease in purine catabolism coincides with inosine and hypoxanthine decrease and the adenosine increase. Liver. Levels of inosine and hypoxanthine remain low during the 24 hours. Adenosine starts to increase at 8 P.M. reaching a 4 A.M. maximum. Brain. The purine pattern was similar to that in the liver. The circadian changes found are related to motor activity (dark) or to the rest period

17. PROTECTIVE EFFECT OF PHENTOLAMINE/PHA/ IN MYOCARDIAL HYPOXIA: POSSIBLE ROLE OF ANTI-ADENOSINE ACTION. Agnes Cseppento, A.J. Szentmiklosi and J. Szegi, Dept. of Pharmacol., Univ. Med. School of Debrecen, H-4012 Debrecen, Hungary.

PHA thought to be a beneficial drug during acute myocardial ischemia, but the mechanism of this effect is not clear. It is also known that adenosine/AR/ plays an important role in the adaptation of myocardial functions to hypoxia, as well as an antagonistic interaction exists between AR and alpha-adrenergic blocking agents. The anti-hypoxic activity of PHA was studied in isolated, electrically driven left and spontaneously beating right atria of guinea pigs. PHA/10-30 μmol/l/ antagonized the hypoxia-induced decrease in the isometric tension. Besides, PHA/1-100 μmol/l/ reduced also the AR-induced decrease in the mechanical activity, as well. The anti-AR effect of PHA was competitive/pA_2:5.6/ only at lower concentrations of PHA/ $\leq$10 μmol/l/, while more than 10 μmol/l PHA exhibited a non-competitive antagonistic action. These higher concentrations of PHA/ $>$ 10 mol/l/ decreased the,sinoatrial automaticity, increased the electrical diastolic threshhold and the effective refractory period in a dose-dependent manner indicating a membrane stabilizing property. This action was not observed at lower concentrations of PHA/$<$10 μmol/l/. It is supposed that anti-hypoxic activity of PHA is related, at least in part, to a competitively non-competitive anti-AR action of this alpha-adrenergic blocking agent/competitive anti-AR and a non-specific membrane stabilizing effect.

18. INTERACTIONS OF PURINE NUCLEOSIDES WITH SYMPATHETIC NEUROTRANSMISSION. Michael D. Devous, Sr., Nuclear Medicine Center, The University of Texas Health Science Center, Dallas, Texas 75235.

Recent evidence has suggested that sympathetic stimulation and/or exogenous administration of adrenergic agonists may alter myocardial and coronary vascular responses to purine nucleosides. Conversely, the cardiovascular neuroeffector response to adrenergic stimulation may also be modified by purine nucleosides. For example, alpha-adrenergic vasoconstrictor responses are potentiated by adenosine in several tissues. Adenosine also inhibits both metabolic and inotropic myocardial responses to various catecholamines. Because adenosine and inosine increase their concentration 10-100-fold in ischemic myocardium, they may play a regulatory role in adrenergic activity during ischemia. Therefore, we have undertaken studies to delineate how the purine nucleosides adenosine and inosine alter myocardial responses to adrenergic stimuli in the dog. Isoproterenol is infused at approximately 0.7 ug/kg/min intravenously which leads to approximately a 100% increase in contractile force. Concomitant changes in heart rate and global cardiac function (reflected in measurements of dP/dt) are monitored as well. Repetitive infusions generate consistent responses over a prolonged period of time. However, when isoproterenol-induced myocardial responses are observed in the presence of adenosine or inosine infusions, the responses are altered. For example, a 5 min infusion of inosine at arterial concentrations of 10^{-5}M increase the contractile force response to isoproterenol by 48% for as long as 30 min after the inosine infusion. However, a similar infusion of adenosine inhibits the contractile force response to isoproterenol, and in some cases completely eliminates that response. In general, isoproterenol-induced increases in myocardial contractility are potentiated by inosine and inhibited by adenosine.

19. NIFEDIPINE REDUCES THE RELEASE OF ADENOSINE (CATABOLITES) FROM ISCHEMIC RAT HEART. J.W. de Jong, E. Harmsen, P.P. de Tombe, and E. Keijzer, Cardiochemical Laboratory, Thoraxcenter, Erasmus Univ., Rotterdam, The Netherlands.

Although many authors assume that the calcium antagonist nifedipine (Nif) has an ATP-sparing effect, few studies support this view. We decided to investigate the action of Nif on the myocardial release of adenosine (Ado) and its catabolites, a sensitive probe for adenine nucleotide breakdown in ischemic heart. Nif (10-100 μg/l) caused an increase in flow and decrease in contractility in perfused rat hearts. During ischemia (75% flow reduction of hearts without drugs) contractility was comparable for Nif-treated and untreated hearts. Ado and its catabolites were assayed by HPLC. Ischemia caused considerable release of adenosine, inosine and hypoxanthine. Nif prevented this dose-dependency. The highest dose reduced release of Ado (catabolites) by 90% ($P<0.01$). The drug acted similarly during reperfusion. Ischemia induced a 15% decrease ($P<0.001$) in adenylate energy charge. Nif (100 μg/l) prevented this ($P<0.05$). We conclude that Nif reduces the release of Ado (catabolites) during ischemia and reperfusion, indicating diminished adenine nucleotide breakdown.

20. TIME-COURSE OF CARDIAC ADENOSINE RELEASE DURING NOREPINEPHRINE INFUSION. D.F. DeWitt, R.D. Wangler, and H.V. Sparks, Dept. of Physiology, Michigan State University, East Lansing, Michigan 48824.

This study was designed to examine the time-course of adenosine release in Langendorff-perfused guinea pig hearts during augmented myocardial metabolism induced by norepinephrine (NE) infusion (0.5-1μg/min). The perfusion medium was a modified Krebs-Henseleit solution containing 8 mM glucose and 2 mM pyruvate, was maintained at 37°C, and was equilibrated with 95% O_2 + 5% CO_2. The hearts were paced at 300 beats/min and perfused at 60 cm H_2O). After a 20 minute control period, coronary effluent was collected every two minutes, during a 16-minute infusion of NE, for measurements of coronary flow (CF=ml/min/g (wet weight)) and adenosine release (Rado-nmol/min/g). Control values of CF and Rado in 5 animals were 6.8$\pm$0.4 ($\bar{X}\pm$SEM) and 0.03$\pm$0.2, respectively. CF and Rado increased significantly after NE infusion was started and reached maxima at 4 minutes (10.7$\pm$0.8 ($P<.01$) and 3.7 $\pm$0.4 ($P<.01$), respectively). CF remained elevated and was 11.3$\pm$0.8 at 12' and 11.8$\pm$0.8 at 16'. Rado declined to a nadir of 1.5$\pm$0.3 at 12" ($P<.01$ as compared to Rado at control and at 4') and remained at a steady-state thereafter (Rado at 16' = 1.3$\pm$0.2). To test whether CF was maximal during the NE infusion, an adenosine infusion (10^{-5} moles/min) was started after 12 minutes exposure to NE in 3 animals. CF increased from 13.2$\pm$1.8 to 19.0$\pm$3.9 ml/min/g after 4 minutes. In summary, when cardiac metabolism was augmented by NE infusion, CF and Rado increased rapidly. However, Rado declined thereafter, while CF remained at a submaximal, steady-state level. These results suggest that either (1) Rado does not reflect interstitial adenosine concentration, or (2) at least one other mediator participates in sustained metabolic vasodilation.

This study was suppirted by NIH grants HL 24232 and F32 HL 06277.

21. METABOLISM OF ^{14}C-ADENOSINE IN HUMAN TONSILLAR AND PERIPHERAL BLOOD LYMPHOCYTES. W.N.M. Dinjens, J.P.R.M. van Laarhoven, C.H.M.M. DeBruyn, Department of Immunobiology, Central Laboratory of the Netherland Red Cross and Department of Human Genetics, University Hospital Nijmegen, Nijmegen, The Netherlands.

It is known that purine nucleoside metabolism in certain lymphoid cell subsets differs from that in other subsets. In addition, changes occur during lymphoid cell differentiation and maturation. We have explored differences between intact peripheral blood lymphocytes (PBL) and tonsillar lymphocytes, with respect to ^{14}C-adenosine metabolism. The two lymphoid cell populations differed in relative amounts of T and B cells and state of maturation.

Intact cells were incubated with 10 μM ^{14}C-adenosine; after certain periods of time reactions were terminated by rapid centrifugation of the cells through a silicone oil layer. Labeled intracellular and extracellular metabolites were analyzed by means of thin layer chromatography on PEI cellulose plates followed by liquid scintillation counting. It was shown that ^{14}C-adenosine was metabolized more rapidly by freshly isolated tonsillar lymphocytes than by fresh PBL. In both cell types the intracellular label was almost completely recovered as adenine nucleotides (mainly ATP). Storage overnight at 4° C did not result in a change of the rate of ^{14}C-adenosine uptake. However, dramatic alterations were observed in the case of tonsillar lymphocytes. In PBL no such significant changes were seen. No differences in activities of adenosine metabolizing enzymes were noted in lysates from freshly isolated and stored PBL and tonsillar lymphocytes.

These data suggest that handling of adenosine by lymphoid cells not only differs with respect to various cell types and differentiation stages, but can also depend on the duration of the isolation procedures. This should be taken into consideration when studying the effects of adenosine on lymphoid cell subfractions.

22. POTENTIATION BY ADENOSINE DEAMINASE OF ISOPROTERENOL ELICITED CONTRACTILE RESPONSE IN OXYGENATED AND HYPOXIC RAT ATRIA. J.G. Dobson, Jr. and L. Griffin-Shea, Department of Physiology, University of Massachusetts Medical School, Worcester, MA 01605 USA.

Adenosine has been reported to reduce catecholamine induced enhancement of contractile state in atria (Amer. J. Physiol. 239: H365, 1980). The present study was undertaken to determine if endogenous adenosine is capable of inhibiting the isoproterenol (ISO, 10^{-7}M) elicited increase in contractility (+dF/dt) in isolated oxygenated and hypoxic rat atria. In oxygenated atria ISO produced a maximal increase in +dF/dt of 74% within 1 min. After 6 min of ISO the maximal increase in +dF/dt faded (decreased) by 26%. Adenosine deaminase (AD, dialyzed, 4 U/ml bath) which had no effect by itself, potentiated the maximal increase in +dF/dt by 19% and attenuated the fade after 6 min of ISO by 46%. The +dF/dt of atria made hypoxic for 6 min was reduced by 23% below control and AD did not significantly elevate the reduced hypoxic control value. During the last min of hypoxia ISO produced a maximal increase in the +dF/dt of 45% in the absence of AD, whereas in the presence of AD +dF/dt increased by 84%. Boiling AD eliminated the effects of the enzyme. These results suggest that endogenous adenosine may, in part, be responsible for (1) the attenuation of maximal catecholamine elicited contractile responses, (2) the fade of contractile responses with prolonged catecholamine stimulation in oxygenated cardiac muscle, and (3) the marked reduction of catecholamine induced contractile responses in the hypoxic myocardium.

Supported by Grants 18280 and 22828 from the NIH & BI.

23. PHARMACOLOGICAL CHARACTERIZATION OF ELECTROPHYSIOLOGICAL AND BEHAVIORAL RESPONSES TO ADENOSINE AGONISTS AND ANTAGONISTS. Thomas V. Dunwiddie and Thomas Worth, Department of Pharmacology, University of Colorado Health Sciences Center, Denver, Colorado 80262.

We have compared the pharmcological actions of various adenosine analogs in vitro to their behavioral actions, and have related these types of responses to the relative affinities of these drugs for adenosine A_1 and A_2 receptors. Adenosine, adenine nucleotides, related purines such as inosine, and metabolically stable analogs of adenosine antagonized synaptic transmission and inhibited seizure-like activity in the rat hippocampus in vitro. Analogs which have been reported to be more potent in displacing ligands from A_1 receptors were also more potent in eliciting physiological responses (L-PIA > CHA > NECA). Several non-metabolizable analogs of adenosine were found to be very potent (EC50 values below 300 nM) in reducing the amplitude of extracellularly recorded EPSPs and in inhibiting the rate of interictal discharge in the hippocampus in vitro. This would indicate that these types of responses are mediated via A_1 rather than A_2 receptors. All of the responses to these putative ligands for A_1 receptors could be antagonized by various xanthines; the xanthines appeared to be competitive antagonists.

In vivo, these analogs of adenosine exhibited sedative, hypothermic, and anticonvulsant activity in both mice and rats. However, it did not appear that these effects were the result of the interaction of these analogs with a homogeneous population of adenosine receptors.

The results from both in vitro and in vivo experiments suggest that the receptors mediating these responses are similar to those tentatively characterized as A1 receptors.

This research was supported by DA 02702 to T.V.D.

24. ROLE OF ADENOSINE IN THE REGULATION OF CORONARY BLOOD FLOW IN THE HYPERTENSIVE DOG. Stephen Ely, Robert Knabb, Chi Wen Sun, Jeffrey Gidday, Rafael Rubio, Robert M. Berne, Department of Physiology, University of Virginia, Charlottesville, Virginia 22908.

If adenosine plays a central role in coronary blood flow (CBF) regulation, it must be related to changes in myocardial O_2 consumption (MVO_2) and CBF under normal as well as pathophysiological conditions. To test this, we compared the changes in MVO_2 (ml/min/100g), CBF(ml/min/100g) and cardiac adenosine release (Rado, nmoles/min/100 g) at rest and during isoproterenol infusion (ISO, 0.1 μg/Kg/min, iv) in a group of normotensive control (NTC) dogs to a group with 6 wk. renal hypertension (RHT). Rado (sinus-arterial plasma (ado)x coronary plasma flow) was used as an index of changes in interstitial fluid (adenosine). Mean arterial blood pressure (mmHg) was significantly different for NTC(99+5) vs. RHT(135+5; $p < .05$), while resting CBF (40+3 vs. 45+4), MVO_2(6.9+4 vs. 8.2+1.2) and Rado (.05+1.0 vs. 1.9+2.1) were not different between NTC and RHT, respectively. ISO increased ($p < .05$) CBF (66+6 vs. 61+4), MVO_2 (10.8 +.8 vs. 10.0+1.1) and Rado (9.4+3.6 vs. 5.2+1.3) in NTC and RHT, respectively. The ratio Δ Rado/Δ coronary vascular resistance for NTC (-9.9+3.8) and for RHT (-3.8+9) was significantly different from zero ($p < .05$), but these ratios were not different from each other ($p > .2$). The data suggest that under these experimental conditions, metabolic regulation of the coronary circulation is unimpaired in hypertension, and that adenosine may play a role in the regulation of CBF in both normotensive and hypertensive dogs.

Supported by NIH HL10384 and the American Heart Association VA Affiliate.

25. RIBOSE-SPECIFIC (A_2) ADENOSINE RECEPTORS MEDIATE CYCLIC AMP ACCUMULATION AND CHLORIDE TRANSPORT IN THE ELASMOBRANCH RECTAL GLAND. J.N. Forrest, Jr., E. Poeschla, and R. Reick, Yale University, New Haven, CT, and the Mt. Desert Island Biological Laboratory.

The shark rectal gland is a useful model for cyclic AMP (cAMP) stimulated sodium-coupled electrogenic chloride transport. Since external adenosine (Aden) receptors modulate cAMP levels in several cell types (Londos et al, PNAS 77, 1980), we examined the effects of Aden in the isolated perfused rectal gland. Addition of Aden to the perfusate markedly increased Cl^- secretion from basal levels (87+ 10 μEq/hr/g wet weight) to 231+34, 679+74 and 1077+56 at 5, 10, and 50 μM Aden respectively). Aden increased tissue cAMP levels in parallel with enhanced Cl- secretion (r=0.87, $p < 0.001$, n=21) rising from basal levels (14+2 pmoles/mg protein) to 362+50 at 50 μM Aden. These effects require the integrity of the ribose (R) moiety of Aden since purine intact analogs (9β - D arabinofuranosyl adenine and 2' deoxyadenosine) did not stimulate Cl^- secretion or tissue cAMP. Theophylline, an antagonist of R specific Aden receptors completely (>90%) and reversibly inhibited both Cl^- transport and cAMP accumulation while methylxanthines without effect on Aden receptors in other tissues had no effect. The relative potency of R intact Aden analogs in stimulating Cl^- secretion (N-ethylcarboxamide adenosine > Aden > phenylisopropyl adenosine) was identical to that reported for activation of adenylate cyclase in the turkey erythrocyte. These studies demonstrate that R specific (A_2) external Aden receptors mediate stimulation of cAMP and Cl^- secretion in an epithelia characterized by secondary active Cl^- transport.

26. VASODILATION BY ADENOSINE-RELATED NUCLEOTIDES IS REDUCED AFTER ENDOTHELIAL DESTRUCTION IN BASILAR, LINGUAL, AND PULMONARY ARTERIES. G.W. Frank and J.A. Bevan, Department of Pharmacology, University of California School of Medicine, Los Angeles, California 90024.

Adenosine and other purinegeric transmitters may play a role in vasodilation with sympathetic nerve stimulation and post-occlusion. The vasodilator action of these compounds has been shown to rely on an intact vascular endothelium. We examined this phenomenon *in vitro* using basilar, lingual and intralobar pulmonary arteries isolated from New Zealand white rabbits. Ring segments were mounted on force transducers in Kreb's bicarbonate tissue baths. After constricting the vessel segments with norepinephrine or prostaglandin F_{2a} we tested the responses to adenosine, AMP, ADP and ATP in the concentration range 10^{-8} to 10^{-5}M. All four agents had a dose-dependent relaxing effect on the vessels. In vessels pre-treated by removing the endothelial layer mechanically, the four nucleotides caused relaxation, but the responses (expressed as percent of dilation caused by papaverine 10^{-4}M) were significantly less than those seen in the untreated control vessels. Thus the vasodilating effect of adenosine, AMP, ADP and ATP is in part mediated by the vascular endothelium, either with the release of a second messenger, or through some other potentiating effect on the smooth muscle.

Supported by USPHS HL 15805 and HL 20581. Dr. Frank is a fellow of the Parker B. Francis Foundation.

27. K^+ AND ADENOSINE AS MODULATORS OF MECHANICAL AND ELECTRICAL ACTIVITY IN ISOLATED AURICULAR MUSCLE OF THE RAT. X. Garcia, E. Gijon, and V. Chagoya de Sanchez, Dept. de Fisiologia, Fac. de Medicina and Dept. de Bioenergetica, Centro de Investigaciones en Fisiologia Celular, Universidad Nacional Autonoma de Mexico, Mexico, D.F. 04510, MEXICO.

In a previous study it has been shown that negative inotropic and chronotropic effects of adenosine are dependent on the potassium concentration of the bathing fluid. Low potassium solutions prolong time to muscle arrest with adenosine 7.5×10^{-4} and positive inotropic and chronotropic effects are observed with adenosine in auricles spontaneously beating. The present work deals with the influence of external potassium concentration on adenosine effects on mechanical and electrical activity in isolated rat atrial muscle electrically driven (0.5 Hz). Hyperpolarizing effect of adenosine increases with low potassium. Adenosine also increases amplitude of action potential, shortens its duration and increases reversed potential with low potassium solutions. Low potassium solutions produce a slight hyperpolarization, increase amplitude of action potential, reduce action potential duration (0.25) followed by increase in action potential duration (0.0, 0.25, 0.50) and decrease reversed potential. We consider that this ion participates in the interaction of adenosine with its membrane receptor and that adenosine modulates ionic transport.

28. COEXISTENCE OF A PURINE NUCLEOSIDASE AND ADENOSINE RECEPTOR ON THE MICROSOMAL FRACTION OF THE CORONARY ARTERIES. Geetha Ghai, Dept. of Pharmacology, University of South Alabama, Mobile, Alabama 36688.

Adenosine, a metabolic mediator, vasodilates coronary vessels through a putative receptor(s) on the cell surface. Standard receptor ligand binding studies have indicated the presence of a saturable site. Scatchard analysis of the same data indicated clustering of the data points around the ordinate suggesting the lack of receptor saturation

(Bas. Res. Cardio. (1981) 76, 380). Further analyses revealed the presence of a purine hydrolyzing enzyme (nucleosidase) in the microsomal fraction. The enzyme assay contained 40 mM (pH 7.4); potassium phosphate buffer; unlabelled adenosine with tracer amounts of 2,8 ^{3}H-adenosine (40,000 to 60,000 cpm), and microsomal protein. The reaction was stopped by boiling the assay tubes in a water bath for 2 mins. The product adenine was separated by TLC on a polyethyleneimine cellulose plate. The area corresponding to adenine was scrapped, eluted with water and counted in a scintillation counter. Analysis of the chromatograms indicated the presence of adenine in the experimental and not in the control samples. The enzyme reaction was pH, time and temperature dependent, linear with different substrate (10 to 80 nM), and protein concentrations (15 to 45 μg). Lineweaver-Burk plot indicated a Km of 117.6 nM. Ca^{++} stimulated the activity of this enzyme. For the first time these results suggest the presence of a purine nucleosidase in the mammalian cells.

Supported by a Starter Grant from Pharmaceutical Manufacturers Fdtn.

29. Na+ AND ADENOSINE AS MODULATORS OF MECHANICAL AND ELECTRICAL ACTIVITY IN ISOLATED AURICULAR MUSCLE OF THE RAT. E. Gijon, X. Garcia, and V. Chagoya de Sanchez, Depto. de Fisiologia, Fac. de Medicine and Depto. de Bioenergetica, Centro de Investigaciones en Fisiologia Celular, Universidad Nacional Autonoma de Mexico, Mexico, D.F. 04510, MEXICO.

In results from our laboratory it has been found that adenosine increases amplitude of the action potential and reversed potential with low potassium solutions on rat atria. To investigate this problem more fully, the Na+ dependence of the changes evoked by adenosine with very low Na+ concentrations is clearly of interest, and this has been studied in isolated rat auricles electrically driven (0.5 Hz). Amplitude of action potential in low sodium solutions decreases (0.75) and returns to control height (0.50, 0.25, 0.0). Adenosine increases the amplitude of action potential (1.0, 0.75, 0.50) and leaves it unchanged with lowest sodium solution (0.0). Poor sodium solutions decrease duration of action potential (0.75, 0.50, 0.25, 0.0), adenosine reverses this effect. Low sodium solutions produce negative inotropic effect (0.75, 0.50) followed by positive inotropic effect (0.25, 0.0). Adenosine causes positive inotropic effect with decreasing external sodium (0.75, 0.50, 0.25, 0.0) and with increasing external sodium (0.25) in spite of normal K^+ concentration. These results suggest that adenosine exerts a modulation toward membrane stabilization.

30. MODULATION OF STRIATAL DOPAMINERGIC FUNCTION BY THE LOCAL INJECTION OF 5'-N-ETHYLCARBOXAMIDE ADENOSINE (NECA), Richard D. Green, Herbert K. Proudfit, and Siu-Mei Helena Yeung, Department of Pharmacology, University of Illinois School of Medicine, Chicago, Illinois.

The rotation model for studying dopaminergic function in the striatum was used to determine if the adenosine receptors known to be present in the striatum modulate dopaminergic function. Adenosine analogs were injected directly into the left caudate nucleus (CN) via cannulae inserted into chronically implanted guide sheathes. Apomorphine (APO) and theophylline were administered systemically. Rats were placed in metal bowls and rotational and self-mutilatory behavior recorded manually. No rotation was observed after the injection of NECA (up to

6.5 nmoles in 2 1) into the CN or after the systemic administration of APO (1 mg/kg). Rats rotated toward the NECA-injected side when the NECA injection was followed by APO. This response was increased when the NECA injection preceded the APO injection by 40 min. Rotational responses to NECA:APO were greater than those to L-phenylisopropyladenosine:APO. Theophylline decreased the rotational response to NECA:APO. The response to NECA:APO was also decreased when 2',5'-dideoxyadenosine (DDA) was injected into the CN with NECA. We interpret these results to indicate that NECA stimulates adenosine-sensitive adenylate cyclase in the CN and that this inhibits the effect of APO. This leaves the effect of APO on the contralateral CN unopposed and causes rotation toward the NECA-injected side. Higher doses of APO in combination with the same doses of NECA induced self-mutilatory behavior. This behavior was also inhibited by theophylline. These results suggest that endogenous adenosine may function to modulate dopaminergic function in the striatum.

31. ENDOGENOUS PURINES AS MODULATORS OF AUTONOMIC CHOLINERGIC NEUROTRANSMISSION. Lars E. Gustafsson, Dept. of Physiology, Karolinska Institutet, Stockholm, Sweden.

The release of endogenous purines from isolated preparations of plexus-containing longitudinal muscle of guinea pig ileum was measured by HPLC. Adenosine was found at concentrations of 20 nM in the perifusates, and levels were increased both by field stimulation of intrinsic nerves and by addition of inhibitors of adenosine elimination. Combination of field stimulation and inhibitors gave bath concentrations of adenosine of 60 nM, levels sufficient for modulation of cholinergic responses. The increase of adenosine overflow was accompanied by inhibition of responses to nerve stimulation, most likely due to inhibition of acetylcholine release, since acetylcholine overflow can be shown to be inhibited by adenosine application. Furthermore, ATP and especially ADP were found at 20-30 nM concentrations in perifusates. Thus the modulating action of purines can only be assessed by measuring the total outflow of adenine nucleosides and nucleotides, and seems to be marked as judged from the enhancing effect by adenosine antagonists lacking effect on tissue cAMP.

In rabbit bronchi, isolated stomach circular muscle and iris sphincter, but not in bovine iris, an enhancing action was found, during adenosine application, on responses to cholinergic nerve stimulation. This could be shown to be accompanied by a sensitization to acetylcholine and was thus due to a postjunctional enhancing effect by adenosine. This effect was also antagonized by alkylxanthines. Thus, purines might exert both inhibitory and facilitatory actions on cholinergic responses, with important tissue and species variations. Adenosine antagonism might contribute to bronchial relaxation induced by alkylxanthines.

32. INHIBITION OF THE SITE-SPECIFIC BINDING OF NITROBENZYLTHIOINOSINE, AN INHIBITOR OF NUCLEOSIDE TRANSPORT, TO CNS MEMBRANES. J.R. Hammond and A.S. Clanachan, Department of Pharmacology, University of Alberta, Edmonton, Alberta, Canada, T6G 2H7.

We have estimated the affinities of several benzodiazepines (BDZs) and recognized nucleoside transport inhibitors for the nucleoside transport system of CNS membranes through inhibition of the binding of (G-^{3}H)nitrobenzylthioinosine (NEMPR), a potent and specific inhibitor of nucleoside transport. Site-specific binding of NEMPR to guinea pig cortical synaptosomes (K_D = 0.25+0.01nM, B_{max}=300+10 fmoles/mg protein) was competitively inhibited by BDZs (K_i values (μM): Ro5-4864, 5.4+0.6; diazepam, 16+2; clonazepam, 34+5; lorazepam, 52+14) and by inhibitors of nucleoside transport (K_i values (nM):nitrobenzylthioguanosine, 0.36+0.03; hexobendine, 6.7+1.4; dipyridamole, 11.2+1.7). These results demonstrate the existence of a high affinity, saturable binding site for NBMPR on CNS membranes which displays characteristics similar to those of the high affinity NBMPR site located on human erythrocytes and cultured cells, the occupation of which is associated with the inhibition of nucleoside transport. Differences between the affinities and order of potency of the above agents for the NBMPR site and the BDZ site indicate that these sites are unrelated. The low affinity of the BDZs for the nucleoside transport system suggests that significant effects due to nucleoside transport inhibition are unlikely to occur with concentrations ($<1\mu$M) achieved after normal anxiolytic doses.

Supported by the Alberta Heart Foundation and the Alberta Heritage Foundation for Medical Research.

33. SYNERGIC EFFECT OF NIFEDIPINE AND PROPRANOLOL ON ADENOSINE RELEASE FROM ISCHEMIC RAT HEART? E. Harmsen, P. Ph. de Tombe, and J.W. de Jong, Cardiochemical Laboratory, Thoraxcenter, Erasmus University of Rotterdam, The Netherlands.

We studied the combined effects of the calcium antagonist nifedipine (Nif) and the β-blocker propranolol (Prop) on ATP-breakdown in the isolated ischemic rat heart. Myocardial nucleotide content was monitored by the release of adenosine and its catabolites inosine, hyopxanthine and uric acid. Ischemia was induced with a roller pump (flow: 2.5 ml/min). Neither Nif nor Prop (both 30 μg/l) showed negative inotropic properties during ischemia. Nif prevented the release of adenosine and total purines from the ischemic heart by 82% ($P<0.001$) and 60% ($P<0.001$), respectively, whereas Prop was without effect on the release of adenosine (catabolites). The combination of Nif and Prop caused a 90% ($P<0.001$) decrease of adenosine release, and myocardial ATP and creatine phosphate breakdown was blocked ($P<0.05$). We conclude from this preliminary work that nifedipine (+ propranolol) prevents catabolism of high-energy phosphates.

34. ADENOSINE UPTAKE AND METABOLISM IN THE PORCINE ISOLATED PERFUSED LUNG. P.G. Hellewell and J.D. Pearson, A.R.C. Institute of Animal Physiology, Babraham, Cambridge, CB2 4AT, U.K.

Isolated neonatal lungs were perfused with Krebs' solution containing 4.5% Ficoll 70 and in 40 sequential samples of venous effluent, collected in 60s following bolus arterial injection of 0.15 nmol (^{3}H) adenosine (ADO) and (^{14}C) sucrose, 46.9+8.2% of ^{3}H and 85.4+4.3% of ^{14}C were recovered (17 lungs). Dilution curves indicated that (^{3}H) ADO was taken up by the lung, but there was no subsequent ^{3}H release. Overall uptake, U = 1-($^3H/^{14}C$), was 45.2+9.2%. Analysis of pooled venous effluent showed 11.0+4.1% formation of inosine (11 lungs). This was not due to leakage of enzymes from the lungs into perfusate. Dipyridamole inhibited U by 85% at 1 μM and by 97% at 10 μM, but the inosine present was not significantly reduced (10.0+5.1% and 7.3+0.7% respectively, 4 observations) suggesting that inosine is produced by ecto-enzymic activity; inosine formation was reduced to 0.8% (2 lungs) in the presence of 10 μM 2'deoxycoformycin. Two minutes after arterial bolus injection of 2.5 nmol (^3H) ADO, 94% of ^3H retained by the lung was in nucleotides. 5-Iodotubericidin (1 μM) inhibited nucleotide formation 78%, and the dilution curves indicated that ^3H efflux now occurred after uptake; this ^3H was found mostly in inosine and hypoxanthine, indicating that under these conditions intracellular deamination of ADO predominates. Similar dilution curves to these were obtained when tracer analysis was performed in lungs perfused with 50-100 μM unlabelled ADO; U was inhibited by 55-70%, confirming that ADO is taken up by a high affinity saturable process.

35. PRESYNAPTIC MODULATION OF SYNAPTIC EFFICACY BY ADENOSINE IN MAMMALIAN SYMPATHETIC NEURONS. Barbara K. Henon and Donald A. McAfee, Division of Neurosciences, City of Hope Research Institute, 1450 E. Duarte Road, Duarte, California 91010.

Since adenosine has been shown to reduce the spontaneous firing of central neurons (Phillis and Wu, 1981), to reduce the amplitude of evoked EPSPs (Siggins and Schubert, 1981) and to activate presynaptic receptors at axon terminals to limit the release of neurotransmitters (Daly et al., 1981), adenosine is generally considered to be an inhibitor of neuronal synaptic transmission. We have found that the adenosine analog, 2-chloroadenosine, inhibits single* evoked cholinergic EPSPs recorded intracellularly from postganglionic neurons. The average EPSP amplitude declined by 42+9% of control in 10 μM (N=5) and 27% in 5 μM 2-chloroadenosine (N=2). Peak inhibition occurred in about 10 minutes, and the effects were reversed in normal Locke solution within 30 min. There was no detectable change in the input resistance of the postsynaptic neuron nor was there any change in the sensitivity of the postsynaptic neuron to bath-applied carbachol (0.5 μM, N=3). These results indicate that the reduction of the EPSP amplitude by 2-chloroadenosine occurs by a presynaptic mechanism.

However, adenosine does not necessarily always act as an inhibitor of synaptic transmission. Our results indicate that adenosine may actually have a facilitory effect during repetitive activity. The same concentration of 2-chloroadenosine (10 μM) that inhibits EPSP amplitude was found to increase the number of synaptically evoked action potentials recorded intracellularly in response to a preganglionic stimulus train (20 Hz for .75 sec). The pulse train was delivered every 20 sec. The response to the presynaptic stimulus train shows considerable synaptic depression with a few stimuli eliciting spikes and the remainder subthreshold EPSPs. In the presence of 2-chloroadenosine, more of the EPSPs exceeded threshold and the number of action potentials gradually increased

* Single stimulus pulses were delivered to the preganglionic nerve at 20 second intervals.

to 181$\pm$12% of control over a period of 10-12 minutes (8 trials on 4 cells). The number of spikes/train returned to control levels by washing in normal Locke solution for 20-30 min.

Adenosine has been shown to inhibit the voltage-dependent Ca^{++} current in the somata of postganglioic neurons (Henon and McAfee, in press) by activation of adenosine receptors on the cell surface. If adenosine also inhibits the Ca^{++} channels on the presynaptic axon terminals, it would reduce the EPSP amplitude as 2-chloroadenosine has now been shown to do.

The facilitative effects of adenosine during repetitive activity in the SCG also seem to involve Ca^{++} since low external Ca^{++} concentrations were shown to reduce the amount of synaptic depression during extracellularly recorded spike trains. The effects of low Ca^{++} on synaptic depression were mimicked by adenosine (Henon et al., 1980). The capability of inhibiting both the postsynaptic and presynaptic Ca^{++} channels would allow adenosine to play a powerful modulatory role in synaptic transmission. It will be interesting to establish the relative contributions of pre- and postsynaptic mechanisms to adenosine modulatory action.

36. MODE OF ACTION OF ADENOSINE-POTENTIATING VASODILATORS. K. J Henrichs, H. Matsuoka, and W. Schaper, Max-Planck Institute, Bad Nauheim, FRG.

To test the hypothesis that adenosine is involved in the action of coronary vasodilators and in the phenomenon of myocardial reactive hyperemia, intraatrial infusions of adenosine deaminase (from calf intestinal mucosa, 22 U/kg/min) were performed in the open-chest canine experimental model.

The dose-response of adenosine on the CX-flow was shifted to the right by 10^2 mM during the infusion of adenosine deaminase. The dilatory response to dipyridamole was completely abolished and the dilatory response to lidoflazine was partly inhibited during the administration of adenosine deaminase.

After coronary occlusions (5 sec, 10 sec, 20 sec), adenosine-deaminase reduced the debt overpay during reactive hyperemia at 25% after 5 sec occlusions, 29% after 10 sec occlusions, and 30% after 20 sec occlusions ($p < 0.05$).

During the control period, the increase in coronary conductance after occlusions was 11.6 ml/min/mmHg after 5 sec occlusions, 15 ml/min/mmHg after 10 sec occlusions, and 19.9 ml/min.mmHg after 20 sec occlusions. This increase of coronary conductance after occlusions was reduced by 23% for all occlusions (5, 10, 20 sec) during the infusion of adenosine deaminase ($p < 0.01$). The effect of adenosine deaminase on the coronary blood flow regulation was reversible.

L-homocysteine infusions (1 mM/min) were also employed in the same experimental model, and it was found to have no influence on the actions of coronary vasodilators and on myocardial reactive hyperemia, either.

Our data favor the notion that adenosine is involved in myocardial reactive hyperemia and the action of dipyridamole and lidoflazine.

37. EFFECTS OF ADENOSINE AND ANOXIA ON ISOLATED VASCULAR SMOOTH MUSCLE (VSM). J.T. Herlihy, Department of Physiology, University of Texas Health Science Center, San Antonio, Texas 78284.

The contractile response of vascular strips of the hog carotid artery (HC), rat aorta (RA) and rat portal vein (RPV) to hypoxia and adenosine were examined to assess the possible role of adenosine as mediator of hypoxia-induced relaxation of VSM. RA exhibited the same sensitivity to hypoxia as that shown by the HC, but was much less sensitive to adenosine. The relationship between the active stress developed by hypoxic muscle was related to the pre-hypoxic stress developed by the normoxic muscle and this relationship was similar for the two vessels. For the HC (N=77), hypoxic stress=0.79 normoxic stress -0.15 (r=0.82); for the RA (N=159), hypoxic stress=0.85 normoxic stress -0.10 (r=0.93). In contrast, the ED_{50}s for adenosine-induced relaxation were significantly different ($p < 0.05$):67$\pm$29 μM and 7$\pm$2 μM for the RA and HC, respectively. The low sensitivity of rat VSM to adenosine necessitated the use of high concentrations to relax the strips to levels obtained with hypoxia. One mM adenosine is required to relax by 50% RA contracted with maximal doses of norepinephrine or K^+-depolarization. Further, relaxation is obtained when the strips are made hypoxic. The same dose of adenosine will reduce the mean contractile response of spontaneously active RPV, but will not completely stop it. In contrast, the activity of this tissue ceases when the strips are made hypoxic. Adenosine deaminase 2 U/ml) reversed the effects of exogenously added adenosine on RA and RPV, but was without effect in either vessel with regard to hypoxia-induced relaxation. The present results do not support a role for adenosine as the mediator of hypoxia-induced relaxation of isolated rat VSM.

Supported in part by NIH Grant HL-19082.

38. 5'-N-ETHYLCARBOXAMIDOADENOSINE (NECA) - A POTENT STEREOSPECIFIC INHIBITOR OF HUMAN PLATELET AGGREGATION. S.M.O. Hourani and N.J. Cusack, Department of Pharmacology, University of London, King's College, Strand, London WC2R 2LS, U.K.

The effects of NECA, an adenosine analogue modified in the ribose moiety, were tested on human platelet function. NECA, like adenosine, noncompetitively inhibited ADP-induced aggregation. After 30 sec incubation, NECA was about 2 times more potent than adenosine both at inhibiting ADP (5 μM)-induced aggregation and at causing increases in levels of platelet cyclic AMP, an action that was competitively inhibited by theophylline (100 μM, K_i 10 μM), an adenosine receptor antagonist. After 3 min incubation, NECA was about 10 times more potent than adenosine as an inhibitor of aggregation induced by ADP (5 μM) and 11α, 9α-epoxymethano PGH_2 (5 μM). The effects of NECA, like those of adenosine, were stereospecific in that the enantiomer of NECA (L-NECA) was completely inactive, but unlike adenosine were not decreased by prolonged (60 min) incubation. NECA is the most potent analogue of adenosine yet tested on human platelets, and the only example of a ribose modification to retain affinity for the human platelet adenosine receptor. The greater potency of NECA relative to adenosine appears at least in part to be due to its resistance to both deamination and to uptake by platelets.

39. THE EFFECT OF ADENOSINE ON POTASSIUM PERMEABILITY OF SINUS VENOSUS OF HEART, STUDIED USING TRACER FLUX MEASUREMENTS. O.F. Hutter and A.C. Rankin, Departments of Physiology and Medical Cardiology, University of Glasgow, Scotland.

The negative chronotropic effect of adenosine is associated with a hyperpolarization, probably due to an increase in potassium permeability (Hartzel, J. Physiol., 293, 23-49, 1979). To obtain more direct evidence we have studied the effect of adenosine on the permeability of tortoise heart sinus venosus to ^{43}K or ^{86}Rb. Adenosine produces an increase in the efflux of both tracers. The dose response curve was determined, and the concentration for half maximal response was approximately 6 x 10^{-6}M. The magnitude of the response varied greatly from animal to animal, and for different parts of the same heart. The largest responses were obtained from the muscular strip around the mouth of the sinus venosus. In one such preparation adenosine (10^{-4}M) increased the efflux of ^{86}Rb to 6.2 times the resting rate. The right horn of that, and some other hearts, also gave large responses, with lesser increases in efflux from the left horn, and little or no response from the interauricular septum. In contrast, acetylcholine (10^{-6}M) produced large increases of ^{86}Rb efflux in all areas of sinus venosus and also interauricular septum, indicating that the distribution of the purinoceptors is more confined than that of the muscarinic receptors.

The adenine nucleotides, ATP and ADP, also produced an increase in ^{86}Rb efflux. The dose response curve for ATP was similar to that of adenosine.

40. NUCLEOSIDE TRANSPORT DIFFERS IN ERYTHROCYTES FROM FRESH AND OUTDATED HUMAN BLOOD. S.M. Jarvis*, J.R. Hammond+, A.R.P. Paterson*, and A.S. Clanachan+, Cancer Research Unit (McEachern Laboratory)* and Department of Pharmacology+, University of Alberta, Edmonton, Alberta, Canada, T6G 2H7.

This report describes major differences in the kinetic parameters for uridine transport between erythrocytes from fresh and outdated human blood. Kinetic constants (mean $\pm$ SEM (n)) for *zero-trans* influx and efflux of (5-^{3}H)uridine in fresh human erythrocytes (at 22° C) were equivalent (K_m values of 0.17$\pm$0.02 (3) and 0.14$\pm$0.04 (4) mM for influx and efflux, respectively, and V_{max} estimates of 116$\pm$14(10) and 135$\pm$16 (4) mmol.1 H_2O/hr). In contrast, uridine transport in erythrocytes from outdated blood (21-35 days) exhibited directional asymmetry in that *zero-trans* influx and efflux were not comparable (K_m values of 0.13$\pm$0.03(4) and 0.38$\pm$0.10(4) mM for influx and efflux, respectively, and V_{max} estimates of 35.4$\pm$3.9(4) and 117$\pm$18(4) mmol/1 H_2O/hr). Inward fluxes of uridine in equilibrium exchange in erythrocytes from fresh and outdated blood were faster than inward *zero-trans* fluxes (K_m values of 0.76$\pm$0.07(6) and 1.1$\pm$0.1(4) mM for cells from fresh and outdated blood, respectively, and V_{max} estimates of 462$\pm$60(6) and 355$\pm$41(4) mmol/l H_2O/hr). The numbers of functional nucleoside transport sites present on erythrocytes, estimated by site-specific NBMPR binding, were similar (~ 12,000 sites/cell). These results demonstrate that uridine transport can be described by a simple carrier model which appears to operate with directional symmetry in fresh erythrocytes but asymmetrically in cells from outdated stored blood.

Supported by NIC (Canada) and the Alberta Heart Foundation.

41. MOLECULAR MODEL FOR THE ERYTHROCYTE NUCLEOSIDE TRANSPORTER. S.M. Jarvis* and J.D. Young+, *Cancer Research Unit, University of Alberta, Edmonton, Alberta, Canada T6G 2H7 and +Department of Biochemistry, Faculty of Medicine, The Chinese University of Hong Kong, Shatin, NT, Hong Kong.

This report describes a possible model for nucleoside transport in erythrocytes. The kinetic characteristics of the single broad specificity saturable nucleoside transport system that exists in human erythrocytes are consistent with a carrier model involving one substrate binding site and two forms of a free carrier, one outward-facing (A) and the other inward-facing (B). Translocation depends on the interconversion of the two carrier forms and is effected by a conformational change of the transporter. Reorientation of the unloaded carrier is the rate-determining step in the translocation process.

The carrier forms (A and B) differ in their inhibitor binding and chemical properties. Nitrobenzylthioinosine (NBMPR) inhibits nucleoside influx into nucleoside permeable sheep erythrocytes competitively (apparent K_i = 1 nM), whereas inhibition of efflux is noncompetitive, indicating that NBMPR binds preferentially to the outward-facing form of the carrier. For simplicity, NBMPR is proposed to occupy both the nucleoside binding site and a second hydrophobic binding site which accommodates the nitrobenzene ring and stabilizes inhibitor binding to the carrier. The latter site is only freely exposed in conformation A. However, since A + B are interconvertible, all carrier sites will eventually bind inhibitor. Finally, the effects of thiol reagents on NBMPR binding to intact cells and disrupted membranes have provided evidence that the nucleoside transporter exhibits chemical asymmetry.

42. ADENOSINE ACTIVATES A POTASSIUM CONDUCTANCE IN GUINEA-PIG ATRIAL HEART MUSCLE. G. Jochem and H. Nawrath, Dept. Pharmacol., Univ. Mainz, D-6500 Mainz, West Germany.

The effects of adenosine on the force of contraction, transmembrane potential and currents and K-42 efflux were studied in myocardial preparations from guinea-pigs. The duration of the action potential and the force of contraction were concentration dependent depressed in atrial preparations; ventricular preparations were not responsive. A hyperpolarization in response to adenosine was seen in partially depolarized atrial preparations. Under voltage clamp conditions, the steady state outward current was clearly increased at holding potentials positive to the resting potential. The most plausible explanation for the hyperpolarization and the increase in the steady state outward current is an increase in the potassium conductance, since the efflux of K-42 was enhanced in the presence of adenosine. We have shown that adenosine activates a potassium conductance in guinea-pig atrial heart muscle. This mechanism may be responsible for the abbreviation of the action potential and the negative inotropic effect of adenosine in the atrium. In the ventricle, obviously, this mechanism cannot be activated.

Supported by a grant from the Deutsche Forschungsgemeinschaft.

43. ADENOSINE MECHANISMS IN RAT HIPPOCAMPUS. Bror Jonzon and Bertil B. Fredholm, Department of Pharmacology, Karolinska Institutet, Box 60 400, S-104 01 Stockholm, Sweden.

Radioactive purines were continuously released from ^{14}C- or ^{3}H-adenine labelled transversely cut hippocampal slices. The rate of release was enhanced by veratridine, by K^{+}-depolarization and by electrical field stimulation. More intense stimulations were required to release labelled purines than ^{3}H-noradrenaline. The time course was also different. Evoked purine release was inhibited by dipyrimadole as reported earlier for other systems, but it was not significantly affected by diazepam.

Adenosine, PIA, CHA and NECA caused an increased accumulation of ^{3}H-cyclic AMP in the slices, with an absolute and relative potency that differed from that of binding to a membrane fraction of the tissue or of inhibition of field EPSP's. The effect of adenosine was enhanced by diazepam, but not by several other antiepileptic drugs. The effect of diazepam was mimicked by dipyridamole and other uptake inhibitors and blocked by Ro 15-1788. Diazepam and clonazepam inhibited the cyclic AMP accumulation induced by NECA. This effect was not shared by other antiepileptic drugs.

The results suggest that adenosine is continuously released from hippocampal slices and that the rate of release is enhanced by neuronal activity. Diazepam specifically enhances adenosine effects on cyclic AMP by an action not related to receptors but possibly related to uptake inhibition. Whether diazepam also influences adenosine induced depression of pyramidal cell firing is not known.

Supported by SMFR, Wibergs Stiftelse, and the Karolinska Institute.

44. EFFECT OF ALTERATION OF ADENOSINE METABOLISM IN POST-EXERCISE HYPEREMIA IN THE CANINE GRACILIS MUSCLE. Joyce Kille and Richard E. Klabunde, West Virginia University, Morgantown, West Virginia 26506.

In an earlier study (Physiologist 24: 75, 1981), dipyridamole, an inhibitor of the cellular transport of adenosine, was found to potentiate post-exercise hyperemia (PEH) in the free-flowing canine gracilis muscle. We have more recently studied the effects of erthro-9-(2-hydroxy-3-nonyl) adenine (EHNA), an adenosine deaminase (ADA) inhibitor, and α, β-methylene adenosine 5'-diphosphate (AOPCP), an inhibitor of 5'nucleotidase (5'N), on PEH in the same preparation. Calculations were made for excess blood flow (EQ, ml/100g), excess oxygen consumption (EVO_2, vol O_2/100g), and the tension-time integral (TT, kg.s/100g) following 1, 3, 5, and 10 sec of sustained, maximal, isometric contraction. The PEH response was analyzed by comparing the slope of the linear regression of EQ vs TT or EVO_2, and EVO_2 vs TT in the absence and presence of an intra-arterial drug infusion. EHNA (50 μM in blood) increased the slope of the regression line for EQ vs EVO_2 by 41% ($p < 0.05$). For the same regression, AOPCP (50 μM in blood) caused a 41% decrease ($p < 0.01$) in the slope. Neither drug had significant effects on EQ or EVO_2 vs TT. Other investigators have shown that 50 μM EHNA and AOPCP can effectively inhibit the deamination of adenosine by ADA and the production of adenosine by 5'N, respectively. The potentiation of PEH by EHNA may result from increased accumulation of interstitial adenosine when adenosine metabolisn by ADA is inhibited. The reduction of PEH by AOPCP suggests that endogenous production of adenosine by 5'N contributes to PEH. These findings are consistent with the hypothesis that accumulation of interstitial adenosine contributes to PEH.

Supported by NHLBI grant # 25412.

45. ADENOSINE BLOCKS RAT ATRIAL SLOW ACTION POTENTIALS GENERATED WITH PHOSPHATIDIC ACID. Maureen T. Knabb, Rafael Rubio, and Robert M. Berne, Department of Physiology, University of Virginia School of Medicine, Charlottesville, Virginia 22908.

Previous work suggests that phosphatidic acid (PA) plays a physiological role in Ca^{2+} influx. We investigated the effects of PA on rat atrium partially depolarized with 22 mM of K^+. PA (10 μg/ml) induced slow action potentials upon electrical stimulation with a mean amplitude of 49.7 mV and maximum rate of rise of 13.2 V/sec. Increase in (Ca^{2+}) from 0.5 to 5.0 mM caused the amplitude of the slow action potential to rise linearly with the logarithm of the (Ca^{2+}). The calculated slope of the line was 21.1 mV for a ten-fold change in $(Ca^{2+})_0$. Mn^{2+} and verapamil blocked the slow response, but a higher concentration was necessary to block PA-induced action potentials as compared to isoproterenol-induced action potentials. Phospholipase D (PLD), an enzyme which breaks down phospholipids into PA, also was able to induce slow action potentials. However, phospholipase C, an enzyme which generates 1, 2-diacylglycerol, did not cause a slow response. Adenosine blocked PA-induced and PLD-induced slow action potentials and aminophylline reversed the effects. The data further support a role for adenosine in the regulation of calcium influx, possibly through control of phospholipid turnover or interference with PA-induced channel formation.

Supported by grants #HL 10384 and 19242.

46. "MICRO" ADENOSINE DEAMINASE (MADA) AND THEOPHYLLINE POTENTIATE ISOPROTERENOL-INDUCED SLOW ACTION POTENTIALS DURING NORMOXIA. Maureen T. Knabb, Rafael Rubio, and Robert M. Berne, Department of Physiology, University of Virginia School of Medicine, Charlottesville, Virginia 22908.

Previous work has shown that adenosine is an effective blocker of atrial slow action potentials. However, it is not known whether endogenous adenosine inhibits the catecholamine-induced slow response during normoxia. We investigated the effects of "micro" adenosine deaminase (MADA molecular weight between 4,000 and 25,000) and theophylline on slow action potentials in an attempt to minimize the action of endogenously-release adenosine. Various concentrations of isoproterenol were used to induce the slow response in partially depolarized rat atrium. The threshold for isoproterenol-induced slow action potentials was 10^{-8} M and maximal stimulation occurred at 5 x 10^{-7} M. Both theophylline and MADA markedly shifted the isoproterenol dose-response curve to the left. In the presence of MADA or theophylline, the threshold for isoproterenol was approximately 10^{-9} M and maximal effects occurred at 5 x 10^{-8} M. These data suggest that denosine exerts important inhibitory effects during catecholamine stimulation under normoxic conditions.

Supported by grants #HL 10384 and 19242.

47. EFFECTS OF DIPYRIDAMOLE ON RELATIONSHIPS AMONG MYOCARDIAL OXYGEN CONSUMPTION (MVO_2)- CORONARY BLOOD FLOW (CBF), AND PERICARDIAL INFUSATE (PCI) ADENOSINE. R.M. Knabb, S.W. Ely, R. Rubio, and R.M. Berne, Department of Physiology, University of Virginia School of Medicine, Charlottesville, Virginia 22908.

Dipyridamole, a vasodilator which blocks adenosine uptake, has been shown to elevate tissue adenosine content, but it is not known if this is due to increased adenosine in the interstitial fluid (ISF) or within cells. Using PCI as an index of ISF adenosine, we have previously demonstrated parallel relationships among MVO_2, CBF, and PCI adenosine. The current series of experiments sought to establish the effects of dipyridamole on ISF adenosine and on the relationships among MVO_2, CBF, and PCI adenosine. Measurements were made at two or more levels of MVO_2. Dipyridamole elevated CBF and decreased oxygen extraction, which resulted in an unchanged MVO_2, and a higher slope for the relationship between CBF and MVO_2. The slope of PCI adenosine vs MVO_2 was similarly elevated by dipyridamole, indicating that at each level of MVO_2 ISF adenosine was greater after dipyridamole. The relationship between CBF and PCI adenosine was not altered by dipyridamole, suggesting that changes in ISF adenosine result in proportional changes in CBF, regardless of whether the changes in adenosine concentration are induced by alterations of MVO_2 or by dipyridamole. Thus, these experiments are highly suggestive of a cause and effect relationship between changes in ISF adenosine and changes in CBF.

48. 5'-NUCLEOTIDASE IN THE NERVOUS SYSTEM. Georg Kreutzberg, Peter Schubert, Martin Reddington, Kevin Lee and Dagmar Heymann. Max-Planck-Institute for Psychiatry, Munich, FRG.

Considerable evidence has accumulated indicating that adenosine is a neuromodulator in the central nervous system. The formation of adenosine probably occurs from nucleotides released from nerve endings as a result of nerve cell activity. An enzyme strategically placed for the hydrolysis of 5'-AMP to adenosine is 5'-nucleotidase, an integral membrane protein which expresses its enzymatic activity at the cell surface. 5'-Nucleotidase has a highly specific distribution in the CNS. Cytochemical studies reveal its presence on oligodendroglia, on astrocytes surrounding nerve endings and in direct association with some synapses. Biochemical investigations suggest that the availability of 5'-nucleotidase is an important determinant of the local extracellular concentration of adenosine. Taken together, these observations support the proposal that 5'-nucleotidase plays a key role in purinergic neuromodulation.

49. SYNAPTIC DEPRESSION AND POST-TENTANIC POTENTIATION (PTP) CAN BE MEDIATED BY ADENOSINE DERIVATIVES ACCUMULATED DURING THE TETANUS. Yoichiro Kuroda, Department of Neurochemistry. Tokyo Metropolitan Institute for Neurosciences, Fuchu-Shi, Tokyo 183, Japan.

ATP and its metabolites, adenosine derivatives (AD), are released into synaptic clefts and removed by re-uptake under moderate pre-synaptic stimulation. However, tetanic stimulation may accumulate AD in the synaptic clefts, since AD overflowed to the extracellular medium from olfactory cortex slices when lateral olfactory tracts (LOT) were tetanically stimulated. Simulateously, synaptic depression during the tetanus and PTP a few minutes after the tetanus were observed on the post-synaptic potentials (PSP) evoked by the stimulation of LOT in the slices. Extracellular application of AD appeared to mimic the synaptic plasticities. AD did cause two effects on the PSP. One was direct inhibition of PSP in the presence of AD, and the other was indirect facilitation which appeared after the removal of AD from the medium. The effects are suggested to be presynaptic and via two distinct adenosine receptors, ADO-1, ADO-2, respectively. A high affinity receptor, ADO-1, mediates the inhibition by the decrease of depolarization-dependent Ca influx, which results in the reduction of transmitter release from LOT terminals. A low affinity receptor, ADO-2, mediates the facilitation by the increase of cyclic AMP in the LOT nerve terminals. Cyclic AMP analogues increase the depolarization-dependent CA influx, therefore the increased cyclic AMP may facilitate the transmitter release from LOT terminals. Thus, physiological significance of AD on neurotransmission in mammalian brain is proposed as a modulator of synaptic depression, PTP and may be heterosynaptic facilitation. The latter two are considered to be basic electrophysiological models of learning and memory.

50. SORTING OUT THE REGULATOR(S) OF CORONARY BLOOD FLOW. J.D. Laird, J.A.E. Spaan, Department of Physiology and Physiological Physics, Leiden University, Leiden, The Netherlands.

The correlation between coronary blood flow and oxygen consumption is often cited in support of a role for local metabolites in the control of flow. The variance in published correlations is substantial, in part, because different interventions in different dogs are compared. In five anesthetized dogs we have measured coronary blood flow and A - V O_2 difference following metabolic vasodilation and increases in isolated coronary artery perfusion pressure in the same dog under comparable hemodynamic conditions. Heart rate was varied between 60 and 180 bpm while perfusion pressure was parametrically varied between 60 and 140 mmHg. Increases in heart rate lead to a positive linear relationship between coronary blood flow and oxygen consumption: metabolic vasodilation. Increasing perfusion pressure (autoregulation) at constant heart rate and ventricular pressure leads to vasoconstriction despite a 20 percent increase in oxygen consumption. Thus vascular conductance is not simply related to oxygen consumption. Several theoretical models based on homogeneous compartmentalization have been used to interpret these results. Attempts to predict these results with a model based on adenosine have been unsuccessful. However, a model assuming a link between tissue oxygen tension and vascular conductance works quite well. A mechanism presumed to affect metabolic vasodilation cannot be ignored in other interventions such as autoregulation. It is easy to conclude that more than one factor is involved in flow regulation; however, the tools to analyze such complex systems are still wanting.

51. RELEASE OF CYCLIC AMP IN CORONARY BLOOD INDUCED BY ADENOSINE.
Jacques Lammerant and Thanh Huynh-Thu, Department of Physiology, School of Medicine, Namur, Belgium.

In 30 closed-chest anesthetized dogs, infusion of adenosine (ADO) into the left anterior descending coronary artery (LAD) at a rate of 150 nmole/min for 10 min induced a significant increase in cAMP release in the great cardiac vein, from a control value of 15.2 ± 3.0 to 104.8 ± 29.5 pmole/min.100g g myocardium ($p < 0.01$). In 22 propranolol-treated, vagotomized dogs, ADO was infused into the left ventricle (LV) at rates of 25, 50 and 100 nmoles/kg.min during three successive 15 min periods, while blood pressure in the ascending aorta was prevented from falling by appropriate inflation of a balloon in the thoracic aorta. The infusion induced a significant dose-dependent increase in both cAMP release in the coronary sinus, from a control value of 48.2 ± 7.1 up to 206.5 ± 56.2 pmole/min.100 g myocardium ($p = 0.001$), without changing the myocardial cAMP content in the biopsies of the apex of the heart (n=10) obtained at the highest rate of ADO infusion. The estimated ADO concentration achieved in the elevated LAD inflow or LV outflow were lower than needed to induce cAMP release from coronary blood platelets, as indicated by the in vitro experiments. The results suggest that increased cAMP generation in coronary cells is involved in the metabolic regulation of coronary flow by ADO.

52. ADENOSINE-INDUCED IMMUNOSUPPRESSION OF NORMAL HUMAN LYMPHOCYTES: A MODEL OF IMPAIRED CELLULAR IMMUNE FUNCTION.
David L. Larson, Karen P. Cannon, Keith E. Jensen, and Leland D. Loose, Pfizer Inc., Groton, CT 06340.

Total E-rosette (E-R) formation has been found to be a use-diagnostic measure of depressed cell-mediated immunity (CMI). We report that the addition of 32 μM adenosine to mononuclear cells (MNC) obtained from healthy donors causes depressed E-R formation. These effects can be reversed by levamisole (LV) or indomethacin (I). Normal donors (N=33) had an E-R level of $54 \pm 5\%$ ($\bar{X} \pm SD$) which was depressed ($p < .01$) after incubation with adenosine to $33 \pm 6\%$. In the presence of monocytes, this immunosuppressive effect was completely abolished when the MNC were coincubated with adenosine and LV or I; dose response studies showed each drug to have an ED_{50} of 0.03 μM. Adenosine treatment of MNC in the absence of monocytes (N=5) resulted in depression ($p < .01$) of E-R formatiom from $54 \pm 4\%$ to $32 \pm 2\%$; coincubation of these cells with adenosine and LV resulted in normal E-R levels of $52 \pm 2\%$. In contrast, coincubation of adenosine and I did not normalize ($33 \pm 2\%$) the adenosine-depressed E-R levels. Since the action of LV and I in this model mimics the immunorestorative effects of LV and I on MNC derived from Hodgkin's disease patients (Ramot et al., 1981), we suggest that adenosine-induced immunosuppression may play a role in the molecular pathology of Hodgkin's disease as well as immunodeficiency diseases associated with adenosine deaminase deficiency.

53. PROBLEMS ENCOUNTERED IN THE CLASSIFICATION OF PURINERGIC PRESYNAPTIC RECEPTORS ON ADRENERGIC AND CHOLINERGIC NERVES.
H.J. Leighton and L.L. Parmeter, Department of Pharmacology, Wellcome Research Labs, Research Triangle Park, North Carolina 27709.

Adenosine (Ad) and 2-chloroadenosine (2ClAd) inhibited twitch contractions, induced by field-stimulation, of the rat vas deferens and guinea pig ileum. In both tissues, theophylline antagonized the inhibitory actions of the purines qualitatively suggesting a P-1 receptor mechanism. In the guinea pig ileum, Schild analysis gave pA_2 values for theophylline of 4.6 (slope 0.79) and 4.9 (slope 1.08) for Ad and 2ClAd, respectively. After dipyridamole (D) and with Ad as agonist, the pA_2 for theophylline was 5.3 (slope 0.95). The improvement of slope in the presence of D indicates the importance of eliminating uptake and metabolism prior to receptor analysis. In the rat vas deferens, pA_2 values obtained with theophylline against 2ClAd varied between 4.2 and 5.2, depending on the frequency of stimulation. In this tissue the slopes of the Schild regression differed from unity. Although the IC50 for 2ClAd should vary with frequency employed to induce contractions, the pA_2 should remain constant. Thus the antagonist actions of theophylline in the rat vas deferens are in part due to a physiologic antagonism (e.g. calcium mobilization and/or phosphodiesterase effect) rather than pharmacologic and serves to demonstrate the need for careful receptor analysis in classifying purinergic receptors on adrenergic nerve endings.

54. AN HPLC-FLUORESCENCE ASSAY FOR MEASUREMENT OF ENDOGENOUS ADENYL PURINE COMPOUNDS. B. Levitt, R.J. Head and D.P. Westfall, Department of Pharmacology and Toxicology, West Virginia University Medical Center, Morgantown, West Virginia.

Previous studies in this laboratory have provided evidence for the simultaneous release of ^{3}H-adenyl purines and ^{3}H-norepinephrine from sympathetic nerves in the guinea-pig vas deferens and the rabbit portal vein following transmural nerve stimulation. A major disadvantage in using a ^{3}H-precursor of ATP for these experiments is the predominant labelling of extraneuronal sites, particularly smooth muscle cells. For a more direct approach, we investigated the use of high performance liquid chromatography (HPLC) in conjunction with fluorescence detection for the measurement of endogenous ATP and its analogues. Incubation of ATP with chloracetalydehyde at 80° C for 40 min yields the fluorescent derivative 1,N^6-etheno-ATP (ϵ-ATP). ϵ-ADP, ϵ-AMP, and ϵ-adenosine (ϵ-Ado) are prepared in the same manner, and these compounds are well resolved by reverse-phase HPLC (Bondapak C18 cartridge, 5μ). This system involves a 15 min gradient elution utilizing a .1 M potassium-phosphate buffer (pH 6.0) and 25% methanol (v/v) as the mobile phase. Under these conditions, the limits of detection for ATP, ADP, AMP and adenosine are approximately 3.5, 2, 1.5 and .50 picomoles, respectively, and this detection is linear in the range of 2 to 500 picomoles for all four ϵ-derivatives. Thus, this method provides a sensitive assay for the detection of endogenous adenyl purines and based on preliminary studies, appears to be useful for measuring release of ATP and its metabolites from autonomically innervated tissues.

55. STUDIES WITH 5'-NUCLEOTIDASE INHIBITORS. G. Lutoslawska and H.P. Baer, Department of Pharmacology, University of Alberta, Edmonton, Alberta, Canada, T6G 2H7.

A rapid micro-screening assay for 5'-nucleotidase, using ^{14}C-labelled AMP as a substrate, was developed allowing a comparative study of potential inhibitors of this enzyme. Inhibitors were compared by testing their ability to reduce the extent of hydrolysis of 0.1 M AMP relative to a control conversion of 50% of total substrate, using microsomal enzyme preparations from rat, guinea-pig and dog ventricles. The inhibitors tested so far included unlabelled AMP as a competing substrate, ATP, adenosine-5'- α,β-methylenediphosphate (APCP) and inosine-5'-α,β-methylene diphosphate (IPCP). IC_{50} values in the absence of Mg^{++} for AMP, ATP, APCP and IPCP were as follows (μM): rat heart: 85, 30, 0.75, 9. Guinea-pig heart: 30, 17, 0.18, 1.3. Dog heart: 80, 30, 22, 14. In the presence of 5 mg Mg^{++} the IC_{50} values for AMP were not much affected, but values for ATP were up to 20-fold larger, and values for APCP and IPCP were several 100-fold larger. Thus, the efficiency of the inhibitors was strongly modified by Mg ion. These and other observations indicate that considerable differences exist with respect to the kinetic properties of 5'-nucleotidases from different species and that the mode of inhibition is difficult to assess. Creatine phosphate in concentrations as high as 16mM did not show any inhibitory effects. It is not yet clear whether the enzyme activities tested are identical to those studied by other authors after partial purification or in perfused organs.

Supported by the Alberta Heart Foundation and the Alberta Heritage Foundation for Medical Research.

56. ON THE MECHANISM OF ADENOSINE EFFECT INCREASING RAT ERYTHROCYTE GLYCOLYSIS. V. Madrid-Marina, H. Martinez-Valdez, R. Gutierrez-Juarez, V.M. Espinosa, C.V. Sanchez, and E. Pina, Department of Biochemistry, Sch. Med. National University of Mexico, P.O. Box 70-159, Mexico D.F. Mexico.

It has been shown that adenosine increased adenine nucleotide pools in rat erythrocytes. Hence, attempts were made to test the effect of the nucleoside on glycolytic pathway, major source of energy in the red cell.

A 2.5 fold increase in the level of fructose 1,6 diphosphate and lactate production, together with changes in other glycolytic intermediates were observed, 30 min after adenosine administration. Activation of phosphofructokinase in the erythrocyte is produced as a consequence of the effects of the injected adenosine. It is concluded that the nucleoside increases glycolysis rate.

57. SEPARATE RECEPTORS FOR ATP AND ADENOSINE IN THE GUINEA PIG TAENIA COLI. M. Helen Maguire and David G. Satchell, Department of Pharmacology, Toxicology and Therapeutics, University of Kansas Medical Center, Kansas City, Kansas 66103 and Department of Zoology, University of Melbourne, Australia 3052.

The question of separate inhibitory receptors for ATP and adenosine in the guinea-pig taenia coli has been addressed by comparing the effects of similar structural modifications to ATP and adenosine on their inhibitory actions in the isolated guinea-pig taenia coli preparation, both in the absence and presence of dipyridamole. The related 5'-monophosphates were also studied. The structural modifications were 8-bromo substitution, 2'-hydroxyl reduction and 2'-hydroxyl epimerization. 8-Bromoadenosine and ara-adenosine were inactive at 1 mM; 2'-deoxyadenosine was a partial agonist, pD_2 3.34. 8-Bromo-ATP, ara-ATP and 2'-deoxy-ATP induced rapid ATP-like relaxations with pD_2s 5.66, 5.60 and 5.76, respectively. For ATP and adenosine pD_2s were 6.10 and 4.67. Dipyridamole, a known potentiator of ATP and adenosine, did not potentiate the actions of 8-bromo-ATP, ara-ATP or 2'-deoxy-ATP or 2'-deoxyadenosine. If one inhibitory receptor mediated responses to both ATP and adenosine, then similar structural modifications in ATP and adenosine should have parallel effects on their agonist properties. Non-parallel effects were observed, supporting the view that the taenia coli has separate receptors for ATP and adenosine. The 5'-monophosphate analogues elicited relaxation of the taenia coli. 2'-Deoxy-AMP was more potent than 2'-deoxyadenosine, but was also a partial agonist. 8-Bromo-AMP and ara-AMP were full agonists, more potent than 2'-deoxy-AMP. These findings indicated that the 5'-monophosphates act as such to relax the muscle, and that hydrolysis to the parent adenosine analogue is not required for activity.

Supported by grants from NHMRC and ARGC, Australia, and BRSG Fund, University of Kansas Medical Center.

58. TRANSLOCATION AND RECEPTORS IN CEREBRAL ACTIONS OF ADENOSINE AND ITS NUCLEOTIDES. Henry McIlwain, Department of Biochemistry, St. Thomas's Hospital Medical School, London SE1 7EH England.

After electrical stimulation of neocortical tissues their cAMP was increased for some minutes and histone kinase, mainly type II, was activated; subcellular fractionation showed a synaptosomal increase in cAMP and kinase, delayed by 5-10 min in relation to changes in the tissue as a whole. Output of adenosine accompanied the excitation and the induced rise in cAMP was diminished by theophylline; the synaptosomal fraction from unstimulated tissue responded to excitation by output of adenosine, but not by increased cAMP. These findings are consistent with the synaptosomal fraction carrying adenylate cyclase inhibited by adenosine at A_1 receptors, and the adenylate cyclase of postsynaptic regions being stimulated by adenosine A_2 receptors and by other neurohumoral agents liberated concomitantly. The presynaptic regions are concluded to receive cAMP by cytoplasmic transport rather than from local adenosine-activated adenylate cyclase, for this is associated with decrease rather than increase in histone kinase. Presynaptic inhibition is thus occurring during the period when the persisting cAMP, by traversing cell processes, is giving an endogenously patterned activation of successive cAMP-sensitive postsynaptic regions.

59. MYOCARDIAL ADENOSINE CONTENT AND CHRONIC EXERCISE. Jack E. McKenzie, Department of Physiology, Uniformed Services University, Bethesda, Maryland 20814.

The influence of chronic exercise on myocardial adenosine content, catecholamine content, and coronary vascular resistance was studied in male Sprague-Dawley rats. Animals were randomly assigned to sedentary or swimming exercised groups. Exercised rats swam 1 hr/day, 5 days/wk, for a total of 10 wks. Following the 10 wk period rats were randomly assigned to one of four groups: sedentary control(SC), exercise control (EC), sedentary aortic constricted(SAC), and exercise aortic constricted (EAC). For groups SAC and EAC the ascending aorta was constricted proximal to the first aortic branch to produce a two fold increase in intraventricular pressure. Exercise increased the heart weight/body weight ratio above that of the sedentary group, 3.03±0.19 and 2.33±0.24, respectively. Coronary vascular resistances were: 0.59±0.14, 0.55±0.03, 0.42±0.07, and 0.34±0.08 $mmHg \cdot ml^{-1} \cdot min \cdot 100g$ for SC,EC,SAC, and EAC respectively. Myocardial adenosine content was greater in EC and EAC (13.27±2.42 and 14.18±1.20 nmoles/g) as compared to SC and SAC (7.6±1.8 and 11.09±2.45 nmoles/g). Catecholamine content was less in sedentary groups (1012±159 and 1098±108 ng/g for SC and SAC, respectively), than the exercised groups (1380±180 and 1168±191 ng/g for EC and EAC respectively). Myocardial adenosine content significantly correlated with total catecholamine content, r=0.60, $p < 0.05$. These data show increased myocardial adenosine and catecholamine contents in EC and EAC groups as compared to SC and SAC groups. However, coronary vascular resistances are similar between SC and EC groups and between SAC and EAC groups. This implies that with chronic exercise adenosine production increases to elicit similar coronary vascular resistances and that there may be a relationship between myocardial adenosine content and catecholamine content.

Supported by USPHS Grant #CO 7635.

60. EFFECTS OF ADENOSINE, ITS ANALOGS, ADRENERGIC AGENTS, AND PROSTAGLANDINS ON ADENYLATE CYCLASE OF HEART MICROVESSELS. Gunvantrai Mistry and G.I. Drummond, University Biochemistry Group, University of Calgary, Calgary, Alberta, T2N 1N4, Canada.

Microvessels were isolated in a highly purified form from rabbit heart using a modified procedure of Simionescu and Simonescu (Microvascular Research 26:426 (1978)). Stimulation of adenylate cyclase by adenosine and its analogs was examined. Activation of the enzyme by 5'-N-ethylcarboxamideadenosine (NECA) EC_{50}, 0.2 μM) was 2-fold over basal, while adenosine (EC_{50}, 0.6 μM) and N^6-phenylisopropyl adenosine (PIA) (EC_{50}, 1.6 μM) stimulated less than 2-fold. The relative potency NECA adenosine PIA suggest the presence of stimulatory adenosine receptors (R_a). The response to adenosine was biphasic suggesting the presence of a P site. Catecholamines: isoproterenol (INE) (EC_{50}, 0.4 μM) and epinephrine (E) (EC_{50}, 1.4 μM) stimulated the enzyme 4-5 fold over basal, while norepinephrine (NE) (Ec_{50}, 10 μM) was a partial agonist. The catecholamine potency ranking INE > E > NE suggest the presence of $beta_2$ receptors. Various alpha adrenergic agents were found to effect catecholamine-stimulated activity. Ergot alkaloids at 10 μM inhibited INE-stimulated activity with relative potencies ergotamine > bromocryptine > dihydroergotamine > ergocryptine > ergovine. Prostaglandins stimulated the enzyme with the following potency order: $PGE_1 > PGE_2 > PGA_1 = PGB_1 > PGB_2 > PGF_{2\beta} > PGF_{1\alpha}$. Thus, our studies indicate the presence of adenosine, adrenergic and prostaglandin receptors linked to adenylate cyclase of heart microvessels.

Supported by AHFMR and MRC of Canada.

61. EFFECT OF ETHANOL INTOXICATION ON BRAIN ADENOSINE RECEPTORS. S. Jamal Mustafa, Marcia Hobbs, M. Saeed Dar, and Wallace R. Wooles, Department pf Pharmacology, School of Medicine, East Carolina University, Greenville, North Carolina 27834.

Adenosine has been reported to cause a central depressing effect through its specific binding sites in brain. This study was aimed to investigate the effect of chronic ethanol intoxication on adenosine receptor binding in the ICR mice. Individually housed mice were fed (10 or 32 days) a liquid diet fortified with vitamins to which ethanol was added equal to 33% of calories. Control mice received the same volume of diet containing an isocaloric amount of sucrose. The animals were sacrificed, the brains removed and adenosine receptor binding measured using ^{3}H-L-pheynlisopropyl-adenosine. Mean blood alcohol levels after 10 days of ethanol diet were found to be 4.52 mg/ml while at 32 days blood ethanol decreased to 2.99 mg/ml. The Scatchard analysis of the data resulted in a K_d value of 2.08 ± 0.35 nM and 7.48 ± 1.35 nM and a β_{max} of 454 ± 18 and 499 ± 55 fmol/mg protein in 10 day control and ethanol treated group, respectively. The K_d values in 32 day group were 5.90 ± 0.25 and 5.85 ± 0.92 nM and a β_{max} of 494 ± 7 and 532 ± 19 fmol/mg protein for control and ethanol treated animals, respectively. A higher K_d value in 10 day ethanol treated group may indicate a significant deviation in ethanol induced behavioral response. The mouse brain possesses a high affinity A_1 receptor of the R type.

Supported by NIH Grant No. HL 27339.

62. RENAL TRANSPORT OF PURINE NUCLEOTIDES. J. Arly Nelson, John F. Kuttesch, and Morris J. Robins, Pharmacology Laboratory, Department of Experimental Pediatrics, the University of Texas M.D. Anderson Hospital and Tumor Institute, Houston, Texas 77030.

In a child lacking adenosine deaminase and in adults treated with the enzyme inhibitor, deoxycoformycin (DCF), apparent renal secretion of 2'-deoxyadenosine (dAdo) and reabsorption of adenosine (Ado) were observed. That is, the renal clearance of dAdo was approximately 5-fold that of creatinine, whereas the corresponding value for Ado was 1/5 that of creatinine. In mice treated with DCF (5 mg/kg), plasma levels of dAdo and Ado increase to near 1 μM and 15 μM, respectively, whereas the level of inosine decreases from approximately 6 to 2 μM. These altered levels plateau in about 30 min and are maintained relatively-constant for at least 90 min after DCF administration. For renal clearance determinations in male mice, the urethra was ligated 30 min after DCF, and urine was allowed to collect in the bladder *in situ*. Radioactive inulin was administered 20 min prior to urethral ligation in order to obtain an estimate of glomerular filtration rate. Samples of plasma and bladder contents were obtained 30 min after ligation. Purine nucleosides were measured by HPLC and radioactive inulin was measured by liquid scintillation counting. The clearances of dAdo, Ado and inosine relative to that of inulin were 1.71+0.16, 0.42 +0.11, and 1.16+0.30, respectively (mean values+S.E., n$\geq$7). Renal secretion of dAdo was confirmed by administration of radioactive dAdo (to obviate the kidney as a source of urinary Ado) and by producing an alkaline diuresis with methazolamide (to eliminate "ion-trapping"). We were unable to determine the transport maximum for dAdo since adminstration of large doses (100 to 1500 nmoles/g) to DCF treated mice produced a progressive decrease in inulin clearance (renal failure). Thus, dAdo may produce renal vasoconstriction in the mouse.

The mechanism for renal secretion of dAdo was studied in mouse kidney slices using 2'-deoxytubercidin (dTub), an analog which is not significantly metabolized. Uptake characteristic of active transport was demonstrated, i.e., it was metabolically-dependent, concentrative and saturable. Classical substrates for the organic cation system (tetraethylammonium, N^+-methylnicotinamide, choline) inhibited dTub uptake whereas substrates for the organic anion system (p-aminohippurate, probenecid, phenol red) failed to inhibit. Conversely, dTub inhibited tetraethylammonium uptake but failed to inhibit that of p-aminohippurate. Therefore, dTub appears to be a substrate for the organic cation system in mouse kidney, in spite of the fact that it is a rather weak base, pKa = 5.3.

Sugar-modified derivatives of the tubercidin aglycone (i.e., 2'-deoxyribosyl, 3'-deoxyribosyl, xylosyl, arabinosyl) are secreted by the mouse kidney in vivo. The lack of structural specificity for secretion suggests that dAdo secretion may occur via the organic cation system and reabsorption of Ado may only be apparent due to extensive metabolism by the Ado kinase of renal tissue.

Supported by Grants HD-13951 and CA-28034 from the National Institutes of Health and the National Cancer Institute of Canada.

63. INHIBITION OF cAMP ACCUMULATION AND LIPOLYSIS BY ADENOSINE ANALOGS IN ISOLATED HUMAN ADIPOCYTES. Jorma J. Ohisalo and Ilpo Huhtaniemi, Department of Medical Chemistry, University of Helsinki, Siltavuorenpenger 10a, 00170 Helsinki 17, Finland.

Adipocytes were prepared by collagenase treatment of pieces of human subcutaneous adipose tissue excised in the beginning of regular abdominal operations. The cells were incubated in a total volume of 1 ml of a buffer containing 125 mM NaCl, 6 mM KCl, 2.5 mM $MgCl_2$, 1 mM $CaCl_2$, 1 mM KH_2PO_4, 10 mM Tris, 4 mM glucose and 2% of bovine serum albumin at pH 7.4, using air as the gas phase. L-isoproterenol (1 μM) stimulated the accumulation of cAMP in these cells; the effect was observed after 1 min of incubation, and it lasted for at least 60 min. The stimulation was completely inhibited by 100 nM N^6-(phenylisopropyl) adenosine, half-maximal effect being observed at about 10 nM concentration of the nucleoside analog. The corresponding figures for 2',5'-dideoxyadenosine were 1000-fold higher. Lipolysis was also stimulated by isoproterenol; 50% inhibition of the lipolytic effect was observed at N^6-(phenylisopropyl) adenosine concentrations that totally abolished the rise in cAMP concentration. Several adenosine analogs were tested as inhibitors of isoproterenol-stimulated lipolysis and the following order of potency was observed: N^6-(phenylisopropyl) adenosine > N^6-isopentenyladenosine > 2-chloroadenosine >adenosine > N^6-methyladenosine, 2',5'-dideoxyadenosine > 2'-deoxyadenosine. Inosine was without effect in concentrations up to 100 μM. Neither did ribose (10mM) have any effect on catecholamine-stimulated lipolysis or on its inhibition by N^6-(phenylisopropyl)adenosine. 3-Methylisobutylxanthine (100 μM) was also a good stimulator of lipolysis in human fat cells. Stimulation by this agent was also inhibited by N^6-(phenylisopropyl) adenosine and 2',5'-dideoxyadenosine.

This study was supported by the Nutrition Research Foundation of Finnish Sugar Co. Ltd and the Paulo Foundation.

64. ADENOSINE (Ado) INITIATES CORONARY RELAXATION VIA ADENYLATE CYCLASE-ASSOCIATED STIMULATORY (R_α) RECEPTORS. R.A. Olsson and S. Kusachi. USF College of Medicine, Tampa, Florida 33612.

An order of potency Ado-5'-alkylcarboxamide > L(-)-N^6-1-methyl-2-phenethyl) Ado (PIA) characterizes the R_α receptor of isolated cell membranes in vitro (Adv Cyclic Nucleotide Res. 14: 199, 1981). We employed this criterion to test the hypothesis that R receptors mediate coronary relaxation by Ado. In 5 open-chest dogs the molar potency ratio (MPR) of Ado-5'-cyclopropylcarboxamide vs Ado was 90±6 (mean±SEM). In 4 other dogs the MPR of PIA vs Ado was 5.4±0.9. PIA is neither a substrate for Ado deaminase (ADA) nor does intracoronary ADA, 5 U/min per kg, alter the vasoactivity of PIA, suggesting that the activity of PIA is not due to blockade of the uptake of endogenous Ado. Thus, these results support the hypothesis that R_α receptors initiate coronary relaxation by Ado. Further, the criteria for recognizing R_α receptors in vitro seem applicable to studies in the whole animal.

Supported by NIH H1-26611 and a Grant-in-Aid from the Suncoast Chapter, Florida AHA Affiliate.

65. INFLUENCE OF ADENOSINE DEAMINASE ON VASCULAR SMOOTH MUSCLE RELAXED BY HYPOXIA. Roland N. Pittman, Department of Physiology, Medical College of Virginia, Richmond, Virginia 23298.

In vitro strips of contracted blood vessel wall generally relax in response to decreased PO_2 in their environment. We have previously shown that addition of adenosine deaminase (ADA) to the anoxic medium bathing hog carotic artery strips leads to a reversal of the relaxation brought about by anoxia. This observation has been interpreted as evidence indicating the involvement of endogenously produced adenosine in the mediation of hypoxic vascular smooth muscle relaxation. Thus, ADA removes the strips from the relaxant influence of adenosine. Further tests have been performed to delineate the action of ADA on hypoxic vascular strips. Adenosine deaminase denatured by boiling had no effect on anoxic contractile force. If ADA produces the reversal of relaxation by converting adenosine to inosine and ammonia, it is necessary to rule out a direct effect of these latter substances on contractile force. Inosine (1 to 100 μm) produced only relaxation of anoxic hog carotid strips. Addition of NH_4OH to the anoxic bathing medium in concentrations up to 1mM did not affect contractile force. Experiments were performed in which ADA was added to anoxic bathing medium, reversing the anoxic relaxation. After contractile force attained a steady value, 100 m erythro-9-(2-hydroxy-3-nonyl) adenine (EHNA), an inhibitor of ADA, was added to the bathing medium. In most cases, EHNA reversed the action of ADA. However, in some experiments EHNA caused further contraction which could not be relaxed by 10 mM aodium nitrite, a concentration that otherwise always completely relaxed contracted strips. Pretreatment of contracted normoxic strips with the same concentration of ADA as above failed to prevent the relaxation produced by anoxia. The time course of relaxation in the presence of ADA was, however, slower than that observed in its absence. A possible explanation of this finding is that the inosine produced by deamination of adenosine accumulated during the normoxic ADA incubation period and to an even greater extent during anoxia. Dose-response relations for inosine on hog carotid arteries showed that inosine relaxed contractions with an ED^{50} (32 μM) about 18 times higher than the ED_{50} for adenosine (1.8 μM). Thus, the progressive accumulation of inosine within the strips is consistent with the slower time course of anoxic relaxation. In summary, the magnitude of the relaxation of hog carotid artery strips due to hypoxia can be accounted for on the basis of endogenously produced adenosine. The action of adenosine deaminase on hypoxic strips is consistent with this explanation.

Supported in part by PHS grants HL 18292 and HL 25383 from the NHLBI.

66. RAPID METABOLISM OF 5'-DEOXYADENOSINE BY LINES OF ANIMAL CELLS IN CULTURE. Peter G.W. Plagemann and Robert M. Wohlhueter, Department of Microbioloby, University of Minnesota, Minneapolis, Minnesota 55455.

5'-Deoxyadenosine (5'-Ado) has been reported not to be a substrate for adenosine deaminase or adenosine kinase (AK) and therfore a non-metabolizable substrate suitable for nucleoside transport measurements (Kessel, J. Biol. Chem. 253, 400, 1978). However, 5'-dAdo has been found to be an efficient substrate for 5'-methylthioadenosine phosphorylase from Ehrlich ascites tumor cells (Savarese et al., Biochem. Pharmac, 30, 189, 1981). We find that 5'-Ado is also rapidly cleaved to adenine by cell-free dialyzed extracts of Chinese hamster ovary (CHO) and Novikoff rat hepatoma cells in a phosphate dependent reaction. Radioactivity from $(5'-^3H)5'$-dAdo is also incorporated into the acid soluble pool (uptake) by whole cells as rapidly as from labeled adenosine or adenine (all at 5 μM extracellular concentration). Radioactivity in the acid soluble pool is mainly associated with a component tentatively identified as 5'-deoxyribose-1-phosphate. A second component, probably 5'-deoxyribose, is formed more slowly and also accumulates in the medium. Cleavage of 5'-Ado by whole cells is dependent on the continuous removal of the product adenine, since uptake is greatly reduced in cells deficient in adenine phosphoribosyltransferase (APRT) and 50 μM adenine strongly inhibits 5'-Ado cleavage.

Adenosine (Ado) and hypoxanthine are slightly inhibitory, presumably by competing with adenine for PRPP required for phosphoribosylation. Uridine strongly inhibits 5'-dAdo uptake, but its mode of action is less obvious. Ado is also cleaved in cell free extracts and whole cells, but less efficiently than 5'-dAdo. AK-deficient CHO or Novikoff cells incorporate (^{3}H-8)Ado into intracellular ATP at a significant rate. Incorporation is strongly inhibited by adenine, but not hypoxanthine, and is greatly reduced in double mutants lacking AK and APRT.

67. REDUCTION OF INTERSTITIAL ADENOSINE AS A TEST FOR THE ROLE OF ADENOSINE IN EXERCISE HYPEREMIA. K.G. Proctor and B.R. Duling, Department of Physiology, University of Virginia, Charlottesville, Virginia.

The purpose of this study was to evaluate the importance of adenosine (ADO) as a regulator of striated muscle blood flow during exercise. Hamster cremaster muscle small arterioles (resting diameter $< 30\mu m$) were observed with intravital microscopy during muscle contraction caused by electrical stimulation (15V; 0.04 msec; 1 Hz; 1.5 min). ADO was applied to the interstitial space surrounding arterioles by pressure ejection from micropipettes. The dilation induced by ADO was significantly cecreased when adenosine deaminase (ADA) was added to the superfusate (11 μg/ml). ADA had no observable effect on: resting arterioles; the vasoconstriction in response to 95% superfusate O_2 (minimum diameter 55% of resting); or the vasodilation in response to topical 2-chloroadenosine (10^{-6}M), a synthetic analogue which binds to the ADP receptor (maximum diameter 200% of resting). In contrast, ADA rapidly, but reversibly eliminated the vasodilation caused by topical ADO (10^{-5}M). These data suggest that ADA diffused to the vascular site of ADO action without producing nonspecific alteration of vascular responsiveness or sensitivity. Vasodilation during muscle contraction with 0% superfusate O_2 (minimum tissue PO_2 = 10 mm Hg) was 21% lower, and with 10% superfusate O_2 (minimum tissue PO_2 = 20 mm Hg) was 26% lower with ADA, compared to a control solution. ADO thus appeared to be released during exercise by a process independent of hypoxia. The quantitative relation was unchanged by decreasing the dose of ADA by one order of magnitude. Thus, the dose of ADA was probably supramaximal. The fractional reduction in the hyperemic response is likely to be a low estimate of the maximum contribution of ADO to blood flow regulation because ADO receptor affinity is probably high relative to the K_m of ADA, and because other factors, which mediate short periods of submaximal exercise hyperemia, are less prominent during prolonged or more intensive exercise.

Supported by PHS #12792 and #06234.

68. ADENOSINE AND PANCREATIC ENDOCRINE FUNCTION. G. Raberger, S. Bacher, W. Conca, and O. Kraupp. Pharmakologisches Institut der Universität Wien, A-1090 Wein, Währinger Str. 13a, Austria.

NECA (adenosine-5'-N-ethylcarboxamide) a long acting adenosine analogue with pronounced vasoactivity leads to hyperglycemia and hyperglucagonemia in intact dogs. Therefore the effect of NECA was investigated on glucagon and insulin release from the <u>in situ</u> isolated blood perfused pancreas in anesthetized dogs. NECA (10^{-9} to 10^{-5} mol/l) led to a dose-dependent increase in glucagon release. Insulin release was inhibited by NECA at low concentrations, but significantly increased at higher concentrations of the adenosing analogue. Similar effects were observed with infusion of adenosine at 10^{-7} and 10^{-6} mol/l. Aminophylline (10^{-4} mol/l) produced a 10-fold attenuation of the actions of NECA. It is concluded that adenosine receptors, which regulate hormone release, do exist in the pancreas. Physiological concentrations of adenosine should lead to a preponderance of glucagon release from the pancreas and hence to glucose mobilization from the liver.

69. BIOCHEMICAL AND ELECTROPHYSIOLOGICAL CHARACTERIZATION OF ADENOSINE RECEPTORS IN THE RAT HIPPOCAMPUS. Martin Reddington, Kevin Lee, Peter Schubert, and Georg Kreutzberg. Max-Planck-Institute for Psychiatry, Munich, FRG.

Studies on the interaction of nucleosides with adenylate cyclase have allowed two types of extracellular adenosine receptor, designated A_1 and A_2 to be distinguished. They differ pharmacologically in several respects, e.g., in relative affinities and in stereospecificity. In the central nervous system adenosine depresses synaptic transmission. Investigations of the depressive effects of various nucleoside derivatives on evoked potentials in a hippocampal slice preparation indicated participation of a receptor of the A_1 type in this response. Further, the pharmacological properties of a high affinity ^{3}H-cyclohexyladenosine binding site in rat brain membranes are similar to those of the response of hippocampal evoked potentials. Adenosine was found to be less effective in its ability to depress evoked potentials in slices from the dorsal hippocampus than in those cut ventrally, a difference which was correlated with a difference in the density of ^{3}H-cyclohexyladenosine binding sites in those regions. The results suggest that the density of A_1-adenosine receptors is important in determining the strength of adenosine action in synaptic transmission.

70. DOES THE S-ADENOSYLHOMOCYSTEINE PATHWAY CONTRIBUTE TO THE PHYSIOLOGICAL EFFECTS OF ADENOSINE IN THE CNS? Martin Reddington and Renate Pusch, Max-Planck-Institute for Psychiatrie, Munich, FRG.

The incorporation of 14C-adenosine into various metabolites was studied in a rat hippocampus slice preparation in order to assess the extent of adenosine metabolism via synthesis of S-adenosylhomocysteine (SAH), a potent inhibitor of transmethylation reactions. Highest incorporation of 14C occurred into nucleotides, with only a few percent being recovered in inosine+hypoxanthine, SAH and the free adenosine pool. Labeling of SAH did not significantly increase with higher concentrations of added adenosine despite greater accumulation of free 14C-adenosine in the tissue. Addition of L-homocysteine significantly increased the labeling of SAH. The results indicate that SAH synthesis is a minor pathway of adenosine metabolism in brain tissue under steady state conditions. Further, changes in adenosine concentration, without a concomitant increase in homocysteine availability, are unlikely to lead to a significanty accumulation of SAH. SAH is therefore not likely to play a significant role in mediating the biological effects of adenosine in the CNS via inhibition of transmethylations.

71. A NEW MECHANISM FOR ADENOSINE-INDUCED MECHANICAL UNCOUPLING. H.H. Rotmensch, C.P. Bianchi, T. Grau, S. Narayan, R.K. Ferguson, Clinical Pharmacology, Departments of Medicine and Pharmacology, Jefferson Medical College, Philadelphia, Pa.

Adenosine causes inhibition of contraction of the isolated, spontaneously beating carp atrium. Its response is characterized by a dose-dependent inhibition of contraction within seconds, spontaneous recovery of most of the contraction amplitude within 10-15 minutes and rebound potentiation of contraction on washout. To evaluate the relation between the negative inotropic effect of adenosine and calcium exchange, the residual (R) and strontium extractable (Sr Ex) Ca^{2+} in the atrial tissue were determined at peak suppression of contraction. In contrast to previous reports, peak suppression of contraction (adenosine 10^{-4}M) coincided with significant increases in both R Ca^{2+} from 0.50 ± 0.07 to 0.80 ± 0.09 μmol/g tissue and Sr Ex Ca^{2+} from 1.69 ± 0.15 to 2.57 μmol/g ($n=7$, $p < 0.05$). We, therefore, conclude that in the carp atrium adenosine-induced negative inotropy is not due to inhibition of calcium entry but rather to inhibition of "triggered" calcium release. The spontaneous recovery of inotropy and rebound on washout could be explained by the accumulation of intracellular Ca^{2+} and subsequent reversal of the inhibition of "triggered" Ca^{2+} release.

72. EFFECTS OF AMP DEPENDENT ADENOSINE TRANSPORT INHIBITION ON MYOCARDIAL POSTANOXIC MECHANICAL FUNCTION AND ADENINE NUCLEOTIDES. Michael J. Rovetto and David O. Williams, II, Department of Physiology, University of Missouri School of Medicine, Columbia, Missouri 65212.

Adenosine (ADO) uptake into myocardial cells occurs via two processes: one, a free-adenosine facilitated carrier system and the second, a system which transports Ado derived from AMP (AMP-Ado carrier). The latter system has a higher velocity and lower Km than the first. The two carriers can be distinguished by use of specific inhibitors of each. The studies determined whether anoxic cellular Ado efflux could be decreased by inhibitors of AMP-Ado carrier mediated Ado uptake. The inhibitor studied was concanavalin A which is not thought to cross cell membranes. Concanavalin A and Ado inhibited Ado efflux in anoxic tissue. ATP levels were higher and recovery of post-anoxic mechanical function was greater than in hearts perfused with Ado during anoxia and control anoxic hearts. After the longest anoxic period studied (2 hr) 100% of the hearts perfused with concanavalin A recovered, and mechanical function and ATP concentrations were about 78% of the pre-anoxic values. In contrast, no hearts perfused with Ado recovered, and ATP concentrations were reduced 60% from aerobic values. The data support the hypothesis that the facilitated carrier system is primarily for extracellular Ado salvage, whereas the AMP-Ado carrier system is utilized for rapid cellular Ado efflux. Such efflux can occur without formation of free intracellular Ado. Furthermore, inhibition of AMP-dependent Ado efflux during anoxia facilitates recovery of ATP and mechanical function following anoxia.

Supported in part by American Heart Association, NIH Grant #HL27336 and HL00922.

73. ADENOSINE AND THE CONTROL OF RECTAL GLAND SECRETION DURING VOLUME REGULATION IN ELASMOBRANCH FISHES. R. Rubio and D. Erlij, Department of Physiology, University of Virginia, Charlottesville, Virginia, and MDIBL, Salsbury Cove, Maine, and SUNY, Downstate Medical Center, Brooklyn, New York 11203.

Extracellular volume expansion markedly increases the rate of secretion by the rectal gland of the dogfish. We have analyzed the role of adenosine in this response in studies involving anesthetized dogfish or perfused isolated glands. The isolated perfused gland is markedly stimulated to secrete by adenosine. Threshold concentrations are between 10^{-7} and 10^{-6}M. An analysis of the effects of a number of adenosine analogues indicates that adenosine acts by combining with an extracellular receptor of type Ra which is usually associated with activation of adenylate cyclase. In further agreement with this notion is the finding that incubation with adenosine increases cyclic AMP content in gland slices. In experiments with anesthetized fish we have determined that during volume expansion the content of high energy phosphate compounds in the gland is markedly reduced; however no increase in adenosine release into the venous outflow was detected. However, when we stimulated isolated perfused glands with veratridine, an agent that selectively depolarizes nerve terminals, there was an increased rate of adenosine release into the venous effluent during the increase in secretion. These results suggest that adenosine could play an important role in regulating the response of the rectal gland to volume expansion.

Supported by the N.Y. Heart Association.

74. THE USE OF A "MICRO"ADENOSINE DEAMINASE (MADA) TO ELUCIDATE FUNCTIONAL SIGNIFICANCE AND TISSUE DISTRIBUTION OF ADENOSINE. R. Rubio, R.M. Knabb, S.W. Ely, and R.M. Berne, Department of Physiology, University of Virginia, Charlottesville, Virginia 22908.

It is important to demonstrate: 1) A cause and effect relationship for the vasoactive effects of interstitial adenosine (ADO_i) and 2) What fraction of the total tissue adenosine (ADO_t), ADO_i represents. In order to approach these two important issues, we reasoned that if tissues were exposed to a very active MADA it would readily penetrate the interstitium because of its small molecular size. If ADO_i is deaminated as rapidly as it is produced, its vasoactive effects would vanish, and the fraction of ADO_i/ADO_t could be calculated. However, if MADA were not small and active enough, partial effects could be observed. We have prepared MADA (MW $\leq$ 4,000-25,000) by treating commercial ADA with pronase. In experiments on isolated guinea pig hearts perfused at constant pressure and subjected to hypoxia (PO_2=80 torr) coronary flow (CF) increased by 222% but only half this flow increase occurred when the perfusion fluid contained MADA. These results indicate a cause and effect relationship between ADO_i and Cf. The failure to totally abolish the hypoxia-induced flow increment with MADA may be a result of the release of other vasodilators, enhanced release of ADO in MADA-treated hearts (because of a greater degree of hypoxia) or that the MADA is either not small enough or not active enough to inactivate all of the ADO released.

75. ATP AND ADENOSINE CAUSE EXCITATORY OR INHIBITORY RESPONSES IN THE GUINEA-PIG TRACHEALIS MUSCLE: STUDIES ON MECHANISM OF ACTION. D.G. Satchell, Department of Zoology, University of Melbourne, Parkville 3052, Australia.

Transverse strips of the trachea usually relax in response to ATP and adenosine although occasional contractions in response to these compounds were observed.

Spirally cut strips of trachea contract in response to ATP and adenosine, although in preparations contracted with carbachol relaxations were observed. Small but significant reductions in the contractions to both compounds were achieved by treatment with either hyoscine (1.3 μM) or theophylline (4.0 μM). In the presence of dipyridamole (0.5 μM)ATP and adenosine failed to cause contractions and caused marked relaxations. Since dipyridamole inhibits the uptake of adenosine into cells and since ATP is rapidly metabolized to adenosine during its contact time, it is likely that adenosine receptors mediate the contractile responses and are located at an intracellular site. This is in contrast to the receptors mediating relaxation which are located extracellularly.

It is not clear why the usual response to ATP and adenosine is excitatory in the spiral strip and inhibitory in the transverse strip, although it may be related to the tone of the preparation or to differential actions in the transverse oblique or longitudinal components of the trachealis muscle.

76. SEPARATE RECEPTORS FOR ATP AND ADENOSINE IN THE GUINEA PIG TAENIA COLI. David G. Satchell and M. Helen Maguire, Department of Zoology, University of Melbourne, Australia 3052 and Department of Pharmacology, Toxicology and Therapeutics, University of Kansas Medical Center, Kansas City, Kansas 66103.

The question of separate inhibitory receptors for ATP and adenosine in the guinea pig taenia coli has been addressed by comparing the effects of similar structural modifications to ATP and adenosine on their inhibitory actions in the isolated guinea pig taenia coli preparation, both in the absence and presence of dipyridamole. The structural modifications were 8-bromo substitution, 2'-hydroxyl reduction and 2'-hydroxyl epimerization. 8-Bromoadenosine and ara-adenosine were inactive at 1 mM; 2'-deoxyadenosine was a partial agonist, pD_2 3.34. 8-Bromo-ATP, ara-ATP and 2'-deoxy-ATP induced rapid ATP-like relaxations with pD_2s 5.66, 5.60 and 5.76, respectively. For ATP and adenosine pD_2s were 6.10 and 4.67. Dipyridamole, a known potentiator of ATP and adenosine, did not potentiate the actions of 8-bromo-ATP, ara-ATP or 2'-deoxy-ATP. If one inhibitory receptor mediated responses to both ATP and adenosine, then similar structural modifications in ATP and adenosine should have parallel effects on their agonist properties. Non-parallel effects were observed, supporting the view that the taenia coli has separate receptors for ATP and adenosine.

Supported by grants from NHMRC and ARGG, Australia, and BRSG Fund, University of Kansas Medical Center.

77. BEHAVIORAL CORRELATES OF ADENOSINE STIMULATED (A2) CYCLASE ACTIVITY IN RAT CEREBRAL CORTEX. Albert Sattin, Institute of Psychiatric Research, Indiana University School of Medicine and V.A. Medical Center, Indianapolis, Indiana 46223.

Previous work with mice has suggested that strains that accumulate larger amounts of cA in cortex tissue in response to a maximal dose of adenosine manifest lesser degrees of locomotory activity when placed in a strange environment (Sattin, A., in NEUROPHARMACOLOGY OF CYCLIC NUCLEOTIDES, G.C. Palmer, Ed., Urban & Schwartzenberg, 1979). Since electroconvulsive shock (ECS) increases this effect of maximal adenosine in cerebral cortex of rats it was predicted that the locomotory activity of ECS rats would be diminished relative to sham ECS rats. Male S-D rats, 160 g were given 22-30 millicoulombs via ear clips and resultant grand mal seizures at 16:00 hrs on alternate days x 5, then behaviorally tested 1 day after the last ECS and the cortex slices tested with adenosine *in vitro* 2 days after the last ECS, all compared to sham ECS rats. Using a photocell array (Columbus Instruments) total 20 min locomotory scores were 917$\pm$91 (SEM) vs 352$\pm$57 for sham ECS vs ECS, respectively, confirming the prediction. 200 μM adenosine augmented the cA content 25% more in ECS than sham ECS tissue. In parallel groups of rats this augmented response to adenosine following ECS appeared to be lost when the tissue incubations were done later than 16:00 hours.

Supported by MH-31137 and AAMHR&E.

78. EFFECTS OF ADENOSINE (AD), N^6-PHENYLISOPROPYLADENOSINE (PIA) AND 2'-5'-DIDEXOYADENOSINE (DID) ON FORCE OF CONTRACTION (FC) AND CYCLIC NUCLEOTIDES IN THE MAMMALIAN HEART. W. Schmitz, A. Fenner, I. Hackbarth, W. Meyer, T.-M. Nobis and H. Scholz, Abteilung Biochemische Pharmakologie, Medizinische Hochschule Hannover, D-3000 Hannover 61, FRG.

The inhibitory action of AD on adenylate cyclase activity and on lipolysis in fat cells is presumably mediated through two different receptors, classified as external ribose site (R site) and internal purine site (P site). In order to elucidate whether this model holds true for the adenosine mediated mechanical effects on the heart we investigated the effects of AD, the R site agonist PIA and the P site agonist DID on FC, on AC in a broken cell preparation, and on cAMP and cGMP content in intact working left auricles from guinea pigs. AD decreased FC. A similar decrease in FC was seen with PIA. DID had no effect at all. AC in broken cell preparations was inhibited only by

AD and DID, whereas PIA had no effect. In the intact working auricles none of the substances changed the cAMP or cGMP levels.

From the results it is evident that the mechanical and biochemical effects of AD and its derivatives do not at all correspond to each other so that the decrease in FC is obviously not mediated by changes in cyclic nucleotide contents. It seems likely that the proposed cardiac R site adenosine receptor is not coupled to the adenylate cyclase, at least in guinea-pig auricles.

79. EVIDENCE FOR R_a -SITE ADENOSINE RECEPTORS IN RAT BRAIN MICROVESSELS.
W. Schütz, G. Steurer, and E. Tuisl, Department of Pharmacology, University of Vienna, Vienna, Austria.

Adenosine (AR) has been described as an adenylate cyclase (AC) stimulator in guinea pig brain microvessels (Huang and Drummond: Mol. Pharmacol. 16:642, 1979). It was explored in the present study whether this effect fulfills the criteria required for AC activation via an external receptor, recently classified as R_a-site AR receptor. Rat cerebral microvessels were isolated and determined for AC activity, using deoxy ATP as the enzyme substrate. Several AR receptor agonists stimulated AC in a GTP-dependent manner. The rank order of potency 5'-N-(ethylcarboxamide)-AR (NECA, ED_{50}, 0.2 μM) > AR(0.71 μM) $\geq$ 2-chloro-AR (0.72 μM) > N^6-(L-phenylisopropyl)-AR (L-PIA, 1.03 μM) > D-PIA (5.27 μM) is consistent with R_a receptor agonism in other tissues. The stimulatory action of NECA was competitively antagonised by methylxanthines. These findings provide functional evidence for R_a-site AR receptors in rat cerebral microvessels.

However, direct identification of these receptors proved not possible. Binding of ^{3}H-labeled NECA to rat brain microvessels displayed rapid on-off kinetics and saturability, but equivocal specificity.

80. EXTRACELLULAR ADENINE NUCOEOTIDE METABOLISM IN CULTURED ARTERIAL CELLS: RELATIONSHIP TO CELLULAR AND EXTRACELLULAR cAMP.
Linda L. Slakey, Samuel J. Goldman, Ellen Gordon, and Ellen S. Dickinson, Department of Biochemistry, University of Massachusetts, Amherst, Massachusetts 01003.

We have examined the time course of the sequence of reactions, ATP → ADP → AMP → Adenosine in the extracellular medium of cultured endothelial cells and cultures smooth muscle cells from swine aorta, and examined as well the effect of adenosine on accumulation of cAMP by the two cell types.

Both cell types respond to adenosine by elevating cAMP levels. cAMP is found both intracellularly and released into the incubation medium. The rate of release was found to be first order with respect to intracellular cAMP. Adenosine at 160 μM caused an increase in the first order rate constant for release of cAMP, as well as causing an increase in intracellular cAMP. Isobutyl-methylxanthine blocks the adenosine stimulated release.

Since 5'-nucleotidase, the last enzyme in the sequence of hydrolysis of ATP to adenosine is inhibited by ADP and ATP, we predicted that after giving a bolus of ATP there would be a lag in the production of adenosine until most of the ATP and ADP were hydrolyzed. Such a lag occurs with both cell types if the initial concentration of ATP is at least 500 μM. Thus the kinetic properties of the ectonucleotidases act together with specific cellular responses to determine the rate and extent of cellular response to a bolus of extracellular ATP.

81. EVIDENCE THAT A PURINE AND CATECHOLAMINE ACT AS CO-TRANSMITTERS IN THE MOTOR RESPONSE OF THE GUINEA PIG VAS DEFERENS TO FIELD STIMULATION.
P. Sneddon, D.P. Westfall, J.S. Fedan, Department of Pharmacology, WVU Medical Center, Morgantown, WV 26506.

The mechanical response of the guinea-pig vas deferens to field stimulation is biphasic: the second phase is blocked by the classical α-antagonists, such as prazosin, whilst the initial, large twitch is resistant to adrenergic antagonists but is blocked by the antagonist $ANAPP_3$, which has been shown to be a specific antagonist of the P_2 type purine receptor (Hogaboom, et al., Science 208, 1975, 1980). We have performed experiments, using intracellular microelectrodes, to test the hypothesis that $ANAPP_3$ blocks the twitch response because it blocks the underlying E.J.P.s. In control tissues, E.J.P.s of up to 10-15 mV could be recorded when the tissue was stimulated at 0.5, 1.0 or 2 Hz (0.5 m sec, supramaximal voltage). Tissues were then pretreated with $ANAPP_3$ (10^{-4}M) and photolysed for 20 mins, and then washed in normal Krebs solution and transferred to the recording bath. In these tissues the E.J.P.s were much smaller than in control cells, usually being less than 2-5 mV. Tissues photolysed for 20 mins without $ANAPP_3$ served as controls; in these tissues E.J.P.s of 10-15 mVs could be obtained as usual. Prazosin, 10^{-7} or 10^{-6}M, did not antagonize the E.J.P. in any of the cells recorded from. It is concluded that both a purine and a catecholamine act as motor transmitters in this muscle.

82. THE CORRELATION BETWEEN ADENOSINE AND ACETATE DURING SKELETAL MUSCLE HYPEREMIA.
R.P. Steffen, J.E. McKenzie, E.L. Bockman, and F.J. Haddy, Department of Physiology, Uniformed Services University, Bethesda, MD 20814

Several studies support the hypothesis that adenosine participates in skeletal muscle exercise hyperemia. We have proposed that acetate also participates in exercise hyperemia. We studied the effect of simulated exercise on acetate and adenosine content in vascularly-isolated dog gracilis muscles at natural flow. The muscles were stimulated to contract isometrically at 0.5, 1.0, or 2.0 Hz for 10 min (n=6 each). Acetate and adenosine were measured in the same muscles. The change in tissue acetate content and vascular resistance correlated negatively ($r=-.75$, $P < 0.001$) as did tissue adenosine content and vascular resistance ($r=-.57$, $P < 0.001$). During local i.a. acetate infusion in the quiescent naturally-perfused dog gracilis muscle, acetate and adenosine contents increased and vascular resistance decreased. Acetate and adenosine contents correlated inversely with resistance (r=-.83,

$P < 0.001$ and r=-.60, $P < 0.005$, respectively) and correlated positively with one another (r=.66, $P < 0.05$). Adenosine levels were comparable to those during exercise. The increase in skeletal muscle acetate content that occurs during exercise may be involved in the biochemical process triggering the increase in tissue adenosine content associated with exercise hyperemia. Further work is needed to identify the mechanism through which changes in endogenous acetate may be linked to adenosine during exercise hyperemia.

83. AUTORADIOGRAPHIC LOCALIZATION OF ^{3}H-ADENOSINE. C.E. Stirling, Department of Physiology & Biophysics, University of Washington, Seattle, Washington.

The purpose of this study was to examine the cellular distribution of ^{3}H-adenosine and its metabolites in heart and retina. Fifty μCi of ^{3}H-adenosine (10Ci/mmole) was injected into the ascending aorta of frogs. After thirty minutes the retina with pigmented epithelium and choroid were removed and processed for freeze-dry autoradiography (Harms and Stirling, 1977, Am. J. Physiol. 233:E47). Very little activity was observed in the retina. However, a significant accumulation of label was observed in vitreal and choroid capillaries and in the pigmented epithelium.

Similar experiments with hamsters showed that approximately 70%, 10% and 20% of the injected adenosine (100 μCi, left ventricular puncture) was taken up by the lungs, kidneys, and remaining tissues. Autoradiography revealed an uptake pattern identical to that in the frog: little or no entry into the retina and brain and heavy activity in the choroid and vitreal capillaries and pigmented epithelium. Cardiac muscle from the hamsters was examined autoradiographically, also. A very low level of activity was observed over the muscle fibers while a high level of activity was found in the coronary vessels. These results indicate that vascular endothelial cells possess an efficient transport mechanism for removing adenosine and its metabolite from the circulation.

In another series of experiments, excised frog retina was incubated 10 minutes in 1 μM ^{3}H adenosine. Autoradiography showed a heavy uptake of ^{3}H-label over the outer plexiform and proximal photoreceptor layers. Similar experiments with hamsters revealed a different pattern. Accumulation of label was heaviest in the inner plexiform and inner nuclear layers, particularly by amacrine cells.

84. PURINE RELEASE IN THE BLOOD VESSEL WALL. C. Su and T. Katsuragi, Department of Pharmacology, So. Illinois University School of Medicine, P.O. Box 3926, Springfield, Illinois 62708.

Characteristics of drug-induced purine release in the blood vessel well were studied. Helical strip of the rabbit main pulmonary artery was preincubated with 0.1 μM ^{3}H-adenosine for 1 hr and superfused with Krebs solution. KCl (30-70 mM) elicited a ^{3}H efflux and a contraction which were unaffected by phentolamine; only the efflux was greatly reduced by removal of Ca^{2+} and denervation. ℓ-Epinephrine (5 μM) also elicited these responses which were unaffected by denervation but blocked by phentolamine. Thus KCl released purines mainly from adrenergic nerves and epinephrine from extraneuronal sites mediated by α-receptors.

The KCl-induced purine efflux was inhibited by adenosine (1,3 μM). Theophylline (10, 100 μM) and clonidine (30, 100 μM), blockers of adenosine receptors, augmented the efflux caused by KCl but not epinephrine. These receptors may mediate the autoinhibition of neuronal purine release. Clonidine (30 μM) contracted the artery without elevating the efflux. Ouabain (100 μM) evoked a ^{3}H-purine efflux which was reduced by 6-hydroxydopamine or phentolamine, indicative of both neuronal and extraneuronal origins of release. In two-cm intralobular sections of the rabbit pulmonary artery pretreated with ^{3}H-adenosine, acetylcholine (0.3, 1 μM) caused a small efflux increase. However, the purine release was probably insufficient to account for ACh vasodilator effect as the latter was unaltered by theophylline or dipyridamole, an inhibitor of adenosine uptake. These results suggest that the origins of purine release vary with the stimuli, and drugs can initiate extraneuronal purine release by mechanisms not necessarily related to smooth muscle membrane depolarization or contraction.

Supported by NHLBI HL 24683 and AHA-Illinois Affiliate.

85. ON THE POSSIBLE MECHANISM UNDERLYING THE INHIBITORY ACTION OF ADENOSINE /AR/ IN ATRIAL MYOCARDIUM OF GUINEA PIGS. A.J. Szentmiklosi, T. Kovacs, Agnes Cseppento, and J. Szegi, Departments of Pharmacology and Physiology, University Medical School of Debrecen, H-4012, Debrecen, Hungary.

The contribution of changes in K permeability and in activation of intracellularly localized AR receptors / e.g. P sites/ to the negative inotropic action of AR was assessed by applying isolated, electrically driven left atria. AR increased the rate constants for ^{42}K efflux and reduced the tissue K content. 4-Aminopyridine, a specific inhibitor of K channels, antagonized the AR-induced decrease of the isometric tension.

AR deaminase/ADA/ activity could be blocked by 7 μmol/l coformycin /CF; by 92-95%/ or hypoxia / by about 60%/ in intact atrial tissues. 2'-DeoxyAR /P site effector; 100 μmol/l/ decreased the contractile force only slightly by about 8%/, but it strongly reduced the isometric tension in the presence of CF / by about 65%, when intracellular ADA was inhibited. However, the cardiodepressive action of 2-chloroAR /R site effector was not modified by CF.

It is concluded that the increase in K permeability could shorten the duration of action potential, therefore it might prevent the slow inward Ca^{2+} current from running its normal time course. Under hypoxia/ partial inhibition of ADA activity/ a higher quantity of endogenous AR/ which is possibly present also in the intracellular space/ would be available to act on P receptors, the actiation of which might inhibit the phosphorylation of protein constituent of slow channels.

86. POSSIBLE REGULATION OF RENIN RELEASE BY INTRARENAL ADENOSINE. C.I. Thompson and W.S. Spielman, Department of Physiology, Michigan State University, College of Human Medicine, East Lansing, Michigan 44824.

The intra-renal infusion of adenosine results in suppression of renin release. The effects of adenosine on renin release can be blocked by the competitive antagonist of adenosine, theophylline. Similar to the effects of adenosine, the intrarenal infusion of hypertonic saline in sodium-depleted dogs resulted in a significant decrease in renin release (1248±148 to 288 ±74 ngAI/min). During theophylline infusion (5 μmol/min), however, hypertonic saline produced no significant change in renin release (1709±602 to 1907±724 ngAI/min). With either adenosine or hypertonic saline infusion, there was a decrease in GFR which was blocked by theophylline. These results suggest that intra-renal, tubular formation of adenosine during hypertonic saline infusion may be responsible for the decrease in renin release. To test the hypothesis that adenosine acts at a tubular site, we altered the tubular delivery of adenosine using two preparations. In the non-filtering kidney, adenosine infusion (0.3 μmol/min) resulted in a significant decrease in renin release (309±53 to 71±26 ngAI/min), and elicited a typical biphasic flow response (constriction followed by dilation). During acute ureteral pressurization (70 cm H_2O), adenosine again produced a decrease in renin release (5903±1923 to 312.0±146 ngAI/min) but resulted in a sustained vasoconstriction (145±20.7 to 79±18 ml/min). These combined results suggest that tubular delivery of adenosine is not necessary for the decrease in renin release. Furthermore, the typical biphasic effect of adenosine on renal hemodynamics can be separated into a constriction and a dilation and suggests that there are at least two functional receptors mediating the hemodynamic effects of adenosine in the kidney.

Supported by HL01010.

87. EFFECT OF AMINOPHYLLINE ON STEADY STATE EXERCISE OF SKELETAL MUSCLE.
Loren P. Thompson and Harvey V. Sparks, Jr., Department of Physiology, Michigan State University, East Lansing, Michigan 48824.

We observed the effect of aminophylline (AMINO) on steady state exercise hyperemia of hindlimb muscles of dogs. Muscles were vascularly isolated and perfused at constant pressure of 100 mmHg. Arterial inflow (F) and oxygen consumption (VO_2) were measured during rest and sustained twitch rates of 3 Hz produced via nerve stimulation. Hyperemic responses to intraarterial infusions of adenosine (ADO) (≃10 μM) and ATP (≃ 1 μM) were measured during rest and sustained exercise. We infused AMINO (≃ 10 mg/kg) into the muscle until we had blocked the flow response to ADO. Following the AMINO infusions resting flow returned to control value. We then repeated the above measurements during rest and exercise periods along with ADO and ATP infusions. AMINO reduced the flow response to ADO by 92±1% during rest and 90± 7% during exercise but did not significantly reduce the flow response to ATP. After blockade of ADO receptors, resting F and VO_2 were not significantly different from control. Before AMINO, resting F and VO_2 were 12.7±3.3 and 0.3±0.1 (ml/min/100g) whereas during AMINO, F and VO_2 were 11.1±4.4 and 0.4±0.1 (ml/min/100g), respectively. F and VO_2 during exercise in the presence of AMINO were significantly different from control values. Before AMINO, F and VO_2 during 3Hz exercise were 89.2±15.7 and 9.4±1.3 (ml/min/100g), respectively. After AMINO, F during exercise was reduced to 78.2±13.4 and VO_2 was increased to 10.0±1.3. These results indicate that blockade of ADO receptors during 3Hz exercise reduces the hyperemic response even in the presence of an increased VO_2. This suggests that ADO is responsible in part for steady state exercise hyperemia.

Supported by USPHS NIH Grant #25779.

88. BINDING OF ADENOSINE TO INTRACELLULAR S-ADENOSYLHOMOCYSTEINE HYDROLASE IN ISOLATED RAT HEPATOCYTES. P.M. Ueland and S. Helland, Department of Pharmacology, University of Bergen, School of Medicine, MFH-bygget, 5016 Haukeland sykehus, Bergen, Norway.

Binding of adenosine to S-adenosylhomocysteine (AdoHcy) hydrolase (EC 3.3.1.1) and partial conversion of bound adenosine to a substance liberating adenine have been demonstrated under conditions of enzymatic synthesis and hydrolysis of AdoHcy (JBC 255:772, 1980). Gel filtration of cytosol from isolated rat hepatocytes treated with (^{14}Cl) adenosine on a HPLC protein column showed that labeled adenine/adenosine eluted as a peak which co-chromatographed exactly with AdoHcy hydrolase. Formation of this peak was inhibited by exposure of the cells to compounds (ara-A, 3-deazaadenosine or homocysteine) interacting with the catalytic site of the enzyme. On this basis it was concluded that labeled adenosine formed a stable complex with AdoHcy hydrolase. A substantial portion (about 50%) of endogenous adenosine in rat hepatocytes seemed to be associated with AdoHcy hydrolase, and this portion equaled the amount of cellular adenosine which was not mobilized by high level of extracellular adenosine deaminase. Exposure of the hepatocytes to compounds which block the formation of the adenosine-AdoHcy hydrolase complex (ara-A, 3 deazaadenosine or homocysteine) for 1 to 2.5 hours only slightly reduced the amount of adenosine associated with the enzyme, indicating a slow turnover of the complex under the conditions of the experiment. It was concluded that adenosine is sequestered in rat hepatocytes through the interaction with AdoHcy hydrolase, and this enzyme functions as an adenosine binding protein in intact rat hepatocytes.

89. (-)N^6-(^{3}H)PHENYLISOPROPYLADENOSINE BINDING IN RAT FAT CELLS: DIFFERENCES BETWEEN INTACT CELLS AND CELL MEMBRANE PREPARATIONS. D. Ukena and U. Schwabe, Institut für Pharmakologie und Toxikologie der Universität Bonn, 5300 Bonn 1, Fed. Rep. Germany.

Binding of (-)N^6-(^{3}H)phenylisopropyladenosine ((^{3}H)PIA) in rat fat cell membrane preparations is not affected by the adenosine uptake inhibitor dipyridamole (100 μM), but inhibited by 63% in intact isolated fat cells. Therefore, all experiments in intact cells were performed in presence of dipyridamole. Specific binding of 5 nM (^{3}H)PIA in intact cells was rapid and more rapidly reversible than in membrane preparations. Steady state binding dissociated completely within 2 minutes after addition of 100 μM PIA, whereas in membrane preparations less than 50% dissociated within 15 minutes. Radioligand binding was stereospecific since the (-)isomer of PIA was 300-fold more potent than the (+) isomer. In contrast to membrane preparations saturability of specific binding in intact cells was not achieved at concentrations up to 1 μM (^{3}H)PIA. The IC_{50}-value for (-)PIA was 51 nM in intact fat cells and 13 nM in cell membrane preparations. (^{3}H)PIA was displaced in intact cells by 2-chloro-adenosine (IC_{50} 95 nM) and theophylline (IC_{50} 9 μM), whereas adenine, inosine and GTP did not substantially inhibit binding. The results allow to describe several characteristics of (^{3}H)PIA binding in intact fat cells but a considerable component of nonreceptor binding impedes a detailed study of physiologically important adenosine receptors.

90. CARDIOPROTECTIVE EFFECT OF ADENOSINE, LIDOFLAZINE, AND R 51 469.
H. Van Belle, R. Xhonneux, M. Borgers, and W. Flameng*, Janssen Pharmaceutical Research Laboratories, B-2340 Beerse, Belgium and *Department of Cardiovascular Surgery, Catholic University Leuven, B-3000 Leuven, Belgium.

Adenosine, lidoflazine and R 51 469, a new analogue of lidoflazine, significantly protect (at 1 x 10^{-6}M) isolated rat hearts against the damage induced by a submaximal calcium paradox. Inosine and dipyridamole were ineffective. Analysis of the results of 63 analogues of lidoflazine shows no correlation between the protective effect against the calcium paradox and the inhibition by these drugs of the uptake of adenosine in human erythrocytes. The latter effect is strongly correlated with the *in vivo* increase in coronary flow in dogs.

When given orally (2.5 mpk - 2 hr before the experiment), R 51 469 is equiprotective to lidoflazine (1.25 mpk i.v.) against the damage produced by 1 hr of normothermic global ischemia. The mechanism(s) involved in the protection remain to be elucidated.

91. CHRONIC CAFFEINE TREATMENT ENHANCES CARDIOVASCULAR RESPONSIVENESS TO CIRCULATING ADENOSINE *IN VIVO*, R.W. von Borstel and R.J. Wurtman, Laboratory of Neuroendocrine Regulation, MIT, Cambridge, MA 02139.

Adenosine (Ado) administration elicits cardiovascular effects such as bradycardia, vasodilatation, and hypotension. These actions of Ado are blocked by the methylxanthines caffeine and theophylline, and are potentiated by agents that block tissue uptake of Ado; thus, the actions of Ado on the cardiovascular system are presumably mediated by specific extracellular receptors similar to those found in brain. Ado is present in plasma and tissues, but few reliable data are available on actual Ado levels *in vivo*, and it is not known how much Ado is needed to exert significant physiological effects in unanesthetized animals. We have developed a method for assessing plasma Ado, involving rapid mixing of samples with dipyridamole and estimation of Ado by HPLC. Basal levels in arterial plasma are 3.0 $\pm$ 0.8 μM), with a range of 1.6-4.6 μM. To assess their physiologic significance, we correlated arterial Ado during intravenous Ado infusion with mean arterial pressure (MAP) in awake, unrestrained rats fitted with arterial and venous catheters. Steady-state Ado concentrations of 4, 6, and 8 μM were associated with 5, 17, and 33% drops in MAP. Acute caffeine treatment (15 mg/kg i.a. 5 min before testing) markedly attenuated the effect of Ado infusion on MAP. However, among rats drinking a caffeine solution (0.1%) for 3 weeks then caffeine-free water for 24 hours, an enhancement of cardiovascular sensitivity to Ado was seen; plasma Ado levels of 4, 6, and 8 μM caused MAP drops of 32, 41, and 45%. Thus, the rat cardiovascular system seems to respond to long-term Ado receptor blockade (by caffeine) by undergoing a compensatory increase in sensitivity to Ado. Normal variations in circulating Ado may thus have physiologic effects. Moreover, some sequelae of caffeine withdrawal may reflect enhanced responsiveness to Ado.

Supported in part by ILSI.

92. PURINERGIC RECEPTORS FOR ATP IN PERIPHERAL CIRCULATION. Richard P. Watts, Department of Physiology, University of Arizona College of Medicine, Tucson, Arizona 85724.

It is thought that the actions of adenosine triphosphate (ATP) and adenosine are mediated by purinergic receptors. It is questionable whether separate receptors exist for ATP or if its actions are due to its breakdown to adenosine. Three approaches to this problem have been taken. In one series of experiments, adenosine, ATP, β,γ--methyleneadenosine triphosphate (ACP), or β,γ-imidoadenosine triphosphate (ANP), when applied to the surface of an exteriorized hamster cremaster muscle, proved to be potent vasodilators with nearly identical effects. Because ACP and ANP are not broken down, this suggests that ATP breakdown is not necessary for its action. In a second series of studies, the binding constant (K_D) of ATP to isolated femoral arteries was determined at 4°C. K_D values of 29nM, 21nM, and 11 nM for hamster, rat, and cat, respectively, were determined by the equilibrium method whereas the rate method gave a K_D=1.3nM for cat. Despite this discrepancy, these data indicate a high affinity receptor for ATP does exist. A third approach using isolated mesentary arteries mounted on a tension transducer was also tried. Added to the bath, adenosine caused no change in tension. However, ATP, ACP, or ANP caused a rapid phasic contraction. When the arteries were precontracted with norepinephrine, adenosine caused relaxation while ATP, ACP, or ANP caused a further contraction followed by relaxation. These opposing actions argue strongly for separate receptor mechanisms.

This work supported by a Grant-in-Aid from the American Heart Association, Arizona Affiliate.

93. EFFECTS OF PURINES ON THE A-V NODE OF ISOLATED GUINEA-PIG HEARTS. A. West, L. Belardinelli, and R.M. Berne, Cardiology Division and Department of Physiology, University of Virginia, Charlottesville, VA.

Adenosine (ADO) is known to cause atrioventricular (A-V) block by slowing conduction through the A-V node, thereby lengthening the atria to His (A-H) interval. It was tested whether or not ATP and its less hydrolysable analog β,γ-methylene ATP also cause A-V block and prolong the A-H interval. Isolated guinea-pig hearts (n=9) were perfused at a constant flow (4-5 ml/min per g) with Krebs-Hensleit solution gassed with 95% O_2+5%CO_2 (35°C, pH 7.4). Extracellular electrograms were recorded from left atrium, His bundle and left ventricle. The overall control (pre- and post-intervention) for the A-H (msec) was 32.6$\pm$1.0 msec ($\bar{X}\pm$SEM).

	A-H Prolongation (msec)		
Concentration (M)	ATP Analog	ATP	ADO
$7x10^{-6}M$	4	8	27
$2x10^{-5}M$	7	20	A-V block

In three out of eight hearts $2x10^{-5}M$ ATP caused second-degree A-V block. ADO ($7x10^{-6}M$) caused second degree A-V block in two out of seven hearts and complete A-V block in all hearts at $2x10^{-5}M$.

In addition, the effects of ATP were markedly potentiated by nitrobenzylthioinosine, a nucleoside transport blocker. Conversely, aminophylline, a competitive antagonist of ADO-induced A-H prolongation, antagonized the ATP-induced A-H prolongation. In summary, these preliminary data show that in contrast to ATP, the β,γ-methylene ATP which is more resistant to hydrolysis than ATP has essentially no effect on the A-V transmission. Thus, it is reasonable to suggest that ATP-induced prolongation of A-H interval and A-V block is a consequence of its degradation to ADO.

94. ACETYLCHOLINE RELEASES ATP FROM ISOLATED MYENTERIC VARICOSITIES.
Thomas D. White and M. Al-Hummayd, Department of Pharmacology, Dalhousie University, Halifax, N.S., Canada B3H 4H7.

It has been suggested that ATP might be a non-adrenergic inhibitory neurotransmitter in vertebrate intestinal smooth muscle. Recently, we described a Ca^{2+}-dependent release of ATP when varicosities isolated from the myenteric plexus of guinea pig small intestine were depolarized with veratridine or elevated extracellular K^+. We report here that nicotinic agonists also release ATP from myenteric varicosities. A P_2 preparation of myenteric varicosities was isolated from the longitudinal muscle of the small intestines of 3 guinea pigs and preincubated for 30 min at 37°C in Krebs-Henseleit-bicarbonate medium. Release of ATP was detected directly and continuously by monitoring the light produced when ATP reacted with firefly luciferin-luciferase present in the medium. Acetylcholine and nicotine released ATP from isolated myenteric varicosities, whereas the muscarinic agonist, bethanecol, did not. The presence of bethanecol had no effect on the subsequent release of ATP by acetylcholine. Release of ATP by acetylcholine was diminished when Ca^{2+} was omitted from the medium and blocked when 1 mM EGTA was added to chelate residual Ca^{2+}. The EC_{50}'s for acetylcholine and nicotine were $4x10^{-6}M$ and $8x10^{-6}M$ respectively. The dose-response curve for nicotine was bell-shaped, suggesting that, at higher concentrations, nicotine desensitized its receptors. d-Tubocurarine ($5x10^{-5}M$) competitively antagonized the release of ATP by either acetylcholine or nicotine. Atropine sometimes also diminished the release of ATP induced by acetylcholine, but this effect was not consistently observed in all experiments. Maximal release of ATP by acetylcholine ($10^{-3}M$) was approximately 23% that observed with the depolarizing drug, veratridine ($5x10^{-5}M$), suggesting that not all of the varicosities capable of releasing ATP possess nicotinic receptors. Tetrodotoxin ($8x10^{-7}$M) produced a slight but significant (16%) reduction in ATP release induced by acetylcholine. This suggests that some of the nicotinic release of ATP from myenteric varicosities involves the opening of fast Na^+-channels in their membranes. Finally, preliminary experiments with 6-hydroxydopamine pretreatment indicate that 60-70% of the ATP released by acetylcholine might originate from nonadrenergic varicosities in the preparation. The residual non-adrenergic release of ATP could be "purinergic".

Supported by the MRC of Canada.

95. EFFECTS OF THEOPHYLLINE ON CEREBRAL BLOOD FLOW (CBF) IN THE RAT. H.R. Winn, S. Morii, A.C. Ngai, D.D. Weaver, R.M. Berne, Departments of Neurosurgery and Physiology, University of Virginia, Charlottesville, VA

We treated rats with theophylline (0.2 μmol/g, IP) to determine if this adenosine receptor blocker would alter CBF during hypoxia. Adult rats (300-420 gms) were initially anesthetized with halothane, tracheostomized, and mechanically ventilated to maintain their PaO_2 $>$100 mmHg and their $PaCO_2$ between 30 and 40 mmHg. Blood pressure and rectal temperature were continuously monitored. Transient hypoxia (30 sec) was induced by switching the animals' tracheostomy tube to a second respirator containing almost 100% nitrogen. Cerebral blood flow was measured by the retroglenoid venous outflow technique. Cerebral blood flow increased 160% in the untreated animals within 15 sec. of the onset of hypoxia. In contrast, in the theophylline treated animals, CBF increased by only 75% ($p < 0.01$). Cerebral vascular resistance was significantly higher in the theophylline treated animals. A similar response was noted in animals treated with 0.05 μmol/g (IP) of theophylline, whereas 0.02 μmol/g did not result in any significant alteration in the CBF response to hypoxia. The significant attenuation of the increase in cerebral blood flow during hypoxia in the theophylline treated animals supports the hypothesis that adenosine is a critical factor in regulation of cerebral blood flow.

96. LOCATION OF ADENOSINE RELEASE AND ADENOSINE RECEPTORS TO RAT STRIATAL NEURONS. W.J. Wojcik and N.H. Neff, Lab. Preclin. Pharmacol., NIMH, St. Elizabeth's Hospital, Washington, D.C. 20032.

We found an uneven distribution of adenosine in various brain regions in rats killed by decapitation. The striatal adenosine content was one of the highest of the brain regions studied. It was also of interest that the striatal homogenate contained an adenosine receptor which stimulated adenylate cyclase activity. We did not find this adenosine sensitive adenylate cyclase in the cortex, hippocampus, or cerebellum. Thus, these observations of striatal adenosine receptors coupled to adenylate cyclase and the elevated concentration of adenosine in the striatum, indicated that the striatum could be the location of a possible adenosine neuronal synapse. In order to determine the source of the adenosine synthesis and the site of the receptors, three separate striatal lesions were performed. The lesions were unilateral and involved decortication to remove the cortical-striatal projections, an injection of 6-hydroxydopamine (6-OHDA) into the medial forebrain bundle to destroy the dopaminergic pathway or an injection of kainic acid directly into the striatum which lesioned the neuronal cell bodies. Decortication and the 6-OHDA lesion did not greatly affect the synthesis of adenosine, nor did these lesions alter the adenosine receptor activation of cyclase. However, the kainic acid lesion prevented the elevation of striatal adenosine and the adenosine sensitive adenylate cyclase. The decrease of adenosine content after kainic acid may not reflect the destruction of an adenosine releasing neuron. To test whether this lesion affected adenosine release, we measured the amount of endogenous adenosine released into the superfusate from normal and kainic acid lesioned striatal slices after a 50 mM K^+ depolarization. This experiment provided evidence that the kainic acid lesion destroyed the neuron which releases adenosine, since a lower adenosine content was observed in the superfusate from the lesioned slices. We conclude from these studies that the synthesis and release of adenosine and the adenosine sensitive adenylate cyclase are associated with neurons contained in the striatum.

97. ALKYLXATHINES AD ADENOSINE RECEPTOR ANTAGONISTS AND MEMBRANE PHOSPHODIESTERASE INHIBITORS IN CENTRAL NERVOUS TISSUE: EVALUATION OF STRUCTURE-ACTIVITY RELATIONSHIPS. P.H. Wu and J.W. Phillis, Department of Physiology, Wayne State University School of Medicine, Detroit, Mi48201.

A series of alkylxanthines were examined as antagonists of the adenosine A_1-receptor in rat brain synaptosomal membranes and as inhibitors of brain membrane phosphodiesterase. The structure-activity relationships suggest that 8-(p-bromophenyl) theophylline and 8-(p-chlorophenyl) theophylline are more selective in their action as adenosine receptor antagonists, whereas 1,3-dimethyl-2,6-dithioxopurine and methylxanthines are selective as phosphodiesterase inhibitors.

N^6-cyclohexyl-(^{3}H)-adenosine (^{3}H-CHA) was used as a ligand to label the adenosine A_1-receptor on brain synaptosomal membranes. The K_D and B_{max} values for the binding sites were 2.14 nM and 578 femto mole/mg protein, respectively. Competition studies showed that the methylxanthines inhibit ^{3}H-CHA binding in a competitive manner with potency order as adenosine receptor antagonists of 8-(p-bromophenyl)theophylline, 8-(p-chlorophenyl)theophylline, > 8-(dimethylaminophenyl)theophylline, > 8-benzyltheophylline > theophylline. IC_{50} values for these compounds to inhibit ^{3}H-CHA binding are 8-(p-bromophenyl)theophylline, $2.9x10^{-7}$ M; 8-(p-chlorophenyl)theophylline, $4.1x10^{-7}$M; 8-methylphenyltheophylline, $3.2x10^{-7}$M; 8-phenyltheophylline, $3.5x10^{-7}$M, 8-(methoxyphenyl)theophylline, $5.8x10^{-7}$M; 8-(dimethylaminophenyl)theophylline, $1.3x10^{-6}$M; 8-benzyltheophylline, $1.1x10^{-5}$M; 3-isobutyl-1-methylxanthine, $2.1x10^{-5}$M and theophylline, 6.0 x 10^{-5}M. Other methylxanthines tested were less active with IC_{50} values ranging $5x10^{-5}$M to $4x10^{-4}$M. Purified ^{3}H-cyclic AMP (1 μM) was used as a substrate for membrane phosphodiesterase. The reaction product, ^{3}H-adenosine, was isolated following treatment of reaction mixture with snake 5'-nucleotidase. The order of potency for inhibition of brain membrane phosphodiesterase was: 1,3-dimethyl-2,6-

dithioxopurine (IC_{50}:8.6x10^{-6}M) > methylxanthines (IC_{50}:3x10^{-5}M)>> 8-substituted theophyllines (IC_{50}: > 5x10^{-5}M).

The structure-activity relationships showed that the addition of certain substituting groups on the 8th position of the theophylline molecule produced molecular structures which were generally in favor of adenosine receptor antagonism whereas an increase in lipid solubility produced by altering the methylxanthine molecule at the 1,2,3 and 6 positions of the purine ring may result in an increase in phosphodiesterase inhibitor activity. Theophylline is approximately equipotent in its action as an adenosine receptor antagonist and a phosphodiesterase inhibitor.

98. CIRCADIAN VARIATIONS OF ADENOSINE METABOLIZING ENZYMES IN THE BLOOD, BRAIN, AND LIVER OF THE RAT. L. Yanez, R. Hernandez-Munoz, M. Diaz-Munoz, J. Suares, S. Vidrio, and V. Chagoya de Sanchez, Depto. Bioenergetica, Ctro. Inv. Fisiologia Celular, Universidad Nacional Autonoma de Mexico, 04510 Mexico, D.F., Mexico.

Adenosine concentration circadian variations in the blood, brain and liver suggest that this nucleoside could be a signal for certain physiological events. The regulation of adenosine concentration should be related to its metabolic system. This possibility was explored by measuring the activity of adenosine deaminase, adenosine kinase, 5'-nucleotidase, and S-adenosyl homocysteine hydrolase in the blood, brain and liver of adult male Wistar rats adapted to natural dark-light cycle two weeks prior to the experiment. The animals were sacrificed every 2 or every 4 hours during the 24-hour period.
Blood. No variation in the activity of adenosine demainase and adenosine kinase were observed during the 24 hours, but the activity of 5'-nucleotidase decreased almost to zero levels from midnight to 5 A.M. suggesting that the adenosine found in the blood at that time is not formed *in situ* but probably stems from the liver.
Liver. The activity of the studied enzymes remained constant during twelve hours (8 A.M.-8 P.M.). After this period a decrease in adenosine deaminase and an increase in adenosine kinase and 5'-nucleotidase was observed, in agreement with the adenosine concentration observed at that time.
Brain. The activity of adenosine kinase and 5'-nucleotidase is oscillatory, showing a peak at the dark period: adenosine deaminase presented low activity during twelve hours (8 A.M.-8 P.M.) but a marked increase after this time occurs corresponding to the activity period. A good correlation between adenosine concentration and adenosine metabolizing enzymes was found.

99. ANALYSIS OF ADENOSINE RECEPTORS IN RAT STRIATUM AND HIPPOCAMPUS WITH (^{3}H)-N-ETHYLCARBOXAMIDE ADENOSINE ((^{3}H)-NECA). Siu-Mei Helena Young and Richard D. Green, Department of Pharmacology, University of Illinois College of Medicine, Chicago, IL 60680.

NECA stimulated and inhibited adenylate cyclase in preparations of rat striatum and hippocampus, respectively. Both of these effects were dependent on GTP and were antagonized by methylxanthines. Binding of (^{3}H)-NECA was studied at 37°C using N^6-(L-phenylisopropyl)adenosine (L-PIA) to define non-specific binding. (^{3}H)-NECA bound to two sites in striatum with K_D values of approximately 4 and 40 nM. The K_D was $\simeq$ 200 nM with G_{pp}(NH)p present. Competitive binding curves (20$_6$ nM (^{3}H)-NECA) for theophylline, MIX, 8-phenyltheophylline, N^6-benzyladenosine, and 5'-ethylcarboxylate adenosine were reasonably steep ($n_H > 0.8$) while those of L-PIA, N^6-cyclohexyladenosine (CHA) and 2-chloroadenosine (2-ClAd) were shallow and in many cases clearly biphasic. (^{3}H)-NECA also bound to two sites in the hippocampus; the K_D values were 2-4 nM and 50-250 nM. Competitive binding curves (20 nM(^{3}H)-NECA) for L-PIA, CHA, and 2-ClAd were different from those in striatum in that they were steeper and to the left of the corresponding curves in striatum. As only the inhibition of adenylate cyclase in hippocampus has been demonstrated we suggest that the two sites in hippocampus are two affinity states of the R_i site. R_i site mediated inhibition of adenylate cyclase in striatum has not been demonstrated. The two sites in striatum may be R_a and R_i sites or may be two affinity states of R_a sites.

100. UREOGENESIS STIMULATION BY ADENOSINE. A. Zentalla, Depto. de Bioquimica Fac. de Medicina UNAM 04510 Mexico, D.F., Mexico.

Intraperitoneal injection of adenosine increases seric urea concentration and enhances incorporation of radioactive glutamate into hepatic N-acetylglutamate fraction.

Adenosine not only produces a rise in ATP content and energy charge, but also provides free NH_4^+ through adenosine deaminase present in the membrane of hepatocytes.

These previous findings indicate the modification by adenosine, of several elements, potentially responsible for ureogenic stimulation.

Experiments trying to evaluate the stimulatory role of these elements are discussed.

INDEX

GPSR Compliance

The European Union's (EU) General Product Safety Regulation (GPSR) is a set of rules that requires consumer products to be safe and our obligations to ensure this.

If you have any concerns about our products, you can contact us on ProductSafety@springernature.com

In case Publisher is established outside the EU, the EU authorized representative is:

Springer Nature Customer Service Center GmbH
Europaplatz 3
69115 Heidelberg, Germany

Batch number: 10755248

Printed by Printforce, the Netherlands